UNEP

United Nations Environment Programme

Environmental Data Report 1993–94

Prepared for UNEP by the

...nd Assessment Research Centre, London, UK

in co-operation with the

World Resources Institute, Washington DC
UK Department of the Environment, London

D0231010

BLACKWELL
Reference

For bibliographic and reference purposes this publication should be referred to as:
UNEP 1993 United Nations Environment Programme
Environmental Data Report, 1993–94
Blackwell Publishers, Oxford.

1st edition published 1987.
2nd edition published 1989.
3rd edition published 1991.

Blackwell Publishers
108 Cowley Road, Oxford OX4 1JF, UK

238 Main Street, Cambridge, Massachusetts 02142, USA

British Library Cataloguing-in-Publication Data
   Environmental Data Report.
   1. Environment
   I. United Nations. Environment Programme.   II. GEMS Monitoring and
   Assessment Research Centre.   III. World Resources Institute.   IV. Great Britain.
   Department of the Environment
   333.7021
   ISBN 0–631–19043–0
   ISSN 0956–9324

Library of Congress Cataloging-in-Publication Data
   Environment data report/prepared for UNEP by the GEMS Monitoring and
   Assessment Research Centre London UK, in co-operation with the World Resources
   Institute, Washington DC, UK Department of the Environment, London.—1993–94.
   At head of title: United Nations Environment Programme.
   1. Pollution.   2. Environmental monitoring.   I. United Nations Environment
   Programme.   II. GEMS Monitoring and Assessment Research Centre.
   III. World Resources Institute.
   TD 174.E576 1993   363.7—dc20   91–9393 CIP
   ISBN 0–631–19043–0
   ISSN 0956–9324

Typeset in Palatino and Helvetica Narrow by Imogen Bertin, Cork
Printed in Great Britain at The Alden Press, Oxford

This book is printed on acid-free paper.

Contents

iv   Contents

Foreword

Today we are more aware than ever of the widespread effects of human activities on our environment. All around us we see the environment changing as a result of our actions, whether from the wasteful exploitation of natural resources, or from inefficient consumption patterns, particularly of energy. Building and reshaping of landscapes, land degradation, forest damage, hazardous wastes and loss of natural resources can clearly be seen as consequences of our actions in the industrialized countries. Today environmental problems have, however, acquired new global dimensions as shown by the threats of climate change and sea-level rise, depletion of the stratospheric ozone layer and the loss of biodiversity.

An evaluation of the state of the environment at the global level requires an understanding of the pressures and responses contributing to environmental change. The link between environment and development was recognized by the new global partnership that arose from the United Nations Conference on Environment and Development (UNCED) held in Rio de Janeiro in June 1992. One major agreement – The Rio Declaration on Environment and Development – is of particular importance because of the acceptance of principles of shared but differentiated responsibility. Together they constitute a significant first step in the transition to sustainable development.

The most significant output of the Rio conference is Agenda 21. It is a broad prescriptive document agreed by governments which outlines programmes of actions covering, in an integrated way, environment and development in the 21st century. The responsibility to act lies with governments but also with United Nations (UN) organizations, intergovernmental bodies, municipalities, companies and non-governmental organizations. Agenda 21 also highlighted the need for an enhanced and strengthened role for the United Nations Environment Programme (UNEP) and recognized its continuing co-operative and collaborative role with partner organizations. In the area of environmental economics, for example, action has already been taken to build up the capacity within UNEP to handle natural resource accounting and impacts of consumption patterns on resource depletion.

The implementation of Agenda 21 can only be successful when it is based on sound evaluation and assessment of the issues using the best available knowledge. At the national level recognition is growing of the need for further action on all environmental issues to identify, study and evaluate current conditions and future prospects arising from human activities. While many data already exist, their quality, comparability and accessibility can be short of the standard required for informed decision-making. A large-scale strengthening of the scientific capacity of developing countries is still required to enable them effectively to sense their environment and assess human effects.

For many years UNEP has stated that in order to avoid long-term environmental degradation, all planning, policy- and decision-making have to be based on reliable information. Consequently, the 'UNEP Environmental Data Report' series has been produced to bring together in single volumes the best available data and information on a wide range of environmental topics. The biennial editions provide from most countries updated core data sets of common environmental topics including pollution, health, natural resources, population and settlements, energy, wastes and disasters.

Although the 'UNEP Environmental Data Report' series contains the best available data, gaps in essential knowledge are still recorded. These gaps may arise from inabilities of scientists in earlier years to gather data in a sufficiently systematic way, or to sense the environment in a particular region. Gaps may also arise through new issues that are only now being recognized as important for our understanding of inter-relationships between natural cycles and human actions. One such gap currently being addressed by UNEP and its partners is the need for a reliable set of environmental indicators, or indices, which can help governments identify environmental problems and enable decision makers to act on them. The need for indicators and indices to strengthen data gathering – in effect "bridging the data gap" – has been highlighted in the Introduction to this 1993/94 volume.

This report documents data and their sources; what is available, why it was collected, who collected it, where it is held and how it was measured – all in one reliable source. Collection and evaluation of data are not ends in themselves. Data must be used to develop management plans for action by decision makers. Data are also needed to ensure that solutions are found for the issues of today and to help focus thoughts on resolution of the problems of tomorrow. I consider that this report is a major contribution towards the provision of the data necessary for national, regional and world-wide action.

Elizabeth Dowdeswell
Executive Director

Notes to Users

The following notes serve as a guide to the 'UNEP Environmental Data Report'. They outline the structure and will assist readers in locating the information they seek.

Environmental and environment-related data in this report is organized as follows:

○ *The state of the environment*: The first section describes the state of the global environment in terms of environmental contamination, climate change and the status of the natural resource base (Parts 1 to 3).

○ *Human and economic activity*: These chapters describe significant trends in the pressures on the global environment. These include population growth and economic development (Part 4), energy production and use (Part 6), industrial activity and transport (Part 7). The problems associated with the generation and disposal of wastes are outlined in Part 8. The impacts of natural and man-made hazardous environmental events (Part 9) and human health issues (Part 5) are also addressed.

○ *Policy responses*: The final section outlines some of the regional and global measures taken to protect the global environment and assesses their effectiveness (Part 10).

Each chapter comprises tabular material (presented as a continuous block at the end of the chapter) plus supporting text and graphics. The supporting texts identify the principal data sources, discuss the quality of the existing data sets and summarize key environmental trends, and the footnotes to the data tables provide information of a more technical nature and full details of the original data sources.

Although each edition of the 'UNEP Environmental Data Report' contains a central or "core" set of data tables which are updated each time (these are identified by a green background), many of the data and supporting information presented in successive editions are new. This material is carefully selected to reflect topical and emerging issues. Owing to space restrictions it is not possible to provide full coverage of every topic in each edition; some issues are thus covered in alternate editions. However, within the supporting texts extensive cross-referencing is made to earlier editions and to other chapters. For convenience references in the text are given after each individual sub-section and the key findings are highlighted in a summary paragraph at the beginning of each chapter. In the present edition, inset boxes have been introduced in the text to provide more detailed discussions of specific issues or a particular study and, where appropriate, to provide examples of environmental indicators.

Notes to all Tables

Symbols and Conventions

Throughout this publication the following notation scheme has been adopted in the data listings (deviations from this scheme are noted in the footnotes to individual tables):

Blank space	Data not available or not applicable
-	Negligible quantity, i.e., usually less than one-half the unit indicated
0	None, magnitude nil or zero
(.)	A point is used to indicate decimals
(-)	A dash used between years, e.g., 1984-86, indicates the calendar period inclusive of beginning and end years, i.e., 1 January 1984 - 31 December 1986
(/)	A slash between years, e.g., 1984/85 indicates a crop or financial year as opposed to a calender year.

Regional and world totals and percentage distributions may not necessarily tally due to rounding. Regional and world totals refer to the sum of values for the countries listed in the table unless otherwise stated. When the aggregates are not considered to be representative of the global or regional situation the aggregate is labelled 'Total'. For further details relating to country nomenclature the reader is referred to Appendix 2 where recent changes in designation are listed.

Units

As far as possible, SI units (the International System of Units) are used throughout this publication. Multiplying prefixes are used to indicate decimal multiples or fractions:

10^{12}	tera	T	10^{-1}	deci	d
10^{9}	giga	G	10^{-2}	centi	c
10^{6}	mega	M	10^{-3}	milli	m
10^{3}	kilo	k	10^{-6}	micro	μ
10^{2}	hecto	h	10^{-9}	nano	n
10	deka	da	10^{-12}	pico	p

One billion is used to refer to 10^{9}.

Introduction

Environmental information (whether as data, statistics, appropriately packaged experience or qualitative materials), is becoming increasingly recognized as an essential component of national development plans. National activities, such as transportation, agriculture, energy production and use, waste management, forestry or fisheries have a potential for direct and indirect impacts on the environment. The United Nations Conference on Environment and Development (UNCED) held in Rio de Janeiro in June 1992 stressed the need to integrate environment and development in future policy formulation. Reliable environmental data are therefore necessary for use by governments, as well as by industrial, public and private sectors, to frame policy, to set priorities and to assess results. The improvement and use of available environmental information is a major requirement for the development of national and international policy.

The need for sound information at all levels, from that of the senior decision maker at the international level to the field worker at the national level, was also highlighted at UNCED. The action plan, commonly known as Agenda 21, which was adopted at the Conference identified two programme areas needing better implementation to ensure that decisions are based increasingly on sound information: bridging the data gap, and improving information availability (UN, 1992).

The absence of data from a number of developing countries, especially in a regional context, hinders national environmental protection and national development and planning, as well as the evaluation of regional and global effects. To achieve cost-effective data collection and assessment, users and their requirements need to be more clearly identified. In addition, ways of transforming existing information into forms more useful for decision-making are essential, together with targeting the information at different user groups.

Environmental Sensing and Statistics

Environmental sensing is the basic approach to assessment of environmental quality and the gathering of information for decision makers. The need to use such data for early warnings, i.e., giving advanced information of an adverse environmental event or situation before it happens, is increasingly recognized as an institutional requirement for national development plans.

Since inception, the United Nations Environment Programme (UNEP) has stimulated national monitoring activities and developed regional and global monitoring programmes. The Global Environment Monitoring System (GEMS), with its United Nations (UN) partners, the World Health Organization (WHO), the World Meteorological Organization (WMO), the Food and Agriculture Organization of the United Nations (FAO), and the United Nations Educational, Scientific and Cultural Organization (UNESCO) amongst others, was established and mandated to meet these needs. Extensive monitoring networks are now in operation around the world. These networks gather data according to defined measurement and monitoring methods, thereby providing comparable information for regional and global assessments of environmental quality.

In addition to direct environmental quality measurements, the process of assessment of environmental quality and potential environmental and health hazards requires knowledge of emissions from industrial activities and natural processes. Emission inventories, whether measured and compiled for point and diffuse sources or conceptually based on emission factors, provide data on the potential for effects. When used in conjunction with dispersion models and data on critical loads or human responses, emission inventories can be used to provide early warnings of potential hazardous situations.

Environmental statistics are compiled, stored and disseminated by central statistical services, government departments, research institutes, local authorities and international organizations. They are collected through censuses, surveys, the use of administrative records and monitoring networks. These statistics describe the state and trends of the environment, covering the media of the natural environment, the biota found in those media and human settlements. Many national and international organizations now periodically review their environmental statistics and use them to assess environmental quality and trends. These assessments are commonly known as 'State of the Environment' (SOE)-type reports. A standard approach to SOE reporting has been produced by UNEP to help to standardize presentation and this has to some extent been adopted by several countries (UNEP, 1985).

As a run-up to UNCED, governments were asked to report on the state of the environment at the country level using a set format. The reports produced varied in content from generalized information on the environment and national perceptions, to detailed data reports containing many fundamental monitoring and statistical data (see Part 10: International Co-operation). The SOEs have been mainly produced by national governments and by their statistical departments, although several non-governmental organizations (NGOs) and international agencies have also produced their own reports, including UNEP (Tolba and El-Kholy, 1992) and other UN organizations.

Over the past several decades the UN has strongly supported efforts by member states and specialized organizations to gather basic quantitative and qualitative data necessary for national planning and policy formulation. The United Nations Statistical Office (UNSTAT) of the Department of Economic and Social Development has primary responsibilities for the development of international statistical systems (UNSTAT, 1991). The annual statistical yearbook provides basic national data collections on a vast range of topics. Environmental data are now being included in the statistical yearbooks.

The UNSTAT Framework for the Development of Environmental Statistics (FDES) provides an extensive description of the scope, coverage and contents of environment statistics, based on a world-wide survey of national and international agencies. In addition, UNSTAT is currently developing methodologies for a Satellite System of Integrated Environmental and Economic Accounting (SEEA) in response to requests by UNCED. Integrated environmental and economic accounting attempts to broaden the scope for interlinkages leading to the developing of indicators of sustainable development. Such issues have been addressed at the three Intergovernmental Working Group meetings on the Advancement of Environment Statistics (UNSTAT, 1993). In addition to the work of UNSTAT, all the UN agencies, departments and specialized organizations also collect sectoral data needed for planning purposes.

With the increasing number of sectoral and national data bases, the concept of the meta-data base has become an important vehicle to help identify "who is doing what and where". Co-operative developments in establishing comparable meta-data bases are under way within UNEP at the Global Resource Information Database (GRID), the UNEP Harmonization of Environmental Measurement Office (UNEP-HEM), the GEMS Monitoring and Assessment Research Centre (MARC), the World Conservation Monitoring Centre (WCMC) and the World Resources Institute (WRI) in addition to co-operative activities with UNSTAT, the Statistical Office of the European Communities (EUROSTAT) and national agencies.

Improving access to information within the UN system is one of the aims of the inter-agency Advisory Committee for the Co-ordination of Information Systems (ACCIS). The Advisory Committee maintains a data base of computerized data bases and information systems, including systems of environmental interest, operated by UN organizations. Information from the data base is published in the Directory of UN Data bases and Information Services (DUNDIS) (Walker, 1991). Co-operation between ACCIS and the International Environmental Information System (INFOTERRA) has given rise to a guide focusing on UN information concerning the environment.

At the non-governmental level other meta-data base activities are also in progress. For example, the Consortium for an International Earth Science Information Network (CIESIN) is being established to facilitate access to, and use and understanding of, global change information world-wide. The International Geosphere-Biosphere Programme's (IGBP) Data and Information System (IGBP-DIS) is responsible for data management within the programme and for facilitating access to global data sets collected by other research groups and agencies, especially those obtained through remote sensing.

Standardization and Harmonization

An examination of data availability and quality, especially on a global scale, shows that they can be extremely uneven. Quality variations, differences in compilation methods and poorly documented procedures all lead to difficulties of interpretation and poor management decisions. Moreover, recognition of the regional and global scope of environmental problems has increased demand for internationally comparable data sets. The past lack of universally accepted definitions and standardized classification systems and emission inventory methodologies has, however, limited the inter-comparability of environmental statistics generated to date.

Global monitoring programmes, such as those operated by UNEP GEMS, have been based on the application of standardized sampling and analytical methodologies. These practices, often accompanied by the use of the same commercial equipment, have enabled comparable data sets to be compiled. However, as different measurement methods are increasingly being developed and applied even within the one monitoring network, the concept of the harmonization of measurements and analysis has come to the fore.

In recent years some progress has been made in addressing some of these data quality issues at the international level. The work of UNSTAT and the International Council for Scientific Unions' (ICSU) Committee on Data for Science and Technology (CODATA) are particularly noteworthy in this regard. The establishment of the HEM office has been UNEP's response to the perceived requirement for improved comparability or harmonization of environmental data. This office is currently evaluating ways of harmonizing sectoral environmental measurements, focusing initially on urban air quality measurements. A second activity is the development of a directory and data base of standard analytical reference materials; standard reference materials are an essential component of quality assurance and analytical quality control procedures for all environmental measurements. The work of the International Standards Organization (ISO), which provides information and guidance on a wide range of analytical techniques for environmental monitoring, is also relevant in this context.

The development of standardized nomenclatures for environmental statistics within the European region is an ongoing activity of the UN Economic Commission for Europe (UN ECE), and is conducted in conjunction with the Organisation for Economic Co-operation and Development (OECD) and the Statistical Office of the European Communities (EUROSTAT). To date, standardized definitions and classifications have been developed for a range of sectors including air and water quality, water use, land use, fauna and flora, and wastes; these have provided the basis for the construction of the UN ECE environmental data compendia (UN ECE, 1992). A manual providing full details of the UN ECE standard classification systems and definitions, together with information on methodologies for the construction of emission inventories, is due to be published at the end of 1993 (UN ECE, 1993).

Development of Environmental Indicators and Indices

Decision makers at all levels are facing increasingly complex policy choices. Therefore, the need for rapid measures to integrate environmental and economic factors into strategic advice is becoming paramount. At present, the information needs for some decision makers cannot easily be met from existing data collections and current analytical techniques. This has led to the development of the concept of environmental indicators.

Environmental indicators have been conceived as a shorthand method for examining environmental situations in a manner readily understandable by experts and the public alike. However, the task is a complicated one because not only do the indicators have to be scientifically valid and reflect environmental trends, but they have to be able to respond to changes brought about by management actions. Moreover, inadequate data severely limit the use of even those indicators that are conceptually sound.

Environmental indicators can be of various types designed to fulfil particular functions, for example:

○ SOE indicators to reflect the quality of the environment;
○ Impact or stress indicators to reflect effects;
○ Environmental-economic indicators to evaluate cost-benefits;
○ Sustainable development indicators or performance indicators to evaluate long-term achievements;
○ Environmental health indicators to reveal inter-relationships and links between environmental hazards and human health.

The development of environmental indicators is likely to become a key component of environmental reporting systems in the future. Many individual countries as well as UNEP, WRI, the OECD, the UN ECE, the World Bank and other organizations, are all actively developing policies and procedures for a core set of indicators of different types (e.g., for policy, sectoral issues and other purposes). As complex issues and value judgments are involved, it has proved to be difficult to agree on the conceptual framework for indicators. The different types of potential users – the public, policy makers, specialists and the number of agencies and organizations wishing to devise indicators has broadened indicator discussion yet slowed down their immediate construction and implementation. Nevertheless, examples of some proposed indicators have been highlighted in the appropriate sections of this report.

In addition to environmental indicators, some environmental indices have been developed. Indices incorporate two or more variables to produce a single value. Several excellent examples of indices are known and already established as tools for scientists and policy makers (US EPA, 1990). The Human Development Index devised by UNDP incorporates both social and economic factors (UNDP, 1992) (see Part 4: Population and Development). Great care will have to be expressed in the compilation, interpretation and use of environmental indicators and indices, not just at the national level but when comparisons are made between nations.

UNEP's 'Environmental Data Report' Series

Governments must have at their disposal not only their own national data upon which to assess the impact of their policies, but also international data to aid their development of sound environmental action plans. In order to disseminate better environmental data at regional and international levels of relevance to management needs, UNEP established, in the mid-1980s, its own biennial 'Environmental Data Report' series (UNEP 1987, 1989, 1991). The 'UNEP Environmental Data Report' is a source volume of carefully checked, verified and annotated data that provides a basis for detailed environmental assessments at the national and international level. The reports also cover environmentally-related sectors such as transport, energy, human health and industry to satisfy the needs of decision makers. Effective linkages between the different sectors or environmental concerns are made in each report. It is not possible for any one of the reports in this ongoing series to contain all the data and information which are available for a particular sector or topic. Therefore, in each report, data are chosen to illustrate selected topical issues, geographical concerns, transboundary issues and global data collections.

The 'UNEP Environmental Data Report' series is designed as a critical review of currently available environmental data and related information. More specifically, through its mix of tabular and textual presentation of material, it serves a number of functions, namely:

○ Identification of the principal sources and compilers of environmental data sets,
○ Provision of extracts or summaries of key data sets in tabular form,
○ Highlighting the deficiencies in the available information base, not only in terms of information gaps and the quality of the existing data, but also in terms of the relevance to policy making,
○ Highlighting the ongoing efforts to correct the deficiencies in the information base,
○ Provision of an overview of the significant and emerging environmental trends, and
○ Attempting to relate environmental conditions to human activities or pressures on the environment.

Examples of data included in this edition come from the UN agencies and their co-operating organizations, particularly WHO, WMO and FAO, as well as other international and regional organizations, economic institutions and national SOE reports.

Future Challenges

It is widely acknowledged that environmental data gathering and its relevance for sustainable development has to be supported by the international community. The recent International Forum on Environmental Information for the Twenty-First Century (EIFS, 1992) has highlighted a number of general and specific requirements concerning environmental information for policy making. These include:

○   As information is a strategic resource, it should be made readily available, both as raw data and in an interpreted form for appropriate user groups.
○   Any information which is made available should have adequate peer review and the data sets clearly specified as they relate to a particular issue.
○   The best possible use should be made of existing data before large new data-collection efforts are undertaken.
○   Co-ordination of data collection amongst partners is necessary to ensure that the information gathered can be aggregated to provide a larger picture.
○   Environmental monitoring and data gathering must be strengthened in developing countries.
○   The scientific base concerning the links between the environment and economics and human health should be broadened and deepened.

The special needs of developing countries were also recognized by the ministers from 11 developing countries who met in Beijing in June 1991. The Beijing Declaration, as it became known, stressed the special situation pertaining to the needs of developing countries and that a strengthening of their scientific and technological capabilities was of major importance. There was wide recognition that the critical natural resources and the environments of developing countries are subject to stresses of unprecedented magnitude. Since the health, nutrition and general well-being of the world's poor depend directly on the integrity of these resources, a government's capacity to manage them effectively over both the short and long term is one of the important prerequisites to the eradication of poverty, the protection of the environment and achievement of sustainable development.

Acquiring better data and using them for management and policy formulation is only part of the task. Research is still required on how ecosystems function and how human activities affect them. The links between the environment and sustainable development and economic progress need to be better understood. The understanding of these links will enable UNEP and UN agencies to support more effectively the gathering of national environmental data necessary for local, regional and global management.

References

EIFS 1992 Environmental Information for the Twenty-First Century, International Forum, May 21–24, Montreal, Proceedings of Conference, Ottawa, Canada.

Tolba, M. K. and El-Kholy, O. A. (Eds) 1992 The World Environment 1972-1992: Two Decades of Challenge, UNEP and Chapman and Hall, London.

UN 1992 *Report of the United Nations Conference on Environment and Development* (Rio de Janeiro, 3–14 June 1992) A/CONF.151.26 (Vols I, II, III and IV), United Nations, New York.

UN ECE 1992 *The Environment in Europe and North America, Annotated Statistics*, United Nations, New York.

UN ECE 1993 *Readings in International Environment Statistics*, United Nations Economic Commission for Europe, Geneva. In press.

UNDP 1992 *Human Development Report*, United Nations Development Programme and Oxford University Press, Oxford.

UNEP 1985 *Guidelines for the Preparation of National State of Environment Reports*, United Nations Environment Programme, Nairobi.

UNEP 1987 *United Nations Environment Programme Environmental Data Report 1987/88*, Basil Blackwell, Oxford.

UNEP 1989 *United Nations Environment Programme Environmental Data Report 1989/90*, Basil Blackwell, Oxford.

UNEP 1991 *United Nations Environment Programme Environmental Data Report 1991/92*, Basil Blackwell, Oxford.

UNEP-HEM 1992 User Requirements for the Harmonization of Environmental Measurement Information System, HEMIS, United Nations Environment Programme, Nairobi.

UNSTAT 1991 *Concepts and Methods of Environment Statistics: Statistics of the Natural Environment*, A technical report ST/ESA/STAT/Ser.F./57, Statistical Office, Department of International Economic and Social Affairs, United Nations, New York.

UNSTAT 1993 Intergovernmental Working Group on the Advancement of Environmental Statistics, Report of the Third Meeting, Report ESA/STAT/AC.45, Department of International Economic and Social Affairs, United Nations, New York.

US EPA 1990 *Ecological Indicators for the Environmental Monitoring and Assessment Programme*, Atmospheric Research and Exposure Assessment Laboratory, Research Triangle Park, North Carolina.

Walker, C. 1991 United Nations information on environmental concerns, *UNEP Industry and Environment*, **14**, 21–22.

Part 1

Environmental Pollution

Chemical contamination of the environment due to human activities has now reached global proportions. The potential threats of greenhouse gas-induced climate change and stratospheric ozone depletion, concerns which epitomize the global scope of chemical contamination, are universally recognized. However, despite the catalogue of evidence suggesting trends towards worsening environmental pollution, some findings of a more positive nature are reported in this edition of the 'UNEP Environmental Data Report'. Key trends highlighted here include:

○ In the past two years there has been a small decline in the atmospheric growth rate of some greenhouse gases, notably methane, CFC-11, CFC-12 and the halons H-1301 and H-1211. This is attributed, at least in the cases of the CFCs and halons, to the impact of the Montreal Protocol.

○ There is a growing body of evidence for contamination of all environmental media, especially wildlife, in remote regions of the world, once thought to be pristine.

○ Arctic ecosystems in particular are becoming increasingly impacted by industrial pollutants; some organochlorine substances, for example, appear to be preferentially accumulating in the colder, polar regions.

○ Contamination of the environment by lead is showing a downward trend in some world regions where lead additives in petrol have been phased out.

○ Enhanced lake eutrophication is becoming recognized as a global water quality issue.

○ Urban air quality in terms of sulphur dioxide and particulate pollution has steadily improved in some cities.

Since the advent of industrial times, human activities have resulted in the release of chemical contaminants into the biosphere and have thus become major agents of environmental change on global, regional and local scales. Truly global phenomena such as greenhouse warming and stratospheric ozone depletion need no introduction, as these issues have dominated the environmental agenda in the recent past. More localized environmental pollution problems, for example trace metal and organochlorine contamination, are reported with increasing regularity and have reached such an extent that they can be described as having almost global occurrence.

For logistical reasons, the material on environmental pollution presented in this chapter of the 'UNEP Environmental Data Report' is grouped according to environmental compartments or media (i.e., air, soil, water and biota). Data sets are selected so as to document key and emerging 'state of the environment' issues and concerns. Where appropriate, attempts are made in the accompanying texts to highlight the pertinent data issues and developments in approaches to environmental monitoring, assessment, and regulation of environmental pollution. Recent advances in the development of 'integrated' approaches to environmental monitoring are singled out in a separate section (see Integrated Monitoring). The use of indicators in environmental monitoring also appears to be gaining popularity; a selection of approaches for soils, sediments, water quality and natural ecosystems are outlined in some of the inset boxes in this chapter.

International concerns such as global warming, stratospheric ozone depletion and acid deposition, have led to the establishment of some of the most successful collaborative environmental monitoring programmes, the results of which form the basis of much of the material presented here. The discussions of greenhouse gases, for example, draws heavily on the scientific work of the Intergovernmental Panel on Climate Change (IPCC), a joint activity of the World Meteorological Organization (WMO) and the United Nations Environment Programme (UNEP). Throughout the chapter, particular reference is also made to those programmes which are co-ordinated by UNEP's Global Environment Monitoring System (GEMS). Such programmes are typically carried out according to standardized methodologies and therefore are able to provide internationally comparable data sets. Results of national monitoring programmes and individual studies have, however, also been used to expand subject coverage.

Over the past 20 years or so considerable progress has been made in assessing and managing the environmental and human health risks posed by chemicals in the environment. However, the sheer number of commercially available chemicals, currently estimated to be over the 100,000 mark, means that there is still a huge number of chemicals that are without detailed toxicity and hazard assessment profiles. Moreover, internationally agreed guidelines for harmonized reporting and testing of potentially toxic chemicals are still lacking. Attempts are being made at the international level to remedy this situation. In 1976 UNEP established the International Register of Toxic Chemicals (IRPTC) to improve the availability and dissemination of information on toxic chemical contaminants. IRPTC's computerized files now contain data profiles for over 800 chemicals. Additional files contain information on waste management and disposal options for individual chemicals, on chemicals currently being tested for toxic effects and on national regulations for over 8,000 substances. Risk evaluation of chemicals, together with the development of methodologies for harmonized testing and evaluation is the remit of the International Programme on Chemical Safety (IPCS), an activity jointly established by UNEP, the International Labour Organization (ILO) and the World Health Organization (WHO) in 1980.

Atmosphere

In this section of the 'UNEP Environmental Data Report' recent trends in emissions and atmospheric concentrations of selected gaseous and particulate pollutants are reviewed. The selection covers the complete range of major air pollution concerns in the sense that it considers problems of a global nature (i.e., greenhouse gases and stratospheric ozone depletion) as well as those of a more regional and local nature. This is in contrast to the approach adopted in previous reports in this series where deposition of acidic pollutant gases was discussed as a separate section on 'Atmospheric Deposition'.

As trends in atmospheric concentrations, sources and sinks of greenhouse gases have been extensively reviewed in earlier editions of this report, the present edition directs attention towards new information which has advanced the understanding of the global budgets of the principal greenhouse gases during the past two years. Owing to space restrictions and the fact that it is the least important of the principal greenhouse gases, the section on nitrous oxide is omitted on this occasion. The section on stratospheric ozone depletion has, however, been comprehensively updated following the publication in 1992 of WMO's latest scientific assessment of this issue. This present edition also incorporates the main findings of a recently completed WHO/UNEP review of urban air pollution problems in megacities. As previously, the issue of climate change detection is covered separately (see Part 2: Climate).

Greenhouse Gases

Routine monitoring of atmospheric concentrations of carbon dioxide (CO_2) and the other principal greenhouse gases, methane (CH_4), the chlorofluorocarbons (CFCs) and nitrous oxide (N_2O) is conducted primarily in remote regions of the world, i.e., removed from the distorting influences of major industrial centres. Global-scale networks are now well established and have been described in previous editions of this report.

More recently the WMO has established the Global Atmosphere Watch (GAW) as an "umbrella" programme for integrating and co-ordinating many of the existing monitoring and research activities involved in the measurement of atmospheric composition (WMO, 1990). This new system, approved in June 1989, will take up the responsibility for preparing scientific assessments and advice to governments on matters relating to the chemical composition of the atmosphere. Information gathered by the GAW programme will contribute to the newly established Global Climate Observing System (see Part 2: Climate).

When fully implemented, GAW will comprise approximately 30 "global" stations and 200–300 "regional" stations. The programme of monitoring at global stations will include the routine measurement of all the principal greenhouse gases; data obtained as a result of this activity will be centrally stored at the newly established WMO World Data Centre for Greenhouse Gases (located at the Japanese Meteorological Agency in Tokyo) and other data centres.

In view of the importance attached to the modelling and prediction of greenhouse gas-induced climate change, considerable effort has been expended in recent years in improving emissions inventories for all the major greenhouse gases. For scientific assessment and modelling purposes, the primary requirements are for inventories at the global and regional levels. National inventories of anthropogenic greenhouse gas emissions are, however, equally as important, especially in the policy-making context; national inventories provide the basis for assessing and then setting priorities for controlling the emissions of gases from different sources and sectors. In addition, more accurate national estimates of greenhouse gas emissions help to provide a more complete picture of global-scale sources and sinks.

With the possible exception of industrial emissions of CO_2 and the halocarbons, anthropogenic greenhouse gas source strengths are subject to considerable uncertainty at both the global and national level, partly because sources are difficult to characterize. Natural sources of CH_4 and N_2O are also poorly documented. Calculation of emissions for many of the greenhouse gas sources thus involves making a number of major assumptions.

Gaseous emissions are generally computed as the product of an "activity" level (i.e., a measure of the type and scale of an anthropogenic source, for example, fossil fuel combustion or deforestation) and an "emissions" factor, the quantity of gas emitted per unit of activity. Using a combination of currently available international data sets for various human "activities" and global average or regionally-specific emission factors, a number of national inventories of greenhouse gas emissions have been derived; some examples are given in this report (see Carbon Dioxide and Methane). Although this approach achieves a degree of inter-country comparability, accuracy of such estimates is hampered by the limitations of the international data sets and the representativeness of the global or regional emission factors at the country level. National emissions estimates of greenhouse gases generated in this way have been utilized in the formulation of various "greenhouse gas indexes" which attempt to reflect national contributions to global warming. Some of these approaches are outlined in Box 1.1.

In recent years a number of countries have independently developed their own inventories of greenhouse gas emissions. Attempts to harmonize the method of calculation for sources and sinks of greenhouse gases at the national level are currently being made under the guidance of Working Group I of the

BOX 1.1 Accountability for Greenhouse Gas-Induced Climate Change

The threat of global climate change has led to increased efforts in recent years to develop appropriate policy-making tools for assigning responsibility for climate change at the national level and for setting greenhouse gases emission reduction targets. Measures which attempt to assess accountability for greenhouse gas-induced change include the IPCC's Global Warming Potential and the World Resources Institute (WRI)'s Greenhouse Gas Index; these and other alternative approaches are briefly reviewed below.

The IPCC's Global Warming Potential

By using simple one-dimensional models, it is possible to derive numerical values which represent the strength of the radiative forcing exerted by individual greenhouse gases and other forcing factors which have the capacity to affect global climate. This radiative forcing is defined as the net change in the flux of radiative energy induced at the tropopause. To date emissions of greenhouse gases from human activities have caused an additional radiative forcing of about 2.5 Wm^{-2}.

The IPCC have developed the concept of radiative forcing into a simple and convenient means for evaluating the relative climatic effects of individual greenhouse gases. They have thus devised the Global Warming Potential (GWP) as a measure of the relative globally-averaged warming effect arising from the emission of a given greenhouse gas. The GWP is more accurately defined as the time-integrated change in radiative forcing due to an instantaneous release of 1 kg of a trace gas relative to that from the release of CO_2.

The derivation of GWPs is outlined in detail in IPCC's First Scientific Assessment Report (IPCC, 1990). On the basis of new information, the IPCC have since revised their original GWP values for several of the major greenhouse gases; these revised values for selected species are listed in the accompanying table.

The IPCC acknowledges that its GWP concept suffers from a number of limitations. The GWP provides a measure of the global effect of a given greenhouse emission; it does not take into account latitudinal and seasonal variations in radiative fluxes and forcings. Secondly, the GWP values are sensitive to uncertainties in atmospheric residence times; moreover, they assume a constant composition, contemporary atmosphere. Thirdly, the GWP definition only reflects radiative forcing – it does not reflect a particular response of the climatic system, for example, surface temperature (IPCC 1990, 1992). Nevertheless, by providing a simple measure of the relative warming

---

GWPs for key greenhouse gases

Gas	Direct GWP	Indirect GWP
CO_2	1	None
CH_4	11	Positive
N_2O	270	Uncertain
CFC-11	3,400	Negative
CFC-12	7,100	Negative
HCFC-22	1,600	Negative
HFC-134a	1,200	None

Note that the GWP has two components, the direct and the indirect component. The direct component refers to the first order radiative effects of greenhouse gas increases, i.e., it ignores any radiative effects due to the products of chemical transformation in the atmosphere. These are accounted for by the indirect component of the GWP. Owing to incomplete understanding of the chemical processes involved, estimates of the indirect effects are more difficult to obtain and thus only the sign of the indirect effects is known with any confidence. The GWPs given here are calculated over a 100-year time horizon.
Source: IPCC, 1992

---

effects of key greenhouse gases, the IPCC hopes to provide policy-makers with a useful tool for evaluating the options that affect the emissions of various greenhouse gases.

Greenhouse Gas Indexes

Several attempts have been made to construct indexes that represent the contribution of individual nations to the global burden of greenhouse gas emissions. These rely on national inventories of greenhouse gas emissions of the type presented in Tables 1.3 and 1.5 for CO_2 and CH_4, respectively. Although these inventories provide the basis for assigning responsibility for emissions, they in themselves do not describe nations' relative responsibility for increases in atmospheric concentrations of greenhouse gases. This requires defining responsibility measures e.g., per capita emissions or emissions per unit land area, GDP or energy consumption) and an equivalence index for comparing the relative warming effect of the greenhouse gases (e.g., the IPCC's GWPs) which can be used to weight national emissions. Furthermore, criteria other than current emissions – such as historical emissions – may be more appropriate when judging national contributions (Subak, 1990).

Indexes which use this type of approach include that compiled by the Washington-based WRI. This index is based on national emission

estimates for CO_2, CH_4, CFC-11 and CFC-12; weighting factors employed include the IPCC's GWPs and the WRI's own weighting factor which is the product of the airborne fraction and the radiative forcing (WRI, 1992). Smith (1991) has proposed a "natural debt" index which invokes the notion of cumulative emissions. In effect this index represents the cumulative surviving human-induced emissions of greenhouse gases per capita; it takes into account the differing lifetimes and potencies of each gas. Other approaches involve allocating carbon emissions or quotas on a per capita basis (Fujii, 1990; Agarwal and Narain, 1991). Agarwal and Narain (1991), for example, advocate the use of an index which allocates the natural sinks for CO_2 and CH_4 to each nation in proportion to its population, and then calculates each country's "excess" emissions of each gas, i.e., that which is beyond its share of the global sinks.

The approaches outlined above all suffer from limitations of one kind or another and in recent years have been the subject of intense debate. Most significantly, the ranking of countries according to the various indexes varies markedly, casting a different light on the issue of accountability with each accounting system. These differences in national rankings limit the usefulness of the accountability concept in a political sense. If accountability is to be used as a basis for setting future restrictions on emissions, criteria that are perceived as fair by the majority of the world's nations will need to be found.

References

Agarwal, A. and Narain, S. 1991 *Global Warming in a Global World*, Centre for Science and Environment, Delhi.
Fujii, Y. 1990 CO_2: a balancing of accounts, *Options*, December issue, 10–13.
IPCC 1990 *Scientific Assessment of Climate Change*, Report prepared for the Intergovernmental Panel on Climate Change by Working Group I, Cambridge University Press, Cambridge.
IPCC 1992 *Climate Change 1992: The Supplementary Report to the IPCC Scientific Assessment*, Report prepared for Intergovernmental Panel on Climate Change by Working Group I, J. T. Houghton, B. A. Callander and S. K. Varney (Eds), Cambridge University Press, Cambridge.
Smith, K. R. 1991 Allocating responsibility for global warming: the Natural Debt Index, *Ambio*, **20**(2), 95–96.
Subak, S. 1990 *Accountability for Climate Change*, Research Memorandum, The Stockholm Environment Institute, Boston. Draft report.
World Resources Institute 1992 *World Resources 1992–93*, Oxford University Press, New York.

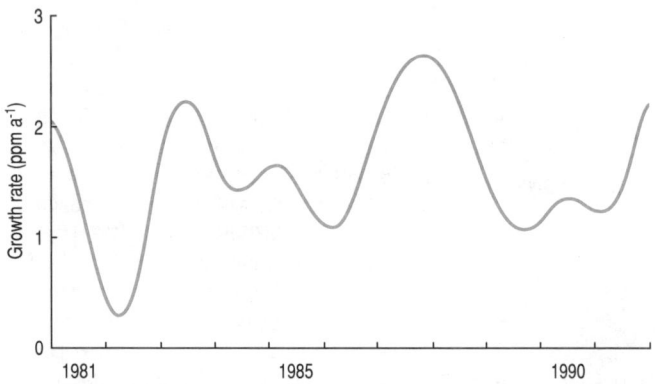

FIGURE 1.1 Globally averaged growth rates for atmospheric CO_2 as a function of time, 1981–1991 (based on NOAA/CMDL flask sampling network data)
Source: Figure provided by T. Conway, NOAA/CMDL, Boulder, USA

IPCC. This effort, which draws on the earlier work of the Organisation for Economic Co-operation and Development (OECD), aims to have developed an internationally-agreed methodology for calculating and reporting of national net emissions of greenhouse gases by the end of 1993 (IPCC, 1992a).

The Framework Convention on Climate Change, developed at the 1992 United Nations Conference on Environment and Development (UNCED), and discussed in more detail in Part 10: International Co-operation, includes a requirement for the preparation of inventories of greenhouse gas emissions and sinks. Until the framework convention comes into force, provision for assisting individual countries conducting this type of work has been made through a Global Environment Facility (GEF)-funded project on Sources and Sinks of Greenhouse Gases. Although the Convention is, in its present form, limited in scope, it nevertheless represents an important step towards achieving the goal of reduction in greenhouse gas emissions.

Carbon Dioxide Routine monitoring for CO_2, established in 1958 and expanded significantly during the 1980s, has provided a reasonably comprehensive picture of the temporal and spatial trends in atmospheric CO_2 in the recent past. Analysis of air trapped in polar ice cores has extended this information yet further backwards in time, to about 100,000 years BP (before present) (UNEP 1989, 1991). On the basis of this observational evidence, it is concluded that global average concentrations of atmospheric CO_2 have increased from pre-industrial levels of around 270 parts per million by volume (ppmv) to a 1991 value of 355 ppmv (IPCC, 1992b).

Of the ongoing monitoring programmes for atmospheric CO_2, that operated by the Climate Monitoring and Diagnostic Laboratory (CMDL) of the US National Oceanic and Atmospheric Administration (NOAA) is one of the most comprehensive; results obtained from the 32-site flask sampling network are summarized in Table 1.1. On the basis of these data a global average growth rate for atmospheric CO_2 over the 1981–1991 period of 1.55 ± 0.54 ppm per year is calculated (Conway, 1993). Trends in the growth rate of atmospheric CO_2 are illustrated in Figure 1.1. More extensive

monitoring data for atmospheric CO_2 and other trace gases, including those generated by NOAA/CMDL, are available from the Carbon Dioxide Information Analysis Center (CDIAC) in Oak Ridge, USA in both published form (CDIAC, 1991) and computerized format (CDIAC, 1992).

The observed accumulation of CO_2 in the atmosphere is a result of an excess of anthropogenic emissions over uptake by the CO_2 sinks – the terrestrial biosphere and the oceans. Manmade sources of CO_2 emissions comprise fossil fuel combustion, cement manufacture and land-use changes, primarily deforestation. Global and national CO_2 emissions estimates for fossil fuel combustion sources and cement manufacturing, as prepared by the CDIAC, are given in Tables 1.2 and 1.3, respectively. These data indicate that, globally, industrial CO_2 emissions have more than trebled since 1950 and are currently of the order of 6 Gt per year (as carbon).

Increases in fossil fuel CO_2 emissions on a regional basis are shown in Figure 1.2. Whereas there are signs of a stabilization in CO_2 emissions from the more developed regions (North America, western Europe), emissions are continuing to rise in some of the less developed world regions (South America, the Far East, centrally planned Asia and Africa). Nevertheless, the more developed regions (USA, Canada, Japan, Europe, former USSR, Australia and New Zealand) still account for approximately 70 per cent of the global total and have significantly higher per capita emission rates. Regionally averaged per capita emission rates vary from over 5 t a^{-1} (as carbon) in North America to less than 0.5 t a^{-1} (as carbon) in the Far East and Africa (CDIAC, 1992).

Relative to those from industrial sources, estimates of global and national CO_2 emissions from land-use changes are less precisely known and still subject to considerable uncertainty (IPCC, 1992b). Deforestation in the tropical regions is, however, thought to account for much of the present-day net biotic flux of carbon to the atmosphere.

Several groups have attempted to develop inventories for CO_2 emissions from land-use changes, at both the global and national levels. Two such inventories for national emissions, which differ in their methods of derivation and selection of raw data sets and "emission factors", are given in Table 1.3.

The IPCC has recently reviewed available estimates of CO_2 releases at the global level and has concluded that a value of 1.6 ± 1.0 Gt a^{-1} (as carbon) represents the best estimate of emissions from land-use changes for the decade 1980–1989. Although this value represents an average for the 1980s it is, however, considered probable that annual fluxes of CO_2 increased in the early part of the decade, peaked in the middle years and subsequently declined towards 1990. This pattern reflects the most likely changes in tropical deforestation rates during this time (IPCC, 1992b). It is noted that both the national emission inventories presented in Table 1.3 sum to global totals which lie within the IPCC range of 1.6 ± 1.0 Gt a^{-1} (as carbon).

During the past few years some progress has been made in attempting to balance the carbon budget, i.e., to account for the apparent discrepancy between net emissions of CO_2 (around 7 Gt a^{-1} as carbon) and the known sinks – the atmosphere (in which about 3.5 Gt of carbon accumulate each year) and the oceans (which absorb approximately 2 Gt of carbon annually).

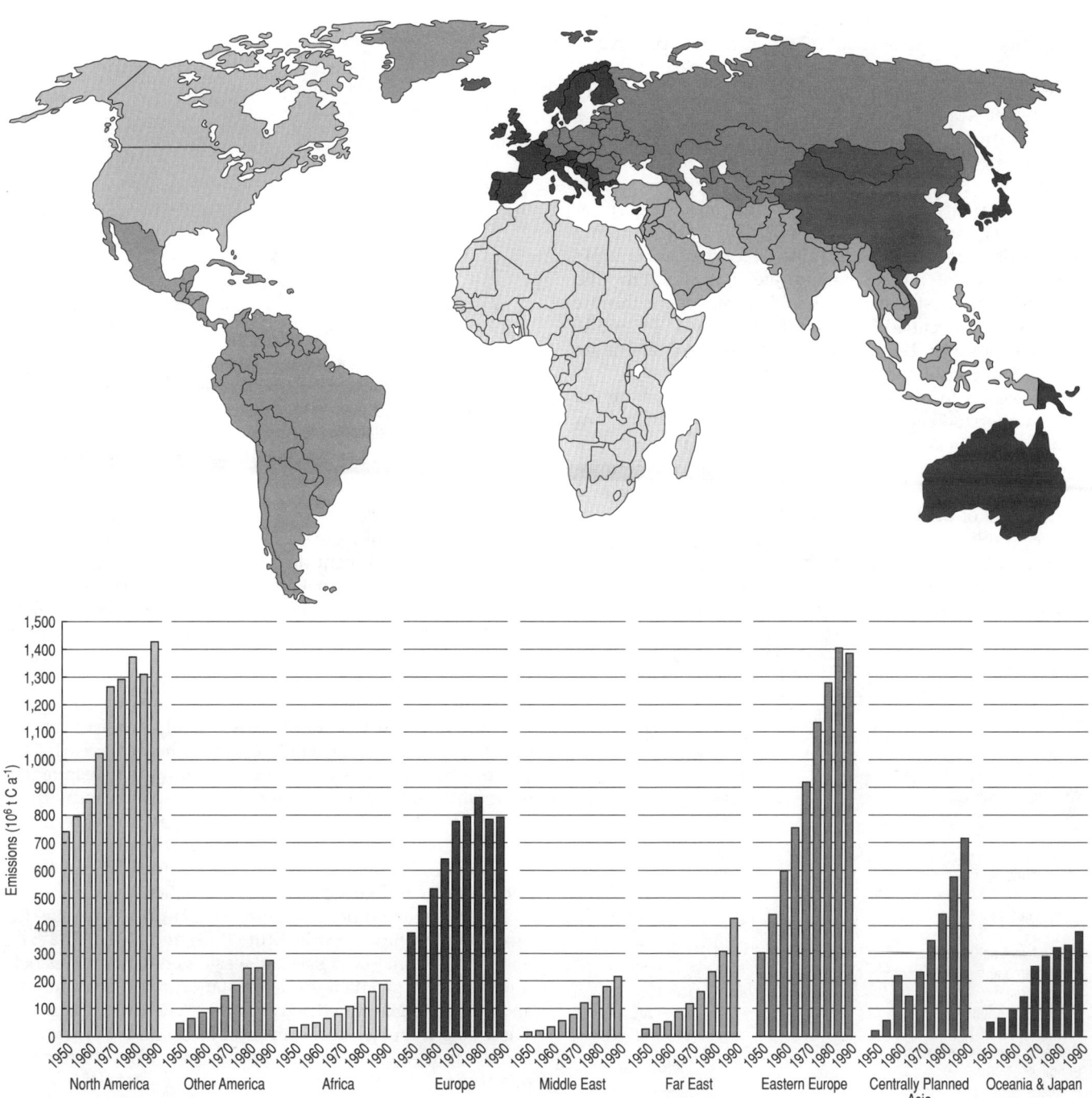

FIGURE 1.2 Trends in emissions of CO_2 from industrial sources in major world regions, 1950–1990
Source: Data from Table 1.3

The terrestrial biosphere, through forest regrowth, productivity enhancement due to atmospheric CO_2 increases, and nitrogen fertilization from atmospheric deposition, has been identified as far the most likely candidate for the so-called missing carbon sink (some 1.6 Gt of carbon). However, it has been difficult to quantify the magnitude of terrestrial sinks, particularly the last two, in terms of carbon storage. New analyses of forest timber inventories support the notion that forest area and timber volume have been expanding in recent decades in the Northern Hemisphere temperate latitudes (see Part 3: Natural Resources). For example, it has been estimated that, despite potentially adverse impacts of air pollution, biomass has actually increased in European forests during the 1970s and 1980s. The annual average rate of carbon storage during this time was estimated to be in the range 0.09–0.12 Gt (as carbon), or just under 10 per cent of the "missing" carbon (Kauppi et al., 1992). In a similar type of study, Sedjo (1992) estimated that the Northern Hemisphere temperate forests

(Europe, former USSR, USA and Canada) are currently sequestering carbon at a rate of about 0.7 Gt (as carbon) per year.

Methane   As in the case of CO_2, modern measurements and measurements of air trapped in ice cores have shown convincingly that atmospheric levels of CH_4 have risen considerably since the onset of industrial times. Present day concentrations average some 1.7 ppmv, more than double the pre-industrial value of 0.8 ppmv (IPCC, 1992b). Annual mean concentrations of CH_4 recorded at NOAA/CMDL flask sampling network sites are presented in Table 1.4 for the period 1984–1991. Analysis of these data indicate that although CH_4 concentrations are still increasing in the atmosphere, the rate of increase has declined from about 13.5 ppbv a^{-1} in 1983 to about 9.5 ppbv a^{-1} in 1991 (Figure 1.3). These observations are discussed in considerable detail by Steele et al. (1992).

The reasons for this decline in CH_4 growth rates are not clearly understood at this time. The most likely explanations are a decrease in emission rates from either natural or man-made sources, an increase in CH_4 loss rate due to an increase in tropospheric hydroxyl radical (OH) concentrations, or a combination of the two (IPCC, 1992b).

Since publication of the previous edition of this report, a number of revisions to the global CH_4 budget have been made. The most recent estimates of the individual source strengths and sinks for CH_4 are thus as listed below:

Sources/sinks	Best estimate (10^6 t a^{-1})	Range (10^6 t a^{-1})
SOURCES	515	
Natural	155	
Wetlands	115	(100–200)
Termites[a]	20	(10–50)
Oceans	10	(5–20)
Fresh water	5	(1–25)
CH_4 hydrate	5	(0–5)
Anthropogenic	360	
Coal mining, natural gas and petrochemical industries[a]	100	(70–120)
Rice paddies[a]	60	(20–150)
Enteric fermentation	80	(65–100)
Animal wastes[a]	25	(20–30)
Domestic sewage treatment[a]	25	?
Landfills[a]	30	(20–70)
Biomass burning	40	(20–80)
SINKS	500	
Atmospheric (tropospheric and stratospheric) removal[a]	470	(420–520)
Removal by soils	30	(15–45)
ATMOSPHERIC INCREASE	32	(28–37)

[a] Indicates a recently revised estimate.
Source: IPCC, 1992b

With respect to emissions, new information has revised downwards the CH_4 emissions estimate from rice paddies,

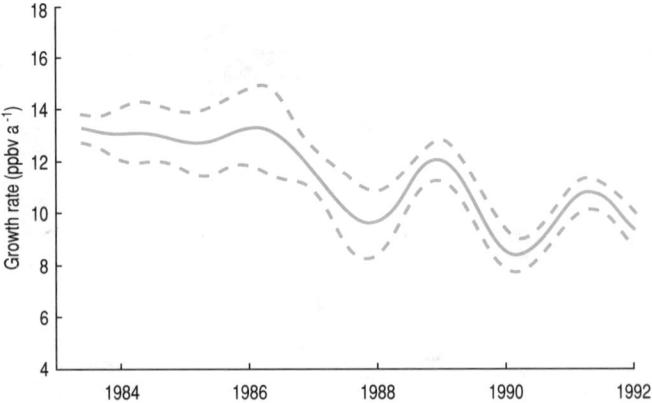

FIGURE 1.3   Trends in the growth rate of atmospheric CH_4, 1983–1991 (the solid line represents variations in the globally averaged CH_4 mixing ratios and the dashed lines reflect the level of uncertainty in this result as determined by a non-parametric statistical technique)
Source: Figure provided by E. Dlugokencky and P. Lang, NOAA/CMDL, Boulder, USA

landfills and termites. Two new emission sources have been quantified, i.e., emissions from animal wastes and from domestic sewage treatment processes. Annual emissions from industrial sources, i.e., coal mining (25–47 x 10^6 t), natural gas production, transmission, distribution and use (25–42 x 10^6 t) and the oil industry (5–30 x 10^6 t) have been revised slightly upwards (IPCC, 1992b). Furthermore, studies of the C^{14} content of atmospheric CH_4 suggest that about 20 per cent of the total annual CH_4 emissions (i.e., 100 x 10^6 t) are from fossil fuel carbon sources, a result which is broadly consistent with independent estimates of emissions from these sources (IPCC, 1992b). The table presented here also shows that despite the uncertainty in the emission rates from most of the individual sources detailed above, there is nevertheless good agreement between the sum of the estimates for individual sources and the global CH_4 emission rate as deduced from the size of the known sinks and atmospheric accumulation.

A national emissions inventory for CH_4, developed by the Stockholm Environment Institute (SEI) and based on available international data sets for waste generation, livestock numbers, rice production, deforestation rates, coal mining and other energy-related activities, is presented as Table 1.5. These data are subject to the uncertainties outlined previously and are thus likely to be less accurate than inventories compiled by individual countries. However, for most of the source categories, the world totals are within the range of the 1992 IPCC global estimates. According to the SEI inventory, rice cultivation and livestock collectively account for almost 60 per cent of the man-made flux of CH_4 to the atmosphere. Energy-related sources contribute the bulk of the remainder (23 per cent), with just three countries, China, USA and the former USSR accounting for 60 per cent of the emissions from this source (Subak et al., 1992).

Halocarbons   Routine global-scale monitoring programmes for a number of the more abundant halocarbons, including CFC-11 and CFC-12, have operated since the late 1970s. Recent trends in the atmospheric abundance of such gases are thus relatively well documented and have been extensively reported in previous editions of this report (UNEP 1989, 1991). Estimates of present-day atmospheric concentrations and growth rates for a range of halocarbons are given below (data refer to the situation in 1989 as assessed by the WMO and approximate to global means):

Trace gas		Annual mean concentration (pptv)	Increase (pptv a⁻¹)
CCl_3F	(CFC-11)	255–268	9.3–10.1
CCl_2F_2	(CFC-12)	453	16.9–18.2
$CClF_3$	(CFC-113)	64	1.0
CCl_4		107	1–1.5
CH_3CCl_3		135	4.8–5.1
$CBrClF_2$	(H-1211)	1.6–2.5	0.2–0.4
$CBrF_3$	(H-1301)	1.8–3.5	0.4–0.7

Source: WMO, 1992

Although the available evidence suggests that the atmospheric concentrations of the halocarbons are continuing to grow in the background troposphere of both hemispheres, a slow down in the rate of growth of not only the most abundant CFCs, CFC-11 and CFC-12, but also the halons, H-1301 and H-1211, has been recently reported. These findings are based on measurements made at selected NOAA/CMDL sites and have been discussed in detail by Butler et al. (1992) and Elkins et al. (1993). The reduction in rate of increase of atmospheric abundances of CFC-11 and CFC-12, which is apparent in the observational record since 1989, is illustrated in Figure 1.4. Calculated growth rates for CFC-11 and CFC-12 at the end of 1991 are 4.1 ± 1 pptv a⁻¹ and 14 ± 2 pptv a⁻¹, respectively which compare with average growth rates of about 10 pptv a⁻¹ for CFC-11, and around 18 pptv a⁻¹ for CFC-12, during the 1980s. It is further calculated that if the observed slow-down in growth rates continues at the 1990–1991 levels, global atmospheric CFC-11 and CFC-12 mixing ratios will reach a maximum before the turn of the century and thereafter begin to decline (Elkins et al., 1993).

The observational record for the halons H-1301 and H-1211 is considerably shorter than that for the CFCs – only six years. Nevertheless, the analysis of available data from both fixed monitoring sites and ship cruises clearly points to a decrease in the more recent years in the atmospheric growth rate for H-1301 and to a lesser extent for H-1211. In the case of H-1301, annual average growth rates in the period 1989–1992 of 0.14 ± 0.3 pptv a⁻¹ are approximately half those calculated for the period 1987–1989 (i.e., 0.29 ± 0.4 pptv a⁻¹) (Butler et al., 1992).

In the past, atmospheric concentrations of the most widely used of the hydrochlorofluorocarbons (HCFCs), namely chlorodifluoromethane or HCFC-22 (a widely-used replacement for some of the CFCs), have been subject to some uncertainty as a consequence of the discrepancies in the results obtained by different measurement techniques for this particular gas. Concentrations in Northern Hemisphere air determined by ground-based chromatographic measurements have been between 10 and 40 per cent higher than those

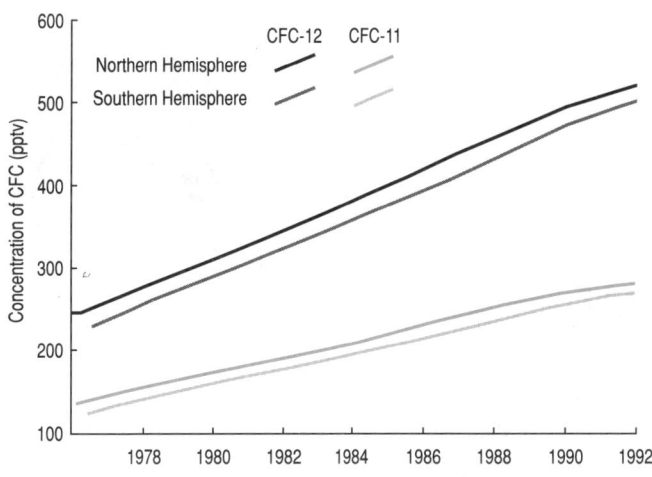

FIGURE 1.4 Trends in mean atmospheric concentrations of CFC-11 and CFC-12 in the Northern and Southern Hemispheres, 1978–1992
Source: After Elkins et al. (1993)

measured within the troposphere by other techniques.

According to new analysis of surface-based measurements, a global mean concentration for HCFC-22 of 102 pptv in 1992 is derived (Montzka et al., 1993). This result is believed to be accurate to within ± 2.5 per cent, which represents a substantial improvement compared with previous observations. Moreover, these new results are substantially lower than those reported previously for ground-based chromatographic measurements but are consistent with calculated estimates of atmospheric HCFC-22 concentrations based on emission data and estimated lifetimes (CMDL, 1992). A re-analysis of archived air samples suggests a mean growth rate for atmospheric HCFC-22 of 7.3 ± 0.3 per cent per year over the past five years (Montzka et al., 1993).

These recently reported trends in atmospheric abundances of the major CFCs (CFC-11 and CFC-12), the halons (H-1301, H-1211) and HCFC-22, reflect recent changes in the use patterns of these halocarbons. The world-wide production and consumption of CFC-11 and CFC-12 – compounds which have traditionally found widespread application as aerosols, refrigerants and foam-blowing agents – have fallen dramatically relative to 1986. This is attributed to the effect of the Montreal Protocol, first adopted in 1987, and which committed signatory nations to a reduction in their consumption of eight controlled substances – including CFC-11 and CFC-12 – by 50 per cent of their 1986 level by 1 July 1999. Continuing fears for the stability of the stratospheric ozone layer (see next section) have led to a progressive tightening of the controls for CFC-11 and CFC-12; in 1990 industrialized countries agreed to phase out the controlled CFCs entirely by the year 2000 (with a 10-year time lag for developing nations) and in November 1992, the deadline for the complete ban on CFC-11 and CFC-12 was brought forward to January 1996 (Kirwin, 1993). The status of the ratification of the Montreal Protocol and its subsequent amendments is described in Part 10: International Co-operation.

Under the terms of the Montreal Protocol, parties are required to submit data concerning their levels of consumption,

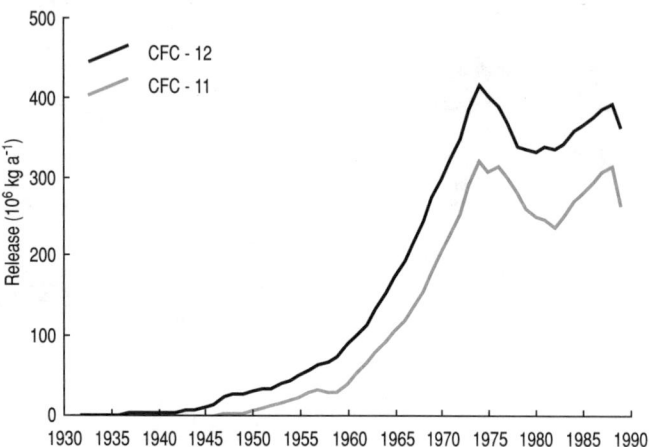

FIGURE 1.5 Trends in releases of CFC-11 and CFC-12 to the atmosphere, 1931–1989
Source: CDIAC, 1991

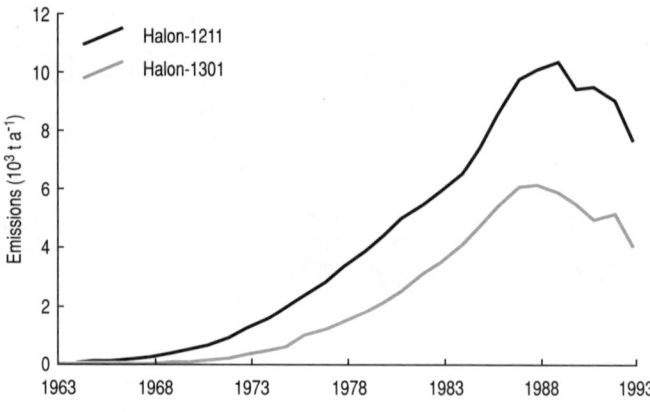

FIGURE 1.6 Estimated global emissions of H-1211 and H-1301, 1960–1990
Source: Based on McCulloch, 1992 with revisions by the Ozone Secretariat, UNEP, Nairobi.

production and trade of the controlled substances to UNEP. Available consumption data (i.e., as of August 1993) for the eight substances originally included in the Protocol – five CFCs and three halons – are presented in Table 1.6, for the reporting years 1986, 1990 and 1991.

Production, consumption and trade data for selected CFCs are also reported by member companies of the Chemical Manufacturers Association Fluorocarbon Program Panel (FPP). From these data the FPP calculates release estimates to the atmosphere for selected CFCs, broken down by end-use category. Trends in CFC-11 and CFC-12 releases, as calculated by the FPP, are illustrated in Figure 1.5 for the period 1931–1989. These data do not take account of CFC production in India, China, eastern Europe or the former USSR. However, it is considered that these omissions do not represent a major source of uncertainty in the release estimates as these countries are only minor sources of CFCs at the present time (CDIAC, 1991). Trend data indicate an exponential increase in the use and release of CFCs during the 1960s and early 1970s. Thereafter, CFC production and release declined following a US ban on the use of CFCs in aerosols. The trend reversed in the 1980s as CFCs continued to be used in aerosols in countries not participating in the ban and as use of CFCs in other applications grew rapidly (CDIAC, 1991). A sharp decline in emissions of these compounds in 1989, which is coincident with the global slow-down of atmospheric growth rates mentioned earlier, is also noteworthy (Figure 1.5).

The recent reduction in the atmospheric growth rates of H-1211 and H-1301 is likewise paralleled by a marked drop in the production, consumption and release of these chemicals. Since the early 1960s the halons, especially H-1211 and H-1301, have been widely used as fire extinguishants. However, in view of their high ozone-depleting potentials (ODPs) (see Box 1.2), the halons have also been singled out for control under the Montreal Protocol. According to the latest amendment to the Protocol, production and consumption of H-1211, H-1301 and H-2402 are scheduled to be phased out by 1994.

Using industry data provided by individual manufacturers, McCulloch (1992) has derived global estimates for the

production, consumption and emission to the atmosphere of H-1211 and H-1301. These data indicate that annual production of both H-1211 and H-1301 peaked in 1988 at 20.2×10^3 t and 12.6×10^3 t, respectively. By 1990, production of H-1211 had already fallen by 26 per cent (to 14.9×10^3 t) and production of H-1301 by 27 per cent (to 9.1×10^3 t). It is estimated that to date 127.9×10^3 of H-1211 and 70.2×10^3 t of H-1301 have been released to the atmosphere. These quantities, which are broadly consistent with measured atmospheric concentrations, represent about 74 per cent of the cumulative production of H-1211 and around 58 per cent of that for H-1301 (the remainder is contained in existing equipment). Trends in emissions of both halons are plotted in Figure 1.6.

As one of the major replacement chemicals for the CFCs for refrigeration, air conditioning and foam-blowing purposes, the use of HCFC-22 has increased considerably in the past few years. Modelling studies have, however, suggested that HCFC-22 could have a significant impact on the ozone layer if its use is allowed to continue unabated (see also Box 1.2). Consequently in 1990 HCFC-22 was added to the list of substances controlled by the Montreal Protocol; under the terms of the 1992 amendment a timetable for the phase out of HCFC-22 by 2030 was established (Kirwin, 1993). Recent increases in the production and release of HCFC-22 are discussed by Midgley and Fisher (1993). The development of CFC replacements and other technologies aimed at reducing the use of controlled substances is discussed further in Part 8: Wastes and Waste Management.

References

Butler, J. H., Elkins, J. W., Hall, B. D., Cummings, S. O. and Montzka, S. A. 1992 A decrease in the growth rates of atmospheric halon concentrations, *Nature* (London), **359**, 403–405.

CDIAC 1991 *Trends '91: A Compendium of Data on Global Change*, Carbon Dioxide Information Analysis Center, Oak Ridge National Laboratory, Oak Ridge.

CDIAC 1992 *Estimates of CO₂ Emissions from Fossil Fuel Burning and Cement Manufacturing, Based on the United Nations Energy Statistics and the US Bureau of Mines Cement Manufacturing Data*, Numeric Data

BOX 1.2 The Ozone Depleting Potential Index

In order to assist the policy-making process, the stratospheric research community has developed a number of measures which reflect the relative impact of selected halocarbons on the ozone layer. These comprise:
a) the chlorine loading potential (CLP), and
b) the ozone depleting potential (ODP).

Conceptually both of these measures are similar to the Global Warming Potential (GWP) that has been derived to represent the relative impact of the climatically important trace gases (see Box 1.1).

Of the two, the CLP is the simplest; it represents the amount of total chlorine delivered from the troposphere to the stratosphere due to the emission of a given halocarbon, relative to that from a reference molecule (generally CFC-11). Calculation of steady-state CLPs relies on two-dimensional atmospheric models which are used to estimate the atmospheric lifetime of the source gas in question. Best estimates of CLPs for selected halocarbons are provided here.

The second measure, the ODP, is the more widely known and is used as a guide for selecting halocarbons for control by the Montreal Protocol. The steady-state ODP is simply defined as the amount of O_3 destroyed by emission of a gas over its entire atmospheric life-time relative to that due to the emission of the same mass of CFC-11. Thus the ODP reflects not only the transport of halocarbons to the stratosphere but also the breakdown of the halocarbon in the stratosphere (i.e., the process which releases reactive chlorine and bromine atoms) and subsequent participation in ozone-depleting chemical reactions.

Traditionally ODPs have been calculated using two-dimensional numerical models. Their reliability is thus constrained by the limitations of such models with respect to their treatment of atmospheric transport and chemical processes. More recently, semi-empirical methods have been used to estimate steady-state ODPs. This approach relies on atmospheric measurements of the source gases and observations of O_3 loss, and is considered to provide more realistic ODPs, especially for the polar regions. Best estimates of the steady-state ODP for a range of halocarbons, as recommended by the scientific assessment panel are also given in the accompanying table.

Although the steady-state ODP index does not provide information about the absolute amount of O_3 destruction, it nevertheless provides a useful

Chlorine loading potentials (CLPs) and ozone-depleting potentials (ODPs) for selected halocarbons

Compound	CLPs	Steady-state ODP	Time-varying ODP[a]
CFCs			
CFC-11	1.0	1.0	1.0
CFC-12	1.60	≈1.0	
CFC-113	1.47	1.07	0.59
CFC-114	2.14	≈0.8	
CFC-115	2.96	≈0.5	
HCFCs			
HCFC-22	0.15	0.055	0.14
HCFC-142b	0.19	0.065	0.14
HCFC-225ca	0.02	0.025	0.10
HCFC-225cb	0.07	0.033	0.11
Halons			
H-1301		≈16	10.5
H-1211		≈4	9.0
H-2402		≈7	11.0

[a] Time-dependent ODPs are based on a 20-year time horizon and employ CFC-11 as the reference gas.
Sources: WMO, 1992; Solomon and Albritton, 1992

guide as to which halocarbons are likely to be less environmentally damaging on long-term time scales, i.e., of centuries or more. On the basis of the fact that the ODPs for the HCFCs are generally lower than those for the CFCs and halons, a number of these compounds (e.g., HCFC-22, HCFC-142, HCFC-225aa, HCFC-225cb) have found increasing application as replacements for some of the CFCs and halons currently controlled by the Montreal Protocol. However, a number of researchers have recently highlighted a possible shortcoming in this approach. Whereas the CFCs and halons typically have very long atmospheric lifetimes, the HCFCs (due to tropospheric breakdown by reaction with hydroxyl radicals) are relatively short-lived. Thus while the steady-state ODP for HCFCs may suggest a fairly modest impact on O_3 concentrations on longer time scales, ODPs calculated as a function of time indicate that the HCFCs have relatively large ODPs for shorter time scales. Solomon and Albritton (1992) have calculated time-dependent ODPs for a range of halocarbons and conclude that some of the

proposed replacement HCFCs may induce significant O_3 destruction in the short-term (i.e., within the next 20 years). It is further concluded that HCFC-141b and CH_3Br (methylbromide) have some of the largest relative impacts on the O_3 layer in both the long and short term. These molecules have only recently been added to the list of substances to be phased out under the terms of the Montreal Protocol amendments.

For a more detailed account of the derivation of CLPs and the ODPs, the reader is referred to the most recent WMO scientific assessment of the depletion of stratospheric ozone (WMO, 1992).

References

Solomon, S. and Albritton, D. L. 1992 Time-dependent ozone depletion potentials for short- and long-term forecasts, *Nature* (London), **357**, 33–37.
WMO 1992 *Scientific Assessment of Ozone Depletion: 1991*, Global Ozone Research and Monitoring Project, Report No. 25, World Meteorological Organization, Geneva.

Package – 030/R4, Carbon Dioxide Information Analysis Center, Oak Ridge National Laboratory, Oak Ridge.

CMDL 1992 *Summary Report 1991*, Climate Monitoring and Diagnostics Laboratory, National Oceanic and Atmospheric Administration, Boulder, USA.

Conway, T. 1993 Personal communication, National Oceanic and Atmospheric Administration, Climate Monitoring and Diagnostics Laboratory, Boulder, USA.

Elkins, J. W., Thompson, T. M., Swanson, T. H., Butler, J. H., Hall, B. D., Cummings, S. O., Fisher, D. A. and Raffo, A. G. 1993 Slowdown in the growth rates of atmospheric chlorofluorocarbons 11 and 12, *Nature* (London). In press.

IPCC 1992a *Progress Report on the IPCC/OECD Programme on the Development of a Methodology for National Inventories of Net Greenhouse Gas Emissions*, Report submitted to the Eighth session of the Intergovernmental Panel on Climate Change, 11–13 November 1992, Harare, Zimbabwe.

IPCC 1992b *Climate Change 1992: The Supplementary Report to the IPCC Scientific Assessment*, Report prepared for the Intergovernmental Panel on Climate Change by Working Group I, J. T. Houghton, B. A. Callender, S. K. Varney (Eds), Cambridge University Press, Cambridge.

Kauppi, P. E., Mielikainen, K. and Kuusela, K. 1992 Biomass and carbon budget of European forests, 1971 to 1990, *Science*, **256**, 70–74.

Kirwin, J. 1993 Ozone update, *Our Planet*, **5**(1), 14–18.

McCulloch, A. 1992 Global production and emissions of bromochlorodifluoromethane and bromotrifluoromethane (halons 1211 and 1310), *Atmospheric Environment*, **24A**(7), 1325–1329.

Midgley, P. M. and Fisher, D. A. 1993 The production and release to the atmosphere of chloro-difluoromethane (HCFC-22), *Atmospheric Environment*. In press.

Montzka, S. A., Myers, R. C., Butler, J. H., Cummings, S. C. and Elkins, J. W. 1993 Global tropospheric distribution and calibration scale of HCFC-22, *Geophysical Research Letters*. In press.

Sedjo, R. A. 1992 Temperate forest ecosystems in the global carbon cycle, *Ambio*, **21**(4), 274–277.

Steele, L. P., Dlugokencky, E. J., Lang, P. M., Tans, P. P., Martin, R. C. and Masarie, K. A. 1992 Slowing down of the global accumulation of atmospheric methane during the 1980s, *Nature* (London), **358**, 313–316.

Subak, S., Raskin, P. and von Hippel, D. 1992 *National Greenhouse Gas Accounts: Current Anthropogenic Sources and Sinks*, Stockholm Environment Institute, Boston.

UNEP 1989 *United Nations Environment Programme Environmental Data Report 1989/90*, Basil Blackwell, Oxford.

UNEP 1991 *United Nations Environment Programme Environmental Data Report 1991/92*, Basil Blackwell, Oxford.

WMO 1990 *The Global Atmosphere Watch*, Fact Sheet No. 3, World Meteorological Organization, Geneva.

WMO 1992 *Scientific Assessment of Ozone Depletion: 1991*, Global Ozone Research and Monitoring Project, Report No.25, World Meteorological Organization, Geneva.

Stratospheric Ozone

The 1985 discovery of the large springtime loss of lower-stratospheric ozone (O_3) over Antarctica focused the attention of the international community on the possibility that rising atmospheric concentrations of chlorine and bromine-containing chemicals may be causing potentially serious damage to the earth's ultraviolet (UV) shield, the stratospheric O_3 layer. Since that time, the growing concern for the stability of the stratospheric O_3 layer on a global scale has culminated in the adoption of the Montreal Protocol to control the production and consumption of the O_3-depleting substances. Trends in production/consumption and atmospheric abundances of the CFCs and the halons are outlined in the previous section (see Greenhouse Gases); this section outlines the observational evidence for depletion of the O_3 layer and for increased incidence of UV-radiation.

Available scientific evidence for depletion of stratospheric O_3 has been periodically reviewed by the WMO (WMO 1985; 1989; 1992). These scientific assessments, together with parallel works on the environmental impacts of ozone depletion (UNEP, 1991) and on the technology and economic implications (Technology and Economic Assessment Panel, 1991), collectively form the information base which supports the formulation of the Montreal Protocol and its subsequent amendments. The concept of the ozone-depleting potential (ODP), developed during the assessment process as a measure of the relative impact of various halocarbons on stratospheric O_3, has been shown to be a useful policy-making tool. The derivation of ODPs is outlined in Box 1.2.

As the main findings of the 1985 and 1989 scientific assessments were covered in earlier editions of this report series (UNEP, 1989), the discussion here will focus on the advances in the understanding of the impact of human activities on stratospheric O_3 levels made in more recent years, i.e., those made since publication of the 1989 review, and which are summarized in the 1991 assessment (WMO, 1992).

Current capability for routine monitoring of O_3 concentrations in the atmosphere comprises a ground-based network of Dobson spectrophotometers and satellite-based instrumentation, namely the Total Ozone Monitoring Spectrometer (TOMS) on board Nimbus-7. Both the Dobson spectrophotometer and the TOMS instrument measure total column ozone, i.e., the amount of O_3 contained in a vertical column of air of basal area 1 cm^2 at standard temperature and pressure. Concentrations are expressed in units of milli-atmosphere centimetre (matm.cm), otherwise known as Dobson units (DU). Typical total O_3 amounts are in the range 230–500 DU, with a global average of around 300 DU. Ozone is not distributed uniformly throughout the atmospheric column; about 90 per cent is found in the stratosphere with maximum concentrations occurring at altitudes of 25 km over the equator and at 15 km of the poles. Thus measurements of total column O_3 provide a good indication of changes in stratospheric O_3.

The network of Dobson spectrophotometers, operated by the WMO as part of their Global Ozone Observing System (GO$_3$OS), currently comprises some 140 stations world-wide. Measurements of total O_3, recorded as daily means, are reported to the WMO World Ozone Data Centre in Toronto, Canada and subsequently published on a regular basis as 'Ozone Data for the World'. Records for some stations extend back to the late 1950s and thus provide the longest records of atmospheric O_3 available to date. The main deficiency of the existing Dobson network is its limited geographical coverage (it is biased towards the middle latitudes of the Northern Hemisphere); Dobson data are thus best suited to characterizing O_3 trends for specific regions of the globe. Assessment of trends in total O_3 over the period 1979–1991 at individual Dobson stations, reveals predominantly downwards trends (WMO, 1992).

In contrast, satellite-derived observations of total O_3 provide global coverage. However, the length of the available records (approximately 14 years) is presently inadequate for the derivation of long-term trends. In addition, data sets obtained by satellites suffer from problems of instrument drift arising from a gradual degradation of the reflective properties of the TOMS's diffuser plate. A correction for instrument drift, estimated by comparison with the ground-based Dobson data, is therefore required. Regionally averaged trends in total O_3, derived from satellite and ground-based instrumentation, are compared in the following table:

	Trend 1979–1991 (% per decade)		
	Dec.–Mar.	May–Aug.	Sept.–Nov.
Satellite-derived			
45°N	−5.6 ± 3.5	−2.9 ± 2.1	−1.7 ± 1.9
Equator	+0.3 ± 4.5	+0.1 ± 5.2	+0.3 ± 5.0
45°S	−5.2 ± 1.5	−6.2 ± 3.0	−4.4 ± 3.2
Ground-based			
26°N–64°N	−4.7 ± 0.9	−3.3 ± 1.2	−1.2 ± 1.6

Source: WMO, 1992

These data indicate that, between 65°N and 65°S, total column O_3 abundance has decreased, on average, by around 3 per cent during the 1980s; this rate of O_3 loss is thought to be significantly greater than that observed in the 1970s. Ozone losses are not, however, distributed uniformly throughout this latitude band; the greatest losses occur over Antarctica (i.e., the now well-known phenomenon of the "ozone hole"). Whereas no significant changes in O_3 levels are indicated for the tropical regions, statistically significant depletion is apparent throughout the year in the mid- to high latitudes of both hemispheres. In the Northern Hemisphere, the largest depletion appears to take place over mid-latitudes, i.e., between 40°N–50°N during the winter months (WMO, 1992).

Measurements of the vertical distribution of O_3 through the atmosphere, such as those obtained by ground-based Dobson Umkehr, ballonsondes and more recently from the Stratospheric Aerosol and Gas Experiment (SAGE) satellite instruments, have provided confirmation that the decreases in total column O_3 observed during the last few decades are in fact largely due to concentration decreases in the lower stratosphere (McCormick et al., 1992; WMO, 1992).

The springtime reduction of O_3 over Antarctica is now well established observationally and the processes involved relatively well understood (Solomon, 1990). Indeed, scientific investigations conducted in the last few years only strengthen the view that polar O_3 losses are the result of heterogeneous chemical processes involving reactive forms of chlorine and bromine, acting in combination with atmospheric circulation conditions unique to the Antarctic Austral Spring (i.e., a closed atmospheric circulation vortex which develops in the spring and effectively isolates Antarctic air masses from south-north meridional exchanges) (WMO, 1992).

In more recent years (1990–1993), "holes" have continued

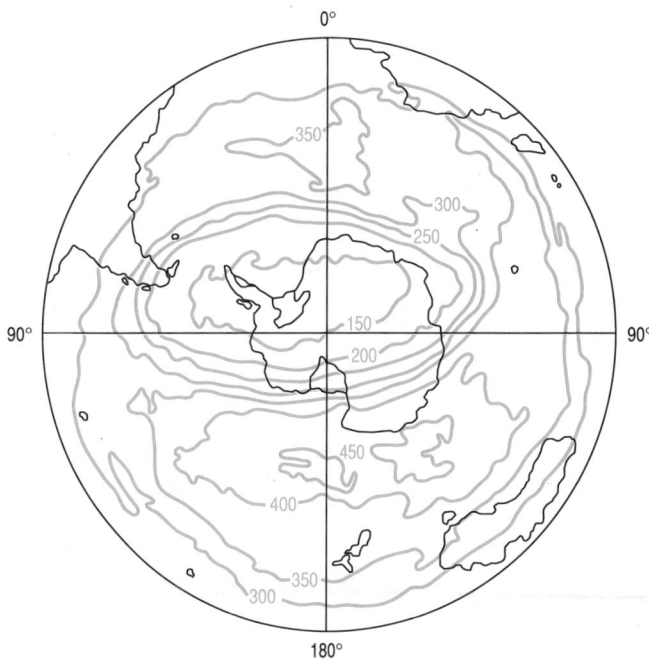

FIGURE 1.7 Ozone concentration isopleths over the southern polar regions recorded on 5 October 1992 by satellite instrumentation (Dobson units)
Source: WMO, 1993a

to occur with increasing depth and spatial extent during the Antarctic spring. In 1992, the total amount of O_3 over Antarctica fell to below 200 DU (i.e., a 30 per cent loss relative to normal values) at the end of August, two weeks earlier than in previous years. Record lows of O_3 were recorded at several stations, including Halley Bay (111 DU) on 2 October, and the South Pole (105 DU) on 11 October. Moreover, during October values fell below 170 DU over South America (Tierra del Fuego); this is the first time depletion has been recorded over a permanently populated region (WMO, 1993a). The spatial extent of the O_3 hole in 1992 is illustrated in Figure 1.7. Preliminary data for 1993 suggest that record-breaking O_3 lows have again occurred over Antarctica (Mendonca, 1993).

In the wake of the eruption of Mount Pinatubo in June 1991, renewed consideration has been given to the possibility that volcanic eruptions can influence stratospheric O_3 abundance. Hofmann et al. (1992), for example, attribute the unexpected high altitude O_3 depletion observed during the 1991 Antarctic hole to sulphate aerosols derived from the eruption of Mount Hudson, Chile in August 1991. On the basis of this observation it is postulated that sulphate aerosols injected into the stratosphere as a result of the Mount Pinatubo eruption will be trapped into the vortex which develops over Antarctica during the spring and will thus have the capacity markedly to affect the morphology of the O_3 hole in the years immediately following the eruption (Hofmann et al., 1992).

The middle and higher latitudes of the Northern Hemisphere have similarly experienced record low values of total O_3. In 1993, the January average total O_3 values were 18.2 per cent below the average for the 45°–65°N latitude band. At the end of January, during February and part of March, total O_3

over North America and parts of Europe and Siberia fell to more than 20 per cent below normal, a deviation greater than that in 1992, which itself had been a record (WMO, 1993b). Although these data set new minimum records for this part of the world, total O_3 seldom fell below 240 DU; in contrast, levels in Antarctica have fallen to near 100 DU. The reason for the disparity between the hemispheres in terms of O_3 hole formation is the differences in meteorological conditions between the North and South Poles; the more frequent exchanges of air masses in the Northern Hemisphere compared with that in the south limit the severity of O_3 depletion, despite the fact that the concentration of chlorine appears to be similar over both polar regions (WMO, 1993b).

It has been recognized for some time that changes in the vertical distribution of O_3, i.e., depletion in the stratosphere and increases in the troposphere (see next section) are likely to have climatic implications (IPCC, 1992, Wang et al., 1993). Indeed one of the most significant findings to emerge from research of recent years, is the suggestion that stratospheric O_3 depletion may actually be offsetting global warming. This situation arises because O_3 depletion tends to cool the stratosphere which, in turn, decreases the radiative-forcing of the surface-troposphere system due to increases of greenhouse gases. Recent attempts to quantify the magnitude of this effect suggest that the effect of the stratospheric O_3 depletions observed in the 1980s could be sufficient, when averaged over the globe, to offset the warming induced by the increases in atmospheric concentrations of the CFCs (IPCC, 1992). The IPCC stresses the preliminary nature of these findings.

Although the science behind the theory that reduced levels of O_3 in the stratosphere will increase the amount of incident solar UV radiation has been well established for some time, until very recently there has been very little observational evidence to support this assertion. This is partly because of the lack of direct measurements of incident UV on adequate spatial and temporal time scales for trend analysis. It is also possible that increases in tropospheric O_3 and aerosol loadings (see below) may be offsetting the consequences of stratospheric O_3 depletion for UV radiation (UNEP, 1991; WMO, 1992).

Of the limited UV data sets available, that obtained by the "global" network of Robertson-Berger (R-B) meters is the most comprehensive. This network comprises 36 stations, 25 of which are located in the USA. Analyses of records from individual stations or groups of stations have been reported in the scientific literature. Although some of these analyses suggest that increases in the amount of UV incident on the earth's surface have occurred over the past decade or so, the reliability of the results is currently a matter of some debate (UNEP, 1991; WMO, 1992).

It has become increasingly apparent that existing UV monitoring capabilities fall short of those required to study the effects of O_3 changes on UV radiation incidence. The R-B meters cannot distinguish UV-B radiation from UV-A, the former being of particular concern in view of its potentially damaging effects on living tissues (see Human Exposure). Moreover, many of the existing monitors are located in urban areas where the effects of increased tropospheric O_3 and aerosol concentrations referred to above are likely to be at their maximum.

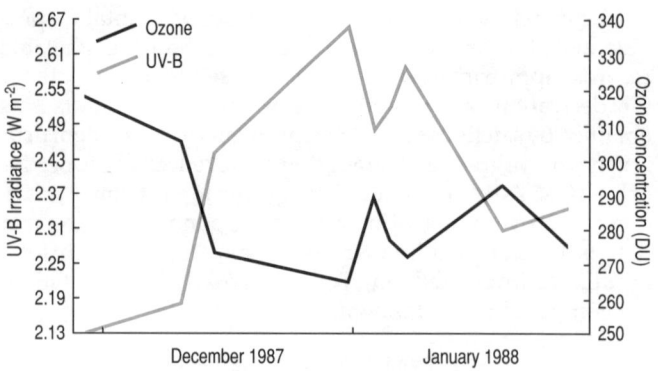

FIGURE 1.8 Relationship between the incidence of UV-B radiation and O_3 levels at Melbourne, Australia, 1987
Source: WMO, 1992

In response to the recognition of the need for more detailed spectral measurements of incident UV radiation, four sites have been established in the high-latitude regions of the Southern Hemisphere for this type of monitoring. Since O_3 losses are more severe in the Southern Hemisphere, and increases in tropospheric O_3 and aerosol concentrations are not as pronounced, it is assumed that O_3 depletion and UV-B incidence reduction associations are more likely to be detected in this region. Since becoming operational in 1988, the Antarctic stations (Palmer and McMurdo) have provided evidence of enhancement of UV radiation under the springtime O_3 holes of 1989 and 1990 (UNEP, 1991; WMO, 1992). Evidence of associations of high UV-B with low O_3 conditions in the mid-latitudes of the Southern Hemisphere has been reported by some researchers. Figure 1.8 shows the strong anti-correlation between UV-B and O_3 recorded at Melbourne, Australia during the intrusion of O_3-poor air in December 1987.

References

Hofmann, D. J., Oltmans, S. J., Harris, J. M., Solomon, S., Deshler, T. and Johnson, B. J. 1992 Observation and possible causes of new O_3 depletion in 1991, *Nature* (London), **359**, 283–287.

McCormick, M. P., Veiga, R. E. and Chu, W. P. 1992 Stratospheric ozone profile and total ozone trends derived from the SAGE I and SAGE II data, *Geophysical Research Letters*, **19**(3), 269–272.

Mendonca, B. 1993 Personal communication, National Oceanic and Atmospheric Administration, Climate Monitoring and Diagnostics Laboratory, Boulder, USA.

IPCC 1992 *Climate Change 1992: The Supplementary Report to the IPCC Scientific Assessment*, Report prepared for the Intergovernmental Panel on Climate Change by Working Group I, J. T. Houghton, B. A. Callender, S. K. Varney (Eds), Cambridge University Press, Cambridge.

Solomon, S. 1990 Progress towards a quantitative understanding of Antarctic ozone depletion, *Nature* (London), **347**, 347–354.

Technology and Economic Assessment Panel 1991 *Montreal Protocol 1991 Assessment*, Report of the Technology and Economic Assessment Panel, Technology and Economic Assessment Panel.

UNEP 1989 *United Nations Environment Programme Environmental Data Report 1989/90*, Basil Blackwell, Oxford.

UNEP 1991 *Environmental Effects of Ozone Depletion: 1991 Update*, Panel report pursuant to Article 6 of the Montreal Protocol on Substances that Deplete the Ozone Layer, United Nations Environment Programme, Nairobi.

Wang, W. C., Zhuang, Y. C. and Bojkov, R. D. 1993 Climate implications of observed changes in ozone vertical distributions at middle and high latitudes of the Northern Hemisphere, *Geophysical Research Letters*, August, 1567–1571.

WMO 1985 *Atmospheric Ozone 1985: Assessment of Our Understanding of the Processes Controlling its Present Distribution and Change*, Global Ozone Research and Monitoring Project, Report No. 16, World Meteorological Organization, Geneva.

WMO 1989 *Scientific Assessment of Stratospheric Ozone: 1989*, Global Ozone Research and Monitoring Project, Report No. 20, World Meteorological Organization, Geneva.

WMO 1992 *Scientific Assessment of Ozone Depletion: 1991*, Global Ozone Research and Monitoring Project, Report No. 25, World Meteorological Organization, Geneva.

WMO 1993a Ozone hole bigger and deeper, *World Climate News*, Issue No. 2, 8–9.

WMO 1993b Ozone deficit in Northern Hemisphere, *World Climate News*, Issue No. 3, 12–13.

Tropospheric Ozone

In the troposphere, O_3 not only behaves as a greenhouse gas, but is also a respiratory irritant and potentially damaging to plants. It is generated by two processes: the downwards mixing of stratospheric O_3 and, more significantly, as a result of chemical reactions involving the absorption of solar radiation by nitrogen dioxide (NO_2) in the presence of volatile organic compounds (VOCs) and carbon monoxide (CO).

As photochemical O_3 formation was first observed in cities, surface O_3 has traditionally been viewed as a predominantly urban pollutant. Research in both America and Europe has since established that extensive O_3 formation frequently occurs throughout the Northern Hemisphere mid-latitudes during the summer months, most noticeably downwind of major cities and industrial regions. In recognition of this long-range transport of O_3 and its precursor emissions (NO_2, CO and VOCs), O_3 is increasingly being viewed as a regional pollutant, particularly in Europe. Moreover, there is evidence to suggest that tropospheric levels of O_3 have roughly doubled in rural parts of Europe since the turn of the century. This increase is widely attributed to human activities (more specifically to the rise in precursor emissions of NO_x, CO and the VOCs from man-made combustion sources). In recent years this observation has been the subject of increased concern in view of its climatic implications. Thus it is in this context of regional pollution that surface O_3 trends are reviewed here; trends in urban areas are briefly considered in the subsequent section on Urban Air Pollution.

Due to the sparsity of reliable long-term records (i.e., >15 years), the evidence for large-scale increases of surface O_3 concentrations is somewhat fragmentary (Penkett, 1991; Lefohn et al., 1992). Evidence for increases over the past 100 years rely heavily on the analysis of records from a handful of rural sites in Europe and North America, the reliability of which is impaired by uncertainties in the accuracy of older O_3 measurement techniques.

The evidence for increases in surface O_3 on more recent timescales, i.e., over the past 2–3 decades, is similarly flawed. One of the longest surface O_3 data sets is that recorded at Hohenpeissenberg, Germany. Data recorded since 1971 indicate an increase of 1–2 per cent per year (Lefohn et al., 1992). However, there is still some debate regarding the validity of the earlier measurements made at this site; in particular, there is some question as to the effects of interference from SO_2 and NO_2 in the older iodometric measurements (Low et al., 1991). Tropospheric O_3 measurements made by ozonesonde also suggest that concentrations over continental Europe have increased by 1–1.5 per cent per year during the last 20–25 years (Staehelin and Schmid, 1991). However, an analysis of surface O_3 data from 20 European stations of differing character (remote, rural and urban) by Low et al. (1992) revealed few statistically significant trends among the station records for the period 1978–1988. Moreover, no dominant region-wide trend was reported.

Surface O_3 has been routinely measured at NOAA/CMDL observatory sites since the early 1970s. The most significant trend observed in these data is a 0.7 per cent per year increase observed over the past 20 years at the Barrow, Alaska monitoring station. This is largely attributed to the large summer (May–September) increases of over 25 per cent since the start of measurements. A small but significant increase has also taken place at the Mauna Loa Observatory. Elsewhere, i.e., in the Southern Hemisphere, weakly negative trends in surface O_3 concentrations are reported (CMDL, 1992).

The concern over rising surface O_3 concentrations has increased the level of basic monitoring for this pollutant. For example, measurements of O_3 are now included as part of the routine monitoring programme at sites participating in the United Nations Economic Commission for Europe's (UN ECE) Co-operative Programme for Monitoring and Evaluation of the Long Range Transmission of Air Pollutants in Europe (EMEP). Surface O_3 data are also collected and disseminated as part of the WMO's newly expanded and strengthened GAW programme (see Greenhouse Gases). Other initiatives include the Tropospheric Ozone Research (TOR) project, a sub-project of EUROTRAC, started in 1986 with the aim of collecting high quality O_3 and other oxidant measurements at sites throughout Europe. An analogous programme is in operation in the USA – the Southern Oxidant Study (SOS).

Modelling activities, which seek to emulate the formation of O_3 in the boundary layer, are also now well advanced. Figure 1.9a shows the distribution of daily maximum O_3 concentrations during April–September 1989 as calculated by EMEP's photo-oxidant model. The model, which reproduces observed O_3 concentrations quite well at most locations for which data exist, indicates that during the summer months peak surface O_3 concentrations range from a maximum of around 70 ppb in central and southern Europe to around 40 ppb in more northerly regions. Given that hourly-average O_3 concentrations of 75 ppb represent the threshold for potential vegetation damage, it has been useful to define and map what is known as "excess ozone". This statistic, mapped in Figure 1.9b, represents the sum over all hours in a given time period (in this case April–September) of all O_3 concentrations in excess of 75 ppb.

Concern over enhanced formation of tropospheric O_3 has also led to the development of more detailed inventories for the precursor emissions in order to support modelling activities and formulation of appropriate control strategies for

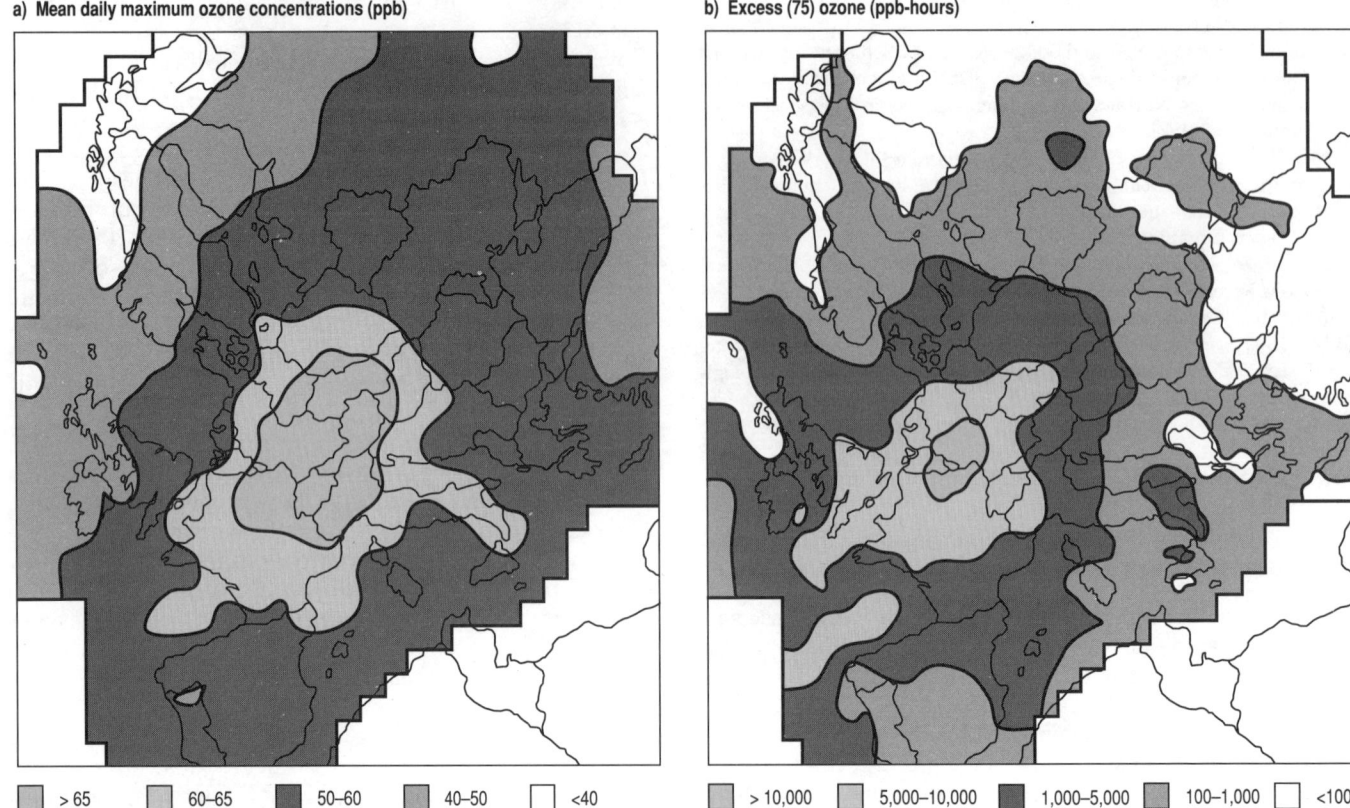

a) Mean daily maximum ozone concentrations (ppb)

b) Excess (75) ozone (ppb-hours)

| > 65 | 60–65 | 50–60 | 40–50 | <40 |

| > 10,000 | 5,000–10,000 | 1,000–5,000 | 100–1,000 | <100 |

FIGURE 1.9 Distribution of a) mean daily maximum tropospheric O₃ concentrations and b) "excess ozone" (75 ppb) in Europe, April–September 1989 (as calculated by the EMEP photo-oxident model)
Source: Simpson and Styve, 1992

VOCs and NO$_x$ precursors. Those for VOCs are briefly reviewed here. In view of their additional role as acid deposition precursors, NO$_x$ emissions are quantified in a subsequent sub-section on 'Acidic Deposition'.

The US Environmental Protection Agency (EPA) has attempted to construct a global anthropogenic emissions inventory for the VOCs (US EPA, 1991). Figure 1.10 shows the major sources of VOCs according to this inventory. Of the global total of 120.6×10^6 t a^{-1}, about one fifth of emissions are attributed to fuelwood utilization and a further 16 per cent to savanna burning, mainly in tropical and sub-tropical areas. It is estimated that petrol storage, consumption, transportation and marketing also accounts for 16 per cent of VOC emissions. As petrol consumption is also a major source of NO$_x$ this makes motor vehicles one of the most important anthropogenic sources of tropospheric O₃ precursors.

National anthropogenic emissions estimates of non-methane VOCs are shown in Table 1.7. It is important to note that VOC emissions inventories are extremely difficult to compile due the huge number of industrial sources which have to be quantified. Moreover, emissions estimates for individual sources are subject to considerable uncertainty. Recent research conducted in the USA suggests that the emissions from mobile sources may be subject to the greatest uncertainties, and that existing inventories may underestimate

the contributions from this sector by a factor of between two and four (Committee on Tropospheric Ozone Formation and Measurement, 1991). In recognition of the role played by natural VOC emissions in regional O₃ formation, inventories of biogenic emissions of VOCs are also being developed.

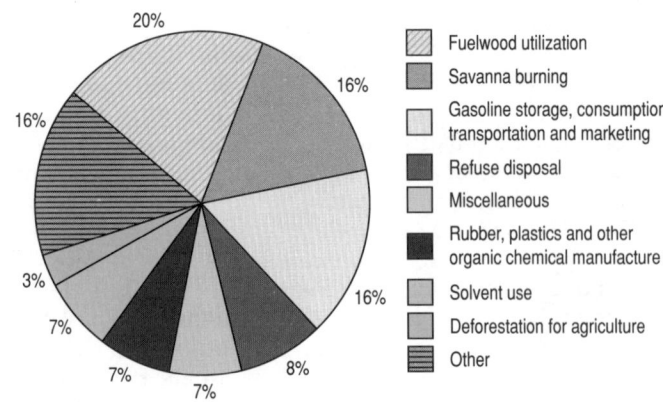

- ▨ Fuelwood utilization
- ▨ Savanna burning
- ▨ Gasoline storage, consumption, transportation and marketing
- ▨ Refuse disposal
- ▨ Miscellaneous
- ■ Rubber, plastics and other organic chemical manufacture
- ▨ Solvent use
- ▨ Deforestation for agriculture
- ▨ Other

FIGURE 1.10 Global emissions inventory for VOCs by major source category, 1980s
Source: US EPA, 1991

References

CMDL 1992 *Summary Report 1991*, National Oceanic and Atmospheric Administration, Climate Monitoring and Diagnostics Laboratory, Boulder.

Committee on Tropospheric Ozone Formation and Measurement 1991 *Rethinking The Ozone Problem in Urban and Regional Air Pollution*, National Academy Press, Washington D.C.

Lefohn, A. S., Shadwick, D. S., Feister, U., Mohnen, V. A. 1992 Surface-level ozone: climate change and evidence for trends, *Journal of the Air and Waste Management Association*, **42**(2), 136–144.

Low, P. S., Davies, T. D., Kelly, P. M., and Reiter R. 1991 Uncertainties in surface ozone trend at Hohenpeissenberg, *Atmospheric Environment*, **25A**(2), 511–515.

Low, P. S., Kelly, M. and Davies, T. D. 1992 Variations in surface ozone trends over Europe, *Geophysical Research Letters*, **19**(11), 1117–1120.

Penkett, S. A. 1991 Changing ozone: evidence for a perturbed atmosphere, *Environmental Science and Technology*, **25**(4), 631–635.

Simpson, D. and Styve, H. 1992 *The Effects of the VOC Protocol on Ozone Concentrations in Europe*, EMEP MSC-W Note 4/92, Meteorological Synthesizing Centre-West, The Norwegian Meteorological Institute, Oslo.

Staehelin, J. and Schmid, W. 1991 Trend analysis of tropospheric ozone concentrations utilizing the 20-year data set of O_3 balloon soundings over Payerne (Switzerland), *Atmospheric Environment*, **25a**, 1739–1749.

US EPA 1991 *Global Inventory of Volatile Organic Compound Emissions from Anthropogenic Sources*, Project Summary EPA/600/88–91/002, United States Environmental Protection Agency, Air and Energy Engineering Research Laboratory, Research Triangle Park.

Acidic Deposition

Acidic gases, such as sulphur dioxide (SO_2) and NO_x, are significant atmospheric pollutants on both local and regional scales. Both species are associated with respiratory morbidity and mortality in humans (WHO, 1987). Long-range transport and the subsequent atmospheric deposition of these gases has caused widespread acidification of terrestrial and aquatic ecosystems in selected world regions, most notably in Europe and eastern North America. In addition, gaseous SO_2 and NO_x, as precursors to aerosol particle formation, have been implicated in offsetting global warming (see Atmospheric Aerosols).

Revised estimates of global emissions of sulphur published by the IPCC in 1992 and listed below suggest that man-made sources account for 55–80 per cent of the combined total (anthropogenic and natural):

Source	Emissions (10^6 t a^{-1} as S)
Anthropogenic emissions (mainly SO_2)	70–80
Biomass burning (SO_2)	0.8–2.5
Oceans (DMS)	10–50
Soils and plants (DMS and H_2S)	0.2–4
Volcanic emissions (mainly SO_2)	7–10

Source: IPCC, 1992

Whereas natural emissions of sulphur are approximately equally divided between the North and South Hemispheres, it is estimated that over 90 per cent of emissions from anthropogenic sources originate in the Northern Hemisphere alone. The dominance of anthropogenic emissions is especially pronounced in the 35°–50°N region, where only 8 per cent of sulphur emissions are considered to be derived from natural sources (Bates et al., 1992).

The main natural sources of NO_x are soils and lightning; some estimates place emissions from each of these sources on a similar level to those from fossil fuel burning by man. Estimates of global NO_x emissions are in the range 35–79 x 10^6 t a^{-1} (as nitrogen), with fossil fuel combustion accounting for some 24 x 10^6 t a^{-1} (as nitrogen):

Source	Emissions (10^6 t a^{-1} as N)
NATURAL	
Soils	5–20
Lightning	2–20
Transport from stratosphere	~1
ANTHROPOGENIC	
Fossil fuel combustion	24
Biomass burning	2.5–13
Tropospheric aircraft	0.6

Source: IPCC, 1992

Increasing demand for energy has resulted in a continuing increase in the emissions of acidic gases over the past century. Long-term trends in global and regional emissions of both nitrogen and sulphur oxides from fossil fuel combustion have been estimated by Dignon and Hameed (1989). These emission estimates are based on derived statistical relations between emissions and fuel consumption for areas with quasi-uniform emission controls.

More recently, the same authors have calculated SO_2 and NO_x source strengths (fossil fuel sources only) on a national basis and on a latitude-longitude grid for the period 1970–1986 (Hameed and Dignon, 1992). On the basis of this more recent work it is estimated that global emissions of SO_2 increased by approximately 18 per cent between 1970 and 1986, from 57 x 10^6 t a^{-1} (as sulphur) to 67 x 10^6 t a^{-1} (as sulphur). The highest rate of increase in SO_x emissions occurred in Asia, where emissions have almost doubled, from around 11 x 10^6 t a^{-1} (as sulphur) in 1970 to approximately 20 x 10^6 t a^{-1} (as sulphur) in 1986. Results for Latin America and Africa show a much slower rate of increase: in contrast, emissions in North America have fallen significantly from over 17.5 x 10^6 t a^{-1} (as sulphur) in 1970 to approximately 14 x 10^6 t a^{-1} (as sulphur) in 1986, while those in Europe have remained more or less the same (Hameed and Dignon, 1992). Emissions estimates for NO_x suggest an increase of nearly one-third between 1970 (18 x 10^6 t a^{-1} (as nitrogen)) and 1986 (24 x 10^6 t a^{-1} (as nitrogen)). Steady increases in NO_x emissions have been observed in all continents except North America where emissions have remained relatively constant (Hameed and Dignon, 1992).

The latitudinal distribution of NO_x and SO_2 emissions from fossil fuel combustion sources (derived from Hameed and Dignon's projection of emissions on a 1° latitude by 1° longitude grid) is shown in Figure 1.11. This presentation clearly demonstrates the marked Northern Hemisphere bias in source strengths for NO_x and SO_2 mentioned earlier. Although the North-South profile is broadly similar for both pollutant gases there are nevertheless some significant differences in the latitudinal distribution of emissions. For example, European emissions of SO_2 produce a large peak at around 50°N; in the case of NO_x emissions, however, the profile is dominated by North American emissions which produce the peak at around 40°N. Furthermore the broad feature near 30°S, which represents South African and Australian emissions, is more pronounced in the sulphur profile (Dignon, 1992).

At the national level, emission inventories for SO_2 and NO_x have been made by the application of emission factors to source-specific fuel consumption statistics, this being the traditional approach for compiling emission inventories for SO_2 and NO_x. The statistical models developed by Hameed and Dignon allow emission estimates to be made when information about fuel types, source categories and emission factors is not readily available. The use of this approach on an individual country basis has generated a reasonable data base of emissions estimates for most industrialized countries spanning the 20-year period 1970–1990. Available data of this type, which have been extracted from 'State of the Environment'(SOE)-type publications and as such represent official estimates, are summarized for SO_2 in Table 1.8, and for NO_x in Table 1.9. It is important to note the procedures for calculating the emission inventories can vary markedly between countries. Moreover, the emission factors employed and the raw "activity" statistics are subject to varying degrees of uncertainty. Thus inter-country comparisons should be made with caution. Given a consistent methodology, trends within a country are, however, likely to be more reliable.

In recent years, considerable effort has been made to standardize or harmonize the calculation of national emission inventories for SO_2 and NO_x in order to improve the comparability of national estimates for regulatory purposes. In the European region, where work of this nature is the most advanced, the initiatives of the UN ECE EMEP programme and the Commission of the European Communities' (CEC) CORINAIR project are particularly noteworthy. As part of this project, detailed SO_2 and NO_x emissions inventories for 1985 have been compiled from raw data supplied by European Community (EC) member countries. The CORINAIR inventory is the most comprehensive of its kind, providing emissions broken down by 41 individual source categories; a 1990 inventory is in preparation (EUROSTAT, 1992).

Through a series of international workshops, the EMEP Task Force on Emission Inventories has developed a set of agreed technical guidelines for the calculation and reporting of national emissions of SO_2, NO_x, VOCs and NH_3. These guidelines have been harmonized with those used by the CORINAIR system. Under the terms of the Protocols to the UN ECE Convention on Long-Range Transboundary Air Pollution, signatory nations are required to submit data on their emissions of SO_2, NO_x and more recently on VOCs to EMEP according to these guidelines. As far as possible, these estimates, i.e., estimates submitted to EMEP, have been used in Tables 1.8 and 1.9 for the European countries; however, these values may differ from estimates provided in national

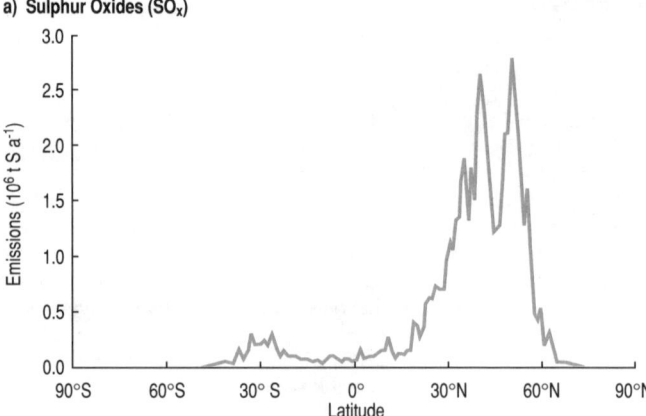

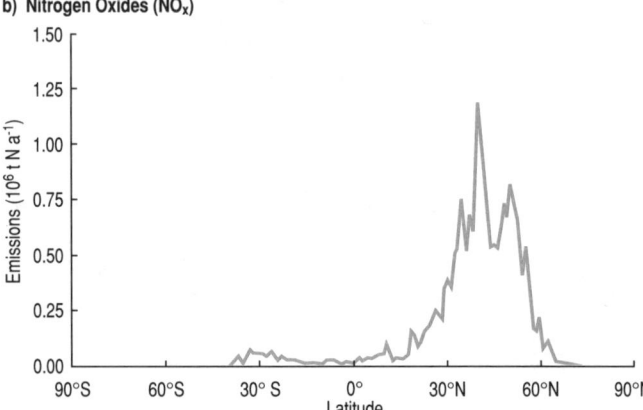

FIGURE 1.11   Latitudinal distribution of a) sulphur dioxide and b) nitrogen oxide emissions from fossil fuel combustion sources, 1980
Source: Dignon, 1992

SOE reports. Also included in Tables 1.8 and 1.9 is a set of emissions estimates for 25 Asian countries compiled by Kato and Akimoto (1992) which are based on international data sets for fuel consumption and a mix of country and regionally-specific emission factors. Data derived in this way will have a greater degree of inter-country comparability, but again, are likely to differ from official estimates provided in national SOE reports.

In addition to listing total emissions, Tables 1.8 and 1.9 calculate the percentage change in emissions between 1980 and 1990, per capita emissions in 1990 and emissions per unit of economic output, i.e., gross domestic product (GDP) in 1990. The latter measures, per capita emissions and emissions per unit GDP, have been proposed as possible indicators of environmental stress by several groups, for example the OECD (OECD, 1991). In the case of SO_2 substantial declines in emissions between 1980 and 1990 are evident in a number of countries, from both the developed and developing world. Decreases in SO_2 emissions have been achieved in most cases despite an increase in fuel consumption and economic output through the adoption of a mix of pollution control strategies, energy conservation measures and through fuel switching (UNEP/WHO, 1988). Structural reform in the industrialized countries may also have played a role (see Part 6: Energy and Part 7: Industry and Transport). A decrease in Kuwait's emissions from 1987 to 1990 of 26 per cent is an artifact caused by the interruption of industrial production during the Kuwait-Iraq conflict. One of the most significant rises in SO_2 emissions are those from China which have increased by 50 per cent between 1980 and 1987, overtaking the USSR to become the second largest emitter of SO_2 after the USA (Table 1.8).

The highest per capita SO_2 emissions occur in Qatar, the former German Democratic Republic, Kuwait and the former Czechoslovakia. Oil refining is the most likely source of high per capita SO_2 emission rates in central Asian countries while heavy reliance on high-sulphur coal for industry and domestic purposes is a major source of SO_2 in many eastern European countries. Of the countries included in Table 1.8, the southeast Asian countries of Cambodia, Laos, Bangladesh, Viet Nam and Myanmar have the lowest per capita emissions; this is ascribed to the heavy dependence on biomass fuels which have much lower sulphur contents than fossil fuels (Kato and Akimoto, 1992).

High SO_2 emissions per unit GDP are, to some extent, characteristic of those countries undergoing rapid economic and industrial development or which have high industrial output in relation to population. High emissions per unit GDP may also reflect a lack of pollution control for SO_2 and/or a reliance on high-sulphur coal. The former Czechoslovakia, China and Poland had the highest emissions of SO_2 per unit GDP around 1990.

Recent trends in NO_x emissions (1980–1990) amongst countries for which data are available are predominantly upwards or at best stationary (Table 1.9). The greatest percentage increases in NO_x emissions between 1980 and 1990 occurred in Brunei, Nepal, Turkmenistan, Kazakhstan, Israel and Ireland, presumably as a result of greater industrialization and an increased level of motorization. However, significant reductions (>10 per cent) in NO_x emissions are noted in some countries, including Estonia, Georgia, Moldova in the former USSR, Luxembourg, Portugal and Hungary, probably because of reduced industrial output.

Motor vehicles are a major source of NO_x emissions in most countries. However, as shown in Table 1.9, the actual ratio of mobile to stationary NO_x varies greatly between countries being dependent upon the level of traffic, the type and level of industry and the climate (which dictates the need for domestic heating). In some countries, the proportion of national NO_x emissions arising from motor vehicles is increasing, partly because of reductions in emissions from stationary sources and partly because of continuing growth in vehicle numbers and traffic volume (see also Part 7: Industry and Transport).

In response to the threat of ecosystem damage due to enhanced rates of acid deposition, monitoring networks for the measurement of precipitation quality and acid deposition have been established at global, regional and national scales. Global monitoring of precipitation chemistry is conducted as part of the WMO's Background Air Pollution Monitoring Network (BAPMoN). Established in 1969 the network currently comprises 152 precipitation chemistry stations throughout the world. Summary data from BAPMoN stations have been presented in previous editions of the 'UNEP Environmental Data Report' (UNEP 1987, 1989).

Continuing the regional focus approach adopted in the previous edition of this report (the work of the UN ECE EMEP programme was covered in the third edition), the status of acid deposition monitoring in the former USSR is briefly reviewed here. Figure 1.12 shows the spatial pattern of sulphate deposition (wet only) over the territory of the former USSR. This information is derived from measurements made by the 80-station strong network for background air pollution monitoring currently operated by the republics of the former USSR. Stations within this network contribute data to both the EMEP and BAPMoN programmes.

Data indicate that sulphur deposition is greatest to the west of the region. It is suggested that deposition in these areas is chiefly due to long-range transport of sulphur from European sources. Elsewhere, elevated sulphur deposition occurs in a localized manner close to major population centres. In addition, it is reported that wet deposition is greatest during the winter and is greatly influenced, both temporally and spatially, by the amount and frequency of precipitation. Nitrate deposition follows a similar distribution pattern to sulphate. Sulphate and nitrate deposition has increased over the European part of the former USSR, west Siberia and Kazakhstan in recent years. The level of sulphate and nitrate deposition is currently of concern in northern Estonia, Lithuania, Belarus, the Ukraine, parts of the central economic region, the central Chernozen economic region, the Volga region, central and southern Urals, and around Kuzbass and Altai (Izrael and Rovinsky, 1991).

The impetus for controlling emissions of SO_2 and NO_x over the past decade or so has stemmed as much from the threat of acidic deposition as from concerns for human health risks in the urban environment. However, while attempts have been made to reduce emissions of SO_2 and NO_x through international agreement (for example, the UN ECE Convention on Transboundary Air Pollution and its related Protocols), legislation has been aimed at a step-wise reduction of emissions from large combustion sources rather than at achieving specific targets based on environmental benefits. In more recent years, governments and scientists have thus sought a more effective means of assessing the magnitude of acid deposition effects and of planning SO_2 and NO_x emission controls. To this end the critical loads approach, which attempts to identify the threshold of damage for a biological receptor (e.g., individual species or whole ecosystems), has proved to be particularly useful. The use of the critical loads approach and its application to setting emission reduction targets in Europe is discussed in more detail in Box 1.3.

References

Bates, T. S., Lamb, B. K., Guenther, A., Dignon, J. and Stoiber, R. E. 1992 Sulphur emissions to the atmosphere from natural sources, *Journal of Atmospheric Chemistry*, **14**, 315–337.

Dignon, J. 1992 NO_x and SO_2 emissions from fossil fuels: a global distribution, *Atmospheric Environment*, **26A**(6), 1157–1163.

Dignon, J. and Hameed, S. 1989 Global emissions of nitrogen and sulphur oxides from 1860 to 1980, *Journal of the Air Pollution Control Association*, **39**, 180–186.

EUROSTAT 1992 *Environment Statistics 1991*, Statistical Office of the European Communities, Luxembourg.

Hameed, S. and Dignon, J. 1992 Global emissions of nitrogen and sulphur

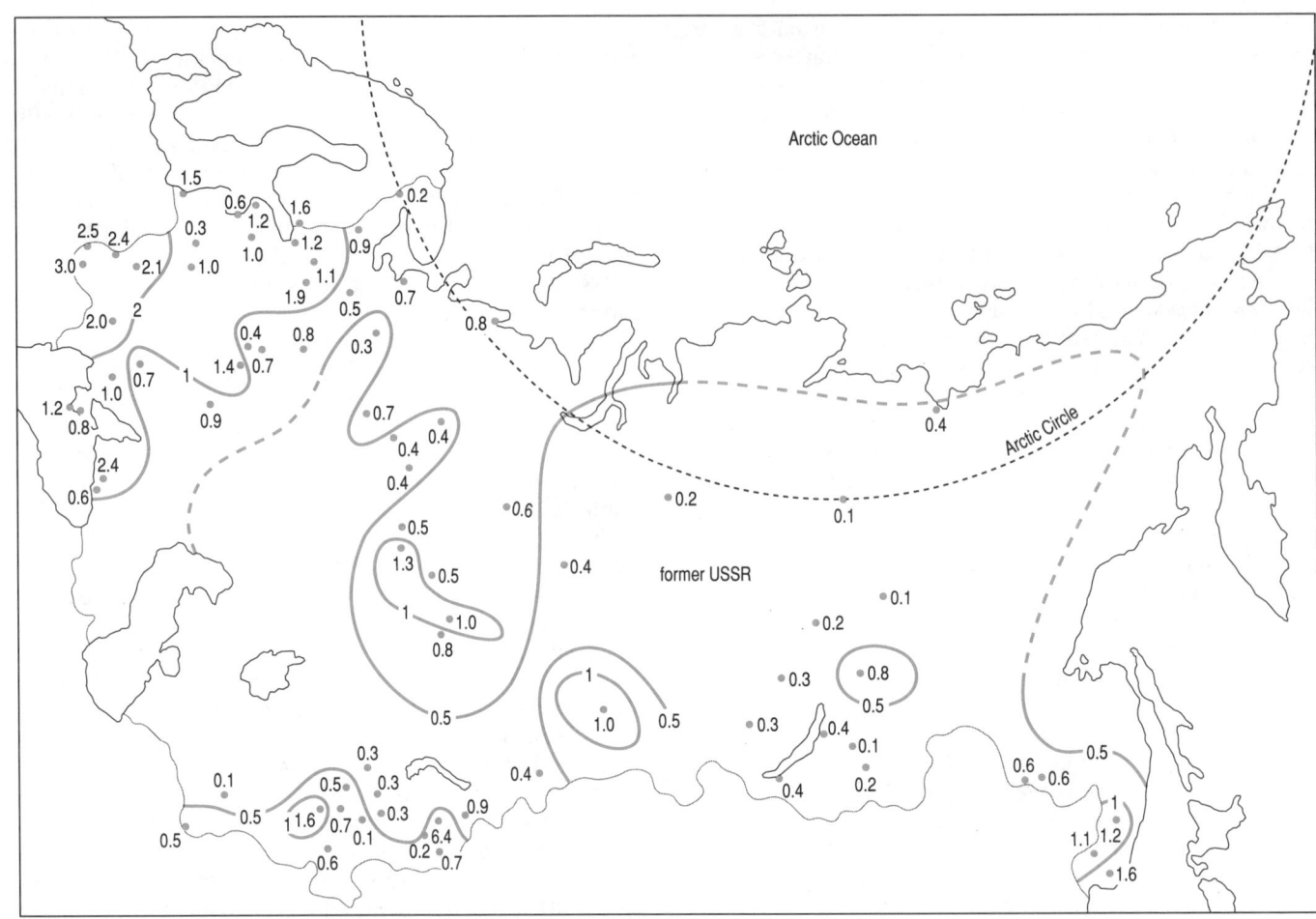

FIGURE 1.12 Wet deposition of sulphate in the former USSR, 1990 (in t m^{-2})
Source: Figure provided by F. Ya. Rovinsky, Institute of Global Climate and Ecology, Moscow, Russia

oxides from fossil fuel combustion 1970–1986, *Journal of the Air and Waste Management Association*, **42**, 159–163.

IPCC 1992 *Climate Change 1992: The Supplementary Report to the IPCC Scientific Assessment*, Report prepared for the Intergovernmental Panel on Climate Change by Working Group I, J. T. Houghton, B. A. Callender and S. K. Varney (Eds.), Cambridge University Press, Cambridge.

Izrael, A. Yu. and Rovinsky, F. Ya. (Eds.) 1991 *Review of the State of the Natural Environment in the USSR*, Natural Environment and Climate Monitoring Laboratory, Moscow.

Kato, N. and Akimoto, H. 1992 Anthropogenic emissions of SO$_2$ and NO$_x$ in Asia: emission inventories, *Atmospheric Environment*, **26A**(16), 2997–3017.

OECD 1991 *Environmental Indicators: A Preliminary Set*, Organisation for Economic Co-operation and Development, Paris.

UNEP/WHO 1988 *Assessment of Urban Air Quality*, United Nations Environment Programme, Nairobi, and World Health Organization, Geneva.

UNEP 1987 *United Nations Environment Programme Environmental Data Report*, Basil Blackwell, Oxford.

UNEP 1989 *United Nations Environment Programme Environmental Data Report 1989/90*, Basil Blackwell, Oxford.

WHO 1987 *Air Quality Guidelines for Europe*, WHO Regional Publications – European Series No. 23, World Health Organization, Regional Office for Europe, Copenhagen.

Atmospheric Aerosols

Airborne particles or aerosols, which are emitted directly into the atmosphere by emission sources or may be formed in the atmosphere by the conversion of natural or man-made gaseous constituents to particulate form, have been the subject of considerable scientific interest for a number of decades. For example, the atmospheric chemistry of sulphate aerosols, derived from the gas-to-particle conversion of precursor emissions of SO$_2$ have been extensively studied in relation to the acid rain issue (see preceding section). More recently, attention has focused on the possible implications of increases in sulphate aerosol concentrations on global climate. It is the latter concern that is the subject of the discussion here.

The presence of aerosol particles in the atmosphere, or more specifically in the troposphere, can affect the radiative balance of the earth (and thus global climate) in a number of

BOX 1.3 The Critical Loads Approach and its Application to Policy Making

The concept of a critical load is essentially that of a dose-response relationship; the threshold of a harmful response triggered by a certain pollutant load is known as the critical load. If values for both the critical load of an ecosystem and the incident pollutant deposition are known, it is possible to calculate if – and by how much – a critical load is exceeded. This exceedance of the critical load may give an indication of the level of damage to be expected. Further, given suitable models critical loads can be used to determine by how much emissions must be reduced to prevent damage or to protect specified ecosystems or communities and thus provide a suitable basis for planning emission controls.

Through a series of international workshops the UN ECE has developed methods of calculating and mapping critical loads of actual acidity, sulphur and nitrogen for the European region (Hettelingh et al., 1991). Work is currently being co-ordinated by the appointed Co-ordination Centre for Effects located at the National Institute of Public Health and Environmental Protection in the Netherlands. The Centre's main role is to generate Europe-wide maps of critical loads from national data for subsequent use in integrated assessment models. The maps of critical loads and critical load exceedances rely on the UN ECE EMEP monitoring data, models and emission inventories and therefore use the EMEP 150 x 150 km grid as their basis.

To date critical loads for two major categories of receptors have been defined, namely soils and fresh waters. In each case there are a variety of methods available for the calculation of critical loads for specific soil and fresh-water receptors, the choice of which is dependent upon national priorities and data availability. The accompanying figure shows the exceedances of critical loads of sulphur; this exceedance map combines national critical loads data for a variety of ecosystems, calculated by a variety of methods. On the basis of this work it is estimated that about three-quarters of the European region currently receives levels of sulphur deposition in excess of the critical loads. The greatest exceedances of critical leads for sulphur occur in the central European region, i.e., eastern Germany, Poland and Czechoslovakia.

For the purposes of formulating emission abatement strategies for sulphur and nitrogen in the European region, it has been proposed that critical loads be used to set "target loads". Target loads are defined as the permitted pollutant load "determined by political agreement" and, as such, allow governments to take account of social and economic considerations in addition to environmental factors when deciding on emission control policies. Thus target loads may be higher or lower than the critical load for a given area; in the former case this would provide a safety margin and the latter would tolerate some damage where critical loads are unattainable. It is considered that this approach will help to direct emission abatement strategies towards protection of those "priority" areas which are known to be particularly sensitive to the effects of acid deposition and where critical loads are exceeded.

In summary the critical loads approach provides a direct link between the protection of the environment and abatement strategies. It is intended that this approach will be used as a basis for the planned 1993 revisions to the Sulphur Protocol and the 1995 revisions to the Nitrogen Protocol.

Reference

Hettelingh, J-P., Downing, R. J. and de Smet, P. A. M. (Eds) 1991 *Mapping Critical Loads for Europe*, CCE Technical Report No 1, Co-ordination Centre for Effects, National Institute of Public Health and Environmental Protection, Bilthoven.

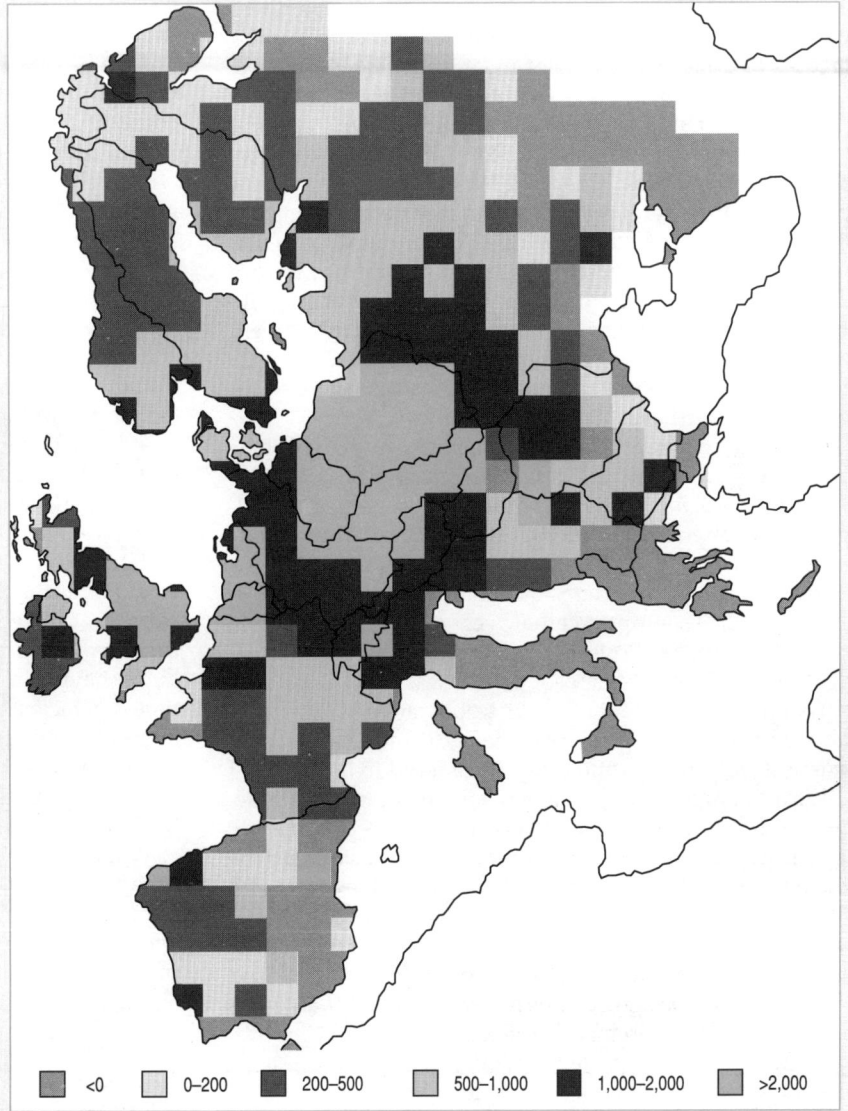

<0	0–200	200–500	500–1,000	1,000–2,000	>2,000

Exceedance of defined critical loads for sulphur deposition in Europe (in equivalents ha^{-1} a^{-1})
Source: UN ECE, 1991

ways. The dominant effect is that of backscattering and absorption of incident solar radiation, leading to cooling at the earth's surface. Of particular importance in this regard are the smaller particles (i.e., <1 μm diameter) formed by chemical reactions involving sulphur, nitrogen and carbon-containing gases. It is highly probable that the increases in emissions of SO_2 compounds from man-made combustion sources that have occurred since pre-industrial times may be exerting a cooling influence on the present climate. These effects are likely to be particularly significant since the 1950s and in the Northern Hemisphere where the SO_2 emissions from fossil fuel combustion sources have been more marked (Figure 1.11).

In the absence of observational records of sulphate aerosol concentrations, Langner et al. (1992) have employed a global transport-chemical model to estimate the changes in the distribution of tropospheric sulphate aerosol concentrations since pre-industrial times. Results of this exercise confirm that sulphate aerosol increases have been small over the Southern Hemisphere, but over the most polluted parts of Europe, increases of up to two orders of magnitude, have occurred.

Model simulations have also been used to estimate the cooling effect of sulphate aerosols at current levels of emissions; this has been estimated to correspond to a negative radiative forcing of the order of 1 Wm^{-2} (averaged over the Northern Hemisphere), which is comparable in magnitude with the positive radiative forcing effect of non-CO_2 greenhouse gases. Thus it is currently believed that the presence of sulphate aerosols in the atmosphere may have offset a significant part of the expected greenhouse warming in the Northern Hemisphere during the past several decades (IPCC, 1992). The climatic implications of increased aerosol concentrations are discussed further in Part 2: Climate.

In order to be able to quantify the climatic role of sulphate aerosols better, an improved data base of aerosol observations is required. Basic measurements of aerosol properties, including aerosol scattering extinction co-efficients, have been made on a routine basis at NOAA's four observatory sites (Barrow, Mauna Loa, Samoa and the South Pole) since the mid-1970s (CMDL, 1992). To support climate research, however, these basic observations need to be supplemented with more detailed measurements of aerosol properties.

In contrast to the situation in the troposphere, where man-made emissions of precursor gases dominate aerosol formation, volcanic eruptions are considered to be the most significant source of sulphate aerosols in the stratosphere. Satellite and lidar observations made over the past decade suggest that stratospheric aerosol concentrations during the 1980s have remained consistently higher than those recorded in 1979. This has been largely attributed to a series of volcanic eruptions, in particular the 1982 El Chichon eruption. However, the increase may also be in part due to increased concentrations of background or non-volcanic sulphate aerosols, derived from aircraft emissions of SO_2 (IPCC, 1992). Hoffman (1991) has reported evidence of increases in non-volcanic aerosol mass of about 5 per cent per year during the 1980s over the Northern Hemisphere mid-latitudes. Increases in stratospheric aerosols have implications for stratospheric ozone (see previous section) and climate (see Part 2: Climate).

References

Hofmann, D. J. 1991 Increase in the stratospheric background sulphuric acid aerosol mass in the past 10 years, *Science*, **248**, 996–1000.

IPCC 1992 *Climate Change 1992: The Supplementary Report to the IPCC Scientific Assessment*, Report prepared for the Intergovernmental Panel on Climate Change by Working Group I, J. T. Houghton, B. A. Callender and S. K. Varney (Eds), Cambridge University Press, Cambridge.

Langner, J., Rodhe, H., Crutzen, P. J. and Zimmermann, P. 1992 Anthropogenic influence on the distribution of tropospheric sulphate aerosol, *Nature* (London), **359**, 712–716.

CMDL 1992 *Summary Report 1991*, National Oceanic and Atmospheric Administration, Climate Monitoring and Diagnostics Laboratory, Boulder.

Urban Air Pollutants

Exposure to air pollution is now an almost inescapable feature of urban living throughout the world. Although the relationship is still poorly defined, increased mortality and morbidity believed to be caused by urban air pollution are of great concern.

Over the past 20 years there has been a significant shift in the type of air pollution affecting the urban areas of developed countries. Traditional pollutants from stationary sources such as SO_2 and suspended particulate matter (SPM) have been effectively controlled by the introduction of "clean air" legislation. In Europe and North America, a change from domestic coal burning to electricity and natural gas for heating and cooking purposes has led to a huge reduction in low-level urban emissions of SO_2 and SPM and concomitant improvements in air quality (UNEP/WHO, 1988).

However, as emissions and concentrations of these traditional urban air pollutants decreased, further economic development (and increasing personal wealth) has resulted in a dramatic increase in motor vehicle traffic, both in terms of vehicle numbers and distances travelled. This in turn means that emissions of those pollutants associated with motor vehicle transport, most notably NO_x, CO and hydrocarbons, have also increased in many cities of the developed world over the past 20 years. Attempts to control emissions from motor vehicles, mainly through the introduction of three-way catalytic converters and more fuel-efficient engines, have to a large extent been outstripped by growth in motor vehicle traffic. Motor vehicle emissions are discussed in greater detail in Part 7: Industry and Transport.

In many developing countries, rapid urbanization has resulted in many of the problems faced by developed countries being repeated. In certain countries, heavy reliance on coal and oil for domestic heating and cooking means that urban SO_2 and SPM levels are high. In addition, rapid economic development has meant that emissions from industry and motorized vehicles are increasingly causing air quality problems.

Many countries, in both the developed and less-developed world, have now established nation-wide urban air quality networks for the monitoring and assessment of pollution levels in their major cities. Standardized data on SO_2 and SPM concentrations from selected cities are submitted to the World Health Organization (WHO) as part of the WHO/UNEP GEMS Urban Air Quality Monitoring Project (known as

GEMS/Air). Participation in this project has expanded significantly since its inception in 1974; it now extends to around 80 cities in over 50 countries. Typically, cities report data to GEMS/Air from three sites; these are selected from an existing city air quality monitoring network according to predefined selection criteria.

A number of detailed assessments of urban air quality problems, which draw on data collected by the GEMS/Air project, have been produced (WHO, 1984; UNEP/WHO, 1988). The most recent of these is a review of urban air pollution in megacities of the world (WHO/UNEP, 1992). This latter study defines a megacity as one which has, or will have, a population of 10 million by the year 2000; these comprise Bangkok, Beijing, Bombay, Buenos Aires, Cairo, Calcutta, Delhi, Jakarta, Karachi, London, Los Angeles, Manila, Mexico City, Moscow, New York, Rio de Janeiro, São Paulo, Seoul, Shanghai and Tokyo. Key findings and summary data for selected cities which reflect the approach adopted for the assessment are included as Box 1.4.

In order to make its monitoring and assessment capabilities more relevant to current air quality problems, the GEMS/Air programme has latterly undergone a major review. The review process culminated in the publication of a report which outlined proposals for a revised global programme for urban air pollution monitoring and assessment (WHO, 1992). One recommendation of this report was that additional pollutants be included in the GEMS/Air project (with carbon monoxide as priority); thus, in the future, countries will be asked to submit data on all the pollutants monitored at their participating GEMS/Air sites. It was further proposed that the number of participating countries and cities be expanded to give greater global coverage. Detailed consideration was also given to harmonization of monitoring methodologies and quality control procedures within GEMS/Air.

A major finding of the megacities study was the widespread prevalence of SPM pollution followed, in terms of prevalence and severity of air quality problems, by SO_2 and O_3 (WHO/UNEP, 1992). In view of this finding, attention here is focused on urban air quality problems with respect to particulates; O_3 concerns are mentioned in Part 7: Industry and Transport in the context of motor vehicle emissions.

A summary of annual mean total suspended particulate (TSP) and smoke concentrations as measured at selected GEMS/Air sites is given in Table 1.10. It should be noted that TSP and smoke, although both representing measurements of SPM, are not directly comparable as they are derived from different measurement techniques. A SO_2 data summary was included in the 1991 edition of this report (UNEP, 1991).

It is difficult to make any generalizations about regional or global trends in SPM pollution from the data given in Table 1.10. However, it can be seen that concentrations in the cities of many developing countries are conspicuously higher than those of developed nations. It is important to note that geography and climate play an important role in determining ambient SPM concentrations. Cities in, or close to, arid areas tend to have much higher concentrations than those in tropical

or temperate climates. In Beijing, for example, it is estimated that in the summer 60 per cent of SPM arises from natural sources, that is, fall-out from dust storms originating in the western plains.

Health effects of SPM are usually associated with those of SO_2 as traditionally these pollutants have been attributed to the same source(s) (chiefly domestic coal combustion). Particles in the respirable range, i.e., smaller than 10 µm are of particular interest (WHO, 1987). The chemical composition of airborne particulate matter also plays an important role in determining effects on health. Elemental carbon, polynuclear aromatic hydrocarbons (PAHs) and toxic base metals are amongst the constituents of particulate matter which are of particular interest in this regard. The decline in domestic coal use in many countries over the past 10–20 years has led to a progressive reduction in the emission of black elemental carbon particulates (black smoke). Indeed, diesel vehicles have now overtaken domestic coal use as the major source of carbonaceous particles in many cities. A detailed inventory of organic and elemental carbon emissions in Los Angeles, for example, attributed approximately 25 per cent of emissions to petrol-engined vehicles and a further 25 per cent to diesel vehicles. Black elemental carbon emissions alone were dominated by diesel exhaust (Spengler et al., 1990).

The particulate fraction may also contain potentially toxic trace metals such as arsenic, lead, cadmium and mercury. Trace metal emission rates have generally decreased over the past decade in line with the overall reduction in the emission of particulates by industry (UNEP, 1991). Polynuclear aromatic hydrocarbons, which may be absorbed or adsorbed on particulate matter, are of great concern in terms of human health because many are carcinogens. Generally PAH concentrations have fallen in most developed countries along with domestic and industrial smoke (UNEP, 1992).

References

Spengler, J. D., Brauer, M. and Koutrakis, P. 1990 Acid air and health, *Environmental Science and Technology*, **24**, 946–956.

UNEP/WHO 1988 *Assessment of Urban Air Quality*, United Nations Environment Programme, Nairobi, and World Health Organization, Geneva.

UNEP 1991 *United Nations Environment Programme Environmental Data Report 1991/92*, Basil Blackwell, Oxford.

UNEP 1992 *Chemical Pollution: A Global Overview*, United Nations Environment Programme, Geneva.

WHO 1984 *Urban Air Pollution 1973–1980*, World Health Organization, Geneva.

WHO 1987 *Air Quality Guidelines for Europe*, WHO Regional Publications, European Series No. 23, World Health Organization, Regional Office for Europe, Copenhagen.

WHO 1992 Urban Air Pollution Monitoring – Report of a Meeting of UNEP/WHO Government-Designated Experts, Geneva, 5–8 November 1991, WHO/PEP/92.2, UNEP/GEMS/92.A.1, World Health Organization, Geneva.

WHO/UNEP 1992 *Urban Air Pollution in Megacities of the World*, World Health Organization, United Nations Environment Programme, Blackwell, Oxford.

BOX 1.4 Urban Air Pollution in Megacities of the World

In order to assess the problems of urban air pollution in a global context, the WHO and UNEP initiated a detailed study of air quality in 20 of the world's megacities (cities which already have or will have 10 million inhabitants by the year 2000). Megacities were chosen for special study as the majority have air quality problems; they encompass large land areas and many people. Moreover, urbanization is a global phenomenon and therefore the lessons of today's megacities may have great relevance to all urban areas and, in particular, those with rapidly growing populations. The major findings of this study have since been published as a WHO/UNEP report, 'Urban Air Pollution in Megacities of the World' (WHO/UNEP, 1992).

As part of the study, available air quality data on SO_2, SPM, Pb, CO, NO_x and O_3 were compiled and evaluated. However, in order to assess fully the air quality situation in these 20 cities, it was necessary to examine not only air quality monitoring data but also information on pollutant sources and emissions. On the basis of these data a subjective assessment of the air pollution problems in each of the 20 cities was made, the results of which are shown in the figure presented here. It is concluded that at least one WHO health guideline is exceeded in all 20 cities; fourteen cities have two pollutants, and seven cities have three or more pollutants whose concentrations exceed the guidelines. Other key findings of the study are listed below:

○ High SPM concentrations are the most prevalent form of pollution (WHO guidelines exceeded in 17 megacities);
○ High concentrations of SO_2 and SPM combined occur in five cities;
○ The upward trend in SO_2 has been reversed in 10 megacities, largely due to fuel switching policies;
○ Motor vehicle traffic is the most important source of air pollution in 10 of the cities;
○ Domestic use of coal or biomass causes a problem in 5 of the 20 megacities, resulting in high human exposure to SPM, SO_2 and PAHs;
○ Only 6 of the 20 megacities have satisfactory monitoring networks and data handling capabilities; and
○ There is no systematic collection of information on the health risks and effects of air pollution in most of the megacities.

Data from five of the megacities included in the study are highlighted here – Beijing, Calcutta, Los Angeles, Mexico City and São Paulo. These

Air pollution status and key socio-economic data for selected megacities, 1990

	Beijing China	Calcutta India	Los Angeles USA	Mexico City Mexico	São Paulo Brazil
Population					
1990 (estimated)	9.74	11.83	10.47	19.37	18.42
2000 (projected)	11.47	15.94	10.91	24.44	23.6
Climate	CCS	S	M	MWC	TRF
Economic status	LIE	LIE	HIE	LMIE	UMIE
Major energy source	Coal	Coal	Hydro-electric	Oil	Oil
Motor vehicles (10^3)	200	500	7,900	2,400	4,000
Air quality: annual mean and maximum concentrations ($\mu g\ m^{-3}$)					
SO_2[a]	25–130 (682)	47–59 (529)	0–10 (60)	80–200 (550)	35–62 (180)
SPM[c]	250–410 (1,286)	236–453 (2,091)	46–115 (1,770)	100–500 (1,000)	50–85 (285)[b]
N_2O[c]	14–54 (76)	30–52 (176)	39–104 (526)	113–207 (714)	35–100
O_3[c]	(320)	nm	(660)	100–400 (900)	(370–600)
Pollutant emissions ($10^3\ t\ a^{-1}$)					
SO_2	526	25	50	206	107
SPM	116[d]	200	400[b]	451	68
CO		177	1,800	2,951	1,400
NO_x		40	440	177	245

a = Daily maxima.
b = PM_{10}.
c = Hourly maxima.
d = Industry only

CCS = Continental cool summer.
S = Savanna.
M = Mediterranean.
MWC = Marine west coast.
TRF = Tropical rain forest.

LIE = Lower income economy.
LMIE = Lower-middle income economy.
UMIE = Upper-middle income economy.
HIE = High income economy.
nm = Not routinely measured.

Most data refer to 1990. The climatic descriptions refer to the classification of global climates according to the Köppen system. The economic classifications are those of the World Bank. Annual mean concentrations of selected pollutants are given as a range and reflect the range of concentrations measured at different sites within the city. The maximum concentration is given in parenthesis and represents the highest daily or hourly concentration measured at any one site within the city.

Source: WHO/UNEP, 1992

cities vary greatly in terms of their location, climate and socio-economic status (see table). Air pollutant emissions and, ultimately, measured concentrations broadly reflect these differences.

In Beijing, for example, high SO_2 levels are concurrent with the main heating period (winter) when large amounts of coal are used for domestic heating. In contrast, in Mexico City high SO_2 concentrations are a result of high industrial emissions. On the basis of industrial and domestic coal consumption, SO_2 levels in Calcutta might also be expected to be high; however, local coal has a low sulphur content and so annual

average concentrations are relatively low. São Paulo has greatly reduced its SO_2 levels in recent years through the implementation of stringent emissions controls on industrial point sources together with fuel switching. Similarly, the strict Californian emission regulations applicable in Los Angeles have reduced SO_2 levels to well below WHO guidelines in recent years.

The factors influencing SPM concentrations are particularly complex. Extremely high SPM concentrations dominate in Beijing, Calcutta and Mexico City. These high concentrations can, to

Box continued

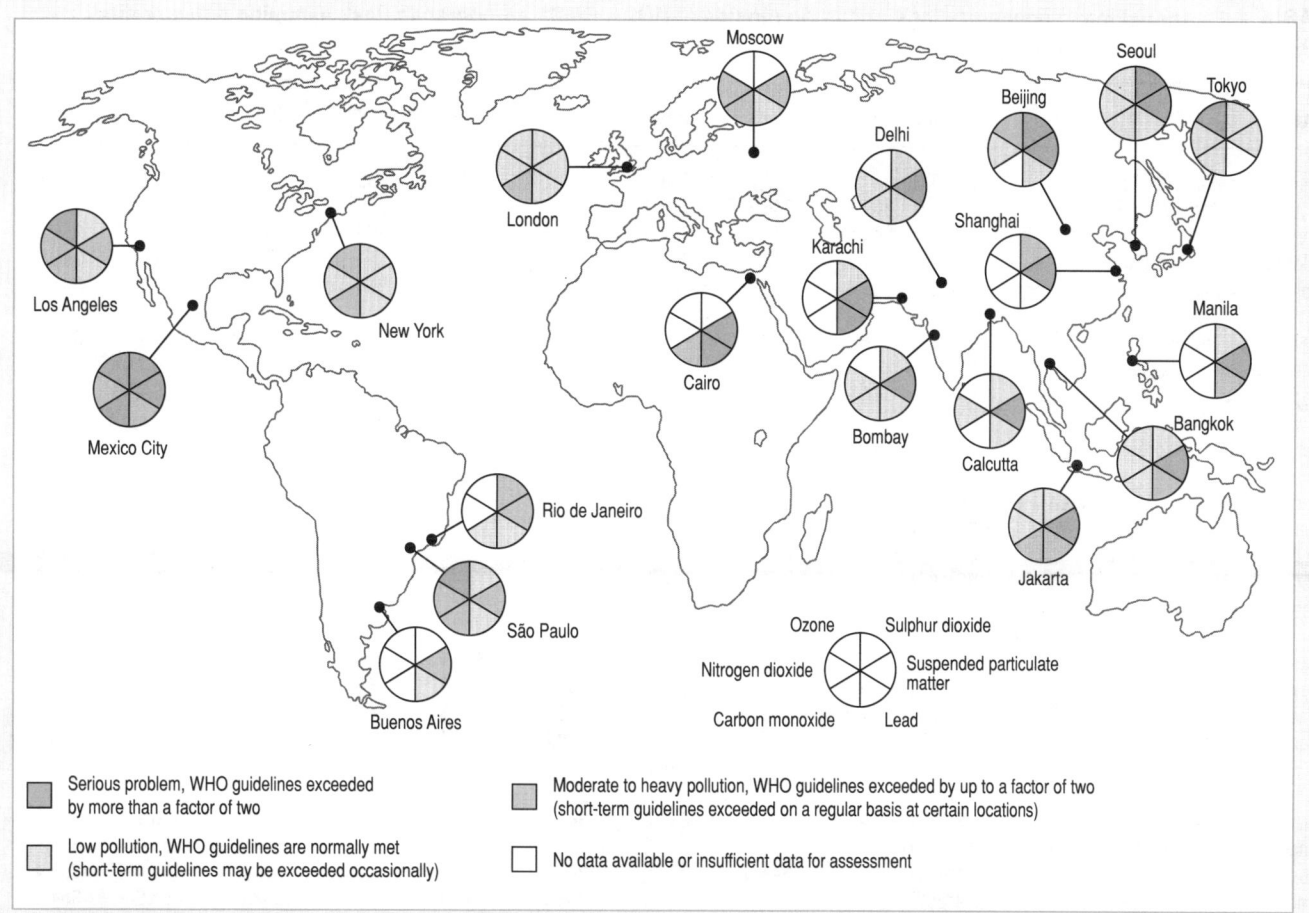

Air pollution problems in 20 "megacities", late 1980s
Source: Adapted from WHO/UNEP, 1992

some extent, be attributed to the large amount of natural dust and erosion products which are entrained in the atmosphere. The high SPM maximum of Los Angeles is similarly largely due to natural dust originating from the warm climate and desert-like surroundings.

The combustion of coal is also a major contributor to SPM emissions in both Beijing and Calcutta. These emissions are of great concern with respect to human health effects. In those cities with long-term elevated SO_2 and SPM levels (Beijing), chronic respiratory illness is common. Reductions in SPM levels have been achieved in São Paulo over recent years but not, however, to the same degree as SO_2. Mortality studies in São Paulo have demonstrated an increase in the number of deaths during high pollution episodes.

Owing to their large spatial variability, it is difficult to compare CO concentrations between cities. Generally speaking concentrations are highly correlated with motor vehicle numbers. In Los Angeles, despite strict emission controls, CO emissions and concentrations are still high due to the extremely high vehicular population. Mexico City is unique among the megacities in its altitude; the high altitude, and thus low oxygen conditions, favours the formation of CO rather than CO_2 (i.e., incomplete combustion) in the internal combustion engine. The low oxygen also exacerbates the health effects attributed to CO.

Nitrogen dioxide and O_3 are regarded as secondary, photochemical pollutants, both relying on NO_x emissions as a precursor. As with CO, the principal sources of NO_x in urban areas are motor vehicles. High NO_2 and O_3 concentrations occur in those cities which combine high precursor emissions and insolation (sunshine), namely Mexico City, Los Angeles and São Paulo. In Los Angeles it has been estimated that each resident, on average, suffers ozone-related symptoms on 17 days per year and that attaining air quality standards (at a cost of US$ 10 billion) would save approximately 1,600 lives each year.

The photochemical pollutants pose great problems in terms of control, not only for those cities with problems now, but for virtually all mid-latitude cities where motor vehicle traffic is increasing. The control of photochemical precursors has already proved to be very difficult and will undoubtedly create a much greater challenge to pollution control engineers and legislators than that posed by the more traditional pollutants, such as SO_2, which have been successfully controlled in many countries (see also Part 7: Industry and Transport).

Reference

WHO/UNEP 1992 *Urban Air Pollution in Megacities of the World*, World Health Organization, United Nations Environment Programme, Blackwell, Oxford.

TABLE 1.1 Annual mean concentrations of carbon dioxide at NOAA/CMDL co-operative flask sampling network sites, 1984–1992 (ppmv in dry air)

Site	Latitude	1984	1985	1986	1987	1988	1989	1990	1991	1992
Alert, North West Territories	82°27'N			348.4	349.7	353.0	355.1	355.5	357.2	357.5
Mould Bay, North West Territories	76°15'N	345.6	346.8	348.9	350.1	353.5	355.5	356.0	357.6	357.4
Barrow, Alaska	71°19'N	345.5	346.8	348.9	349.9	353.4	355.0	356.0	357.5	357.5
Atlantic Ocean (Polarfront)	66°00'N	344.7	346.1	347.6	349.1	352.4	354.0	354.8	356.5	356.5
Cold Bay, Alaska	55°12'N	345.6	347.1	348.5	350.1	352.3	354.4	355.3	357.3	357.3
Mace Head, Ireland	53°20'N								–	356.1
Shemya Island, Alaska	52°43'N			349.3	350.4	353.1	354.5	355.0	356.7	357.1
Cape Meares, Oregon	45°29'N	344.9	347.3	348.2	351.2	352.6	354.4	355.4	356.7	356.3
Niwot Ridge, Colorado	40°03'N	344.6	346.1	346.8	349.1	351.9	353.5	354.7	356.1	356.8
Terceira Island, Azores	38°45'N	344.4	346.0	348.8	348.6	351.6	–	–	355.8	–
Tae-ahn Peninsula, South Korea	36°44'N								359.8	360.6
Qinghai Province, China	36°16'N								–	356.4
St David's Head, Bermuda	32°22'N							355.2	356.1	357.0
Southhampton, Bermuda	32°16'N							355.1	356.6	356.3
Tenerife, Canary Islands	28°18'N								–	356.2
Sand Island, Midway	28°13'N			347.9	349.9	352.9	354.1	355.2	357.0	356.8
Key Biscayne, Florida	25°40'N	345.3	346.7	348.1	350.4	352.7	354.4	355.9	356.5	357.3
Mauna Loa, Hawaii	19°32'N	344.2	345.4	346.6	348.7	351.3	352.8	354.1	355.5	356.4
Cape Kumukahi, Hawaii	19°31'N	344.3	345.7	346.9	348.8	351.4	352.9	354.3	355.8	356.3
St Croix, Virgin Islands	17°45'N	343.4	345.5	346.8	348.5	351.5	353.5	–		
Guam, Mariana Islands	13°26'N	344.5	346.1	347.6	349.6	352.1	353.6	354.2	356.1	356.3
Ragged Point, St Phillip's Parish	13°10'N					351.5	352.9	354.7	355.9	355.9
Christmas Island, Pacific Ocean	1°42'N		346.0	347.0	348.8	351.3	353.3	354.3	355.4	356.4
Mahe Island, Seychelles	4°40'S			346.4	348.7	350.3	352.2	353.3	353.9	354.8
Ascension Island, Atlantic Ocean	7°55'S	343.9	345.1	346.2	348.5	350.4	351.6	352.9	353.9	355.1
Tutuila, American Samoa	14°15'S	342.9	344.5	345.7	347.5	350.2	351.7	352.8	354.2	354.9
Amsterdam Island, Indian Ocean	37°57'S	342.4	344.2	345.6	347.9	349.5	350.5	–	–	
Cape Grim, Tasmania	40°41'S		343.5	344.8	346.5	348.9	350.3	351.6	352.8	353.5
Palmer Station, Antarctica	64°55'S	342.7	344.0	–	346.9	349.6	350.9	351.9	353.2	354.2
Syowa Station, Antarctica	69°00'S			345.5	–	349.3	–	352.2	353.3	354.2
Halley Bay, Antarctica	75°40'S	342.7	344.3	345.4	347.5	349.6	–	–	–	–
South Pole, Antarctica	89°59'S	342.2	343.7	345.0	346.9	349.0	350.6	351.8	353.1	354.1

A dash in this table signifies insufficient values for the calculation of a valid annual mean.

The NOAA/CMDL flask sampling network currently comprises 32 globally distributed sites at which air samples are taken approximately weekly in pairs of 0.5 litre pyrex glass flasks. In 1991 four new sites were added to the network, at Tae-ahn Peninsula, Qinghai Province, Mace Head and Izaña Observatory, Tenerife. Sites are listed according to latitude, starting with the most northern.

Air samples are returned to the NOAA/CMDL laboratory in Boulder, USA for CO_2 analysis by non-dispersive infrared gas analysers. All samples, including calibration gases are air dried prior to analysis. Results are reported relative to the Scripps Institution of Oceanography (SIO) X85 mole fraction calibration scale. Data for 1992 are provisional, pending recalibration of standard gases. These flask samples are also subject to analysis for CH_4 (see Table 1.4).

To be considered valid, results from flask pairs must agree to within 0.50 ppmv. When CO_2 concentrations in the two flasks differ by more than this amount, the higher result of the pair is rejected. Additional validation procedures identify contaminated samples (e.g., those resulting from incorrect sampling and/or flask stopcock failures) and those that are not considered representative of the background conditions at the site. Remaining data are subsequently subjected to a statistical analysis to obtain weekly values fitted to a smoothed curve function. The monthly mean values reported here are calculated from weekly values derived from the curve. Further details of the raw data selection and processing procedures are given by Conway et al. (1988).

These data are available in electronic format, (i.e. as a numerical data package) from the Carbon Dioxide Information Analysis Center in Boulder, USA. Data are also archived at the WMO World Data Centre for Trace Gases in Tokyo, Japan.

Sources and references:
Data supplied by T.J. Conway, National Oceanic and Atmospheric Administration, Climate Monitoring and Diagnostics Laboratory, Boulder, Colorado, USA.
Conway, T. J., Tans, P., Waterman, L. S., Thoning, K. W., Masarie, K. A. and Gammon, R. H. 1988 Atmospheric carbon dioxide measurements in the remote global troposphere, 1981-1984, *Tellus*, **40B**, 81-115.

TABLE 1.2 Global emissions of carbon dioxide from fossil fuel combustion and cement manufacture, 1950–1990

Year	Solid fuels 10^6 t C a^{-1}	%	Liquid fuels 10^6 t C a^{-1}	%	Natural gas 10^6 t C a^{-1}	%	Cement manufacture 10^6 t C a^{-1}	%	Natural gas flaring 10^6 t C a^{-1}	%	Total emissions 10^6 t C a^{-1}	Total emissions per capita t C a^{-1}
1950	1,077	66	423	26	97	6	18	1	23	1	1,638	0.65
1951	1,137	64	479	27	115	6	20	1	24	1	1,775	0.69
1952	1,127	63	504	28	124	7	22	1	26	1	1,803	0.69
1953	1,132	61	533	29	131	7	24	1	27	1	1,848	0.70
1954	1,123	60	557	30	138	7	27	1	27	1	1,871	0.69
1955	1,215	59	625	30	150	7	30	1	31	2	2,050	0.74
1956	1,281	59	679	31	161	7	32	1	32	1	2,185	0.78
1957	1,317	58	714	31	178	8	34	1	35	2	2,278	0.80
1958	1,344	57	732	31	192	8	36	2	35	1	2,338	0.80
1959	1,390	56	790	32	214	9	40	2	36	1	2,471	0.83
1960	1,419	55	850	33	235	9	43	2	39	2	2,586	0.86
1961	1,356	52	905	35	254	10	45	2	42	2	2,602	0.85
1962	1,358	50	981	36	277	10	49	2	44	2	2,708	0.86
1963	1,404	49	1,053	37	300	11	51	2	47	2	2,855	0.89
1964	1,442	48	1,138	38	328	11	57	2	51	2	3,016	0.92
1965	1,468	47	1,221	39	351	11	59	2	55	2	3,154	0.95
1966	1,485	45	1,325	40	380	11	63	2	60	2	3,314	0.97
1967	1,455	43	1,424	42	410	12	65	2	66	2	3,420	0.98
1968	1,456	40	1,552	43	445	12	70	2	73	2	3,596	1.01
1969	1,494	39	1,674	44	487	13	74	2	80	2	3,809	1.05
1970	1,571	38	1,838	45	516	13	78	2	87	2	4,091	1.11
1971	1,571	37	1,946	46	554	13	84	2	88	2	4,242	1.12
1972	1,587	36	2,056	47	583	13	89	2	94	2	4,409	1.15
1973	1,594	34	2,240	48	608	13	95	2	110	2	4,648	1.18
1974	1,591	34	2,244	48	618	13	96	2	107	2	4,656	1.16
1975	1,686	36	2,131	46	623	13	95	2	93	2	4,629	1.13
1976	1,723	35	2,313	47	647	13	103	2	109	2	4,895	1.18
1977	1,786	35	2,390	47	646	13	108	2	104	2	5,034	1.19
1978	1,802	35	2,384	47	674	13	116	2	107	2	5,082	1.18
1979	1,899	35	2,535	47	714	13	119	2	100	2	5,366	1.23
1980	1,921	36	2,409	46	725	14	120	2	89	2	5,264	1.18
1981	1,930	38	2,272	44	735	14	121	2	72	1	5,129	1.13
1982	1,993	39	2,178	43	734	14	121	2	69	1	5,094	1.11
1983	1,998	39	2,163	43	736	14	125	2	63	1	5,085	1.09
1984	2,088	40	2,186	42	785	15	128	2	57	1	5,243	1.10
1985	2,243	41	2,169	40	819	15	131	2	55	1	5,416	1.12
1986	2,300	41	2,274	41	836	15	137	2	52	1	5,600	1.13
1987	2,345	41	2,285	40	876	15	143	3	48	1	5,698	1.13
1988	2,403	41	2,387	40	916	15	150	3	55	1	5,912	1.16
1989	2,443	41	2,423	40	951	16	154	3	54	1	6,024	1.16
1990	2,431	40	2,476	41	981	16	154	3	54	1	6,097	1.15

Note that emissions are in units of 10^6 t a^{-1} as carbon (C); for emissions in 10^6 t a^{-1} as carbon dioxide (CO_2), multiply by 3.67.

Global annual emissions of CO_2 from fossil fuel combustion sources are derived from emission factors applied to fuel production data (UN 1992 *Energy Statistics Yearbook 1990*, United Nations, New York) and to gas flaring data compiled by Rotty (for 1950–1970) and the US Department of Energy (after 1970). Emissions from cement manufacturing are calculated from cement production data given in the US Bureau of Mines Yearbooks.

The uncertainty of the total emissions estimates is ± 8 per cent. Full details of the procedures for calculating emissions are given in the source document.

Percentage distribution of CO_2 emissions may not necessarily total 100 per cent due to rounding.

These data are available in electronic format (i.e., as a numerical data package) from the Carbon Dioxide Information Analysis Center in Boulder, USA.

Source:
CDIAC 1992 *Estimates of CO_2 Emissions from Fossil Fuel Burning and Cement Manufacturing, Based on the United Nations Energy Statistics and the US Bureau of Mines Cement Manufacturing Data*, Numeric Data Package-030/R4, Carbon Dioxide Information Analysis Center, Oak Ridge National Laboratory, Oak Ridge.

TABLE 1.3 National emissions of carbon dioxide from anthropogenic sources, 1960–1990

	Industrial sources						Land-use changes	
	Total (10^3 t C a^{-1})				Per capita (t C a^{-1})	Per unit GDP (kg C a^{-1})	WRI (10^3 t C a^{-1})	SEI (10^3 t C a^{-1})
Region/country	1960	1970	1980	1990	1990	1990	1989	1985
WORLD			5,066,855	5,824,051			1,743,869	855,938
AFRICA								
Algeria	1,717	4,126	18,063	18,585	0.74	0.44		–3,619
Angola	151	966	1,449	1,369	0.14	0.18	8,992	12,397
Benin	44	77	133	182	0.04	0.10	2,589	2,646
Botswana	–	6	272	465	0.36	0.17	708	529
Burkina	12	39	116	147	0.02	0.05	4,632	3,109
Burundi		14	31	55	0.01	0.06	144	–167
Cameroon	74	174	1,057	1,518	0.13	0.14	16,349	22,388
Cape Verde	6	10	33	23	0.06			
Cent. African Rep.	24	55	29	54	0.02	0.04	3,542	2,844
Chad	15	35	58	59	0.01	0.05	4,087	2,545
Comoros	3	7	11	18	0.03			
Congo	61	149	105	542	0.24	0.19	3,270	9,458
Côte d'Ivoire	126	663	1,282	2,332	0.19	0.31	95,368	20,562
Djibouti	11	28	83	103	0.25			
Egypt	4,388	5,783	12,329	21,842	0.42	0.66		–245
Equatorial Guinea	6	10	15	31	0.09		490	
Ethiopia	96	428	493	791	0.02	0.14	8,174	4,304
Gabon	36	318	1,298	1,695	1.45		2,534	8,002
Gambia	5	13	43	51	0.06		518	588
Ghana	399	716	662	983	0.07	0.16	8,447	6,833
Guinea	112	208	257	272	0.05	0.10	10,082	11,177
Guinea-Bissau	5	18	37	55	0.06		4,905	5,910
Kenya	663	760	1,691	1,590	0.07	0.21	3,542	3,636
Liberia	45	388	554	127	0.05		10,627	11,822
Libya	189	8,807	7,347	11,667	2.57	0.51		–1,568
Madagascar	109	248	436	265	0.02	0.10	32,698	23,604
Malawi		114	194	145	0.02	0.09	15,804	5,210
Mali	33	55	107	115	0.01	0.05	2,098	1,203
Mauritania	10	117	167	712	0.35	0.75		663
Mauritius	49	135	161	314	0.29	0.15		–63
Morocco	995	1,928	4,354	6,237	0.25	0.25		–2,073
Mozambique	518	790	874	275	0.02	0.21	8,174	6,492
Niger	8	59	156	279	0.04	0.11	2,016	2,568
Nigeria	934	5,855	18,588	22,786	0.21	0.66	73,569	38,284
Réunion	20	75	188	250	0.42			–47
Rwanda	29	16	70	109	0.02	0.05	572	610
São Tomé & Príncipe	3	4	11	18	0.15			
Senegal	226	340	762	707	0.10	0.12	2,997	1,466
Seychelles	–	8	26	45	0.66			
Sierra Leone	195	197	161	183	0.04	0.22	1,253	1,515
Somalia	23	59	166	229	0.03	0.26	1,417	1,316

Continued

TABLE 1.3 Continued

Region/country	Industrial sources Total (10^3 t C a^{-1}) 1960	1970	1980	1990	Per capita (t C a^{-1}) 1990	Per unit GDP (kg C a^{-1}) 1990	Land-use changes WRI (10^3 t C a^{-1}) 1989	SEI (10^3 t C a^{-1}) 1985
South Africa	26,907	40,795	58,230	75,824	2.15	0.84		−1,075
Sudan	372	1,302	900	941	0.04		26,703	18,551
Swaziland	9	117	7,880	13,454	1.80			−703
Tanzania	226[a]	482	511	591	0.02	0.29	5,722	6,352
Togo	18	71	161	159	0.05	0.10	790	850
Tunisia	471	994	2,580	3,766	0.46	0.34		−755
Uganda	114	385	176	225	0.01	0.08	2,725	3,806
Western Sahara		19	44	54·	0.30			
Zaire	644	750	946	1,132	0.03	0.15	35,422	32,869
Zambia		1,036	967	662	0.08	0.21	7,357	11,223
Zimbabwe	2,960	2,247	2,652	4,193	0.71	0.15	4,360	2,287
NORTH AMERICA								
Antigua & Barbuda	10	126	39	82	1.08			
Bahamas	112	703	2,179	364	1.44			
Barbados	47	114	184	275	1.08			
Belize	12	33	52	69	0.37			
Bermuda	43	62	116	161	2.77			
Br. Virgin Is	–	5	7	13	1.03			
Canada	52,700	90,610	115,820	115,260	4.35	0.20		−79,050
Cayman Is	3	10	44	68	2.71			
Costa Rica	134	339	672	896	0.30	0.16	7,084	15,215
Cuba	3,751	5,075	8,369	9,540	0.90		243	−733
Dominica	3	7	10	16	0.19			
Dominican Rep.	284	848	1,738	1,716	0.24	0.23	354	532
El Salvador	168	384	581	700	0.13	0.13	436	904
Greenland	61	104	152	151	2.69			
Grenada	6	12	13	33	0.38			
Guatemala	368	618	1,225	1,117	0.12	0.14	11,172	17,474
Haiti	78	106	205	209	0.03	0.08	234	231
Honduras	168	358	560	533	0.10	0.23	11,444	12,414
Jamaica	401	1,362	2,304	1,265	0.52	0.32	221	410
Martinique	46	99	213	375	1.10			
Mexico	17,223	28,914	71,001	89,396	1.01	0.38	54,496	55,968
Neth. Antilles	3,025	4,524	2,863	137	2.29			
Nicaragua	146	378	553	564	0.15	0.17	16,076	25,415
Panama	345[b]	710[b]	851	729	0.30	0.16	5,177	9,177
Puerto Rico	1,972	3,179	3,832	5,447	1.57			
St Kitts & Nevis	3[c]	7[c]	14[c]	18	0.40			
St Lucia	4	18	30	44	0.30			
St Pierre & Miquelon	10	10	10	25	4.19			
St Vincent & Grenadines	3	8	10	22	0.19			
Trinidad & Tobago	704	2,285	4,552	4,087	3.19	0.86	90	−2

Continued

TABLE 1.3 Continued

Region/country	Industrial sources Total (10³ t C a⁻¹) 1960	1970	1980	1990	Per capita (t C a⁻¹) 1990	Per unit GDP (kg C a⁻¹) 1990	Land-use changes WRI (10³ t C a⁻¹) 1989	SEI (10³ t C a⁻¹) 1985
USA	799,544	1,165,477	1,259,281	1,310,341	5.26	0.24	5,995	−60,306
US Virgin Is	33	1,597	3,475	1,440	12.41			
SOUTH AMERICA								
Argentina	13,325	22,157	29,337	29,961	0.93	0.32		22,672
Bolivia	274	672	1,232	1,335	0.18	0.30	10,082	11,493
Brazil	12,808	23,622	48,178	54,379	0.36	0.13	258,856	212,986
Chile	3,702	6,703	7,356	9,309	0.71	0.33		−7,035
Colombia		6,952	10,734	14,534	0.44	0.35	114,441	60,783
Ecuador	492	1,132	3,666	4,638	0.44	0.43	43,597	25,394
French Guiana	6	26	100	147	1.50			
Guyana	180	431	483	173	0.22		300	410
Paraguay	83	197	402	476	0.11	0.09	18,256	43,626
Peru	2,223	4,798	6,423	5,788	0.27	0.16	38,147	30,388
Suriname	118	439	648	523	1.24		300	
Uruguay	1,179	1,556	1,576	1,068	0.35	0.13		−875
Venezuela	15,755	22,033	24,456	27,544	1.40	0.57	16,076	8,799
ASIA								
Afghanistan	117	460	486	1,711	0.10			−105
Bahrain	157	248	1,795	3,574	6.93			
Bangladesh			2,084	4,240	0.04	0.19	2,371	−248
Bhutan		1	6	35	0.02	0.13	234	131
Brunei	91	2,238	1,871	1,421	5.34			521
Cambodia	64	320	78	123	0.01		2,997	4,987
China	215,295	211,607	406,440	678,016	0.61	1.86		−105,571
Cyprus	242	465	873	1,195	1.70			−18
Hong Kong	807	2,263	4,489	7,401	1.26	0.12		9
India	33,193	53,261	95,547	183,547	0.22	0.72	32,698	25,659
Indonesia	5,844[d]	9,050[d]	25,825[d]	38,506[d]	0.21	0.36	237,057	123,659
Iran	10,199	23,991	31,682	49,417	0.90	0.43		−115
Iraq	2,254	6,352	11,997	14,173	0.75			470
Israel	1,764	4,521	5,768	9,564	2.08	0.18		72
Japan	63,997	202,973	254,881	289,288	2.34	0.10		4,391
Jordan	203	403	1,290	2,774	0.69	0.83		−42
Korea	3,455	14,230	34,312	65,884	1.54	0.28		−241
Korea, Dem.	6,520	20,173	34,269	42,606	1.96			−58
Kuwait	2,128	6,950	6,752	7,027	3.45	0.30		
Laos	22	152	50	62	0.01	0.07	65,395	−25
Lebanon	705	1,078	1,684	2,509	0.93			−52
Macau	14	56	136	280	0.59			
Malaysia	1,112[e]	4,092	7,764	16,216	0.90	0.38	76,294	57,654
Mongolia	348	750	1,829	2,753	1.26			−134
Myanmar	744	1,120	1,369	1,378	0.03		103,542	88,503

Continued

TABLE 1.3 Continued

Region/country	Industrial sources Total (10^3 t C a^{-1}) 1960	1970	1980	1990	Per capita (t C a^{-1}) 1990	Per unit GDP (kg C a^{-1}) 1990	WRI (10^3 t C a^{-1}) 1989	SEI (10^3 t C a^{-1}) 1985
Nepal	22	63	148	186	0.01	0.06	8,719	9,067
Oman		64	1,603	3,361	2.24	0.44		
Pakistan			8,648	17,754	0.14	0.50	1,090	−264
Philippines	2,303	6,733	9,971	11,752	0.19	0.27	51,771	28,846
Qatar	48	2,063	3,518	3,852	10.47			
Saudi Arabia	731	9,831	35,685	51,427	3.64	0.64		14
Singapore	380	4,964	8,217	10,279	3.77	0.30		−47
Sri Lanka	618	981	930	1,052	0.06	0.15	5,995	5,503
Syria	880	1,663	5,274	8,271	0.66	0.56		−87
Thailand	1,012	4,190	10,921	25,535	0.46	0.32	79,019	53,109
Turkey	4,587	11,504	20,741	38,301	0.69	0.40		−2,079
United Arab Em.	3	4,158	9,897	14,374	9.05	0.51		
Viet Nam	2,061[f]	7,696[f]	4,633	6,337	0.10		40,872	39,282
Yemen	16	46	326	1,022	0.11	0.39		
Yemen, Dem.	975	639	579	1,590	0.64			

EUROPE

Region/country	1960	1970	1980	1990	Per capita	Per unit GDP	WRI	SEI
Albania	277	932	2,639	2,656	0.82			−252
Austria	8,394	13,729	14,247	14,821	1.95	0.09		−2,671
Belgium	25,033	34,396	34,860	28,241	2.87	0.15		125
Bulgaria	5,956	21,777	30,332	24,872	2.76	1.25		−891
Czechoslovakia	35,394	54,341	66,094	56,715	3.62	1.28		−736
Denmark	8,141	16,920	17,254	13,940	2.71	0.11		−627
Faeroe Is	16	70	115	167	3.56			
Finland	4,132	10,997	15,036	14,023	2.82	0.10		−8,274
France	74,791[g]	116,176[g]	132,129[g]	97,432[g]	1.74	0.08		−7,741
German Dem. Rep.	71,940	73,845	83,678	82,023	5.05			
Germany, Fed. Rep.	148,614	200,858	208,021	180,554	2.94	0.18		−6,833[h]
Gibraltar	13	12	18	16	0.53			
Greece	2,545	6,559	14,031	18,886	1.88	0.33		−381
Hungary	12,208	19,654	23,282	15,748	1.49	0.48		−1,367
Iceland	331	379	509	614	2.43	0.15		
Ireland	3,039	4,915	6,845	8,427	2.27	0.20		−2,110
Italy	30,143[i]	77,955[i]	101,564[i]	103,649[i]	1.82	0.10		−67
Luxembourg	3,175	3,755	2,883	2,650	7.10	0.38		
Malta	93	178	269	455	1.29	0.23		
Netherlands	20,173	34,813	41,364	37,916	2.54	0.14		516
Norway	3,582	6,846	10,948	10,449	2.48	0.10		−2,476
Poland	55,049	82,783	125,436	100,040	2.60	1.57		−3,475
Portugal	2,248	3,716	7,396	11,189	1.09	0.20		−1,945
Romania	14,585	32,974	55,395	49,378	2.12	1.42		−2,101
Spain	13,423	30,194	54,546	55,426	1.41	0.11		−6,214
Sweden		25,179	19,416	13,488	1.60	0.06		−8,315
Switzerland	5,346[j]	10,780[j]	11,169[j]	11,439[j]	1.72	0.05		−305

Continued

TABLE 1.3 Continued

	Industrial sources						Land-use changes	
	Total (10^3 t C a^{-1})				Per capita (t C a^{-1})	Per unit GDP (kg C a^{-1})	WRI (10^3 t C a^{-1})	SEI (10^3 t C a^{-1})
Region/country	1960	1970	1980	1990	1990	1990	1989	1985
UK	160,770	175,397	160,551	152,773	2.65	0.16		419
Yugoslavia	9,368	18,616	29,205	35,669	1.50	0.43		−4,421
USSR	**396,016**	**628,209**	**895,504**	**1,055,499**	**3.66**			**−140,371**
OCEANIA								
American Samoa	9	38	116	78	2.05			
Australia	24,060	38,884	55,348	72,913	4.32	0.25		−2,576
Fiji	53	142	213	204	0.27			84
French Polynesia	10	50	77	167	0.81			
Guam	35	343	556	408	3.46			
Nauru		18	34	36	4.00			
New Caledonia	238	651	540	428	2.56			
New Zealand	3,167	3,884	4,802	7,012	2.07	0.16		−826
Pacific Is. Tr. Tr.	4	37	42	64	0.36			
Papua New Guinea		189	499	623	0.16	0.19	3,270	42,669
Solomon Is	3	11	28	44	0.14			
Tonga	3	7	11	20	0.21			
Vanuatu	−	11	17	18	0.12			

[a] Estimate comprising the sum of emissions from the former Tanganyika and Zanzibar which united to form Tanzania in 1969.

[b] Estimate comprising the sum of emissions from Panama (excluding the Canal Zone) and the Former Panama Canal Zone.

[c] Includes Anguilla.

[d] Includes East Timor.

[e] Estimate comprising the sum of emissions from Peninsula Malaysia, Sabah and Sarawak.

[f] Estimate comprising the sum of emissions from former South Vietnam Republic and the former Democratic Republic of Vietnam, which unified to form Viet Nam in 1970.

[g] Includes Monaco.

[h] Data refer to the unified Federal Republic of Germany.

[i] Includes San Marino.

[j] Includes Liechtenstein.

Note that emissions are in units of 10^3 t a^{-1} and kg a^{-1} as carbon; for emissions in 10^3 t a^{-1} and kg a^{-1} as carbon dioxide multiply by 3.67.

National emissions of CO_2 from industrial sources are based on UN consumption data for gas, liquid and solid fuels plus cement manufacturing statistics to which appropriate emission factors have been applied. Full details of the procedures for calculating these emissions are given in the source document. Per capita emissions use UN population data for 1990 and emissions per unit GDP are based on GDP values compiled by the World Bank for 1990 (in US$).

Two independently derived sets of CO_2 emissions estimates for land-use changes are presented here. The first data set refers to 1989 and has been compiled by the World Resources Institute (WRI). These estimates rely on the methodology developed by Houghton et al. (1987) for the calculation of net CO_2 emissions from land-use changes (mainly deforestation) in tropical regions. The original approach has, however, been modified by WRI to take account of new estimates of carbon densities, reported by Houghton (1991). The WRI has also subtracted the weight of carbon contained in national log production from the estimated CO_2 release from land-use changes in each country to approximate the quantity of carbon sequestered as durable wood products. Full details of the calculations can be found in the source document and in the additional references cited here.

The second set of estimates given here have been compiled by the Stockholm Environment Institute (SEI) and are similarly based on land clearing data, estimates of carbon densities on cleared land, and assumptions about carbon release rates. These are detailed in the source document (Subak et al., 1992). These calculations take account of the amount of carbon sequestered as long-lived wood products. However, unlike the WRI calculations, the SEI approach also incorporates estimates of the magnitude of possible carbon sinks in the form of afforestation. (These are likely to be significant in the temperate regions of Europe, the former USSR and China). A negative sign in this data column thus implies a net uptake of carbon. On a global scale the SEI estimate that CO_2 emissions from land clearing alone are in the range 1.3–2.1 x 10^9 t a^{-1} (as C). This is, however, offset by an uptake of carbon by afforestation and forest products of about 0.5 x 10^9 t a^{-1} (as C), yielding a net release of CO_2 to the atmosphere of about 0.9 x 10^9 t a^{-1} (as C).

Sources and references:
Houghton, R. A. 1991 Tropical deforestation and atmospheric carbon dioxide, *Climate Change*, **19**, 99–118.
Houghton, R. A., Boone, R. D., Fruci, J. R., Hobbie, J. E., Melillo, J. M., Palm, C. A., Peterson, B. J., Shaver, G. R., Woodwell, G. M., Moore, B., Skole, D. L., and Myers, N. 1987 The flux of carbon from terrestrial ecosystems to the atmosphere in 1980 due to changes in land use: geographic distribution of the global flux, *Tellus*, **39B**, 122–139.
CDIAC 1992 *Estimates of CO2 Emissions from Fossil Fuel Burning and Cement Manufacturing, Based on the United Nations Energy Statistics and US Bureau of Mines Cement Manufacturing Data*, Numeric Data Package – 030/R4, Carbon Dioxide Information Analysis Center, Oak Ridge National Laboratory, Oak Ridge.
Subak, S., Raskin, P. and von Hippel, D. 1992 *National Greenhouse Gas Accounts: Current Anthropogenic Sources and Sinks*, Stockholm Environment Institute, Boston.
World Resources Institute 1992 *World Resources 1992–93*, Oxford University Press, New York.

TABLE 1.4   Annual mean concentrations of methane at fixed sites of the NOAA/CMDL co-operative flask sampling network, 1984–1991 (ppbv in dry air)

Site	Latitude	1984	1985	1986	1987	1988	1989	1990	1991
Alert, North West Territories	82°27'N			1,743.5	1,751.6	1,765.9	1,777.2	1,783.7	1,793.8
Mould Bay, North West Territories	76°15'N	1,714.7	1,730.7	1,745.7	1,759.5	1,770.3	1,779.3	1,790.1	1,801.4
Barrow, Alaska	71°19'N	1,724.2	1,733.3	1,755.1	1,758.7	1,773.8	1,788.7	1,797.0	1,801.4
Atlantic Ocean (Polarfront)	66°00'N	1,708.5	1,725.7	1,735.1	1,750.4	1,763.8	1,765.2	1,779.3	1,789.4
Cold Bay, Alaska	55°12'N	1,707.4	1,722.5	1,736.5	1,746.1	1,757.7	1,763.9	1,779.0	1,787.8
Shemya Island, Alaska	52°43'N			1,740.1	1,748.3	1,761.4	1,771.6	1,783.8	1,790.5
Olympic Peninsula, Washington	48°15'N		1,713.3	1,719.0	1,732.9	1,745.0	1,749.6		
Cape Meares, Oregon	45°29'N	1,698.4	1,707.6	1,717.5	1,733.6	1,747.1	1,756.0	1,760.2	1,767.8
Niwot Ridge, Colorado	40°03'N	1,673.1	1,682.2	1,689.2	1,707.1	1,720.1	1,726.2	1,733.1	1,749.6
Terceira Island, Azores	38°45'N	1,689.2	1,703.3	1,714.7	1,728.2	1,732.4	1,750.3	1,762.9	1,768.9
St David's Head, Bermuda	32°22'N							1,757.3	1,761.7
Southampton, Bermuda	32°16'N							1,745.2	1,760.5
Sand Island, Midway	28°13'N			1,700.5	1,719.4	1,717.0	1,729.7	1,741.8	1,753.0
Key Biscayne, Florida	25°40'N	1,667.1	1,679.5	1,691.6	1,706.6	1,719.8	1,721.9	1,742.0	1,736.2
Mauna Loa, Hawaii	19°32'N	1,635.6	1,649.0	1,661.9	1,673.0	1,683.4	1,697.7	1,710.1	1,720.2
Cape Kumukahi, Hawaii	19°31'N	1,654.7	1,670.6	1,677.8	1,694.5	1,697.9	1,705.4	1,723.4	1,732.3
St Croix, Virgin Islands	17°45'N	1,650.5	1,668.6	1,674.3	1,685.3	1,697.1	1,704.0		
Guam, Mariana Islands	13°26'N	1,624.8	1,648.4	1,667.7	1,680.0	1,680.1	1,694.2	1,709.3	1,717.9
Ragged Point, St Phillip's Parish	13°10'N					1,693.9	1,702.1	1,716.3	1,723.5
Christmas Island, Pacific Ocean	1°42'N		1,616.0	1,627.6	1,639.2	1,648.8	1,662.6	1,667.9	1,677.9
Mahe Island, Seychelles	4°40'S	1,601.4	1,609.8	1,625.0	1,642.0	1,649.3	1,659.7	1,670.7	1,673.8
Ascension Island, Atlantic Ocean	7°55'S	1,588.7	1,602.3	1,615.2	1,627.6	1,635.5	1,647.4	1,657.3	1,669.5
Tutuila, American Samoa	14°15'S	1,585.2	1,599.7	1,609.5	1,617.3	1,629.9	1,643.2	1,647.1	1,660.4
Amsterdam Island, Indian Ocean	37°57'S	1,578.4	1,587.9	1,601.6	1,612.1	1,620.8	1,635.3		
Cape Grim, Tasmania	40°41'S		1,588.5	1,599.8	1,612.3	1,621.3	1,635.8	1,644.1	1,651.9
Palmer Station, Antarctica	64°55'S	1,577.1	1,587.9	1,599.4	1,610.3	1,622.4	1,636.0	1,641.4	1,652.1
Syowa Station, Antarctica	69°00'S				1,611.1	1,623.0	1,634.8	1,642.7	1,652.4
South Pole, Antarctica	89°59'S	1,575.9	1,586.8	1,599.4	1,610.4	1,622.0	1,634.2	1,642.3	1,652.9

Data in the above table are obtained from the fixed stations of the NOAA/CMDL co-operative flask sampling network at which air is sampled on approximately a weekly basis.

Flask samples (taken in pairs) are returned to the NOAA/CMDL laboratory at Boulder, Colorado for analysis for CH₄ by gas chromatography with flame ionization detection. Up until October 1991, two aliquots of air were extracted from a single flask for CH₄ analysis, prior to the extraction of a single aliquot (from both flasks) for CO₂ analysis (see Table 1.1). However, now that the precision of the measurements has been established, the procedure has since been altered; both flasks, one aliquot from each, are now analysed for CH₄. This change allows pair agreement between flasks to be assessed; this can yield useful information relating to sample quality. All samples, including calibration gases, are air dried before analysis.

All CH₄ measurements have been calibrated against the original pair of standards obtained at the time the programme was started in 1983. The annual means represent 12-month running means, determined from smoothed curve fits to the

flask sampling data for 1984–1991. Blank spaces in the above table indicate that there are insufficient data to calculate a representative annual mean. Full details of the sampling, measurement and data analysis procedures are given by Lang et al. (1990a,b).

Sources and references:
Data supplied by E. J. Dlugokencky and P. M. Lang, National Oceanic and Atmospheric Administration, Climate Monitoring and Diagnostics Laboratory, Boulder, Colorado, USA.
Lang, P. M., Steele, L. P., Martin, R. C. and Masarie, K. A. 1990a *Atmospheric Methane Data for the Period 1983–1985 from the NOAA/CMDL Global Cooperative Flask Sampling Network, Technical Memorandum ERL CMDL-1,* National Oceanic and Atmospheric Administration, Boulder.
Lang, P. M., Steele, L. P. and Martin, R. C. 1990b *Atmospheric Methane Data for the Period 1986–1988 from the NOAA/CMDL Global Cooperative Flask Sampling Network, Technical Memorandum ERL CMDL-2,* National Oceanic and Atmospheric Administration, Boulder.

TABLE 1.5 Estimates of national emissions of methane from man-made sources, late 1980s (10^3 t a^{-1})

Region/country	Total	Energy production and use	Landfills	Biomass burning	Livestock	Rice cultivation
WORLD	352,398	79,188	35,726	35,819	103,170	98,495
AFRICA	33,642	4,646	1,758	11,680	12,625	2,932
Algeria	805	434	90		281	
Angola	334	99	21	57	145	13
Benin	95	10	14	6	61	5
Botswana	117	8	2	1	105	
Burkina	221	16	6	9	175	15
Burundi	50	8	3		31	9
Cameroon	436	53	42	96	240	5
Cent. African Rep.	140	7	11	10	109	4
Chad	276	7	13	5	233	18
Congo	82	24	7	43	6	2
Côte d'Ivoire	288	23	42	92	3	59
Egypt	1,035	261	129		338	306
Ethiopia	1,809	73	48	15	1,673	
Gabon	85	40	4	36	4	
Gambia	39	2	1	3	20	13
Ghana	190	1	19	30	97	43
Guinea	419	8	13	47	84	267
Guinea-Bissau	91	1	2	23	18	46
Kenya	853	69	39	21	712	12
Lesotho	61	4	3		54	
Liberia	94	9	8	56	8	12
Libya	390	266	24		100	
Madagascar	1,695	14	22	117	446	1,096
Malawi	110	26	9	10	48	17
Mali	506	10	13	2	351	130
Mauritania	201	0	6	2	185	7
Mauritius	10	4	4		3	
Morocco	532	12	96		415	8
Mozambique	229	30	26	20	71	82
Namibia	185	0	6	2	177	
Niger	342	9	10	6	306	11
Nigeria	2,213	509	301	187	944	273
Réunion	7	1	3		2	
Rwanda	63	11	4	4	43	2
Senegal	247	8	23	3	168	44
Sierra Leone	146	6	10	7	19	103
Somalia	944	13	17	6	905	3
South Africa	3,760	2,223	344		1,192	1
Sudan	1,426	41	44	39	1,301	1
Swaziland	37	5	2		30	
Tanzania	1,049	59	53	20	651	267
Togo	56	1	7	3	32	13
Tunisia	176	29	42		105	
Uganda	294	24	13	15	225	16

Continued

TABLE 1.5 Continued

Region/country	Total	Energy production and use	Landfills	Biomass burning	Livestock	Rice cultivation
Zaire	408	70	109	128	76	24
Zambia	231	28	32	51	116	4
Zimbabwe	363	91	20	5	246	
NORTH AMERICA	**38,357**	**12,899**	**11,059**	**2,400**	**11,224**	**775**
Canada	3,218	1,041	1,093		1,083	
USA	32,739	11,858	9,966		10,140	775
LATIN AMERICA	**40,722**	**3,909**	**2,684**	**12,988**	**18,600**	**2,541**
Argentina	4,662	890	250	116	3,326	80
Bolivia	502	23	30	48	342	60
Brazil	12,574	748	959	1,006	8,866	995
Chile	382	45	99		205	34
Colombia	1,972	224	201	238	1,102	207
Costa Rica	187	7	12	63	77	28
Cuba	543	45	71	2	246	180
Dominican Rep.	217	7	35	3	105	68
Ecuador	629	74	48	117	210	180
El Salvador	94	9	20	4	52	9
Guatemala	238	19	31	72	93	23
Guyana	100	1	2	2	13	82
Haiti	165	11	16	1	94	44
Honduras	213	11	17	51	125	9
Jamaica	32	2	11	2	17	
Mexico	3,625	986	532	266	1,787	54
Nicaragua	238	7	18	103	83	26
Panama	164	6	11	35	65	46
Paraguay	567	11	16	186	333	21
Peru	845	56	127	136	302	224
Trinidad & Tobago	106	91	7	0	5	2
Uruguay	701	6	24		586	85
Venezuela	1,465	630	148	38	565	85
MIDDLE EAST	**6,237**	**2,026**	**1,340**	**900**	**1,580**	**392**
Afghanistan	409	19	69		209	113
Bahrain	75	67	7		1	
Cyprus	17	1	7		10	
Iran	1,877	424	473		733	246
Iraq	670	231	232		174	33
Israel	113	3	79		31	
Jordan	55	1	35		19	
Kuwait	189	148	33		7	
Lebanon	56	2	44		10	
Oman	94	70	2		21	
Qatar	112	105	5		2	
Saudi Arabia	891	620	175		97	
Syria	260	29	107		125	

Continued

TABLE 1.5 Continued

Region/country	Total	Energy production and use	Landfills	Biomass burning	Livestock	Rice cultivation
United Arab Em.	327	292	22		13	
Yemen	134	13	32		90	
Yemen, Dem.	58	2	18		38	
SOUTH EAST ASIA	**90,484**	**4,576**	**3,325**	**5,151**	**17,737**	**59,695**
Bangladesh	8,290	93	117	4	1,089	6,988
Bhutan	58	6	1	1	19	32
Brunei	65	60	1	3	1	1
Cambodia	1,493	10	8	23	113	1,340
China, Taiwan	617	26	123		43	424
Hong Kong	91	3	85		4	
India	42,079	2,543	1,443	208	11,854	26,030
Indonesia	8,876	408	462	610	679	6,717
Korea	1,720	471	261	1	119	868
Malaysia	935	196	58	250	59	372
Myanmar	5,111	52	87	416	538	4,017
Nepal	1,373	36	13	44	498	783
Pakistan	3,714	254	279	4	1,756	1,421
Papua New Guinea	204	12	5	177	10	
Philippines	2,614	82	215	140	280	1,896
Singapore	61	10	42		9	
Sri Lanka	811	17	26	30	124	614
Thailand	9,372	299	100	240	540	8,192
CENTRALLY PLANNED ASIA	**60,655**	**18,818**	**2,356**	**1,200**	**8,174**	**30,107**
China	52,369	17,921	2,085		7,369	25,021
Korea, Dem.	1,367	761	127		74	406
Laos	331	7	6		85	233
Mongolia	375	35	16		324	
Viet Nam	4,986	95	122		322	4,448
EUROPE	**38,025**	**12,579**	**8,361**	**300**	**16,533**	**253**
Albania	127	26	17		82	2
Austria	380	67	27		286	
Belgium	633	122	236		274	
Bulgaria	599	207	101		284	7
Czechoslovakia	1,404	692	171		540	
Denmark	453	22	137		294	
Finland	257	21	107		129	
France	3,706	444	836		2,415	
Germany[a]	6,314	2,608	1,312		2,393	11
Greece	508	95	148		254	
Hungary	903	298	39		557	11
Iceland	23	0	5		17	9
Ireland	572	31	50		492	
Italy	2,133	250	710		1,069	105
Luxembourg	57	4	7		46	
Netherlands	1,122	362	171		589	

Continued

TABLE 1.5 Continued

Region/country	Total	Energy production and use	Landfills	Biomass burning	Livestock	Rice cultivation
Norway	312	117	78		117	
Poland	5,653	3,841	706		1,106	
Portugal	278	7	66		205	
Romania	1,658	595	188		843	32
Spain	2,003	357	767		842	38
Sweden	281	21	86		174	
Switzerland	261	30	33		197	
Turkey	2,029	152	575		1,268	34
UK	4,934	2,023	1,521		1,391	
Yugoslavia	1,125	186	266		668	5
USSR	**33,350**	**17,887**	**3,085**	**300**	**11,724**	**354**
OCEANIA & JAPAN	**10,925**	**1,848**	**1,758**	**900**	**4,974**	**1,445**
Australia	5,040	1,321	967		2,676	77
Fiji	17	2	3		9	4
Japan	3,539	464	721		989	1,365
New Zealand	1,429	61	68		1,300	

[a] Data refer to the unified Federal Republic of Germany.

Country-level estimates of anthropogenic emissions of methane (CH_4) are assembled from separate inventories for each of the sources given in the above data table. For each source, emissions are computed as a product of an "activity level" (i.e., a measure of the type and scale of the anthropogenic source) and an "emissions factor" (i.e., the quantity of gas emitted per unit of activity).

Emissions of CH_4 from energy sources comprise emissions from fossil fuel and wood fuel combustion sources, coal mining operations, plus fugitive emissions from oil and natural gas extraction operations and distribution systems. CH_4 losses from natural gas transportation and distribution are calculated from "loss" data provided by OECD/IEA Energy Balance Statistics coupled with recent studies conducted in the USA, Germany and Australia.

Emissions of CH_4 from landfills are computed as the product of solid waste generated, fraction landfilled, carbon content and carbon fraction converted to gaseous form. The estimates draw on the work of Bingemer and Crutzen (1987); more recent country-specific data on waste generation rates and disposal from selected countries (Europe, North America, Japan and some developing countries) have, however, been incorporated where possible. The calculation assumes that 50 per cent of degradable organic carbon in landfills is converted to biogas containing about 50 per cent CH_4 by volume (Bingemer and Crutzen, 1987).

CH_4 emissions from rice cultivation are computed on a country-specific basis as the product of the area under wet rice cultivation, the growing period (i.e., number of days rice is grown in a given year) and an average daily CH_4 emission rate. The latter are based on detailed studies of CH_4 emissions from Italian rice paddies reported by Schuetz et al. (1989).

Emissions of CH_4 from biomass burning refer to releases to the atmosphere resulting from deforestation and other forms of biomass burning such as the burning of grasslands and agricultural residues, shifting cultivation and prescribed burning. Emissions of CH_4 from deforestation are calculated on a country-specific basis; country data in the above table thus refer to emissions from this source only. Emissions from biomass burning sources other than deforestation have been estimated on a regional basis only due to the lack of country-specific data on these activities. Regional and world totals thus include an estimate of CH_4 emissions from grassland burning, shifting cultivation, agricultural residue burning and prescribed forest fires. Emissions from other biomass burning sources are estimated to be approximately 30×10^6 t CH_4 per year globally, with grassland burning, shifting cultivation and agricultural residue burning, each contributing 30 per cent (or about 9×10^6 t CH_4) to this total and prescribed fires the remaining 10 per cent. These estimates are based on the work of Crutzen and Andreae (1990). The grassland fraction has been ascribed to Africa, emissions from shifting cultivation to Latin America and those from prescribed forest fires to North America, Japan, Australia and New Zealand. It is assumed that 90 per cent of the agricultural residue burning takes place in the developing countries.

Emissions of CH_4 from livestock comprise emissions from enteric fermentation (about 75 per cent) plus those from decomposing animal wastes (about 25 per cent). CH_4 emissions from enteric fermentation are computed for each animal type as the product of the livestock population and a regionally-average CH_4 emission factor according to the methodology proposed by Crutzen et al. (1986). FAO estimates for national livestock populations for 1988 are used in the present study. Emissions from decomposing animal wastes (under anaerobic conditions) are also calculated by applying an emission factor to national livestock populations. The regionally-average emission factors used in this case are themselves a product of manure production, the amount of degradable organic material and the CH_4 conversion rate, and are based on the work of Casada and Safely (1990). For further details of the procedures employed in the calculation of CH_4 emissions, please refer to the source document and the additional references provided.

Sources and references:
Bingemer, H. G. and Crutzen, P. J. 1987 The production of methane from solid wastes, *Journal of Geophysical Research*, **90**(D2), 2181–2187.
Casada, M. E. and Safely, L. M. (Jr) 1990 *Global Methane Emissions from Livestock and Poultry Manure*, A report submitted to the US Environmental Protection Agency by the Biological and Agricultural Engineering Department, North Carolina State University, USA.
Crutzen, P. J. and Andreae, M. O. 1990 Biomass burning in the tropics: impact on atmospheric chemistry and biogeochemical cycles, *Science*, **250**, 1669–1677.
Crutzen, P. J., Aselmann, I. and Seiler, W. 1986 Methane production by domestic animals, wild ruminants, other herbivorous fauna, and humans, *Tellus*, **38B**, 271–284.
Schuetz, H., Holzapfel-Pschorn, A., Conrad, R., Rennenberg, H. and Seiler, W. 1989 A three-year continuous record on the influence of daytime, season fertilizer treatment on methane emission rates from an Italian rice paddy, *Journal of Geophysical Research*, **94**(D13), 16405–16416.
Subak, S., Raskin, P. and von Hippel, D. 1992 *National Greenhouse Gas Accounts: Current Anthropogenic Sources and Sinks*, Stockholm Environment Institute, Boston.

TABLE 1.6 Consumption of chlorofluorocarbons and halons in selected countries, 1986, 1990 and 1991

Region/country	CFCs[a] 1986 (t a⁻¹)	1990 (t a⁻¹)	1991 (t a⁻¹)	Halons[b] 1986 (t a⁻¹)	1990 (t a⁻¹)	1991 (t a⁻¹)	Total 1986 (t a⁻¹)	1990 (t a⁻¹)	1991 (t a⁻¹)	Per capita 1990 (kg a⁻¹)
WORLD							**1,549,933**	**1,027,748**		**0.72**
AFRICA	**16,620**	**8,736**		**3,303**	**47**		**26,168**	**8,782**		
Algeria							2,200[c]			0.10[d]
Côte d'Ivoire	168		258	5		19	173		277	0.02
Egypt[e]	2,803		544	0		36	2,803		580	0.16
Gabon							115[c]			0.10[d]
Ghana[e]	90	0.4	107	5		4	95	0.4	111	0.01
Kenya[e]	136	230			16		136	246		0.01
Madagascar	49			0			49			–
Morocco							2,200[c]			0.10[d]
Nigeria[e]		934						934		0.01
Senegal							600[c]			0.09[d]
South Africa[e]	12,790	6,841	4,795	3,290	28	1,121	16,080	6,870	5,916	0.19
Togo[e]							300			0.10[d]
Tunisia[e]	584	730	1,055	3	3	20	587	733	1,075	0.13[f]
Zimbabwe							830[c]			0.10[d]
NORTH AMERICA	**337,714**	136,532		61,250	40,361		400,711	176,920		
Belize							16[c]			0.09[d]
Canada[e]	19,958	13,174	8,819	3,218	2,128	1,642	23,176	15,302	10,461	0.58
Cuba[e]	884	778		4	3		888	781		0.06[d]
Dominica							2			0.03[d]
Dominican Rep.							620[c]			0.10[d]
El Salvador							480[c]			0.10[d]
Guatemala	1,800			80			1,880			0.23[d]
Honduras							160[c]			0.04[d]
Jamaica	196	424		27			196	451		0.46[d]
Mexico[e]	8,805	11,117		117	3,676		8,922	14,793		0.17
Nicaragua							300[c]			0.09[d]
Panama[e]	303			1			304			0.14[d]
USA[e]	305,964	111,039	173,062	57,803	34,554	33,453	363,767	145,593	206,515	0.88
SOUTH AMERICA	**24,750**	14,995		**1,496**	450		29,216	**15,445**		
Argentina[e]	5,015			1,076			6,091			0.20[d]
Bolivia							650[c]			0.10[d]
Brazil[e]	10,974	8,539			7		10,974	8,546		0.06
Chile[e]	730	684		61	239		791	923		0.07
Colombia	2,611	2,026	1,599		38	27	2,611	2,064	1,626	0.05[f]
Ecuador[e]	385	709					385	709		0.07
Paraguay							320[c]			0.08[d]
Peru	832			0			832			0.04[d]
Uruguay[e]	323		416	35		10	358		426	0.15[f]
Venezuela[e]	3,880	3,037	3,037	324	166	244	4,204	3,203	3,280	0.17[f]

Continued

TABLE 1.6 Continued

Region/country	CFCs[a] 1986 (t a^{-1})	1990 (t a^{-1})	1991 (t a^{-1})	Halons[b] 1986 (t a^{-1})	1990 (t a^{-1})	1991 (t a^{-1})	Total 1986 (t a^{-1})	1990 (t a^{-1})	1991 (t a^{-1})	Per capita 1990 (kg a^{-1})
ASIA	**173,384**	150,118		35,051	**25,075**		**236,736**	192,985		
Bahrain[e]	113		122	11		25	124		147	0.28[f]
Bangladesh[e]	_[c]			_[c]			_[c]			_[d]
China[e]	19,418	35,478	43,252	12,250	17,790	19,569	31,668[c]	53,268	62,821	0.05[f]
Cyprus[e]							500[c]			0.74[d]
India[e]	4,600			700			5,300			0.01[d]
Indonesia[e]	2,489			5			2,494			0.01[d]
Iran[e]							4,400[c]			0.10[d]
Iraq							1,590[c]			0.10[d]
Israel[e]							5,000			1.16[d]
Japan[e]	118,134	97,723	88,436	16,955	22,351		135,089	120,074	88,436	0.97
Jordan[e]	302	540	545	48	255	210	350	795	755	0.18
Korea[e]	9,244			150			9,394			0.23[d]
Kuwait	981			84			1,065			0.59[d]
Malaysia[e]	2,190	3,384	3,829	1,650	809	268	3,840	4,194	4,098	0.24[f]
Pakistan							10,000[c]			0.10[d]
Philippines[e]	4,299	2,957	2,023	60	185	89	4,359	3,142	2,112	0.05
Saudi Arabia							5,181			0.43[d]
Singapore[e]	4,052	3,167		2,439	1,151		6,491	4,318		1.44
Sri Lanka[e]	215	209	203	30	1	3	245	210	206	0.01[f]
Syria[e]	925			484			1,409			0.13[d]
Thailand[e]	2,300	6,660	7,904	60	324	420	2,360	6,984	8,324	0.16[f]
Turkey[e]	4,122		3,223	125		201	4,247		3,424	0.06[f]
United Arab Em.[e]							1,630			1.18[d]
EUROPE	**426,017**	249,794		94,504	**65,170**		**524,520**	315,138		
EEC Members[e,g]	302,477	170,331	150,640	40,993	37,783	35,420	343,470	208,114	186,060	0.60
Austria[e]	7,760	1,802	1,738	1,650	57	14	9,410	1,858	1,752	0.24
Bulgaria	2,180	2,434	1,556		22	16	2,180	2,456	1,592	0.17[f]
Czechoslovakia[e]	6,650	5,870		139	160		6,788	6,030		0.39
Denmark[e]	5,528	2,473		1,425	803		6,953	3,276		0.64
Finland[e]	3,301	1,859	1,199	598	516	362	3,899	2,375	1,561	0.31[f]
France[e]	71,018	38,989		34,465	23,776		105,483	62,765		1.12
Germany[e,h]	131,046	78,470		19,749	15,910		150,795	94,380		1.19
Hungary[e]	5,468	4,390	2,055	1,883	2,628	1,030	7,351	7,018	3,085	0.68
Iceland[e]	195	133	93	81	33	26	276	166	120	0.46[f]
Luxembourg[e]	136			4			140	136		0.36
Malta[e]	287	179	85	18	15	17	305	195	102	0.28[f]
Netherlands[e]	42,331	16,249			1,360		42,331	17,609		1.18
Norway[e]	1,313	722	414	1,411	1,332	879	2,724	2,054	1,293	0.30[f]
Poland[e]	6,656	4,939	2,562	3,900	330	828	10,556	5,269	3,390	0.14
Romania							4,000			0.17[d]
Spain[e]	19,832	23,596		600			20,432	23,596		0.61

Continued

TABLE 1.6 Continued

Region/country	CFCs[a] 1986 (t a⁻¹)	1990 (t a⁻¹)	1991 (t a⁻¹)	Halons[b] 1986 (t a⁻¹)	1990 (t a⁻¹)	1991 (t a⁻¹)	Total 1986 (t a⁻¹)	1990 (t a⁻¹)	1991 (t a⁻¹)	Per capita 1990 (kg a⁻¹)
Sweden[e]	4,962	1,818	1,119	1,831	396	259	6,793	2,214	1,378	0.16[f]
Switzerland[e]	7,960	2,920	2,186	1,050	473	352	9,010	3,394	2,538	0.40
UK[e]	102,014	58,081		16,500	15,036		118,514	73,117		1.27
Yugoslavia[e]	7,380	4,870		9,200	2,360		16,580	7,230		0.30
USSR[e]	110,654	110,654		28,752	28,752		139,406	139,406		0.50
OCEANIA	16,378	7,762		4,760	655		21,208	8,417		
Australia[e]	14,290	7,204	6,812	4,270	18	446	18,560	7,222	7,257	0.42[f]
Fiji[e]							70[c]			0.10[d]
New Zealand[e]	2,088	558	752	490	637	3	2,578	1,195	756	0.22[f]

[a] Chlorofluorocarbons refer to the Group I compounds: CFC-11; CFC-12; CFC-113; CFC-114; and CFC-115.

[b] Halons refer to the Group II compounds: Halon-1301; Halon-1211 and Halon-2402

[c] UNEP estimate.

[d] 1986.

[e] Parties to the Montreal Protocol.

[f] 1991.

[g] The EEC member states (Belgium, Denmark, France, Germany, Greece, Ireland, Italy, Luxembourg, Netherlands, Portugal, Spain and the UK) reported consumption data collectively. Some member countries also provided national data; these data are included where available.

[h] Data refer to the unified Federal Republic of Germany.

Data on consumption of CFCs (Group I) and halons (Group II) are based on official reports submitted to the Ozone Secretariat, UNEP, under the terms of the Montreal Protocol.

Data for 1986 (the baseline year for reductions in consumption) comprise official estimates for 60 countries, the EEC plus UNEP Secretariat estimates for 23 countries. As of August 1993, only 39 countries plus the EEC had reported complete consumption data for 1990; a further 6 countries reported incomplete data. As of August 1993, 46 countries had reported complete data for 1991. Regional totals include listed countries only; world totals represent UNEP estimates of the global total.

Source:
Data supplied by the Ozone Secretariat, United Nations Environment Programme, Nairobi in August 1993.

TABLE 1.7 Emissions of non-methane volatile organic compounds from man-made sources in selected countries, 1970–1990

Region/country	Total emissions (10^3 t a^{-1})					Per capita (kg a^{-1}) 1990	M/S ratio 1990	Source(s)
	1970	1975	1980	1985	1990			
NORTH AMERICA								
Canada[a]	2,017	2,168[b]	2,099	2,315	2,256	85.1	0.58	1
USA	26,200	22,000	22,300	20,000	18,500[c]	74.2	0.49	1
ASIA								
Israel			34	45	61	13.2		2
Malaysia					37			3
EUROPE								
Albania					33[d]			4
Austria			382	441	466[c]	62.2	0.43	1
Belgium					335[d]			4
Bulgaria					167[d]			4
Czechoslovakia				107	295[d]	18.8	0.89	4
Denmark			197	176[e]	176[d]	34.4	0.64[f]	1,4
Finland			163	181	162[d]	32.6	0.89	1,4
France			1,972	1,972[e]	1,972[d,e]	34.0	1.35	1,4
German Dem. Rep.				940	1,050[d]	63.1		4
Germany, Fed. Rep.	2,881	2,808	2,754	2,624	2,536[d]	41.9	1.01	1
Greece				657			0.32[f]	5
Hungary					205	19.4		4
Ireland[g]		48	62	64	108	29.0	1.51	1
Italy				1,521[e]	1,642[d,e]	28.6		4
Netherlands	540	555	502	416	381[h]	25.8	1.10	1
Norway		149	158	224	226[h]	53.7	0.41	1
Poland					1,280	33.3	0.55	4,5
Portugal			55	134	156[c]	15.2	0.66	1
Sweden[a]		432	410	446	440[c]	52.8	1.40	1
Switzerland	289	261	311	339[i]	297	45.5	0.27	1
UK	1,750	1,733	1,887	1,926	2,066[d]	36.3	0.62	1
USSR								
European USSR[j]				6,639	10,411			4
Belarus			174	150	164[c]			5
Ukraine				1,626	1,604[c]		2.04	5

[a] Data refer to total VOCs, i.e., including methane.
[b] 1976.
[c] 1988.
[d] 1989.
[e] EMEP estimate.
[f] 1985.
[g] Data for 1990 refer to 1987. A revised method for calculating emissions was introduced in 1987. This may account for at least part of the observed increase in emissions.
[h] Provisional estimate.
[i] 1984.
[j] Data refer to the European part of the USSR only, i.e., region within the EMEP area of calculation.

Emissions of non-methane volatile organic compounds (VOCs) presented here generally represent official country estimates and have been extracted from 'state of the environment'-type reports. Methods of estimation of emissions may vary between countries and therefore inter-country comparisons should be made with caution. The M/S ratio is the ratio of VOC emissions from mobile sources (i.e., transportation) to those from stationary sources.

Sources:
1. OECD 1991 Environmental Data Compendium 1991, Organisation for Economic Co-operation and Development, Paris.
2. Ministry of the Environment 1992 The Environment in Israel: National Report to UNCED, Ministry of the Environment, Jerusalem.
3. Department of the Environment 1991 Environmental Quality Report 1990, Ministry of Science, Technology and the Environment, Kuala Lumpur.
4. Simpson, D. and Styve, H. 1992 The Effects of the VOC Protocol on Ozone Concentrations in Europe, EMEP Meteorological Synthesizing Centre-West, The Norwegian Meteorological Institute, Oslo.
5. UN ECE 1992 The Environment in North America and Europe: Annotated Statistics 1992, United Nations, New York.

TABLE 1.8 Emissions of sulphur dioxide from man-made sources in selected countries, 1970–1990

Region/country	Total emissions (10^3 t a^{-1} as SO$_2$)					% change 1980–90	Per capita 1990 (kg a^{-1})	Per unit GDP 1990 (kg a^{-1})	Source(s)
	1970	1975	1980	1985	1990				
NORTH AMERICA									
Canada	6,677	5,319[a]	4,643	3,704	3,800[b,c]	−18	143.3	6.7	1
USA	28,400	25,900	23,400	21,100	21,100[d]	−10	84.7	3.9	1,2
ASIA									
Afghanistan		8.1	8.5	8.6	11[b]	26	3.9	2.7	3
Bangladesh		40	57	46	49[b]	−14	0.5	2.1	3
Brunei		0.4	0.9	1.1	1.1[b]	22	4.4	0.3	3
Cambodia		1.2	1.3	2.8	2.9[b]	123	0.3		3
China		10,180	13,370	17,260	19,990[b]	50	17.7	54.8	3
China, Taiwan		609	1,040	693	605[b]	−42	29.8	7.0	3
Hong Kong		109	166	144	150[b]	−10	25.9	2.5	3
India		1,650	2,010	2,830	3,070[b]	53	3.7	12.1	3
Indonesia		201	329	435	485[b]	47	2.7	4.5	3
Israel			308	250	273	−11	58.6	4.8	4
Japan		2,570	1,600	1,180	1,140[b]	−29	9.2	0.4	3
Korea		234	271	324	333[b]	23	7.9	1.4	3
Korea, Dem.		1,160	1,920	1,370	1,290[b]	−33	59.3	48.5	3
Kuwait			450[e]	627b	465	133	222.5	19.8	5,6
Laos		1.3	1.4	1.6	1.7[b]	21	0.4	2.0	3
Macao		0.9	3.0	6.2	8.4[b]	180	18.9		3
Malaysia		193	272	271	263[b]	−3	14.7	6.2	3
Maldives			−	0.3	0.3[b]	−	1.5		3
Mongolia		39	65	90	101[b]	55	49.4	42.5	3
Myanmar		17	31	30	30[b]	−3	0.7	4.1	3
Nepal		3.7	4.9	7.6	11.0[b]	124	0.6	3.8	3
Pakistan		148	198	351	381[b]	92	3.4	10.7	3
Philippines		807	1,040	510	370[b]	−64	6.0	8.4	3
Qatar					159		430.8	21.0	6
Saudi Arabia				1,150[f]			99.2[f]	14.2	7
Singapore		85	122	147	155[b]	27	51.7	4.5	3
Sri Lanka		22	30	24	28[b]	−6	1.7	3.9	3
Thailand		224	420	507	612[b]	46	11.2	7.6	3
Turkey			276	322	398[b]	44	7.2	4.1	8
United Arab Em.				165[g]					6
Viet Nam		40	34	38	39[b]	14	0.6	4.2	3
EUROPE									
Albania			50[h]	50[h]	50[h]			16.3	8
Austria			390	190	98	−75	13.1	0.6	1,8
Belgium			828	452	420	−49	42.3	2.2	1,8
Bulgaria			1,034	1,094	1,030	−	114.6	45.3	8
Czechoslovakia			3,100	2,782	2,564[h]	−17	177.1	57.7	8

Continuod

TABLE 1.8 Continued

Region/country	Total emissions (10^3 t a^{-1} as SO_2)					% change 1980–90	Per capita 1990 (kg a^{-1})	Per unit GDP 1990 (kg a^{-1})	Source(s)
	1970	1975	1980	1985	1990				
Denmark	574	418	448	339	266	−41	52.0	2.0	1,8
Finland	515	535	584	382	256	−56	51.5	1.9	1,8
France	2,966	3,328	3,338	1,470	1,206	−64	21.5	1.0	1,8
German Dem. Rep.			4,264	5,340	5,242[h]	23	314.9	34.8	8
Germany, Fed. Rep.	3,750	3,350	3,194	2,396	1,002	−69	16.6	0.7	1,8
Greece			400	500	500[h]	25	50.3	8.6	8
Hungary			1,632	1,404	1,010	−38	95.7	30.7	8
Iceland			6	6	6[h]	–	24.8	1.5	8
Ireland[i]		186	222	140	168	−24	45.2	4.0	1,8
Italy			3,800	2,504	2,406[h]	−37	42.0	2.2	8
Luxembourg			24	16	10[h]	−58	32.7	1.6	1,8
Netherlands	772	386	466	276	238	−49	16.1	0.9	8
Norway	171	137	142	98	60	−58	14.2	0.6	1,8
Poland			4,100	4,300	3,210	−22	83.5	50.5	8
Portugal	116	178	266	198	212	−20	20.6	3.7	1,8
Romania			1,800	1,800[h]	1,800[h]	–	45.6	28.1	8
Spain			3,250	2,190	2,190[h]	−33	56.7	4.5	8
Sweden	930	690	514	270	204	−60	24.5	0.9	1,8
Switzerland	125	109	126	96	62	−51	9.5	0.3	1,8
UK	6,330	5,310	4,898	3,724	3,774	−23	66.3	3.9	8,9
Yugoslavia			1,300	1,500	1,480	14	62.1	18.0	1,8
USSR[j]			**20,051**	**19,746**	**16,488**	**−18**	**57.3**		**10**
Armenia			141	100	73	−48	20.9[k]		10
Azerbaijan			119	140	90	−24	12.4[k]		10
Belarus			740	699	562	−24	54.6[k]		10
Estonia			462	205	192	−58	121.3[k]		10
Georgia			59	123	76	28	13.9[k]		10
Kazakhstan			1,607	1,612	1,484	−8	87.0[k]		10
Kyrgyzstan			62	69	56	−10	12.4[k]		10
Latvia			60	65	54	−11	20.0[k]		10
Lithuania			228	223	143	−37	38.0[k]		10
Moldova			289	272	231	−20	52.9[k]		10
Russia			12,123	11,945	10,166	−16	68.2[k]		10
Tajikistan			7.0	6.5	17	136	3.0[k]		10
Turkmenistan			5	15	22	315	5.8[k]		10
Ukraine			3,850	3,663	2,782	−28	53.3[k]		10
Uzbekistan			298	608	542	82	25.2[k]		10

[a] 1976.
[b] 1987.
[c] OECD Secretariat estimate.
[d] 1988.
[e] 1983.
[f] 1986.
[g] 1984.
[h] EMEP estimate.
[i] A revised method for calculating emissions was introduced in 1987.
[j] Data refer to emissions from 46,684 industrial and power generating combustion sources in 611 cities throughout the USSR. Emissions estimates are considered to be more reliable after 1985.

[k] Per capita estimates based on UN population data for 1992.

Note that emissions are given in units of 10^3 t a^{-1} as sulphur dioxide (SO_2); to convert to emissions in 10^3 t a^{-1} as sulphur (S) divide by 2.0.

Data presented in the above table generally represent official country emissions estimates (with the exception of selected countries in Asia) as reported in 'state of the environment'-type reports or as reported to EMEP. As methods of estimation may vary between countries, inter-country comparisons should be made with caution. Trends observed within each country are more reliable than comparisons between countries.

Notes continued

TABLE 1.8   Continued

Data for the majority of countries in Asia listed here are derived from an independent study of emissions reported by Kato and Akimoto (1992). Estimates of emissions are made by applying country-specific emission factors to IEA/OECD fuel consumption statistics (IEA/OECD 1990 *World Energy Statistics and Balances 1971–1988*, International Energy Agency, Organisation for Economic Co-operation and Development, Paris).

Per capita emissions of SO_2 in 1990 are based on UN population statistics for 1990 unless otherwise indicated. The calculation of emissions of SO_2 per unit GDP is based on World Bank estimates of GDP in 1990 in US$ as reported in the 'World Development Report 1992'. Emissions of SO_2 per unit GDP are given in kg a^{-1} per 10^3 US$.

Sources:

1. OECD 1991 *Environmental Data Compendium 1991*, Organisation for Economic Co-operation and Development, Paris.
2. US EPA 1991 *National Air Quality and Emissions Trends Report 1989*, Office of Air Quality Planning and Standards, US Environmental Protection Agency, Research Triangle Park.
3. Kato, N. and Akimoto, M. 1992 Anthropogenic emissions of SO_2 and NO_x in Asia: emission inventories, *Atmospheric Environment*, **26A**(16), 2997–3017.
4. Ministry of the Environment 1992 *The Environment in Israel: National Report to UNCED*, Ministry of Environment, Jerusalem.
5. UNEP/WHO 1988 *Assessment of Urban Air Quality*, United Nations Environment Programme, Nairobi and World Health Organization, Geneva.
6. ROPME 1992 *Urban Air Quality Monitoring Capabilities in the Middle East*, Report prepared for the forthcoming GEMS/Air study on air quality monitoring capabilities, Regional Organization for Protection of the Marine Environment, Kuwait. Unpublished report.
7. Ahmed, A. F. M. 1990 SO_2 and NO_x emissions due to fossil fuel combustion in Saudi Arabia: a preliminary inventory, *Atmospheric Environment*, **24A**(12), 2917–2926.
8. Sandes, M. and Styve, H. 1992 *Calculated Budgets for Airborne Acidifying Components in Europe, 1985, 1987, 1988, 1989, 1990 and 1991*, EMEP Meteorological Synthesizing Centre-West, The Norwegian Meteorological Institute, Oslo.
9. Department of the Environment 1992 *The Digest of Environmental Protection and Water Statistics, No. 14: 1991*, HMSO, London.
10. Data supplied by the Global Institute of Ecology and Climate, USSR Academy of Sciences, Moscow, Russia.

TABLE 1.9 Emissions of nitrogen oxides from man-made sources in selected countries, 1970–1990

Region/country	Total emissions (10^3 t a^{-1} as NO_2)					% change 1980–90	M/S ratio 1990	Per capita 1990 (kg a^{-1})	Per unit GDP 1990 (kg a^{-1})	Source(s)
	1970	1975	1980	1985	1990					
NORTH AMERICA										
Canada	1,364	1,756[a]	1,959	1,959	1,943[b]	−1	1.59	73.3	3.4	1
USA	18,300	19,200	20,400	19,800	19,800[c]	−3	0.69	79.8	3.7	1,2
ASIA										
Afghanistan		20	22	24	30[d]	36		1.8	7.6	3
Bangladesh		46	58	61	66[d]	13		0.6	2.9	3
Brunei		2	4	8	11[d]	175		44.2	2.9	3
Cambodia		9	9	12	12[d]	30		1.4		3
China		3,730	4,910	6,360	7,370[d]	50		6.5	20.2	3
China, Taiwan		124	225	261	325[d]	44		16.0	3.7	3
Hong Kong		51	88	95	99[d]	12		17.1	1.7	3
India		1,380	1,670	2,310	2,560[d]	53		3.1	10.1	3
Indonesia		331	465	561	639[d]	37		3.6	6.0	3
Israel			80	112	148	85		31.8	2.8	4
Japan		2,330	2,130	1,950	1,940[d]	−9		15.7	0.7	3
Korea		220	365	464	555[d]	52		13.1	2.3	3
Korea, Dem.		325	383	456	468[d]	22		21.5	17.6	3
Laos		8	8	9	9[d]	14		2.2	10.5	3
Macao		2	3	4	5[d]	72		11.3		3
Malaysia		90	126	167	177[d]	40		9.9	4.2	3
Maldives				1	1[d]			2.9		3
Mongolia		31	49	66	72[d]	46		35.3	30.4	3
Myanmar		38	47	50	45[d]	−3		1.1	6.2	3
Nepal		18	21	34	50[d]	137		2.6	17.3	3
Pakistan		101	164	193	231[d]	41		2.1	6.5	3
Philippines		172	184	173	184[d]	0		3.0	4.2	3
Qatar					51			138.2	6.7	6
Saudi Arabia				775[e]			0.14	66.8[f]		7
Singapore		43	67	81	88[d]	31		29.3	2.5	3
Sri Lanka		23	31	34	37[d]	20		2.2	5.1	3
Thailand		182	255	327	384[d]	51		7.0	4.8	3
Turkey			175[g]	175[g]	175[g]			3.0	1.8	8
United Arab Em.				148[e]						6
Viet Nam		120	88	95	99[d]	12		1.5	10.6	3
EUROPE										
Albania			9[g]	9[g]	9[g]			2.8	2.9	8
Austria			233	232	209	−10	2.28	27.9	1.3	1,8
Belgium			317	281	300	−5	0.79	30.2	1.6	1,8
Bulgaria			150	150	150			16.8	6.6	8
Czechoslovakia			1,204	992	1,122[h]	−7	0.19	71.3	39.2	8
Denmark		178	245	258	254	4	0.56	49.6	1.9	1,8
Finland		160[i]	264	251	290	10	1.07	58.3	2.1	1,8
France	1,322	1,608	1,823	1,615	1,742	−4	3.50	31.0	1.5	1,8

Continued

TABLE 1.9 Continued

Region/country	Total emissions (10^3 t a^{-1} as NO$_2$)					% change 1980–90	M/S ratio 1990	Per capita 1990 (kg a^{-1})	Per unit GDP 1990 (kg a^{-1})	Source(s)
	1970	1975	1980	1985	1990					
German Dem. Rep.			630	670	705[h]	12	0.74	42.3	4.7	8
Germany, Fed. Rep.	2,381	2,571	2,980	2,959	2,707,[h]	−9	2.11	44.7	1.8	1,8
Greece			746[g]	746	746[g]		2.97	75.1	12.9	8
Hungary			273	262	238	−13	0.97	22.6	7.2	8
Iceland			13	12	12[g]	−		49.8	3.0	8
Ireland[j]		60	73	91	135	85	0.89	36.3	3.2	1,8
Italy			1,480	1,595	1,755[g]	19	1.09	30.6	1.6	1,8
Luxembourg			23	19	15	−35	1.75	40.9	2.4	1,8
Netherlands[k]	456	464	553	544	529	−4	1.73	35.9	1.9	1,8
Norway	159	176	185	203	212	15	3.44	50.1	2.0	1,8
Poland			1,500[g]	1,500	1,280		0.46	33.3	20.1	8
Portugal	72	104	166	96	142	−14	1.67	13.8	2.5	1,8
Romania			390[g]	390[g]	390[g]			16.8	11.2	8
Spain		625	950	950	950	0	1.14	24.2	1.9	1,8
Sweden	302	310	398	394	373	−6	4.12	23.2	1.6	1,8
Switzerland	149	162	196	214	184	−6	2.12	28.2	0.8	1,8
UK	2,510	2,427	2,442	2,402	2,690	10	1.19	47.3	2.8	1,8
Yugoslavia			350	400	420	20	0.65	17.6	5.1	1,8
USSR[l]			5,486	5,670	6,729	23	0.45	23.4[m]		10
Armenia			16	23	23	38		6.5[m]		10
Azerbaijan			50	66	59	19		8.1[m]		10
Belarus			98	84	101	3		9.8[m]		10
Estonia			46	28	20	−56		12.6[m]		10
Georgia			40	43	24	−40		4.3[m]		10
Kazakhstan			169	221	330	95		19.4[m]		10
Kyrgyzstan			11	11	12	16		2.7[m]		10
Latvia			10	11	14	37		5.3[m]		10
Lithuania			30	38	35	18		9.4[m]		10
Moldova			51	42	39	−25		8.9[m]		10
Russia			2,578	2,498	3,050	18		20.5[m]		10
Tajikistan			6	8	8	42		1.5[m]		10
Turkmenistan			15	34	35	135		9.0[m]		10
Ukraine			841	754	761	−9		14.6[m]		10
Uzbekistan			80	111	117	47		5.5[m]		10

[a] 1976.
[b] OECD Secretariat estimate.
[c] 1988.
[d] 1987.
[e] 1986.
[f] Data approximate to per capita emissions in 1985.
[g] EMEP estimate.
[h] 1989.
[i] 1977.
[j] A revised method for calculating emissions was introduced in 1987.
[k] Emissions from industrial processes are included after 1987.
[l] Data refer to emissions from 46,684 Industrial and power generating combustion sources located in 611 cities throughout the USSR. Emissions estimates are considered to be more reliable after 1985.

[m] Per capita estimates are based on UN population data for 1992.

Note that emissions are given in units of 10^3 t a^{-1} as nitrogen dioxide (NO$_x$); to calculate the emissions in 10^3 t a^{-1} as nitrogen (N), divide by 3.29.

Per capita emissions of NO$_x$ in 1990 are based on UN population statistics for 1990 unless otherwise indicated. The calculation of emissions of NO$_x$ per unit GDP is based on World Bank estimates of GDP in 1990 in US$ as reported in the 'World Development Report 1992'. Emissions of NO$_x$ per unit GDP are given in kg a^{-1} per 10^3 US$. The M/S ratio is the ratio of NO$_x$ emissions from mobile (i.e., transportation) sources to those from stationary sources.

The remaining notes and sources are as those given in Table 1.8

TABLE 1.10 Concentrations of suspended particulate matter at selected GEMS/Air sites, 1980–1990 (mean annual values, $\mu g\ m^{-3}$ and number of observations)

Total suspended particulate concentrations (high-volume gravimetric or equivalent method)

Region/country	City	Site	1980	1981	1982	1983	1984	1985	1986	1987	1988	1989	1990
AFRICA													
Ghana	Accra	SR	92 (18)	93 (35)				95 (3)	107 (26)	119 (38)	147 (38)	92 (26)	124 (17)
		SI	109 (21)	129 (34)					123 (27)	129 (38)	152 (40)	117 (35)	150 (52)
NORTH AMERICA													
Canada	Hamilton	SR[a]	112 (55)	86 (54)	95 (58)	91 (56)	106 (52)	75 (53)	85 (57)	83 (52)	93 (55)		
		CCC	99 (56)	87 (46)	102 (58)	95 (54)	94 (57)	76 (51)	92 (60)	89 (56)	88 (55)		
	Montreal	SR	75 (49)	44 (57)	46 (56)	44 (53)	42 (49)	37 (56)	34 (47)	41 (44)	29 (56)	32 (45)	34 (45)
		SR	84 (60)	81 (57)	60 (60)	68 (59)	68 (51)	58 (51)	55 (56)	66 (56)	59 (55)		
		CCC	69 (56)	71 (59)	64 (61)	56 (58)	59 (58)	56 (55)	50 (56)	63 (58)	59 (55)		
	Toronto	SI	99 (58)	103 (60)	87 (59)	59 (53)	68 (58)	60 (56)	66 (55)	72 (61)	75 (59)		
		SR	76 (57)	67 (52)	62 (59)	58 (58)	69 (58)	54 (58)	59 (56)	60 (58)	53 (57)	77 (59)	53 (57)
		CCC		59 (34)	61 (61)	65 (59)	66 (60)	53 (61)	57 (56)	59 (58)	62 (56)	66 (60)	62 (48)
	Vancouver	CCR	62 (51)	47 (53)	42 (51)	38 (49)	35 (45)	44 (45)	38 (4)	90 (59)	58 (55)	51 (57)	40 (57)
		SI	79 (52)	75 (38)	63 (53)	50 (57)	46 (56)	65 (56)	63 (54)	43 (55)	41 (58)	31 (51)	29 (54)
		CCC	76 (51)	73 (60)	61 (55)	45 (54)	49 (50)	56 (56)	51 (58)	36 (52)	32 (54)		
USA	Birmingham	CCC	95 (290)	82 (249)	69 (229)	78 (292)	74 (304)	71 (302)	76 (285)	96 (250)	91 (137)	89 (61)	94 (59)
		CCI	121 (290)	120 (209)	99 (67)	82 (219)	89 (304)	87 (297)	92 (283)	127 (61)	120 (58)	125 (56)	114 (61)
	Azusa	SI					115 (57)	105 (61)	106 (60)	92 (60)	93 (61)	87 (61)	88 (61)
	Long Beach	SC	129 (52)										
	Chicago	CCI		52 (56)	96 (59)	107 (56)	98 (54)	104 (50)	87 (54)	71 (55)	82 (50)	74 (52)	79 (55)
		CCR		65 (56)	72 (61)	86 (57)	77 (54)	73 (51)	71 (57)	72 (53)	68 (42)		
		CCI		68 (55)	95 (60)	99 (57)	82 (60)	84 (56)	78 (60)	77 (54)	86 (56)	87 (60)	
	New York City	SR	51 (56)	64 (61)	45 (55)	43 (59)	50 (54)	48 (53)	44 (55)	50 (58)	61 (57)	65 (47)	56 (61)
		CCR	62 (50)	56 (59)	58 (58)	54 (58)	62 (60)	68 (56)	59 (56)	60 (60)	68 (60)	69 (53)	67 (28)
		CCI	63 (53)	75 (53)	60 (54)	63 (57)	67 (60)	74 (57)	64 (59)	73 (56)	59 (61)	50 (58)	48 (59)
	Chattanooga	CCC	62 (43)		58 (58)	63 (57)	60 (61)	60 (57)	80 (62)	62 (60)	52 (25)		
		CCR			46 (61)	50 (60)	50 (58)	43 (58)	51 (60)	47 (59)	48 (18)		
	Houston	CCC	82 (52)		81 (56)	69 (57)	66 (54)	59 (54)	53 (60)	51 (58)			
		SR	93 (31)	104 (33)	85 (44)	72 (51)	67 (55)	62 (56)	54 (57)				
SOUTH AMERICA													
Brazil	São Paulo	CCR	162 (28)	129 (50)	110 (55)	92 (58)	101 (58)	100 (56)					

Continued

TABLE 1.10 Continued

Region/country	City	Site	1980	1981	1982	1983	1984	1985	1986	1987	1988	1989	1990
Venezuela	Caracas	NA		132 (27)	132 (37)	96 (43)	95 (25)	50 (57)					
		NA		77 (33)	66 (36)	61 (37)	109 (36)	105 (57)					
ASIA													
China	Beijing	SI		479 (75)	454 (162)	421 (146)	512 (88)		385 (164)	400 (152)	414 (152)	418 (161)	430 (168)
		CCC		422 (76)	527 (165)	504 (160)	595 (84)		402 (163)	417 (157)	415 (168)	407 (173)	418 (179)
		SR		252 (62)	277 (165)	220 (151)	308 (76)		234 (156)	250 (165)	268 (171)	247 (174)	371 (174)
		CCR		415 (79)	409 (163)	384 (163)	400 (81)		356 (153)	367 (164)	375 (164)	369 (175)	371 (175)
	Guangzhou	SR		375 (38)	335 (151)	343 (150)	305 (148)	301 (147)	308 (145)	297 (145)	324 (167)	282 (165)	114 (155)
		CCC		117 (39)	174 (158)	178 (163)	206 (150)	216 (144)	234 (145)	240 (148)	242 (148)	232 (159)	176 (157)
		CCC		260 (38)	235 (150)	234 (163)	197 (144)	160 (149)	199 (146)	174 (150)	164 (150)	159 (160)	142 (157)
		SR		96 (24)	119 (151)	103 (158)	92 (145)	109 (147)	131 (144)	143 (145)	129 (145)	129 (167)	269 (162)
	Shanghai	CCI		330 (89)	335 (174)	271 (182)	249 (88)	233 (175)	274 (176)	280 (177)	340 (177)	291 (163)	269 (166)
		CCR		235 (87)	245 (178)	205 (178)	212 (180)	225 (180)	270 (171)	290 (175)	305 (170)	289 (166)	280 (163)
		CCC		205 (88)	243 (177)	200 (180)	198 (174)	155 (176)	232 (177)	249 (178)	278 (176)	231 (167)	211 (167)
		SR				151 (178)	151 (177)		185 (173)	195 (179)	215 (179)	192 (167)	161 (166)
	Shenyang	SR		465 (72)	476 (144)	517 (144)	507 (144)	463 (144)	437 (147)	372 (144)	532 (144)	492 (144)	393 (144)
		CCR		523 (72)	520 (144)	545 (144)	496 (144)	554 (144)	507 (144)	489 (144)	536 (145)	406 (144)	447 (144)
		CCI		406 (72)	412 (144)	464 (144)	504 (72)		457 (144)	428 (144)	456 (144)	422 (144)	394 (145)
		CCC		225 (72)	213 (144)	245 (144)	317 (72)		254 (145)	256 (144)	260 (144)	232 (145)	209 (144)
	Xian	SR		235 (97)	315 (199)	273 (176)	359 (178)	363 (154)	395 (144)	356 (132)	472 (132)	455 (132)	361 (144)
		CCR		387 (120)	414 (209)	387 (182)	477 (174)	493 (155)	584 (144)	490 (144)	592 (143)	658 (144)	487 (144)
		CCC		350 (115)	447 (211)	395 (184)	497 (178)	526 (150)	642 (144)	518 (144)	641 (144)	526 (144)	444 (144)
		SI		463 (117)	439 (211)	384 (185)	531 (175)	528 (156)	575 (144)	489 (144)	623 (144)	552 (144)	444 (144)
Hong Kong	Hong Kong	CCC						72 (29)	89 (101)	90 (97)	95 (107)		
		CCI						95 (26)	102 (107)	120 (105)	143 (103)		
		CCR						76 (28)	89 (106)	90 (93)	102 (107)		
India	Bombay	CCC	142	141 (32)	356 (29)		104 (3)	175 (32)					
		SR		203 (32)	431 (26)		244 (30)	289 (29)					
		CCC		207 (34)	378 (31)		230 (26)	227 (30)					
	Calcutta	CCC	462 (31)	384 (12)	344 (31)		412 (33)	374 (31)					
		SI	356 (30)	575 (15)	270 (31)		443 (19)	405 (26)					
		SR	393 (27)	604 (16)	374 (31)		323 (29)	297 (30)					
	Delhi	CCC	535 (35)	486 (35)			481 (29)	439 (27)					
		CCR	322 (39)	349 (32)			308 (32)	294 (25)					
		CCI	453 (40)	441 (33)			498 (30)	488 (27)					

Continued

TABLE 1.10 Continued

Region/country	City	Site	1980	1981	1982	1983	1984	1985	1986	1987	1988	1989	1990
Indonesia	Jakarta	CCR	274 (57)	273 (53)	215 (40)	271 (46)	201 (47)	204 (43)	204 (21)	183 (53)	188 (51)	237 (53)	
		SI	167 (50)	172 (56)	18 (20)								
		CCI			359 (21)								
		NA				262 (36)	180 (44)	175 (28)	212 (22)	156 (54)	188 (55)		
Iran	Tehran	CCC	292 (80)	129 (86)	161 (57)	278 (61)	243 (62)	271 (38)	200 (22)	199 (16)	216 (35)	287 (58)	241 (22)
		SI	370 (60)	188 (74)	259 (52)	355 (62)	318 (62)	372 (32)	275 (19)	254 (11)	299 (14)	299 (42)	277 (13)
		SR	291 (65)	123 (76)	160 (45)	277 (64)	230 (58)	258 (36)	237 (22)	204 (10)	189 (20)	246 (52)	220 (8)
Japan	Osaka	CCC	49 (361)	51 (364)	47 (363)	39 (363)	41 (366)	40 (364)	43 (358)	43 (360)	41 (363)	41 (365)	42 (338)
		CCI	60 (362)	61 (358)	60 (354)	47 (325)	49 (366)	48 (325)	47 (365)	54 (365)	54 (362)	54 (365)	56 (365)
		SR	58 (362)	60 (364)	57 (363)	48 (361)	50 (360)	50 (364)	49 (345)	54 (346)			
	Tokyo	CCI	53 (364)	54 (360)	52 (352)	36 (275)	39 (366)	36 (363)	41 (360)	46 (365)	44 (360)	45 (361)	43 (304)
		CCC	59 (354)	63 (365)	57 (365)	54 (364)	64 (334)	56 (363)	58 (355)	60 (350)	55 (366)	47 (356)	56 (365)
		SR	52 (315)	53 (363)	53 (365)	51 (361)	56 (362)	51 (290)	47 (365)	55 (365)	48 (359)	46 (365)	49 (350)
		CCI					52 (314)	51 (349)	61 (361)	62 (365)	55 (356)	58 (357)	63 (353)
Malaysia	Kuala Lumpur	SI	182 (137)	159 (86)	120 (53)	104 (99)	188 (121)	139 (109)	124 (29)	138 (70)	136 (49)		
		SR		247 (41)	96 (50)		104 (20)		162 (42)	120 (65)	121 (47)		
Pakistan	Lahore	SR	690 (106)	796 (59)						619 (36)	400 (125)	468 (92)	405 (64)
Thailand	Bangkok	SI	174 (79)	244 (111)	265 (124)	247 (114)	218 (81)	204 (104)	320 (51)	241 (45)	292 (8)	198 (67)	
		SR	115 (55)	118 (136)	151 (78)	163 (156)	205 (75)	156 (131)	127 (110)	92 (80)	107 (85)	115 (105)	
		SR	232 (11)		169 (93)	188 (108)	237 (118)	236 (84)	261 (125)	242 (69)	204 (104)	194 (109)	
EUROPE													
Belgium	Brussels	CCC	25 (366)	20 (362)	20 (364)	17 (358)	23 (326)	24 (362)	24 (364)				
Denmark	Copenhagen	SI	46 (343)	46 (364)	69 (120)	49 (347)	60 (361)	55 (335)	55 (341)				
Germany, Fed. Rep.	Frankfurt	CCC	25 (318)	21 (251)	22 (324)	44 (94)	31 (314)	46 (214)	35 (358)	39 (325)	39 (91)	43 (86)	36 (89)
Finland	Helsinki	CCC	75 (167)	67 (175)	80 (178)	83 (174)	78 (163)		75 (170)	74 (163)	83 (172)	87 (175)	
		SI	76 (161)	56 (198)	62 (172)	61 (175)	73 (169)		69 (159)	59 (163)	66 (176)	60 (171)	
Greece	Athens	SI	202 (41)	186 (28)	125 (4)	240 (33)	183 (53)	167 (16)	136 (29)				
		CCC	218 (57)	218 (85)	195 (20)	192 (38)	193 (94)	173 (62)	155 (44)				
Portugal	Lisbon	CCR		99 (128)	98 (89)	94 (97)	89 (88)	122 (107)	83 (16)	85 (75)	75 (49)	144 (32)	91 (12)
		CCR		105 (50)	104 (56)	118 (64)	105 (93)	128 (100)	105 (30)	120 (39)	103 (42)	159 (19)	138 (10)
		SR			100 (42)	106 (104)	86 (79)	110 (67)	76 (92)	81 (84)	76 (56)	99 (43)	156 (9)

Continued

TABLE 1.10   Continued

Smoke concentrations (smoke shade reflectance method)

Region/country	City	Site	1980	1981	1982	1983	1984	1985	1986	1987	1988	1989	1990
Yugoslavia	Zagreb	CCC	119 (197)	105 (179)	110 (177)	122 (181)	118 (181)	138 (162)	130 (149)	116 (173)	136 (172)	127 (137)	130 (131)
		SR	130 (201)	138 (193)	123 (228)	122 (202)	105 (215)	124 (185)	115 (201)	115 (193)	83 (201)	81 (196)	64 (196)
		CCI	122 (232)	111 (231)	106 (228)	111 (187)	107 (189)	113 (206)	121 (144)	129 (198)	118 (154)		
OCEANIA													
Australia	Melbourne	CCC	73 (46)	66 (36)	74 (53)	58 (55)	58 (24)						
	Sydney	CCC	118 (55)	89 (56)	91 (57)	91 (57)	113 (58)	138 (59)					
		SI	85 (56)	63 (57)	71 (59)	69 (58)	55 (57)	51 (54)					
New Zealand	Christchurch	NA									34 (211)	29 (291)	22 (356)
AFRICA													
Egypt	Cairo	CCC	101 (44)						73 (216)	56 (238)	68 (202)	118 (213)	49 (171)
		SI							130 (144)	77 (109)	76 (171)	45 (194)	42 (207)
		SR	83 (17)						63 (205)	54 (305)	78 (173)	31 (198)	38 (180)
SOUTH AMERICA													
Brazil	São Paulo	CCR	80 (365)	92 (364)	75 (363)	66 (365)	74 (365)	72 (362)	42 (61)	56 (59)	66 (57)		
		CCM	126 (361)	134 (363)	112 (361)	103 (362)	111 (362)	97 (363)	76 (61)	71 (56)	65 (59)		
		CCR	56 (362)	72 (363)	62 (365)	60 (365)	69 (366)	63 (363)	49 (61)	50 (57)	50 (60)		
Chile	Santiago	CCC	100 (313)	149 (337)	203 (287)	232 (343)	97 (335)	75 (362)					
		CCR	40 (246)	48 (335)	74 (328)	108 (347)	64 (359)						
Venezuela	Caracas	CCC	23 (183)	28 (221)	27 (230)	32 (276)	29 (253)	32 (221)	28 (198)	25 (258)	20 (273)		
ASIA													
Hong Kong	Hong Kong	CCC	49 (366)	46 (365)	57 (356)	45 (364)	43 (207)						
		SI	25 (366)	22 (365)	21 (355)	21 (364)	21 (208)						
Iran	Tehran	CCC	175 (80)	121 (91)	112 (71)	219 (54)	142 (71)	114 (48)	113 (28)	73 (19)	113 (34)	93 (62)	155 (19)
		SI	222 (69)	146 (88)	218 (76)	197 (65)	138 (59)	118 (32)	187 (19)	159 (13)	269 (15)	150 (37)	181 (11)
		SR	128 (76)	68 (90)	71 (69)	83 (61)	61 (61)	59 (32)	63 (25)	31 (10)	63 (21)	48 (41)	56 (5)
EUROPE													
Denmark	Copenhagen	CCC	23 (366)	26 (365)	36 (120)	29 (365)	33 (317)	28 (356)	28 (365)				
		SI	16 (366)	14 (365)	22 (118)	19 (361)	25 (365)	22 (363)	18 (334)				

Continued

TABLE 1.10 Continued

Region/country	City	Site	1980	1981	1982	1983	1984	1985	1986	1987	1988	1989	1990
Greece	Athens	SI				34 (175)	38 (204)	34 (197)	25 (342)				
		CCC				107 (174)	122 (340)	129 (360)	91 (346)				
Ireland	Dublin	CCR	49 (348)	51 (357)	51 (364)	45 (365)	60 (362)	50 (362)	45 (354)	77 (341)	67 (358)	43 (361)	
		CCI	17 (361)	31 (363)	40 (364)	25 (349)	29 (230)	39 (357)	30 (312)	56 (346)	50 (366)	39 (364)	
		SR	19 (364)	34 (362)	44 (321)	22 (338)	30 (349)	28 (351)	31 (356)	59 (324)	43 (361)	47 (265)	
Luxembourg	Vianden	NA	7 (331)	6 (365)	6 (351)			7 (364)	4 (348)	4 (358)	3 (317)		
Poland	Warsaw	CCR	44 (188)	37 (118)	57 (145)	48 (194)	56 (212)	69 (213)	60 (198)	47 (194)	45 (198)	44 (212)	46 (103)
		CCI	47 (205)	43 (172)	65 (193)	45 (195)	51 (292)	76 (276)	71 (280)	73 (248)	58 (263)	54 (260)	55 (136)
		CCC	56 (193)	44 (173)	64 (202)	54 (227)	60 (297)	88 (297)	76 (293)	77 (294)	63 (295)	63 (279)	57 (143)
	Wroclaw	CCC	76 (293)	79 (281)	97 (297)	85 (289)	91 (294)	89 (288)	113 (278)	140 (277)	109 (285)	119 (289)	86 (145)
		CCI	53 (291)	54 (290)	67 (282)	62 (283)	61 (285)	62 (246)	75 (248)	77 (256)	67 (268)	81 (279)	78 (137)
		CCR	51 (291)	52 (277)	63 (284)	84 (276)	55 (276)	53 (279)	63 (259)	57 (264)	53 (258)	70 (262)	63 (140)
Spain	Madrid	CCC	129 (281)				112 (197)	84 (303)	120 (339)	113 (334)	113 (317)	135 (349)	104 (334)
		SR	56 (330)				47 (286)	56 (296)	45 (350)	25 (341)	34 (350)	26 (308)	20 (358)
		SI							57 (281)	45 (365)	61 (339)	75 (346)	62 (363)
UK	Glasgow	CCC	30 (366)	32 (365)	21 (364)	18 (365)	18 (366)	20 (365)					
	London	SR	27 (366)	28 (365)	18 (365)	15 (365)	11 (349)	15 (339)					
		CCC	16 (366)	19 (365)	20 (365)	33 (365)	33 (366)	32 (307)					
		SI	27 (346)	26 (353)	19 (365)	17 (361)	12 (366)	12 (361)					
OCEANIA													
New Zealand	Auckland	SI	6 (331)	7 (357)	7 (365)	5 (357)	6 (361)	8 (350)	9 (345)	6 (358)	4 (344)	4 (333)	3 (342)
		SR	41 (325)	26 (333)	33 (344)	20 (336)	23 (345)	24 (353)	31 (319)	16 (184)			
	Christchurch	SI	21 (354)	16 (339)	18 (346)	9 (341)	13 (346)	17 (340)	26 (212)				
		SC	32 (316)	21 (331)	20 (349)	17 (342)	13 (340)	22 (339)	30 (210)				

a Site location changed in 1985.

The GEMS/Air site codes are as follows:

CCC = City Centre Commercial
CCI = City Centre Industrial
CCR = City Centre Residential
CCM = City Centre Mobile
SI = Suburban Industrial
SR = Suburban Residential
NA = Not Applicable, i.e., the site has not been assigned a GEMS/Air site code.

Station data presented above have been extracted from the GEMS/Air data base and represent a summary of the data received by the GEMS/Air programme as of September 1991. Stations reporting less than three annual mean values during the period 1980–1990 have not been included.

The number of observations made during the year are given in parentheses in order to provide an indication of the level of completeness of the data set. Although many sites make observations on a daily basis, at some locations measurements are made less frequently, for example, once every six days. As long as these measurements are made regularly throughout the year, this level of sampling will generate a representative annual mean. Thus the use of the number of observations to assess data quality must be viewed with some caution.

Cities have been grouped according to the method employed for the measurement of suspended particulate matter in air. Measurement of particles collected on a filter by the high-volume sampler is based on a mass or gravimetric determination. The Black Smoke Method involves drawing a known volume of air through a filter; the density or reflectivity of the resulting stain is then related to a weight of smoke particles by means of an established calibration curve. Measurements of airborne particles by the two different methods are not directly comparable.

Source:
Data supplied by the Division of Environmental Health, World Health Organization, Geneva, Switzerland.

Soils and Sediments

Soils represent a major natural resource that serves a variety of interconnecting and vital functions – for biomass production (in the form of crops and timber), as a habitat for living creatures, as a store of biodiversity, and as a relatively stable reservoir for the whole ecosystem. Yet increasingly it is becoming apparent that soils are not an unlimited resource, but a finite one that can be lost or degraded by both natural forces and human activities.

This direct threat to the world's soils has been recognized for many years at the international level by both the Food and Agriculture Organization of the United Nations (FAO) which adopted a World Soils Charter and UNEP which has developed a World Soils Policy. More recently, UNEP has commissioned a Global Assessment of Soil Degradation (GLASOD). This project, conducted through the International Soil Reference and Information Centre (ISRIC), Wageningen in association with the Winand Staring Centre, the International Society of Soil Science, the FAO and the International Institute for Aerospace Survey and Earth Sciences was completed in 1991 (Oldeman et al., 1991).

The GLASOD work defined four broad categories of soil degradation: wind erosion, water erosion, physical deterioration (e.g., compaction, waterlogging and subsidence), and chemical deterioration. This latter category – which comprises pollution by chemical contaminants, acidification, salinization and nutrient depletion – will, as in previous editions of this report, form the focus of the discussion here. Other forms of soil degradation and the wider issues of land degradation and desertification are discussed in greater depth, and with particular reference to the GLASOD study, in Part 3: Natural Resources. This section of the 'UNEP Environmental Data Report' also reviews the issue of marine and lake sediment contamination, focusing in this edition on the problems of the Irish and Baltic seas.

The GLASOD work, while providing important decision-support tools for policy makers at the global level, has highlighted the lack of detailed data on soil quality at the national level. Keeping track of a nation's soil resources or, more specifically, monitoring changes in soil properties in relation to human activities, has hitherto received relatively low priority in many national environmental monitoring programmes (Young, 1991). There is thus an almost universal need for more reliable, quantitative data on a wide range of soil properties on appropriate temporal and spatial scales. Given such data it should then be possible not only to establish which land-use systems on which soil types are causing soil degradation and loss of fertility, but also to assess the effects of acidification, salinization and other types of pollution. Data requirements, with particular reference to the development of soil quality indexes, are discussed in more detail in Box 1.5.

Within the context of soils the foci of attention, and thus of information requirements, differ markedly between the less-developed and developed countries. In the tropics, physical forms of soil degradation, nutrient losses and wind and water erosion caused, at least in part, by over-cultivation and poor land-use management are the major concerns. In the industrialized countries the preoccupation is with chemical pollution and associated off-site effects on water bodies, especially on ground waters (Young, 1991). However, that is not to say that wind and water erosion do not occur in some of the more developed countries. Soil maps for Hungary, for example, show that 6 per cent of the country is strongly eroded and 25 per cent eroded to some degree (Varallyay, 1989). Similarly, in the UK it has been estimated that one third of the arable land of England and Wales is at risk from either water or wind erosion or both (Brown, 1992). In Canada soil erosion by water and wind has been identified as the most widespread soil degradation problem (Environment Canada, 1991). For example, in the Saskatchewan and Alberta Provinces soil erosion by water affects almost 5×10^6 ha of agricultural land.

Strong evidence exists for the transboundary, and even transcontinental, transport of large quantities of soil dust by wind erosion. A well-known example is the transport of reddish Saharan desert dust across the Atlantic Ocean to Bermuda, and northward to Europe. Elsewhere, wind erosion in China is considered responsible for the transport of dust to Hawaii (Parrington et al., 1983). Dust and airborne particles typically contain trace elements and selected organic compounds. Data show that heavy metals (such as lead), hexachlorocyclohexane (HCH), polychlorinated biphenyls (PCBs) and chlordanes are transported from the former USSR, Eurasia and North America to the Arctic in this way (Oehme, 1991).

Soil Contamination

At the same time as being an essential resource for human development, soils also act as important sinks for pollutants released through industrial discharges and other human activities. This function is irrespective of whether the pollutants have been released into the air, to rivers or directly to soil.

The areal extent of chemical deterioration of soil on a global scale, as determined as part of the GLASOD study, is summarized below:

Region	Degraded land area (10^6 ha)	Percentage distribution			
		Cn	Cs	Cp	Ca
WORLD	239.1	57	32	9	2
Africa	61.5	73	24	–	2
North America	0.1	0	0	0	100
Central America	6.9	61	33	6	0
South America	70.3	97	3	0	0
Asia	73.2	20	72	2	6
Europe	25.8	12	15	72	1
Australasia	1.3	31	69	0	0

In the above table "degraded land area" refers to the total land area affected by all forms of chemical soil degradation; the areal extent of soils affected by nutrient losses (Cn), salinization (Cs), pollution (Cp) and acidification (Ca) are expressed as a percentage of this total. Asia includes the Asian part of the former USSR and Europe includes the European part of the former USSR. For further details pertaining to these data and the GLASOD study please refer to Table 3.2a–e (Part 3: Natural Resources).
Source: Oldeman et al., 1991

BOX 1.5 Soil Quality Indicators and Indices

The concept of soil quality is a complex issue. In some senses the quality of soil must be evaluated in relation to its use. However, "use" is only part of the quality issue. The soil quality concept ideally must be able to respond to threats to, and provide protection for, the environment and at the same time allow flexibility in the future use of the land.

Current knowledge about soil is based primarily on scientific studies which isolate a particular fraction or property for quantitative measurement. These studies tend to be either chemical, physical or biological in focus. However, it is the interaction of these three broad aspects which gives soil its unique character and which we attempt to encompass in the term "soil quality". Hitherto soil quality has been viewed as a rather qualitative term; it has not been as well defined as air or water quality.

In view of the growing global concern about the various forms of soil degradation, the need for better ways of evaluating soil quality has become more acute. Moreover, indices or indicators of soil quality which identify and document the changes in soil quality in relation to human activities are widely perceived as being essential tools in the policy-making process (Haberern, 1991). At the present time, however, the development of such tools is severely constrained by a number of factors. In the first instance there is, as yet, no universally accepted set of criteria for evaluating changes in soil quality. Nor is the science for standardizing methods of soil investigations as advanced as that for air and water (Hortensius and Nortcliff, 1991). Thirdly, the detailed analytical data necessary for an evaluation of soil quality at the national level are often lacking. Without these raw data and standard methods of description and analysis, attempts to formulate policies for protection of the soil environment will necessarily be based on flawed information.

A number of countries have responded to this perceived need for an improved soil data base at the national level. For example, Agriculture Canada has recently established a Soil Quality Evaluation Program, as part of its National Soil Conservation Program, in order to determine soil quality from a National perspective and to assess how soils respond to

environmental influences and land-use practices. Its intention is to provide periodic reports on the state of the soil (Environment Canada, 1991).

It is generally recognized that there are three basic soil quality issues which need to be addressed (Haberern, 1991; Varallyay et al., 1992; Wood and Horvath, 1992):

❍ Productivity (e.g., soil fertility, hydrology, toxicity);
❍ Environment (e.g., contaminant leaching and water quality); and
❍ Health (e.g., animal and human requirements and effects, bioavailability, toxicology, and human exposure with emphasis on children).

In the past the quality of soil was mainly considered from an agricultural viewpoint; that is to say, soil productivity has been the main focus of concern (Haberern, 1991). The need to broaden this perspective to include environmental quality, human and animal health and food safety is not in dispute. However, many questions remain to be answered when evaluating soil quality in terms of environment and health. What can usefully be measured as an indicator? Are pesticide use or chemical fertilizer use useful indicators of soil quality? How can animal and human health be related to soil quality when many other factors are involved in the determination of health?

Haberern (1991) and others have proposed a soil health index which comprises three components: nutritional quality (NQ), management (M) and toxicants (T), i.e.,

$$Index = NQ + M - T$$

The emphasis or weight of each of these components in the equation would be different depending on the agro-ecosystem and the use for which the soil will be put. Further development of this concept is anticipated and will be reported in future editions of the 'UNEP Environmental Data Report'.

A more specific step-wise approach to evaluating the hazards of chemicals in soils in terms of human health· and the environment has been developed by the European Chemical Industry Ecology and Toxicology Centre (ECETOC, 1990) and outlined by Poels et al. (1991). In essence, it is based on human exposure and exposure commitment concepts (Bennett, 1981). The approach

estimates the maximum tolerable exposure level (MTEL) required to protect human health and compares it with the environmental exposure level (EEL). If the EEL is lower than the MTEL no hazard will exist, consequently there is no risk to human health. If the EEL exceeds the MTEL, a potentially significant hazard exists and remedial action is required.

Whatever approach is adopted to develop soil quality indicators or indices, whether contaminant migration models, exposure assessment models or risk models, one point is clear: unfavourable soil quality has major human and economic consequences. A method is urgently required to evaluate vulnerability of natural areas to contamination and degradation.

References

Bennett, B. G. 1981 *Exposure Commitment Assessments of Environmental Pollutants, Vol 1*, MARC Report Number 23, Monitoring and Assessment Research Centre, London.

ECETOC 1990 *Hazard Assessment of Chemical Compounds in Soil*, Technical Report Number 40, European Chemical Industry Ecology and Toxicology Centre, Brussels.

Environment Canada 1991 *The State of Canada's Environment 1991*, Government of Canada, Ottawa.

Haberern, J. 1991 Report of the International Conference on the Assessment and Monitoring of Soil Quality, 11–13 July 1991, Emmaus, USA, Rodale Institute, Emmaus.

Hortensius, D. and Nortcliff, S. 1991 International standardization of soil quality measurement procedures for the purpose of soil protection, *Soil Use and Management*, **7**(3), 163–166.

Poels, C. C. M., Gruntz, U., Isnard, P., Riley, D., Spitellier, M., ten Berge, W., Veerkamp, W. and Bontinck, W. J. 1991 Hazard assessment of chemical contaminant in soil, *Chemosphere*, **23**(1), 3–24.

Várallyay, G., Salomons, W. and Csikós, I. 1992 *Report of the Workshop on Long-term Environmental Risks for Soils, Groundwaters and Sediments in the Danube Catchment Area*, 13–15 December 1992, Budapest, Chemical Time Bombs Project Workshop.

Wood, B. and Horvath, D. 1992 *Australia and New Zealand Guidelines for the Assessment and Management of Contaminated Sites*, Australia and New Zealand Environment and Conservation Council, National Health and Medical Research Council, Ministry for the Environment, Wellington.

On the basis of these GLASOD data, it is estimated that globally chemical soil deterioration currently affects 239×10^6 ha or approximately 2 per cent of the earth's land surface. Of this total 25.8×10^6 ha are located in Europe, where the dominant form of chemical soil degradation is pollution, i.e., soils are contaminated by pesticides, trace metals, industrial wastes, oil, mineral acids (from atmospheric deposition) and other substances. Note that according to GLASOD terminology, the category "acidification" refers to the impacts of over-application of acidifying fertilizers and the drainage of pyrite-containing soils only. More detailed information on soil contamination by trace metals, selected organic compounds and acids (i.e., the process of acidification), is given in the sub-sections below.

Elsewhere, and particularly in Asia, salinization is the predominant form of chemical soil degradation. Areas which are severely affected by salinization include Iraq and Syria in the Middle East, Senegal and the Sudan in Africa, the Aral Sea Basin, the Euphrates Valley and the Indus Basin in Asia, and the lower Colorado River in the USA (UNEP, 1992). Soil salinization is often linked to irrigation, an issue which is covered in more detail in Part 3: Natural Resources. Although soil degradation by chemical impacts may be significant locally, on a global scale it is a relatively minor factor in comparison with wind and water erosion which collectively account for $1,642 \times 10^6$ ha (or 83.5 per cent) of degraded soils world-wide (Oldeman et al., 1991).

As noted above, problems of soil contamination by chemical pollutants are most pronounced in the industrialized regions, notably Europe and North America. Despite this, and the fact that Europe has the longest history of chemical pollution, knowledge about the state of Europe's soils is extremely fragmentary (CEC, 1992). Present-day concerns have largely arisen from the discovery of contaminated soil in the late 1970s. This has resulted in legislative action and in major clean-up programmes in a number of countries. Although comprehensive assessments are lacking, the scale of the problem is believed to be large. Estimates of the number of contaminated sites in selected countries are presented in Part 8: Wastes and Waste Management.

In Europe the presence of contaminated sites has given rise to the programme with the colloquial name "Chemical Time Bombs". A chemical time bomb is a concept "that refers to a chain of events resulting in the sudden occurrence of harmful effects due to the mobilization of chemicals stored in soils and sediments in response to slow alterations of the environment" (Stigliani et al., 1989). As part of the Chemical Time Bombs project, several international workshops dealing with the long-term risks of contaminated soils have been held. One such workshop on Soil Vulnerability Mapping for Europe (SOVEUR) recognized the need to protect soils and by implication water, food crops and human health (Batjes and Bridges, 1991). Participants drew attention to the paucity of information on effects of specific chemical compounds on soil, soil properties and soil vulnerability throughout European countries.

Heavy Metals Metals may accumulate in rural and urban soils as a result of a variety of human activities. Potential pathways include the deposition of metal-containing particulates from coal-fired plants, municipal incinerators, metal processing industries and from motor vehicle exhausts (lead); the application of phosphate fertilizers to agricultural land (cadmium); the re-use of waste water for irrigation; the application of sewage sludge and animal manure (often high in copper) to agricultural land and, in urban areas, the deposition of paint flakes from buildings (lead). With the exception of lead, which tends to remain adsorbed to organic matter in the surface layers, most metals gradually move down the soil profile. The rate of this downward movement depends on the chemical characteristics of the soil and the intrinsic properties of the element.

Instances of soil contamination by radionuclides, which have been released into the environment as a result of atmospheric nuclear weapons testing in the 1950s and from nuclear accidents, have been widely reported in the scientific literature. Of the radionuclides, caesium-137 (^{137}Cs) and strontium-90 (^{90}Sr) are the most significant in view of their long half-lives and potential interactions with the plant-animal system. However, little routine monitoring of radionuclide contamination is undertaken other than by a few nations which monitor known contaminated sites.

In general, metal concentrations in soil increase with increasing proximity to human habitation and industrial centres. Despite the plethora of scientific publications reporting metal data in soils, they "are not very representative in global terms" (Davies, 1992). Most case studies have taken place in the industrialized countries and have focused on polluted or geochemically anomalous areas. Metal pollution of soils in a range of European countries has been reviewed by Batjes and Bridges (1991). In general, it is reported that metal concentrations are largely dependent upon the site factors (for example, geology, climate, land use and topography), proximity to industrial sources and on traffic density. Contamination of soils in Poland, particularly in the "Black Triangle" area of southern Poland, has been described as particularly serious with long-term ecological and health implications (Smal and Salomons, 1991).

As an example of a national data set, concentrations of acid extractable cadmium and nickel in topsoils and subsoils from Switzerland are shown in Figure 1.13. National guideline values for soils, as set by the Swiss Ordinance Relating to Pollutants in Soil (VSBo), are also included on the graphs for comparison. These data show that in the case of both cadmium and nickel, metal contents lie largely below the guideline level. It is noteworthy that cadmium concentrations tend to be higher in the topsoils, a pattern which is consistent with a man-made source for this metal. In contrast the distribution for nickel, i.e., roughly equal subsoil and topsoil contents, is indicative of pedogenous migration (Vogel et al., 1992).

It is generally accepted that since the advent of industrial times human activities have had a major impact on the global biogeochemical cycles of many trace metals (Nriagu, 1991). The biogeochemical cycle of lead has possibly been affected by man to a greater degree than that of most other heavy metals. In a study of historical emission patterns, Bergbäck et al. (1992) estimate that human activities over the 100-year period 1880–1980 have resulted in a doubling of lead concentrations in soils in some parts of Sweden.

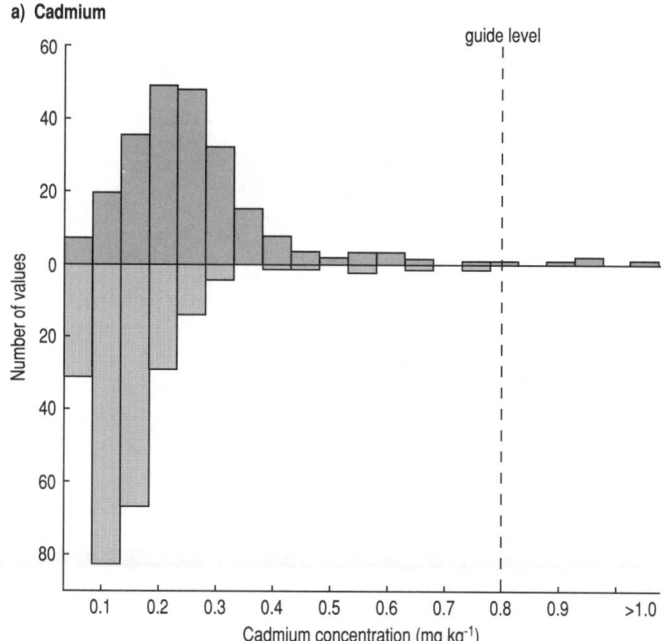

a) Cadmium

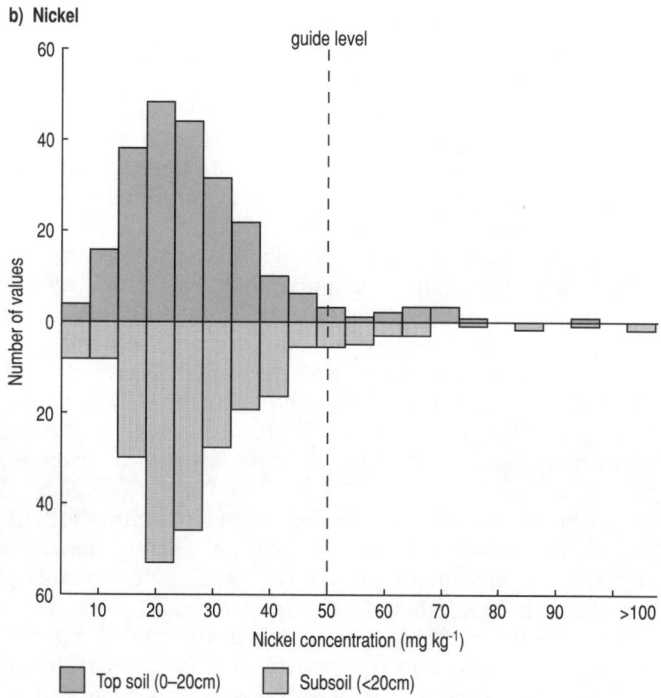

b) Nickel

Top soil (0–20cm)   Subsoil (<20cm)

FIGURE 1.13   Distribution of measured concentrations of a) cadmium (extractable in 2 M HNO₃), and b) nickel (extractable in 2 M HNO₃) in topsoils and subsoils, Switzerland (a total of 194 sites were sampled)
Source: Vogel et al., 1992

In more recent years, with the reduction in emissions of lead to the atmosphere following the phasing out of lead in petrol in many countries, decreases in lead concentrations in various environmental compartments are considered likely. Indeed, decreases in lead contents of air, snow and soils have been reported (Görlach et al., 1991; Friedland et al., 1992). For example, a 17 per cent decrease in lead concentrations between

1980 and 1990 was observed in the top 4 cm of forest soils from the north-eastern USA (Craig et al., 1991; Friedland et al., 1992) (Figure 1.14). Given that lead is believed to have a long residence time in organic soils, this relatively rapid decrease in soil surface lead concentrations is somewhat surprising. The authors conclude that lead appears to be leaving the forest floor but does not appear to be entering streams, it must be moving into the underlying mineral soil horizon as organic complexes (Friedland et al., 1992).

Organic Compounds There are many thousands of organic substances of varying degrees of toxicity which are emitted from industrial processes and which end up in the soil. The use of pesticides – a number of which are persistent chemicals – in agriculture adds yet more organic substances to ecosystems. Farmyard manure, sewage sludge and emissions from motor vehicles also add organic substances to the soil environment.

At present there is little systematic monitoring of organic compounds in soil, even at the national level. However, individual studies, particularly of chlorinated compounds, have been undertaken in a number of countries, some of which have been reported in previous editions of this report (UNEP 1989, 1991). Results of such studies have been used in conjunction with knowledge of basic physico-chemical properties to develop models to explain the behaviour and fate of such compounds in soils.

Until recently, few studies have examined the occurrence of organic substances in soils over historical times. One such study, based on archived soil samples from a semi-rural plot in south-east England which were collected between 1846 and 1986, has shown that the concentrations of polychlorinated dibenzo-*p*-dioxins and -furans (PCDDs and PCDFs) have increased in soil and associated herbage over time (Figure 1.15). Concentrations started to increase around the turn of the century, rising from 31 to 92 ng total PCDD and PCDF per kg of soil (Kjeller et al., 1991). Net rates of increase in the soil over the last century are calculated as approximately 190 ng m⁻² a⁻¹.

Dioxins and furans have both natural and man-made sources; natural sources include forest and bush fires, and man-made sources comprise the various industrial processes which involve the combustion of organic materials, e.g., municipal incinerators, and coal- and oil-fired power stations. The historical soil analysis reported here, however, clearly demonstrates that man-made activities far outweigh the contribution from natural combustion sources and that there is an imbalance between the contemporary rates of PCDD/F release into the environment and their destruction.

As noted above, a diverse range of organic compounds enters agricultural soils from sewage sludge applications, however few systematic investigations on the behaviour and fate of such compounds have been made. Wild and Jones (1992) have reviewed the fate of 46 organic contaminants known to occur in sewage sludge, and have developed a screening approach to ascertain their potential transfer from soil to ground waters, crop plants and grazing livestock.

Acidification Acidification of soils occurs when hydrogen and/or aluminium ions replace base cations, thereby depleting the

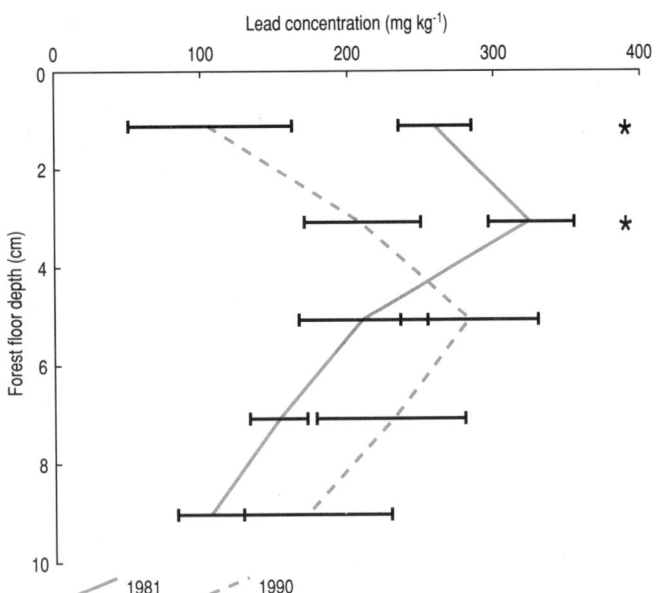

FIGURE 1.14 Changes in lead concentrations in forest floor soil samples taken from spruce-fir stands in Vermont, USA, 1980–1990 (the asterisk indicates a significant difference ($p < 0.05$) between the two sampling periods)
Source: Craig et al., 1991

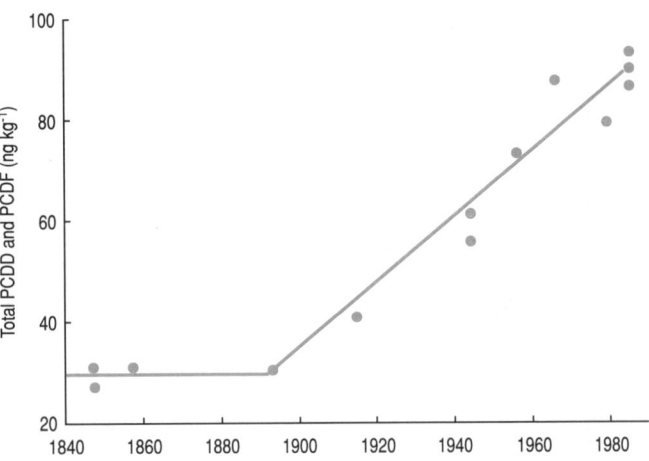

FIGURE 1.15 Trends in concentrations of total PCDDs and PCDFs in soil samples taken at Broadbalk, UK, 1846–1986 (all soil samples were taken from the cultivated plough layer, i.e., the 0–23 cm layer)
Source: Kjeller et al., 1991

store of exchangeable base cations in the soil. A more strict description of the process defines acidification as a decrease in acid neutralizing capacity. The decrease in exchangeable base cations (calcium, magnesium, sodium and potassium) and elevated levels of exchangeable aluminium can markedly reduce soil fertility, root function and crop growth. The magnitude of such effects depends on the characteristics of the soil and on the intensity of the acidification process, but generally adverse impacts are noted in soils having a pH <4.5.

Soil acidification can be a natural process. In areas of high rainfall, for example, leaching of base cations may lead to increasing soil acidity. More importantly, however, increases in emissions of sulphur and nitrogen oxides (largely from coal-fired power plants and smelters) which, following transformation in the atmosphere, are returned to the earth's surface as acidic deposition have resulted in the acidification of soils and fresh waters in a number of countries. The effects of acid deposition are particularly noticeable in Europe and North America (UNEP 1989, 1991).

Acidification of soils in the Nordic countries is mostly attributed to sulphur deposition although nitrogen, emitted as ammonia or as nitrogen oxides, is estimated to contribute up to 30 per cent of the acidifying load in this region (Brodin and Kuylenstierna, 1992). Significant decreases of soil pH over several decades, or longer in some countries, have been recorded at a number of locations in Scandinavia and Western Europe. Selected examples have been highlighted in earlier editions of this report (UNEP, 1991).

In Sweden, spatial and temporal trends in soil acidification have been shown to follow closely the pattern of acid deposition (Eriksson et al., 1992). In south-west Sweden acidification of the soil could be detected well below 2 m depth in the soil, while soils near the east coast were less seriously affected. In north

Sweden the level of acid deposition has no noticeable affect on the soils. Other studies have shown that losses in exchangeable base cations of about 50 per cent over a period of 35 years (1945–1980) have taken place in southern Sweden; moreover, losses of base cations have continued at the same rate in topsoils during the last decade. Decreases in base cations have also been shown to occur in the C-horizon of soils under deciduous forests in this part of Sweden (Falkengren-Grerup and Eriksson, 1990; Falkengren-Grerup and Tyler, 1992).

It is probable that soil acidification may have occurred in other areas of the Nordic countries but reliable evidence is lacking (Brodin and Kuylenstierna, 1992). Acidification effects on surface-water chemistry are, however, well recognized in these countries as a result of the Norwegian SNSF project (Acid Precipitation – Effects on Forest and Fish), the UK/Norway/Sweden SWAP project (Surface Water Acidification) and the Finnish HAPRO programme (Research Project on Acidification).

Despite clear indications of changes in soil chemistry, there is still no concrete evidence linking soil acidification with decreases in tree-stand productivity and/or deteriorating forest health. Throughout Europe, but especially in the coniferous forests in mountainous areas of central Europe (Germany, Poland and the former Czechoslovakia), forest decline – as measured by needle loss – has been widely observed. (See also Part 1: Environmental Pollution, Biological Monitoring). Forest decline is believed to be linked to soil acidification and also to high concentrations of ozone and sulphur dioxide in ambient air (Johnson et al., 1991).

The problems encountered in establishing a causal relationship here are well illustrated by a study of oak and beech stands in southern Sweden (Falkengren-Grerup and Eriksson, 1990). Although decreases in base cation concentrations and increases in exchangeable cations were observed during the period of study, 1947–1988, many field layer species actually increased in cover over this time. Moreover, yields of beech, but not of oak, also increased more than expected. The authors suggest that increased beech yields are a consequence of a fertilization effect of increased nitrogen deposition which more than compensates for the loss of the other macronutrients (i.e., the base cations).

The Arctic region, in this context defined as the area north of the Arctic Circle, is characterized as an ocean surrounded by land masses of the eight circumpolar countries – the former USSR, USA, Canada, Norway, Iceland, Finland, Denmark and Sweden. Small increases in acidity and conductivity have been measured in Greenland and Iceland glaciers and in the snowpack but, for the most part, acidified snow and soils in the Arctic region are found near industrial centres and large cities (Reiersen, 1991). For example, acidic emissions from the Kola Peninsula are considered to have contributed to adverse impacts on soil and forests in northern Finland, Lapland and northern Norway (Tikkanen and Raitio, 1990). However, observed symptoms in forests of northern Sweden and northern Canada are considered to be caused by climatic factors rather than acidic deposition. Extensive research is required in this area as the processes of acidification in the varied Arctic environments are not yet well understood (Nenonen, 1991).

Acidic deposition is also believed to be involved in soil changes in eastern North America and Canada but, as with European studies, results are inconsistent (Johnson and Ball, 1990/91; Johnson et al., 1991). Recent work reported by Joslin et al. (1992) and based on studies of red spruce (*Picea rubens*) forests, suggests that acidic deposition could account for more than half of the current leaching losses of base cations from the soil in north-eastern parts of the USA. Attempts to link soil chemistry changes and foliar patterns with red spruce declines have, however, proved inconclusive. In western North America, on the other hand, acidic deposition is not thought to be a major factor in soil acidification. Rather, the acidification is considered to be due to nitrification associated with excessive nitrogen fixation of red alder (*Alnus rubra*).

There is now the growing realization that, owing to rapid industrialization, there is a risk of substantial increases in acidification of soils and potential for acidification damage in many countries around the world, but particularly in Asia and South America (Rodhe and Herrera, 1988).

In recent years the concept of critical loads has been developed in order to assess both the impact of acidic deposition on soils and lakes and the capacity of particular ecosystems to buffer the effects. The concept is based on the assumption of a threshold response in terms of the onset of harmful effects and is thus soil, site and ecosystem specific. The application of critical load approaches to soil mapping has been adopted in a number of countries and is discussed in more detail elsewhere in this chapter (see Atmosphere).

References

Batjes, N. H. and Bridges, E. M. 1991 Mapping of soil and terrain vulnerability to specified chemical compounds in Europe at a scale of 1:5M. In: Proceedings of an International Workshop held at Wageningen, The Netherlands (20–23 March 1991), International Soil Reference and Information Centre, Wageningen.

Bergbäck, B., Anderberg, S. and Lohm, U. 1992 Lead load: historical pattern of lead use in Sweden, *Ambio*, **21**, 159–165.

Brodin, Y-W. and Kuylenstierna, J. C. I. 1992 Acidification and critical loads in Nordic countries: a background, *Ambio*, **21**, 332–336.

Brown, A. (Ed.) 1992 *The UK Environment*, Government Statistical Service, HMSO, London.

CEC 1992 *The State of the Environment in the European Community, Overview*, Commission of the European Communities, Brussels.

Craig, B. W., Friedland, A. J., Herrick, G. T., Siccama, T. G. and Johnson, A. H. 1991 Changes between 1981 and 1990 in lead, copper and zinc in organic horizon profiles from New England, USA. In: *Heavy Metals in the Environment* 1, J. G. Farmer (Ed.), CEP Consultants, Edinburgh, 302–305.

Davies, B. E. 1992 Trace metals in the environment: retrospect and prospect. In: *Biogeochemistry of Trace Metals*, D. C. Adriano (Ed.), Lewis Publishers, Boca Raton, 1–17.

Environment Canada 1991 *The State of Canada's Environment*, Government of Canada, Ottawa.

Eriksson, E., Karltun, E. and Lundmark, J-M. 1992 Acidification of forest soils in Sweden, *Ambio*, **21**, 150–154.

Falkengren-Grerup, U. and Eriksson, H. 1990 Changes in soil, vegetation and forest yield between 1947 and 1988 in beech and oak sites of southern Sweden, *Forest Ecology and Management*, **38**, 37–53.

Falkengren-Grerup, U. and Tyler, G. 1992 Changes since 1950 of mineral pools in the upper C-horizon of Swedish deciduous forest soils, *Water, Air, and Soil Pollution*, **64**, 495–501.

Friedland, A. J., Craig, B. W., Miller, E. K., Herrick, G. T., Siccama, T. G. and Johnson, A. H. 1992 Decreasing lead levels in the forest floor of the Northeastern United States, *Ambio*, **21**, 400–403.

Görlach, U., Candelone, J-P. and Boutron, C. 1991 Changes in heavy metal concentrations in Greenland snow during the past 23 years. In: *Heavy Metals in the Environment* 1, J. G. Farmer (Ed.), CEP Consultants, Edinburgh, 74–77.

Johnson, D. W. and Ball, J. T. 1990/91 Environmental pollution and impacts on soils and forests nutrition in North America, *Water, Air, and Soil Pollution*, **54**, 3–20.

Johnson, D. W., Cresser, M. S., Nilsson, S. I., Turner, J., Ulrich, B., Binkley, D. and Cole, D. W. 1991 Soil changes in forest ecosystems: evidence for and probable causes, *Proceedings of the Royal Society of Edinburgh*, **97B**, 81–116.

Joslin, J. D., Kelly, J. M. and Van Miegroet, H. 1992 Soil chemistry and nutrition of North American spruce-fir stands: evidence for recent change, *Journal of Environmental Quality*, **21**, 12–30.

Kjeller, L-O., Jones, K. C., Johnston, A. E. and Rappe, C. 1991 Increases in the polychlorinated dibenzo-p-dioxin and -furan content of soils and vegetation since the 1840s, *Environmental Science and Technology*, **25**(9), 1619–1627.

Nenonen, M. 1991 Report on acidification in the Arctic countries: man-made acidification in a world of natural extremes. In: *The State of the Arctic Environment Reports*, Arctic Centre Publications Number 2, University of Lapland, Rovaniemi, 7–81.

Nriagu, J. O. 1991 Human influence on the global cycling of trace metals. In: *Heavy Metals in the Environment* 1, J. G. Farmer (Ed.), CEP Consultants, Edinburgh, 1–5.

Oehme, M. 1991 Further evidence for long-range air transport of polychlorinated aromates and pesticides: North America and Eurasia to the Arctic, *Ambio*, **20**, 293–297.

Oldeman, L. R., Hakkeling, R. T. A. and Sombroek, W. G. 1991 *World Map of the Status of Human-induced Soil Degradation: An Explanatory Note*, 2nd Edition, International Soil Reference and Information Centre, Wageningen, and United Nations Environment Programme, Nairobi.

Parrington, J. R., Zoller, W. H. and Aras, N. K. 1983 Asian dust: seasonal transport to the Hawaiian Islands, *Science*, **220**, 195–197.

Reiersen, L-O. 1991 *State of the Arctic Environment: Updated Draft Proposals for Arctic Monitoring and Assessment Programme (AMAP)*, State Pollution Control Authority, Norway.

Rodhe, H. and Herrera, R. (Eds) 1988 *Acidification in Tropical Countries*, SCOPE Report Number 36, J. Wiley and Sons, Chichester.

Smal, H. and Salomons, W. 1991 Heavy metals and acidification in

Poland: a chemical time bomb? Paper presented at the International Workshop on Long-Term Environmental Risks for Soils, Sediments and Groundwater in the Baltic Catchment Area, 2–6 July, 1991, Tuczno, Italy.

Stigliani, W. M., Brouwer, F. M., Munn, R. E., Shaw, R. W. and Antonovsky, M. 1989 *Future Environments for Europe: Some Implications of Alternative Development Paths*, International Institute for Applied Systems Analysis, Executive Report 15, Laxenburg.

Tikkanen, E. and Raitio, H. 1990 On the occurrence and causes of needle loss observed in northern Finland in summer 1987. In: *Effects of Air Pollution and Acidification in Combination with Climatic Factors on Forests, Soils and Waters in northern Fennoscandia*, K. Kinnunen and M. Varmola (Eds), Nordic Council of Ministers, Oslo.

UNEP 1989 *United Nations Environment Programme Environmental Data Report 1989/90*, Basil Blackwell, Oxford.

UNEP 1991 *United Nations Environment Programme Environmental Data Report 1991/92*, Basil Blackwell, Oxford.

UNEP 1992 *World Atlas of Desertification*, United Nations Environment Programme, Nairobi and Edward Arnold, London.

Varallyay, G. 1989 Soil degradation processes and their control in Hungary, *Land Degradation and Rehabilitation*, **1**, 171–188.

Vogel, H., Desaules, A. and Häni, H. 1992 Heavy metals contents in the soils of Switzerland, *International Journal of Environmental Analytical Chemistry*, **46**, 3–11.

Wild, S. R. and Jones, K. C. 1992 Organic chemicals entering agricultural soils in sewage sludges: screening for their potential to transfer to crop plants and livestock, *Science of the Total Environment*, **119**, 85–119.

Young, A. 1991 Soil monitoring: a basic task for soil survey organization, *Soil Use and Management*, **7**, 126–130.

Sediments

Aquatic sediments, comprising fine-, medium-, and coarse-grained minerals and organic particles, originate largely from weathering and erosion of soils. These particulates can be transported by rivers or deposited as wind-blown particulates and dispersed into lakes, reservoirs, estuaries, coastal embayments and the open sea. Contaminants such as heavy metals and organics derived from anthropogenic and natural sources can be adsorbed during the transport process or removed from the water column by settling particulates.

In recent years, protecting sediment quality has been increasingly viewed as a logical and needed extension of water quality protection. The basic premise that is used to protect water quality, i.e., to restrict chemicals from occurring in water at concentrations above known "safe" limits, is now being considered for sediments. However, in order to protect sediment quality and thus public health, water supplies and natural ecosystems, agreed measures of the quality of sediments, either *in situ*, or when dredged and applied as land fill, are urgently required. Such measures are important, both locally and nationally, for shared rivers, lakes and for inland seas subjected to multiple use. The information and data needs for the adequate assessment of sediment quality are discussed further in Box 1.6.

The analysis of sediment samples and cores can offer

BOX 1.6 Indices for Assessment of Sediment Quality

The contamination of marine, estuarine and fresh-water sediments poses a potential risk to aquatic resources and also to human health via the consumption of edible fish and shell-fish. Disposal of contaminated sediments dredged from waterways and placed in upland or wetland environments also raises questions of chemical hazard, ecosystem effects and potential health risks. Sediments show large variations in contaminant concentrations because of major differences in mineralogy, grain size, organic matter content, water content and sources of anthropogenic input. This makes the interpretation and assessment of sediment contamination difficult.

Assessments of sediment quality have, in the past, involved a range of approaches. Long-term evaluation of contaminated estuarine dredged material, using indicator plants and animals, has been undertaken for a number of years by the US Army Corps of Engineers (Brandon et al., 1992). Other approaches have involved methodologies based on sediment-water partition co-efficients and invertebrate bioassays (National Research Council, Marine Board, 1989).

In a recent review of approaches to sediment quality assessment, Adams et al. (1992)

have listed and described 11 currently available methods. They conclude that, owing to the inherent complexities of the nature of sediments, no single approach offers the flexibility, reliability and scientific credibility that is required to provide policy-relevant sediment criteria on a national scale. In other words, it is considered that sediment contaminant assessment has not yet reached the point where a single value can be generated and which can form an appropriate basis for regulation. However, the combination of existing methodologies into a tiered approach to sediment quality assessment was viewed as a possible way forward.

Adams et al. (1992) envisage a three-tier approach to sediment quality assessment. The first step in the process – Tier I – is a screening step in which sediment contaminant concentrations are compared against a predetermined value or Sediment Assessment Value (SAV). In cases where the SAV is exceeded, further investigative work is required, i.e., Tier II. In this part of the assessment the determination is made whether the sediment contains chemicals in amounts toxic to aquatic organisms or whether chemicals with a high tendency to bioaccumulate are above levels of concern. Testing as part of Tier II would also establish the magnitude of the area that is

impacted. If, for example, the zone of impact is determined to be large, additional testing would be required, i.e., Tier III. Tier III is thus the part of the assessment process that would provide in-depth testing of the sediments in the zone of impact to confirm the significance of contaminants to aquatic life and their potential to move through the food web. This integrated biological and chemical approach to sediment assessment thus provides a comprehensive means of evaluating sediment contamination that could be used in a range of regulatory programmes.

References

Adams, W. J., Kimerle, R. A. and Barnett, J. W. Jr 1992 Sediment quality and aquatic life assessment, *Environmental Science and Technology*, **26**(10), 1865–1875.

Brandon, D. L., Lee, C. R. and Simmers, J. W. 1992 *Long-term Evaluation of Plants and Animals Colonizing Contaminated Dredged Material Placed in Upland and Wetland Environments*, Environmental Effects of Dredging, Vol D-92-4, US Army Corps of Engineers, Vicksburg.

National Research Council, Marine Board 1989 *Contaminated Marine Sediments – Assessment and Remediation*, National Academy Press, Washington DC.

certain advantages over water chemistry monitoring. In particular, sediment data can provide an integrative measure of water pollution over time; a single water sample only represents the pollution at the moment of sampling. Secondly, substances, such as PCDDs and PCDFs, which tend to occur in very low concentrations in the surrounding waters will often accumulate in sediments, thereby easing the analytical difficulties associated with measuring low concentrations. Finally, if sediment cores are available, a chronology or history of past water contamination can be established (Vogel and Chovanec, 1992). Moreover, sediment cores can sometimes provide a historical record of past emissions or pollutant discharges to water bodies. Biological activity and chemical conditions within the sea-bed may, however, re-mobilize portions of the pollutants or natural substances thus modifying the historical records of pollutant inputs.

Sediments represent an important component of aquatic ecosystems because of the niche and food source they provide for benthic aquatic organisms. Sediments provide a substrate for a wide variety of organisms to live in, or on; some such organisms, for example, shrimp, crayfish, lobster and crab, are of economic importance. As these organisms live in intimate contact with sediments, they have been widely used as biological indicators of chemical pollutants. This approach to environmental monitoring is discussed in a subsequent section of this chapter (see Biological Monitoring).

Previous editions in the 'UNEP Environmental Data Report' series have focused on different aspects of the sediment contamination issue. The third edition, for example, reviewed the status of marine sediment contamination with an emphasis on the organics such as PCBs, PAHs and tributyltin (UNEP, 1991). In this present edition, radionuclide contamination of Irish Sea and heavy metals in sediments of the Baltic Sea are highlighted. These examples serve to illustrate the spectrum of results available and their use for historical monitoring of industrial emissions.

Radionuclides Since the early 1950s, the Irish Sea has been a sink for low-level radioactive wastes in liquid form discharged from the Sellafield nuclear reprocessing facility in Cumbria, UK. The liquid wastes are a mixture of α-, β-, and γ-emitting fission and neutron activation products and include transuranium elements. These radionuclides are predominantly adsorbed onto particulates which sediment over time.

In 1988 the UK Ministry of Agriculture, Fisheries and Food (MAFF), as part of its programme of monitoring for chemical, biological and radiological contaminants, collected sediment cores from an undisturbed site at Senhouse Dock, Maryport Harbour, on the coast north of Sellafield for detailed analysis. The radiological analysis of a suite of 16 Sellafield-derived radionuclides and four naturally-occurring radionuclides profiled down the cores, and corrected for radioactive decay, has been reported by Kershaw et al. (1990). A chronology was established by comparing a variety of radionuclide concentrations and isotope ratios in the sediment core with available information on effluent discharges. On the basis of the data obtained a comprehensive record of Sellafield discharges dating back to the early 1960s has been obtained. This included estimates of releases of radionuclides for which conventional discharge data are incomplete. Data illustrating one of the radioactive chronologies, in this case the two transuranium radionuclides Pu[239] and Pu[240], together with the decay-corrected discharges, are shown in Figure 1.16.

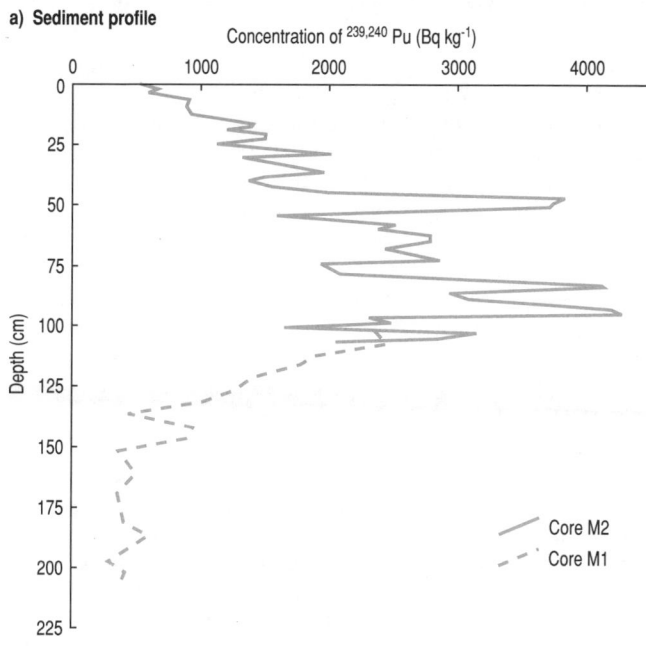

a) Sediment profile

Concentration of ^{239,240}Pu (Bq kg^{-1})

Core M2
Core M1

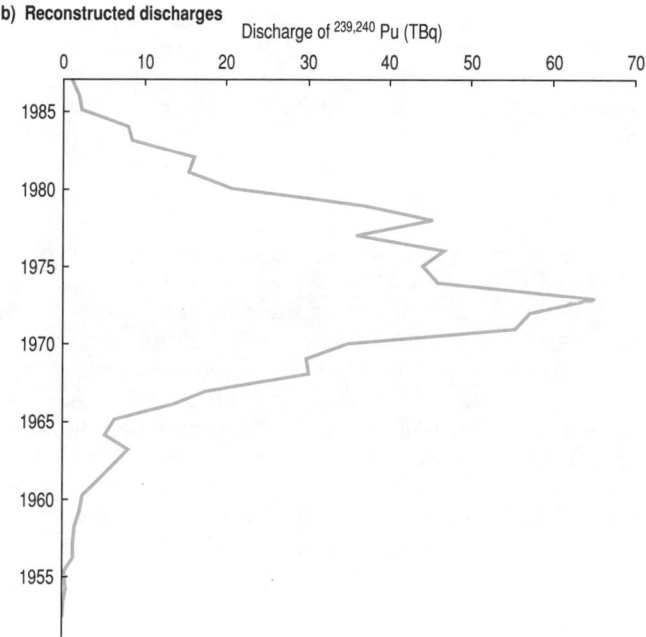

b) Reconstructed discharges

Discharge of ^{239,240}Pu (TBq)

FIGURE 1.16 (a) Vertical sediment profile of the Sellafield-derived radionuclides, ^{239,240}Pu and (b) estimates of the decay-corrected discharges of ^{239,240}Pu from Sellafield, 1952–1987 (based on the analysis of a core taken at Senhouse Dock, Maryport)
Source: Kershaw et al., 1990

Trace Metals The Baltic Sea, one of the largest brackish seas in the world, is almost an enclosed sea and is characterized by very limited water exchange through the Danish sounds. Many studies have evaluated the inputs of anthropogenic substances to the Baltic Sea from a wide range of industries via rivers and atmospheric deposition (Niemistö, 1986; Hallberg, 1991). An international monitoring system for the Baltic Sea has been in place for a number of years under the Helsinki Commission (Krutikow, 1991).

Given that water has a residence time of about 22 years in the Baltic Sea and metals a residence time of a year or less, sedimentation is assumed to be the predominant process for removal of metals from the water. Sediment inventories of 12 metals – zinc, copper, nickel, cadmium, cobalt, iron, arsenic, molybdenum, manganese, vanadium, lead and mercury – have been compiled from samples taken at 59 stations in the Baltic proper (Hallberg, 1991). In line with earlier work (Rodhe et al., 1980) this study confirmed that atmospheric input is the most important source of metals in the Baltic area. The burning of fossil fuels is generally considered to be the most important source of metals to the atmosphere.

An examination of changes of metal concentrations over the 100 years represented in the sediment cores showed that, on average, metal concentrations in the Baltic proper have increased fivefold over the past 50 years. For some elements, cadmium for example, the increase has been even more marked – approximately tenfold (Hallberg, 1991).

An earlier analysis of a longer sedimentary record of metal concentrations in the Baltic showed that, in the case of zinc, there has been very little variation in the concentrations in the layers representing the period 1600–1935. After that the concentration of zinc in the core increased from about 150 µg g^{-1} to about 600 µg g^{-1} at the top of the core (Hallberg, 1974). This result has since been confirmed by the more recent work; the earlier profile for zinc was reproduced from data obtained from a core sampled at the same site 10 years later (Hallberg, 1991). The close resemblance of the two profiles supports the conclusion that metal increases in sediments are a result of increased inputs.

A detailed examination of the onset of increased zinc concentrations in sediment cores in the Baltic area showed a south to north trend which is in accordance with atmospheric deposition patterns described by Hallberg (1991) and the work of Niemistö (1986). Figure 1.17 illustrates the latitudinal variations in the onset of increases in zinc concentration. Had river inputs of zinc assumed a greater importance than atmospheric inputs, the reverse south-north pattern would have been established since zinc inputs from rivers are highest in the north.

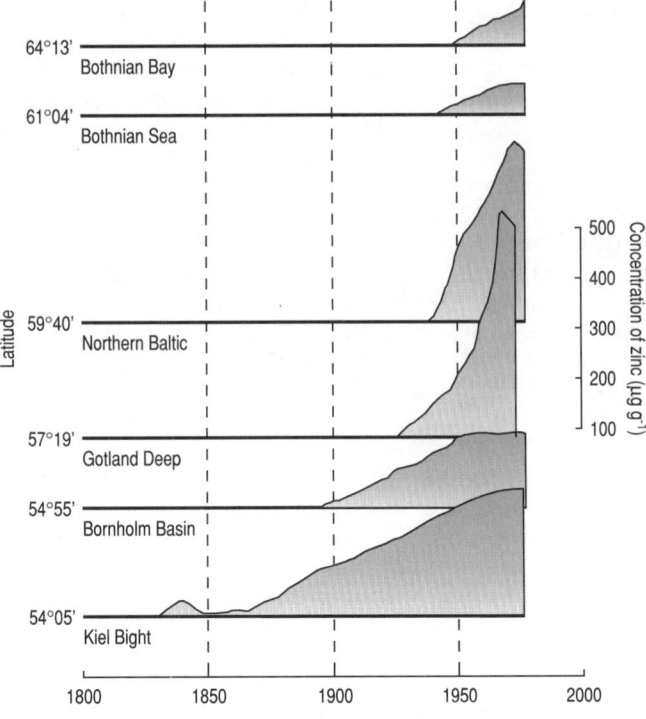

FIGURE 1.17 Variations in the onset of increased zinc concentrations in Baltic Sea sediments with latitude (zinc concentrations below the background of 100 µg g^{-1} are not shown)
Source: Niemistö, 1986

References

Hallberg, R. O. 1974 Paleoredox conditions in the eastern Gobland basin during recent centuries, *Merentutkimuslatt Julk*, **238**, 3–16.

Hallberg, R. O. 1991 Environmental implication of metal distribution in Baltic Sea sediments, *Ambio*, **20**, 309–316.

Kershaw, P. J., Woodhead, D. S., Malcolm, S. J., Allington, D. J. and Lovett, M. B. 1990 A sediment history of Sellafield discharges, *Journal of Environmental Radioactivity*, **12**, 201–241.

Krutikow, M. 1991 Paper presented at of the International Workshop on Long-Term Environmental Risks for Soils, Sediments and Groundwater in the Baltic Catchment Area, July 2–6 1991, Tuczno, Italy.

Niemistö, L. 1986 Monitoring sediments in the Baltic Sea, *Baltic Sea Environment Proceedings*, **19**, 175–180.

Rodhe, H., Söderlund, R. and Ekstedt, J. 1980 Deposition of air-borne pollutants on the Baltic, *Ambio*, **9**, 168–173.

UNEP 1991 *United Nations Environment Programme Environmental Data Report 1991/92*, Basil Blackwell, Oxford.

Vogel, W. R. and Chovanec, A. 1992 Sediment analysis as a method of monitoring industrial emissions, *Hydrobiologia*, **235/236**, 723–730.

Water Quality

Fresh water is important to human health: for drinking, cooking and hygiene. Adequate supplies of clean fresh water are also crucial to many aspects of sustainable development, including agriculture and industry. The quality of marine waters also affects the health and development of those populations living close to the coast who depend on them for sources of food (principally fish and shellfish) and who use marine waters for recreation.

Water quality can be measured in terms of physical, chemical and biological variables, all of which show local and regional variations depending on the geological, biological and climatological conditions of the area. For the purposes of monitoring, water quality is usually considered in relation to specific uses of the water or anticipated impacts on a water body, or in relation to the preservation of the quality of the aquatic ecosystem itself. In addition, monitoring is occasionally carried out to determine natural or background water quality, although in practice it is difficult to find sampling locations which are totally unaffected by human activity. These programmes, when carried out over long time periods, may also provide early indications of the spread of human impacts.

Specific uses of water, such as for drinking water supplies or irrigation, usually have a minimum acceptable quality which can be defined in terms of selected and measurable variables. When monitoring is designed to assess the effect of particular impacts (such as agricultural activities or industrial discharges) on a water body, the variables are also selected according to the known or anticipated effects on specific physical, chemical or biological variables (Chapman, 1992). Thus many of the currently available water monitoring data reflect the aims of objectives of specific programmes or activities and are mainly local or national in extent.

Fresh Waters

Fresh-water quality and quantity are inextricably linked. Although there is sufficient fresh water to meet human demands at present and in the foreseeable future, uneven distribution of ground water, surface water and rainfall means that many arid and semi-arid parts of the world are without reliable sources of fresh water (see Part 3: Natural Resources). To add to this problem, human impacts on fresh-water systems have increased dramatically during past decades while global demands for good quality fresh water have also increased. Water used for domestic purposes, industry or agriculture is frequently returned to its original source contaminated with chemicals or substances not previously present, or contaminated with much elevated concentrations of natural constituents. As a result, the amount of water of acceptable quality available for human use can be further reduced. Problems of water scarcity resulting from poor co-ordination and planning in land and water development, along with water quality degradation, were major foci at UNCED in June 1992. In view of continuing and increasing pressures on the aquatic environment, and the dependence of human health and ecosystem health on good quality water, there is a fundamental need for monitoring of fresh-water quality on a world-wide basis and for scientific assessment leading to improved and sustainable management of national water resources.

The UNEP/WHO/United Nations Educational, Scientific and Cultural Organization (UNESCO)/WMO Programme on Global Water Quality Monitoring and Assessment (GEMS/Water), which was established in 1977, is the first programme to attempt global monitoring of surface and ground-water quality. The GEMS/Water network, which has been described in earlier editions of the 'UNEP Environmental Data Report' (UNEP 1987, 1989, 1991b) and in which 59 countries participate, is currently undergoing a restructuring which will lead to a proposed global network of between 40 and 50 baseline monitoring stations, between 300 and 400 trend monitoring stations and between 60 and 70 global river flux monitoring stations. Monitoring activities at these GEMS/Water stations serve various assessment goals:

○   Determination of natural fresh-water qualities in the absence of significant direct human impact;
○   Determination of long-term trends in the levels of critical water quality indicators in fresh-water resources; and
○   Determination of the fluxes of toxic chemicals, nutrients, suspended solids and other pollutants from major river basins to the continent/ocean interfaces.

The current restructuring of GEMS/Water also aims to place greater emphasis on the monitoring and assessment of ground-water quality and on the application of biological monitoring, including the use of biological indices. Water quality indices of different types (see Box 1.7) are assuming greater importance in the assessment of water quality.

In order to assist in the statistical and analytical presentation of water quality data, the National Water Research Institute (NWRI) of Environment Canada has developed an integrated computer software system called RAISON (Regional Analysis by Intelligence Systems ON a micro-computer). This system uses recent developments in expert systems and artificial intelligence to include the non-numeric and judgmental information, which water managers gain through years of experience, in scientific models. As part of the GEMS/Water programme, RAISON is now used to produce data summaries and to provide statistical interpretation and synthesis of the extensive GEMS/Water data holdings.

The first global assessments of water quality (UNEP/WHO 1988; Meybeck et al., 1989), which used data from the GEMS/Water data base, were able to highlight certain issues of present or increasing importance at the global scale, but also drew attention to the difficulty in making global assessments from widely distributed monitoring stations in diverse natural conditions. Amongst those water quality issues covered in the global assessments, contamination with pathogens, metals and synthetic organic chemicals (including pesticides), as well as eutrophication, salinization, acidification and high concentrations of nitrates, were particularly highlighted. Such issues are often best illustrated by national monitoring programmes, or special surveys of specific water bodies.

Rivers, lakes and ground waters can all be affected by water

BOX 1.7 The Use of Indices in International Water Quality Assessment Programmes:River Danube

A water quality index is obtained by aggregating several water quality measurements into one value (Chapman, 1992). This simplified expression of a complex set of variables allows easy communication of water quality information to non-specialists as well as to specialists who can then use it for a variety of purposes, such as decision-making, planning, identifying water quality problems and assessing changes in water resources.

The UN ECE has a draft scheme for the presentation of water quality of surface water bodies, which is known as the Standard Statistical Classification of Ecological Freshwater Quality (UN ECE, 1992). This classification scheme is based largely on water quality criteria for the protection and maintenance of aquatic life. The criteria are derived from concentrations of substance that do not adversely affect aquatic life in all its forms and life stages.

Water quality variables used in the classification scheme are assigned to one of the following six groups: oxygen regime, eutrophication,

acidification, heavy metals and cyanides, harmful substances (organic micropollutants) and microbial pollution. For example, the group "eutrophication" contains the variables: Total-P, Total-N and chlorophyll a, and the group "acidification" contains the variables: pH and alkalinity. Concentration ranges of these variables are also allocated to one of five quality classes ranging from excellent (class I) to bad (class V). Annual mean or median values of monitoring data for the variables in all the groups are taken and a quality class is assigned to each of the six groups of variables. The assigned quality class is equivalent to the worst quality class associated with any of the individual variables in the group concerned. With respect to heavy metals and organic micropollutants, water quality class III allows "normal" aquatic life, whereas classes IV and V characterize a pollution situation in which harmful substances may be present at chronic or acute toxic concentrations.

The UN ECE classification scheme has been tested on a number of major transboundary rivers

using data available in relevant international data bases such as GEMS/Water, together with data from the national monitoring authorities concerned. The results for selected stations on the Danube for the year 1988 are given in the figure. It was concluded that the changes in water quality classes along the River Danube (particularly acidification, eutrophication, heavy metals and persistent organic substances at the national borders) were influenced by the quality of the data used. Harmonization of monitoring methods and data handling at the international level would improve the value of the classification of the scheme in such situations (UN ECE, 1992).

References

Chapman, D. (Ed.) 1992 *Water Quality Assessments: A Guide to the Use of Biota, Sediments and Water in Environmental Monitoring*. Chapman & Hall, London.
UN ECE 1992 *The Environment in Europe and North America: Annotated Statistics 1992*. United Nations Economic Commission for Europe, New York.

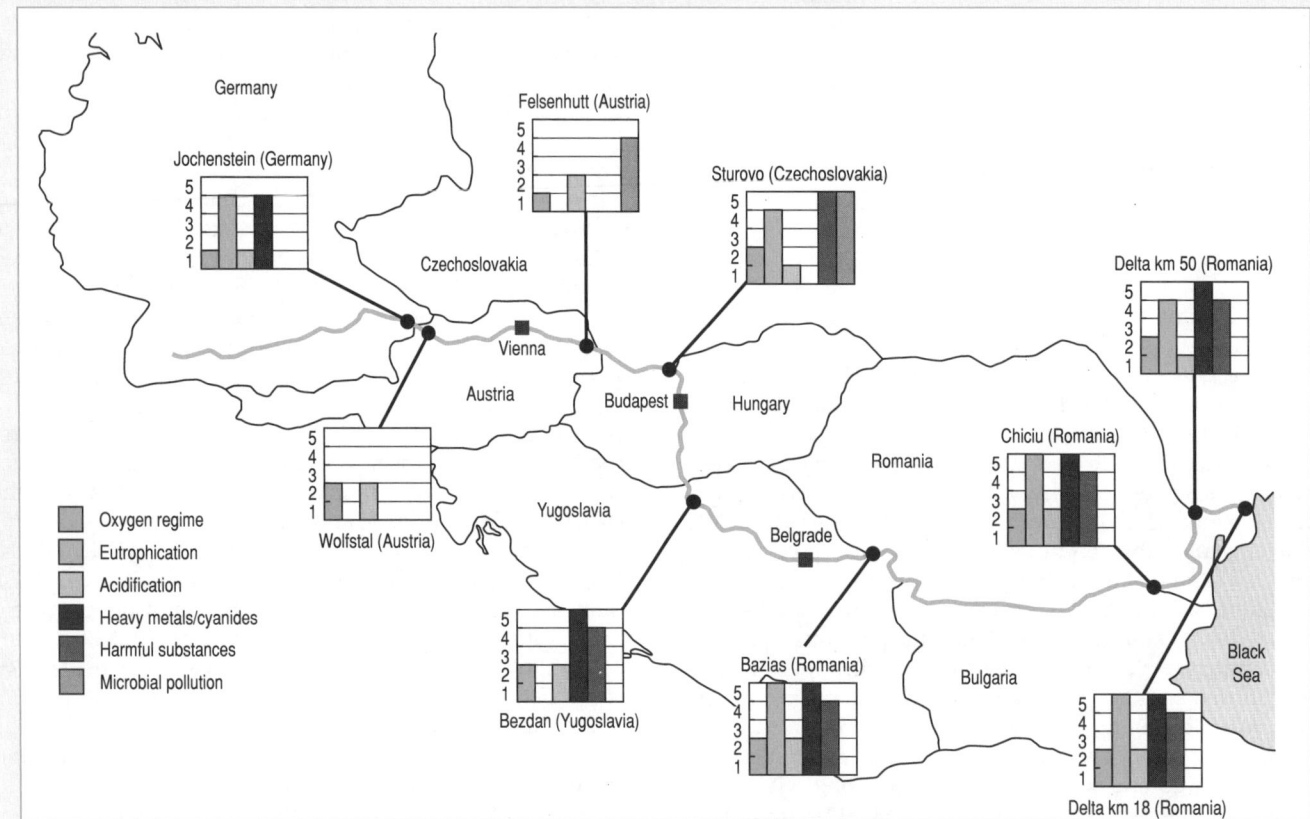

Water quality classes at selected monitoring stations of the River Danube, 1988. Water quality ranges from class 1 (excellent) to class 5 (bad). For further explanation of the groups and classes see text.
Source: After UN ECE, 1992

quality problems to some extent, although certain types of water bodies are more prone to particular problems. For example, eutrophication is particularly associated with lakes (see below) and contamination with pesticides and synthetic chemicals are of particular significance in ground waters. This edition of the 'UNEP Environmental Data Report' illustrates a few examples of these major water quality issues, concentrating on acidification, eutrophication, increasing nitrate concentrations, and contamination with pesticides. Eutrophication is an example of an impact of agriculture on water quality. The important developments in agricultural production over recent decades have led to a wide range of pollution problems including increased sedimentation from erosion, animal wastes, salinization, waterlogging and herbicides. These, and other issues, such as contamination arising from industrial activities, will be addressed in future editions of this report. An example in which contamination of river water can lead to potential health risks to human populations is given in Box 1.8, which describes the problem of mercury contamination of the River Amazon and its tributaries as a result of gold mining activities.

Owing to the complexity of river dynamics and the spatial and temporal variability of rivers, relatively little is known with much certainty about global river-water quality (Dunnette, 1992a). However, one of the ways to achieve a better understanding of river water quality is to study monitoring data on a regional basis with consideration given to environmental pressures in that region. Table 1.11 presents annual data for 1987–1991 for GEMS/Water stations in the Asia-Pacific region. This region contains some of the world's largest rivers such as the Ganges and the Yangtze, and has great biogeochemical diversity, and includes both desert and tropical regions. In some areas of the Asia-Pacific region, the human impacts on water quality have been found to be less than problems arising from naturally occurring concentrations, such as high salinity, high fluoride, and low iodine or selenium (WHO/UNEP, 1991). Acidification of surface waters from long-range transboundary air pollution is assuming greater significance in the region, particularly in south-east Asia (Rodhe et al., 1992).

In addition to acidification and eutrophication (which are highlighted below), other water quality issues have been identified in lakes (Meybeck et al., 1989). These issues, arising principally from industrial and agricultural activities, have mostly been studied and described in industrialized regions, such as the North American Great Lakes (Meybeck et al., 1989). Water quality issues in these regions have progressed from concerns over eutrophication to concerns over toxic chemicals, in a sequence parallel to social and industrial development (Chapman, 1992). Contamination of lake water by industrial pollutants can be monitored by chemical analyses of water samples and sediments. However, for a better understanding of ecosystem effects, such contamination has often been a focus for biological monitoring (see Part 1: Environmental Pollution, Biological Monitoring).

Acidification Acidification of surface and some ground waters is a slow process which is principally caused by increased atmospheric deposition of inorganic acids. Atmospheric deposition in the form of "acid rain" occurs around the world naturally through the chemical reaction of rainwater with carbon dioxide as it falls through the atmosphere. However, significantly accelerated acidification has been recognized as a major issue for some time, particularly in Scandinavia and north-eastern North America where the land surfaces lack sufficient acid-neutralizing soils. More recently, acidification has become a matter of great concern throughout northern Europe, most of eastern North America and parts of southeast Asia. In addition, there is now evidence to suggest that acidification is likely to cause problems in other regions of the world including South America and extended parts of Asia (WHO/UNEP, 1991) (see also Part 1: Environmental Pollution, Atmosphere).

Natural lakes, ponds, rivers and streams with mean pH values below 5.0 are relatively rare, but there is evidence that where these conditions occur the habitats will contain few, if any, fish (UNEP, 1992). It is known that increased acidic deposition in susceptible regions of North America and Europe has reduced the pH of some lakes so that they no longer support fish or any other animal life. In addition, the release of metals to lakes and streams from acidified soils and sediments presents possible risks to human health and to fish communities (Meybeck et al., 1989).

There have been a number of national studies of the extent and effects of acidification of fresh waters in recent years. In the USA, the US Environmental Protection Agency (EPA) performed a National Surface Water Survey (NSWS) between 1984 and 1986 in order to enable quantitative estimates of stream and lake characteristics to be compared. Data from the NSWS for streams and lakes, as well as data from national surveys of acidification in Norwegian lakes and Swedish ground waters, were included in the second and third editions of the 'UNEP Environmental Data Report' (UNEP 1989, 1991b).

In 1987, the International Co-operative Programme (ICP) on Assessment and Monitoring of Acidification of Rivers and Lakes was set up under the auspices of the UN ECE 1979 Convention on Long-range Transboundary Air Pollution. In 1990, 200 sampling sites in 66 locations in Austria, Belgium, Canada, Denmark, Finland, Germany, UK, Ireland, Norway, Sweden, The Netherlands, USA and the former USSR reported data to the Programme Co-ordinating Centre, the Norwegian Institute for Water Research (NIVA). These data are summarized in Table 1.12 and suggest that the effects of acidification are present in a large variety of aquatic systems (NIVA, 1992a). Results are now available for the first three years of this programme and the utility of this data base will increase as more years are added, since long-term data are especially important as a baseline to assess the consequences of climatic changes in the future (NIVA, 1992b). The use of biological monitoring methods in this programme is discussed in Part 1: Environmental Pollution, Biological Monitoring.

Eutrophication Eutrophication is the natural phenomenon of ageing in lakes, where organic material gradually accumulates in the lake basin during the geological history of the lake. However, human activities around lakes can lead to an enrichment of lake waters with nutrients, especially phosphorus and nitrogen, which then results in accelerated eutrophication (also known as cultural eutrophication). Increased concentrations of nutrients in lakes have been attributed to the discharge

BOX 1.8 Gold mining and mercury contamination of rivers in the Amazon Basin, Brazil

Gold mining has become a major activity in the Brazilian Amazon region involving about 650,000 people (Nriagu et al., 1992). At gold mining sites, mercury is used to amalgamate fine gold particles and, due to the large quantities of mercury involved together with its known toxicity, there are potential risks to the health of the local populations.

Some attempts have been made to quantify the problem by estimating mercury losses during the gold mining process based on the quantities of gold produced. For example, Pfeiffer and de Lacerda (1988) estimated that during the gold mining process in the Amazon basin, 1.32 kg of mercury are released to the environment for every 1 kg of gold which is produced; although in some field conditions the proportion of mercury released may be greater. Estimates of mercury fluxes to the environment based on the official and the estimated gold production and mercury use in the Amazon basin suggest that between 16,034 and 128,273 kg a^{-1} of mercury are released to the Amazon environment. Approximately 45 per cent of this is released to rivers as metallic mercury and the remainder is lost to the atmosphere as mercury vapour.

Most gold mining activity in the Amazon is concentrated on the Madeira River in the south-western Amazon basin, in the State of Rondonia. Actual gold production rose from less than 2 t a^{-1} in 1980 to over 20 t a^{-1} in 1983. However, production in 1985 had fallen to only half that of 1983. Consequently, the amount of mercury released to the Madeira River showed a similar trend, reaching a peak of between 14 t and 15 t in 1983 (de Lacerda et al., 1989). The calculation of mercury inputs is based on the estimates of mercury released per kg of gold produced and the estimates of actual gold production (as opposed to official production data).

Water quality monitoring in the Madeira River in response to the mercury problem has been performed on an ad hoc basis and, therefore, no reliable and comparable time-trend data are available. The table presents reliable data from a 1991 study of mercury concentrations in water of the Madeira River and its tributaries. These results are many times lower than previously reported mercury concentrations in the Madeira River which

range from <40 to >9,000 ng l^{-1}. Nevertheless, the average value for the Madeira River is about 17-fold higher than the 0.82 ng l^{-1} estimated to be the mean dissolved mercury concentration in rivers of the world (Nriagu et al., 1992).

The severity of mercury contamination in the Madeira River is illustrated by the elevated mercury concentrations found in fish from the river and its tributaries. Methylmercury is the major mercury species that bioaccumulates in fish. Carnivorous fish species, in particular, accumulate extremely high concentrations of methylmercury in their body tissues because they consume large numbers of smaller contaminated fish. Reported concentrations of 1.01±0.64 µg g^{-1} for carnivorous fish, 0.13±0.08 µg g^{-1} for omnivorous fish and 0.12±0.06 µg g^{-1} for detrital feeding fish, show that some species (particularly carnivorous ones) exceed the limits for mercury in edible fish tissue (0.5 µg g^{-1}) that have been established in many countries (WHO, 1990).

Many of the fish with potentially elevated mercury concentrations form an important component of the diet of the local populations in the Madeira River area. Regular consumption of contaminated fish, particularly by children, could lead to a daily intake of mercury in excess of the WHO recommended daily intake limit of

0.43 µg g^{-1} body weight for prevention of adverse health effects (WHO, 1990). Pregnant women and their unborn children are particularly at risk from the neurological effects of mercury poisoning.

Although there is a lack of good quality monitoring data for mercury in water, sediments and biota from the Amazon region, the information available to date suggests that the potential health risks from mercury inputs to the Madeira River are significant. Further monitoring would help to quantify this risk and highlight those communities most likely to suffer from the toxic effects of widespread mercury contamination in the Amazon region.

References

Nriagu, J. O., Pfeiffer, W. C., Malm, O., de Souza, C. M. M. and Mierle, G. 1992 Mercury pollution in Brazil, *Nature* (London), **356**, 389.

de Lacerda, L. D., Pfeiffer, W. C., Ott, A. T. and de Silveira, E. G. 1989 Mercury contamination in the Madeira River, Amazon - Hg inputs to the environment, *Biotropica*, **21**(1), 91–93.

Pfeiffer, W. C. and de Lacerda, L. D. 1988 Mercury inputs into the Amazon region, Brazil. *Environmental Technology Letters*, **9**, 325–330.

WHO 1990 *Environmental Health Criteria No. 101: Methylmercury*, World Health Organization, Geneva.

Mercury concentrations in the Madeira River and its tributaries (ng l^{-1})

Sample location	Dissolved	Particulate	Total
Mutum Parana (junction of Madeira River)	11.3	14.5	25.8
Mutum Parana (30 m from highway)	9.8	17.5	27.3
Prainha	15.0	18.2	33.2
Humaita, mid-river	15.8	6.4	22.2
Humaita, nearshore	12.5	8.5	21.0
Porto Velho	13.7	10.4	24.1
Cachoreira Teotonio (a fishing village)	13.3	6.6	19.8
2 km above Teotonio Falls	10.7	9.1	19.9
4 km above Teotonio Falls	16.3	6.2	22.5
8 km above Teotonio Falls	17.2	8.5	25.6
0 km above Teotonio Falls	12.7	9.1	21.9
12 km above Teotonio Falls	17.1	15.3	32.4
Mean	13.8±2.5	10.8±4.4	24.6±4.5

Source: Nriagu et al., 1992

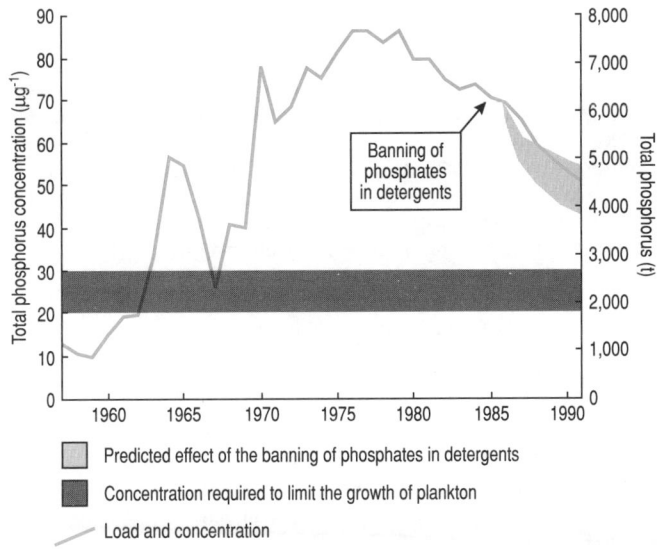

Predicted effect of the banning of phosphates in detergents

Concentration required to limit the growth of plankton

Load and concentration

FIGURE 1.18   Concentrations of total phosphorus in Lake Léman, 1957–1991
Source: After CIPEL, 1992

of waste water into lakes, the use of fertilizers (via run-off from agricultural land) and changes in land use which increase run-off. Data showing annual mean concentrations of phosphorus and nitrogen in selected lakes in OECD countries for 1970 to 1987 were presented in the third edition of this report (UNEP, 1991b). Although eutrophication is commonly associated with lakes it may also be a problem in some rivers and coastal areas (see below).

It is difficult to assess the extent of eutrophication on a global scale, but it is an established problem in many lakes and reservoirs in highly populated, industrialized countries and is probably the most pervasive water quality problem on a global scale (UNEP, 1991a). Nevertheless, eutrophication is not confined to industrialized regions because it can be accelerated by intensified agriculture. In drier climatic regions, or wherever water resources are stored in artificial impoundments, eutrophication can be a serious problem, necessitating extensive treatment of the water before it is suitable for human use. One third of the 800 reservoirs in Spain are highly eutrophic and serious eutrophication problems are also reported in reservoirs from South America, South Africa, Mexico and Australia (UNEP 1991a, 1992).

Lake Léman (Lake Geneva) is one of several lakes which have been well studied with respect to eutrophication. Monitoring has been carried out by the International Surveillance Commission of Lake Léman (CIPEL) since 1986 (CIPEL, 1992). Tables 1.13 and 1.14 show annual mean loads and annual mean concentrations of selected variables in Lake Léman from 1970 to 1991. The water quality of the lake, which was relatively good at the end of the 1950s, rapidly deteriorated due to an increase in nutrient inputs, primarily point source discharges of phosphorus. Eutrophication of the lake reached a critical stage during the period 1976 to 1979 before showing signs of steady recovery (Rapin et al., 1989).

For over 25 years, phosphorus has been recognized as the element most responsible for limiting primary production of fresh-water algae in temperate regions. In the late 1970s, it became clear that the degree of eutrophication in water bodies may often be predicted by phosphorus concentration data (Dunnette, 1992b). The phosphorus concentration increased in Lake Léman from about 15.4 µg l^{-1} in 1960 to 89.5 µg l^{-1} in 1979 (Figure 1.18). Since 1980, a steady decline in phosphorus concentrations has been observed. During this period, concentrations of dissolved reactive phosphorus in the lake have decreased by 56 per cent and this has been attributed to the introduction of tertiary waste-water treatment and to the ban, in 1986, on phosphates in detergents in Switzerland (Rapin et al., 1989).

Nitrates Nitrate concentrations in surface and ground waters have been a focus of studies and monitoring programmes in western countries since the 1960s and results have been well reported. Nitrate concentrations in excess of the WHO drinking water guideline value of 10 mg NO$_3$-N l^{-1} are now widespread in European and some North American aquifers. However, although known to be a problem, nitrate concentrations are less well reported in many other countries. High nitrate concentrations in drinking water may lead to methaemoglobinaemia in bottle-fed infants. Owing to the potential health risks of nitrate in drinking water, measurement of nitrate concentrations has been included in a major survey of ground waters used for drinking-water supplies in the USA (see below).

Elevated nitrate concentrations in ground waters are most commonly related to the application of fertilizers to soils. In recent decades there has been an exponential increase in the use of nitrogen fertilizers in most countries (see Part 3: Natural Resources). The gradual increase in nitrate concentrations in the Slapy reservoir, former Czechoslovakia, between 1960 and 1990 follows a similar trend to the increase in the application of nitrates in fertilizer to agricultural land in the river basin during the same period (Figure 1.19). Table 1.15 presents monitoring data from an aquifer in the former USSR. The data show dramatic decreases in the concentrations of nitrate, nitrite and ammonium (all of which are associated with the application of fertilizers) in the ground waters of the Vakhsh area during the period 1976 to 1987. These decreases are due to the cessation of agricultural activities in the area in preparation for the flooding of the valley after the construction of a hydro-electric dam. This example illustrates the direct influence of some agricultural practices on water quality.

Pesticides The aquatic environment is now exposed to growing numbers and quantities of pesticides. Monitoring for such a wide range of compounds, often present only in very low concentrations, is both costly and complex. Since detection of some of these compounds in water is difficult, many countries now recognize the need for alternative approaches to their monitoring, such as by measuring their concentrations in biota and particulate matter (see Part 1: Environmental Pollution, Biological Monitoring and Soils and Sediments). Some of the pesticides originally manufactured and used in the 1950s and 1960s have low solubility in water and have

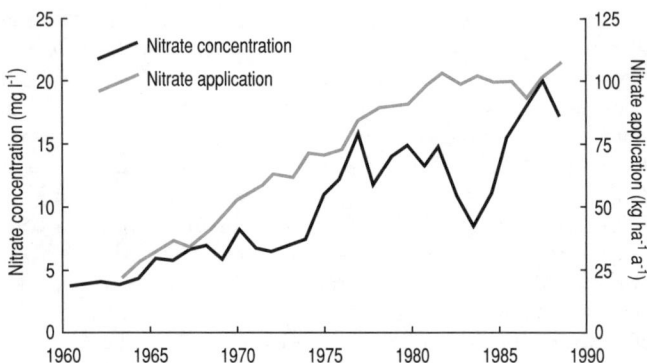

Figure 1.19 Nitrate concentrations in the Slapy reservoir in the Vltava River and the application of nitrates in fertilizers to agricultural land in the river basin, 1960–1990
Source: After Czechoslovak Report to UNCED, 1991

demonstrated great persistence in the environment. Recognition of this in the 1970s led to the banning of the manufacture and use of chemicals such as DDT (dichlorodiphenyltrichloroethane) (although evidence of the continued use of DDT and other organochlorine pesticides of the same generation remains in some world regions, see also Part 1: Environmental Pollution, Biological Monitoring).

Human exposure to pesticides occurs principally through food and studies of the effects of pesticides on human health are mostly focused on acute effects (OECD, 1986). Nevertheless, human exposure through drinking water in which concentrations exceed health-based standards presents health risks through chronic, long-term effects. Table 1.16 presents selected summary data from pesticide monitoring of surface waters used for public drinking supplies in some parts of the USA and Canada. These data show that in some areas certain herbicides were found in very high proportions of both raw and treated water supplies. Maximum concentrations of compounds in treated water used for public supply were often as high as for raw water (Table 1.16).

In response to concerns about pesticide and nitrate pollution of ground-water resources for drinking supplies, the US Environmental Protection Agency has recently published preliminary results of the National Survey of Pesticides and Nitrates in Drinking Water Wells, which was conducted between 1988-1990 (US EPA, 1992). During this survey, 1,349 community water system (CWS) wells and rural domestic drinking water wells (RD) were sampled and analyzed for the presence of 101 pesticides, 25 pesticide degradates and nitrate. Results of the survey showed that less than 0.8 per cent of CWS wells and 0.6 per cent of RD wells contained mean concentrations of pesticides above health-based limits; and 1.2 per cent of CWS wells and 2.4 per cent of RD wells contained mean concentrations of nitrate above health-based limits.

References

Chapman, D. (Ed.) 1992 *Water Quality Assessments: A guide to the use of biota, sediments and water in environmental monitoring.* Chapman & Hall, London.

CIPEL 1992 *Rapports sur les Etudes et Recherches Enterprises dans le Bassin Lémanique: Campagne 1991,* Commission Internationale pour la Protection des Eaux du Léman, Lausanne.

Czechoslovak Report to UNCED 1991 Prague.

Dunnette, D. A. 1992a *Assessing global river water quality: overview and data collection.* In: D. A. Dunnette and R. J. O'Brien (Eds.), The Science of Global Change: The Impacts of Human Activities on the Environment, ACS Symposium Series 483, American Chemical Society, Washington DC.

Dunnette, D. A. 1992b The aquatic component: An overview. In: D. A. Dunnette and R. J. O'Brien (Eds.), *The Science of Global Change: The Impact of Human Activities on the Environment,* ACS Symposium Series 483, American Chemical Society, Washington DC.

Meybeck, M., Chapman, D. and Helmer R. [Eds] 1989 *Global Freshwater Quality: A First Assessment,* Basil Blackwell, Oxford.

NIVA 1992a *International Co-operative Programme on Assessment and Monitoring of Acidification of Rivers and Lakes: Data Report 1990.* Norwegian Institute for Water Research, Oslo.

NIVA 1992b *Evaluation of the International Co-operative Programme on Assessment and Monitoring of Acidification of Rivers and Lakes.* Norwegian Institute for Water Research, Oslo.

OECD 1986 *Water Pollution by Fertilizers and Pesticides.* Organisation for Economic Co-operation and Development, Paris.

Rapin, F., Blanc, P. and Corvi, C. 1989 Influence des apports sur le stock de phosphore dans le lac Léman et sur son eutrophisation, *Revue des Sciences de L'eau,* **2,** 721–737.

Rodhe, H., Galloway, J. and Dianwu, Z. 1992 Acidification in Southeast Asia - prospects for the coming decades, *Ambio,* **21**(2), 148–150.

UNEP 1987 *United Nations Environment Programme Environmental Data Report 1987/88,* Basil Blackwell, Oxford.

UNEP 1989 *United Nations Environment Programme Environmental Data Report 1989/90,* Basil Blackwell, Oxford.

UNEP 1991a *Freshwater Pollution.* UNEP/GEMS Environment Library Series No. 6, United Nations Environment Programme, Global Environment Monitoring System, Nairobi.

UNEP 1991b *United Nations Environment Programme Environmental Data Report 1991/92,* Basil Blackwell, Oxford.

UNEP 1992 *Chemical Pollution: A global overview.* United Nations Environment Programme, Geneva.

UNEP/WHO 1988 *Assessment of Freshwater Quality.* United Nations Environment Programme, Nairobi and World Health Organization, Geneva.

US EPA 1992 *Another Look: National Survey of Pesticides in Drinking Water Wells, Phase 2 Report.* EPA 570/9-91-020, US Environmental Protection Agency.

WHO/UNEP 1991 *Water Quality: Progress in the implementation of the Mar del Plata Action Plan and a strategy for the 1990s.* GEMS Monitoring and Assessment Research Centre (Ed.), World Health Organization and the United Nations Environment Programme.

Estuaries, Coastal Waters and Seas

The open ocean is still relatively clean (GESAMP, 1990) and most monitoring of marine waters is concentrated in the coastal zones, where the impacts of human activities are most noticeable and can potentially affect the use or exploitation of the marine waters and their resources by local populations. Major contaminants discharged directly to the sea or transported with rivers become dispersed and diluted. Measurement of very low concentrations of many contaminants is technically difficult and, therefore, many studies of marine pollution have concentrated on measuring contaminants in particulate matter and biological tissues or have studied the effects of pollution on marine organisms (see Part 1: Environmental Pollution, Biological Monitoring and Soils and Sediments).

Marine pollution either originates on land and reaches the

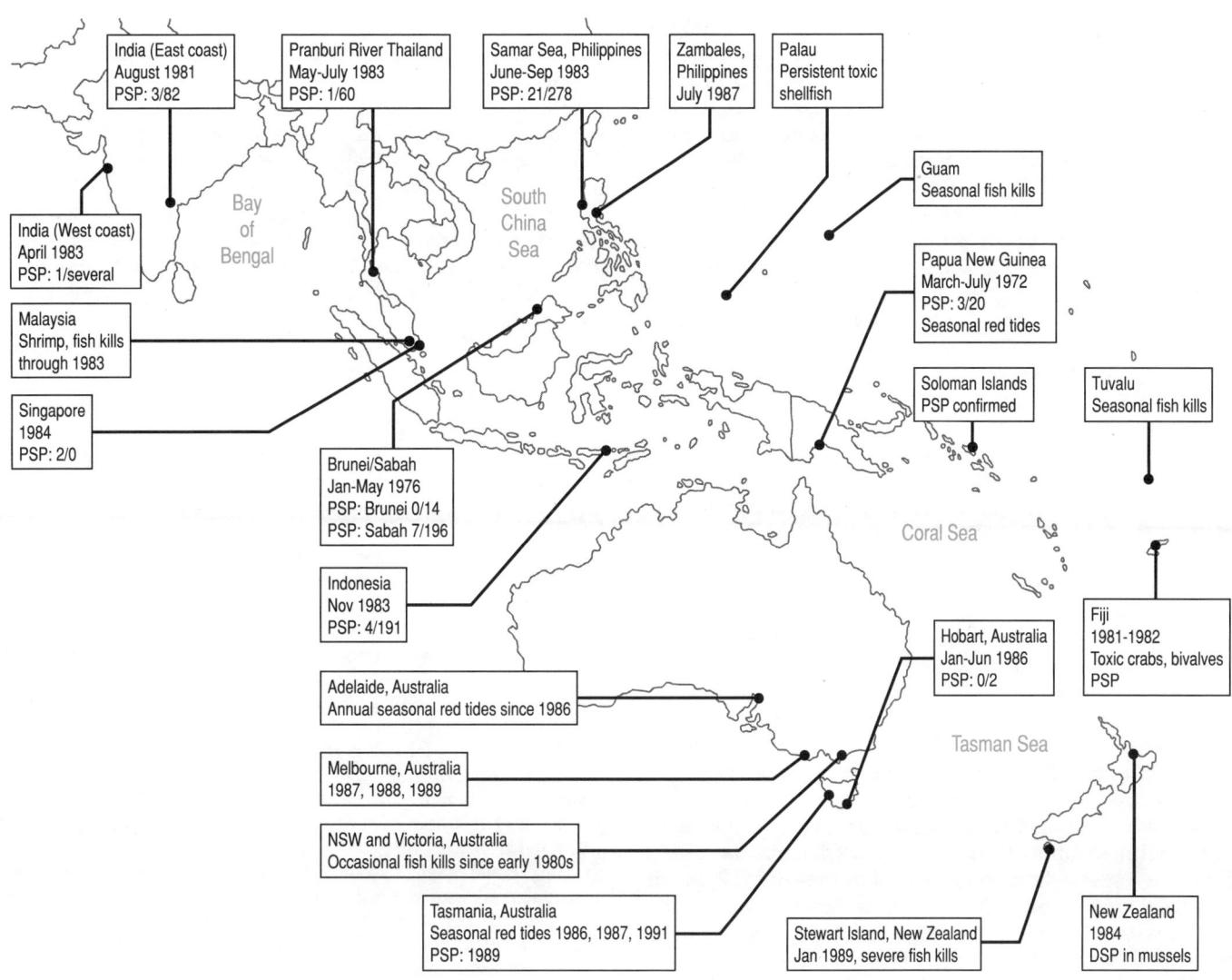

FIGURE 1.20 Locations of red tides, including those responsible for fish kills and incidents of shellfish poisonings in Asia and Oceania, 1980–1991. (numbers and dates in boxes refer to numbers of deaths/illnesses and the time of the first reported incident)
Source: After Maclean, 1989

sea at, or near, the coast from rivers, pipelines or ground-water seepage, or it arises from atmospheric depositions. Non-degradable contaminants, and those bound to particulate matter, can be transported by rivers and ultimately deposited in the seas. Poor land-use practices, such as deforestation, and other factors, such as increasing populations, uncontrolled urbanization and inappropriate agricultural practices have led to a dramatic rise in sediment and nutrient loads of rivers around the world. In view of this, the second phase of the GEMS/Water programme, mentioned above, recognizes the increasing need to determine fluxes of pollutants to continent/ocean interfaces and is in the process of designating 60–70 major river basins which will be monitored to meet this objective (WHO, 1991).

Table 1.17 presents data for 1983 and 1987 comparing fluxes of salts and nutrients from the three largest rivers in Turkey to the Black Sea. The Black Sea is virtually enclosed and is, therefore, particularly prone to pollution from land-based sources.

There is growing scientific evidence to suggest that it has suffered serious ecological damage from eutrophication and contamination by pathogenic microbes and toxic chemicals (Mee, 1992). In response to these problems, UNEP has taken the first initiative towards protection of the Black Sea through international co-operation. Following the signing by the riparian countries, in April 1992, of a treaty for the protection of the Black Sea, UNEP has designated it the twelfth Regional Sea in its Regional Seas Programme.

The Baltic Sea is also a partially enclosed sea which is affected by many of the same problems as the Black Sea. During the 1980s inputs of nitrogen and phosphorus to the Baltic Sea increased by a factor of approximately four to six (Rosenberg et al., 1990). Data on the amounts of nitrogen and phosphorus supplied to different sub-basins of the Baltic Sea in the 1980s were presented in the third edition of the 'UNEP Environmental Data Report' (UNEP, 1991). Long-term changes in loads of selected variables (including nitrogen and phosphorus) discharged to the Baltic

Sea by major Latvian rivers, from 1947 to 1988, are given in Table 1.18. Major increases in discharges occurred for most variables (except silicon) in all the rivers monitored from 1947 to 1988.

Enclosed seas such as the Black Sea and the Baltic Sea, as well as coastal zones around the world, that experience increased inputs of nutrients from land-based sources combined with limited mixing of the contaminated water with open ocean water, are vulnerable to marine eutrophication. Increases in nutrients can contribute to blooms of phytoplankton, particularly dinoflagellates and diatoms, and several other groups of photosynthetic micro-organisms. These blooms can give rise to a number of other environmental and possible human health problems in their localized areas (although such problems usually only last for a short duration, e.g., weeks). Severe algal blooms cause visible coloration of the water which has led to the common term "red tides" (although the coloration may be red, green, yellow or brown). In the Black Sea, successive algal blooms and high organic production have led to deoxygenation of most of the water column as the algae and organic matter are biologically degraded by oxygen-consuming micro-organisms.

Changes in the frequency and extent of algal blooms seem to be increasing in many locations around the world (Sarokin and Schulkin, 1992). In some locations, frequent blooms are being experienced consisting of organisms not known to have bloomed there in the past, or which were not indigenous to that area (e.g., a toxin-producing diatom bloom at Monterrey Bay, California in 1991). In addition, the spread of species responsible for red tides seems to be increasing and toxins are being identified from algal species that had previously been thought harmless. Consequently, the number of reports of catastrophic mass deaths of marine mammals, fish and invertebrates associated with red tides has also increased (Smayda, 1992). However, observed increases in algal blooms may be due to improved monitoring of such events, for example by satellite remote sensing, rather than to a genuine increase in incidents of red tides on a global scale (Smayda, 1992).

Blooms of certain algal species are associated with toxic effects on fish and shellfish and on humans which consume shellfish from areas affected by these algal species. Human intoxication, which can sometimes be fatal, may result from consumption of contaminated shellfish in which toxins have bioaccumulated. These cases include paralytic shellfish poisonings (PSP), diarrhoeic shellfish poisonings (DSP) and neurotoxic shellfish poisonings (NSP). A recent PSP outbreak in Guatemala resulted in 26 deaths and 187 cases of illness (Smayda, 1992). Figure 1.20 indicates the sites of some red tides, fish kills and incidents of PSP in Asia and Oceania, 1980-1991.

The continuing and increasing use of the sea for recreation has led to increased human exposure to pollutants discharged into coastal waters. The bacteriological quality of bathing waters receiving sewage discharges particularly gives cause for concern in many countries. The bacteriological quality of bathing waters in OECD countries was reported in the third edition of the 'UNEP Environmental Data Report' (UNEP, 1991). In addition, the European Community (EC) publishes the results of monitoring for compliance with the EC bathing water quality directive (76/160/EEC) (CEC, 1992). This programme will be highlighted in future editions of this report.

References

CEC 1992 *Quality of Bathing Water 1991*. Directorate-General Environment, Nuclear Safety and Civil Protection, Report EUR 14210 EN, Commission of the European Communities, Luxembourg.

GESAMP 1990 *The State of the Marine Environment*. GESAMP Reports and Studies No. 39, Joint Group of Experts on the Scientific Aspects of Marine Pollution, United Nations Environment Programme, Nairobi.

Maclean, J. L. 1989 Indo-Pacific Red Tides, 1985–1988, *Marine Pollution Bulletin*, **20**(7), 304–310.

Mee, L. D. 1992 The Black Sea in crisis: A need for concerted international action, *Ambio*, **21**(4), 278–286.

Rosenberg, R., Elmgren, R., Fleischer, S., Jonsson, P., Persson, G. and Dahlin, H. 1990 Marine eutrophication case studies in Sweden, *Ambio* **19**(3), 102–108.

Sarokin, D. and Schulkin, J. 1992 The role of pollution in large scale population disturbances, part 1: aquatic populations, *Environmental Science and Technology*, **26**(8), 1476–1484.

Smayda, T. J. 1992 A phantom in the ocean, *Nature* (London), **358**, 374–375.

UNEP 1991 *United Nations Environment Programme Environmental Data Report 1991/92*, Basil Blackwell, Oxford.

WHO 1991 *GEMS/Water 1990–2000: The Challenge Ahead*, World Health Organization, Geneva.

TABLE 1.11 Physical characteristics, nutrient levels and pollution indicators in rivers of the South-east Asian and Pacific regions as reported to the GEMS/Water Programme, 1987–1991 (median values)

Variable/country	Station	1987	1988	1989	1990	1991
PHYSICAL CHARACTERISTICS: Suspended solids (mg l^{-1})						
Australia	La Trobe River	53.5				
	Mitta Mitta River	4.0				
	Yarra River	11.0				
China	Changjiang (Yangtze River)	158.5	153.5	173.0	174.5	
	Huanghe (Yellow River)	1,825.0	2,567.0	3,354.5	6,822.0	
	Zhujiang (Pearl River)	77.5	27.5	15.0	30.5	
Indonesia	Musi River at New Intake	95.0	61.5	89.0	NR	NR
	River Banjir Kanal	44.0	45.6	45.2	48.0	NR
	River Citarum	192.0	77.0	169.0	70.0	NR
	River Garang	76.0	75.5	131.0	245.0	
	River Sunter	86.0	73.5	121.0	140.0	NR
	River Surabaya	48.0	82.0	370.0	90.0	
Japan	Kiso River at Asahi	4.0	5.0	4.5	5.5	
	Kiso River at Inuyama	5.0	3.0	3.0	3.0	
	Kiso River at Shimo-Ochiai	4.0	6.0	5.5	4.0	
	Ohta River at Hesaka	3.0	3.0	4.0	3.0	
	Sagami River at Samukawa	5.5	6.5	9.5	8.5	
	Shinano River at Zuiun Bridge	15.5	23.0	20.0	16.5	
	Tone River at Tone-Ozeki	6.5	11.0	6.5	5.0	NR
	Toyohira River at Shiraikawa	2.5	4.0	5.5	4.5	
	Yodo River at Hirakata Bridge	18.5	17.5	16.5	19.0	
Malaysia	Kelantan River	63.5				
	Klang River	95.5	106.5	156.0		
	Linggi River	NR	82.5	154.0		
	Muda River	62.5				
	Sekudai River	14.5	34.0	16.0	NR	
New Zealand	Waikato River at Mercer Bridge	14.6	16.1	17.4	18.8	18.2
	Waikato River at Taupo Gates	2.0	0.5	1.35	1.25	1.55
Philippines	Cagayan River		29.0	52.0	36.0	6.0
Korea	Han River	3.0	2.9	3.4	3.4	
PHYSICAL CHARACTERISTICS: Instantaneous discharge (m^3 s^{-1})						
Australia	La Trobe River	17.00				
	Mitta Mitta River	10.45				
	Murray River	77.10				
	Yarra River	6.65				
China	Changjiang (Yangtze River)	17,900.00	16,050.00	25,750.00	21,150.00	
	Huanghe (Yellow River)	399.00	400.00	860.00	880.00	
	Zhujiang (Pearl River)	6,260.00	3,705.00	3,260.00	4,510.00	
India	Cauveri River near Musiri	14.34	41.55		24.54	57.44
	Krishna River at Gadwal	23.20				
	Mahi River near Sevalia	8.44				158.00
	Periyar River near Kaladi	436.40	418.00	274.00	192.00	53.30

Continued

TABLE 1.11   Continued

Variable/country	Station	1987	1988	1989	1990	1991
Indonesia	River Banjir Kanal	21.97	15.31	14.43	21.37	NR
	River Citarum	23.11	49.51	86.29	31.62	NR
	River Sunter	8.00	7.58	7.94	9.25	NR
Japan	Kiso River at Asahi	100.98	131.50	167.50	170.00	
	Kiso River at Inuyama	133.82	146.00	205.50	141.00	
	Kiso River at Shimo-Ochiai	83.49	119.00	125.00	85.00	
	Ohta River at Hesaka	56.77	40.33	60.85	69.90	
	Sagami River at Samukawa	19.12	28.37	44.30	37.55	
	Shinano River at Zuiun Bridge	231.60	225.10	224.00	195.50	
	Tone River at Tone-Ozeki	130.00	180.00	205.00	170.00	NR
	Toyohira River at Shiraikawa	13.85	13.00	16.00	18.10	
	Yodo River at Hirakata Bridge	129.45	118.30			
New Zealand	Waikato River at Mercer Bridge	236.00	384.00	428.50	372.00	391.50
	Waikato River at Taupo Gates	238.00	NR	217.00	220.00	131.50
Korea	Han River	173.62	40.11	19.84	145.16	

NUTRIENTS: Nitrogen, Nitrate + Nitrite (mg l^{-1} as N)

Variable/country	Station	1987	1988	1989	1990	1991
Australia	La Trobe River	0.08				
	Mitta Mitta River	0.01				
	Murray River	0.01				
	River Murray, Mannum	0.06				
	Yarra River	0.54				
India	Bhima River near Takali	0.18	0.314	0.164	0.281	0.348
	Cauveri River near Musiri	0.20	0.040	0.195	0.085	0.185
	Chaliyar River at Kalpalli	0.31	0.315	0.255	0.220	0.340
	Godavri River near Dhalegaon	0.28	0.377	0.153	0.381	0.335
	Kallada River at Panamthottam Kad	0.35	0.295	0.410	0.395	0.350
	Krishna River at Gadwal	0.15	0.260	0.200	0.230	0.135
	Mahi River near Sevalia	0.13	0.210	0.255	4.148	0.185
	Narmarda River at Sethani Ghat	0.02	0.035	0.033	0.021	0.035
	Periyar River near Kaladi	0.36	0.295	0.310	0.300	0.320
	Sabarmati River in Ahmedabad	0.54	0.350	0.240	0.300	0.228
	Tapti River near Burhanpur	0.60	0.545	0.478	0.404	0.586
	Wainganga River near Ashti	0.28	0.241	0.148	1.120	0.264
Malaysia	Klang River	5.40	3.70			
	Linggi River	NR	1.65			
	Sekudai River	1.75	1.92	2.16	NR	
New Zealand	Waikato River at Mercer Bridge	0.38	0.35	0.32	0.35	0.13
	Waikato River at Taupo Gates	0.01	0.01	<0.01	0.01	<0.01

NUTRIENTS: Nitrate (mg l^{-1} as N)

Variable/country	Station	1987	1988	1989	1990	1991
China	Changjiang (Yangtze River)	0.42	0.88	0.73	1.15	
	Huanghe (Yellow River)	2.30	2.76	2.44	2.48	
	Zhujiang (Pearl River)	0.67	0.67	0.84	0.70	
Japan	Kiso River at Asahi	0.40	0.44	0.37	0.36	
	Kiso River at Inuyama	0.32	0.37	0.32	0.31	
	Kiso River at Shimo-Ochiai	0.23	0.25	0.24	0.24	

Continued

TABLE 1.11 Continued

Variable/country	Station	1987	1988	1989	1990	1991
Japan	Ohta River at Hesaka	0.42	0.44	0.42	0.41	
	Sagami River at Samukawa	1.83	1.89	1.74	1.97	
	Shinano River at Zuiun Bridge	0.75	0.85	0.78	0.84	
	Tone River at Tone-Ozeki	1.95	NR			
	Toyohira River at Shiraikawa	0.14	0.16	0.14	0.18	
	Yodo River at Hirakata Bridge	0.81	0.80	0.93	0.96	
Malaysia	Kinta River	NR	0.05	NR		

NUTRIENTS: Total Orthophosphate (mg l^{-1} as P)

Variable/country	Station	1987	1988	1989	1990	1991
China	Changjiang (Yangtze River)	<0.01	<0.01	<0.01	<0.01	
	Huanghe (Yellow River)	0.01	0.01	0.01	0.01	
	Zhujiang (Pearl River)	<0.01	<0.01	<0.01	<0.01	
Indonesia	Musi River at New Intake	0.02	0.01	0.04	NR	NR
	River Banjir Kanal	0.17	0.19	0.13	0.12	NR
	River Citarum	0.03	0.03	0.03	0.05	NR
	River Garang	0.07	0.06	0.10	0.08	
	River Sunter	0.11	0.16	0.08	0.07	NR
	River Surabaya	0.15	0.12	0.10	0.11	
Japan	Kiso River at Asahi	0.02	0.01	0.01	0.01	
	Kiso River at Inuyama	0.01	0.01	0.01	NR	
	Kiso River at Shimo-Ochiai	0.01	NR	NR	NR	
	Ohta River at Hesaka	0.01	0.01	0.01	0.01	
	Sagami River at Samukawa	0.08	0.08	0.06	0.06	
	Shinano River at Zuiun Bridge	0.04	0.05	0.04	0.05	
	Tone River at Tone-Ozeki	0.04	0.03	0.04	0.05	NR
	Toyohira River at Shiraikawa	0.02	0.02	0.02	0.03	
	Yodo River at Hirakata Bridge	0.08	0.08	0.05	0.04	
Malaysia	Kelantan River	0.19				
	Muda River	0.04				
New Zealand	Waikato River at Mercer Bridge	0.03	0.02	0.02	0.02	0.02
	Waikato River at Taupo Gates	<0.01	NR	<0.01	<0.01	<0.01
Korea	Han River	0.01	0.01	0.01	0.02	

NUTRIENTS: Total Phosphorus (mg l^{-1} as P)

Variable/country	Station	1987	1988	1989	1990	1991
Australia	Yarra River	0.11				
China	Changjiang (Yangtze River)	0.10	0.09	0.12	0.05	
	Huanghe (Yellow River)	1.15	1.76	2.27	5.13	
	Zhujiang (Pearl River)	0.05	0.05	0.05	0.05	
Indonesia	Musi River at New Intake	0.08	0.08	0.09	NR	NR
	River Banjir Kanal	0.40	0.45	0.14	0.30	NR
	River Citarum	0.13	0.16	0.08	0.10	NR
	River Garang	0.12	0.11	0.16	0.09	
	River Sunter	0.33	0.35	0.14	0.18	NR
	River Surabaya	0.24	0.16	0.13	0.16	
Japan	Kiso River at Asahi	0.03				

Continued

TABLE 1.11 Continued

Variable/country	Station	1987	1988	1989	1990	1991
Japan	Ohta River at Hesaka	0.02	0.02	0.03	0.02	
	Yodo River at Hirakata Bridge	0.18	0.17	0.13	0.13	
Malaysia	Kinta River	NR	0.01			
New Zealand	Waikato River at Mercer Bridge	0.10	0.08	0.09	0.09	0.07
	Waikato River at Taupo Gates	0.01	0.01	0.01	0.01	0.01
Thailand	Chao Phrya R. D/S Nakhon Sawan	NR	NR	NR	0.45	

POLLUTION INDICATORS: BOD$_5$ (mg l^{-1})

Variable/country	Station	1987	1988	1989	1990	1991
China	Changjiang (Yangtze River)	0.70	0.80	1.05	0.80	
	Huanghe (Yellow River)	1.70	1.55	0.93	1.18	
	Zhujiang (Pearl River)	0.50	0.60	0.65	0.85	
Fiji	Waimanu River	0.90	0.95			
India	Bhima River near Takali	4.00	4.50	4.10	4.40	3.80
	Cauveri River near Musiri	1.74	1.10	1.00	1.00	2.00
	Chaliyar River at Kalpalli	0.67	0.30	0.60	0.90	1.00
	Godavri River near Dhalegaon	5.30	4.10	4.00	4.00	3.70
	Kallada River at Panamthottam Kad	0.73	0.30	0.40	0.60	0.50
	Krishna River at Gadwal	3.55	2.90	3.20	2.50	2.30
	Mahi River near Sevalia	1.50	1.60	1.30	2.30	2.00
	Narmarda River at Sethani Ghat	2.75	2.20	2.50	2.30	2.90
	Periyar River near Kaladi	0.61	0.20	0.30	0.60	0.50
	Sabarmati River in Ahmedabad	92.00	101.00	111.00	84.00	72.00
	Tapti River near Burhanpur	2.55	1.60			2.20
	Wainganga River near Ashti	4.30	3.70	4.00	6.00	3.70
Indonesia	Musi River at New Intake	1.40	1.60	1.10	NR	NR
	River Banjir Kanal	8.10	8.80	5.70	9.20	NR
	River Citarum	7.00	6.80	7.80	7.40	NR
	River Garang	1.40	2.65	NR	2.30	
	River Sunter	10.00	8.55	6.20	7.90	NR
	River Surabaya	11.70	15.00	10.00	5.90	
Japan	Kiso River at Asahi	0.95	0.60	0.75	0.90	
	Kiso River at Inuyama	0.80	1.00	0.70	0.65	
	Kiso River at Shimo-Ochiai	0.60	0.75	0.55	0.30	
	Ohta River at Hesaka	0.55	0.70	0.60	0.60	
	Sagami River at Samukawa	1.95	2.15	1.70	1.45	
	Shinano River at Zuiun Bridge	1.60	1.40	1.25	1.50	
	Tone River at Tone-Ozeki	1.35	1.45	1.20	1.25	NR
	Toyohira River at Shiraikawa	1.15	1.05	1.25	1.50	
	Yodo River at Hirakata Bridge	3.85	2.90	2.85	2.35	
Malaysia	Kinta River	NR	0.80	NR		
	Klang River	7.00	7.00	4.80		
	Linggi River	NR	1.95	3.10		
	Sekudai River	2.20	1.35	2.20	NR	
New Zealand	Waikato River at Mercer Bridge	0.90	1.30	1.10	0.90	
	Waikato River at Taupo Gates	0.58	NR	0.55	0.75	0.55

Continued

TABLE 1.11 Continued

Variable/country	Station	1987	1988	1989	1990	1991
Philippines	Cagayan River		0.70	0.30	1.00	0.44
Korea	Han River	1.80	1.60	1.20	1.30	
Thailand	Chao Phrya R. D/S Nakhon Sawan		NR	1.20	0.50	

POLLUTION INDICATORS: Faecal coliforms (No. per 100ml)

Variable/country	Station	1987	1988	1989	1990	1991
China	Changjiang (Yangtze River)	475	640	945	865	
	Huanghe (Yellow River)	730	1,025	1,450	2,950	
	Zhujiang (Pearl River)	240	155	95	205	
India	Cauveri River near Musiri	230	600	1,350	2,900	2,250
	Chaliyar River at Kalpalli	240	270	270	700	500
	Kallada River at Panamthottam Kad	720	900	900	335	700
	Krishna River at Gadwal	11	4			
	Mahi River near Sevalia	52,500	2,400	4,600	1,700	430
	Periyar River near Kaladi	270	240	240	135	260
	Sabarmati River in Ahmedabad	150,000	1.3×10^6	24.0×10^6	24.0×10^6	35.0×10^6
	Tapti River near Burhanpur	130				
Indonesia	Musi River at New Intake	5,164	5,674	5,284	NR	NR
	River Banjir Kanal	$> 1 \times 10^6$	$> 1 \times 10^6$	820,000	$> 1 \times 10^6$	NR
	River Citarum	460,000	400,000	$> 1 \times 10^6$	890,000	NR
	River Garang			NR	82,500	
	River Sunter	$> 1 \times 10^6$	$> 1 \times 10^6$	310,000	560,000	NR
Japan	Kiso River at Asahi	265	120	185	155	
	Kiso River at Inuyama	710	490	170	120	
	Kiso River at Shimo-Ochiai	490	215	380	220	
	Ohta River at Hesaka	1,010	330	155	102	
	Sagami River at Samukawa	1,500	350	1,400	2,850	
	Tone River at Tone-Ozeki	745	490	435	1,045	NR
	Toyohira River at Shiraikawa	205	200	135	34	
New Zealand	Waikato River at Mercer Bridge	250	350	425	150	160
Korea	Han River	7	11	5	23	

TRACE METALS: Total Cd (mg l^{-1})

Variable/country	Station	1987	1988	1989	1990	1991
Australia	River Murray, Mannum	<0.01				
Indonesia	River Banjir Kanal	0.001	NR	0.004	0.004	NR
	River Citarum	0.003	NR	0.020	0.011	NR
	River Garang	NR	0.002	NR	NR	
	River Sunter	0.001	NR	0.009	0.010	
	River Surabaya	0.016	NR	NR	NR	
Japan	Kiso River at Asahi	0.001	0.005	0.001	0.001	
	Kiso River at Inuyama	0.001	0.001	0.001	0.001	
	Kiso River at Shimo-Ochiai	0.001	0.001	0.001	0.001	
	Sagami River at Samukawa	0.001	0.001	0.001	0.001	
	Shinano River at Zuiun Bridge	0.005	0.005	0.005	0.005	
	Yodo River at Hirakata Bridge	0.010	0.010	0.010	0.010	
	Sekudai River		0.100			

Continued

TABLE 1.11 Continued

Variable/country	Station	1987	1988	1989	1990	1991
Thailand	Chao Phrya R. D/S Nakhon Sawan	NR	NR	0.002	NR	

TRACE METALS: Total Hg (μg l-1)

Japan	Kiso River at Asahi	NR	NR	NR	0.500	
	Kiso River at Inuyama	0.250	0.500	0.500	0.500	
	Kiso River at Shimo-Ochiai	0.250	0.500	0.500	0.500	
	Ohta River at Hesaka	0.500	0.500	0.500	0.500	
	Sagami River at Samukawa	0.100	0.100	0.100	0.100	
	Shinano River at Zuiun Bridge			0.200	0.200	
	Tone River at Tone-Ozeki	0.500	0.500	0.500	0.500	NR
	Toyohira River at Shiraikawa	0.500	0.500	0.500	0.500	
	Yodo River at Hirakata Bridge	0.500	0.500	0.500	0.500	

TRACE METALS: Total Pb (mg l^{-1})

Australia	River Murray, Mannum	0.006				
Indonesia	Musi River at New Intake	0.001	0.010	NR		
	River Banjir Kanal	0.010	NR	0.025	0.325	NR
	River Citarum	0.009	NR	0.030	0.155	
	River Sunter	0.007	NR	0.035	0.268	NR
	River Surabaya	0.080	0.090	0.050	NR	
Japan	Kiso River at Asahi	0.002	0.002	0.002	0.002	
	Kiso River at Inuyama	0.003	0.003	0.002	0.002	
	Kiso River at Shimo-Ochiai	0.003	0.003	0.002	0.002	
	Shinano River at Zuiun Bridge	0.020	0.020	0.005	0.005	
	Yodo River at Hirakata Bridge	0.010	0.010	0.010	0.010	
Malaysia	Klang River		NR	0.020		
	Linggi River		NR	0.020		
	Sekudai River		0.100			
Thailand	Chao Phrya R. D/S Nakhon Sawan	NR	NR	NR	0.020	

ORGANOCHLORINES: Total DDT (μg l^{-1})

China	Changjiang (Yangtze River)	0.058	0.058	0.058	0.058	
	Huanghe (Yellow River)	0.040	0.040	0.264	0.264	
	Zhujiang (Pearl River)	0.070	0.070	0.070	0.070	
Japan	Ohta River at Hesaka	0.100	0.100	0.100	0.100	
	Shinano River at Zuiun Bridge	0.005	0.005	0.005	0.005	

ORGANOCHLORINES: PCBs (μg l^{-1})

Japan	Ohta River at Hesaka	0.500	0.500	0.500	NR	
	Shinano River at Zuiun Bridge	0.500	0.400	0.300	0.300	

NR in this table indicates that fewer than four measurements were made during the year.

Data in the above table are from all GEMS/Water river stations where data are available for one or more of the selected variables. In the case of India, not all stations which submit data to the data base are listed. The number of decimal places to which data are presented varies according to the variable and may differ from the reporting accuracy of the GEMS/Water data base. For further information please refer to the GEMS/Water data summaries for 1982–84 and 1985–87. Recommended sampling frequencies and analytical methods are given in the 'GEMS/Water Operation Guide'. Further information on the GEMS/Water network coverage and reporting experience can be obtained from WHO, Geneva.

Source:
Data supplied by the National Water Research Institute, Canada Centre for Inland Waters, Burlington, Ontario.

TABLE 1.12 Annual mean concentrations of selected water quality variables in rivers and lakes in countries contributing to the UN ECE International Co-operative Programme for Assessment and Monitoring of Acidification of Rivers and Lakes, 1990

Region/country	Location of river or lake	N	pH	SO$_4^{2-}$ (mg l^{-1})	Ca^{2+} (mg l^{-1})	Mg^{2+} (mg l^{-1})	Al^{3+} (μg l^{-1})	Alkalinity[a]
NORTH AMERICA								
Canada	Nova Scotia, Beaverskin Lake	6	5.40	4.2	0.4	0.4	56.8	4.2
	Ontario, Algoma region, Batchawana Lake	70–73	6.00	5.2	2.7	0.4	114.0	44.5
	Ontario, Algoma region, Little Turkey Lake	57–75	6.74	5.7	4.7	0.5	52.2	133.1
	Ontario, Algoma region, Turkey Lake	57–74	6.80	5.7	5.6	0.5	41.2	180.3
	Ontario, Algoma region, Wishart Lake	57–75	6.56	5.3	3.9	0.4	74.4	92.4
	Quebec, Laflamme Lake	38–51	6.24	4.1	2.4	0.3	106.4	91.6
	Quebec, Parc de la Jaques–Cartier, Lac Bonneville	1–6	5.00	4.2	0.9	0.2	200.3	–380.0
USA	Colorado, Summit Lake	5		1.0	1.0	0.2		60.5
	New York, Adirondack Mountain, Arbutus	11		6.2	3.2	0.6	31.2	68.6
	New York, Adirondack Mountain, Constable	11		6.0	2.0	0.3	221.8	–2.1
	New York, Adirondack Mountain, Dart Lake	11		5.6	2.0	0.3	161.6	1.0
	New York, Adirondack Mountain, Heart Lake	11		4.5	2.3	0.3	28.3	39.6
	New York, Adirondack Mountain, Lake Rondaxe	11		5.5	2.3	0.4	107.8	26.5
	New York, Adirondack Mountain, Moss Lake	11		5.8	2.8	0.5	67.2	55.8
	New York, Adirondack Mountain, Otter Lake	10		5.7	1.7	0.4	159.7	4.8
EUROPE								
Austria	Kaernten, Gradenbach	5	7.54	15.7	13.6	3.2		552
	Kaernten, Wangenitzbach	5	7.40	13.6	11.4	<2.8		490
	Tirol, Piburger Bach	10	7.42	6.3	8.1	1.8		325
Belgium	Eupen, Getz River, small dam	44–48	4.50	14.6	3.2	1.0	758.7	
	Eupen, Vesedre River, Bellesfort	43–48	4.72	17.6	4.2	1.3	740.5	
	Eupen, Vesedre River, dam (0.5m)	39–48	4.40	14.6	3.2	1.1	840.8	
	Jalhay, Gileppe River, Chemin des Charbonniers	45–48	4.64	15.5	3.5	1.3	690.7	
	Jalhay, Helle River, Schornstein	44–48	4.25	13.4	2.3	1.0	720.7	
	Jalhay, Lake Gilleppe (0.5m)	43–47	4.90	16.0	4.1	1.4	539.1	
	Jalhay, Louba River, Les Hes	46–48	6.41	16.8	6.4	2.0	251.0	
	Jalhay, Soor River, small dam	44–48	4.26	16.1	3.0	1.1	971.6	
Denmark	Sestrup Sande, Skaerbaek, Station A	9–13	5.18	10.9	3.1	1.8		0.034
	Sestrup Sande, Skaerbaek, Station B	13–25	5.50	13.6	5.8	2.2		0.038
	Sestrup Sande, Skaerbaek, Station C	9–13	5.62	14.0	6.1	2.2		0.049
	Sestrup Sande, Skaerbaek, Station D	13–23	6.42	17.4	10.3	2.9		0.121
	Sestrup Sande, Skaerbaek, Station F	12–23	6.58	17.4	10.3	2.9		0.122

Continued

TABLE 1.12 Continued

Region/country	Location of river or lake	N	pH	SO$_4^{2-}$ (mg l^{-1})	Ca^{2+} (mg l^{-1})	Mg^{2+} (mg l^{-1})	Al^{3+} (µg l^{-1})	Alkalinity[a]
Finland	Hirvilampi	4	5.12	9.0	1.9	0.5	270	−6.7
	Maekilampi	4	5.12	7.4	1.8	0.4	220	−13.7
	Vuorilampi	6	5.52	8.1	2.3	0.6	297	6.5
Germany, Fed. Rep.	Hunsrueck, Graefenbach	10	4.47	29.4	7.1	2.7	1,000.0	
	Hunsrueck, Schwollbach	10	5.29	10.1	2.7	1.5	560.0	
	Hunsrueck, Traunchbach, Station A	11	4.46	10.2	2.1	1.5	873.6	
	Hunsrueck, Traunchbach, Station B	10–11	5.41	8.1	2.2	1.3	520.9	
	Rothaargebirge, Elberndorfer Bach	20	6.85	14.3	3.9	3.2	<55.0	
	Rothaargebirge, Zinse	18 18–20	6.73	13.4	3.7	2.7	<27.0	
Netherlands	Achterste Goorven, Station A	5	5.26	24.2	3.0	1.8	370.0	0.049
	Achterste Goorven, Station B	5	5.10	33.8	3.6	2.1	570.0	<0.041
	Achterste Goorven, Station E	5	5.26	33.4	3.5	2.1	660.0	<0.076
	Gerritsfles	12	4.54	10.2	1.4	0.8	131.7	<0.015
Norway	Buskerud, Langtjern outflow	51	4.77	3.0	1.0	0.1	175.4	
	Aust Agder, Birkenes	51	4.66	5.2	1.1	0.3	429.6	
	Aust Agder, Tovdalselva, Boen Bruk	11	4.92	3.6	0.8	0.2	168.3	
	Finmark, Dalelva, Jarfjord	56	5.95	6.0	1.6	1.0	32.0	
	Oppland, Aurdoela, Aurdalsfjorden	11	6.15	2.5	1.1	0.2	49.7	
	Rogaland, Vikedalselva, Vindatfjorden	21	5.26	2.3	0.6	0.4	60.1	
	Sogn og Fjordane, Gular, Eldalen	17	5.28	1.3	0.3	0.2	61.7	
	Sogn og Fjordane, Nausta, Naustdal	19	5.57	1.3	0.4	0.2	33.9	
	Sogn og Fjordane, Trodoela	43	5.34	1.3	0.3	0.2	<31.9	
	Telemark, Storgama outflow	51	4.54	3.2	0.5	0.1	157.9	
Sweden	Alsteraan, Getebo	12	6.48	17.3	6.9	1.8		111.4
	Alsteraan, Stroemsborg	9	6.26	17.6	6.4	1.9		102.7
	Anraasen, Hoersvatn	16	4.44	8.8	1.0	1.0		83.0
	Brunnsjoen (0.5m)	7	5.54	19.3	5.5	2.0		14.5
	Delaangeraan	12	6.77	5.7	3.7	1.1		172.3
	Fiolen (0.5m)	6	6.15	10.9	3.8	1.2		31.7
	Fraecksjoen (0.5m)	7	6.22	11.1	4.5	1.5		44.1
	Haersvatn (0.5m)	7	4.37	8.8	1.0	1.0		
	Stensjoen (0.5m)	6	6.28	2.8	1.3	0.3		43.7
	Storasjoen	6	5.32	7.4	1.8	0.7	95.0	
	Tvaeringen (0.5m)	6	6.76	3.1	2.6	0.6		122.3

Continued

TABLE 1.12 Continued

Region/country	Location of river or lake	N	pH	SO_4^{2-} (mg l^{-1})	Ca^{2+} (mg l^{-1})	Mg^{2+} (mg l^{-1})	Al^{3+} (μg l^{-1})	Alkalinity[a]
UK	England, Scoat Tarn	4	4.91	3.2	0.7	0.7		-0.5
	Scotland, Loch Coire nan Arr	4	6.28	2.0	0.7	0.8		1.5
	Scotland, Lochnager	4	5.33	2.8	0.6	0.4		0.0
	Scotland, Round of Glenhead	4	4.83	3.3	0.7	0.6		-0.6
	Wales, Llyn Llagi	4	5.20	3.2	1.4	0.7		1.5

[a] In Austria, Canada, Finland, Sweden and USA alkalinity is measured using the Gran titration method and data are given in μeqv l^{-1}. Alkalinity is measured using different methods in other countries and data are given in mg CaCO$_3$ l^{-1} for Denmark and UK, and in mmol l^{-1} for the Netherlands.

N = Number of samples.

Only locations where at least four observations were made during 1990 are included in the above table.

Please refer to the manual for the International Co-operative Programme on Assessment and Monitoring of Acidification of Rivers and Lakes for further details relating to sampling and analytical procedures.

Source:
UN ECE 1992 *International Co-operative Programme on Assessment and Monitoring of Acidification of Rivers and Lakes. Data Report 1990*, Norwegian Institute for Water Research, Oslo.

TABLE 1.13 Total loads of selected variables in Lake Léman, 1970–1991 (t)

Year	O_2	P_{total}	PO_4^-	N_{total}	NH_4^+	NO_2^-	NO_3^-	Cl^-
1970	830,600	6,918			245	109	32,450	
1971	830,500	5,790	3,910		770	138	31,760	237,000
1972	792,400	6,090	4,830		1,255	141	33,030	251,000
1973	801,900	6,900	5,660	49,180	1,185	201	33,920	271,000
1974	781,700	6,700	5,240	50,350	1,180	164	36,200	295,000
1975	767,500	7,200	5,670	51,970	905	167	37,220	314,000
1976	716,800	7,670	6,200	53,820	1,000	116	37,800	329,000
1977	712,100	7,660	6,340	52,140	960	155	39,000	347,000
1978	732,300	7,440	6,290	52,860	595	153	39,810	358,000
1979	765,500	7,670	6,340	54,970	470	129	39,330	372,000
1980	776,200	7,070	6,130	56,270	635	159	40,780	376,000
1981	798,600	7,080	6,130	58,970	680	90	42,720	388,000
1982	791,600	6,640	5,950	57,830	720	104	44,450	394,000
1983	787,600	6,460	5,760	59,360	875	105	47,020	403,000
1984	810,200	6,550	5,790	60,500	965	83	47,440	418,000
1985	817,600	6,260	5,570	62,970	1,205	50	47,600	439,000
1986	842,600	6,150	5,300	61,500	580	72	47,160	454,000
1987	824,200	5,800	5,000	61,130	570	69	48,530	462,000
1988	799,940	5,290	4,665	60,750	460	74	50,350	480,000
1989	741,520	4,995	4,430	61,020	480	74	51,220	486,000
1990	714,200	4,740	4,145	59,000	495	77	49,890	496,000
1991	727,600	4,480	3,880	56,540	510	81	49,080	514,000

Source: CIPEL 1992 *Rapports sur les Etudes et Recherches Entreprises dans le Bassin Lémanique: Campagne 1991*, Commission Internationale pour la Protection des Eaux du Léman, Lausanne.

TABLE 1.14 Annual mean concentrations of selected variables in Lake Léman, 1970–1991

Year	Diss. O_2 ($mg\ l^{-1}$)	P_{total} ($\mu g\ P\ l^{-1}$)	PO_4^--P ($\mu g\ l^{-1}$)	N_{total} ($\mu g\ N\ l^{-1}$)	NH_4^+-N ($\mu g\ l^{-1}$)	NO_2^--N ($\mu g\ l^{-1}$)	NO_3^--N ($\mu g\ l^{-1}$)	Cl^- ($mg\ l^{-1}$)	Transparency (m)
1970	9.69	80.5	50.5		2.9	1.3	379		8.6
1971	9.69	67.6	45.6		9.0	1.6	371		9.5
1972	9.25	71.1	56.3		14.6	1.6	385		8.5
1973	9.36	80.5	66.1	574	13.8	2.4	396		9.3
1974	9.12	78.2	63.2	588	13.8	1.9	422		8.5
1975	8.96	84.0	66.1	606	10.6	1.9	434		7.3
1976	8.36	89.6	72.3	628	11.7	1.4	441		8.2
1977	8.31	89.4	74.0	608	11.2	1.8	455		8.0
1978	8.55	86.8	73.4	617	7.0	1.8	465		7.3
1979	8.93	89.5	74.0	641	5.5	1.5	459		10.4
1980	9.06	82.5	71.5	657	7.4	1.9	476		8.9
1981	9.32	82.6	71.6	688	8.0	1.1	498		8.1
1982	9.24	77.5	69.5	675	8.4	1.2	519		7.5
1983	9.19	75.4	67.3	693	10.2	1.2	549		8.2
1984	9.46	76.4	67.6	706	11.2	1.0	554		7.6
1985	9.54	73.1	65.0	734	14.0	0.6	556		8.4
1986	9.83	71.8	61.9	718	6.8	1.0	550	106.70	7.5
1987	9.62	67.7	58.3	713	6.6	0.8	566	72.90	8.0
1988	9.33	61.7	54.5	709	5.4	0.9	588	115.30	7.2
1989	8.65	58.3	51.7	712	5.6	0.9	598	93.30	8.9
1990	8.33	55.3	48.3	689	5.8	0.9	582	101.70	7.8
1991	8.49	52.3	45.3	660	5.9	0.9	572	91.50	7.8

Data shown are annual mean values, weighted so as to accommodate changes in monitoring methods.

Source:
As for Table 1.13.

TABLE 1.15 Quality of ground water near Kamensk and Vakhsh in the former USSR, 1976–1987

Location	Year	Concentration (mg l⁻¹)			Location	Year	Concentration (mg l⁻¹)		
		Annual mean	Mean monthly minimum	Mean monthly maximum			Annual mean	Mean monthly minimum	Mean monthly maximum
Kamensk[a]		*Aluminium (Al)*			Vakhsh[b]		*Ammonium (NH₄⁺)*		
	1983	2.6	1.10	5.9		1976	91.8	0.02	1,945.0
	1984	2.4	1.01	12.8		1977	32.7	0.01	518.9
	1985	2.1	1.01	4.3		1978	27.3	0.01	525.0
	1986	2.6	1.06	26.7		1979	40.5	0.05	1,520.0
	1987	2.2	1.02	11.9		1980	53.1	0.10	689.0
						1981	54.3	0.05	3,228.0
		Manganese (Mn)				1982	53.2	0.13	2,037.0
						1983	21.1	0.06	503.0
	1983	6.5	0.20	21.5		1984	31.2	0.12	832.7
	1984	2.4	0.10	24.8		1985	20.9	0.04	368.3
	1985	1.5	0.15	19.5					
	1986	2.2	0.16	29.4			*Nitrate (NO₃⁻)*		
	1987	1.6	0.17	24.5					
						1976	44.8	0.48	488.3
		Zinc (Zn)				1977	28.0	1.20	133.8
						1978	27.6	0.50	362.0
	1983	2.5	1.10	27.0		1979	30.2	0.90	128.5
	1984	2.4	1.00	20.4		1980	20.2	1.00	348.0
	1985	1.7	1.05	9.6		1981	18.6	0.50	393.5
	1986	2.4	1.10	13.0		1982	14.1	0.12	197.0
	1987	1.7	1.05	5.2		1983	13.9	0.40	82.9
						1984	6.5	0.03	62.8
		Ammonium (NH₄⁺)				1985	5.1	< 0.01	37.3
	1983	6.9	2.10	39.0					
	1984	3.8	2.00	21.0			*Nitrite (NO₂⁻)*		
	1985	7.0	2.10	30.0					
	1986	4.3	2.10	39.0		1976	4.9	0.01	35.8
	1987	5.3	2.10	14.5		1977	7.5	< 0.01	38.8
						1978	5.8	< 0.01	56.3
						1979	3.8	0.01	17.4
						1980	2.5	0.02	85.5
						1981	1.5	0.01	53.9
						1982	1.8	< 0.01	70.7
						1983	1.3	0.02	27.5
						1984	1.3	0.00	46.6
						1985	1.2	< 0.01	52.0

[a] Near the city of Kamensk-Shakhtinsky in the Rostov District, Russian Federation.
[b] The Vakhsh hydrochemical site in the Kurgan District, Tajikistan.

Samples were taken by local hydrogeologists to determine the effectiveness of nature protection activities with respect to ground-water quality.

Frequency of sampling: six to ten times per month.

Source:
Data supplied by the Institute of Global Climate and Ecology, Russian Academy of Sciences, Moscow.

TABLE 1.16 Summary data from pesticide monitoring of surface waters used for public drinking water supplies in the USA and Canada

Pesticide	State or province	Raw water			Treated water		
		N	Samples with detections (%)	Max. conc. (μg l-1)	N	Samples with detections (%)	Max. conc. (μg l-1)
HERBICIDES							
Alachlor	Illinois	334	54.0	8.50			
	Iowa	15	67.0	9.30	33	52.0	8.80
	Kansas	3	100.0	2.60	4	100.0	2.10
	Ontario[a]	417	0.0				
	Ohio				NR		14.30
Atrazine	Illinois	334	77.0	24.00			
	Iowa	15	93.0	26.00	33	91.0	24.00
	Kansas	7	100.0	4.20	13	92.3	4.80
	Ontario[b]	422	62.3	29.40	150	76.0	37.00
	Ohio				NR		30.00
Butylate	Illinois	334	0.3	0.39			
	Iowa	15	0.0	ND	33	3.0	0.27
Cyanazine	Illinois	334	42.0	28.00			
	Iowa	15	73.0	20.00	33	79.0	17.00
	Ontario	422	8.1	6.80	150	14.0	8.80
	Ohio				NR		2.40
Dicamba	Iowa	14	7.0	1.20	30	3.0	1.40
Linuron	Ohio				NR		0.61
Metolachlor	Illinois	334	52.0	12.00			
	Iowa	15	73.0	10.00	33	64.0	21.00
	Kansas	3	100.0	1.20	3	66.7	0.70
	Ontario	417	9.6	15.00	150	15.3	5.97
	Ohio				NR		24.20
Metribuzin	Illinois	334	15.0	3.70			
	Iowa	15	7.0	0.89	33	12.0	0.45
	Ontario	418	0.7	1.70	150	0.0	ND
Simazine	Ontario[c]	422	0.2	0.15	150	0.7	0.06
	Ohio				NR		1.90
Trifluralin	Illinois	334	4.0	0.73			
	Iowa	15	0.0	ND	33	3.0	0.13
2,4-D	Iowa	14	14.0	0.17	30	7.0	0.30
	Kansas	1	100.0	0.51	2	100.0	3.20
INSECTICIDES							
Carbofuran	Iowa	15	33.0	17.00	33	27.0	14.00
Chlorpyrifos	Illinois	334	0.2	0.05			
	Iowa	15	0.0	ND	33	0.0	ND

[a] Sampled during 1986 while alachlor use was suspended.
[b] Atrazine plus diethylatrazine.
[c] Simazine plus diethylsimazine.

N = Number of samples. ND = Not detected. NR = Not reported.

Source:
UN ECE 1992 *The Environment in Europe and North America: Annotated Statistics 1992*, United Nations Economic Commission for Europe, New York.

TABLE 1.17 Fluxes of suspended matter, salts and nutrients from the three largest rivers in Turkey to the Black Sea, 1983 and 1987 (10^6 kg a^{-1})

Variable	Sakarya 1983	Sakarya 1987	Yesilirmak 1983	Yesilirmak 1987	Kizilirmak 1983	Kizilirmak 1987
Average flow rate (10^9 m^3 a^{-1})	6.00	6.99	5.57	6.27	5.92	5.93
Total dissolved solids	2,410	3,020	1,590	1,806	5,930	14,400
Suspended material	4,190	4,222	630	552		
Chloride (Cl$^-$)	125.0	152.4	61.0	72.1	133.3	
Ammonium (NH$_4^+$-N)	2.70	3.01	1.84	1.76	2.43	2.97
Nitrite (NO$_2^-$-N)	0.138	0.147			0.136	0.308
Nitrate (NO$_3^-$-N)	6.900	7.970		3.574	10.120	4.507
Total organic nitrogen	6.96	7.62				
BOD$_5$	10.81	13.28	6.12	7.52		
Orthophosphate (PO$_4^-$)	0.966	1.188	0.278	0.314	0.474	0.415
Sulphate (SO$_4^{2-}$)	576	706	318	333	1,833	1,849
Iron (Fe^{2+})	10.38	12.16	1.28	1.76		
Sodium (Na$^+$)	187.10	231.91	107.80	128.74		
Potassium (K$^+$)	27.10	31.74	12.10	12.98		
Calcium (Ca^{2+})	371	433	244	297	847	
Magnesium (Mg^{2+})	178	211	116	133	205	

Sources:
Balkas, T., Dechev, G., Mihnea, R., Serbanescu, O. and Unluata, U. 1990 *State of the Marine Environment in the Black Sea Region,* UNEP Regional Seas Reports and Studies No. 124, United Nations Environment Programme, Nairobi.

EPFT 1989 *Environmental Profile of Turkey 1989,* Environmental Problems Foundation of Turkey, Ankara.

TABLE 1.18 Long-term changes in loads of selected variables discharged into the Baltic Sea by major Latvian rivers, 1947–1988

River/site	Period	Mean annual loads (10^3 t a^{-1})								Water flow (km^3 a^{-1})
		SO_4^{2-}	$Na^+ + K^+$	Cl^-	TDS	NO_3^--N	PO_4^--P	NO_2^--N	Total Si	
Daugava Daugavpils	1947–60	112.0	29.3	24.5	2,030	2.890	0.560	0.065	42.60	15.00
	1961–70	144.0	58.6	67.9	2,260	3.230	0.502	0.203	50.20	12.80
	1971–80	175.0	92.2	106.0	2,480	10.800	0.461	0.214	27.60	12.10
	1981–88	341.0	122.0	141.0	3,090	13.700	0.746	0.267	33.00	14.20
	Change, 1947–88	229.0	92.7	116.0	1,060	10.800	0.186	0.202	−9.60	−0.80
Lielupe Elgava	1947–60	169.0	8.1	11.7	791	0.678	0.051	0.021	6.78	2.30
	1961–70	166.0	16.0	25.2	766	0.718	0.084	0.027	7.27	1.76
	1971–80	223.0	29.6	45.4	970	5.590	0.138	0.046	3.40	1.71
	1981–88	260.0	36.4	57.1	1,190	6.010	0.186	0.029	4.13	2.14
	Change, 1947–88	91.0	28.3	45.4	399	5.330	0.135	0.009	−2.65	−0.16
Venta Kuldiga	1947–60	29.8	9.5	10.4	590	0.520	0.030	0.011	4.88	2.06
	1961–70	46.4	11.4	19.4	571	0.468	0.090	0.027	6.83	1.85
	1971–80	94.2	23.4	34.2	888	2.820	0.048	0.027	4.52	2.10
	1981–88	110.0	29.1	40.4	1,030	4.760	0.095	0.027	5.18	2.58
	Change, 1947–88	80.2	19.6	30.0	440	4.240	0.064	0.016	0.30	0.52
Gauya Valmiera	1947–60	17.7	3.7	4.0	329	0.518	0.039	0.005	5.02	1.61
	1961–70	26.6	6.2	8.1	326	0.446	0.051	0.018	4.46	1.30
	1971–80	38.0	8.8	10.9	388	1.380	0.036	0.018	2.63	1.25
	1981–88	49.2	10.7	13.6	464	2.270	0.028	0.011	3.63	1.55
	Change, 1947–88	31.5	7.0	9.6	135	1.750	0.012	0.006	−1.39	−0.06
All Latvian rivers[a]	Change, 1947–88	681.0	233.0	317.0	3,208	34.900	0.590	0.366	−21.00	

[a] Estimated values.

TDS = Total dissolved solids.

Source:
Tsirkunov, V. V., Nikanorov, A. M., Laznik, M. M. and Dongwei, Z. 1992 Analysis of long-term and seasonal river water quality changes in Latvia, *Water Resources*, **26**(9), 1203–1216.

Biological Monitoring

Biological monitoring encompasses a range of techniques and strategies which broadly can be defined as the measurement of the responses of living organisms to anthropogenic stresses. Such responses take many forms including changes in cellular biochemistry, physiology, growth or health status, effects on development or reproduction of the individual and, on the larger scale, changes in population, community or ecosystem structure.

Traditionally, biological monitoring has tended to take the form of the measurement of contaminant concentrations in biotic material. In more recent decades, however, its scope has widened markedly and now the term incorporates the measurement of a wide range of genetic, biochemical, physiological and ecological parameters that have been demonstrated by research to be influenced by measured contaminant concentrations. Such biological changes, or to use a term that is becoming more widespread, biomarkers, can provide information on exposure to contaminants and in some cases may also provide information on effects of that exposure. The advantage of this approach to biological monitoring over straight contaminant measurement lies in the fact that it can demonstrate that a change has taken place. Even if, initially, no adverse effect is demonstrated it can be used as a trigger for further investigation. Conversely, if no changes in biomarkers, organisms or community structure in an impacted area are found, there is a sound basis for not investigating further. Chemical residue data are harder to use for this purpose as many persistent pollutants now have a global distribution. Nevertheless, where a substance has a known toxic effect, its presence in organisms is sometimes used as a basis for precautionary regulation.

Whereas some biological changes or biomarkers may demonstrate exposure, others demonstrate effects on the system being studied. One such biomarker is the presence of deoxyribose nucleic acid (DNA) adducts. Although in many ways its measurement is similar to contaminant analysis, it does have the advantage of showing that the chemical contaminant has reached the target. Other biomarkers, such as the induction of mixed function oxidases, have in some circumstances been used to show effects. This has been done for polychlorinated biphenyls (PCBs) by the use of dioxin equivalents.

The degree of specificity of biological changes that can be measured varies greatly. Some changes are contaminant-specific; the inhibition of aminolevulinic acid dehydratase (ALAD), for example, is specific to lead. Other biomarkers are at the other end of the scale, i.e., changes may be induced by a wide variety of man-made and naturally occurring chemicals. In general, however, specificity decreases up the organizational levels from the cellular changes to community structure, and thus biomarkers are most easily used at lower organizational levels. However, biological changes at higher organizational levels, i.e., in population structure and above, are of greater ecological importance.

Most approaches to biological monitoring suffer from the fact that they only give a "snapshot" of the status of the system being studied. The measured concentration of a contaminant in a tissue represents the net product of pollutant exposure, metabolism and excretion. Concentrations may have been caused by chronic low-level exposure, be decreasing after a massive single exposure or any other exposure regime. Thus the interpretation of this type of data is difficult even if detailed toxicological information is available. Similarly, the physiological condition of the animal may be stable, it may be improving, it may be getting worse, but we do not know which as we do not usually know the regime of exposure.

The analysis of biological materials and data is customarily used as a supplement, rather than as an alternative, to the analysis of air, soil and water chemistry. As regulatory controls are generally based on the chemical quality of these environmental media, these forms of chemical monitoring will inevitably assume continuing importance. However, biological monitoring has the advantage of offering greater insight into environmental effects in that it provides a much clearer indication of actual response to changes in environmental quality. On the other hand, this approach does not necessarily give information on which agent has caused the change; hence the complementary nature of biological and other approaches to environmental monitoring.

This section presents data from internationally co-ordinated biological and chemical monitoring programmes where appropriate; these are supplemented by data from individual studies, published in either 'State of the Environment'-type reports or in the scientific literature. An initial discussion of the contamination of Arctic species is given to illustrate the now global nature of environmental pollution which has arisen from long-range transboundary air pollution. This is followed by discussions of examples of biological monitoring using plants and animals from terrestrial, fresh-water and marine ecosystems.

Arctic Ecosystems

During recent decades there has been a growing awareness of the increasing extent of wildlife contamination by environmental pollutants. More recently, remote and, in the case of organochlorine compounds, cold areas of the world, have been the subject of increasing attention. In recognition of the potential role of long-range transboundary air pollution, concentrations of pollutants in air and biota have been widely measured in polar regions in order to evaluate the role of the atmosphere in the transport of pollutants and subsequent contamination of these remote areas (Calamari et al., 1991).

For many years the Arctic region was viewed as being too remote to be subject to any substantial chemical contamination from human activities. However, this region can no longer be considered pristine, since it experiences regular intrusions of polluted air masses and of surface ocean currents from Europe and North America (Figure 1.21). Analyses of Arctic biota as well as of water and air samples for organochlorines, heavy metals and radionuclides are now regularly performed, often on a systematic basis and as part of co-ordinated monitoring programmes.

Concentrations of contaminants in the Arctic ecosystem are generally lower than those in more polluted areas such as the North American Great Lakes or the Baltic Sea (time-trend

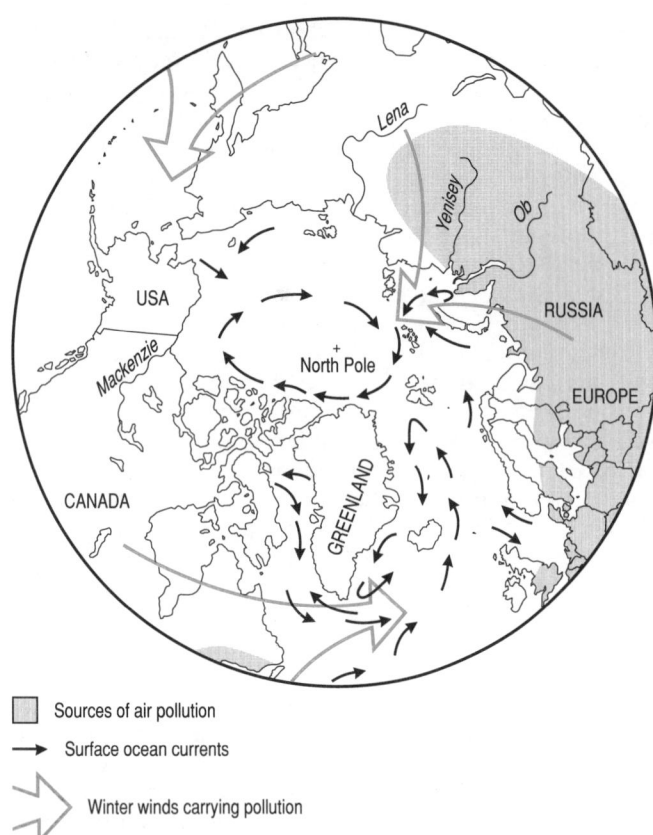

Sources of air pollution

→ Surface ocean currents

Winter winds carrying pollution

FIGURE 1.21 Transport pathways of pollutants into the Arctic
Source: Environment Canada, 1991

contaminant data for these areas were presented in the third edition of this report (UNEP, 1991)). However, as shown below, some of the more volatile compounds, such as hexachlorobenzene (HCB), are often detected in the Arctic at concentrations similar to those in source regions. The generally lower concentrations of most other contaminants in the Arctic region do not necessarily diminish the potential significance of their effects on ecosystem health. Due to the slow rate of breakdown of contaminants, the Arctic region is considered to be particularly vulnerable to environmental pollution.

In recognition of the fact that knowledge of sources and effects of contaminants in the Arctic environment is limited at the present time, a co-operative research effort – the Arctic Monitoring and Assessment Programme (AMAP) – has recently been set up by the Arctic countries. The primary function of AMAP is to co-ordinate existing national programmes and to develop further these programmes where and when necessary. During the initial stages of development of the programme the focus will be on persistent organic contaminants and selected heavy metals and radionuclides. A longer-term objective is the reduction of the output levels of the pollutants concerned.

Organochlorine Compounds Many organochlorine compounds are very persistent in the environment and have a tendency

to bioaccumulate through food chains. Such compounds, including certain pesticides and PCBs are now found in measurable concentrations in many Arctic wildlife species (see Table 1.19). However, the extent to which the measured concentrations actually have biological effects remains poorly understood. Nevertheless, the concentrations of selected organochlorines found in Arctic wildlife tissues are considered to be high enough to trigger concerns about the impact on those indigenous human populations which are heavily dependent on local foods and thus may be exposed to contaminant levels sufficient to cause adverse health effects (Wania and Mackay, 1993).

There are very few local sources of organochlorine contaminants in the Arctic, the major users of such compounds being some mid-latitude countries. Long-range transport from more industrialized regions, particularly via the atmosphere, is the major route of entry of organochlorines into the Arctic environment.

Table 1.19 shows mean organochlorine concentrations measured in Baltic and Arctic wildlife samples collected during 1985–1989. When comparing concentrations, it can be seen that DDT and PCB residues bioaccumulate significantly along the food chain from fish to seals. Similarly DDT, PCB, HCB and chlordanes show a bioaccumulation from fish to the white-tailed eagles. However, toxaphene and HCH (hexachlorocyclohexane) residues show no significant bioaccumulation. It is suggested that eagles accumulate DDT, PCB, HCB and chlordanes from their food, but are able to metabolize HCH and toxaphene. In general, Arctic species contained lower concentrations of organochlorines than the same species from the Baltic, with the exception of toxaphene which occurs at similar concentrations at both locations. The Baltic:Arctic concentration ratios, which are also presented in Table 1.19, may be taken as measures of the local:global nature of the pollutant (Paasivirta and Rantio, 1991).

It has been proposed that the often surprisingly high concentrations of some organochlorines, particularly those of intermediate volatility, are explained in part by a process of "global distillation and cold condensation" (Wania and Mackay, 1993). As temperature decreases there is a tendency for chemicals present in the atmosphere to "condense" into soil, water, aerosols, snow or ice, and hence to be deposited on to the earth's surface. Chemicals which are non-volatile and thus immobile (e.g., benzo-a-pyrene) will tend to remain in the region of their emission as they "condense" and deposit rapidly. The more volatile the compound, however, the further it may travel via the atmosphere before finally condensing and accumulating at the colder latitudes. In effect, compounds with a given volatility, for example HCB, will preferentially accumulate in colder, polar regions.

Data reported by Calamari et al. (1991) illustrate the above theory of "global distillation". Calamari et al. (1991) have compared levels of HCB, HCH and DDT in plant foliage sampled at various locations world-wide and covering a latitude band from 78°N to 74°S. For HCB, one of the more volatile compounds, they found a strong dependence of foliage concentration on the average annual temperature of the sampling site, with concentrations increasing towards cold

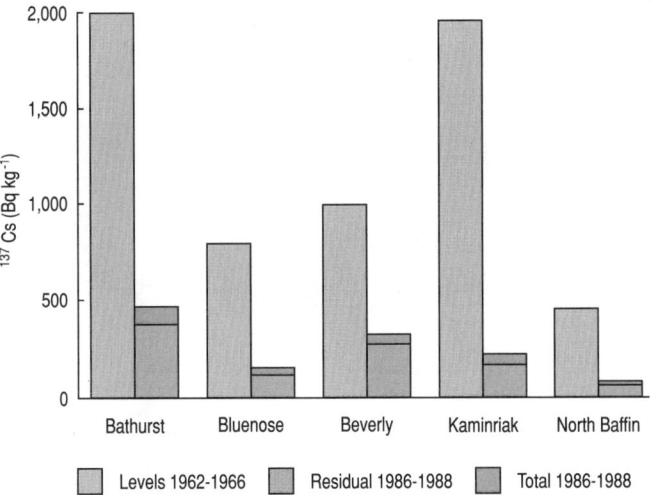

FIGURE 1.22 Average concentrations of ^{137}Cs in caribou meat sampled at five Canadian Arctic locations, 1962–66 and 1986–87
Source: Thomas et al., 1992

regions. However, no such pattern for HCHs and DDT was reported. It is concluded that the observed distribution of HCB in plant foliage is influenced more by its own physico-chemical properties and by temperature than by direct contamination and usage patterns. In contrast, current or recent usage patterns are more likely to control the distribution of the less volatile HCHs and DDT in the environment (Calamari et al., 1991).

Radionuclides Since the 1950s radionuclides have entered the Arctic ecosystem in significant amounts, largely as a result of atmospheric fallout from the nuclear weapons testing that took place between 1952 and 1978 and, more recently, following the accident at the Chernobyl nuclear power plant in 1986. The primary long-lived fission products arising from such events are ^{90}Sr (half-life of 28.1 years) and ^{137}Cs (half-life of 30.1 years). Other radioactive threats to the Arctic environment include accidents connected to the activities of nuclear-powered vessels and satellites, such as Cosmos 954 which crash-landed in the Canadian Arctic in 1978 (see Part 9: Environmental Disasters) and the dumping of radioactive waste in the Arctic ocean (see Part 8: Wastes and Waste Management).

Studies on radioactivity in the Arctic carried out during the past 30 years have tended to focus on events which occurred in two distinct time periods. The first period – the early 1960s – refers to the time of atmospheric nuclear weapons testing when large radio-ecological research projects were started in most Arctic countries. These projects and studies centred on the accumulation of radionuclides up the lichen-reindeer/caribou-human food chain. Data from a selection of such studies were included in the second 'UNEP Environmental Data Report' (UNEP, 1989). The second period of interest has been in the aftermath of the Chernobyl accident. Figure 1.22 compares levels of ^{137}Cs in caribou meat sampled at five locations throughout the Canadian Arctic, initially in the mid-1960s and again during 1986–87. Approximately 25 per cent

of the 1986–87 values are ascribed to the Chernobyl accident; the remainder is residual fallout from weapons testing (Thomas et al., 1992). The precise biological effects of contamination are difficult to determine.

References

Calamari, D., Bacci, E., Focardi, S., Gaggi, C., Morosini, M. and Vighi, M. 1991 Role of plant biomass in the global environmental partitioning of chlorinated hydrocarbons, *Environmental Science and Technology*, **25**(8), 1489–1495.

Paasivirta, J. and Rantio, T. 1991 Chloroterpenes and other organochlorines in Baltic, Finnish and Arctic wildlife, *Chemosphere*, **22**(1–2), 47–55.

Thomas, D. J., Tracey, B., Marshall, H. and Norstrom, R. J. 1992 Arctic terrestrial ecosystem contamination, *The Science of the Total Environment*, **122**, 135–164.

UNEP 1989 *United Nations Environment Programme Environmental Data Report 1989/90*, Basil Blackwell, Oxford.

UNEP 1991 *United Nations Environment Programme Environmental Data Report 1991/92*, Basil Blackwell, Oxford.

Wania, F. and Mackay, D. 1993 Global fractionation and cold condensation of low volatility organochlorine compounds in polar regions, *Ambio*, **22**(1), 10–18.

Terrestrial Ecosystems

Although there are a number of well established, long-term biological monitoring programmes in existence around the world, much biological monitoring of terrestrial environments is undertaken as individual one-off studies. These studies are generally designed to indicate the environmental impact of point sources of pollution or to highlight regional differences in pollutant levels in biota at the time of study. However, these individual studies rarely provide an integrated picture of the effects of pollution on terrestrial ecosystems. It is now becoming increasingly apparent that more comprehensive data bases, produced as a result of long-term monitoring programmes, are needed to manage ecosystems more effectively.

The few long-term national monitoring programmes that exist for terrestrial wildlife have tended to concentrate on the measurement of contaminant concentrations in birds and plants. Extracts of data from programmes assessing concentrations of trace metals in birds were reported in the first two editions of this report (UNEP 1987, 1989) while data showing organochlorine concentrations in adult woodcock (*Philohela minor*) from eastern Canada were presented in the third edition (UNEP, 1991).

Since they have a high capacity for interception and retention of airborne contaminants, mosses and lichens have been the most widely used plants for monitoring atmospheric deposition of heavy metals and radionuclides, and concentrations of gaseous pollutants. Since the late 1960s levels of heavy metals in mosses have been monitored systematically on a nation-wide basis in Sweden in order to assess spatial and temporal variations in trace metal deposition. Extracts of data arising from the Swedish monitoring efforts were presented in the second edition of this report (UNEP, 1989).

In more recent years, this type of monitoring has been adopted by other countries and is now co-ordinated on a European scale. Data from participating countries are now

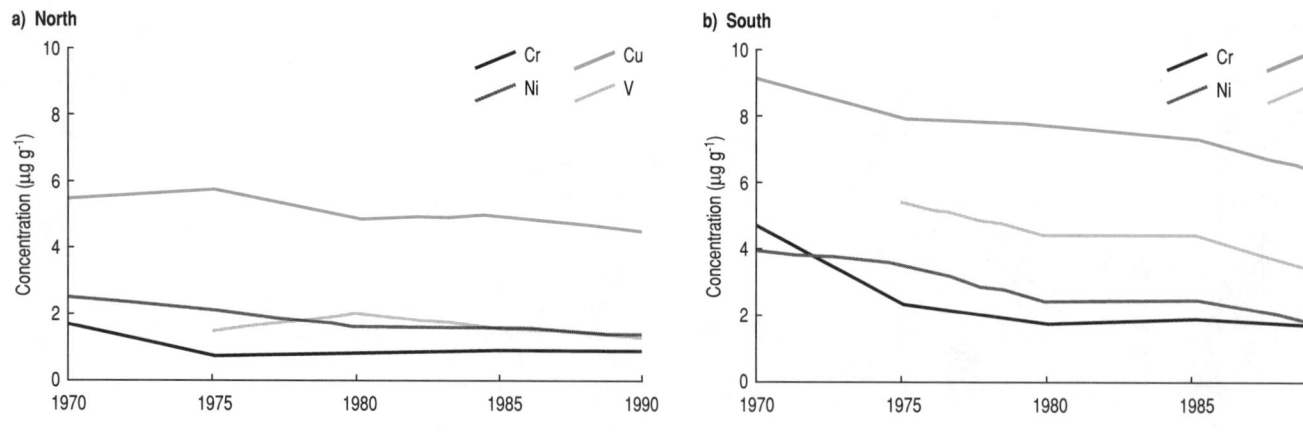

FIGURE 1.23 Variations in concentrations of selected heavy metals in moss samples from (a) northern Sweden and (b) southern Sweden, 1970–1990
Source: Data supplied by Å. Rühling, Department of Plant Ecology, University of Lund, Sweden

contributed to the UN ECE ICP for Monitoring and Evaluation of the Long-range Transmission of Air Pollutants in Europe (EMEP). In 1990 surveys of heavy metal deposition based on the analyses of moss samples were performed in a number of countries including Denmark, Finland, Norway and Sweden. Figure 1.23 shows variations in concentrations of four heavy metals in moss samples from locations in north and south Sweden during the period 1970–1990. Higher concentrations in the south reflect closer proximity to pollution sources; moreover, it can be seen that metal concentrations have decreased significantly in the southern samples (south of 60°N) whereas concentrations in the northern samples (north of 64°N), although lower, have remained much more constant. This may be due to a slower rate of turnover of biomass in the north and hence a slower rate of metal movement to sinks.

In addition, bryophytes and lichens have been widely used as monitors of a number of the traditional gaseous air pollutants. This is an example of biological monitoring where effects on organisms can be linked relatively specifically to certain pollutants. Lichens, for example, are known to be particularly sensitive to sulphur dioxide (SO_2) (Burton 1986, 1990). In this respect London is a classic case; the re-appearance of a number of lichen species in the Greater London area during the 1980s, some of which had not been observed since the 1700s and 1800s, is considered to be indicative of falling SO_2 concentrations (Hawksworth and McManus, 1989). The detailed survey work of Hawksworth and McManus (1989) supports the conclusion that ambient SO_2 levels are a dominant factor in determining the coverage of lichen vegetation on trees and wood in urban areas.

Crop plants may also provide an indication of the level of contamination in the terrestrial environment. Table 1.20 presents data on concentrations of the pesticide HCH in crop plants in selected republics of the former USSR from 1986–1990. The data show that, in most republics, mean HCH concentrations have decreased to such an extent that less than 1 per cent of samples taken exceeded the nationally imposed maximum permissible concentration of 0.1 mg kg⁻¹ by 1989 or 1990. The exceptions are Azerbaijan and Uzbekistan where 18.8 and 4.1 per cent of samples, respectively, exceeded the

maximum permissible concentration. The widespread use of HCH for pest control in southern parts of the former USSR on cotton and tobacco crops is considered to be the most likely explanation for these observations (Rovinsky, 1993).

The study of variations in crop yields in response to environmental stresses, especially air pollution, has become an important area of research (Adams and Crocker, 1992). Under the auspices of the UN ECE Convention on Long-range Transboundary Air Pollution, a co-operative programme for the monitoring of the effects of gaseous pollutants on agricultural crops was established in 1989. A pilot project on research into the effects of ozone on radish plants has recently been completed, but co-ordinated monitoring is yet to begin.

Similarly, the apparent connection of forest damage with air pollution has led to the establishment of a programme of large-scale surveys of forest condition in Europe which have, since 1986, been co-ordinated by the UN ECE (UN ECE/CEC, 1992). This co-operative programme, like that for crops, operates under the auspices of the UN ECE Convention on Long-range Transboundary Air Pollution. Surveys of tree health are conducted in the participating countries on an annual basis by trained observers according to a standardized methodology. The degree of defoliation of sample trees is used as an indicator of tree health and is defined as follows:

Class	Needle/leaf loss (%)	Tree health
0	0–10	no defoliation
1	11–25	slight defoliation
2	26–60	moderate defoliation
3	61–99	severe defoliation
4	100	dead

Table 1.21 presents annual data from the forest damage survey from 1986–1991, and Figure 1.24 illustrates the percentage change in degree of defoliation between 1988 and 1991 in European countries for which national data for both years are available.

In 1991 survey results from 28 European countries encompassing 36,000 sample plots and 700,000 sample trees were

a) Broad leaved trees   b) Coniferous trees

☐ Less than -2.0%   ☐ Between -2.0% and +2.0%   ☐ Between +2.0% and +10.0%   ☐ Greater than +10.0%   ☐ No information available

FIGURE 1.24 Percentage change in the number of sample trees in damage classes 2+3+4 (i.e., with >25 per cent defoliation) among (a) broadleaved and (b) coniferous trees in selected European countries, 1988–1991
Source: Data from Table 1.21

reported. Of the 214×10^6 ha of forests in Europe (including major parts of the forests in the western Russian Federation), around 168×10^6 ha (78 per cent) were included in the 1991 survey, 6×10^6 ha more than in 1990. Of the trees sampled in 1991, the survey revealed that 22.5 per cent of trees suffered greater than 25 per cent defoliation (i.e., were in damage classes 2+3+4) and can thus be classified as damaged. The majority of the highly affected forest areas occur in Bulgaria, the former Czechoslovakia, Germany, Poland and the UK. In these countries defoliation in coniferous forests is particularly high. Amongst the broad-leaved trees, beech in Denmark, birch in Sweden and oak in Portugal also show high levels of defoliation. Although the assessment of survey results with respect to temporal variations in forest condition is constrained by a number of inherent limitations in the available data, it is generally believed that survey results obtained to date indicate a continuing overall deterioration in the vitality of European forests (UN ECE/CEC, 1992).

Within the present format of the survey, it is not easy to distinguish changes in forest health due to air pollution from those caused by other factors, such as pest invasions and extreme climatic conditions. In order to gain further insight into the complex relationship between forest condition and air pollution, more detailed monitoring of soils and forest vegetation at a sub-set of permanent sample plots is being planned for future surveys. These data will be combined with further data on ecological parameters and on air quality collected by other international programmes and thus provide the range of observations which detailed integrative assessments of forest damage and air pollution will require (UN ECE/CEC, 1992).

References

Adams, R. M. and Crocker, T. D. 1992 The economic impact of air pollution on agriculture: an assessment and review. In: *Towards Sustainable Agricultural Development*, M. D. Young (Ed.), Belhaven Press, 265–319.

Burton, M. A. S. 1986 *Biological Monitoring of Environmental Contaminants: Plants*, MARC Report Number 32, Monitoring and Assessment Research Centre, London.

Burton, M. A. S. 1990 Terrestrial and aquatic bryophytes as monitors of environmental contaminants in urban and industrial habitats, *Botanical Journal of the Linnean Society*, **104**, 267–280.

Hawksworth, D. L. and McManus, P. M. 1989 Lichen recolonization in London under conditions of rapidly falling sulphur dioxide levels, and the concept of zone skipping, *Botanical Journal of the Linnean Society*, **100**, 99–109.

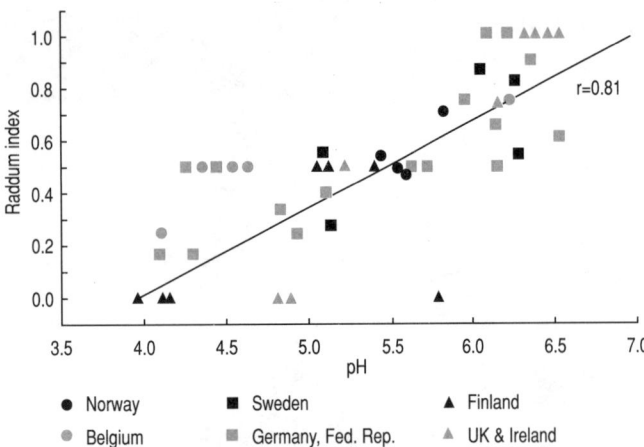

Norway   Sweden   Finland
Belgium   Germany, Fed. Rep.   UK & Ireland

FIGURE 1.25 Average Raddum Index scores and annual mean pH in river basins in various European countries, 1987–1989
Source: NIVA, 1992

Rovinsky, F. Ya. 1993 *Personal communication*, Institute of Global Climate and Ecology, Moscow, Russia.

UN ECE/CEC 1992 *Forest Condition in Europe: 1992 Report*, United Nations Economic Commission for Europe, Geneva and Commission of the European Communities, Brussels.

UNEP 1987 *United Nations Environment Programme Environmental Data Report 1987/88*, Basil Blackwell, Oxford.

UNEP 1989 *United Nations Environment Programme Environmental Data Report 1989/90*, Basil Blackwell, Oxford.

UNEP 1991 *United Nations Environment Programme Environmental Data Report 1991/92*, Basil Blackwell, Oxford.

Fresh-water Ecosystems

Techniques for biological monitoring for chemical contaminants or for ecosystem or organism responses are relatively advanced for fresh waters. Species distribution, growth rates and other developmental responses, together with bioaccumulation of contaminants in food chains are now studied by an increasing number of national and international biological monitoring programmes in fresh-water environments.

The North American Great Lakes have been particularly well studied in this regard. Biological monitoring programmes have been conducted in the Great Lakes since the early 1970s and have been instrumental in the identification of various metal and organochlorine contaminants within the Great Lakes ecosystem and have enabled managers to assess the effects of remedial measures taken to reduce pollutant loads. Contaminant data for various fish species and herring gulls (*Larus argentatus*) were presented in the third edition of this report (UNEP, 1991). Box 1.9 links the use of pesticides to organochlorine contamination of wildlife in a tropical agricultural river basin in India.

Biological information is considered to be a useful indicator of overall water quality, and biological monitoring programmes that assess the impact of general environmental quality on fresh-water ecosystems are widely used in selected countries for water quality management. With increasing

concern about the environmental effects of long-range transboundary air pollution, co-ordinated international efforts are now beginning to be made in this field, particularly with regard to the effects of fresh-water acidification.

In order to relate changes in water quality due to acidification to biological and ecosystem responses, a score technique which relies on the presence or absence of acid sensitive species for its calculation has been developed as part of the UN ECE International Co-operative Programme on Assessment and Monitoring of Acidification of Rivers and Lakes. This score technique, the Raddum Index, has gained widespread acceptance, and has led to a better understanding of the relationship between water quality and biota (NIVA, 1992). Similar scores or indices have been developed for specific national situations but so far have been rarely used in international programmes.

Figure 1.25 shows the relationship between the score given by the UN ECE ICP's Raddum Index and annual mean pH for a selection of catchments in different countries. The index system is based on the tolerance of species to acidic conditions found at participating sites; species with the lowest tolerance to acidification and which are the first to die when pH starts to fall get the score or index value 1. The most tolerant species which survive in strongly acidic conditions receive the index value 0. The average Raddum Index will therefore be a number between 0 and 1.

References

NIVA 1992 *Evaluation of the International Co-operative Programme on Assessment and Monitoring of Acidification of Rivers and Lakes*, Report prepared by the Programme Centre in co-operation with an expert panel appointed by the Programme Task Force, Norwegian Institute for Water Research, Oslo.

UNEP 1991 *United Nations Environment Programme Environmental Data Report 1991/92*, Basil Blackwell, Oxford.

Marine Ecosystems

Heavy metals, organochlorine compounds and petroleum hydrocarbons have long been recognized as the most deleterious contaminants to biota in the world's marine and estuarine waters (Martin and Richardson, 1991). During the past two decades various biomonitoring strategies have been developed to monitor and evaluate the adverse impacts of these compounds on marine ecosystems. One of the most successful efforts has involved the use of bivalve shellfish as sentinel organisms. Mussels (and oysters) have become widely used for monitoring contamination in coastal and estuarine ecosystems because, as filter feeders, they bioaccumulate contaminants. This approach has become popularly known as the "Mussel Watch".

The primary goals of Mussel Watch programmes are to provide long-term information on the concentrations of trace contaminants in coastal marine ecosystems, and to identify areas where these compounds may be accumulated by living organisms in higher than expected concentrations. In short, Mussel Watch programmes are intended to be indicator systems for particular substances in waters of the coastal zone.

A considerable data base now exists for a variety of contaminants in mussels and oysters in Northern Hemisphere

BOX 1.9 Pesticide Use and Persistent Organochlorine Contamination in Wildlife: A Case Study

To meet the food requirements of rapidly growing populations, some countries have increased their use of selected pesticides, some of which are banned or restricted in other countries. For example, DDT is still used in some Asian countries and some forms of HCH are widely produced and used in the Far East. In India persistent organochlorines such as HCH and DDT have been used in significant quantities for agricultural purposes and malaria control respectively, where their annual use amounts to 45 x 10^6 kg and 19 x 10^6 kg, respectively. Together these compounds account for about 60 per cent of the pesticides (in quantity) currently used in the Indian sub-continent.

The table given here shows mean concentrations of HCH and DDT in selected fish and bird species measured as part of a case study of an agricultural watershed in the Vellar River basin near Parangipettai in southern India. Levels of four organochlorine compounds were measured; of these, concentrations of HCH ranked the highest, followed by total DDT, PCBs and HCB (Ramesh et al., 1992). Concentrations of HCH were higher than those of total DDT in most of the samples. The overall pattern of HCH residues can be explained by the fact that this insecticide is widely applied to a variety of crops resulting in significant contamination of the local environment. Lower concentrations of total DDT in wildlife samples can be attributed to the relatively smaller use of DDT in the Vellar River basin for agricultural purposes. Under the National Malaria Eradication Programme of the Government of India, about 85 per cent of the DDT produced in India is used for malaria control. Similarly, sources of PCBs and HCB in the Vellar River basin are industrial rather than agricultural, although HCB is also used as a fungicide.

The feeding habits of wildlife species in the Vellar River basin are considered to be an important factor in determining the extent of contamination and the bioaccumulation of contaminants. Among the wild birds, the data show that mean concentrations of organochlorines are higher in the carnivores, such as the kites, which are at the top of the food chain. However, concentrations of HCHs and total DDT in cattle egret and pond heron are almost as high as those in kites; these species forage mainly in agricultural fields, i.e., they reside in close proximity to sources of contamination. Among the fish, concentrations of HCHs in estuarine species (mullet, catfish

Mean concentrations of persistent organochlorines in wildlife from Parangipettai, South India, 1987–1991

Species	Sample date Month	Year	No. of samples	Mean concentration (ng g^{-1} wet wt.) ΣHCHs	ΣDDTs
FISH					
Seer fish	Jan.	1988	5	1.7	31
Sole fish	Jan.	1989	5	4.8	12
Grey mullet	July	1988	5	63	3.8
Catfish	July	1989	5	61	4.6
Pearlspot	Dec.	1989	3	84	1.8
WILD BIRDS					
Cattle egret	Nov.	1988	2	1,100	470
Pond heron	Jan.	1989	8	1,400	550
Chicken	Dec.	1989	3	29	24
Brahminy kite	Feb.	1991	4	1,500	570
Kentish plover	Feb.	1991	4	130	220
Pariah kite	Feb.	1991	3	1,300	260
Redwattled lapwing	Feb.	1991	2	59	1.5
Roseringed parakeet	Feb.	1991	3	17	1.0
White-breasted kingfisher	Feb.	1991	4	950	150
White-breasted waterhen	Feb.	1991	2	46	4.0

ΣHCHs = Total hexachlorocyclohexane
ΣDDTs = Total dichlorodiphenyltrichloroethane

Seer fish = *Scomberomorous commersoni*; sole fish = *Cynoglossus paraplagusia*; grey mullet = *Mugil cephalus*; catfish = *Arius maculatus*; pearlspot = *Etroplus suratensis*; cattle egret = *Bubulcus ibis*; pond heron = *Ardeola grayii*; chicken = *Gallus* sp.; brahminy kite = *Haliastur indus*; kentish plover = *Charadrius alexandrinus*; pariah kite = *Milvus migrans*; redwattled lapwing = *Vanellus indicus*; roseringed parakeet = *Psittacula krameri*; white-breasted kingfisher = *Halcyon smyrnensis*; white-breasted waterhen = *Amaurornis phoenicurus*.
Fish samples consisted of flesh (muscularculture). Breast muscle was sampled in birds.
This study was conducted as part of a wider research programme aimed at understanding the comprehensive behaviour and fate of organochlorine insecticides in the tropical environment.

Source:
Ramesh, A., Tanabe, S., Kannan, K., Subramanian, A. N., Kumaran, P. L. and Tatsukawa, R. 1992 Characteristic trend of persistent organochlorine contamination in wildlife from a tropical agricultural watershed, South India, *Archives of Environmental Contamination and Toxicology*, **23**, 26–36.

and pearlspot) are much higher than those in oceanic species (seer fish and sole fish), suggesting that significant amounts of HCHs enter the coastal water bodies through agricultural drainage (Ramesh et al., 1992).

It is noticeable that HCH residues in birds from India and other developing countries in the tropics are higher than those detected elsewhere. On the other hand, in spite of the extensive use of HCH in tropical countries, residues in fish species are comparable to values reported from other countries. On the basis of these results,

Ramesh et al. (1992) concluded that much of the HCH applied to the Vellar River is removed to the air and only a relatively small proportion drains into the sea.

Reference

Ramesh, A., Tanabe, S., Kannan, K., Subramanian, A. N., Kumaran, P. L. and Tatsukawa, R. 1992 Characteristic trend of persistent organochlorine contamination in wildlife from a tropical agricultural watershed, South India, *Archives of Environmental Contamination and Toxicology*, **23**, 26–36.

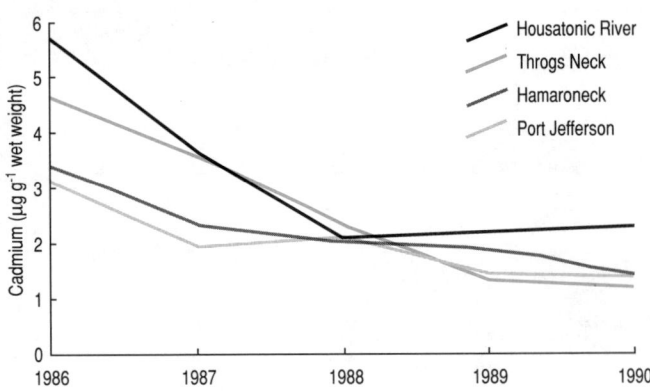

FIGURE 1.26 Concentrations of cadmium in mussels at sites in Long Island Sound, USA, 1986–1990
Source: O'Connor, 1992

locations. The Mussel Watch approach is also assuming greater importance in Southern Hemisphere regions, most notably in Australia (Martin and Richardson, 1991) and Antarctica (Berkman and Nigro, 1992). However, monitoring in these regions remains relatively unco-ordinated and the lack of baseline data reduces the usefulness of the data sets for assessment purposes.

One of the most developed marine biological monitoring programmes is the Mussel Watch Project of the National Status and Trends Program co-ordinated by NOAA in the USA. The NOAA Mussel Watch Program was initiated in 1984 and now incorporates some 200 coastal and estuarine sampling sites in the USA. Figures 1.26 presents data from the NOAA Mussel Watch Program and shows recent trends in cadmium in mussels from Long Island Sound. The decreases in contamination observed at these sites during the second half of the 1980s are representative of many sites in the USA and imply that some benefits have resulted from the improved management of polluting discharges. However, for most chemical compounds data for more years will be required before the effects of human activity and natural influences can be clearly distinguished (O'Connor, 1992).

The Mussel Watch approach has been adopted by the International Council for the Exploration of the Sea (ICES), a body which is engaged in regular marine biomonitoring programmes in the North Atlantic region. However, ICES places greater emphasis on the use of fish as biomonitors. Sixteen countries bordering the North Atlantic, North Sea and Baltic Sea areas, including the USA in the west and the former

USSR in the east, participated in the ICES 1985 baseline study of contaminants in fish and shellfish species. The results of this study for cod, herring and blue mussel were included in the third edition of this report (UNEP, 1991). In more recent years ICES activity has been mainly confined to monitoring of "hot spots", i.e., areas which are known to receive substantial inputs of contaminants. These ocean and coastal areas are also monitored as part of the Joint Monitoring Programme (JMP) of the Oslo and Paris Commissions. The major purpose of these programmes is to check that concentrations of contaminants in organisms do not exceed levels which would be hazardous to either human health or to the fish population (UK Department of Environment, 1992).

Tables 1.22 and 1.23 present data showing trends in contamination of fish and shellfish by heavy metals and organochlorine compounds in Dutch and UK estuaries. These data has been contributed to ICES through the JMP. A series of North Sea Conferences (London, 1987 and The Hague, 1990) have resulted in international agreement on the reduction of micropollutant loads to the North Sea by 1995. The ultimate goal is the reduction of contaminant levels in the North Sea environment, especially in organisms. These data sets may, therefore, serve as an interim indicator of progress towards the achievement of these goals.

In 1992, based on the work of ICES, the North Pacific Marine Sciences Organization established an equivalent international programme of co-operative investigations. This programme is called the Pacific International Council for the Exploration of the Sea (PICES) and it is intended that existing national monitoring activities will be co-ordinated and expanded. It is anticipated that data reporting will begin in 1993 (Wooster, 1992).

References

Berkman, P. A. and Nigro, M. 1992 Trace metal concentrations in scallops around Antarctica, *Marine Pollution Bulletin*, **24**(6), 322–323.

Martin, M. and Richardson, B. J. 1991 Long term contaminant biomonitoring: views from southern and northern hemisphere perspectives, *Marine Pollution Bulletin*, **22**(11), 533–537.

O'Connor, T. 1992 *Recent Trends in Coastal Environmental Quality: Results from the First Five Years of the NOAA Mussel Watch Project*, National Oceanic and Atmospheric Administration, Rockville.

UK Department of Environment 1992 *Digest of Environmental Protection and Water Statistics No.14, 1991*, Her Majesty's Stationery Office, London.

UNEP 1991 *United Nations Environment Programme Environmental Data Report 1991/92*, Basil Blackwell, Oxford.

Wooster, W. S. 1992 Personal communication. Department of Marine Sciences, University of Washington, Seattle, USA.

TABLE 1.19 Mean concentrations of selected organochlorines in Baltic and Arctic wildlife samples, 1985–1989

Compound or group of compounds	Species	Tissue	Baltic (ng g⁻¹)	Arctic (ng g⁻¹)	Baltic/Arctic ratio
Total DDT	Baltic Ringed Seal	Blubber	26,500		
	Cod	Liver	1,480	147	10.10
	Grey Seal	Blubber	27,200		
	Herring	Whole body	754		
	Salmon	Muscle	3,225	516	6.25
	Whitetailed Eagle	Eggs	68,000	19,000	3.60
		Muscle	110,800		
γ-HCH	Grey Seal	Blubber	20		
	Herring	Whole body	71.50		
	Salmon	Muscle	50	14	3.55
	Whitetailed Eagle	Muscle	<10		
HCB	Cod	Liver	61	33	1.85
	Grey Seal	Blubber	230		
	Herring	Whole body	57		
	Salmon	Muscle	123.25	56	2.20
	Whitetailed Eagle	Eggs	2,320	1,380	1.70
		Muscle	2,565		
Chlordanes	Cod	Liver	168	128	1.31
	Grey Seal	Blubber	1,100		
	Herring	Whole body	43.59		
	Salmon	Muscle	566.75	139	4.33
	Whitetailed Eagle	Eggs	4,800	1,060	4.50
		Muscle	12,560		
Toxaphene	Cod	Liver	640	540	1.19
	Grey Seal	Blubber	80		
	Salmon	Muscle	3,105.50	2,870	1.08
	Whitetailed Eagle	Muscle	<10		
PCBs	Baltic Ringed Seal	Blubber	40,900		
	Cod	Liver	5,600	570	9.80
	Grey Seal	Blubber	227,000		
	Herring	Whole body	1,030		
	Salmon	Muscle	4,793.50	572	8.38
	Whitetailed Eagle	Eggs	362,000	116,000	3.10
		Muscle	722,000		
TCDF	Baltic Ringed Seal	Blubber	0.12		
	Cod	Liver	0.16	0.026	6.30
	Grey Seal	Blubber	0.19		
	Salmon	Muscle	0.22	ND	
	Whitetailed Eagle	Muscle	<1.00		

ND = Not detected.

Baltic ringed seal = *Phoca hispida*; cod = *Gadus morhua*; grey seal = *Halichoerus grypus*; herring = *Clupea harengus*; salmon = *Salmo salar*; whitetailed eagle = *Haliaeetus albicilla*.

Total DDT = Dichlorodiphenyltrichloroethane and its metabolites DDE and TDE.
γ-HCH = Hexachlorocyclohexane; trade name Lindane.
HCB = Hexachlorobenzene.
Chlordanes = 1,2,4,5,6,7,8,8-octachloro-2,3,3a,4,7,7a-hexahydro-4,7-methano-1H-indene and related compounds.
Toxaphene = A complex mixture of over 177 C_{10} polychloro derivatives with an overall formula $C_{10}H_{10}Cl_{18}$.

PCBs = Polychlorinated biphenyls.
TCDF = 2,3,7,8-tetrachloro-dibenzofuran.

Data represent mean concentrations of organochlorines in samples collected between 1985 and 1989 at single sites in the Baltic and Arctic regions, except in the case of salmon (average of four Baltic and two Arctic sites) and eagle muscle (average of two Baltic sites). All results are given on a lipid weight basis.

Source:
Paasivirta, J. and Rantio, T. 1991 Chloroterpenes and other organochlorines in Baltic, Finnish and Arctic wildlife, *Chemosphere*, **22**, 47–55.

TABLE 1.20 Concentrations of HCH in crop plants in Republics of the former USSR, 1986–1990

Country	Year	Number of samples	Mean concentration (mg kg^{-1})	% of samples in which HCCH was detected	% of samples which exceeded the MPC
Azerbaijan	1986	529	0.062	51.5	20.5
	1987	575	0.180	64.0	18.0
	1988	269	0.250	71.0	44.6
	1989	671	0.075	38.7	0.0
	1990	750	0.190	51.3	18.8
Belarus	1989	13	<0.001	0.0	0.0
Kazakhstan	1986	4,504	0.135	46.6	15.5
	1987	5,426	0.090	40.0	12.0
	1988	5,420	0.040	28.3	0.0
	1989	6,610	0.120	10.0	2.0
	1990	2,558	0.026	10.1	0.4
Kyrgyzstan	1986	60	0.191	41.8	11.0
	1987	634	0.030	32.0	0.0
	1988	1,184	0.020	18.5	0.0
	1989	588	<0.001	0.0	0.0
	1990	4,388	0.030	31.9	0.0
Latvia	1988	7	0.012	0.0	0.0
Tajikistan	1986	395	0.049	58.0	0.0
	1987	13	0.059	78.0	0.0
	1988	189	0.040	54.5	0.0
	1989	637	0.023	47.1	0.0
	1990	472	<0.001	12.0	0.0
Turkmenistan	1986	372	0.068	78.0	7.0
	1987	4	<0.001	0.0	0.0
	1988	376	0.005	3.5	0.0
	1989	324	<0.001	0.0	0.0
	1990	4,681	0.130	14.9	0.1
Ukraine	1986	224	0.044	53.0	21.0
	1987	75	0.062	42.0	19.0
	1988	251	0.008	40.0	0.0
	1989	212	0.025	6.1	0.0
	1990	488	0.030	9.0	0.0
Uzbekistan	1986	1,113	0.050	67.0	3.7
	1987	103	0.050	71.0	4.1
	1988	622	0.012	37.3	0.0
	1989	1,050	0.040	55.7	4.1

HCH = Hexachlorocyclohexane.
MPC = Maximum permissible concentration (0.1 mg kg^{-1}).

Crop plants are sampled once a year before harvesting.

Source:
Data supplied by Institute of Global Climate and Ecology, Russian Academy of Sciences, Moscow (1992).

TABLE 1.21   European forest damage survey results, 1986–1991

Region/country	Conifers (all ages) Percentage of trees with >25% defoliation[a]							Broadleaves (all ages) Percentage of trees with >25% defoliation[a]						
	1986	1987	1988	1989	1990	1991	% change 1988–1991	1986	1987	1988	1989	1990	1991	% change 1988–1991
EUROPE														
Austria			12.0	10.1	8.3	7.0	−5.0			16.6	15.7	14.9	11.1	−5.5
Belgium		4.7	10.8	15.0	10.7	23.4	12.6		16.0	10.0	8.1	5.2	13.5	3.5
Bulgaria	4.7	3.8	7.6	32.9	37.4	26.5	18.9	4.0	3.1	8.8	16.2	17.3	15.3	6.5
Czechoslovakia	16.4	15.6	27.0	32.0	50.3	46.0	19.0			29.1	37.0	33.9	23.7	−5.4
Denmark		24.0	21.0	24.0	18.8	31.4	10.4		20.0	14.0	30.0	25.4	27.3	13.3
Finland		13.5	17.0	18.7	18.0	17.2	0.2		4.7	7.9	12.6	11.6	7.7	−0.2
France	12.5	12.0	9.1	7.2	6.6	6.7	−2.4	4.8	6.5	5.3	4.8	7.7	7.4	2.1
Germany						24.8							26.5	
German Dem. Rep.	9.2	10.3	15.5	17.5	31.5			1.6	2.8	9.0	12.9	49.0		
Germany, Fed. Rep.	19.5	15.9	14.0	13.2	15.0			16.8	19.2	16.5	20.4	23.8		
Greece			7.7	6.7	10.0	7.2	−0.5			28.5	18.4	26.5	28.5	0.0
Hungary			9.4	13.3	23.3	17.8	8.4			7.0	12.5	21.5	19.9	12.9
Ireland			4.8	13.2	5.4	15.0	10.2							
Italy						13.8			3.6	2.9	9.5	16.7	17.1	14.2
Liechtenstein	22.0	27.0	23.0	12.4	7.1			10.0	7.0	5.0	9.0			
Luxembourg	4.2	3.8	11.1	9.5				5.6	10.1	12.3	13.9		33.9	21.6
Netherlands	28.9	18.7	14.5	17.7	21.4	21.4	6.9	13.2	26.5	25.4	13.1	11.5		
Norway			20.8	14.8	17.1	19.0	−1.8				18.2	25.1		6.9
Poland			24.2	34.5	40.7	46.9	22.7			7.1	17.7	25.6	34.8	27.7
Portugal			1.7	9.8	25.7	19.8	18.1			0.8	8.6	34.1	36.5	35.7
Romania						6.9							10.4	
Slovenia					34.6	31.3	−3.3					4.4	5.8	1.4
Spain	18.2	10.7	7.3	3.5	3.1	7.3	0.0	13.7	13.7	6.8	3.2	4.4	7.4	0.6
Sweden	11.1	5.6	12.3	12.9	16.1	12.3	0.0			5.2		22.1	9.1	3.9
Switzerland	16.0	14.0	15.0	14.0	19.0	21.0	6.0	8.0	15.0	7.0	6.0	12.0	13.0	6.0
UK		23.0	27.0	34.0	45.0	51.5	24.5		20.0	20.0	21.0	28.8	65.6	45.6
Yugoslavia[b]	23.0	16.1	17.5	39.1	34.6	15.9	−1.6		7.3	9.0	8.2	4.4	8.2	−0.8
USSR														
Belarus				76.0	57.0						33.4	45.0		
Estonia			9.0	28.5	20.0	28.0	19.0							
Latvia				43.0								27.0		
Lithuania		14.8	3.0	24.0	22.9	27.8	24.8			1.0	16.0	15.8	14.9	13.9
Russia						4.2								
Ukraine				1.4	3.0						1.4	2.7		

[a] Percentage of sampled trees in defoliation classes 2–4 inclusive; i.e., those with > 25 per cent defoliation ("unhealthy" trees).

[b] Excluding Croatia and Slovenia.

The UN ECE forest damage survey assesses defoliation (the loss of leaves or needles) as an indicator of tree health. These results pertain to all broadleaved and coniferous species.

The survey specifies that defoliation of sample trees is classified as follows:
Class 0 = 0–10% (no effect on tree health)

Class 1 = 11–25% (slight defoliation, warning stage)
Class 2 = 26–60% (moderate defoliation)
Class 3 = 61–99% (severe defoliation)
Class 4 = 100% (dead tree).

Source:
UN ECE/CEC 1992 *Forest Condition in Europe – Report of the 1991 Forest Damage Survey Results*, prepared by the Programme Co-ordinating Centres with the assistance of the Commission of the European Communities and the Secretariat of the United Nations Economic Commission for Europe.

TABLE 1.22 Concentrations of heavy metals and organic compounds in blue mussel and flounder from two Dutch estuaries, 1985–1990

Species/estuary	Pollutant	1985	1986	1987	1988	1989	1990
MUSSEL							
Ems Dollard	Cd[a]	0.098	0.068	0.062	0.186	0.135	0.078
	Zn[a]	16.116	18.368	12.743	19.796	25.387	21.478
	As[a]	1.249	1.115	0.654	1.106	1.273	1.204
	Pb[a]	0.395	0.360	0.175	0.172	0.423	0.381
	Cu[a]	1.471	1.784	1.351	2.172	1.708	1.465
	Ni[a]	0.341	0.302	0.224	0.544	0.323	0.294
	Hg[a]	0.030	0.034	0.017	0.043	0.043	0.046
	PCBs[b]	2.429	1.159	1.144	2.210	1.729	1.597
	PCB[138 b]	0.636	0.318	0.314	0.604	0.466	0.413
	PAH[b]	4.162	1.338	1.672	8.520	3.756	3.557
	p,p'-DDE[b]		0.061	0.056	0.153	0.098	0.101
	Lindane[b]	0.068	0.061	0.056	0.153	0.098	0.067
	Dieldrin[b]	0.136	0.077	0.056	0.153	0.085	0.116
Western Scheldt	Cd[a]	0.869	0.802	0.330	0.513	0.694	0.589
	Zn[a]	29.174	31.365	21.987	36.236	39.861	34.837
	As[a]	1.290	1.288	0.937	0.973	1.100	0.869
	Pb[a]	0.711	0.516	0.394	0.545	0.435	0.394
	Cu[a]	2.216	1.872	1.619	3.476	2.252	1.614
	Ni[a]	0.503	0.500	0.435	0.604	0.529	0.546
	Hg[a]	0.040	0.047	0.025	0.038	0.059	0.038
	PCB[b]	8.020	4.317	4.446	6.410	6.348	5.033
	PCB[138 b]	1.854	1.025	1.064	1.652	1.504	1.215
	PAH[b]	16.241	4.885	5.060	5.088	5.459	7.652
	p,p'-DDE[b]		0.226	0.562	0.400	0.358	0.315
	Lindane[b]	0.272	0.121	0.107	0.145	0.130	0.120
	Dieldrin[b]	0.304	0.243	0.200	0.223	0.199	0.150
FLOUNDER							
Ems Dollard	Cd[a]		0.066	0.061	0.077	0.090	0.097
	Hg[a]		0.092	0.092	0.119	0.196	0.130
	PCB[138 b]		0.062	0.056	0.039	0.043	0.039
Western Scheldt	Cd[a]	0.073	0.027	0.043	0.048	0.080	0.084
	Hg[a]	0.109	0.101	0.144	0.160	0.110	0.129
	PCB[138 b]	0.112	0.098	0.130	0.128	0.114	0.105

[a] Concentrations in mg^{-1} kg^{-1} wet weight
[b] Concentrations in mg^{-1} kg^{-1} fat weight

Mussel = *Mytilus edulis*, flounder = *Platichthys flesus*.

PCBs = Polychlorinated biphenyls; sum of seven congeners.
PCB[138] = PCB congener 138.
PAH = Polycyclic aromatic hydrocarbons.
p,p'-DDE = 1,1'-(2,2-dichloroethenylidene)bis(4-chlorobenzene).
γ-HCH = 1,2,3,4,5,6-hexachlorohexane; otherwise known as Lindane.
Dieldrin = (1aα,2β,2aα,3β,6aα,7β,7aα)-3,4,5,6,9,9-hexachloro-1a,2,2a,3,6,6a,7,-
7a-octahydro-2,7:3,6-dimethanonaphth(2,3-b)oxirene.

Sample collection and analyses were carried out in conformity with procedures stipulated by the International Council for the Exploration of the Sea (ICES).

These data were collected as the Dutch contribution to the Joint Monitoring Programme (JMP) of the Oslo and Paris Commissions and were also submitted to ICES.

Sources:
Stronkhorst, J. 1992 Trends in pollutants in blue mussel *Mytilus edulis* and Flounder *Platichthys flesus* from two Dutch estuaries, 1985–1990, *Marine Pollution Bulletin*, **24**(5), 250–258.
Data supplied by J. Stronkhorst, Ministry of Transport, Tidal Waters Division, Middelburg.

TABLE 1.23 Mean heavy metal concentrations in the muscle of two common fish species in three coastal areas of the UK, 1980–1990 (mg kg^{-1} wet weight)

Coastal area	Species	Pollutant	1980	1981	1982	1983	1984	1985	1986	1987	1988	1989	1990
Thames	Cod	Hg	0.15	0.11	0.10	0.13	0.12	0.08[a]		0.07[a]	0.11		0.09[b]
		Cd	0.10	0.20									
		Pb	0.20	0.20									
		Cu	0.20	0.20				0.20		0.30	0.20		0.10
		Zn	4.00	3.30				3.50		3.30	4.20		3.50
	Plaice	Hg	0.09	0.08	0.08	0.06	0.09	0.04	0.04		0.06		0.06[b]
		Cd	0.10	0.10									
		Pb	0.20	0.20									
		Cu	0.40	0.30				0.20	0.30		0.20		0.20
		Zn	6.20	7.00				5.50	5.60		4.90		4.30
Liverpool Bay	Cod	Hg		0.37	0.26	0.30	0.28	0.27	0.15	0.25	0.17	0.15	0.11
		Cd	0.10	0.20									
		Pb	0.60	0.20									
		Cu	0.30	0.40				0.20	0.20	0.20	0.30	0.10	0.20
		Zn	3.70	3.60				3.60	3.50	3.20	3.20	3.20	3.10
	Plaice	Hg	0.26	0.23	0.26	0.29	0.26	0.15	0.20	0.18	0.15	0.13	0.11
		Cd	0.10	0.10									
		Pb	0.20	0.20									
		Cu	0.20	0.60				0.30	0.30	0.20	0.30	0.30	0.10
		Zn	6.30	6.00				5.40	5.10	4.40	4.40	4.00	3.90
North Sea (Southern Bight)	Cod	Hg	0.04		0.10	0.10	0.08	0.07	0.08	0.07	0.08	0.10	0.07
		Cd	0.10										
		Pb	0.20										
		Cu	0.30					0.30	0.20	0.20	0.20	0.20	0.20
		Zn	3.60					3.30	3.10	3.50	3.30	3.30	3.40
	Plaice	Hg	0.04		0.07	0.08	0.06	0.05	0.04	0.05	0.06	0.05	0.06
		Cd	0.10										
		Pb	0.20										
		Cu	0.20					0.30	0.20	0.20	0.30	0.30	0.30
		Zn	6.60					4.10	4.00	4.40	4.50	3.80	

[a] Only small specimens caught.
[b] Few specimens collected some distance offshore from the estuary.

Cod = *Gadus morhua*; plaice = *Pleuronectes platessa*.

Sample collection and analyses were carried out in conformity with procedures stipulated by the International Council for the Exploration of the Sea (ICES). Data were subsequently contributed to ICES.

Source:
Department of Environment 1992 *Digest of Environmental Protection and Water Statistics* No. **14**, 1991, HMSO, London.

Integrated Monitoring

Integrated monitoring may be defined as the repeated measurement of a range of related environmental variables or indicators in the living and non-living compartments of the environment, and the investigation of the transfer of substances or energy from one environmental compartment to another (UNEP, 1980). Monitoring becomes truly integrated when the measurements of different variables in different compartments are co-ordinated in time and space to provide a comprehensive picture of the system under study. In short, integrated monitoring incorporates integrated physical, chemical and biological measurements at the same site.

The integrated monitoring approach generally aims to identify the environmental status of ecological areas which are isolated from direct human activity and influences. The information gathered from the monitoring areas may be used for three main purposes: to act as a background reference with which anthropogenic influences in other areas can be compared; to give better insights into the functioning of natural systems and the interaction of environmental compartments, especially with respect to pollutants; and to act as a possible indicator of emerging environmental problems (Söderman, 1990; Wiersma, 1990).

Since publication of the second edition of this report, considerable progress has been made in the development of existing international integrated monitoring programmes; furthermore, a number of new programmes have been established. This is in part a result of the growing recognition of the need for long-term monitoring programmes to support the study of terrestrial ecosystem responses to global change, in particular greenhouse gas-induced climate change. Integrated approaches to monitoring are also being used on an increasing scale by national monitoring programmes. It is estimated that there are currently about 160 monitoring sites world-wide that could be counted as integrated monitoring sites (Wiersma, 1993).

While the various existing networks for integrated monitoring have broadly common aims, the failure to use a set of standard concept definitions has hindered international co-operation (Söderman, 1990). Nevertheless, there are today two international networks that can be said to begin to meet these requirements. On a national level, it is hoped that the development of several established networks will lead to further strengthening of data networking and international co-operation towards integrated monitoring.

International Networks

Of the two best examples of international integrated monitoring networks, one is co-ordinated by UNEP GEMS and the other by the UN ECE. The programmes are based on monitoring stations which are mostly located in small water catchment areas.

The Global Environment Monitoring System has been helping to co-ordinate an integrated monitoring programme in countries which formerly comprised the Council for Mutual Economic Assistance (CMEA). Monitoring of the atmosphere, soil, fresh and marine water and biota for priority pollutants is undertaken at remote sites (Rovinsky, 1990). At present, 17 stations (7 in the Russian Federation and 6 in other former USSR republics) are in operation and a further 30–40 are planned (Rovinsky, 1992). The measurement of air and precipitation quality is supplemented by regular observations made at BAPMoN stations and at UN ECE EMEP stations. Table 1.24 summarizes the results of monitoring of lead, cadmium and mercury concentrations in all environmental compartments at an integrated monitoring site in the former Czechoslovakia between 1983 and 1989.

The UN ECE Pilot Programme on Integrated Monitoring began in 1990 and is one of five monitoring programmes operated under the Convention on Long-range Transboundary Air Pollution. The main aims of this programme are:

❍   To monitor long-term changes in selected catchments in Europe;
❍   To identify levels of pollutants; and
❍   To quantify fluxes of pollutants between environmental compartments.

Information gathered can then be used to produce input and output budgets of pollutants in the catchments, providing the basis for more detailed understanding of environmental processes and pollution influences. Evaluation is done with the aid of statistical analyses and mathematical models. The pilot period of the project has now ended. In August 1992 it was recommended that the programme should continue under the name "International Co-operative Programme on Integrated Monitoring of Air Pollution Effects on Ecosystems". At present 37 designated sites contribute data to the programme.

The Environment Data Centre (EDC) in Helsinki, Finland is responsible for the collation, storage, and assessment of data provided by national focal centres in the 17 European and North American countries that participate in the UN ECE programme. The Finnish centre has published two 'Annual Synoptic Reports' in which the results of the UN ECE programme are presented (EDC 1990, 1991). Since the programme is only in its fourth year, integrated, time-series input and output data exist for only a few stations. Table 1.25 shows concentrations and fluxes of selected variables at three sites in Finland, Norway and Sweden during 1990.

In the Nordic countries, where interest in integrated monitoring has expanded rapidly in recent years, monitoring efforts are co-ordinated by the Nordic Council of Ministers. The Council has been particularly active in the development of harmonized methodologies in integrated monitoring in order to enable intercalibration between institutes and laboratories of different countries (Nordic Council of Ministers, 1989).

Integrated monitoring is a term which has traditionally been applied to multiple media and pollution monitoring, but there is now an increasing tendency to expand the integration of environmental monitoring data with other data to increase understanding of the anthropogenic impacts on the environment, such as those arising from an enhanced greenhouse effect and depletion of the ozone layer. The inclusion of social and economic data, to facilitate the study of economic and social impacts on the environment, is an approach which has been used by the European Community in its Co-ordination of Information on the Environment (EC CORINE) network and by the UNESCO Man and Biosphere Programme (MAB).

The potential usefulness of integrated monitoring data to assess global change is reflected by the proposed Global Terrestrial Observing System (GTOS) which has five parent international organizations: UNESCO, FAO, WMO, International Council for Scientific Unions (ICSU) and UNEP. The role of GTOS will be to provide a global network of sites at which responses of terrestrial and fresh-water ecosystems to environmental processes will be monitored. It is proposed that GTOS will be based upon a set of hypotheses about global change and terrestrial ecosystems and that the location of monitoring sites and measurements made will be dependent on questions being posed. An important, though not exclusive, aim of GTOS will be to provide the terrestrial component of the Global Climate Observing System (GCOS) in a similar way to which the already established Global Ocean Observing System (GOOS) provides the marine component. A more detailed description of GCOS is given in Part 2: Climate.

The International Geosphere Biosphere Programme (IGBP) also incorporates the main principles of integrated monitoring into its core projects. Most notable in this respect are the International Global Atmospheric Chemistry (IGAC) project, the Global Change and Terrestrial Ecosystems (GCTE) project, and the Joint Global Ocean Flux Study (Izrael and Rovinsky, 1991a). The IGBP is also discussed in Part 2: Climate.

National Programmes

A few national integrated monitoring networks are now operational. In Europe, Sweden, Finland and Norway each operate national networks which, in part, contribute to both the UN ECE Programme and the co-operative network of the Nordic Council of Ministers. Similarly, the integrated background monitoring activities in the former USSR have aided 'State of Environment Reports' for the USSR, independently of other former CMEA countries (Izrael and Rovinsky, 1991b). In the UK, the Environmental Change Network was launched in 1992 by the Natural Environment Research Council. Physical, chemical and biological data are being collected from a number of sites according to rigorous standard protocols, to provide comparable data over a long time-span. This standardization in long-term monitoring is fundamental to the programme, and is aimed at providing suitable and reliable information for analysis in order to gain insight into the occurrence and mechanisms of environmental change (Lane, 1992).

In North America, on-going integrated monitoring efforts have been established by the US EPA Environmental Monitoring and Assessment Programme (EMAP), and by Environment Canada's National Integrated Monitoring Programme (NIMP) (EPA, 1992; Söderman, 1990). Both these

programmes monitor environmental changes and status and aim to anticipate future problems. The EMAP programme, for example, aims to identify status and trends for each category of ecosystem by measuring a range of ecological, physical, chemical and human variables. In addition to gathering background information, it examines the effects of different levels of human activity on selected ecosystems, much in common with the international EC CORINE and UNESCO MAB programmes. The EMAP programme is therefore considered necessary to evaluate the effectiveness of environmental policy for protecting ecological resources (Bromberg, 1990).

References

Bromberg, S. M. 1990 Identifying ecological indicators: an environmental monitoring and assessment program, *Journal of the Air and Waste Management Association*, **40**(7), 976–978.

EDC 1990 *Pilot Programme on Integrated Monitoring: Annual Synoptic Report 1990*, Environmental Data Centre, National Board of Waters and the Environment, Helsinki.

EDC 1991 *Pilot Programme in Integrated Monitoring: Annual Synoptic Report 1991*, Environmental Data Centre, National Board of Waters and the Environment, Helsinki.

EPA 1992 An overview of the Environmental Monitoring and Assessment Program, *EMAP Monitor*, March issue, US Environmental Protection Agency, Washington DC.

Izrael, Yu. A. and Rovinsky, F. Ya. 1991a *Integrated Background Monitoring of Environmental Pollution in Mid-Latitude Eurasia*, Report No. 72, Global Atmosphere Watch, World Meteorological Organization, Geneva.

Izrael, Yu. A., and Rovinsky, F. Ya. (Eds) 1991b *Review of the State of the Natural Environment in the USSR*, Natural Environment and Climate Monitoring Laboratory, USSR State Committee for Hydrometeorology, USSR Academy of Sciences, Moscow.

Lane, M. 1992 The Environmental Change Network (ECN), *The Globe*, **10**, 1–2.

Nordic Council of Ministers 1989 *Methods for Integrated Monitoring in the Nordic Countries*, Report of the Working Group for Environmental Monitoring, NORD 1989:68, Miljorapport 1989:11, Nordic Council of Ministers, Oslo.

Rovinsky, F. Ya. (Ed.) 1990 *1989 Bulletin of the Background State of the Environment in CMEA Member Countries*, Issue 8/3, CMEA Co-ordinating Centre for Scientific Co-operation, Natural Environment and Climate Monitoring Laboratory, Moscow.

Rovinsky, F. Ya. (Ed.) 1992 *1990 Bulletin of the Background Pollution of the Natural Environment in Some East-European Countries*, Issue 9/4, Institute of Global Climate and Ecology, Moscow.

Söderman, G. 1990 Integrated monitoring as defined by existing concepts and networks. Background document for UNEP GEMS Expert Meeting on Integrated Monitoring, Gothenberg, Sweden, 31 May–1 June 1990. Unpublished paper.

UNEP 1980 *Selected Works on Integrated Monitoring*, Global Environment Monitoring System Information Series No. 2, United Nations Environment Programme, Nairobi.

Wiersma, G. B. 1990 Conceptual basis for environmental monitoring programmes, *Toxicology and Environmental Chemistry*, **27**(4), 241–249.

Wiersma, G. B. 1993 Personal communication, College of Forest Resources, University of Maine, USA.

TABLE 1.24 Concentrations of lead, cadmium and mercury in environmental samples at the GEMS integrated monitoring station near Kosetice, Czechoslovakia, 1983–1989 (ppm dry weight)

Substrate	Year(s)	No. of samples	Lead Mean	Lead Min.	Lead Max.	Cadmium Mean	Cadmium Min.	Cadmium Max.	Mercury Mean	Mercury Min.	Mercury Max.
AIR AND ASSOCIATED INDICATORS											
Wet precipitation[a]	1987–89	33	9.4	0.7	53.0	2.5	0.005	12.0	2.5	0.1	5.0
SPM[b,c]	1987–89	32	30.3	2.0	67.0	0.9	0.2	3.4			
Mosses											
Pleurozium schreberi	1984–87	4	38.4	0.8	129.0				0.8	0.005[d]	2.2[d]
Rhytideadelphus squarosus	1984–87	4	19.2	0.8	32.0				0.2	0.003[d]	0.6[d]
Hylocomium splendens	1984–87	4	68.6	42.6	133.0				2.1	0.004[d]	7.9[d]
Lichen											
Hypogymnia physodes	1984–87	26	21.3	0.7	74.2	0.2	0.003	2.2	0.6	0.003[d]	4.1
SOIL AND ASSOCIATED INDICATORS											
Humus layer (0–5 cm)	1986–89	15	18.1	1.3	47.1	2.4	0.01	6.7	0.8	0.1	1.8
Wood clearing (0–5 cm)	1986		24.7			1.7			1.3		
Wood clearing (5–20 cm)	1986		1.8			3.2			0.8		
Spruce stand (0–5 cm)	1986		28.7			6.7			0.7		
Spruce stand (5–20 cm)	1986		1.3			2.2			0.9		
Spruce needles (Picea excelsa)	1988	3	17.0	2.7	25.8	4.3	0.9	8.5	0.5	0.4	0.7
Earthworms (Lumbricus terrestris)											
Wood clearing	1984–88	15	9.1	0.8	31.5	1.4	0.01	5.3	0.2	0.009[d]	1.2
Border of pond	1984–88	15	13.2	2.6	36.4	1.6	0.01	7.6	0.3	0.006[d]	0.7
Hare hair (Lepus europaeus)											
Young (< 2 years)	1983,1985–86	5	6.7	5.1	11.0				3.1	0.1	8.0
Old (> 2 years)	1983,1985–86	5	25.1	22.0	32.9				5.7	0.2	12.0
Vole hair (Microtus arvalis)	1984	8	21.5	1.4	110.0	1.3	0.01	5.2	1.9	0.1	12.1
WATER INDICATOR											
Sediment	1983–89	30	12.4	1.7	52.7	1.1	0.005	13.5	1.8	0.005[d]	4.8

a   Concentrations are in units of $\mu g\ dm^{-3}$.
b   Suspended particulate matter; concentrations in units of $ng\ m^{-3}$.
c   Concentrations of total SPM in the same time period ranged between 7.3 and 39.0 $ng\ m^{-3}$ with a mean of 22.7 $ng\ m^{-3}$.
d   Limit of detection.

The monitored area comprises the Anansky Brook catchment and is covered with a mosaic layout of fields (53%), woods (43%) and grassland (4%). The original communities of wood species no longer exist. Human-planted monocultures now predominate. Tree age varies from 15 to 130 years, with 40–60 year old stands prevailing. The altitude of the sampling area ranges from 464 m to 633 m above sea level. Samples were collected from a number of sites within an area of 2.84 km^2 throughout the stated period.

Source:
Skácel, F. and Pekárek, J. 1992 Monitoring of lead, cadmium and mercury in environmental samples at the regional station of the integrated background monitoring network of GEMS in Czechoslovakia, The Science of the Total Environment, 115, 261–276.

TABLE 1.25 Concentrations and fluxes of selected variables at three integrated monitoring sites in Scandinavia, 1990

Country/site	Year	Month	Variable	Input data		Output data	
				Concentration[a] (mg l^{-1})[e]	Flux[b] (meqv m^{-2} month^{-1})	Concentration[c] (mg l^{-1})	Flux[d] (meqv m^{-2} month^{-1})
Finland	1990	January	SO_4^{2-}-S	0.400	1.257	0.900	1.332
Hietajarvi			NO_3^--N	0.420	1.511	0.015	0.025
			NH_4^+-N	0.120	0.432	0.025	0.042
			pH	4.380	2.101	6.390	0.010
			Ca^{2+}	0.080	0.201	1.500	1.776
			Na^+	0.120	0.263	1.300	1.342
			K^+	0.050	0.064	0.500	0.303
			Mg^{2+}	0.020	0.083	0.400	0.781
			Cl^-	0.260	0.370	0.250	0.167
Norway			SO_4^{2-}-S	0.130	0.908	1.040	3.000
Karvatn			NO_3^--N	0.030	0.240	0.150	0.495
			NH_4^+-N	0.100	0.800		
			pH	5.440	0.407	6.300	0.023
			Ca^{2+}	0.090	0.503	0.760	1.754
			Na^+	1.480	7.210	1.580	3.179
			K^+	0.140	0.401	0.150	0.177
			Mg^{2+}	0.170	1.566	0.250	0.951
			Cl^-	2.730	8.624	2.850	3.718
Sweden			SO_4^{2-}-S	1.050	10.479	2.820	26.349
Berg			NO_3^--N	0.640	7.311	0.173	1.850
			NH_4^+-N	0.630	7.196	0.011	0.118
			pH	4.320	7.658	4.300	7.508
			Ca^{2+}	0.160	1.277	1.400	10.465
			Na^+	2.850	19.835	5.200	33.882
			K^+	0.160	0.655	0.430	1.647
			Mg^{2+}	0.350	4.607	1.170	14.418
			Cl^-	5.320	24.009	8.830	37.308
Finland	1990	June	SO_4^{2-}-S	0.540	1.162	6.620	0.006
Hietajarvi			NO_3^--N	0.200	0.493	0.009	0.016
			NH_4^+-N	0.030	0.074	1.100	1.187
			pH	4.470	1.170	0.003	0.005
			Ca^{2+}	0.090	0.155	0.012	0.033
			Na^+	0.070	0.105	0.300	0.612
			K^+	0.100	0.088	0.400	0.280
			Mg^{2+}	0.020	0.057	0.400	0.254
			Cl^-	0.220	0.214	1.200	1.486
Norway			SO_4^{2-}-S	0.260	0.555	0.690	58.057
Karvatn			NO_3^--N	0.120	0.293	0.080	7.704
			NH_4^+-N	0.050	0.122		
			pH	4.750	0.608	5.900	1.698
			Ca^{2+}	0.040	0.068	0.290	19.521
			Na^+	0.170	0.253	1.020	59.849
			K^+	0.080	0.070	0.100	3.450
			Mg^{2+}	0.020	0.056	0.130	14.426
			Cl^-	0.260	0.251	1.780	67.727

Continued

TABLE 1.25   Continued

Country/site	Year	Month	Variable	Input data		Output data	
				Concentration[a] (mg l^{-1})[e]	Flux[b] (meqv m^{-2} month^{-1})	Concentration[c] (mg l^{-1})	Flux[d] (meqv m^{-2} month^{-1})
Sweden							
Berg			SO_4^{2-}-S	1.750	11.571	2.790	5.671
			NO_3^--N	0.560	4.328	0.086	0.200
			NH_4^+-N	0.960	7.265	0.009	0.021
			pH	4.140	7.679	5.100	0.259
			Ca^{2+}	0.100	0.529	2.100	3.415
			Na^+	0.180	0.830	6.670	9.454
			K^+	0.060	0.163	0.350	0.292
			Mg^{2+}	0.030	0.262	1.390	3.726
			Cl^-	0.240	0.716	12.300	11.306

[a]   Mean concentration of ions in precipitation at sampling point.
[b]   Input flux, calculated as wet deposition plus an estimation of dry deposition per m^2 per month.
[c]   Mean concentration in run-off at sampling point.
[d]   Output flux, calculated as run-off plus seepage per m^2 per month.
[e]   Except for pH.

Where output concentration values exist but corresponding flux values do not, precipitation or run-off data were lacking.

Data were produced as part of the UN ECE Pilot Programme on Integrated Monitoring which was set up in 1990 under the Convention on Long-range Transboundary Air Pollution. The programme now comprises 36 integrated monitoring areas within 16 countries. However, at the time of writing, data have been reported only by Czechoslovakia, Finland, Norway and Sweden. Only sites reporting monthly data, showing between year variations, for the three-year period 1988–1990 have been included in the above table. Sampling and analytical procedures were conducted according to agreed methodologies as directed by the Environment Data Centre, National Board of Waters and the Environment, Helsinki, Finland.

Sources:
Environment Data Centre 1990 *Pilot Programme on Integrated Monitoring: Annual Synoptic Report 1990*, National Board of Waters and the Environment, Helsinki.
Environment Data Centre 1991 *Pilot Programme on Integrated Monitoring: Annual Synoptic Report 1991*, National Board of Waters and the Environment, Helsinki.
Data supplied by S. Kleemola, Environment Data Centre, National Board of Waters and the Environment, Helsinki, 1992.

Human Exposure

Traditionally the monitoring and assessment of human exposure to environmental contamination has involved the measurement of levels of pollutants in food and the ambient environment. It was generally assumed that data on urban air quality, water quality and food contamination would provide estimates of the potential for exposure to the contaminants of concern. Human exposure to particulate matter and ultraviolet radiation have, for example, been assessed this way.

It is now increasingly recognized that actual human exposure to contaminants may differ significantly from estimates of exposure based on measured concentrations in food, air and/or water. This is partly a consequence of the fact that people spend a significant proportion of their time indoors, either in residences, in the workplace or in shops. Thus alternative measurement techniques for exposure assessment based on more direct human contact with pollutants, by either personal monitoring, or by determining the level of substances in tissues, secreta or excreta, have been developed in order to provide more meaningful data of relevance to human health. For example, exposure to lead and to cadmium (i.e., heavy metals with multiple exposure pathways via food, water and air) have been usefully measured by adopting this latter technique.

To improve exposure monitoring and assessment internationally, WHO and UNEP through GEMS have been co-operating since 1986 on an internationally co-ordinated research and monitoring effort – the Human Exposure Assessment Locations (HEALs) project. Previous editions of this report have described some of the HEALs pilot phase studies which have involved the assessment of exposure to nitrogen dioxide, lead, cadmium, DDT and HCB. Results of ongoing studies will be highlighted in future editions.

Earlier editions of this report have also reviewed selected studies and monitoring efforts related to the assessment of human exposure to selected heavy metals, aflatoxins, radionuclides, dioxins and a range of organochlorines, including the PCBs. In the present edition indoor air pollution arising from the burning of biomass in the home, ultraviolet radiation and two heavy metals – lead and cadmium – have been selected for discussion. Whereas biomass burning in the home is particularly prevalent in developing countries, exposure to lead and cadmium is predominantly an issue for concern in developed countries. With the possible increases of exposure associated with reduction in stratospheric ozone, exposure to ultraviolet radiation has emerged as an issue for concern with world-wide environmental health implications.

Indoor Air Pollution

It is estimated that world-wide some 2,000 million people depend on fuelwood as their principal source of domestic energy; a further 800 million people rely on other forms of unprocessed biomass, such as agricultural residues, twigs, straw and dung, as a source of fuel to meet their daily energy needs (see Part 6: Energy). The combustion of biomass fuels in the home gives rise to significant indoor air pollution and probably represents the largest energy-related source of air pollutant exposure (WHO, 1992a,b,c).

In developing countries, and especially in rural areas of Africa, Latin America and South and South-East Asia, an open fire inside the dwelling is commonly used for cooking and heating. Often there is no chimney and poor ventilation. Owing to the inefficiency of combustion, large amounts of incompletely combusted materials are produced as smoke and irritating gases. The combination of inefficient stoves, the absence of a chimney and poor ventilation, can lead to severe indoor air pollution and concomitant adverse effects on human health (de Koning, 1988; WHO, 1991; WHO, 1992c).

During the combustion of biomass fuels a range of potentially toxic pollutants are emitted; these include SPM, carbon monoxide (CO), oxides of nitrogen (NO_x) and SO_2. Available data, albeit limited, indicate that WHO air quality guidelines are sometimes exceeded by several orders of magnitude in homes using biomass fuels. The table below reflects the magnitude of average and maximum indoor concentrations of SPM that have been recorded at selected locations in various world regions:

		Concentration ($\mu g\ m^{-3}$)	
Region	Fuel	Mean	Maximum
Africa	Wood	2,300	6,200
Asia	Wood	15,800	30,000
	Wood/dung	18,400	30,000
Latin America	Wood		30,000
Oceania	Wood	1,300	4,800

The above values may be loosely compared with the WHO air quality guidelines for particulates; these are 120 $\mu g\ m^{-3}$ for a 24h averaging time and 60–90 $\mu g\ m^{-3}$ for an annual average (WHO, 1987).
Source: WHO, 1992b

The principal cause of harm to health from the domestic combustion of biomass arises from inhalation of the pollutant emissions. Acute effects include carbon monoxide poisoning and smoke inhalation with excessive acute respiratory infection (particularly among children). Smoke and irritant gases can penetrate into the lower respiratory tract producing inflammatory effects which in turn can impair the ability of tissues to deal with invading micro-organisms and lead to the accumulation of secretions that clog the airways. The result is chronic bronchitis. Decreased pulmonary function and stress on the heart (cor pulmonale) can also occur.

Women and children, who spend a large proportion of their day inside the home, are especially heavily exposed to biomass fuel emissions. It is suggested that such exposure represents the largest single "occupational health" problem facing women in the world today; as many as 700 million women in developing countries may be at risk from developing chronic respiratory diseases. Exposure and resultant health effects may be particularly pronounced in populations residing at high altitudes where indoor fires are used to heat sleeping areas, although data on prevalence of respiratory diseases and mortality in such populations are sparse.

The use of more efficient stoves and processed forms of biomass (charcoal, biogas or methanol), together with the adoption of simple ventilation measures, help to reduce indoor air pollution and to mitigate the risks to health associated with the combustion of biomass fuels. It has been estimated, for example, that emissions of SPM could be reduced by 60 per cent, and those of CO by 86 per cent, by the introduction of more fuel-efficient stoves equipped with a ventilating chimney (WHO, 1992b). The use of more suitable stoves has the added advantage of increasing the cooking temperature and thus improving the elimination of food pathogens. The adoption of such stoves has, however, been slow and has encountered a number of social and cultural problems.

References

de Koning, H. W. 1988 *Air Pollution in African Villages and Cities*, Report WHO/PEP/88.8, World Health Organization, Geneva.

WHO 1987 *Air Quality Guidelines for Europe*, WHO Regional Publication, European Series No. 23, World Health Organization Regional Office for Europe, Copenhagen.

WHO 1991 *Indoor Air Pollution from Biomass Fuel*, Working Papers from a WHO Consultation, June 1991, Report No. WHO/PEP/92.3B, World Health Organization, Geneva.

WHO 1992a *Our Planet, Our Health: Report of the WHO Commission on Health and the Environment*, World Health Organization, Geneva.

WHO 1992b *WHO Commission on Health and Environment, Report of the Panel on Energy*, Report No. WHO/EHE/92.3, World Health Organization, Geneva.

WHO 1992c *Indoor Air Pollution from Biomass Fuel*, Report No. WHO/PEP/92.3A, World Health Organization, Geneva.

Cadmium and Lead

Pollution of the environment by cadmium arises from the mining and smelting of zinc and cadmium, the application of phosphate fertilizers containing appreciable cadmium concentrations, and the application of cadmium-containing sewage sludge to agricultural land. Historical trends in cadmium emission to the atmosphere and subsequent deposition have been described in the previous edition of this report (UNEP, 1991).

Food is the major source of exposure to cadmium for non-occupationally exposed people (Vahter and Slorach, 1990). In general, average daily cadmium intakes from food in areas not significantly polluted with cadmium have been estimated at 10 to 40 µg, whereas in polluted areas the estimated daily intake is believed to be in the range 140–200 µg (WHO, 1992). However, the uptake of cadmium by heavy smokers may equal the intake from food in non-polluted areas, i.e., 10–40 µg day^{-1}. Based on recommendations made by the Joint FAO/WHO Expert Committee on Food Additives and Food Contaminants, the provisional tolerable weekly cadmium intake has been set at 400–500 µg (for adults) (WHO, 1989a).

Cadmium is a nephrotoxin with a long biological half-life in humans. Thus, with respect to cadmium exposure, the kidney is the main organ for concern. An association between cadmium exposure and its nephrotoxic action can be detected on the basis of various urinary biomarkers (e.g., low and high molecular weight proteins, enzymes and tubular antigens) or blood-borne biomarkers (e.g., serum ß$_2$-microglobulin or creatinine) (WHO 1991, 1992). The usefulness of such biomarkers of early renal changes for the assessment of cadmium exposure has recently been tested in a series of European studies involving

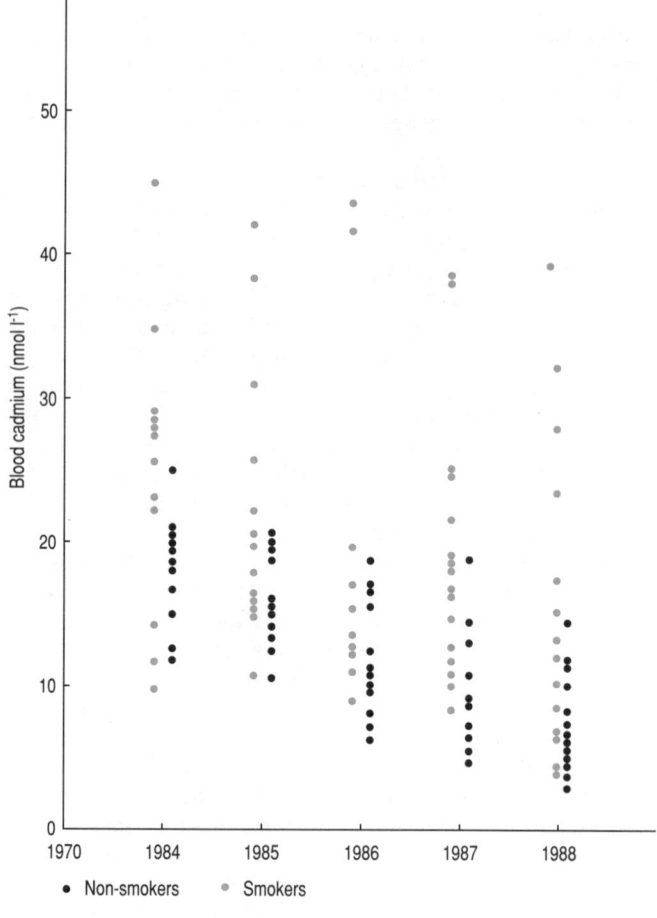

FIGURE 1.27 Blood cadmium concentrations in 31 male individuals from an urban district of Liege, Belgium, 1984–1988 (1 nmol = 0.112 µg l^{-1})
Source: Ducoffre et al., 1992

occupationally exposed workers, the results of which have been published by Roels et al. (1993). Blood cadmium levels were found to be significantly higher in the group of exposed workers, averaging 5.52 µg l^{-1} (range 1.6–14.6 µg l^{-1}) compared with only 0.84 µg l^{-1} (range 0.3–2.8 µg l^{-1}) in the group of "control" workers (Roels et al., 1993).

Belgium has been one of the principal producers of zinc and cadmium in Europe for many decades although, since the early to mid-1980s, production has declined markedly largely due to economic factors. A number of investigations of cadmium exposure amongst the general population of Belgium have been carried out over the years. A group of 31 males, aged between 24 and 58 years and living in a polluted urban district of Liege once dominated by zinc and cadmium factories, have formed the focus of one such study (Ducoffre et al., 1992). Continuous monitoring of blood cadmium levels in these individuals over a five-year period (1984–1988) revealed a steady decline in blood cadmium: the geometric mean blood cadmium level fell by about 56 per cent from a 1984 value of 2.25 µg l^{-1} (range 1.10–6.77 µg l^{-1}) to only 1.00 µg l^{-1} (range 0.30–4.38 µg l^{-1}) in 1988. Figure 1.27 illustrates these trends in blood cadmium levels for both non-smokers and smokers. The decrease in blood cadmium is attributed to the closure of many of the metal producing factories in this area in 1982 (Ducoffre et al., 1992). A second cross-sectional study, involving 527 subjects residing in a nearby rural area, was conducted over a similar time span. These results also showed a decline in blood cadmium; average blood cadmium concentrations in the test population were 40 per cent lower in 1988 than in 1985. The estimated annual decrease rate was 13 per cent (Ducoffre et al., 1992).

Other forms of environmental monitoring in Belgium support the suggestion from human exposure studies that cadmium pollution has decreased substantially during the 1980s. For example, cadmium in SPM decreased, on average, from 50 to 20 ng m^{-3} in a non-industrial urban area and from 40 to 10 µg m^{-3} in a rural area (Ducoffre et al., 1992). Similar patterns in cadmium contamination are observed elsewhere in Europe. A decrease in blood cadmium was reported in individuals from North Rhine (Westphalia) between 1974 and 1988; values declined from 1.20 to 0.40 µg l^{-1}, representing an annual decrease rate of 10 per cent per year (MURL, 1989).

Lead is another potentially toxic heavy metal emitted into the atmosphere during smelting of non-ferrous metals and, more importantly, from motor vehicles using leaded petrol. Exposure of the population takes place through inhalation of lead emitted into the air from motor vehicles, from drinking water reticulated in lead piping, and via food, especially canned food (Vahter and Slorach, 1990). Other sources include soils contaminated with lead, lead-based paint and house dust. These latter sources represent major exposure routes of bioavailable lead to children, because they have high hand-to-mouth contact and are particularly susceptible to lead poisoning (WHO, 1989b).

Lead additives in petrol have been progressively phased out in much of North America, the European Community, Japan and a number of other countries (WHO/UNEP, 1992) (see also Part 7: Industry and Transport) resulting in marked declines in levels of lead in a range of environmental media, including air. For example, in the USA an 87 per cent decrease in ambient concentrations of lead in air was recorded between 1980 and 1989 (see Figure 1.28); a comparable decrease was also noted in Canada (Hilborn and Still, 1990; US EPA, 1991).

Lead in blood has been widely measured in order to indicate exposure and to assess the likelihood of health effects arising from exposure to lead. Reductions in blood lead values have been observed in both adults and children from a number of countries following reductions of lead in air, in drinking water and in food. Selected data of this type have been summarized by the US Centre for Disease Control (CDC), and as part of the National Health and Nutrition Examination Survey (NHANES) for the USA and are shown in Figure 1.28; reductions in blood lead are associated with the introduction of lead-free petrol and concomitant reduction in national emissions of lead.

References

Ducoffre, G., Claeys, F. and Sartor, F. 1992 Decrease in blood cadmium levels over time in Belgium, *Archives of Environmental Health*, **47**, 354–356.

Hilborn, J. and Still, M. 1990 *Canadian Perspectives on Air Pollution, A State of the Environment Report, Number 90–1*, Environment Canada, Ottawa.

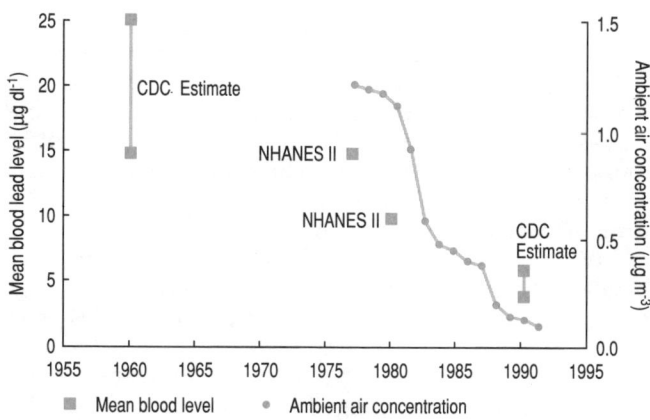

FIGURE 1.28 Trends in average concentrations of lead in blood and in ambient air in the USA, 1960–1991 (data are the results of the ongoing NHANES and estimates prepared by the CDC)
Source: Figure provided by S. Binder, US Centre for Disease Control, Research Triangle Park, USA

MURL 1989 *Luftreinhaltung in Nordrhein-Westfalen: eine Erfolgsbilanz der Luftreinhalte Planung (1975-1988)*, Minister für Umwelt, Raum Ordnung und Landwirtschaft, Düsseldorf.

Roels, H. et al. 1993 Markers of early renal changes induced by industrial pollutants, III Application to workers exposed to cadmium, *British Journal of Industrial Medicine*, **50**, 37–48.

UNEP 1991 *United Nations Environment Programme Environmental Data Report 1991/92*, Basil Blackwell, Oxford.

US EPA 1991 *National Air Quality and Emissions Trends Report*, EPA 450/4-91-003, United States Environmental Protection Agency, Washington DC.

Vahter, M. and Slorach, S. 1990 *Exposure Monitoring of Lead and Cadmium: An International Pilot Study within the WHO/UNEP Human Exposure Assessment Locations (HEALs) Programme*, World Health Organization, Geneva and United Nations Environment Programme, Nairobi.

WHO 1989a Cadmium. In: *Evaluation of Certain Food Additives and Contaminants*, Thirty-third Report of the Joint FAO/WHO Expert Committee on Food Additives, Technical Report Series 776, World Health Organization, Geneva, 28–31.

WHO 1989b *Lead: Environmental Aspects*, Environmental Health Criteria, Number 85, World Health Organization, Geneva.

WHO 1991 *Principles and Methods for the Assessment of Nephrotoxicity Associated with Exposure to Chemicals*, Environmental Health Criteria, Number 119, World Health Organization, Geneva.

WHO 1992 *Cadmium*, Environmental Health Criteria, Number 134, World Health Organization, Geneva.

WHO/UNEP 1992 *Urban Air Pollution in Megacities of the World*, World Health Organization, United Nations Environment Programme, Blackwell, Oxford.

Ultraviolet Radiation

Incident solar radiation comprises mainly ultraviolet (UV), visible and infrared radiation (IR), although both shorter wavelength (ionizing) and longer wavelength (microwaves and radiowaves) radiation are present in the solar spectrum. Ultraviolet radiation (UVR) is generally taken to be that having wavelengths between 10 and 400 nm. However, different methods of classification of component spectral bands of solar radiation have been devised, largely to suit particular branches of science and technology. In terms of photobiological properties, the following spectral regions are defined: UVA (315–400 nm); UVB (280–315 nm); and UVC (100–280 nm).

These approximate to spectral regions in which certain biological absorption properties and biological interactions may dominate (IARC, 1992). The UV component of terrestrial radiation received from the sun comprises about 95 per cent UVA and 5 per cent UVB; UVC is removed by stratospheric ozone. It has been stressed by IARC (1992) that this distinction of UVR into UVA, UVB and UVC ranges has no formal biological basis, and the potential of UVR for causing damage to biomolecules, cells, tissues and organisms varies enormously over the entire spectral range. However, radiation in the wavelength band 295–320 mm (i.e., UVB) is considered to be the most biologically significant; this part of the UVR spectrum has been shown in laboratory studies to damage DNA, the genetic basis of living organisms (IARC, 1992).

Before the beginning of this century, the sun was essentially the only source of UVR but, with the advent of artificial sources, the opportunity for additional exposure has increased. Emissions of UVA and UVB arise from medical lamps and those used for cosmetic purposes; UVB radiation is also present in certain lamps used for general illumination, such as unshielded fluorescent and tungsten-halogen lamps. Exposure to UVC is uncommon and related mainly to use of certain lamps and welding arcs.

Despite the introduction of artificial UVR sources, most human exposure to UV radiation is still from the sun. The most highly exposed individuals are those living nearest to the equator and, at any given latitude, those whose work or recreation takes them outdoors.

A number of factors influence terrestrial UVR levels and consequently human exposure to UVA and UVB (UNEP, 1991; IARC, 1992). These comprise:

O   Geographical latitude and altitude;
O   Time of day, cloud cover, season and surface reflection;
O   Variations in stratospheric ozone with latitude and season; and
O   The level of urban air pollution.

Until relatively recently, the monitoring of the levels of incident solar radiation has been a low-key activity. The data base of observations for the separate component bands, UVA, UVB and UVC, is especially weak. Several attempts have been made to compile maps of UVR exposure based on observational data to support epidemiological studies. However, it has been difficult to relate measured radiation amounts to actual human exposure. Three principal factors complicate the interpretation of available monitoring data and contribute to difficulties of developing accurate biological dose-response relationships for UVB. Firstly, as already mentioned, there are marked variations in UVB with season and time of day; secondly, there are considerable variations in exposure

between individuals, and within individuals between various parts of their bodies; and thirdly, there are effects of body geometry on exposure of individuals to consider. Differences in cultural and social behaviour and in the type of clothing worn provide further uncertainties associated with estimates of personal exposure and population exposure.

Although there are difficulties in estimating or measuring personal exposure to solar radiation, data from epidemiological studies have consistently demonstrated a causal relationship between exposure to sunlight and skin and lip cancer, eye cataracts and immunological changes (Tomatis, 1990; IARC, 1992). However, these studies are based on the assessment of exposure to sunlight or UVR rather than to a specific frequency range (i.e., UVA or UVB). Nonetheless, the UVB frequency range is presumed to be the active component of sunlight in view of its known tissue-damaging and mutagenic effects. The range of health effects associated with exposure to UVR radiation has been described in detail by Tomatis (1990) and the effects of increased levels of UVB on plants have been recently reviewed by SCOPE/UNEP (1992). The evidence linking skin cancer and UV exposure is reviewed elsewhere in this report (see Part 5: Human Health).

As a result of reductions in total ozone recorded in recent years, the quantity of UVR reaching the Earth's surface is expected to increase. Evidence for such increases which have been detected in regions where ozone depletion has been large – Antarctica, Australia and New Zealand – are reviewed in Part 1: Environmental Pollution, Atmosphere. It has been calculated that a sustained decrease in the stratospheric ozone layer of 10 per cent would result in a 26 per cent increase in non-melanoma skin cancer world-wide, i.e., more than 300,000 additional cases of non-melanoma skin cancer and 4,500 cases of melanoma (UNEP, 1991).

IARC in association with WHO and UNEP is currently developing an epidemiological research project, titled INTERSUN, to focus on the skin cancer aspects of ozone layer depletion. The project will, however, also cover other effects. The study aims to collect UV radiation data and other information of relevance to human exposure assessment; progress in the work will be outlined in the next edition of the 'UNEP Environmental Data Report'.

References

Tomatis, L. (Ed.) 1990 *Cancer: Causes, Occurrences and Control*, IARC Scientific Publications Number 100, International Agency for Research on Cancer, Lyon.

IARC 1992 *IARC Monographs on the Evaluation of Carcinogenic Risks to Humans: Solar and Ultraviolet Radiation*, Volume 55, International Agency for Research on Cancer, Lyon.

SCOPE/UNEP 1992 *Effects of Increased Ultraviolet Radiation on Biological Systems*, Scientific Committee on Problems of the Environment, Paris.

UNEP 1991 *Environmental Effects of Ozone Depletion: 1991 Update*, United Nations Environment Programme, Nairobi.

Part 2

Climate

The detection of greenhouse gas-induced climate change in the observational record represents an important aspect of climate research and in recent times has become a high priority issue. New analyses of observational records for a range of climatic variables reviewed here reveal the following important trends:

○ Surface air temperature records indicate that the global warmth of the 1980s has continued into the early 1990s; preliminary data for 1992 reflect the effects of the eruption of Mount Pinatubo in June 1991.

○ Average warming over parts of the Northern Hemisphere is characterized by increases in night-time rather than day-time temperatures.

○ Radiosonde data confirm that the mid-troposphere has warmed and the lower stratosphere has cooled over recent decades.

○ Available data indicate a world-wide retreat of mountain glaciers over the last 100 years which provides independent evidence for global warming.

○ New analyses of tide-gauge measurements suggest that global mean sea level has risen, on average by around 1.8 mm per year over the last 60 years.

Although the above findings are broadly in line with model predictions for climate change, it must be stressed that it is still not possible to attribute the observed global warming specifically to increased atmospheric concentrations of greenhouse gases.

To say that the prospect of global climate change due to increasing atmospheric concentrations of greenhouse gases has become a major concern is now something of an understatement. Since the mid-1980s climate issues have in fact dominated the environmental agenda and have been at the forefront of the international political arena. This latter process has culminated with the signing of the Framework Convention on Climate Change (FCCC) at the 1992 United Nations Conference on Environment and Development (UNCED) (see Part 10: International Co-operation).

Although the increase in atmospheric concentrations of greenhouse gases has been well established from observations (see Part 1: Environmental Pollution, Atmosphere), the impact of this increase on the major components of the earth's climatic system – the atmosphere, the oceans, the cryosphere (i.e., snow and ice) and the terrestrial biosphere – is not so easy to determine. Current efforts to detect greenhouse gas-induced climate change in the observational record provides the focus of discussion for the climate section of the 'UNEP Environmental Data Report'. This principal discussion is, however, preceded by a brief synopsis of the current predictions for greenhouse gas-induced warming and an overview of the current capabilities for climate-system monitoring.

Predictions for Climate Change

Comprehensive assessments of the climate change issue are provided by the publications of the Intergovernmental Panel on Climate Change (IPCC). The IPCC was established in 1988 by the World Meteorological Organization (WMO) and United Nations Environment Programme (UNEP) specifically to furnish the world community with the scientific information it needs to support the formulation of policy responses to the threat of global warming. To this end the IPCC formed three Working Groups (WG), which have been assigned the following responsibilities:

○ WG I – Science: to assess the available scientific information on climate change, including the prediction of future climatic change;
○ WG II – Impacts: to assess the environmental and socio-economic impacts of climate change; and
○ WG III – Response: to formulate response strategies.

In 1990 the IPCC adopted its first major assessment report which summarized the more detailed reports prepared by the individual working groups (IPCC, 1990a,b,c). These reports represent the consensus view of the numerous scientists and specialists who participated in their preparation and thus form the most comprehensive assessment of the climate change issue produced to date. In anticipation of the needs of the Intergovernmental Negotiating Committee for the FCCC, the IPCC has since updated its original 1990 assessment and published its findings in a supplementary report in February 1992 (IPCC, 1992). This supplementary report does not address the full range of topics covered by the First Assessment, but focuses on new research, data and analyses that have since become available.

Since UNCED, the IPCC has undergone a slight modification in its structure – whereas the mandate and structure of

Working Group I is unchanged, Working Groups II and III have been amalgamated to form a new Working Group II which will address all aspects of the climate change issue related to impacts, adaptation and mitigation. Within this working group, four sub-groups have been established with responsibilities for issues such as energy, industry and transport; oceans and coastal zones; agriculture and forestry and other managed systems; and finally unmanaged systems. A new Working Group III will address cross sectoral aspects such as emission scenarios and economics. A Second Assessment Report is scheduled for completion in 1995, with an interim report designed to provide the relevant information to support the decision-making at the first conference of the Parties to the UN FCCC planned for 1994.

The IPCC scientific assessment report presents the current consensus amongst the scientific community regarding the most probable responses of the climatic system to increased atmospheric concentrations of greenhouse gases. Earlier conclusions regarding the sensitivity of global mean temperature to a doubling of carbon dioxide (CO_2) concentrations have remained unchanged in the 1992 update; thus the temperature change associated with an instantaneous equilibrium doubling of CO_2 is expected to lie in the range 1.5–4.5°C. However, the IPCC again stresses that predictions concerning the timing, magnitude and regional patterns of climate change are still subject to considerable uncertainty (IPCC, 1992).

With respect to the transient, or time-dependent responses of the climatic system to increased greenhouse gas radiative forcing, two recent findings have caused the IPCC to revise their previous conclusions. In 1990 the IPCC predicted, for a "business-as-usual" emissions scenario, that global mean temperatures would rise, on average, by about 0.3°C per decade during the next century (IPCC, 1990a). However, in light of new research relating to the possible effects of stratospheric ozone (O_3) depletion and increased concentrations of sulphate aerosols on the earth's radiative balance the IPCC concluded that this warming rate is likely to be slightly lower (IPCC, 1992).

Recent losses of O_3 in the stratosphere (caused by man-made increases in the abundance of chlorofluorocarbons (CFCs) are believed to counteract the radiative forcing caused by the increased concentrations of CFCs over the last decade or so. Similarly, the increases in concentrations of tropospheric sulphate aerosols – which have occurred since the 1950s (particularly in the Northern Hemisphere) as a consequence of rising emissions of sulphur dioxide (SO_2) from fossil fuel combustion sources and which exert a cooling effect – are believed to have offset a significant part of the greenhouse warming, at least in the Northern Hemisphere during the past few decades. However, as the current generation of climate models do not simulate the effects of sulphate aerosols or stratospheric O_3 losses, the impacts of these forcing factors on global mean surface air temperature have not been quantified at this time. The climatic implications of stratospheric O_3 depletion and increased sulphate aerosol concentrations are explained in more detail in the appropriate sub-sections of Part 1: Environmental Pollution, Atmosphere.

References

IPCC 1990a *Climate Change: The IPCC Scientific Assessment of Climate Change*, Report prepared for the Intergovernmental Panel on Climate Change by Working Group I, J. T. Houghton, G. J. Jenkins and J. J. Ephraumus (Eds), Cambridge University Press, Cambridge.

IPCC 1990b *Climate Change: The IPCC Impacts Assessment*, Report prepared for the Intergovernmental Panel on Climate Change by Working Group II, W. J. McG. Tegart, G. W. Sheldon and D. C. Griffiths (Eds), Australian Government Publishing Service, Canberra.

IPCC 1990c *Climate Change: The IPCC Response Strategies*, Report prepared for the Intergovernmental Panel on Climate Change by Working Group III, World Meteorological Organization, Geneva and United Nations Environment Programme, Nairobi.

IPCC 1992 *Climate Change 1992: The Supplementary Report to the IPCC Scientific Assessment*, Report prepared for the Intergovernmental Panel on Climate Change by Working Group I, J. T. Houghton, B. A. Callander and S. K. Varney (Eds), Cambridge University Press, Cambridge.

Climate System Monitoring and Research

As a natural consequence of the complex and interdisciplinary nature of the science of climate change, several major international programmes have been established in order to co-ordinate scientific research into the issue of climate change. The World Climate Research Programme (WCRP) – itself a component of the wider World Climate Programme (WCP) which is jointly sponsored by the WMO and UNEP – and the International Geosphere-Biosphere Programme (IGBP) together form the current framework for developing an overall scientific strategy for climate change research.

The WCRP was initially established in 1980, as a joint programme of the International Council of Scientific Unions (ICSU) and the WMO. In January 1993, the Intergovernmental Oceanographic Commission (IOC) of the United Nations Educational, Scientific and Cultural Organization (UNESCO) also became a co-sponsor of this programme. Multi-disciplinary studies carried out under the umbrella of the WCRP are aimed at improving understanding of the physics of the climate system (i.e., the dynamics of the physical interactions between the components of climate system, the atmosphere, the oceans, the cryosphere and the land surface) and the factors which influence climatic variability and change (i.e., radiative forcings). Ongoing WCRP projects or "experiments" include the following (WMO, 1993):

O   The Tropical Ocean and Global Atmosphere project (TOGA);

O   The World Ocean Circulation Experiment (WOCE); and

O   The Global Energy and Water Cycle Experiment (GEWEX).

The programme of work co-ordinated by the IGBP complements that of the WCRP in that it aims to describe and understand the interactive physical, chemical and biological processes that collectively regulate the earth system. Initially established in 1986 by the ICSU, a series of IGBP core projects have now been formulated (IGBP, 1992). These comprise:

O   International Global Atmosphere Chemistry (IGAC);
O   Global Change and Terrestrial Ecosystems (GCTE);
O   Biospheric Aspects of the Hydrological Cycle (BAHC);
O   Joint Global Ocean Flux Study (JGOFS); and
O   Past Global Changes (PAGES).
O   Land-ocean Interactions in the Coastal Zone (LOICZ).
O   Global Ocean Euphotic Zone Study (GOEZS) (potential core project);

Further references to selected WCRP and IGBP projects are made in other sub-sections of this chapter as appropriate.

Despite the intensification of research effort in the past decade, there are still major uncertainties surrounding the science of global change, and greenhouse gas induced warming in particular. The key gaps in understanding are considered to be in the following areas:

O   Sources and sinks of greenhouse gases and aerosols;
O   Behaviour of clouds and other elements of the atmospheric water budget;
O   The role of the oceans;
O   Response of the polar ice sheets; and
O   Land surface processes and feedbacks.

It is generally accepted that a vastly improved observational data base relating to all elements of the climatic system is an essential prerequisite to achieving a better understanding of climate change, improving our capability for detecting climate change and improving predictions of future climates.

At the present time the WMO's World Weather Watch (WWW) provides global data on basic meteorological parameters including surface and upper air temperature, sea-surface temperature (SST), precipitation, cloud cover, wind speed and direction, relative humidity and atmospheric pressure. Observations are made from ground-based meteorological stations, aircraft and merchant ships, with additional information derived from a space-based observing system comprising five geostationary and four polar orbiting satellites. Data are relayed via the WWW's Global Telecommunication System (GTS) to designated meteorological centres for processing by the Global Data Processing System (GDPS) from which short- and medium-range weather forecasts are generated. Summary data (monthly averages) for selected climatic variables, for example, surface air temperature and precipitation that are reported by a sub-set of national meteorological stations operating within WWW via the GTS, are routinely published as the 'World Weather Records'.

The WWW is very much designed for monitoring the weather and as such does not provide the range and quality of climate observations that will furnish the climatic research community with the information they require. Poor global coverage is a major failing of the current system. Expansion

of the current observing system for basic meteorological parameters is thus a priority. Preliminary estimates of the requirements for expansion to the year 2000 are outlined below:

Observation or observing platform	Number of stations		Number of daily observations	
	1990	2000	1990	2000
Surface stations	9,500	10,000	20,000	30,000
Mobile ships	7,000	7,500	3,500	5,000
Drifting buoys	300	600	2,500	5,000
Moored buoys	170	300	600	800
Radiosonde (land-based)	820	900	1,200	1,400
Upper-wind only	600	700	1,000	1,500
Wind profilers	5	100	60	1,200
Automated shipboard upper-air systems	15	50	30	125
Aircraft reports	3,003	4,000	4,650	104,500
Soundings from satellite data			10,000	25,000
Sea surface temperature			50,000	125,000
Cloud motion vectors			7,000	15,000
Satellite imagery			500	1,200

Source: WMO, 1991

In addition to improving the climate observing system in terms of geographical coverage, questions relating to data quality also need to be addressed. Many of the instruments used today for measuring climatic variables are still the traditional types developed in the 19th and early 20th centuries. The quality of data obtained from these traditional instruments is dependent on how well they are maintained and how good are their operators. The siting of the instrument itself can also be a significant factor. For obvious reasons, most established meteorological stations are sited in or near cities. However, many climatic parameters, but especially temperature, are affected by proximity to urban centres; thus locating instruments in or near cities can introduce bias in the readings (see the discussion of Temperature below).

The advent of microelectronics in recent decades has provided the capability of revolutionizing routine ground-based climatic observations. Modern automatic instrumentation is able to monitor the climate to a much higher precision and accuracy than old-style instrumentation. Moreover, as daily operators are no longer required, instrumentation can be more readily located in remote areas. There is currently a drive towards greater automation of meteorological station observations; however, a period of "overlap" during the switch from traditional to automated instrumentation is considered vital in order to ensure continuity of long-term data sets.

The principal systems for routine ocean monitoring currently comprise:

O   The Integrated Global Ocean Services System (IGOSS); and
O   The Global Sea-level Observing System (GLOSS).

The Integrated Global Ocean Services System provides data on selected ocean parameters, namely surface and sub-surface temperature, salinity and currents, via the GTS of the WWW to support weather forecasting. In 1992, approximately 270 ships contributed to the collection of IGOSS data from which 65 sub-surface and 192 surface temperature salinity profiles were prepared; these are described in the IGOSS 'Products Bulletin' which has been published on a regular basis since 1991. The Global Sea-level Observing System, a service of the IOC, provides standardized sea-level data from a global network of 300 tide-gauge stations. Monthly and annual mean sea-level data are routinely submitted to the Permanent Service for Mean Sea Level (based at the Bidston Observatory, UK) for archiving, processing and dissemination.

These existing systems are, however, modest compared with the perceived requirements for ocean monitoring. Despite progress in recent decades, most measurements of basic oceanic parameters (temperature, salinity, current direction and speed) are not made on adequate spatial or temporal scales. Data coverage is especially poor in the southern latitudes, in polar regions and in the deep ocean. Measurements of many chemical and biological species (e.g., dissolved nutrients, CO_2 and plankton) are not made in sufficient detail, even in surface waters, while many observations known to be relevant to our understanding of ocean and sea-ice conditions are not made at all.

In recognition of these needs for climatic data outlined above, the Second World Climate Conference, held in 1990, recommended that urgent consideration be given to establishment of the Global Climate Observing System (GCOS). Framework plans for the GCOS have since been formulated under the guidance of its four sponsor organizations, WMO, the IOC of UNESCO, UNEP and ICSU.

Initial discussions of the data requirements for the study of climate change have reviewed the breadth and diversity of observations that GCOS will need to encompass. Whereas studies of climatic processes typically require fixed-term specialized regional-scale data sets, studies of climate variability and detection work demand homogeneous long-term (on time scales of decades and centuries) data sets. Climate modelling, for the purposes of predicting climatic change, also has very specific data requirements.

In order to meet the needs of climate research, detection and prediction studies, GCOS will seek to improve and expand existing observing systems, and to develop and implement new systems where appropriate. Thus the ongoing operational activities of the WWW, the Global Atmosphere Watch (GAW), the Global Environment Monitoring System (GEMS), IGOSS and GLOSS, together with the various research and projects of the WCP and the IGBP programmes, will form the starting point for the GCOS concept. Observations from space already provide a valuable source of climate data, a role which will expand in the future. Thus GCOS will work with both national and international space agencies, including the Committee of Earth Observation Satellites (GEOS) to ensure effective co-ordination and use of such key resources.

With regard to ocean observations GCOS is developing its requirements in partnership with the Global Ocean Observing System (GOOS) now being established under the aegis of the IOC in co-operation with ICSU, WMO and UNEP. The GOOS is a broad-based observation system which has been designed to address a range of oceanographic issues, including coastal seas, living resources and the health of the oceans. It will, however, include a climate "module", the requirements of which will be formulated in co-operation with GCOS. Similarly, with regard to land surface observations, GCOS will help to develop a climatic component of the wider Global Terrestrial Observing System (GTOS) which is currently being established jointly by UNEP, UNESCO, the Food and Agriculture Organization of the United Nations (FAO), WMO and ICSU. A detailed draft plant which describes the strategy for developing GCOS has recently become available (Joint Scientific and Technical Committee for GCOS, 1993).

References

IGBP 1992 *Global Change: Reducing the Uncertainties*, The International Geosphere-Biosphere Programme, Royal Swedish Academy of Sciences, Stockholm.

Joint Scientific and Technical Committee for GCOS 1993 *Draft Plan for the Global Climate Observing System (GCOS)*, International Council of Scientific Unions, United Nations Environment Programme, Nairobi, United Nations Educational, Scientific and Cultural Organization, Paris and World Meteorological Organization, Geneva.

WMO 1991 *The Global Climate Observing System*, Report of a meeting convened by the Chairman of the Joint Scientific Committee for the World Climate Research Programme, 14–15 January 1991, Winchester, UK, WCRP-56, WMO/TD-No. 412, World Meteorological Organization, Geneva.

WMO 1993 The World Climate Programme, Background paper prepared for the Intergovernmental Meeting on the World Climate Programme, 14–16 April 1993, Geneva.

Detection of Climate Change

In recent years considerable effort has been devoted to the identification of significant changes in a number of climatic variables, particularly global mean temperature. Other parameters which may serve as indicators of climatic change include upper air temperature, sea level, ocean temperature and salinity, sea-ice extent, snow cover, glacier mass balance, and precipitation amount. Analyses of available records with respect to some of these variables are discussed in more detail below. As temperature is the most direct measure of greenhouse gas-induced climatic change, temperature trends are described in some detail. This particular edition of the 'UNEP Environmental Data Report' also covers the oceanic component of the climatic system, with a brief section on the cryosphere. Previous editions have discussed changes in the cryosphere (UNEP, 1991) and in precipitation/cloud cover (UNEP, 1989) in greater depth. Observational records of atmospheric variables, such as precipitation, evaporation, moisture content and circulation are not covered here but will form the focus of the next edition of this report.

In order to claim detection of greenhouse gas-induced climate change it is necessary not only to identify a change in one or more climatic variables, but also to attribute this change to increased atmospheric concentrations of greenhouse gases.

In this respect, the conclusions of the IPCC remain unchanged since the publication of its 1990 assessment. Although analyses of observational records indicate that global mean surface air temperatures have increased by approximately 0.5°C since the late 1800s (see below), there is insufficient evidence to attribute this change specifically to increases in greenhouse gas concentrations in the atmosphere (IPCC, 1992). Moreover, the IPCC considers that unequivocal detection of an enhanced greenhouse effect from observations is not likely for a decade or more (IPCC, 1992). The detection issue is discussed more fully by the IPCC (IPCC, 1990) and more recently by Karoly et al. (1992).

The success of detection studies ultimately relies on the quality of the historical record. However, the analysis of the vast majority of instrumental records is hampered by several major shortcomings inherent in the currently available raw data sets. Although some temperature and precipitation records date back to the mid-1800s, global coverage prior to 1920 is generally poor. Many records suffer from problems relating to discontinuities and inhomogeneity (or bias) introduced by station relocations, changes in observer practices and averaging methods, and changes in measurement technique. Given an adequate level of supporting information relating to individual measurements (or 'metadata') some of the sources of bias can be accounted for and appropriate corrections applied. However, in many cases such information is lacking. Moreover, many potentially useful data are currently inaccessible to the scientific community, being held in unprocessed form by national meteorological stations.

In recent years, however, a number of key projects have been initiated by the WMO's World Climate Data and Monitoring Programme (WCDMP) which will improve the quality and usefulness of existing data sets. These are briefly outlined below; more detailed information can be found in the WMO's series of technical documents (WMO 1991, 1992).

The Climate Change Detection Project (CCD) was originally conceived in 1989 by the WMO Commission for Climatology (CCl) as a world-wide effort to encourage meteorological services to collect more climate data with associated station data (i.e., metadata), and to process data using uniform procedures. Since its approval in 1991, the CCD project has been included in the WCDMP and, in order to achieve its goals as outlined above, has initiated work in three principal areas:

❍ Development of statistical procedures for climate change detection;

❍ Promotion of a network of reference climatological stations (RCS); and

❍ Production of global baseline data sets which will serve as "official" data sets for global climate change studies.

Greater accessibility to climatological records, particularly those held by meteorological services in developing countries, will be achieved though the Data Rescue Project (DARE). The DARE project is currently operational in over 30 countries in Africa and has successfully saved millions of manuscript records, which may otherwise have been lost, by transferring them to microfilm. These records, some of which date back 100 years, will ultimately be converted to computer-readable media and thus be available for scientific research. This project is currently being expanded to the Central American region.

References

IPCC 1990 *Climate Change: The IPCC Scientific Assessment of Climate Change*, Report prepared for the Intergovernmental Panel on Climate Change by Working Group I, J. T. Houghton, G. J. Jenkins and J. J. Ephraumus (Eds), Cambridge University Press, Cambridge.

IPCC 1992 *Climate Change 1992: The Supplementary Report to the IPCC Scientific Assessment*, Report prepared for the Intergovernmental Panel on Climate Change by Working Group I, J. T. Houghton, B. A. Callander and S. K. Varney (Eds), Cambridge University Press, Cambridge.

Karoly, D. J., Cohen, F. A., Meehl, G. A., Mitchell, F. F. B., Oort, A. H., Stouffer, R. J. and Wetherald, R. T. 1992 An example of fingerprint detection of greenhouse gas climatic change, *Climate Dynamics*. In press.

UNEP 1989 *United Nations Environment Programme Environmental Data Report 1989/90*, Basil Blackwell, Oxford.

UNEP 1991 *United Nations Environment Programme Environmental Data Report 1991/92*, Basil Blackwell, Oxford.

WMO 1991 *Report of the Meeting of Experts on the Climate Change Detection Project*, Niagara-on-the-Lake, 26–30 November 1990, WCDP-No. 13, WMO-TD/No. 418, World Meteorological Organization, Geneva.

WMO 1992 *Final Report of the CCl Working Group on Climate Data*, WCDMP-No. 21, WMO-TD/No. 523, World Meteorological Organization, Geneva.

Surface Air Temperature

As the most widely used and accepted measure of the state of the climatic system, global mean surface air temperature is the most important indicator of greenhouse gas-induced climate change. Examination of the historical record for evidence of increases in global mean temperature has thus formed the focus of detection studies during the past decade.

Several independent time series of global (and hemispheric) average temperatures have been compiled by various research groups; these are derived largely from routine observations made at land-based meteorological stations (some of which date back to the middle of the 1800s) and measurements taken by merchant shipping fleets. More recently, satellite-derived data have added to the available data base of temperature observations. Raw meteorological station data (which have generally been archived as monthly average values) are converted to temperature anomalies (i.e., expressed as a deviation from a reference period mean) before aggregating to hemispheric and finally global mean values.

Land-based Records The available land-based records of surface air temperature have been compiled and analysed independently by three different research groups (Hansen and Lebedeff, 1988; Jones et al., 1986a,b, 1991; Vinnikov et al., 1990). These analyses have been reviewed by the original IPCC assessment (IPCC, 1990), and have been updated to 1991 in the 1992 assessment (IPCC, 1992).

As noted above, analyses of this type are constrained by a number of deficiencies in the available records. Although some records date back to the mid-1800s, geographical coverage prior to 1850 was limited to land areas in Western Europe, parts of Asia and North America, and some coastal regions of Africa, South America and Australasia. Between 1850 and 1900 the number of meteorological stations, and thus observations of temperature, increased markedly such that by the 1920s the only land areas without meteorological instrumentation were some interior parts of Africa, South America and Asia, Arctic coasts and Antarctica. Since the 1950s meteorological stations have been established in Antarctica.

In addition to the problems of inadequate geographical coverage, many records suffer from problems of bias. Inhomogeneities in the observational records have been introduced by a number of factors, including changes in observing practices, station relocations and changes in the local environment (in particular, increasing urbanization). Differences in elevation between stations and variations in the techniques employed by different stations for calculating monthly mean temperatures can also introduce bias.

Of the three data sets mentioned above, that compiled by researchers at the Climatic Research Unit (CRU) at the University of East Anglia, UK is considered to be the most comprehensive (IPCC 1990, 1992). This data set comprises monthly mean surface temperatures assembled from the 'World Weather Records' (published since the 1920s by the Smithsonian Institution and since 1959 by the US Weather Bureau) and 'Monthly Climatic Data for the World' (compiled and published by the Climatic Analysis Center, USA); these are updated routinely with data from CLIMAT reports (i.e., monthly data received from national meteorological stations via the GTS) and data supplied directly to CRU by collaborating institutes and scientists (Bradley et al., 1985; Jones et al., 1986a,b).

Time series of global and hemispheric land temperature fluctuations, expressed as deviations from a reference period mean, derived by the CRU from raw station data are presented in Table 2.1 and plotted in Figure 2.1. A brief description of the procedures employed in this analysis, including details of the corrections applied to the raw data to take account of the main sources of bias, is given in the notes to the table.

The CRU analysis of the land-based temperature record indicates that the 1980s was the warmest decade on record. Moreover, the initial indications are that this warmth extends into the early 1990s; 1990 and 1991 are the warmest years on record. However, 1992 was only the 10th warmest year. Although the other analyses of the historical temperature record (i.e., those of Hansen and Lebedeff (1988) and Vinnikov et al. (1990)) differ from that of the CRU in a number of ways, principally in terms of the quantity of raw data used and in the correction procedures employed, a high degree of correlation is reported between the three analyses (IPCC 1990, 1992).

The trend towards increasing urbanization around meteorological stations is still perceived to be the most likely cause of error in the land-based temperature record (IPCC, 1992). As reported in the previous edition of this report (UNEP, 1991), Jones et al. (1990) have estimated that urbanization influences account for at most 0.05°C of the 0.5°C warming observed over land areas during the last 100 years. However,

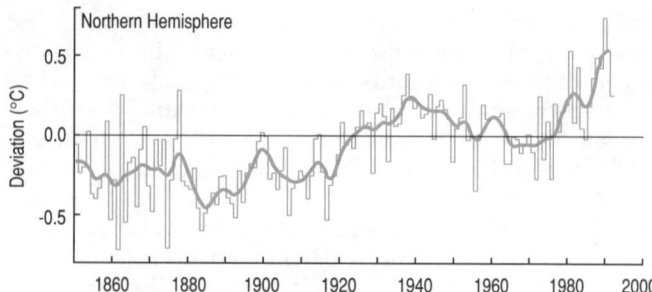

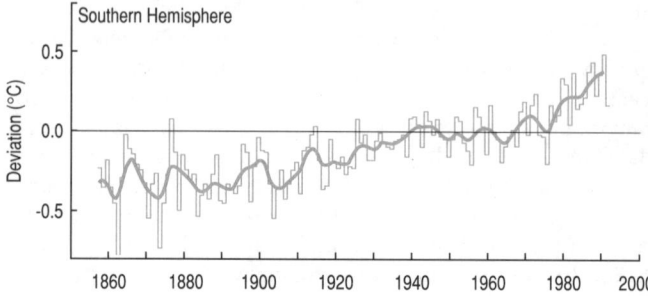

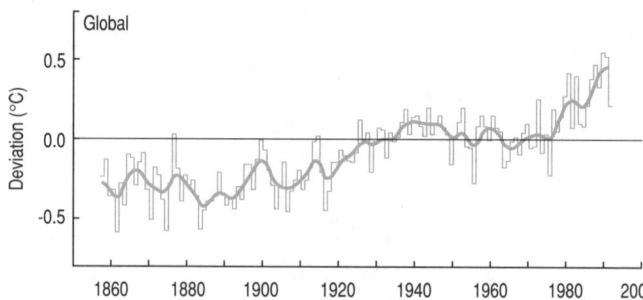

FIGURE 2.1 Trends in mean annual surface air temperatures derived from land-based measurements, 1851–1992 (expressed as deviations in °C from a reference period mean)
Source: Figure provided by P. D. Jones, Climatic Research Unit, University of East Anglia, Norwich, UK

new research indicates that the urbanization problem may be more complex than previously believed. Further investigation is required before the effect of urbanization can be more clearly elucidated (IPCC, 1992).

A detailed analysis of available daily maximum and minimum temperature records for selected land areas of the Northern Hemisphere has recently added a new dimension to the climate detection issue. Karl et al. (1991) have analysed an assemblage of year-month mean maximum and mean minimum temperature records derived from over 750 Northern Hemisphere meteorological stations (including 500 in the USA, 190 in the former USSR and 57 in China). It is concluded that over the last four decades there has been an increase in mean minimum (i.e., night-time) temperatures but little change in mean maximum (i.e., day-time) temperatures. This implies that much of the warming observed to date over land areas of the Northern Hemisphere is due to increases in night-time rather than day-time temperatures.

The origin of the greater warming at night relative to

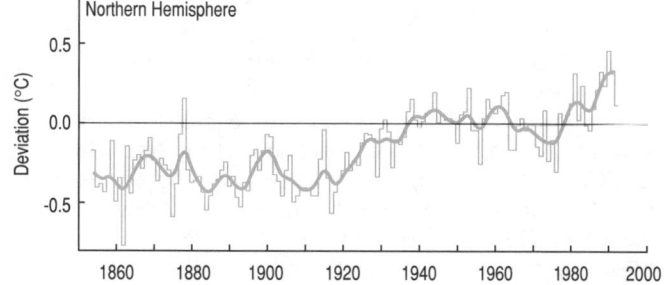

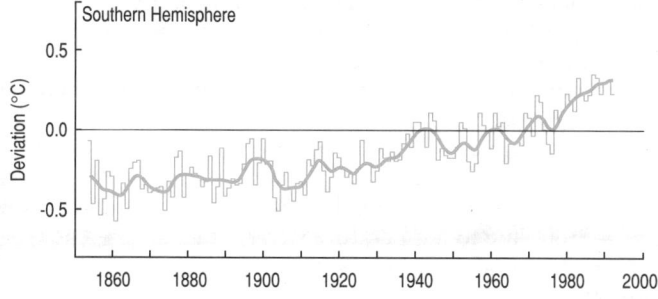

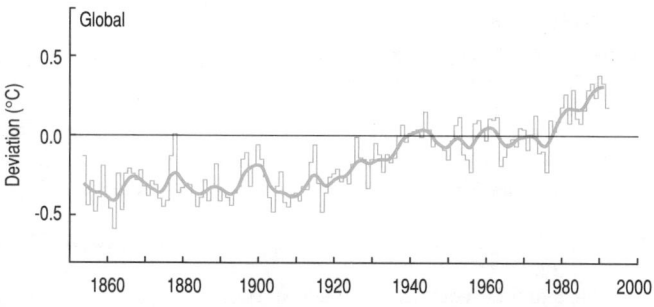

FIGURE 2.2 Variations in mean annual surface air temperatures derived from combined land and SST records, 1854–1992 (expressed as deviations in °C from a reference period mean)
Source: Figure provided by P. D. Jones, Climatic Research Unit, University of East Anglia, Norwich, UK

day is not clearly understood at this stage. The direct radiative impact of greenhouse gases should increase both maximum and minimum daily temperatures. Thus it is likely that factors other than direct radiative forcing by CO_2 are contributing to the observed trends. It has been suggested that the reduced diurnal range could be related to increased cloudiness; increased cloudiness would be expected preferentially to mitigate greenhouse gas-induced warming by day. Moreover, the observational record does in fact suggest that cloud cover has increased over land areas of the Northern Hemisphere (see UNEP, 1989). Other factors which could account for the reduced diurnal range include changes in clear-sky albedo (attributable to increases in sulphate aerosols), and to a lesser extent urbanization effects and changes in atmospheric circulation (IPCC, 1992).

Land plus Marine Records A time series of truly "global" average temperatures has been derived from a combination of land-based data (those of the CRU) and available records of

sea-surface temperature (SST). The analysis of this combined data set represents the major observational evidence in the global warming debate.

In order to increase the volume of data for analysis and avoid certain bias problems, measurements of SST are used as surrogate marine air temperatures. The use of temperature anomalies in analyses of this type means that the difference between air and sea surface temperatures (which can be up to 6–8°C in absolute terms) is of no consequence. Measurements of SST have been made by merchant shipping fleets since the mid-1800s. Thus the geographical coverage is largely determined by shipping routes; in the second half of the 19th century coverage was mainly confined to the Indian and Atlantic oceans north of 40°S. Coverage in the Pacific improved during the 20th century, but is consistently poor in the south-eastern Pacific and everywhere south of 40°S (except near South America). In more recent decades SST measurements have also been made by buoys and bathythermographs. Since the late 1970s, satellite-borne infrared radiometers have added to the data base of SST measurements. To date, however, satellite-derived data have not been used in the analyses reported here.

Two global data bases of *in situ* SST observations have been compiled to date from existing records (mainly ships' logs). These are the Comprehensive Ocean Atmosphere Data Set (COADS) generated by the US National Oceanic and Atmosphere Administration (NOAA) (Woodruff et al., 1987) and the data set produced by the UK Meteorological Office (UKMO), described by Bottomley (Bottomley et al., 1990). The former contains approximately 80 million measurements made between 1854 and the present day. Although the UKMO data set is smaller (about 60 million observations) it contains some unique values. An augmented data set has recently been assembled from these two data bases by the UKMO (IPCC, 1992).

Time series of global and hemispheric temperature anomalies based on an analysis of combined land and sea instrumental records by the CRU are plotted in Figure 2.2; the corresponding data values are provided in Table 2.2. This analysis combines the CRU land data set (Table 2.1) with the COADS SST data (to 1986) and UKMO SST data set (from 1987 onwards). These data update those presented in the previous edition of the 'UNEP Environmental Data Report' (UNEP, 1991). A similar analysis, which combines CRU land data with the new UKMO marine SST data set (the blended COADS and Bottomley (1990) data), is described by the IPCC in its 1992 assessment (IPCC, 1992). The origin and treatment of the main source of bias in the marine data sets, i.e., that arising from changes in the method for sampling sea water over time, are described in the footnotes to Table 2.2.

Updated and improved global and hemispheric temperature series, reported here and by the IPCC (IPCC, 1992), confirm the 1980s as the warmest decade of the past 140 years and 1990 and 1991 as the warmest years on record. There is, however, clear evidence of regional and seasonal diversity in the pattern of warming. Year-to-year and decadal-scale variations are also apparent in the temperature record (Jones and Briffa, 1992; IPCC, 1992). These latest revisions have not, however, altered the earlier conclusion of the IPCC (reported in the previous edition of this report), i.e., that, on average, the surface of the earth has warmed by 0.45 ± 0.15°C since the late 19th century.

Attempts have been made to relate some of the shorter- and medium-term fluctuations in the temperature records (i.e., on time-scales of a few years up to a few decades) to non-greenhouse gas forcing factors. El Niño events, for example, are known to cause temporary increases in hemispheric and global temperatures (the El Niño/Southern Oscillation (ENSO) phenomena are described in greater detail in Part 9: Environmental Disasters). El Niño events occurred in 1982/83, 1986/87 and 1991; these may have contributed to the observed global warmth in those years. When the effects of ENSO events are removed from the temperature record the relative warmth of the 1980s is maintained, but 1989 and 1990 become the warmest years on record (IPCC, 1992).

Volcanic eruptions which inject significant quantities of SO_2 into the upper atmosphere may also affect global and hemispheric mean temperature on time-scales of one to two years. Major volcanic eruptions occurred in 1982 (El Chichon) and in 1991 (Mount Pinatubo); these are likely to exert a short-term cooling effect on global and hemispheric surface temperatures. In particular, the sulphur-rich eruption of Mt Pinatubo (i.e., the type of eruption currently thought to have the maximum climatic effects) is predicted to reduce temporarily global mean surface temperature by 0.3–0.5°C (IPCC, 1992). Thus it is probable that the consequences of the Mt Pinatubo eruption could dominate the global surface temperature records in 1992 or a little beyond. It is interesting to note that preliminary results for 1992 included in Tables 2.1 and 2.2 indicate that 1992 was cooler compared with the late 1980s and early 1990s. Volcanic eruptions also have detectable effects in the low stratosphere; these are discussed in the subsequent section.

As noted earlier in this chapter, it is postulated that increases in atmospheric aerosols, as a result of rising man-made emissions of SO_2, may offset some of the greenhouse gas-induced warming and may already be causing observable effects in the temperature record. In particular, aerosols may be responsible, at least in part, for hemispheric differences in warming rates. Figure 2.2 indicates that, from about 1910 to the 1950s, the Northern Hemisphere surface warmed with respect to the Southern Hemisphere. This differential warming is thought to have a natural, rather than a man-made, origin in view of the small increase of greenhouse gases during this time. Between 1950 and 1980, however, the surface of the Southern Hemisphere warmed relative to the Northern Hemisphere (by nearly 0.3°C). The timing of this reversal of relative warming is consistent with the theory that sulphate aerosols may act to retard the expected warming in the Northern Hemisphere relative to that in the Southern Hemisphere. More recently (i.e., since the early 1980s), reductions in man-made emissions of SO_2 from North America and Western Europe (see Part 1: Environmental Pollution, Atmosphere) may be contributing to the renewed warming that has been observed in the Northern Hemisphere (Figure 2.2). It must be stressed that these results do not necessarily prove the hypothesis of aerosol cooling effects on the observed climate. These changes are within the bounds of natural

variability and many other variables (such as cloudiness changes and ocean circulation fluctuations in the North Atlantic) may influence hemispheric temperature trends (IPCC, 1992).

Supporting evidence for an increase in SST on a regional scale may be provided by observations of coral reef bleaching (IPCC, 1992). Bleaching occurs when given algae are expelled from the cells of the living coral, and is usually a sign of stress. Extremes of SST are known to cause this type of reaction. Thus bleaching can be a manifestation of a SST that is extreme for a particular locality. During the last decade there are indications that coral reef bleaching has become more common in the tropical oceans (Glynn, 1993); some scientists have interpreted these events as indicative of rising SSTs. While some of the most severe bleaching events may indeed be linked to warming caused by the strong 1982–83 El Niño event, a more detailed comparison of coral reef bleaching and historical SST data is needed before these conclusions can be confirmed (IPCC, 1992). Furthermore, the observed bleaching may not be entirely due to temperature stress; other stresses, for example pollution, may be playing a role (see Part 3: Natural Resources).

References

Bottomley, M., Folland, C. K., Hsiung, J., Newell, R. E. and Parker, D. E. 1990 *Global Ocean Surface Temperature Atlas (GOSTA)*, Her Majesty's Stationery Office, London.

Bradley, R. S., Kelly, P. M., Jones, P. D., Diaz, H. F. and Goodess, C. 1985 *A Climate Data Bank for the Northern Hemisphere 1851–1980*, Technical Report Series TR017, US Department of Energy, Carbon Dioxide Research Division, Washington DC.

Glynn, P. W. 1993 Coral reef bleaching: ecological perspectives, *Coral Reefs*, **12**, 1–17.

Hansen, J. and Lebedeff, S. 1988 Global surface temperatures: update through 1987, *Geophysical Research Letters*, **15**, 323–326.

IPCC 1990 *Climate Change: The IPCC Scientific Assessment of Climate Change*, Report prepared for the Intergovernmental Panel on Climate Change by Working Group I, J. T. Houghton, G. J. Jenkins and J. J. Ephraumus (Eds), Cambridge University Press, Cambridge.

IPCC 1992 *Climate Change 1992: The Supplementary Report to the IPCC Scientific Assessment*, Report prepared for the Intergovernmental Panel on Climate Change by Working Group I, J. T. Houghton, B. A. Callander and S. K. Varney (Eds), Cambridge University Press, Cambridge.

Jones, P. D. and Briffa, K. R. 1992 Global surface air temperature variations during the twentieth century: Part 1, spatial, temperal and seasonal details, *The Holocene*, **2**(2), 165–179.

Jones, P. D., Groisman, Ya. P., Coughlan, M., Plummer, N., Wang, W. C. and Karl, T. R. 1990 Assessment of urbanization effects in time series of surface air temperature over land, *Nature* (London), **347**, 169–172.

Jones, P. D., Raper, S. C. B., Bradley, R. S., Diaz, H. F., Kelly, P. M. and Wigley, T. M. L. 1986a Northern Hemisphere surface air temperature variations 1851–1984, *Journal of Climate and Applied Meteorology*, **25**, 161–179.

Jones, P. D., Raper, S. C. B. and Wigley, T. M. L. 1986b Southern Hemisphere surface air temperature variations 1851–1984, *Journal of Climate and Applied Meteorology*, **25**, 1213–1230.

Jones, P. D., Wigley, T. M. L. and Farmer, G. 1991 Marine and land temperature data sets: a comparison and look at recent trends. In: *Greenhouse Gas-Induced Climatic Change: A Critical Appraisal of Simulations and Observations*, M. E. Schlesinger (Ed.), Elsevier Science Publishers, B.V., Amsterdam, 153–172.

Karl, T. R., Kukla, G., Razuvayer, V. N., Changery, M. J., Quayle, R. G., Heim, R. R. Jr, Easterling, D. R. and Fu, C. B. 1991 Global warming: evidence for asymetric diurnal temperature change, *Geophysical Research Letters*, **18**, 2253–2256.

UNEP 1989 *United Nations Environment Programme Environmental Data Report 1989/90*, Basil Blackwell, Oxford.

UNEP 1991 *United Nations Environment Programme Environmental Data Report 1991/92*, Basil Blackwell, Oxford.

Vinnikov, K. Ya., Groisman, P. Ya. and Lugina, K. M. 1990 Empirical data on contemporary global climate changes (temperature and precipitation), *Journal of Climate*, **3**, 662–667.

Woodruff, S. D., Slutz, R. J., Jenne, R. J. and Steurer, P. M. 1987 A comprehensive ocean-atmosphere data set, *Bulletin of the American Meteorological Society*, **68**, 1239–1250.

Upper Air Temperatures

In addition to temperature changes at the surface of the earth, the current generation of general circulation models (GCMs) are capable of predicting greenhouse gas-induced changes in temperature in upper layers of the atmosphere. Increases of CO_2 and other greenhouse gases are expected to cause warming in the troposphere and, more significantly perhaps, cooling in the stratosphere. As stratospheric cooling is characteristic of greenhouse gas forcing (for example, increases in solar irradiance are expected to cause warming in both the troposphere and stratosphere), the analysis of available aerological data sets represents an important aspect of climate change detection work.

At the present time measurements of temperature of the upper layers of the atmosphere are made from the ground, i.e., by radiosondes (a balloon-borne instrument), and also by satellite-borne instrumentation, predominantly microwave and infrared sounders. Radiosonde observations are generally taken twice a day at about 1,000 stations world-wide and the data transmitted to national meteorological centres for processing. Some of these data are exchanged via the WWW's GTS as CLIMATE TEMP reports. Monthly averaged data sets are archived in digitized form at several institutions including the UKMO and NOAA's National Climatic Data Center.

Radiosonde Data To date, two radiosonde data sets have been assembled independently from available station records and subjected to trend analysis (Angell 1988, 1991; Oort and Liu, 1993). Angell's analysis of upper air temperature trends which is based on data from a globally distributed network of 63 stations, was previously reported in this report (UNEP, 1989) and is updated here. Hemispheric and global average annual mean temperature anomalies, relative to a 1958–1977 reference period mean, for the mid-troposphere (850–300 mb layer) and the lower stratosphere (100–50 mb layer) are given in Table 2.3. The footnotes to the data table provide a summary of the data analysis procedures employed; for further details please refer to Angell (1988, 1991). Results of this analysis do indeed suggest a warming of the mid-troposphere (with 1990 as the warmest year on record) and a cooling of the lower stratosphere.

Both radiosonde data sets have been reviewed and compared by the IPCC as part of the 1992 update of its scientific assessment. Figure 2.3 compares the globally averaged

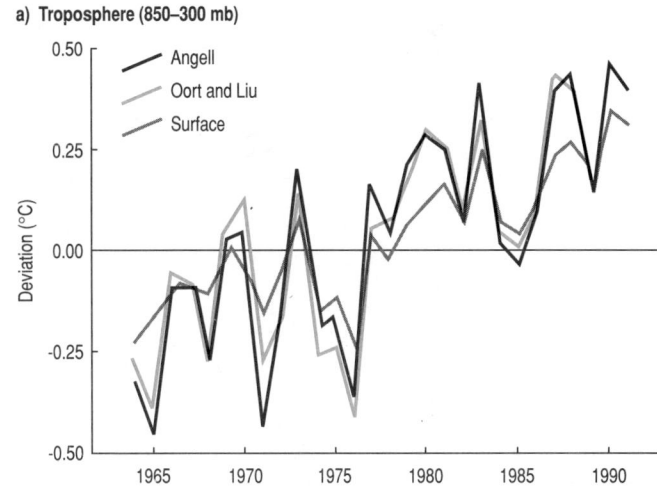

a) Troposphere (850–300 mb)

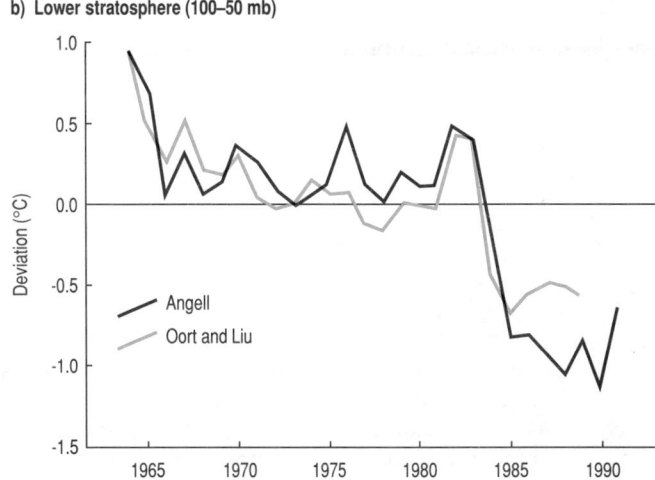

b) Lower stratosphere (100–50 mb)

FIGURE 2.3 Trends in global mean temperatures in (a) the troposphere (850–300 mb layer) and (b) the lower stratosphere (100–50 mb layer), expressed as deviations in °C from a 1964–1989 reference period mean (tropospheric and stratospheric temperatures are derived from radiosonde data and surface temperatures from combined land and SST data)
Source: IPCC, 1992

temperature anomalies for the troposphere (Figure 2.3a) and lower stratosphere (Figure 2.3b) derived by Angell (1964–1991) and by Oort and Liu (1964–1989).

It is noted that for both layers of the atmosphere the correlation between the two independently-derived data sets is high. A high degree of correlation between the radiosonde data sets for the 850–300 mb layer and the combined land and sea-surface temperature is also reported. Moreover, the warming trend in the mid-troposphere layer and the cooling trend in the lower stratosphere are significant in both data sets at better than the 1 per cent level (IPCC, 1992). The IPCC has combined the values from both analyses and has calculated hemispheric and global average temperature trends for three upper atmosphere layers for the period 1964–1991. These results are summarized below (trends are given in degrees Celsius per decade): ·

	Troposphere (850–300 mb)	Tropopause (300–100 mb)	Low stratosphere (100–50 mb)
NH	0.21	−0.05	−0.38
SH	0.23	−0.13	−0.53
Globe	0.22	−0.09	−0.45

NH = Northern Hemisphere
SH = Southern Hemisphere
Source: IPCC, 1992

A notable feature of the upper air temperature time series reported here is the pronounced influence of volcanic eruptions. The injection of aerosols into the atmosphere as a result of a volcanic eruption results in a warming of the stratosphere, the effects of which typically last for 1–2 years. Figure 2.3 clearly shows the influence of the 1982 eruption of El Chichon. The data in Table 2.3 reflect the relative warmth of the lower stratosphere in 1992, an occurrence which is in all likelihood a consequence of the eruption of Mount Pinatubo in June 1991. More significantly, the globally-averaged stratospheric cooling trend of −0.45°C per decade calculated by the IPCC for the period 1964–1991 may be exaggerated as the start of the time series (1964) was a warm year in the tropical stratosphere following the eruption of Agung in 1963 (IPCC, 1992).

Although the above findings appear to be consistent with model predictions of greenhouse gas-induced warming and with the results of analyses of surface air temperature data sets, a degree of caution must be exercised when interpreting these data. Both radiosonde data sets suffer from data-quality problems which may limit their reliability. Detailed studies of possible time-varying biases in the observational records have not yet been made. Moreover, the radiosonde data sets have relatively low spatial and temporal coverage. Angell's data set is based on only 63 stations; the analysis of Oort and Liu has a higher data density in that it relies on records from about 800 stations. Although radiosonde measurements have been made routinely since the 1930s, coverage is biased towards the Northern Hemisphere. Thus truly "global" estimates of upper air temperatures can only really be made after the 1960s (IPCC, 1992).

In an attempt to improve the existing data base of raw radiosonde observations, the US National Data Center (NCDC) is currently compiling and processing available data from over 1,000 meteorological stations world-wide. The Comprehensive Aerological Reference Data Set (CARDS) will initially comprise a data base of observations made from 1970 to 1990. Subsequently, it will be extended to include earlier data (i.e., since the 1930s) and more current data will be added on a periodic basis. The CARDS project will also involve a programme of quality control plus identification of, and correction for, sources of bias. It is hoped that the initial phase of the project will be completed by the middle of 1993 (Doty, 1992).

References

Angell, J. K. 1988 Variations and trends in tropospheric and stratospheric global temperatures, 1958–87, *Journal of Climate*, **1**, 1296–1313.

Angell, J. K. 1991 Changes in tropospheric and stratospheric global temperatures, 1958–1988. In: *Greenhouse Gas-Induced Climatic Change:*

A Critical Appraisal of Simulations and Observations, M. E. Schlesinger (Ed.), Elsevier Science Publishers, Amsterdam, 231–247.

Doty, S. R. 1992 The comprehensive aerological reference data set: a global upper air data base of radiosonde observations, *Earth System Monitor*, December issue, 9–12.

IPCC 1992 *Climate Change 1992: The Supplementary Report to the IPCC Scientific Assessment*, Report prepared for the Intergovernmental Panel on Climate Change by Working Group I, J. T. Houghton, B. A. Callander and S. K. Varney (Eds), Cambridge University Press, Cambridge.

Oort, A. H. and Liu, H. 1993 Upper-air temperature trends over the globe, 1958–1989, *Journal of Climate*, **6**, 292–307.

UNEP 1989 *United Nations Environment Programme Environmental Data Report 1989/90*, Basil Blackwell, Oxford.

Snow and Ice

It has been predicted that greenhouse gas-induced warming will be most pronounced in the high latitude regions, in part due to the snow-ice albedo feedback mechanism (IPCC, 1992). The polar amplification of global warming implies that greenhouse effects may become apparent sooner in the high latitude Arctic and Antarctic environments (Walsh, 1991). Thus climatic variables such as sea-ice extent and thickness, permafrost and snow cover are likely to serve as good indicators of climatic change. In mountain environments, the extent and mass of alpine glaciers are also potentially effective climate change indicators.

Changes in cryosphere parameters, discernible from available observational records and monitoring programmes, are outlined below in varying degrees of detail. The focus is on new data analyses which have become available since the publication of the previous edition of this report; where little new information has come to light, the reader is referred to the third edition of this report for a fuller discussion of the variable in question (UNEP, 1991).

The collection and dissemination of standardized data describing the status of the world's existing perennial ice and snow masses is an on-going task of the World Glacier Monitoring Service (WGMS), which is based at the Swiss Federal Institute of Technology in Zurich. The WGMS is one of the permanent services of the Federation of Astronomical, Geophysical and Data Analysis Services (FAGS/IUSU) and operates under the auspices of the International Commission on Snow and Ice (ICSI/IAHS). The work of the WGMS serves as a contribution to UNEP GEMS and the IHP (International Hydrological Programme) of UNESCO.

In recent years a comprehensive inventory of perennial snow and ice has been compiled by WGMS and published in summary form (Haeberli et al., 1989). On the basis of the results of the World Glacier Inventory (WGI), it is estimated that perennial snow and ice cover 15.6×10^6 km^2 of the earth's surface, 96.6 per cent of which comprises the ice sheets of Greenland and Antarctica. Alpine glaciers and small ice caps account for the remaining 3.4 per cent. The WGI is discussed in greater detail in the previous edition of this report (UNEP, 1991).

Polar Ice Sheets The response of the polar ice caps, which together account for over 96 per cent of the earth's fresh water, to global warming has been the subject of intense speculation for well over a decade. The question of the potential instability of the West Antarctic ice sheet, in particular, has received much attention in view of the not inconsiderable implications for sea-level rise. Predictions made in the early 1980s of a 5 m rise in sea level, following a rapid disintegration of this ice sheet have, however, been revised downwards. It is now considered that for a typical greenhouse gas warming scenario, significant mass outflow would not occur until 100 to 200 years from now. The projected West Antarctic contribution to sea-level rise would be 40 cm after 200 years and 30 cm after 300 years (IPCC, 1990). Indeed, the consensus scientific opinion regarding the response of the Antarctic ice sheets as a whole to climatic change is one of increasing accumulation and a negative contribution to sea-level rise, at least in the shorter term (IPCC, 1990; Jacobs, 1992).

Recent evidence for surface accumulation in Antarctica has been reviewed by Jacobs (1992). One of the most significant studies to date is that of Morgan et al. (1991) involving the analysis of four ice cores from the Wilkes Land sector of east Antarctica. Careful interpretation of the seasonal variations in the oxygen isotope ratio and in ice-crust stratigraphy has enabled Morgan et al. to reconstruct a time series of annual snow accumulation rates at each of the four sites. At one site, Law Dome, the record dates back to 1806 and indicates an overall increase in snow accumulation to the present day, but with large inter-annual variability. The latter part of this record is especially interesting – accumulation decreased rapidly in 1955 reaching a minimum in 1960 and thereafter increased rapidly to the present values, which are the highest on record. Collectively, the cores indicate that recent (1975–1985) accumulation rates are approximately 20 per cent above the long-term mean (1930–1985). The authors attribute this increase in accumulation to higher snowfall caused by enhanced cyclonic activity around Antarctica (Morgan et al., 1991).

At this stage it is unclear how representative the above findings are of the entire continent. Investigations in other regions of Antarctica confirm the increased accumulation rates; other studies, however, indicate decreases or little change in accumulation rates (Jacobs, 1992).

Nor is it possible to say with any confidence whether the Antarctic ice sheet is currently in balance or whether it has been contributing to sea-level rise (by losing mass) or reducing sea-level rise (by gaining mass). On the basis of their observations, Morgan et al. (1991) conclude that since the 1960s Antarctica has had a positive imbalance (i.e., is gaining mass). Before this time the Antarctic ice sheet was close to balance. On the other hand, Doake and Vaughan (1991) have reported that the Wordie ice shelf, which is located on the western side of the Antarctic Peninsula, has shrunk markedly from about 2,000 km² in 1966 to about 700 km² in 1989. The shrinkage has been related to a strong regional warming trend (Doake and Vaughan, 1991). However, Zwally (1991) cautions against extrapolating this result to other parts of Antarctica and notes that substantial further warming would be required to affect the major ice shelves that act to stabilize the west Antarctic ice sheet. Nevertheless, the break up of the Wordie ice shelf serves to highlight the need for continued observation of changes in the Antarctic ice shelves and in grounded ice sheets.

A similar degree of uncertainty surrounds the current balance status of the Greenland ice sheet. Available estimates of the mass balance of the entire Greenland ice sheet are summarized by the IPCC (IPCC, 1990). Studies of recent changes in selected areas are also reviewed. Of particular note is the suggestion of thickening in southern Greenland during the period 1978–1986 as evidenced from satellite altimetry (Zwally, 1989). Although there are some doubts regarding the reliability of the results of this study, the potential usefulness of radar altimetry for monitoring changes in large ice sheets is clearly demonstrated (IPCC, 1990).

Mountain Glaciers Compared with the larger ice masses of Greenland and Antarctica which respond only slowly to temperature and climatic changes, mountain or alpine glaciers typically respond to environmental change on annual and decadal time scales. Indeed, mountain glaciers are highly sensitive integrators of changes in the radiative balance at the surface of the earth. For this reason, variations in glacier parameters are widely perceived as potentially useful indicators of greenhouse gas-induced warming.

Within the complex chain of processes linking glaciers and climate, glacier mass balances are the most directly related to changes in atmospheric conditions. Glacier mass balance, i.e., the net amount of ice mass gained (or lost) through a given year at the surface of the glacier per unit area, is thus of special relevance to climate change detection studies.

Glacier data compiled by the WGMS are wide ranging and include information on the position of glacier fronts, their length, area and volume as well as mass balance. In some instances records date back to the late 19th century. These data are assembled from information supplied by scientists and participating institutes, and are available in published form as the report series, 'Fluctuations of Glaciers'. To date six volumes in this ongoing series have been published, spanning the periods 1959–1965 (Kasser, 1967), 1965–1970 (Kasser, 1973), 1970–1975 (Müller, 1977), 1975–1980 (Haeberli, 1985) 1980–1985 (Haeberli and Müller, 1988) and 1985–1990 (Haeberli and Hoelzle, 1993).

The 'Glacier Mass Balance Bulletin' is a new publication series of the WGMS which provides mass balance data for selected reference glaciers at two-yearly intervals. This publication is designed to complement the more comprehensive 'Fluctuations of Glaciers' series by providing mass balance data on a more regular basis in a simplified, graphical format. To date two issues in the new report series have been published, covering the mass balance years 1987/88 and 1988/89, and 1989/90 and 1990/91 (WGMS 1991, 1993).

Selected glacier mass balance data compiled by WGMS are presented as Table 2.4. These data refer to the set of reference glaciers which form the basis of the 'Glacier Mass Bulletins'; they represent observational series that are both complete and ongoing. It is noteworthy that WGMS glacier mass balance records are available for only a very small proportion (less than one per cent) of the total number of glaciers world-wide; these tend to be biased towards the European Alps and Iceland. More numerous observations, however, exist for variations in the position of glacier fronts or glacier length reductions. Such measurements can be usefully converted to estimates of long-term glacier mass balance by employing a combination of continuity analyses and data on initial glacier length.

Direct measurements of glacier mass balance dating back to the late 19th century are available for a select number of European alpine glaciers. These data, together with ground and aerial surveys, suggest that alpine glaciers are currently considerably smaller than a century ago (IPCC, 1992). Statistical analyses of spatial and temporal variations in mass balance series from glaciers on other continents indicate that the shrinkage of the Alpine glaciers is characteristic of mountain glaciers in general.

This finding, i.e., a world-wide recession of mountain glaciers over the 100 years or more, provides some of the clearest evidence for a change in energy balance at the earth's surface since the turn of the century. It also provides independent evidence of global warming to support that from the temperature records (Haeberli, 1990; IPCC, 1992). In addition it is estimated that the secular rate of glacier mass loss during the 20th century (about 0.3 – 0.4 m a^{-1} water equivalents) when expressed as an energy flux is broadly consistent with the estimated anthropogenic greenhouse forcing (2.5 W m^{-2}) (Haeberli, 1993; WGMS, 1993).

Available evidence indicates that the world-wide mountain glacier retreat has not occurred uniformly throughout the last century. Most noteworthy is the period of net advance in Alpine glaciers during the 1960s and 1970s which coincided with colder average temperatures over much of the North Atlantic and over Western Europe (IPCC, 1992). During this time it is estimated that at least 50 per cent of the European glaciers were advancing (Wood, 1988). More recent data (from the mid- to late 1980s) suggest that the Alpine glaciers are again shifting back to a regime dominated by retreat and shrinkage (Figure 2.4). In fact, it is believed that the melt rates of the European glaciers have accelerated markedly during the past decade to the extent that glaciers now appear to be at their minimum for the past 5,000 years or more (Haeberli, 1993).

Evidence from other world regions, particularly China and Patagonia, suggests that these European trends are typical of mountain glaciers as a whole (IPCC, 1992) (See Figure 2.4). However, there are notable exceptions to these very general secular trends in glacier fluctuations. Some maritime glaciers, especially those in humid coastal locations in the mid-high latitudes (e.g., the Franz Josef glacier in New Zealand and the Wolverine in Alaska) have tended to gain mass since the mid-1970s and early 1980s (WGMS, 1991; IPCC, 1992).

It is probable that the melting of mountain glaciers has contributed to the observed sea-level rise over the last century. Sea-level changes, and the likely contribution of glacier melt waters to the observed changes, are discussed in a subsequent section (see Oceans).

Snow Cover   Until the advent of satellite remote-sensing, relatively little routine monitoring of seasonal changes in snow and ice had been conducted. However, available satellite-derived estimates of snow and sea-ice area now span almost two decades. Satellite-based monitoring of continental snow cover is confined to the Northern Hemisphere where the large-scale variations in snow cover principally occur.

Since 1966, weekly snow and ice cover charts have been

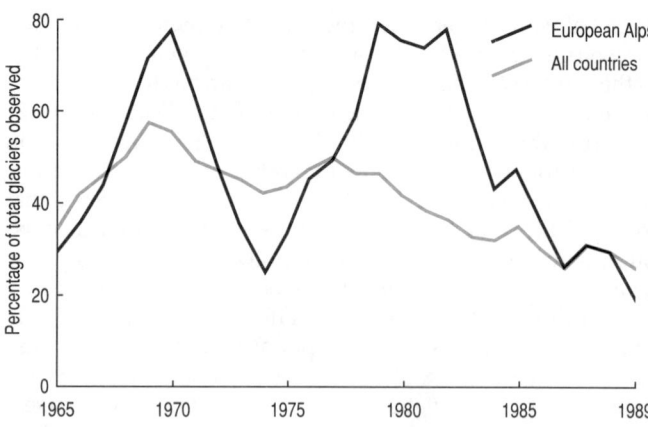

FIGURE 2.4 The percentage of observed glaciers world-wide and in Europe having positive mass balances, 1965–1989 (values are based on five-year moving averages)
Source: IPCC, 1992

compiled for the Northern Hemisphere from visible satellite imagery by NOAA's National Environmental Satellite, Data and Information Service. These charts are subsequently digitized using a 89 x 89 cell Northern Hemisphere grid whereby snow cover area (in km^2) can be computed. This data set, which is available from the World Data Center for Glaciology at Boulder, Colorado, USA, is effectively the only comprehensive observational record of snow cover changes.

The monitoring of snow cover by satellite-borne high resolution radiometers, while producing regular, high resolution and geographically comprehensive observational data sets does, however, suffer from a number of shortcomings. Satellite observations in the visible spectrum are impaired by cloud cover and by conditions of low solar illumination. Secondly, snow cover is often under-estimated in densely forested areas. Monitoring of snow cover in mountainous areas and uniformly lightly-vegetated areas can also present difficulties because of problems of discriminating clouds from snow. Furthermore, satellite monitoring of snow cover does not provide any information about snow depth. It is now recognized that during the earlier years of operation, snow cover was underestimated on the NOAA charts, especially in the autumn season. Since the early 1970s, improvements in the measurement and data processing techniques have, however, improved the reliability of the data set.

Since the publication of the third edition of this report, the methodology for calculating monthly mean snow cover data from weekly digitized charts has been modified (IPCC, 1992; Robinson, 1993). The change in the analysis technique has meant that the pre-1981 values for snow cover that were previously reported are too high. Thus the magnitude of the decrease in Northern Hemisphere snow cover reported in the previous edition of this report for the period of early 1970s to late 1980s needs to be revised downwards. The IPCC now concludes that Northern Hemisphere snow cover has decreased by around 2 x 10^6 km^2 (or 8 per cent of the total area) as opposed to the 3 x 10^6 km^2 estimated previously (IPCC, 1992).

Monthly mean snow cover data, based on the NOAA

a) Annual mean snow cover

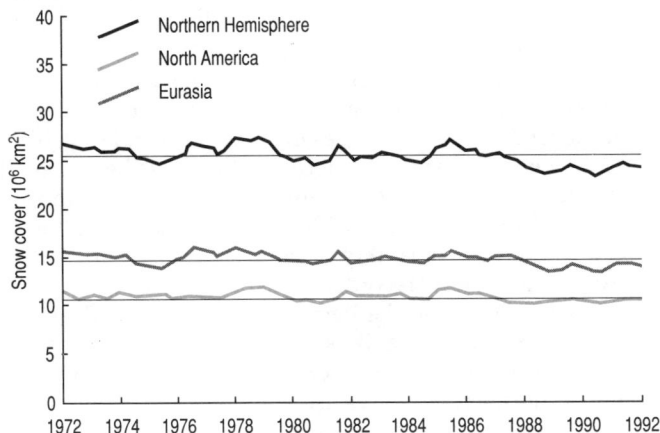

b) Spring mean snow cover

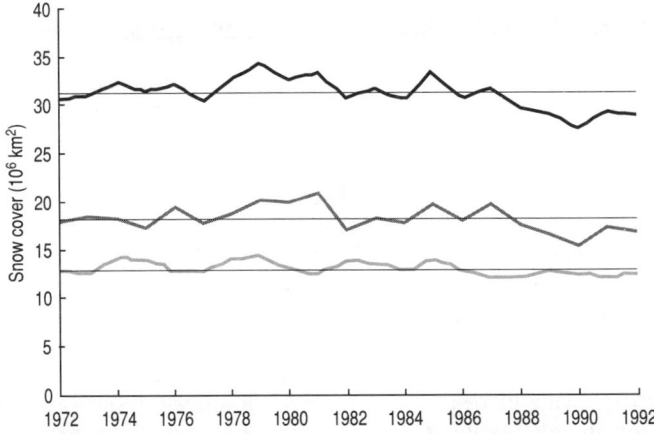

FIGURE 2.5 Trends in (a) annual mean and (b) spring mean snow cover in the Northern Hemisphere, 1972–1992 (areal values are calculated from NOAA weekly charts using the Rutgers Routine – see Table 2.5)
Source: Figure provided by D. Robinson, Rutgers University, New Jersey, USA

charts, are updated to the end of 1992 in Table 2.5; trends in annual and spring mean snow cover since 1973 are illustrated in Figure 2.5. In addition to the overall reduction in Northern Hemisphere snow cover noted above, the record shows that spring snow cover has been particularly low in recent years. The lowest area of spring snow cover in the Northern Hemisphere was recorded in 1990 (Robinson, 1993).

Although the snow cover record is still too short to distinguish a possible greenhouse signal from natural variability (IPCC, 1992), an interesting correlation is noted between snow cover and surface air temperature anomalies. The low areal extent of Northern Hemisphere snow cover in March 1990, noted above, coincides with an anomalously warm period in the Northern Hemisphere; land-based temperature records indicate that parts of Siberia were more than 10°C warmer than normal. It is considered likely that the advection of warm air from the anomalously snow-free areas contributed to the relative warmth in the Northern Hemisphere during March 1990 (Parker and Jones, 1991).

References

Doake, C. S. M. and Vaughan, D. G. 1991 Rapid disintegration of the Wordie Ice Shelf in response to atmospheric warming, *Nature* (London), **350**, 328–336.

Haeberli, W. 1985 *Fluctuations of Glaciers 1975–1980, Volume IV*, International Association of Hydrological Sciences, Paris, and United Nations Educational, Scientific and Cultural Organization, Paris.

Haeberli, W. 1990 Glacier and permafrost signals of 20th century warming, *Annals of Glaciology*, **14**, 99–101.

Haeberli, W. 1993 Accelerated glacier and permafrost changes in the Alps, Paper presented at the International Conference on Mountain Environments in Changing Climates, Unpublished paper.

Haeberli, W., Bösch, H., Scherler, K., Ostrem, G. and Wallén, C. C. (Eds) 1989 *World Glacier Inventory: Status 1988*, International Association of Hydrological Sciences, Wallingford, United Nations Environment Programme, Nairobi and United Nations Educational, Scientific and Cultural Organization, Paris.

Haeberli, W. and Hoelzle, M. 1993 *Fluctuations of Glaciers 1985–1990, Volume VI*, International Association of Hydrological Sciences, Wallingford, United Nations Environment Programme, Nairobi, and United Nations Educational, Scientific and Cultural Organization, Paris.

Haeberli, W. and Müller, P. 1988 *Fluctuations of Glaciers 1980–1985, Volume V*, International Association of Hydrological Sciences, Wallingford, United Nations Environment Programme, Nairobi, and United Nations Educational, Scientific and Cultural Organization, Paris.

IPCC 1990 *Climate Change: The IPCC Scientific Assessment of Climate Change*, Report prepared for the Intergovernmental Panel on Climate Change by Working Group I, J. T. Houghton, G. J. Jenkins and J. J. Ephraumus (Eds), Cambridge University Press, Cambridge.

IPCC 1992 *Climate Change 1992: The Supplementary Report to the IPCC Scientific Assessment*, Report prepared for the Intergovernmental Panel on Climate Change by Working Group I, J. T. Houghton, B. A. Callander and S. K. Varney (Eds), Cambridge University Press, Cambridge.

Jacobs, S. S. 1992 Is the Antarctic ice sheet growing? *Nature* (London), **360**, 29–33.

Kasser, P. 1967 *Fluctuations of Glaciers 1959–1965, Volume I*, International Association of Hydrological Sciences, Paris, and United Nations Educational, Scientific and Cultural Organization, Paris.

Kasser, P. 1973 *Fluctuations of Glaciers 1965–1970, Volume II*, International Association of Hydrological Sciences, Paris, and United Nations Educational, Scientific and Cultural Organization, Paris.

Morgan, V. I., Goodwin, I. D., Etheridge, D. M. and Wookey, C. W. 1991 Evidence from Antarctic ice cores for recent increases in snow accumulation, *Nature* (London), **354**, 58–60.

Müller, F. 1977 *Fluctuations of Glaciers 1970–1975, Volume III*, International Association of Hydrological Sciences, Paris, and United Nations Educational, Scientific and Cultural Organization, Paris.

Parker, D. E. and Jones, P. D. 1991 Global warmth in 1990, *Weather*, **46**, 302–311.

Robinson, D. A. 1993 Recent trends in Northern Hemisphere snow cover. In: *Proceedings of the Fourth Symposium on Global Change Studies*, American Meteorological Society, California, 329–334.

UNEP 1991 *United Nations Environment Programme Environmental Data Report 1991/92*, Basil Blackwell, Oxford.

Walsh, J. E. 1991 The Arctic as a bell-wether, *Nature* (London), **352**, 19–20.

WGMS 1991 *Glacier Mass Balance Bulletin, Bulletin No. 1 (1988–1989)*, International Association of Hydrological Sciences, Wallingford, United Nations Environment Programme, Nairobi, and United Nations Educational, Scientific and Cultural Organization, Paris.

WGMS 1993 *Glacier Mass Balance Bulletin, Bulletin No. 2 (1990–1991)*, International Association of Hydrological Sciences, Wallingford,

United Nations Environment Programme, Nairobi, and United Nations Educational, Scientific and Cultural Organization, Paris.

Wood, F. B. 1988 Global alpine glacier trends, 1960s to 1980s, *Arctic and Alpine Research*, **20**(4), 404–413.

Zwally, H. J. 1989 Growth of Greenland ice sheet: interpretation, *Science*, **246**, 1589–1591.

Zwally, H. J. 1991 Break-up of Antarctic ice, *Nature* (London), **350**, 274.

The Oceans

Ongoing routine ocean observing and monitoring activities have been briefly outlined earlier in this chapter. The major shortcomings of the current ocean observing systems – in terms of the lack of detailed data – were also noted. Consequently, currently available observational records of oceanic parameters, with the exception of those of SST (which are covered elsewhere in this chapter) and possibly global mean sea level (summarized below), have limited application to climate change detection studies. Nevertheless, a number of interesting and potentially significant changes in a range of other oceanic parameters have been reported in recent years, some of which are summarized here. In each case a brief overview of the possible implications for climate change is also provided.

Sea Level The IPCC concluded in 1990 that for a "business-as-usual" emissions scenario, global mean sea level is likely to rise by some 8–29 cm by the year 2030 (with a best estimate of 18 cm) in response to increased greenhouse gas radiative forcing. The prediction to the year 2070 is a rise in the range of 21–71 cm with a best guess of 44 cm. Thermal expansion of the oceans and increased melting of mountain glaciers and small ice caps are considered to be the principal contributors to the increase in global mean sea level (IPCC, 1990).

Since publication of the IPCC's first assessment, there has been considerable speculation in the scientific community regarding the most probable response of the polar ice caps to global warming, in particular with respect to their influence on sea level (Scheider, 1992). It had been generally considered that the Antarctic and Greenland ice sheets would make relatively little contribution to global sea-level rise, at least during the next century. If anything, the likelihood of increased snow accumulation associated with warming in Antarctica is expected to reduce sea level slightly (IPCC, 1990). However, in light of recent evidence of thickening of polar ice caps, especially of the Greenland ice sheet (see Snow and Ice), some workers have predicted that a build-up of land ice in Greenland could exert a major influence on global sea level in a high CO_2 world (Miller and de Vernal, 1992). Furthermore, a detailed study of the sensitivity of mountain glaciers and small ice caps has recently estimated that for a 1 K warming, glacier mass balance is likely to decrease by 0.4 m per year. This corresponds to a sea-level rise of 0.58 mm per year, a value which is significantly less than earlier estimates (Oerlemans and Fortuin, 1992).

In view of the above findings it is possible that predictions of future sea level rise may be revised downwards from the earlier IPCC estimates. However, at the time of writing no scientific consensus regarding this point has been reached.

Relative to other oceanic variables, sea level is fairly well monitored; a global network of around 1,300 tide gauges currently provides data on sea level. Tide gauges actually measure "relative sea level", i.e., they measure sea level in relation to a fixed benchmark and thus record the variations in sea level due to real changes in ocean level and to vertical land movements. In order to derive a globally-coherent trend in sea level, the non-climatic influences must be removed from the observational record of sea level. The difficulties in doing so limit the reliability of the resulting signal.

Available sea level records, some of which date back to the 1880s, are archived at the Permanent Service for Mean Sea Level (PSMSL) located at Bidston Observatory, UK. Records exist for approximately 800 stations world-wide. However, many of these records are discontinuous and less than 20 years in length. In addition, sea-level records are strongly biased towards coastal regions of Europe, North America and Japan; parts of Africa, Asia and the polar regions are particularly poorly represented. These shortcomings limit the usefulness of the existing data base of tide gauge records for determining the global trends in sea-level rise. Efforts to improve the monitoring of sea level through the expansion of IOC's GLOSS under the auspices of the proposed GOOS have been outlined earlier in this chapter.

The PSMSL data set has formed the basis of a number of attempts to quantify the rate of change of global mean sea level over the past 100 years. Published values for global mean sea-level rise over the last 50–100 years, corrected for vertical land movements, vary from about 1–3 mm per year, with formal uncertainties ranging from 0.15–0.90 mm per year (Douglas, 1991). The scatter in results is attributed to the differences in the methods used to analyse the raw data, in particular with respect to station selection, geographical grouping and averaging of station records and corrections for land movements. Several analyses of this type have been reviewed in previous editions of this report, including those of Barnett (1988) and Peltier and Tushingham (1989).

More recent analyses of sea-level rise have been reported by Douglas (1991) and Trupin and Wahr (1990). The former is based on a small, strictly selected sub-set of tide gauge records that yield highly consistent measurements of sea-level trends. Station records are selected on the grounds of length (minimum length of record is 60 years), data completeness (at least 80 per cent) and agreement with other stations. Stations located at or near convergent tectonic plate boundaries are also excluded from the analysis in order to minimize the influence of obvious tectonic effects in the calculation of global mean sea-level change. Only 21 tide gauge station records met these strict selection criteria. When corrected for post-glacial rebound (by applying the ICE-39 model developed by Tushingham and Peltier (1991)), these data yield an estimated rate of global mean sea-level rise of 1.8 ± 0.1 mm a^{-1}.

A similar result is obtained by Trupin and Wahr (1990). Using data from 84 stations having a minimum record length of 37 years, a value of 1.75 ± 0.13 mm a^{-1} is obtained. These more recent results are in agreement with earlier estimates and add support to the 1990 IPCC conclusion that over the past 100 years the rate of sea-level rise has averaged 1–2 mm per year (IPCC, 1990). As a final point, it is interesting to note that contrary to some of the earlier studies, the more recent analysis of Douglas (1991) finds no suggestion of an acceleration of global mean sea-level rise in recent decades. This too is in support of the 1990 IPCC findings.

Ocean Circulation For conceptual convenience, the two forms of ocean circulation are generally distinguished: the deeper ocean or thermohaline circulation (which is driven by changes in water temperature or salinity) and the largely wind-driven horizontal circulation which operates within the upper few hundred metres of the ocean surface.

The main features of the modern-day thermohaline ocean circulation are illustrated in Figure 2.6. In the tropical Atlantic, solar heating and excess evaporation over precipitation create an upper layer of relatively warm, saline water. As this warm, saline surface water flows northwards into the North Atlantic, it loses heat to the atmosphere (through evaporation) and consequently becomes colder and more saline. In parts of the North Atlantic (for example, the Labrador and Norwegian-Greenland Seas) and especially in the winter months, the combination of low temperatures and high salinity is sufficient to make this body of water more dense than the water below it. At this point convection sets in and the water sinks to depths greater than 1,000 m. This cold, dense water then moves southwards as North Atlantic "intermediate" and "deep" water, where it is joined by cold, dense water generated in the Southern Ocean by a similar mechanism. Deep water from these two sources spreads slowly throughout the world's oceans where it is gradually warmed (by mixing with warmer water above) and forced upwards by new supplies of cold water which push under it. Eventually, this body of water is returned to the surface of the ocean where it can once more be warmed directly by the sun and move up into the North Atlantic. Water which is moved into the most distant parts of the ocean, the North Pacific, can take about 1,000 years to complete this cycle. This global system of ocean circulation is sometimes referred to as the "conveyor belt circulation".

The northwards movement of surface warm water from the tropics to the North Atlantic followed by cooling, sinking and then flowing south is the principal mechanism by which the ocean transports heat from the equator to the poles. The quantity of heat involved is estimated to be about 1 PW, i.e., of comparable magnitude to that transported by the atmosphere.

In the recent decade, the North Atlantic thermohaline circulation and the formation of North Atlantic Deep Water (NADW) have been the focus of a great deal of interest. Calculations using simple global ocean models have shown that the overturn in the North Atlantic may be sensitive to relatively small changes in the strength and location of surface fresh-water inputs. Increased precipitation, melting of the Greenland ice cap and/or changes in the way that relatively low salinity water from the Arctic Ocean passes into the North Atlantic all have the capacity to lower surface salinity such that cooling alone will not be sufficient to produce water dense enough to sink to the intermediate and deep parts of the ocean (Stewart, 1991; IPCC, 1992). Effectively, the "NADW conveyor belt" is switched off. If this were to happen warm saline tropical waters would cease to flow into the sub-Arctic North Atlantic and the region would become colder. Moreover,

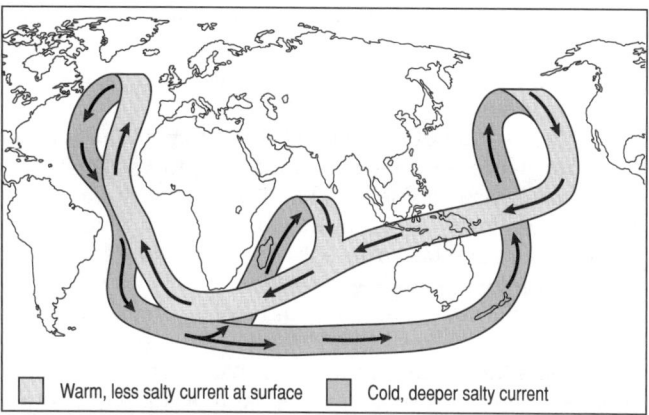

Warm, less salty current at surface   Cold, deeper salty current

FIGURE 2.6 Schematic representation of the thermohaline circulation of the ocean system, present day
Source: Adapted from Broecker, 1987

model results indicate that the breakdown of the circulation can occur relatively quickly, i.e., within a few decades (IPCC, 1992).

The potential instability of NADW formation suggested by ocean models is to some extent mirrored by observations. The palaeoclimatic observational record in particular has provided some useful clues about past fluctuations in North Atlantic ocean circulation and climatic conditions.

Of particular interest is the discovery of sudden oscillations or "flips" in climatic conditions in the North Atlantic region during the period 14,000–8,000 years BP, that is, as the earth was emerging from the grip of the last ice age. Analysis of Greenland ice cores and other palaeo records has suggested that the glacial-interglacial transition was punctuated by episodes of cold conditions, each lasting for a couple of centuries. The most severe and well documented of these is the Younger Dryas event which occurred around 11,000 years BP and during which the North Atlantic returned "briefly" to near glacial conditions. Some scientists have attributed the Younger Dryas event to a temporary breakdown in the North Atlantic thermohaline circulation, possibly triggered off by the increased influx of melt waters from the North American Laurentide Ice Sheet (Broecker et al., 1989).

The observational records find both for and against these theories of on-off modes of North Atlantic thermohaline circulation (Zahn, 1992). Whereas recent analyses of geochemical and plankton data imply a rapid shut-down of circulation during the Younger Dryas in the North Atlantic (Lehman and Keigwin, 1992), benthic isotope data from the same region indicate that North Atlantic convection occurred throughout the period in question and was as vigorous as it is today (Veum et al., 1992). In order to resolve this apparent discrepancy it has been suggested that, rather than switching on and off, the North Atlantic thermohaline circulation merely operates at a shallower depth or the sites of greatest convection shifts between different locations (Zahn, 1992).

The possibility of a collapse of the North Atlantic thermohaline circulation has a number of implications; evidence suggests that such changes can lead to very rapid alterations in North Atlantic climates – this may place stress on the functioning of the biosphere. Moreover, as formation of NADW is an important mixing mechanism in the oceans, any shifts in the rate of overturn may affect the capacity of the oceans to absorb CO_2.

The functioning of the thermohaline circulation is also of direct relevance to modellers attempting to predict future climates. Current GCMs simulate the sinking in the North Atlantic reasonably well; on the basis of such models it is predicted that the effect of an enhanced greenhouse warming is to reduce the rate of deep water formation as a result of a freshening (the dominant factor) and cooling of surface waters in the North Atlantic (IPCC, 1992).

Evidence for a reduction in the formation of deep water in the Greenland sea during the 1980s has indeed been reported (Schlosser et al., 1991). Other recently published data – this time based on the CONVEX-91 survey, a British contribution to the WOCE – has indicated that, relative to 1962, waters at intermediate depths between Greenland and the UK were cooler (by 0.08°C on average) and less saline (i.e., fresher) in 1991 (Read and Gould, 1992). Contrary to model expectations (described above), Read and Gould's measurements indicate that cooling is the dominant factor; if this is the case a renewed convection of the North Atlantic waters would result.

It is clear that a considerable amount of further work is needed in order to resolve the present unknowns in the modes of operation of the deep ocean circulation, especially in the all-important region of the North Atlantic. A more complete knowledge of what the ocean is actually like now will help to generate improved models of the ocean; it will also help to characterize how the ocean is changing. This is the prime objective of the WOCE.

References

Barnett, T. P. 1988 Global sea level change. In: *Climate Variations over the Past Century and the Greenhouse Effect*, A report based on the First Climate Trends Workshop, 7–9 September 1988, Washington DC.

Broecker, W. S. 1987 Unpleasant surprises in the greenhouse? *Nature*, (London), **328**, 123–126.

Broecker, W. S., Kennett, J. P., Flower, B. P., Teller, J. T., Trumbore, S., Bonani, G. and Wolfli, W. 1989 Routing of meltwater from the Laurentide ice sheet during the Younger Dryas episode, *Nature* (London), **341**, 318–322.

Douglas, B. C. 1991 Global sea level rise, *Journal of Geophysical Research*, **96**(C4), 6981–6992.

IPCC 1990 *Climate Change: The IPCC Scientific Assessment of Climate Change*, Report prepared for the Intergovernmental Panel on Climate Change by Working Group I, J. T. Houghton, G. J. Jenkins and J. J. Ephraumus (Eds), Cambridge University Press, Cambridge.

IPCC 1992 *Climate Change 1992: The Supplementary Report to the IPCC Scientific Assessment*, Report prepared for the Intergovernmental Panel on Climate Change by Working Group I, J. T. Houghton, B. A. Callander and S. K. Varney (Eds), Cambridge University Press, Cambridge.

Lehman, S. J. and Keigwin, L. D. 1992 Sudden changes in North Atlantic circulation during the last deglaciation, *Nature* (London), **356**, 757–762.

Miller, G. H. and de Vernal, A. 1992 Will greenhouse warming lead to Northern Hemisphere ice sheet growth?, *Nature* (London), **355**, 244–246.

Oerlemans, J. and Fortuin, J. P. F. 1992 Sensitivity of glaciers and small ice caps to greenhouse warming, *Science*, **258**, 115–117.

Peltier, W. R. and Tushingham, A. M. 1989 Global sea-level rise and the greenhouse effect: might they be connected?, *Science*, **244**, 806–810.

Read, J. F. and Gould, W. J. 1992 Cooling and freshening of the subpolar North Atlantic Ocean since the 1960s, *Nature* (London), **360**, 55–57.

Schlosser, P., Bönisch, G., Rhein, M. and Bayer, R. 1991 Reduction of deep water formation in the Greenland Sea during the 1980s: evidence from tracer data, *Science*, **251**, 1054–1056.

Schneider, S. 1992 Will sea levels rise or fall, *Nature* (London), **356**, 11–12.

Stewart, R. W. 1991 The ocean and climate, *International Marine Science Newsletter*, **55/56**, 4–7.

Trupin, A. and Wahr, J. 1990 Spectroscopic analysis of global tide gauge and sea level data, *Geophysical Journal International*, **100**, 441–453.

Tushingham, A. M. and Peltier, W. R. 1991 ICE-3G: a new global model of late Pleistocene deglaciation based on geophysical predictions of post glacial relative sea level change, *Journal of Geophysical Research*, **96**, 4497–4523.

Veum, T., Jansen, E., Arnold, M., Beyer, I. and Deplessy, J-C. 1992 Water mass exchange between the North Atlantic and the Norwegian Sea during the past 28,000 years, *Nature* (London), **356**, 783–785.

Zahn, R. 1992 Deep ocean circulation puzzle, *Nature* (London), **356**, 744–746.

TABLE 2.1 Trends in hemispheric and global mean annual surface air temperatures derived from land-based records, 1851–1992 (deviation in degrees Celsius from a reference period mean, 1951–1970)

Year	NH	SH	Global	Year	NH	SH	Global
1851	−0.08			1900	0.00	−0.06	−0.03
1852	−0.26			1901	−0.03	−0.14	−0.09
1853	−0.22			1902	−0.29	−0.15	−0.22
1854	0.01			1903	−0.26	−0.35	−0.31
1855	−0.39			1904	−0.36	−0.56	−0.46
1856	−0.42			1905	−0.24	−0.38	−0.31
1857	−0.35			1906	−0.09	−0.26	−0.17
1858	−0.27	−0.25	−0.26	1907	−0.52	−0.44	−0.48
1859	0.08	−0.37	−0.15	1908	−0.35	−0.35	−0.35
				1909	−0.31	−0.26	−0.28
1860	−0.55	−0.20	−0.38	1910	−0.24	−0.21	−0.22
1861	−0.30	−0.37	−0.34	1911	−0.27	−0.41	−0.34
1862	−0.74	−0.47	−0.61	1912	−0.42	−0.14	−0.28
1863	0.24	−0.84	−0.30	1913	−0.27	−0.12	−0.20
1864	−0.57	−0.31	−0.44	1914	−0.04	−0.04	−0.04
1865	−0.19	−0.04	−0.12	1915	−0.01	0.02	0.00
1866	−0.16	−0.13	−0.14	1916	−0.25	−0.20	−0.23
1867	−0.47	−0.16	−0.31	1917	−0.55	−0.38	−0.47
1868	−0.11	−0.23	−0.17	1918	−0.33	−0.36	−0.35
1869	0.04	−0.26	−0.11	1919	−0.27	−0.07	−0.17
1870	−0.34	−0.34	−0.34	1920	−0.14	−0.21	−0.17
1871	−0.50	−0.56	−0.53	1921	0.07	−0.25	−0.09
1872	−0.05	−0.35	−0.20	1922	−0.08	−0.18	−0.13
1873	−0.21	−0.28	−0.25	1923	−0.04	−0.29	−0.16
1874	−0.05	−0.75	−0.40	1924	−0.10	−0.24	−0.17
1875	−0.73	−0.47	−0.60	1925	0.04	−0.25	−0.11
1876	−0.30	−0.29	−0.30	1926	0.14	0.06	0.10
1877	−0.04	0.06	0.01	1927	0.03	−0.09	−0.03
1878	0.27	−0.15	−0.21	1928	0.07	−0.04	0.02
1879	−0.31	−0.51	−0.41	1929	−0.26	−0.20	−0.23
1880	−0.34	−0.17	−0.25	1930	0.13	−0.20	−0.04
1881	−0.36	−0.26	−0.31	1931	0.19	−0.08	0.05
1882	−0.23	−0.32	−0.28	1932	0.11	−0.03	0.04
1883	−0.48	−0.29	−0.38	1933	−0.18	−0.09	−0.14
1884	−0.62	−0.55	−0.59	1934	0.16	−0.12	0.02
1885	−0.51	−0.42	−0.47	1935	0.05	−0.13	−0.04
1886	−0.46	−0.35	−0.41	1936	0.06	−0.08	−0.01
1887	−0.38	−0.44	−0.41	1937	0.24	−0.07	0.09
1888	−0.46	−0.29	−0.38	1938	0.37	−0.04	0.17
1889	−0.28	−0.17	−0.23	1939	0.21	−0.18	0.01
1890	−0.27	−0.45	−0.36	1940	0.16	0.07	0.12
1891	−0.41	−0.47	−0.44	1941	0.18	0.08	0.13
1892	−0.45	−0.35	−0.40	1942	0.10	0.03	0.06
1893	−0.54	−0.38	−0.46	1943	0.12	−0.12	0.00
1894	−0.24	−0.41	−0.32	1944	0.25	0.11	0.18
1895	−0.44	−0.36	−0.40	1945	−0.04	0.05	0.01
1896	−0.26	−0.10	−0.18	1946	0.17	−0.04	0.07
1897	−0.20	−0.15	−0.18	1947	0.21	0.06	0.13
1898	−0.22	−0.46	−0.34	1948	0.15	−0.05	0.05
1899	−0.06	−0.24	−0.15	1949	0.09	−0.07	0.01

Continued

TABLE 2.1 Continued

Year	NH	SH	Global	Year	NH	SH	Global
1950	−0.18	−0.18	−0.18	1970	−0.01	0.17	0.08
1951	0.03	−0.07	−0.02	1971	−0.12	−0.04	−0.08
1952	0.10	0.08	0.09	1972	−0.29	0.15	−0.07
1953	0.31	0.05	0.18	1973	0.23	0.22	0.23
1954	−0.05	−0.09	−0.07	1974	−0.17	−0.04	−0.11
1955	−0.02	−0.14	−0.08	1975	0.08	−0.05	0.01
1956	−0.37	−0.23	−0.30	1976	−0.29	−0.22	−0.25
1957	−0.01	0.14	0.06	1977	0.19	0.15	0.17
1958	0.18	0.08	0.13	1978	0.01	0.05	0.03
1959	0.09	0.02	0.06	1979	0.14	0.09	0.12
1960	0.10	−0.16	−0.03	1980	0.18	0.32	0.25
1961	0.10	0.15	0.13	1981	0.52	0.28	0.40
1962	0.11	−0.02	0.05	1982	0.07	0.03	0.05
1963	0.13	−0.07	0.03	1983	0.42	0.35	0.38
1964	−0.20	−0.21	−0.20	1984	0.03	0.13	0.08
1965	−0.20	−0.11	−0.16	1985	−0.04	0.16	0.06
1966	−0.03	−0.05	−0.04	1986	0.17	0.20	0.19
1967	−0.04	−0.01	−0.01	1987	0.35	0.36	0.36
1968	−0.13	−0.11	−0.12	1988	0.48	0.42	0.45
1969	−0.08	0.11	0.02	1989	0.41	0.21	0.31
				1990	0.73	0.34	0.53
				1991	0.52	0.47	0.50
				1992	0.24	0.15	0.19

NH = Northern Hemisphere.
SH = Southern Hemisphere.

The derivation of mean annual hemispheric and global temperature changes from individual meteorological station records is briefly outlined below. Further details of the analysis procedures employed can be found in the source documents (Jones et al., 1986a,b; Jones and Wigley, 1990; Jones and Briffa, 1992).

Time series of monthly mean temperatures (i.e., raw station data) suffer from inhomogeneity problems which, if uncorrected, can lead to biases in the temperature record. The main sources of discontinuities in temperature records include changes in station locations (from the 1940s onwards many stations moved from old city centre observatories to new airport locations), changes in observing practices and thermometer exposure, and changes in the surrounding station environment (for example, increasing urbanization can lead to warming trends in records of affected sites). Additional problems arise due to variations in site elevation and differences in the techniques used to calculate monthly mean temperatures between stations.

Over 3,000 station records (2,666 from the Northern Hemisphere and 610 from the Southern Hemisphere) were tested for signs of inhomogeneity by comparing each station's record with that of stations situated within a range of a few hundred kilometres. Jumps or trends in the temperatures recorded at one station that were not observed at neighbouring stations were taken to indicate inhomogeneity. Where possible, correction factors were applied to account for such biases; other records were rejected and subsequently excluded from the analysis.

At the remaining stations (1,584 in the Northern Hemisphere and 293 in the Southern Hemisphere), monthly and annual average temperatures were expressed as deviations from their average temperatures over a reference period (the period 1951–1970 was selected as the reference). Expressing temperatures as deviations from a mean overcomes some of the anomalies associated with changes in station numbers throughout the historical record, variations in site elevations and differences in methods used to calculate monthly mean temperatures between stations.

Finally, station data were processed to produce area-averages, initially on a 5° latitude by 10° longitude grid, and subsequently to give hemispheric and global averages. Interpolating station data onto a grid in this way prevents areas with high station densities from unduly influencing the hemispheric averages.

Sources and references:
Data supplied by P.D. Jones, Climatic Research Unit, University of East Anglia, UK.
Jones, P. D. and Briffa, K. R. 1992 Twentieth century global surface air temperature variations, Part I, *The Holocene*, **2**(2), 165–179.
Jones, P. D. and Wigley, T. M. L. 1990 Global warming trends, *Scientific American*, **263**(2), 84–91.
Jones, P. D., Raper, S. C. B., Bradley, R. S., Daiz, H. F., Kelly, P. H. and Wigley, T. M. C. 1986a Northern Hemisphere surface air temperature variations 1851–1984, *Journal of Climate and Applied Meteorology*, **25**, 161–179.
Jones, P. D., Raper, S. C. B. and Wigley, T. M. L. 1986b Southern Hemisphere surface air temperature variations, 1851–1984, *Journal of Climate and Applied Meteorology*, **25**, 1123–1230.

TABLE 2.2   Trends in global and hemispheric mean annual surface air temperatures derived from both land and sea-based records, 1854–1992 (deviations in degrees Celsius from a reference period mean, 1950–1979)

Year	NH	SH	Global	Year	NH	SH	Global
				1900	−0.06	−0.05	−0.05
				1901	−0.07	−0.21	−0.14
				1902	−0.31	−0.19	−0.25
				1903	−0.35	−0.42	−0.38
1854	−0.16	−0.07	−0.12	1904	−0.44	−0.50	−0.47
1855	−0.39	−0.46	−0.43	1905	−0.28	−0.34	−0.31
1856	−0.37	−0.19	−0.28	1906	−0.19	−0.26	−0.22
1857	−0.42	−0.53	−0.47	1907	−0.48	−0.35	−0.41
1858	−0.33	−0.43	−0.38	1908	−0.44	−0.44	−0.44
1859	−0.10	−0.26	−0.18	1909	−0.39	−0.33	−0.36
1860	−0.48	−0.29	−0.38	1910	−0.39	−0.32	−0.35
1861	−0.33	−0.57	−0.45	1911	−0.40	−0.40	−0.40
1862	−0.76	−0.41	−0.58	1912	−0.44	−0.18	−0.31
1863	−0.13	−0.33	−0.23	1913	−0.44	−0.22	−0.33
1864	−0.43	−0.49	−0.46	1914	−0.21	−0.12	−0.16
1865	−0.22	−0.24	−0.23	1915	−0.03	−0.07	−0.05
1866	−0.19	−0.20	−0.20	1916	−0.33	−0.25	−0.29
1867	−0.27	−0.19	−0.23	1917	−0.55	−0.38	−0.47
1868	−0.16	−0.37	−0.27	1918	−0.42	−0.29	−0.35
1869	−0.08	−0.34	−0.21	1919	−0.35	−0.14	−0.25
1870	−0.22	−0.39	−0.31	1920	−0.28	−0.17	−0.23
1871	−0.35	−0.38	−0.37	1921	−0.17	−0.24	−0.20
1872	−0.21	−0.36	−0.28	1922	−0.28	−0.29	−0.28
1873	−0.25	−0.35	−0.30	1923	−0.22	−0.27	−0.25
1874	−0.28	−0.50	−0.39	1924	−0.25	−0.33	−0.29
1875	−0.57	−0.32	−0.44	1925	−0.11	−0.24	−0.17
1876	−0.37	−0.42	−0.40	1926	−0.05	−0.06	0.00
1877	−0.06	−0.17	−0.12	1927	−0.06	−0.20	−0.13
1878	0.17	−0.13	0.02	1928	−0.08	−0.20	−0.14
1879	−0.28	−0.42	−0.35	1929	−0.32	−0.32	−0.32
1880	−0.36	−0.28	−0.32	1930	−0.02	−0.25	−0.14
1881	−0.35	−0.23	−0.29	1931	0.04	−0.11	−0.04
1882	−0.32	−0.27	−0.30	1932	−0.04	−0.17	−0.11
1883	−0.42	−0.29	−0.35	1933	−0.26	−0.19	−0.22
1884	−0.53	−0.35	−0.44	1934	−0.09	−0.13	−0.11
1885	−0.44	−0.32	−0.38	1935	−0.12	−0.19	−0.16
1886	−0.37	−0.16	−0.27	1936	−0.07	−0.18	−0.13
1887	−0.34	−0.45	−0.40	1937	0.09	−0.08	0.01
1888	−0.28	−0.35	−0.31	1938	0.17	−0.02	0.08
1889	−0.23	−0.11	−0.17	1939	0.04	−0.10	−0.03
1890	−0.38	−0.41	−0.40	1940	−0.01	0.06	0.03
1891	−0.32	−0.36	−0.34	1941	0.03	0.06	0.04
1892	−0.45	−0.30	−0.38	1942	0.08	0.02	0.05
1893	−0.51	−0.34	−0.43	1943	0.10	−0.10	0.00
1894	−0.36	−0.33	−0.35	1944	0.21	0.12	0.16
1895	−0.41	−0.21	−0.31	1945	0.02	0.07	0.05
1896	−0.19	−0.08	−0.14	1946	0.07	−0.18	−0.06
1897	−0.15	−0.05	−0.10	1947	0.04	−0.11	−0.04
1898	−0.28	−0.34	−0.31	1948	0.05	−0.15	−0.05
1899	−0.16	−0.20	−0.18	1949	0.02	−0.17	−0.08

TABLE 2.2 Continued

Year	NH	SH	Global
1950	−0.11	−0.17	−0.14
1951	0.07	−0.10	−0.02
1952	0.09	0.06	0.08
1953	0.24	0.02	0.13
1954	−0.03	−0.19	−0.11
1955	−0.03	−0.25	−0.14
1956	−0.24	−0.20	−0.22
1957	0.05	0.12	0.09
1958	0.17	0.04	0.11
1959	0.11	0.00	0.05
1960	0.08	−0.11	−0.02
1961	0.12	0.12	0.12
1962	0.19	0.03	0.11
1963	0.21	0.06	0.13
1964	−0.15	−0.20	−0.18
1965	−0.15	−0.08	−0.12
1966	0.00	−0.07	−0.04
1967	0.05	−0.07	−0.01
1968	−0.03	−0.09	−0.06
1969	0.00	0.12	0.06

Year	NH	SH	Global
1970	0.00	0.09	0.05
1971	−0.13	−0.03	−0.08
1972	−0.19	0.23	0.02
1973	0.10	0.19	0.14
1974	−0.22	0.01	−0.10
1975	−0.09	−0.08	−0.09
1976	−0.29	−0.14	−0.22
1977	0.08	0.14	0.11
1978	0.01	0.08	0.04
1979	0.07	0.13	0.10
1980	0.14	0.24	0.19
1981	0.33	0.20	0.27
1982	0.04	0.13	0.09
1983	0.25	0.34	0.30
1984	0.00	0.24	0.12
1985	−0.03	0.20	0.09
1986	0.11	0.23	0.17
1987	0.23	0.36	0.30
1988	0.34	0.34	0.34
1989	0.25	0.24	0.25
1990	0.47	0.31	0.39
1991	0.35	0.32	0.34
1992	0.13	0.24	0.19

NH = Northern Hemisphere.
SH = Southern Hemisphere.

Sea-surface temperature records (SST) were examined for possible sources of bias and discontinuities in much the same way as land-based temperature records, as outlined in the notes to Table 2.1. In the case of SST the principal problems arise due to changes in the method of sampling of sea water. Before 1940, SST measurements were routinely made by hauling water onto the deck in a canvas bucket and recording the temperature, typically after a few minutes of standing by which time the water had cooled slightly as a result of evaporation. Other types of bucket, for example, wooden and plastic buckets, were also used. Since the early 1940s, most measurements of SST have been made by thermometers inserted in engine-cooling water intake pipes.

Comparative studies have indicated that bucket measurements give temperatures that are typically between 0.3–0.7°C cooler than those recorded in intake pipes. By estimating evaporative cooling rates on a monthly basis and the elapsed time between sampling and reading, corrections are applied to the older bucket measurements. The situation is, however, further complicated by the uncertainty surrounding the type of bucket used to sample water, particularly in the older (pre-1900) data. Further details relating to the allowances made for variations in evaporative cooling rates of different bucket types are given by Jones et al. (1991).

Corrected absolute monthly mean SSTs are then expressed as deviations from a reference period mean. Land and marine data sets are subsequently combined on to a 5° latitude by 5° longitude grid to give area averages from which hemispheric and global average anomalies are then calculated. Where land and sea occur in the same 5° x 5° grid box, the grid value is an average of the land and marine components. In the case of the combined land and marine gridded data sets, deviations are with respect to the 1950–1979 reference period mean.

Sources and references:
Data supplied by P. D. Jones, Climatic Research Unit, University of East Anglia, Norwich, UK.
Jones, P. D., Wigley, T. M. L. and Farmer, G. 1991 Marine and land temperature sets: a comparison and look at recent trends. In: *Greenhouse Gas-Induced Climate Change*, M. E. Schlesinger (Ed.), Elsevier Science Publishers B.V., Amsterdam, 153–172.

TABLE 2.3 Trends in hemispheric and global mean annual tropospheric and stratospheric temperatures, 1958–1992 (deviations in degrees Celsius from a reference period mean)

Year	Troposphere (850–300 mb)			Lower stratosphere (100–50 mb)		
	NH	SH	Global	NH	SH	Global
1958	0.37	0.23	0.30			
1959	0.45	0.22	0.33			
1960	0.12	0.20	0.16			
1961	0.22	0.16	0.19			
1962	0.08	0.02	0.05			
1963	0.04	−0.15	−0.05			
1964	−0.08	−0.36	−0.22		0.99	
1965	−0.36	−0.39	−0.37		0.82	
1966	0.09	−0.05	0.02		−0.43	
1967	0.04	−0.03	0.01		0.43	
1968	−0.20	−0.16	−0.18		−0.11	
1969	0.02	0.23	0.12		−0.13	
1970	0.12	0.16	0.14	0.19	0.02	0.10
1971	−0.45	−0.27	−0.36	0.03	−0.05	−0.02
1972	−0.27	0.13	−0.07	−0.22	−0.09	−0.16
1973	0.16	0.48	0.32	−0.26	−0.28	−0.27
1974	−0.29	0.07	−0.11	−0.11	−0.34	−0.22
1975	−0.09	−0.06	−0.07	−0.05	−0.20	−0.12
1976	−0.43	−0.16	−0.29	0.33	0.11	0.22
1977	0.11	0.40	0.25	0.11	−0.38	−0.14
1978	−0.02	0.27	0.13	−0.46	−0.04	−0.25
1979	0.11	0.47	0.29	0.03	−0.19	−0.08
1980	0.16	0.59	0.37	0.11	−0.43	−0.15
1981	0.25	0.45	0.35	−0.11	−0.16	−0.14
1982	−0.05	0.37	0.16	0.44	−0.02	0.22
1983	0.39	0.63	0.51	0.03	0.29	0.16
1984	−0.04	0.25	0.11	−0.44	−0.38	−0.41
1985	−0.25	0.34	0.05	−0.42	−1.75	−1.08
1986	0.09	0.27	0.18	−0.73	−1.41	−1.07
1987	0.26	0.67	0.47	−0.65	−1.80	−1.22
1988	0.48	0.58	0.53	−1.05	−1.63	−1.34
1989	0.20	0.27	0.23	−0.65	−1.58	−1.12
1990	0.48	0.61	0.55	−1.02	−1.81	−1.41
1991	0.39	0.60	0.49	−0.45	−1.21	−0.83
1992	−0.14	0.05	−0.05	−0.29	−0.55	−0.42

NH = Northern Hemisphere.
SH = Southern Hemisphere.

Upper air temperatures given derived from measurements of the geopotential heights at the 850 mb, 350 mb, 100 mb and 50 mb pressure surfaces; the difference in height of two pressure surfaces (i.e., thickness) is related to the mean temperature of the layer between the two surfaces. The thickness of the 850–300 mb layer provides a measure of tropospheric temperatures, and the 100–50 mb layer, the lower stratosphere. Geopotential heights were obtained from a network of 63 radiosonde stations. Monthly data were averaged to provide seasonal (DJF, MAM, JJA, SON) and 12-month (December to November) means from which seasonal and 12-month mean temperatures were calculated. Seasonal and 12-month mean temperatures were then expressed as deviations from a long-term reference period mean (1958–1977). Individual station deviations were subsequently averaged (with equal weighting) to obtain seasonal and 12-month mean temperatures for seven climatic zones; the North and South Polar (60°–90°), the North and South Temperate (30°–60°), the North and South Sub-tropic (10°–30°) and the Equatorial 10°S–10°N). Hemispheric averages are based on a weighted average of the polar, temperate, sub-polar and equatorial zones and the global mean temperature deviations on the average for the two hemispheres.

An analysis of seasonal data completeness during the period 1970–1988 for the seven climatic zones indicated a reasonable degree of completeness (i.e., more than 90 per cent of data available) for all zones except the Equatorial and South Sub-tropic in the troposphere; in the lower stratosphere only the North Temperate and North Sup-tropic achieved a better than 90 per cent level of data completeness. In the Southern Hemisphere zones up to 25 per cent of the seasonal values were missing at this altitude. Stations with missing seasonal data were excluded from the calculation of seasonal and 12-month mean deviations over the climatic zones, hemispheres and the globe. Owing to the lack of height data for the 50 mb pressure surface from the former USSR prior to 1970, Northern Hemisphere and global average deviations are reported from 1970 onwards only. For further details relating to the these data please refer to Angell (1991).

Sources and references:
Data supplied by J. K. Angell, Air Resources Laboratory, National Oceanic and Atmospheric Administration, Maryland, USA.
Angell, J. K. 1991 Changes in tropospheric and stratospheric global temperatures, 1958–1988. In: *Greenhouse Gas-Induced Climatic Change: A Critical Appraisal of Simulations and Observations*, M. E. Schlesinger (Ed.), Elsevier Science Publishers B.V., Amsterdam, 231–247.

TABLE 2.4 Fluctuations in annual mass balance for selected glaciers, 1960–1991 (mm H$_2$O equivalent)

Region/country	Glacier	1960/1	1961/2	1962/3	1963/4	1964/5	1965/6	1966/7	1967/8	1968/9	1969/70	1970/1	1971/2	1972/3
AFRICA														
Kenya	Lewis													
NORTH AMERICA														
Canada	Helm													
	Place					−680	114	−1,213	−131	−210	−1,510	−343	−344	−299
USA	Gulkana						80	−90	−480	−760	440	−170	−300	50
	South Cascade	−1,100	200	−1,300	1,200	−170	−1,030	−630	10	−730	−1,200	600	1,430	−1,040
	Wolverine					0	−270	−1,670	−290	−90	1,940	770	−980	950
ASIA														
China	Urumqhie S. No. 1	−33	−167	234	2	374	−374	−70	−456	148	−313	102	262	−708
EUROPE														
Austria	Hintereisferner	−205	−696	−603	−1,244	925	344	20	338	−431	−552	−600	−74	−1,229
	Kesselwandferner	271	−416	−406	−537	1,040	594	299	464	−152	0	46	368	−380
	Sonblickkees	148	54	−1,426	−932	1,976	736	160	236	−247	144	−392	128	−721
	Vernagtferner					751	633	83	301	−307	−225	−424	137	−460
France	Sarennes	−390	−910	190	−1,830	30	420	−410	340	−360	−410	−1,100	−370	−870
	Saint Sorlin	−390	−940	50	−1,760	280	660	−730	600	220	−220	−1,280	−590	−1,030
Italy	Careser							−390	260	0	−630	−650	400	−1,280
Norway	Álfotbreen			−1,100	280	480	−1,610	1,280	950	−2,170	−1,230	940	−110	2,180
	Austre Brøggerbreen						−650	−100	−930	−540	−580	−310	−80	
	Engabreen										−990	1,010	−70	2,720
	Gråsubreen		770	−710	320	410	−290	710	−80	−1,300	−660	470	−640	890
	Midtre Lovénbreen								−30	−840	−530	−460	−220	−20
	Nigardsbreen		2,250	−220	950	910	−920	2,160	220	−1,310	−560	820	−140	1,100
	Storbreen	−520	720	−1,180	210	340	−610	720	50	−1,420	−720	180	−310	80
Sweden	Rabotsglaciär													
	Riukojietna													
	Storglaciären	−1,100	320	−190	490	430	−530	−230	−100	−1,040	−1,520	−190	−1,050	50
	Tarfalaglaciären													
Switzerland	Gries		−890	30	−660	510	−280	20	410	470	−540	−970	380	−1,050
	Silvretta	590	−350	−870	−1,260	1,340	1,310	440	650	10	210	−470	−160	−1,130
USSR														
Kazakhstan	Ts. Tuyuksuyskiy[a]	−560	−690	440	520	−50	40	230	−230	210	110	−360	130	−290
Kyrgyzstan	Kara–Batkak[b]	−806	−86	−40	144	−41	−156	10	−652	0	−182	148	53	−753
	Golubin									−130	150	−90	−109	−391
Russia	Djankuat								100	−1,090	410	−230	−1,140	−280
	Kozelskiy													580
	Leviy Aktru													
	Maliy Aktru		−400	−340	−280	−560	−380	290	−20	290	120	250	70	100
	No. 125[c]													
	Praviy Aktru													
Tajikistan	Abramov								−140	840	−110	−890	530	−1,000

[a] Formerly known as the Tuyuksu glacier.
[b] Formerly known as the Karabatkak glacier.
[c] Formerly known as the Vodopadniy glacier.

Data presented in the above table refer to annual specific glacier mass balance which is defined as the difference between glacier accumulation and ablation (i.e., melting) averaged over the area of the glacier. Balances are expressed as water equivalent in mm. A negative value indicates a net loss of glacier mass, and a positive value a net gain.

No distinction has been made between stratigraphic and fixed date systems of measurement of glacier mass balance. Fixed date measurement allows a true

1973/4	1974/5	1975/6	1976/7	1977/8	1978/9	1979/80	1980/1	1981/2	1982/3	1983/4	1984/5	1985/6	1986/7	1987/8	1988/9	1989/90	1990/1
					−70	−1,750	−1,210	−370	−720	−900	−950	−680	−770	−2,030	770	−1,010	−810
								−340	−209	−320	−1,728	−1,334	−788	−150	−1,670	−1,790	−2,239
562	−238	877	−1,227	−433	−2,212	−923	−1,093	−754	−441	−337	−1,882	−1,312	−845	−969	−1,040	−938	−990
−1,090	−370	−1,020	−180	−80	−570	−60	40	−110	30	−310	690	60	−120	−210	−920	−600	−110
1,020	−50	950	−1,300	−380	−1,560	−1,020	−840	80	−770	120	−1,200	−710	−2,560	1,640	−710	−730	−100
−1,180	380	−600	2,020	1,000	−1,020	2,860	1,510	−380	0	−480	340	−180	1,190	1,000	−1,960	−2,510	−410
−125	288	29	180	−110	−84	−335	−652	−45	100	−83	−612	−669	−176	−642	106	52	−706
55	65	−314	760	411	−219	−50	−173	−1,240	−580	32	−574	−732	−717	−945	−637	−995	−1,325
573	369	−37	701	423	66	162	161	−620	−182	178	−8	−494	−243	−265	−151	−242	−849
576	397	79	148	833	224	834	414	−1,282	−535	338	−281	−1,432	−525	−711	252	−561	−818
235	171	76	352	288	44	140	−55	−845	−537	20	−112	−808	−290	−497	−312	−568	−1080
−1,600	110	−2,070	990	550	−110	320	40	−100	−70	−40	−1,210	−1,790	−920	−690	−2,590	−2,140	−1,360
−1,090	−10	−1,910	1,090	950	280	600	310	−1,020	−340	−120	−120	−1,730	−810	310	−2,490	−1,400	−1,270
−320	170	−270	990	80	−180	10	−840	−1,680	−790	−590	−760	−1,140	−1,640	1,010	−820	−1,580	−1,730
1,030	1,210	1,530	−560	−510	−130	−610	220	−130	1,600	1,320	−560	−410	2,070	−2,480	2,930	1,790	790
−920	−310	−450	−110	−560	−710	−520	−550	−40	−270	−730	−550	−320	220	−520	−450	−660	130
800	1,610	2,410	880	−510	420	−500	980	840	1,060	1,050	−900	250	940	−1,790	3,170	850	690
340	950	−1,000	−390	−220	40	−1,890	−190	−510	−50	−370	0	−760	720	−580	450	730	−520
−890	−210	−350	−40	−480	−660	−430	−460	20	−170	−680	−480	−210	240	−490	−240	−510	100
480	270	400	−770	−130	710	−1,220	310	−420	1,090	340	−220	−100	1,480	−890	3,470	1,770	200
240	−150	−90	−540	−440	−100	−1,310	−100	−470	200	−300	−400	−320	320	−950	1,200	1,250	−150
								−70	−110	−510	−1,280	−188	223	1,052	618	−40	−216
												−540	−260	−910	890	210	80
−340	1,170	270	200	−80	−210	−1,270	−190	260	280	120	−720	−60	480	−840	1,240	590	170
												0	900	−1,290	1,230	50	160
−30	280	−990	1,290	970	−880	660	−350	−910	−580	−70	−1,210	−690	−940	−1,100	−1,040	−1,890	−1,480
730	730	−350	600	1,010	−50	1,090	350	−290	−530	360	510	−270	−210	−580	−250	−530	−1,130
−620	−450	−720	−1,100	−1,480	−520	−630	110	−630	−540	−1,250	−550	−520	−340	−610	−460	−960	−1,100
−51	−475	−841	−864	−1,176	−501	−364	−447	−784	−948	−1,572	−1,292	−392	−682	−456	−390	−778	−398
−462	−240	−400	−320	−460	−289	−500	78	−272	−222	−485	−557	−372	152	−432	−422	−590	−722
240	−910	290	−370	440	−310	380	−910	420	−970	210	−380	−500	1,540	520	40	340	−310
1,640	−50	1,330	−880	−940	−200	0	−1,950	60	−250	−340	2,020	−1,660	−300	−1,940	−740	−1,280	460
			240	−370	−680	−20	−320	−410	280	320	150	60	200	370	30	150	
−1,470	400	680	490	−410	−580	110	−310	−660	150	310	240	40	170	470	220	70	
		80	−190	−780	−120	−10	−310	160	290	210	120	180	160	100	150		
							250	−330	−310	330	330	290	150	230	310	190	70
−800	−740	−1,020	−1,470	−1,310	−390	−1,050	70	−770	−410	−1,680	−810	−1,010	240	−10	−220	−540	−420

annual mass balance to be calculated (i.e., annual accumulation and annual ablation) while stratigraphic measurement (sum of winter balance and summer balance) yields a net balance which straddles the calendar year (plotted in the forward year). Although this limits comparability between glaciers for individual years it is less important when considering longer time trends. The absolute accuracy of mass balance observations averaged over the surface of a glacier is difficult to assess, it being dependent on a number of constantly changing physical glacier parameters.

Source:
Data supplied by W. Haeberli, World Glacier Monitoring Service, Swiss Federal Institute of Technology, Zurich, Switzerland.

Table 2.5 Monthly and average snow cover in the Northern Hemisphere, 1972–1993 (10^6 km^2)

Year	Jan.	Feb.	Mar.	Apr.	May	Jun.	Jul.	Aug.	Sep.	Oct.	Nov.	Dec.	Annual
						Northern Hemisphere							
1972	46.6	48.5	40.6	31.4	19.9	11.3	5.6	4.8	7.9	21.6	35.6	44.4	26.5
1973	45.8	46.3	40.1	31.4	21.3	12.3	5.8	4.2	5.9	18.4	37.2	43.9	26.0
1974	45.7	45.1	40.8	32.0	24.1	13.1	5.5	4.1	4.7	19.4	30.5	39.6	25.4
1975	44.8	44.6	41.1	31.8	20.6	11.5	5.3	3.7	5.6	17.0	30.9	42.1	24.9
1976	45.4	45.3	41.6	31.2	23.2	14.4	6.7	4.7	5.6	26.1	34.6	42.0	26.7
1977	48.5	44.1	40.0	30.9	20.4	12.1	6.5	4.3	7.6	19.4	31.5	44.6	25.8
1978	49.4	51.0	43.7	30.5	23.1	15.6	8.0	5.7	7.1	18.5	32.2	43.9	27.4
1979	49.7	47.5	43.6	35.3	23.3	12.7	6.7	4.6	5.6	14.9	28.3	37.6	25.8
1980	45.6	48.2	42.7	33.9	21.3	11.3	6.0	5.1	5.8	14.3	31.7	37.5	25.3
1981	41.7	43.5	42.7	34.1	23.1	13.8	6.4	5.1	4.9	17.0	33.7	43.5	25.8
1982	48.0	46.0	40.7	31.9	19.3	9.3	3.7	3.4	6.0	18.8	34.4	43.0	25.4
1983	46.6	46.9	41.3	31.5	21.9	11.4	5.2	3.9	5.4	17.5	32.8	44.2	25.7
1984	46.2	45.0	41.5	30.9	18.8	8.7	3.9	2.8	4.2	17.6	33.7	42.2	24.6
1985	49.8	48.3	44.1	32.8	22.9	14.0	5.5	3.4	4.8	17.5	37.9	46.0	27.3
1986	46.1	47.8	39.9	30.9	20.9	11.3	4.9	3.5	6.7	17.6	33.9	41.5	25.4
1987	47.4	45.6	43.4	32.2	18.8	12.5	5.8	3.1	5.4	13.5	32.4	42.2	25.2
1988	47.1	45.2	39.7	29.3	19.3	8.4	3.8	2.6	4.4	13.0	31.6	41.0	23.8
1989	45.7	44.0	38.6	30.1	18.4	9.6	4.4	2.6	5.8	16.9	32.7	44.6	24.5
1990	45.4	43.2	37.0	28.2	17.4	7.3	3.4	2.6	3.9	15.4	30.4	43.9	23.2
1991	46.0	45.7	39.6	28.8	19.2	11.0	4.0	3.5	4.4	15.6	34.3	42.4	24.5
1992	45.8	43.2	38.4	28.9	19.0	10.3	3.9	2.6	5.9	16.9	34.0	45.3	24.5
1993	46.5	44.9	39.7										
						Eurasia							
1972	30.0	31.8	25.5	17.9	10.1	4.8	1.1	1.0	3.0	12.4	22.5	27.3	15.6
1973	29.2	29.6	25.5	18.0	11.7	5.3	1.5	0.5	1.9	10.5	22.8	27.4	15.3
1974	28.5	28.6	24.7	17.7	12.3	6.8	1.6	1.1	1.2	9.8	18.3	23.7	14.5
1975	27.4	27.3	24.3	16.8	10.4	5.1	1.9	0.7	1.6	8.5	17.5	25.7	13.9
1976	28.3	28.6	25.8	18.9	13.0	6.9	2.1	1.4	1.8	17.4	21.7	26.1	16.0
1977	31.1	28.2	24.2	18.0	10.8	4.9	1.9	1.1	3.5	11.7	19.2	28.2	15.2
1978	31.1	32.2	26.6	16.9	12.3	7.6	2.3	1.4	2.3	10.4	18.5	26.4	15.7
1979	31.2	28.8	27.1	20.2	12.7	5.6	1.7	1.0	1.7	8.4	16.9	22.8	14.8
1980	28.2	30.1	26.6	20.6	12.1	5.3	1.6	1.2	1.7	7.6	19.1	23.0	14.8
1981	25.8	27.8	28.1	21.2	13.0	6.9	1.9	0.9	1.3	8.8	21.2	26.9	15.3
1982	29.5	28.1	24.5	17.2	9.4	3.6	0.8	0.7	1.9	11.0	20.4	26.8	14.5
1983	29.3	30.0	25.3	17.6	11.3	4.7	1.6	1.2	1.7	9.4	19.4	26.2	14.8
1984	27.9	28.3	25.2	18.4	9.9	3.5	0.9	0.4	0.8	9.6	20.1	25.6	14.2
1985	31.1	30.1	27.9	18.9	11.7	6.1	1.1	0.4	0.9	9.6	23.1	28.1	15.7
1986	29.0	30.0	24.8	17.7	11.1	4.4	1.0	0.4	2.2	8.9	19.6	26.0	14.6
1987	30.2	29.3	27.6	20.5	10.3	6.1	1.5	0.4	1.8	7.1	20.8	26.2	15.1
1988	29.0	28.2	25.3	16.9	10.2	2.9	0.7	0.2	1.1	5.8	18.4	25.4	13.7
1989	29.1	26.7	22.9	16.7	9.5	4.2	1.5	0.4	2.0	9.2	19.3	27.2	14.1
1990	28.7	26.7	22.1	15.6	8.4	2.3	0.5	0.3	0.6	7.7	18.3	27.1	13.2
1991	28.6	29.7	24.7	16.5	10.4	4.8	0.8	0.7	0.6	7.2	19.6	26.5	14.2
1992	29.1	27.5	23.8	16.2	9.3	4.1	0.6	0.1	1.5	8.5	20.0	27.5	14.0
1993	28.7	27.2	23.9										

Snow cover charts are derived on a weekly basis from a visual analysis of satellite imagery (visible wavelengths) by trained meteorologists. Snow cover is plotted on charts using a 1:50,000,000 polar stenographic projection of the Northern Hemisphere. Each map is subsequently digitized; digitized snow charts are used to compute monthly, seasonal and annual average snow cover according to the Rutgers Routine devised by Professor Robinson at Rutgers University.

Source:
Data supplied by D. A. Robinson, Department of Geography, Rutgers University, New Brunswick, New Jersey, USA (May 1993).

Part 3

Natural Resources

It is widely asserted that the world's stock of renewable resources is currently being consumed at an unsustainable rate. Material presented in this chapter of the 'UNEP Environmental Data Report' generally supports this view; in particular, the following trends in the status of natural resources are highlighted:

○ According to preliminary findings, the rate of deforestation in the tropics averaged about 1 per cent per year during the 1980s; this may be sufficient to commit between 2–8 per cent of the planet's species to extinction within 25 years.

○ Approximately 17 per cent of the world's soils are considered degraded as a result of human activities.

○ A dozen or so countries are currently consuming almost all or even more than their total annual supply of renewable fresh waters.

○ On average, increased agricultural productivity over the past three decades has provided for a better fed world; however, in parts of Africa serious problems of food shortages and famine still prevail.

○ Some of the world's major fisheries are currently being fished at unsustainable rates.

○ Although whale catches have been successfully limited in recent years, catches of selected species of smaller cetaceans which are currently unregulated are increasing.

○ Despite rapid growth in the number and extent of nationally and internationally-designated protected areas in recent decades, less than 5 per cent of the world's land area is currently protected.

Human activities are currently placing natural resources under severe stress; furthermore, it is estimated that demands on natural resources could easily double in the next two decades if populations and per capita incomes grow at their projected rates. Paradoxically, it is the overuse of renewable resources such as land, vegetation, water and wildlife, rather than the consumption of non-renewable natural resources, that is perceived as the more immediate threat to the sustainability of economic and human development.

The unsustainability of present levels of natural resource consumption was a major theme of the recent UN Conference on Environment and Development (UNCED). The major outcome of the Conference, known as Agenda 21, lays out in some considerable detail recommendations for moving towards the goal of sustainable natural resource use. This goal requires a shift in thinking from a traditional preservationist approach to resource management, to one that involves a more integrated, holistic perspective. This shift in emphasis is already occurring and is illustrated here in a short discussion of forestry management.

Whereas the following chapter in this report (Part 4: Population and Development) explores some of the linkages between population, development, resource consumption and environment, the aim of this present chapter is to provide an overview of the status of, and trends in, the condition of the natural resource base. Note that this chapter refers to renewable natural resources only, grouped under the headings: land resources, agriculture and food production, forests, fresh-water resources, fisheries and wildlife resources. Non-renewable energy and mineral resources are covered in Part 6: Energy and Part 7: Industry and Transport, respectively.

Two key data sources form the basis of much of the core information on renewable resources presented here. Since its inception, the Food and Agriculture Organization of the United Nations (FAO) has had the responsibility for the compilation of statistics relating to a number of key resources including land cover/use, agriculture, forestry and fisheries, and is now recognized as the principal source of global-scale data of this type. The collection of statistical and related information pertaining to the world's wildlife resources is the remit of the World Conservation Monitoring Centre (WCMC), based in Cambridge, UK. In this edition of the 'UNEP Environmental Data Report', core information is supplemented by a summary of a recently published study sponsored by the United Nations Environment Programme (UNEP) of the status of soil degradation, and the preliminary findings of the FAO's second decadal assessment of forest resources. A review of UNEP's recent work in the assessment of the status of desertification is also provided.

Land Resources

The problem of degradation of land resources has received much attention in recent decades. In particular, UNEP has devoted considerable effort to the study of desertification,

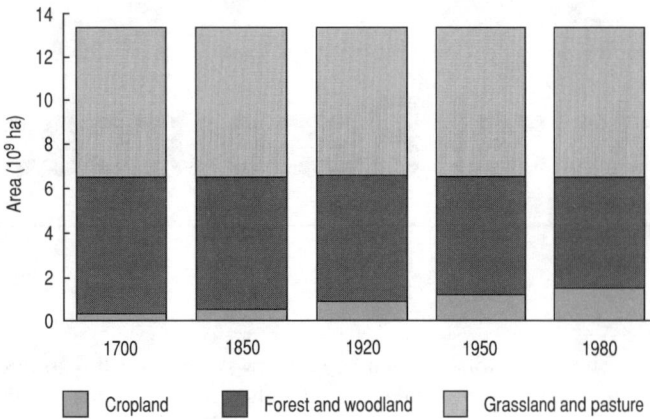

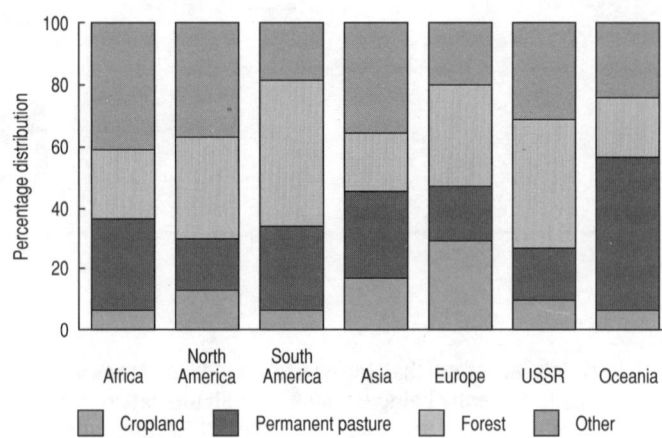

FIGURE 3.1 Long-term changes in global land use, 1700–1980
Source: After Richards, 1990

FIGURE 3.2 Percentage distribution of land use in major world regions, 1988–1990 (mean annual values)
Source: Data from Table 3.1

producing its first assessment of global desertification for the 1977 UN Conference on Desertification (UNCOD). Since that time UNEP has been entrusted with the task of co-ordinating and following up the implementation of the Plan of Action to Combat Desertification (PACD) drawn up at UNCOD. Progress towards PACD goals was reviewed as part of the run-up to UNCED, following which the UN has set up an Intergovernmental Negotiating Committee for the elaboration of an international convention to combat desertification in those countries experiencing drought and/or desertification (see also Part 10: International Co-operation).

Land Cover and Use

The broad patterns of changes in land use and cover that have occurred since the advent of modern industrial times are reasonably well characterized. As illustrated in Figure 3.1, global-scale changes in land use and cover are dominated by the increase in cultivated land; it is estimated that since the 1700s some 1.2×10^9 ha of land have been brought into cultivation (Richards, 1990). This level of expansion has not occurred evenly throughout the world; the Americas, Southeast Asia and the former USSR have experienced greater rates of conversion to agricultural land than the world as a whole. In North America, for example, the cropland area increased by 6,666 per cent during the period 1700–1980 (Richards, 1990). These changes in land cover have largely been driven by real and perceived needs for expanded agricultural production (see also Food and Agricultural Resources below).

The expansion of cropland has largely been at the expense of forested land. According to one estimate, global forest cover has decreased by some 0.9×10^9 ha (of which 0.7×10^9 ha represents loss of closed forest) since pre-agricultural times (Meyer and Turner, 1992). In contrast, the total world area of grassland and pasture has remained virtually constant since the 1700s (Figure 3.1). On a regional basis, however, marked net decreases in grasslands and pastures have occurred between 1700 and 1980 in Europe (27.4 per cent), North America (13.7 per cent) and South-east Asia (26.4 per cent), largely as a result

of conversion to cropland. Such decreases have been offset by increases in other world regions (Latin America and tropical Africa), thereby maintaining a rough balance in the area of grasslands over this period (Richards, 1990). The gains in Latin America have largely been driven by ranching and pasture development (Meyer and Turner, 1992).

On more recent time scales, the collection of detailed land-use statistics by the FAO allows national-level changes in land use to be assessed. Since the 1950s FAO has regularly published land-use data for four categories of land use, namely "cropland", "forest and woodland", "permanent pasture", and "other land" in their 'Production Yearbooks'. The latter category includes urban areas, unmanaged rangelands, wetlands, desert, tundra and all other unclassified land. Selected data of this type are presented in Table 3.1. Regionally aggregated land-use data reveal that at present Europe is the most intensively cultivated continent; almost 30 per cent of the land area is devoted to agriculture (Figure 3.2).

The FAO statistics suggest that the global expansion of cultivated land is continuing; all regions, except Europe, witnessed an increase in cultivated land area from 1968–1970 to 1988–1990. In Europe approximately 5 per cent of cropland has been lost to urban expansion, replaced by forest, or has been abandoned due to low productivity during this time period (Table 3.1).

Although the FAO data given here represent the most widely used global and national land-use/cover statistics, these data have a number of inherent limitations. Firstly, land-use data submitted by individual countries vary in quality and reliability (see also footnotes to Table 3.1). Secondly, the definitions of land-use categories employed by reporting countries vary considerably such that items classified under the same category often relate to differing kinds of land. The FAO class of "other land", for example, combines several important and distinct forms of land cover or use. Thus key ecosystems such as wetlands, and significant changes such as urban encroachment, are not reflected in these data and are therefore impossible to monitor and assess on a global scale.

Thirdly, although the FAO data document land-cover

changes in terms of conversion from one category to another, they provide little information on the modification of land-cover condition within a category. For example, intensification of cultivation (e.g., through the use of synthetic inputs and the planting of hybrid crops), forest thinning and overgrazing of grasslands, which can lead to significant changes in land cover condition, will not be registered as conversion. Furthermore, the data reflect only net changes so that variations in land-use patterns within a country may be concealed.

The scientific community and policy makers are beginning to request better land-use data to serve a myriad of purposes. For example, more precise land-cover data that have spatial congruence with socio-economic variables are required to support the study and assessment of the causes of land-cover change (Meyer and Turner, 1992). In addition, improved global data sets are needed as input for models that predict climate change.

In the past decade, development of satellite-remote sensing techniques has improved the means to monitor land-use changes on global, regional and national scales. Data obtained via Advanced Very High Resolution Radiometer (AVHRR) instruments have been found to be particularly appropriate for global land-cover studies. The strength of AVHRR lies in the fact that its spectral bands are well suited to measurement of terrestrial attributes, especially those relating to vegetation. The previous editions of this report have provided examples of successful use of satellite remote-sensing techniques for land-cover monitoring activities based on AVHRR (UNEP 1989, 1991).

The usefulness of the existing AVHRR data sets for global-scale research is nevertheless still hampered by a number of inherent shortcomings of the system. These include problems relating to instrument drift, poor global average and inadequate resolution. Initiatives are currently under way to address these problems. As part of the International Geosphere-Biosphere Programme (IGBP), researchers are working towards developing a global data set of the land surface with a spatial resolution of 1 km based on remote sensing with AVHRR instrumentation (IGBP, 1992). Through its Harmonization of Environmental Measurement (HEM) office, UNEP is co-ordinating an effort to develop a universal vegetation classification scheme to assist in the process of global mapping and monitoring (UNEP, 1993).

Land Degradation and Desertification

The term "land degradation" is generally used to signify a loss or reduction of land productivity as a result of human activity. Land degradation encompasses degradation of soil and/or vegetation cover and may be caused by a variety of anthropogenic pressures including deforestation, overgrazing, unsustainable agricultural practices and industrial activities. In practice, land degradation typically occurs as a result of a combination of these physical, chemical, biological and socio-economic factors, rather than as a result of any one single factor.

Desertification is a widely used but controversial term; to date more than 100 definitions have been observed in the scientific literature. Although still subject to considerable debate, a perception of desertification as land degradation in arid, semi-arid and dry sub-humid areas resulting mainly from adverse human impact is becoming generally accepted.

Accurate and reliable data on the current extent and severity of land degradation are scarce. Early attempts to assess the extent of desertification on a global scale, such as that produced by UNEP in 1977 and updated in 1984 (UNEP, 1987), have since been subject to some criticism. Nevertheless, reliable identification not only of the spatial extent of land degradation but also of the underlying causes, is essential if viable remedies to the problem are to be reached.

A first step towards this goal has been achieved with the completion of the three-year global study of soil degradation, co-ordinated by the International Soil Reference and Information Centre (ISRIC) with sponsorship from UNEP. The Global Assessment of Soil Degradation (GLASOD) provides the most comprehensive compilation of global information on the amount of soil degraded by human activities based on a consistent methodology that is available to date (Oldeman, 1988; Oldeman et al., 1991; UNEP, 1992). The principal output of the three-year study is a series of maps showing the extent, type and degree (or severity) of human-induced soil degradation world-wide that has occurred since World War II. The maps have since been digitized to enable the calculation of the areal extent of the various types of soil degradation. These data are presented in Tables 3.2a–f; the footnotes to this series of tables provide more detailed information regarding the concepts and definitions invoked in the GLASOD work and also the methodology employed.

A simplified version of the GLASOD world map of human induced soil degradation is given here as Figure 3.3. On the basis of the GLASOD work, it is estimated that 1.2×10^9 ha, representing 10.5 per cent of the world's vegetated surface, have been "moderately", "strongly", and "extremely" degraded since 1945. Of this, 9.3×10^6 ha (approximately the size of Italy) are classified as "extremely" degraded, i.e., the land is considered to be unsuitable for agriculture and impossible to restore. World-wide, an additional area of 749×10^6 ha is categorized as "lightly" degraded, i.e., soils can be restored through appropriate farm preservation practices (Table 3.2a).

On a continental basis, Europe suffers the most soil degradation in percentage terms; 16 per cent (158.3×10^6 ha) of its vegetated area falls into the categories "moderately", "strongly", and "extremely" degraded. The main forms of soil degradation are physical or chemical degradation. Moreover, Europe is the only region where industrial activities contribute significantly to soil deterioration (6 per cent). Chemical forms of soil degradation are discussed in more detail in Part 1: Environmental Pollution, Soils and Sediments.

In absolute terms, Africa and Asia account for the largest areas of degraded soils – 452.5×10^6 ha in Asia and 320.6×10^6 ha in Africa. In Africa almost 25×10^6 ha, farmed under low input agricultural systems, are degraded as a result of nutrient loss. Poor management of irrigation works has led to the salinization of over 10×10^6 ha. Livestock have brought extreme physical deterioration, primarily in parts of the Sahel and southern Africa. In Asia, water erosion is believed to be the principal agent responsible for the removal of nutrient-rich topsoil in western India, throughout the Himalayas, in South-east Asia

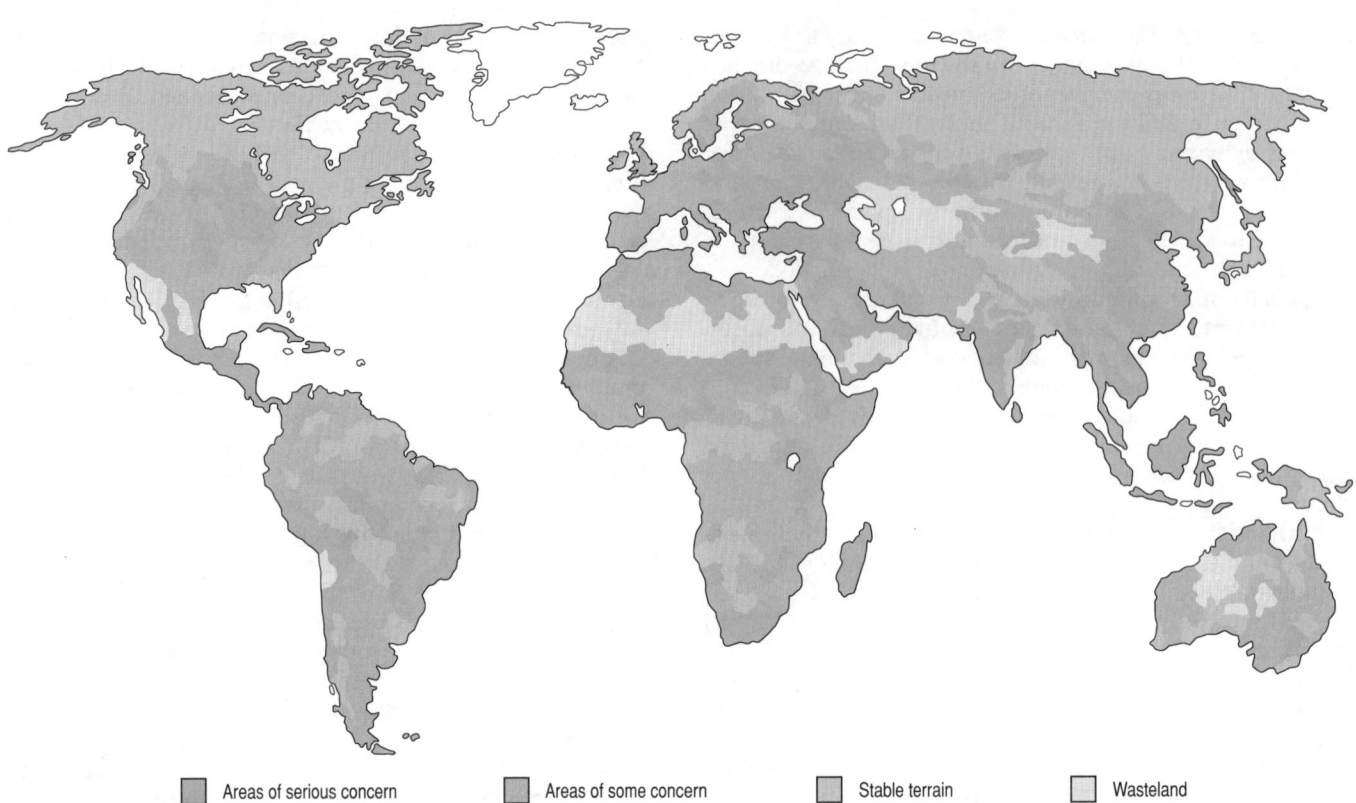

Areas of serious concern   Areas of some concern   Stable terrain   Wasteland

FIGURE 3.3 The extent and severity of human-induced soil degradation, current status (areas of "serious concern" refer to soil units with localized "severe" or "extreme" soil degradation or widespread "moderate" degradation; areas of "some concern" refer to soil units with localized "moderate" or "severe" degradation, or widespread "light" degradation)
Source: Figure provided by WRI, Washington DC

and in large areas of China. Soils in western China and Mongolia are affected by serious wind erosion. In northern India, Bangladesh, Myanmar, Thailand, Malaysia, Indonesia, northern China and the Korean peninsula, soils have become acidified and salinized, and have been losing nutrients over the past 45 years.

It must be stressed that the GLASOD work reported here represents only a first approximation of the status of soil degradation. Although GLASOD specifically deals with human-induced soil degradation, it is often difficult to distinguish between naturally- and human-induced processes. This problem is particularly acute in dryland areas. There thus remains a need for a more quantitative and scientifically acceptable assessment of the problem of soil degradation. The preparation of a more detailed map of the world's soil resources applicable to a national level analysis is currently under way; completion of this data set will facilitate a more accurate assessment of the status and risk of soil degradation. The World Soils and Terrain Digital Database (SOTER), a digitized map of soils and terrain at a scale of 1:1,000,000, is expected to be completed by the year 2000 if sufficient funding is provided.

The GLASOD data base has been used to evaluate the extent of desertification (i.e., land degradation in drylands) in UNEP's most recent global assessment of desertification (UNEP, 1992). In the absence of a suitable data set on

vegetation degradation (another important aspect of desertification), GLASOD data have been integrated with measures of vegetation production (the Global Vegetation Index derived from AVHRR satellite imagery) to show the general relationship between the degree of soil degradation and vegetation production. As the vegetation index does not provide any indication of the quality of vegetation, no causal links between vegetation and soil degradation can be inferred from this exercise. However, the map can be used to pin-point those areas of the world which are likely to be susceptible to land degradation and a number of tentative conclusions are drawn. For example, parts of north Africa, southern Africa, Arabia, Iran, Pakistan and the southern steppes of middle Asia, which have high-severity soil degradation and low biomass, are identified as susceptible to desertification. In addition, low to medium biomass values coupled with an albeit limited degree of soil degradation in parts of North America, Australia and the Argentine Pampas could indicate potential for further degradation by erosion processes, especially if land usage were to intensify (UNEP, 1992).

Despite its shortcomings, coarse resolution and the use of qualitative expert judgement rather than measured data, the GLASOD study and the recent UNEP assessment of desertification have been able to highlight areas of concern and potential problems for decision makers. These assessments send a signal to policy makers to alter short-term oriented land-use practices and to change the perception of land as a disposable resource which can be abandoned when the soils become too severely degraded.

References

IGBP 1992 *Improved Global Data for Land Applications: A Proposal for a New High Resolution Data Set*, Report No. 20, The International Geosphere-Biosphere Programme, Stockholm.

Meyer, W. B. and Turner II, B. L. 1992 Human population growth and global land-use/cover change, *Annual Review of Ecological Systems*, **23**, 39–61.

Oldeman, L. R. (Ed.) 1988 *Guidelines for General Assessment of the Status of Human-Induced Soil Degradation*, International Soil Reference and Information Centre, Wageningen, The Netherlands.

Oldeman, L. R., Hakkeling, R. T. A and Sombroek, W. G. 1991 *World Map of the Status of Human-Induced Soil Degradation: An Explanatory Note*, Rev. 2 Ed., International Soil Reference and Information Centre, Wageningen, The Netherlands.

Richards, J. F. 1990 Land transformation. In: *The Earth as Transformed by Human Action, Global and Regional Changes in the Biosphere over the Past 300 Years*, B. L. Turner II, W. C. Clark, R. W. Kates, J. F. Richards, J. Mathews, W. B. Meyer (Eds), Cambridge University Press with Clark University, Cambridge, USA, 163–178.

UNEP 1987 *United Nations Environment Programme Environmental Data Report 1987/88*, Basil Blackwell, Oxford.

UNEP 1989 *United Nations Environment Programme Environmental Data Report 1989/90*, Basil Blackwell, Oxford.

UNEP 1991 *United Nations Environment Programme Environmental Data Report 1991/92*, Basil Blackwell, Oxford.

UNEP 1992 *World Atlas of Desertification*, Edward Arnold, London.

UNEP 1993 *Vegetation Classification: Report of the UNEP-HEM/WCMC/GCTE Preparatory Meeting, Charlottesville, Virginia, USA, 24–26 January 1993*, GEMS Report Series No. 19, United Nations Environment Programme, Nairobi.

Food and Agricultural Resources

Agriculture, along with forestry and fishing, provides the food and many of the resources upon which the well-being of society depends. Furthermore, these sectors collectively provide a livelihood for approximately one half of the world's population. During the present century the agricultural sector in particular has undergone accelerated change. This agricultural development, i.e., the so-called Green Revolution, has occurred largely in response to world population growth and has operated through processes of intensification of agricultural production and expansion of agricultural land use. The facts that high input farming methods (while sustaining a growth in agricultural production) are undermining the ecological resource base upon which agriculture depends and are contributing to environmental degradation are, however, matters of prime concern at the present time.

In this section of the 'UNEP Environmental Data Report', global- and regional-scale patterns of agricultural and food production are reviewed. Agriculture is considered here more in terms of a resource (i.e., the capability of the agricultural sector to feed the growing world population) rather than as a "stressor" of the environment. Although the environmental impacts of fertilizer use will be covered in this section (see Box 3.1), other environmental implications of agricultural practices are covered in other sections of the report (see Part 1:

Environmental Pollution, Biomonitoring and Water Quality; Part 8: Wastes and Waste Management; and in this chapter, Fresh-water Resources.

Crop Production

The agricultural sector is generally well documented in a statistical sense. The FAO collates national data on crop production and yields, fertilizer use, irrigated land area, livestock numbers, tractor use and other parameters on a routine basis. Many of these data are published as annual yearbooks. Generally speaking, agricultural data are more reliable from the developed countries (i.e., those countries with commercially viable agricultural sectors).

The level of agricultural production may be measured using an index which represents the total value of all crop and livestock products (Table 3.3). On the basis of this index, it is estimated that over the past three decades, world agricultural production has doubled. This growth in agricultural production has been more marked in the developing countries (156 per cent increase) compared with that in the developed countries (56 per cent). Similar trends are evident in the food production index (Table 3.3). The growth in agricultural and food production has generally outpaced population growth; per capita agricultural production has grown in developed countries by 21 per cent, and in developing countries by 30 per cent since the early 1960s (Table 3.3).

Trends in the production of selected food commodities and livestock products in the world, developed and developing countries are compared in Table 3.4. In 1970, developing countries produced just under half of the world's cereals. Since then their cereal production has increased at a faster pace than in developed countries, such that at present, developing countries account for 55 per cent of the global total. Growth in cereal production in developing countries has been fuelled primarily by new varieties of wheat and rice, which are responsive to fertilizer inputs, irrigation and pesticides. Production growth for the coarse grains such as maize, barley, millet and sorghum has, however, been less marked.

Meat production has similarly increased at a faster rate in developing countries than in the more developed countries. Developed countries, however, still dominate meat production (Table 3.4), despite the fact that the largest share of the world's livestock population is found in the developing countries. This is attributed to the fact that much of the livestock in developing countries exist in small-scale and traditional farming systems where they are largely a source of subsistence and transport. Statistics on livestock populations can be found in the previous edition of this report (UNEP, 1991).

Despite the gains in production of cereals and other food crops achieved by developing countries, per capita production in developed countries still exceeds that in developing countries by a wide margin. Moreover, aggregated data for the world, developed and developing regions mask significant regional- and national-scale differences in the level of agricultural and food production. In terms of per capita food production, only Asia of the developing regions sustained gains during the 1970s and 1980s; the increase in per capita food and agricultural production is particularly noticeable in

BOX 3.1 Environmental Indicators for the Agricultural Sector: Fertilizer Use

It is a widely held belief that the adoption of high-input agricultural practices is responsible for many of the adverse environmental impacts currently associated with the agricultural sector. One of the main issues relates to the high application rates of artificial fertilizers. The excessive use of nitrogen-based mineral fertilizers is a particular cause of concern and has thus formed the focus of environmental indicator development for the agricultural sector.

According to one estimate fertilizers currently account for approximately one quarter of the total input of nitrogen to the global land surface (Jenkinson, 1990). Although the long-term implications of excessive fertilization of arable land and pasture have yet to be fully assessed, the short-term environmental effects of the accumulation of nitrates (NO_3^-) which are in excess of crop requirements in soils are well documented. The first and the most obvious is the leakage of excess NO_3^- (through run-off and soil leaching) to ground and surface waters, which in turn can lead to impairment of water quality and even eutrophication. There is little doubt that increased application of fertilizers, whether in the form of mineral fertilizers or manures, have been a major contributor to the rises in NO_3^- levels recorded in numerous fresh-water bodies (see Part 1: Environmental Pollution, Water Quality).

If excess soil NO_3^- is not removed by leaching, then there is potential for loss by denitrification. This process results in the release of nitrous oxide (N_2O), a greenhouse gas. The increasing use of nitrogen fertilizers is likely to have enhanced this otherwise natural process, thereby contributing significantly to the man-made flux of N_2O to the atmosphere (Jenkinson, 1990).

In devising its preliminary set of 24 environmental indicators, the Organisation for Economic Co-operation and Development (OECD) has proposed the inclusion of an indicator which reflects the intensity of use of nitrogenous fertilizers in agriculture; this parameter is expressed as the quantity of commercial fertilizer applied (in tonnes of active ingredient) per km² of cropland (OECD, 1991). Values for this indicator, which are based

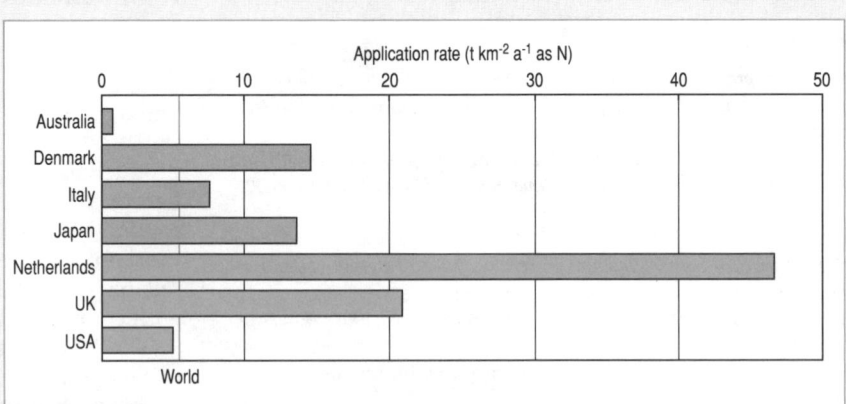

Use of nitrogenous fertilizers in selected OECD countries, 1988
Source: After OECD, 1991

on FAO data, are plotted here for a selection of OECD countries. The world average is also included for comparison.

It is generally accepted that this indicator, as a measure of the environmental pressure exerted by high-input agriculture, has a number of limitations and that further refinements are required. In the first instance, the amount of mineral fertilizer applied as an annual quantity is not necessarily directly related to the environmental impacts or risks outlined above; the magnitude of the associated impacts is heavily dependent on the method and timing of the application, and on the prevailing climatic conditions. Moreover, there is often a considerable time lag between fertilizer application and subsequent leaching (Pain et al., 1991). Secondly, the use of data on application rates of commercial fertilizers ignores possible nitrogen contributions from the application of organic wastes in the form of animal manures or sewage sludges.

The usefulness of this indicator is further impaired by an inherent limitation of the FAO data upon which it is drawn. The current FAO data do not necessarily reflect the actual rate of fertilizer use; the denominator in the calculation, i.e., the area of arable land and permanent cropland, refers to the physical area and not the sum of the areas given to each crop cultivated in

a given year. Thus in countries where double cropping is practised on a large scale, the fertilizer use per hectare of cropland will be overestimated. Overestimation of the fertilizer use rate will also occur in countries where significant quantities of fertilizers are applied to pastures, lawns and ornamental gardens; these do not constitute cropland (Zarqa, 1993). These factors may well account for the relatively high fertilizer application rates recorded for some countries in Table 3.5 (Zarqa, 1993).

Recognizing the limitations of the existing data compilations for fertilizer application rates, the FAO has recently ceased to publish its data in this form. Improved data will be published in a new publication, 'Fertilizer Use by Crops'.

References

Jenkinson, D. S. 1990 An introduction to the global nitrogen cycle, *Soil Use and Management*, **6**(2), 56–61.
OECD 1991 *Environmental Indicators: A Preliminary Set*, Organisation for Economic Co-operation and Development, Paris.
Pain, B., Jarvis, S. and Clements, B. 1991 Impact of agricultural practices on soil pollution, *Outlook on Agriculture*, **20**(3), 153–160.
Zarqa, S. 1993 Personal communication, Statistics Division, Food and Agriculture Organization of the United Nations, Rome.

China and the other centrally planned economies of this region. In Latin America and countries of the near East, food production is, on average, just managing to keep pace with population growth (FAO 1988, 1992).

The African region, especially sub-Saharan Africa, is in stark contrast to the rest of the world and faces a critical problem of producing adequate food supplies. Since the early 1970s, per capita food production has fallen in many countries of this region; this situation – aggravated by severe droughts and unsettled political conditions – reached crisis proportions in the mid-1980s with widespread famine.

Trends in food production are mirrored in regional and national trends in per capita food supplies (i.e., the availability of calories for direct human consumption). In the developing

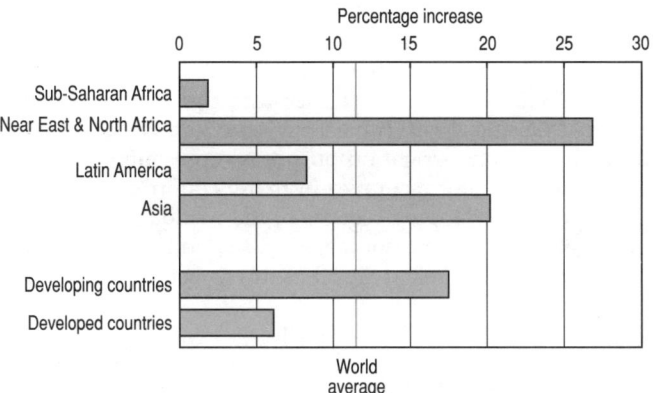

FIGURE 3.4 Percentage increases in per capita food supply (as measured by the availability of food for direct human consumption in 10^3 cal day^{-1}) by major world region, 1969–71 and 1987–89 (mean annual values)
Source: FAO, 1992

world as a whole, per capita food supply steadily increased in the 1980s, albeit at a lower rate than in the 1970s. The same is true of Asia, Latin America and northern Africa; in sub-Saharan Africa, however, food supply has hardly improved at all (Figure 3.4). In some countries of this region, per capita food supply actually declined in the late 1980s (FAO 1988, 1992).

It is encouraging to note that while some countries have experienced declines in per capita food supply, the number of countries which have reached the relatively comfortable level of more than 2,600 calories per day has progressively increased since the 1960s. China entered the "over 2,600 calories" class in the 1980s, so that one half of the world's developing population is now in this class. These changes are reflected in the table below:

Per capita food supply (10^3 cal day^{-1})	No. of countries		% of population	
	1969–71	1987–89	1969–71	1987–89
<2,000	25	12	43	6
2,000–2,600	89	62	52	44
>2,600	16	56	5	50

Source: FAO, 1992

From the above discussion it is evident that considerable progress in food and agriculture, in terms of growth of production, has been made over the past quarter of a century. These increases in crop production have been achieved as a result of the expansion of the area of arable land, higher cropping intensity and higher yields. The latter have been made possible by the use of high-input, intensive farming methods coupled with technological improvements and increased mechanization. Some of these factors are discussed in the next sub-section.

Aside from the production gains, two further important changes in the agricultural sector – which have had significant impacts on agricultural policy and its development – merit a mention here. The expansion of the so-called "livestock economy" in response to greater demands for meat and animal products, represents one such change. In order to raise meat output, livestock producers in the developed countries have adopted intensive, high-input rearing techniques which rely on grains to feed their animals. In fact, livestock currently consume 38 per cent of the global production of grain; in the developed countries this percentage is far higher, rising to as much as 70 per cent in the USA. A comprehensive discussion of the implications of the rise in livestock farming, and the associated environmental impacts (which include the generation and disposal of manure and methane emissions), is given by Durning and Brough (1991).

The increasing reliance on international trade, both as a source of food and as a means for the disposal of surplus output, represents a second fundamental change in the world's food system. In the early 1960s, world food imports amounted to 8 per cent of production; by the mid-1980s this proportion rose to around 12 per cent. The increase took place in both the developed and developing regions and was most pronounced during the 1970s. This surge in demand for food imports was, however, halted in the early 1980s as a consequence of a combination of factors, including reduced import demands by developing countries and increased domestic production by previously food-importing countries (e.g., China and India). As a consequence, the price of agricultural commodities on the international market fell dramatically, largely because of the heavy subsidies and protectionist trade agreements adopted by the developed market economies at this time (FAO, 1988).

The rapid growth in agricultural trade in the 1970s was fuelled, at least in part, by the expansion of agricultural production in the developed countries; this capacity expansion, while allowing the rapid increases in exports in the 1970s, has led to problems of over-production in more recent years. The process of retardation of the growth in agricultural production is currently causing some severe adjustment difficulties in many of these countries.

Developing countries too are suffering in the wake of the agricultural trade expansion of the 1970s. Although the volume of agricultural exports of developing countries continued to grow during the first half of the 1980s, the falling world prices offset most of the economic gains offered by the volume increase. Moreover, the high level of protection has made it increasingly difficult for developing countries to earn foreign income from agricultural exports. Problems have been exacerbated in many developing countries where farmers have been encouraged to grow high-earning cash crops for export. In some cases expansion of cash crops has driven food production and subsistence farming to marginal lands. These policies have led to imbalances in food supply and environmental degradation in some countries (FAO, 1988).

Agricultural Inputs

Although technological, economic and managerial factors have all contributed to the growth in agricultural production, it is the increase in the use of artificial fertilizers that is perceived by many to be the single most influential factor. Increased levels of fertilizer use, together with scientific advances and improvements in other agricultural inputs such

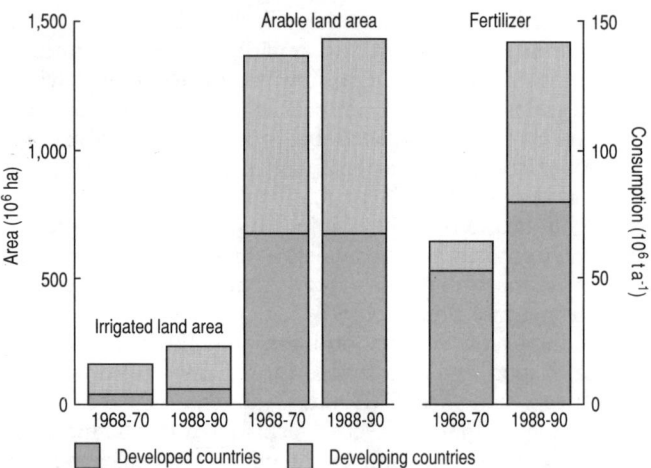

FIGURE 3.5 Increases in selected agricultural inputs in developed and developing countries, 1968–70 and 1988–90 (mean annual values)
Source: Data from Table 3.5

as developments in higher yielding crop varieties, irrigation schemes and pest control practices, are largely responsible for the significant gains in crop yields which have transformed agricultural productivity over the past three decades (FAO, 1988). Changes in selected agricultural inputs in the developed and the developing world are compared in Figure 3.5.

Fertilizers and Pesticides   Table 3.5 summarizes national data on the total consumption and application rate of mineral fertilizer as compiled by the FAO. Globally, fertilizer consumption has doubled over the last 20 years or so, peaking in 1988/89 at 145.7 x 10^6 t. In 1989/90, consumption fell to 143.3 x 10^6 t (FAO, 1992), the decline due to a drop in consumption in developed countries.

The strongest growth rates in fertilizer consumption in recent years have been sustained by the developing countries (Figure 3.5). Fertilizer application per hectare of cropland is also growing at a faster rate in developing countries than in developed countries. Asia saw the largest relative growth (almost 400 per cent), followed by South America (189 per cent) and Africa (100 per cent). Such a comparison, however, masks the dominance of a number of countries in the developing regions. For example – four countries – Egypt, South Africa, Nigeria and Morocco, consume 67 per cent of Africa's fertilizer (Table 3.5). Lack of financial resources, low crop prices, and poor distribution structures have constrained the use of higher rates of fertilizer in many other developing countries.

During the past three decades, the development of pesticides and herbicides has also had a significant role to play in the improvement of crop yields. Pesticide use directly improves crop yields by reducing crop losses due to pest invasions and indirectly improves them by facilitating the planting of larger fields with single crops, thereby increasing efficiency. Data on the application of pesticides in selected countries have been included in previous editions of the 'UNEP Environmental Data Report' (UNEP, 1991); pesticides will be discussed in greater detail in subsequent reports in this series.

Irrigation   Irrigation is considered by some to be the second most important factor responsible for the gains made in agricultural productivity since the advent of the Green Revolution. Globally, irrigated agriculture has expanded by a factor of 2.5 since 1949 (Table 3.5; Figure 3.5). The abstraction of fresh water for irrigation purposes represents the largest single use of water resources in many countries, especially those in the developing regions of the world. The demands of irrigation for water resources are discussed in greater detail in the section on Fresh-water Resources below.

Expansion of Arable Land   Historically, the expansion of agricultural land has been the main source of growth in agricultural production. Since the early 1900s in the developed countries and since the 1950s and 1960s in some developing countries, gains in productivity (i.e., higher yields) have increasingly become the prime source of the growth in agricultural output (Figure 3.5). Nevertheless, although the expansion of agricultural land may no longer be the prime source of growth in agricultural output, the area of land under cultivation continues to grow in most world regions (see also Land Cover/Use).

Currently agricultural production takes place on around 1.4 x 10^9 ha or 11 per cent of the world's land area. It is projected that arable land will need to increase by some 83 x 10^6 ha by the year 2000, two-thirds of which will be achieved by the expansion of irrigation (FAO, 1988).

With a few exceptions most of the land suitable for rainfed agriculture has already been utilized (Meyer and Turner, 1992). Thus there is widespread concern that expansion will increasingly occur in marginal and fragile environments. This concern is particularly apparent in sub-Saharan Africa. Indeed, a number of countries have already reached their potential of arable land and have extended agriculture to marginal areas (Table 3.5). Examples in this category include Burundi, Lesotho, Libya, Mauritius, Morocco, Rwanda, Tunisia, Haiti, India, Iran, Lebanon, Saudi Arabia and Thailand. For these countries increased production can only be reached via more intensive cultivation and higher inputs (FAO, 1988).

In summary, it can be said that currently there is no shortage of food globally nor is the capacity to produce food lacking. During the past 20–30 years the agricultural sector has become increasingly productive and, on the whole, supports a better-fed world; the global average food availability has risen from 2,290 calories per capita per day in the early 1960s to 2,700 in the late 1980s despite an increase in the world's population of 175 per cent during this period (FAO, 1992).

In spite of these overall achievements, the FAO has identified a number of problems that currently face the agricultural sector. As noted here, per capita food supply in the group of low-income countries (excluding China) has not improved significantly since the early 1970s. Secondly, present-day agricultural trade policies which are placing a heavy burden on the economies of both the developed and developing nations are in urgent need of reform. Finally, the threat of environmental damage as production pressures on ecological resources intensifies requires attention. Many countries have already recognized the need to change their agricultural policies. Developed market economies are adopting measures to tackle the problems of environmental pollution from agricultural sources and of surplus and subsidized production. In developing countries, national development plans are increasingly recognizing the importance of agriculture as a path to economic growth whereas, previously, priority was given to the industrial sector.

References

Durning, A. B. and Brough, H. B. 1991 *Taking Stock: Animal Farming and the Environment*, World Watch Paper No. 103, The World Watch Institute, Washington DC.

FAO 1988 *World Agriculture: Toward 2000: An FAO Study*, Alexandratos, N. (Ed.), Belhaven Press, London.

FAO 1992 *The State of Food and Agriculture 1991*, FAO Agricultural Series, Report No. 24, Food and Agriculture Organization of the United Nations, Rome.

Meyer, W. B. and Turner II, B. L. 1992 Human population growth and global land-use/cover change, *Annual Review of Ecological Systems*, **23**, 31–61.

UNEP 1991 *United Nations Environment Programme Environmental Data Report*, 1991/92, Basil Blackwell, Oxford.

Forests and Woodlands

The most widely quoted estimate for the current areal extent of the world's forests is just under 4×10^9 ha, or around one-third of the land surface (Mather, 1990). As noted earlier in this chapter, this represents a decrease of approximately 15 per cent in forest area since pre-industrial times (see Land Cover and Use). This reduction in the areal extent of forest cover, which took place initially in the temperate Northern Hemisphere and in more recent decades has accelerated in tropical regions, has been a source of increasing concern in latter years at both the national and international level.

National estimates of forested areas are compiled on a routine basis by the FAO and published in the 'FAO Production Yearbook'. The FAO data base forms the basis of the data on land cover presented in Table 3.1. Owing to national differences in definitions and data collection methods, these data on forest cover are variable in quality. Moreover, as the same data tend to be quoted over a period of several years for some countries, the apparent degree of comprehensiveness in the FAO statistics is to some extent misleading.

The lack of standard definitions for the terms forest, woodlands and deforestation have hampered attempts to characterize the extent of changes in tree cover and in the past have led to discrepancies between estimates of forest resources and rates of deforestation. Nowadays "Forest and woodland" is generally taken to mean land under natural or planted stands of trees. In the FAO statistics this includes cleared lands which will be reforested in the foreseeable future. "Closed" forest is sometimes interpreted as land with tree crowns covering more than 20 per cent of the land area and "open woodland" as land with tree-crown cover of 5–20 per cent of the surface area (Mather, 1990). However, different descriptions of forested areas are common and care must be exercized when comparing data using differing definitions.

The requirement for a globally consistent monitoring programme to assess periodically the status of the forest resource base was highlighted at the 1972 Stockholm Conference on the Human Environment. Since that time FAO, in association with UNEP and the United Nations Educational, Scientific and Cultural Organization (UNESCO), have developed and advanced methodologies for the compilation and assessment of forest data and in 1982 the first Tropical Forest Resources Assessment was completed (reference year 1980). This assessment provided the first statistically consistent estimates of forest area and deforestation/afforestation rates in tropical countries. It used a single set of concepts and classifications to organize and rationalize the existing information relating to forests and woodlands, including that obtained by remote sensing. Results of this first forest assessment have been presented in earlier editions of this report (UNEP 1987, 1989).

In the early 1990s a second decadal assessment of tropical forest resources (reference year 1990) was initiated, the final results of which are due to be available at the end of 1993. The latter study is supported by two further projects, the joint United Nations Economic Commission for Europe (UN ECE)/FAO assessment of forest resources in temperate zone developed countries and the FAO's survey of forests and woodlands in non-tropical developing countries. Collectively these three studies will provide a comprehensive picture of forest resources world-wide as they were in 1990 (Lanly et al., 1991).

Temperate Forests

At the time of writing, only the results of the UN ECE/FAO 1990 Forest Resources Assessment were available; these are presented in summary form in Table 3.6. On the basis of this study, it is estimated that the developed temperate-zone countries hold 50.7 per cent, or 2.06×10^9 ha of the world's forest resources. The tropics hold a further 42.6 per cent of the world total, with the remainder located in developing temperate-zone countries such as China, Mongolia, Argentina and Chile. Of the 2.06×10^9 ha of forest and other wooded land assigned to the developed temperate region, almost half (46 per cent) is located in the republics of the former USSR and a further 36 per cent in Canada and the USA. Canada has the largest endowment of forest and other wooded land per capita, over 17 hectares per capita, more than 10 times the average for all the assessed temperate countries.

The UN ECE/FAO assessment has, however, been unable to provide a conclusive aggregate estimate of change in forest area in the temperate zone. Among the main reasons for this were incomplete data sets and difficulty in separating afforestation (establishment of forest on previously non-forest land) and reforestation (re-establishment of forest cover on temporarily unstocked forest). The assessment was nevertheless able to provide a relatively firm estimate for the European countries. It is concluded that over the past 40 years, the forest resource base in Europe has expanded in both area and volume; more specifically, between 1981–1990, it is estimated that forest and other wooded area in Europe has increased by about 2×10^6 ha. Elsewhere in the temperate zone, there is evidence to suggest that since 1980 forest and other wooded land has increased in the former USSR, but has declined in the USA, Japan and Canada (UN ECE/FAO, 1992).

To date, assessments of the status of forests have focused on changes in the areal extent of forest cover, as opposed to changes in forest condition. Recognition of the importance of the terrestrial biosphere, and forests in particular, in the context of global carbon cycle is, however, changing this perception. Preliminary evidence, which includes the observation of expanded forest area in Northern Hemisphere temperate latitudes noted here, suggests that at present Northern Hemisphere temperate forests are increasing in biomass and thus acting as a net sink for atmosphere carbon dioxide (CO_2) (see also Part 1: Environmental Pollution, Atmosphere).

The conclusion that temperate forests of the Northern Hemisphere are increasing in biomass is reached despite the observation of widespread "Waldsterben" or forest decline in many European countries. Although the causes and mechanisms of the forest die-back phenomenon are incompletely understood at this time, atmospheric pollutants such as the sulphur and nitrogen oxides (acting directly and indirectly in the form of acidic deposition) are implicated as potential stress factors. The assessment of forest die-back, as measured by degree of defoliation, is reported in Part 1: Environmental Pollution, Biological Monitoring. In terms of implications for forest biomass and productivity, the scale of the impact of this type of forest decline has not been clearly quantified. According to one modelling study conducted by the International Institute for Applied Systems Analysis (IIASA), it is estimated that sulphur emissions may be causing a 15 per cent reduction in potential forest harvests; this translates to economic losses of the order of US$ 36 billion per year (Carrier and Krippl, 1990).

Tropical Forests

Although the final results of the second decadal assessment of the world's tropical forests are not yet available, preliminary results confirm the suspected acceleration in the rate of deforestation in tropical countries during the past decade. These preliminary results, which represent the results of the first, i.e., the statistical phase of the 1990 forest assessment, are presented in Table 3.7. Data are aggregated on a regional basis and cover 87 countries which collectively account for 97 per cent of the world's tropical forests. Details of the definitions and methodologies employed are provided in the footnotes to the table; further discussion is given by Lanly et al. (1991). On the basis of these preliminary findings it is estimated that the annual rate of tropical deforestation during the 1980s averaged 16.9×10^6 ha, or just under 1 per cent per year; this represents about a 50 per cent increase over the estimate for a similar set of countries for the period 1976–1980.

Completion of the second and third phases of the 1990 forest assessment project (i.e., the remote sensing phase and the analytical phase) will provide improved estimates of the areal extent of forest resources in 1990 and changes since 1980. Whereas the 1980 and the first phase of the 1990 assessment are based on available (i.e., published) national estimates of forest area and related data, the second phase of the 1990 assessment utilizes AVHRR satellite data in conjunction with multi-date high resolution LANDSAT imagery to generate what will be the first truly consistent global baseline data set for tropical forest area. Although it is recognized that these data will not be sufficiently detailed for land-use and economic development planning at sub-national levels, this global-scale information will assist policy makers to plan the conservation and development of their forest resources within the regional and world-wide context. It will also help aid agencies and international organizations to formulate their programmes in forestry and related sectors. Thirdly, these data are of value to the scientific community engaged in global change research, especially those concerned with carbon cycle modelling and the generation of emission scenarios. The implications of tropical deforestation with regard to CO_2 emissions is discussed in Part 1: Environmental Pollution, Atmosphere.

As part of its Tropical Forest Conservation Plan, The World Conservation Union (IUCN) has undertaken its own project to assess the extent and location of tropical moist forests. On behalf of the IUCN, WCMC has assembled all available maps and related information pertaining to forests in the tropical regions. These maps are being made available in three separate tropical forest resources atlases. Two such atlases, one covering Asia, the other Africa, have been published already and the third on Latin America is currently in preparation. These maps have been digitized and the data presented in Table 3.8. These data reflect a drastic reduction in forest habitat in some countries since pre-industrial times (IUCN/WCMC 1991, 1992).

Although the forces which drive the trend towards increasing, and even accelerating, tropical deforestation differ considerably across the world regions, some general patterns are recognized. Clearance for cultivation, i.e., expansion of cropland, is probably the most widespread stress factor, the root cause of which, in some areas at least, is the poverty of the people living in and around the forests. In Central and Latin America ranching and pasture development have been identified as significant causes of clearance. In contrast, timber extraction or logging in excess of regrowth typifies the situation in some south-east Asian countries, and to some extent in parts of western Africa. In Africa, the Indian sub-continent and mountainous parts of Latin America, demand for fuelwood is an important cause of the removal of forest cover (Meyer and Turner, 1992). The relationship between forest resources and the demands of the timber industry are explored further in Part 7: Industry and Transport and here in Box 3.2.

Management of Forest Resources

World-wide concern for the deterioration of the forest resource base, in both tropical and temperate locations, has led to a gradual change, at both national and international levels, in the perception of forest resources. The Tropical Forestry Action Plan, launched in 1985, exemplifies the new thinking; its main aim is to mobilize international co-operation for the sustainable development of tropical forests (IUCN/WCMC, 1991). Similarly, in the European region, a series of Ministerial Conferences on the Protection of Forests in Europe has been convened to agree common principles for the protection of European forests based on sustainable development and conservation. Box 3.2 summarizes some of the recent approaches to developing

BOX 3.2 Sustainable Use of Forest Resources: Proposed Indicators

The maintenance of forest resources is widely perceived as an essential component of ecologically sustainable development. In recognition of this fact measures which reflect sustainability of forest resource utilization on a national basis have been included in the majority of sets of environmental indicators that have been proposed to date.

The Organisation for Economic Co-operation and Development (OECD), for example, has proposed an indicator which reflects the intensity of use of commercial forests. This indicator is based on principles of forest resource accounting and is defined as the ratio of the annual harvest (the demand side) to the annual increment or growth in the growing stock of trees (the supply side). A value of unity for this ratio thus indicates a balance in the processes of timber removal (by harvesting) and resource regeneration (replanting and regrowth). Data presented here for selected OECD countries show that in the vast majority of cases the intensity of use ratio is less than one. The OECD concludes that, on a national level at least, OECD countries present a picture of sustainable use of their commercial forest resources (OECD, 1991).

Intensity of use of forest resources in selected OECD countries, 1980–1985

Country	Total harvest/ annual growth
OECD	0.52
Australia	0.39
Austria	0.78
Canada	0.47
Finland	0.82
France	0.62
Germany, Fed. Rep.	0.96
Japan	0.53
Sweden	0.70
Switzerland	0.91
UK	0.44
USA	0.58
Yugoslavia	0.72

Data represent an average for the period 1980–1985
Source: OECD, 1991

In proposing a national set of environmental indicators, Environment Canada has adopted a similar approach to that of the OECD for the forestry sector. Its indicator compares the forest area harvested and regrown, rather than wood volumes. The forest area regrown is more precisely defined as "successfully regenerated areas", i.e., land areas harvested for timber that have been treated so as to establish a new forest that meets minimum forest management criteria (Environment Canada, 1991). Trends in this indicator, i.e., successfully regenerated areas expressed as a percentage of total forest area harvested for timber are shown in the figure. The Canadian indicator has been developed largely in response to the national concern within the forestry sector that, over the years, there has been a gradual build-up of the area of forest land considered to be inadequately restocked, i.e., unable to sustain new forest growth equivalent to that cut.

The Canadian approach, and that of the OECD proposal, refers explicitly to domestic commercial forests and measures the intensity of harvesting pressures for timber products. However, the rising demand for wood products and expansion of international trade mean that countries are increasingly exerting pressures on forest resources overseas. The harvesting of wood from tropical forests to supply demands of the OECD countries is such a particular case.

A second forestry indicator proposed by the OECD reflects this concern and is given as the import of cork and wood from tropical countries, measured in value terms (i.e., in US$ per capita) and as a percentage of total cork and wood imports. Japan (US$ 21.6 per capita) followed by the Netherlands (US$ 15.2 per capita) have the highest imports of cork and wood from tropical countries as measured in value terms; Turkey (US$ 0.3 per capita) and the USA (US$ 0.7 per capita) have the lowest. Portugal derives 61.6 per cent of its imports of cork and wood from tropical countries, the highest percentage contribution in the OECD region. At the other end of the scale Iceland and Sweden obtain less than 2 per cent of their total wood and cork imports from tropical countries (OECD, 1991).

In this context it is interesting to note that several researchers who have conducted statistically-based studies to assess the relative importance of various economic activities (including timber production and trade) as a cause

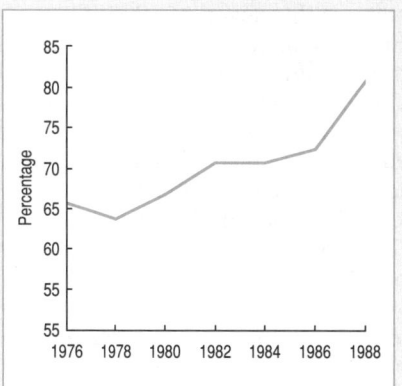

Canada's regeneration success as a percentage of total forest area harvested, 1976–1988
Source: Environment Canada, 1991

of tropical deforestation, have concluded that the international timber trade is not, in fact, the major factor in determining deforestation rates in the majority of tropical countries. Other factors, in particular the conversion of forest land for agriculture and harvesting of trees for fuelwood, are likely to be more important. This finding does not, however, diminish the concern over excessive exploitation and rapid depletion of tropical forest resources (Burgess, 1993).

The basis of the above conclusion is the observation that the volume of tropical timber production that enters the international trade is actually quite small; only 6 per cent of the total tropical non-coniferous roundwood production enters the international trade (Burgess, 1993). Of this, almost three-quarters originates from Malaysia and Indonesia; around 80 per cent of the traded wood is destined for Japan, the EEC and the USA (see also Part 7: Industry and Transport).

References

Burgess, J. 1993 Timber production, timber trade and tropical deforestation, *Ambio*, **22**(2–3), 136–143.
OECD 1991 *Environmental Indicators: A Preliminary Set*, Organisation for Economic Co-operation and Development, Paris.
Environment Canada 1991 *A Report on Canada's Progress Towards a National Set of Environmental Indicators*, SOE Report No. 91–1, Environment Canada, Ottawa.

environmental indicators which reflect the sustainable or non-sustainable use of forest resources.

These, and other similar initiatives, including the Forest Statement developed at UNCED (see Part 10: International Co-operation), are successfully increasing awareness of the multiple values served by forest ecosystems, and are encouraging decision makers to adopt policies which take account of forest values other than wood production. These can include a wide range of non-wood forest products (such as, gums and resins, bamboos, seeds and spices, wildlife products and medicinal plants).

There are some signs that, as well as recreational uses, non-wood forest products are increasingly being exploited in tropical areas (see Part 7: Industry and Transport). A shift in emphasis is also apparent in temperate regions. As part of the UN ECE/FAO forest assessment, national respondents were requested to supply information pertaining to the perceived importance of environmental benefits and non-wood products derived from their forests. Published responses are only qualitative and suggest that for all countries the removal of wood, because of its high economic value, is still seen as the raison d'être for forests. Nevertheless, demands to use forests for environmental and biodiversity protection or recreation were reported to be increasing, sometimes challenging wood production as the priority objective (UN ECE/FAO, 1992).

References

Carrier, J–G. and Krippl, E. 1990 Comprehensive study of European forests assesses damage and economic losses from air pollution, *Environmental Conservation*, **17**(4), 365–366.

IUCN/WCMC 1991 *The Conservation Atlas of Tropical Forests: Asia and the Pacific*, N. M. Collins, J. A. Sayer and T. C. Whitmore (Eds), Macmillan Press Ltd, London.

IUCN/WCMC 1992 *The Conservation Atlas of Tropical Forests: Africa*, N. M. Collins, J. A. Sayer and T. C. Whitmore (Eds), Macmillan Press Ltd, London.

Lanly, J–P., Singh, K. D. and Janz, K. 1991 FAO's 1990 reassessment of forest cover, *Nature and Resources*, **27**(2), 21–26.

Mather, A. S. 1990 *Global Forest Resources*, Belhaven Press, London.

Meyer, W. B. and Turner, B. L. 1992 Human population growth and global land-use/cover change, *Annual Review of Ecological Systems*, **23**, 39–61.

UN ECE/FAO 1992 *The Forest Resources of the Temperate Zones, Main Findings of the UN ECE/FAO 1990 Forest Resources Assessment*, United Nations, New York.

UNEP 1987 *United Nations Environment Programme Environmental Data Report*, 1987/88 Basil Blackwell, Oxford.

UNEP 1989 *United Nations Environment Programme Environmental Data Report 1989/90*, Basil Blackwell, Oxford.

Fresh-water Resources

The growth of the world population and the concomitant expansion of agriculture and industry have placed ever increasing demands on the world's water resources; according to one estimate the consumption of fresh water has increased by a factor of 35 over the past three centuries. Recent decades have witnessed an annual increase in water withdrawal of between 4 and 8 per cent, with the highest rates of growth occurring in developing countries (WRI, 1992).

Although the current world population uses less than one tenth of the earth's reserves of fresh water, inequities in the distribution of fresh-water resources mean that some countries are faced with the prospect of critical water shortages in the near future; indeed some countries are already operating at the limit of their resources (WMO/UNESCO, 1991). Moreover, pollution by chemicals, radioactive materials and suspended solids pose a further threat to the security of water supplies. Whereas aspects of water quality and its degradation are discussed elsewhere in this report (see Part 1: Environmental Pollution, Water Quality), the depletion of water resources is the main issue of concern here.

In recognition of the central importance of proper management of water resources against a background of rising demands, the UN convened a Water Conference in 1977, the recommendations of which have become known as the Mar del Plata Action Plan (MPAP). Since that time periodic reviews of the progress made in implementing the Action Plan have been undertaken, culminating in the late 1980s and early 1990s in the formulation at the interagency level of a strategy for the implementation of the Mar del Plata Action Plan in the 1990s as an input to the UNCED process. This latter work stresses the need for a more integrated and intersectoral approach to water resource management than has hitherto been the case (WMO/UNESCO, 1991; FAO, 1990; WHO/UNEP, 1991).

Water Resources Assessment

Among other things, the MPAP process has highlighted the current inadequacies of the information base pertaining to water resources. There is a clear requirement for more regular and systematic collection of hydrometerological, hydrological and hydrogeological data at the national level which is supported by a comprehensive system for processing information of this type on regional and global scales. A recent review of the national capabilities for monitoring and assessment of water resources has identified technical and operational difficulties, together with the fragmented nature of institutional arrangements for water resource-related matters as the main impediments to the establishment of integrated data bases of water resource information (WMO/UNESCO, 1991).

In the absence of centrally co-ordinated data collection (the gathering of global information on fresh-water resources and use is not the mandate of a specific organization), available estimates of water resources and consumption have been compiled from a range of sources by the World Resources Institute (WRI), Washington DC. Published estimates of water resources and consumption (i.e., withdrawal) have been supplemented by estimates derived from model calculations which use available values of parameters such as cropland under irrigation, livestock numbers and precipitation amount as input. Although the reliability of these estimates varies considerably between countries (and thus limits inter-country comparability), this data set is nevertheless the best available at the present time. As these data have been presented in detail in the previous edition of this report (UNEP, 1991) and only a few changes have since been made to the data base, the data

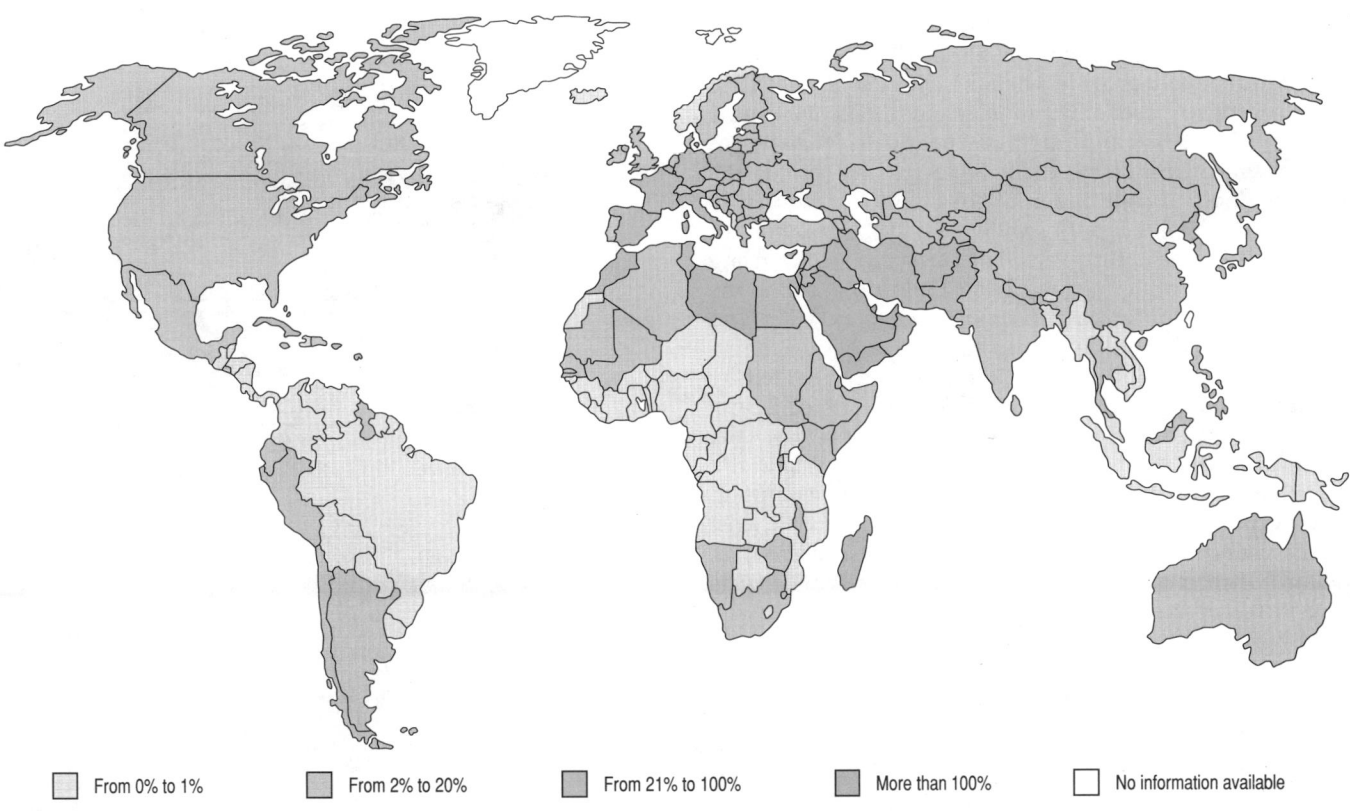

From 0% to 1%   From 2% to 20%   From 21% to 100%   More than 100%   No information available

FIGURE 3.6 Utilization of renewable fresh-water resources, current status (total water withdrawal is given as a percentage of available renewable fresh-water resources; values in excess of 100 per cent indicate utilization of non-renewable resources)
Source: Data supplied by WRI, Washington DC

table has not been included on this occasion. The following discussion on water resources and consumption does, however, draw on these data.

Renewable sources of fresh water comprise rivers, lakes and ground water. Water may also be produced from desalinization plants, or it can be withdrawn from ancient sealed aquifers. The latter are generally referred to as non-renewable fresh-water resources and may represent significant sources of fresh water in some countries. The endowment of renewable water resources varies markedly between countries and regions; this is reflected in the distribution of per capita renewable water resources. In three regions, Oceania, South America and North America, annual per capita availability exceeds the world average of 7.4×10^3 m^3. Asia has the least amount of renewable water available per capita, a consequence of scarce supplies in the arid south-west (WRI, 1992).

Water Use

A number of countries use almost all their total endowment of renewable water supplies; Belgium, Malta, Israel and Egypt use more than 70 per cent of their available resources. Libya, Qatar, Saudi Arabia, United Arab Emirates, the former Yemen Arab Republic and the former People's Democratic Republic of Yemen use more than their annual renewable water resource by satisfying excess demand with ground water (from non-

renewable aquifers) or desalinization plants (Figure 3.6).

Water withdrawal data indicate that almost 70 per cent of global water use is for agricultural purposes; industry uses a further 23 per cent and the remainder is consumed for domestic and municipal purposes. These global patterns of sectoral water use are reflected by many individual countries. European countries, however, tend to be the exception; on average European industry accounts for 54 per cent of the total fresh-water withdrawals (UNEP, 1991).

With respect to agricultural uses of fresh water, irrigation of cropland is by far the most dominant. As noted earlier in this chapter, the area of land under irrigation has increased steadily during recent decades. This trend is likely to continue into the future and thus will place greater demands on existing water supplies. Several rivers of the world, for example, the Nile (Egypt/Sudan), the Euphrates (Syria/Iraq), the Jordan (Jordan/Israel), the Colorado (USA), and the Amu Darya and Syr Darya in the former USSR are already so intensively used for irrigation (and other purposes) that in dry years there is inadequate water in these rivers. The problems of the Aral Sea, for which the Syr and Amu are the main tributaries, have received much attention in recent years; indeed the desiccation of the Aral Sea as a result of over-abstraction has been described as one of the world's worst ecological disasters (WRI, 1992). UNEP has been instrumental in drawing up an Action Plan for the rehabilitation of the Aral Sea (UNEP, 1992).

Inadequate supplies of water for irrigation is, however, only one aspect of the problem. Although development of irrigation generally provides considerable benefits, not all of which are purely economic, adverse environmental effects can accrue in cases of imprudent and poorly designed irrigation schemes. One of the main issues for concern is the waterlogging and salinization of soils, which are typically caused by excessive input of water into systems that have inadequate drainage capacities. The combination of a rising water table (i.e., waterlogging) and a build-up of salt in the soil, in turn, act to reduce crop productivity and yields. The FAO estimates that at the present time some $20–30 \times 10^6$ ha of irrigated cropland world-wide are severely affected by salinity (FAO, 1990). Irrigated cropland in parts of Egypt, India, Iraq, Mexico, Pakistan, the former USSR and the USA are amongst the affected areas.

Data which reflect temporal trends in water consumption are especially limited in their availability. Selected data exist for the countries of the OECD; and these are presented in Table 3.9. In most cases, total and per capita water withdrawals have increased steadily between 1960 and the late 1980s, a trend which is likely to be mirrored in other world regions. All but two countries, Switzerland and Austria, withdraw the majority of their water from surface rather than underground sources (Figure 3.7).

References

FAO 1990 *An International Action Programme on Water and Sustainable Agricultural Development: A Strategy for the Implementation of the Mar del Plata Action Plan for the 1990*, Food and Agriculture Organization of the United Nations, Rome.

OECD 1991 *OECD Environmental Data Compendium 1991*, Organisation for Economic Co-operation and Development, Paris.

UNEP 1991 *United Nations Environment Programme Environmental Data Report, 1991/92*, Basil Blackwell, Oxford.

UNEP 1992 Options for saving the Aral Sea, *Our Planet*, 4(4), 15.

WHO/UNEP 1991 *Water Quality: Progress in the Implementation of the Mar del Plata Action Plan and Strategy for the 1990s*, World Health Organization, Geneva, and United Nations Environment Programme, Nairobi.

WMO/UNESCO 1991 *Water Resources Assessment: Progress in the Implementation of the Mar del Plata Action Plan and a Strategy for the 1990s*, World Meteorological Organization, Geneva, and United Nations Educational, Scientific and Cultural Organization, Paris.

World Resources Institute 1992 *World Resources 1992–93*, Oxford University Press, New York.

Fisheries

The location of the world's marine fisheries is largely governed by the distribution of phytoplankton production, which in turn is dependent on nutrient supplies. These are often greatest in areas of upwelling. Thus, areas of upwelling sustain some of the world's largest commercial fisheries; these are located in the North-west Pacific, the South-east Pacific, the North-east Atlantic and the Western Central Pacific.

Owing to the complexity of the mechanisms that affect biological capacity and fish populations, estimates of commercial fish stocks are uncertain at the present time. For example, small changes in temperature, salinity, ocean currents, strength of upwelling and predator populations can

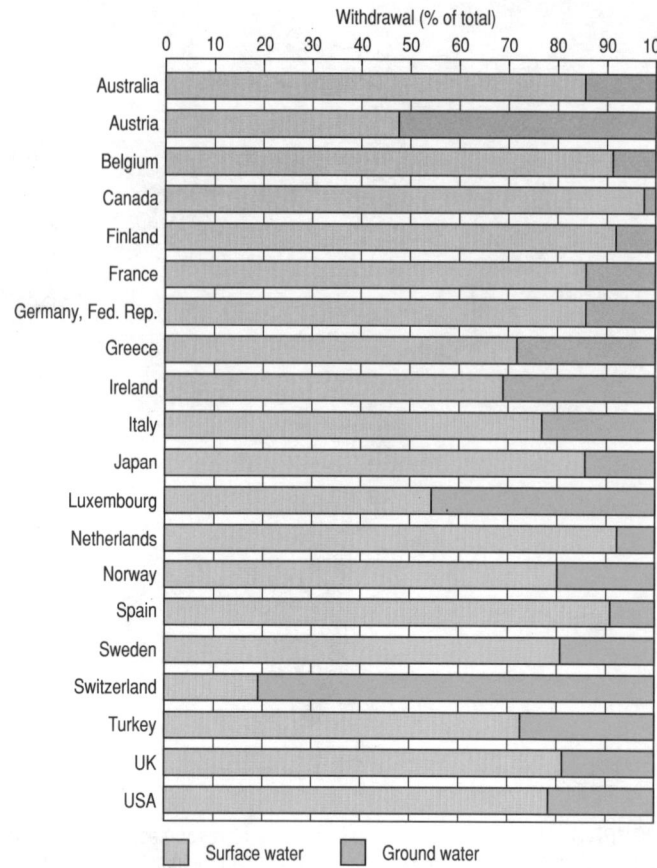

FIGURE 3.7 Relative contributions of surface and ground fresh-water sources to total water withdrawal in OECD countries, latest available year
Source: Based on OECD, 1991

sharply alter the abundance and distribution of fish populations. Human activities can also affect fish stock renewal through over-exploitation and habitat degradation.

In recent years the FAO has voiced concerns about the current level of marine fish catches. It is estimated that stocks of some marine fisheries are now close to their maximum catch limits or sustainable yields. Moreover, some fisheries show signs of biological degradation (FAO, 1990). Figure 3.8, which compares estimates of maximum harvests that can be sustained by a fishery with annual average catches for the 1987–1989 period, reflects this concern.

Given that there appears to be little scope for increased catches of many of the traditionally fished marine species, rising demands for fishery products are increasingly likely to be met by exploitation of open seas resources and aquaculture. Open-sea fishery resources (i.e., those over 200 miles from shore), being more dispersed, are typically harvested by pelagic driftnetting techniques. This practice causes large-scale accidental mortalities of non-target species, an issue which has also given rise to some concern in recent years.

In light of the above concerns, increasing attention has been devoted to the establishment of international agreements relating to shared fishery resources and to the adoption of sound fishery management practices. The establishment of the 200-mile fishing exclusion zones by most coastal states has

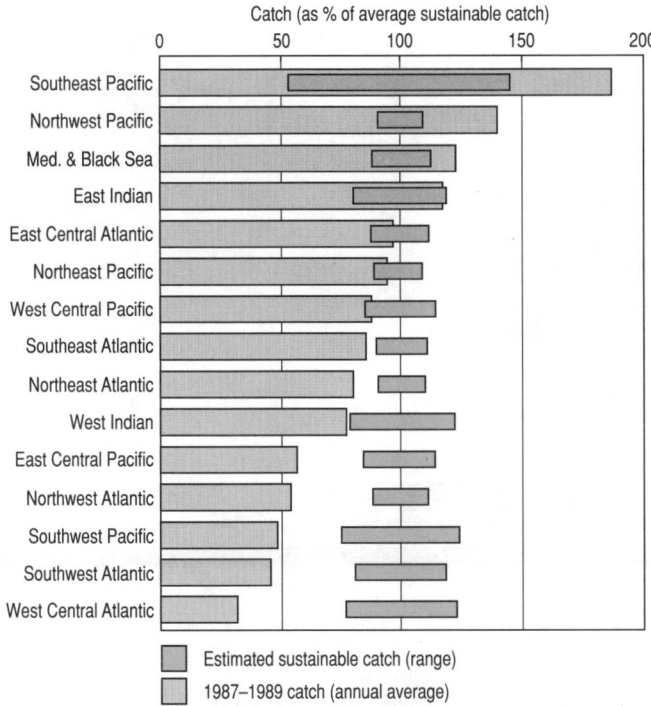

Catch (as % of average sustainable catch)

Southeast Pacific
Northwest Pacific
Med. & Black Sea
East Indian
East Central Atlantic
Northeast Pacific
West Central Pacific
Southeast Atlantic
Northeast Atlantic
West Indian
East Central Pacific
Northwest Atlantic
Southwest Pacific
Southwest Atlantic
West Central Atlantic

☐ Estimated sustainable catch (range)
☐ 1987–1989 catch (annual average)

FIGURE 3.8   Marine fish catch as a percentage of average sustainable yields, 1987–89 (data are provided according to the 17 fishing areas as defined by the FAO for statistical purposes; the Arctic and Antarctic fisheries are, however, excluded from this chart as estimates of sustainable yields are not available for these regions)
Source: After WRI, 1992

been an important step in this process. In accordance with the recommendations of the 1982 UN Convention on the Law of the Sea (UNCLOS), several world regions have set up organizations to foster co-operation between member countries with respect to their shared marine resources. The strength and weaknesses of such efforts have been reviewed by WCMC (1992).

Fish Catch and Production

In contrast to data on fish stocks, statistics on fish catch or harvest are widely available for both developed and developing countries. It should be stressed, however, that with respect to a discussion of fishery resources such data should be interpreted with care; catch statistics do not necessarily reflect the size or health of fish stocks. At best fish harvest data provide a measure of the level of fishing activity upon commercial stocks.

The fishery division of FAO is responsible for collecting international data on world fish catch and landings, and on the production and trade of fish commodities. The FAO organizes its data by 23 marine statistical areas and 8 inland catch areas, by country level, and also by 840 species items. National data on marine and fresh-water fish catches are presented in Table 3.10 for selected countries. Asian countries currently account for the majority of world fish production, a share which has been growing over the last ten years.

Global trends in marine and fresh-water fish catch since 1950 are illustrated in Figure 3.9. The overwhelming trend is one of a substantial growth in the annual world landings of aquatic resources. During the 1960s and early 1970s total landings increased steadily as a result of discoveries of new stocks, increased fishing effort and improvements in fishing technology. In the latter half of the 1970s, growth in the fish harvest slowed following the collapse of the Peruvian anchovy fishery in the early 1970s. The 1980s, however, saw a renewed surge in the growth of world landings, fuelled mainly by the increases in catches of shoaling pelagics in the north-east and south-east Pacific regions (FAO, 1990).

In 1990 the world fish harvest fell for the first time since 1974 from a peak of 100.3×10^6 t to 97.2×10^6 t (Figure 3.9). This 4 per cent decline is attributed to two factors, a 12 per cent decrease in the catch of small pelagic fish in the south-east Pacific and a reduced catch of selected North Atlantic species following the imposition of lower quotas for this region (FAO, 1992). Trends in the catch of selected fish species were reviewed in the previous edition of this report (UNEP, 1991).

Aquaculture

Since 1984, FAO's collection of aquaculture statistics has become a regular feature of its annual survey of world fishery statistics. At present, aquaculture accounts for approximately 13 per cent of the world's total fish production, with almost 6×10^6 t a^{-1} arising from inland aquaculture and a further 5×10^6 t a^{-1} from marine aquaculture. Each year an additional 4×10^6 t of "other" products (i.e., aquatic plants, frogs and turtles), is generated but is not included in the total world fish production. The previous edition of this report provided detailed aquaculture statistics for the top 20 producing countries, ranked by their volume of production (UNEP, 1991). These data are updated and summarized here in Figure 3.10. China is by far the world's principal producer of aquaculture products, both by volume (46 per cent of the world total) and value (35 per cent of the world total). The bulk of its production comprises fresh-water species such as carp and tilapia.

Figure 3.10 clearly illustrates the growing importance of aquaculture production on a global scale; both intensive and

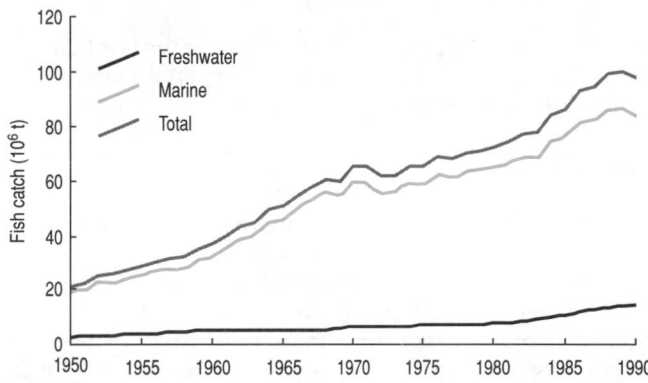

FIGURE 3.9 Trends in the global catch of marine and fresh-water fish, 1950–1990
Source: Data supplied by FAO, Rome

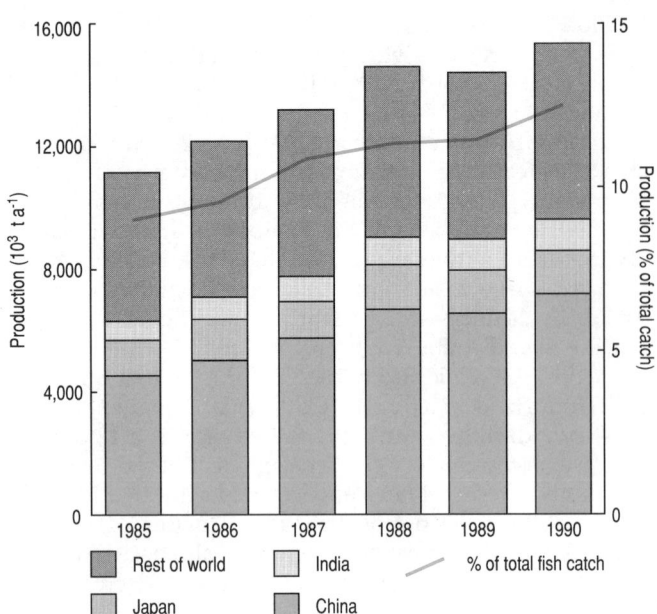

FIGURE 3.10 Global aquaculture production, tonnage and as a percentage of total fish production, 1985–1990
Source: Data provided by FAO, Rome

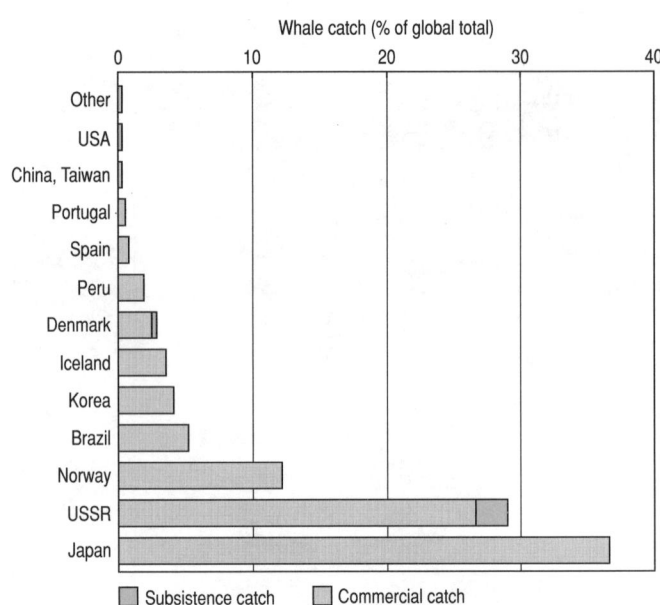

FIGURE 3.11 The proportion of whales caught during 1980–1991 by the main whaling nations
Source: Data provided by R. Gambell, IWC, Cambridge, UK

extensive mariculture have grown at a faster rate than capture fisheries in the past few years (FAO, 1990). It is likely that this trend will continue into the future.

The practice of aquaculture, despite offering potential benefits, is nevertheless associated with a number of adverse impacts on the environment. Iwama (1991) provides a comprehensive review of the issues and identifies the output of organic particulate wastes (i.e., suspended solids derived from feed and faecal material) as the principal cause for concern. Impacts range from a mild enrichment of the receiving waters to the smothering of benthic organisms and the production of anoxic sediment layers. At the extreme end of the spectrum, aquaculture is implicated in the now widespread degradation of mangrove areas in tropical regions.

Marine Mammals

It is likely that the practice of whaling over the last hundred years has greatly reduced the populations of many whale species (Perrin, 1989). Of the 80 recognized species of dolphins, porpoises and whales (i.e., the cetaceans), stocks of some 20 species are sufficiently diminished so as to fall within the IUCN's "threatened" category and at least 14 other species require careful monitoring. Although the rest may not qualify as "threatened species", several species have localized conservation problems that need attention (IUCN/UNEP, 1989).

The International Whaling Commission (IWC) was set up in 1946 in order to regulate commercial whaling. The notion of banning commercial whaling was first raised in 1972 at the UN Conference on the Human Environment. In 1982 the IWC decided that there should be a suspension of commercial whaling commencing with the 1985/86 whaling season. Although this proposal met with some objections, it has

nevertheless been in effect since 1988. However, whales continue to be taken for subsistence purposes and under special permit for scientific research. Furthermore, at the present time there is no agreement on the legal competence of the IWC to regulate the takes of other marine mammals.

The 38-member nation IWC continues to meet regularly and to discuss a possible resumption of commercial whaling. Some whaling nations, including Iceland, Norway and Japan, believe that a number of whale stocks are large enough to support new limited hunts. The most recent annual IWC session, held in Tokyo (10–14 May 1993), ended with the decision to maintain the existing commercial whaling moratorium for another year. It was agreed that some whales would still be killed for scientific purposes.

In addition to its regulatory role, the IWC also collects statistics relating to whale populations and catches. Data collection activities also cover the smaller cetaceans, i.e., dolphins and porpoises. Current estimates of whale (large cetacean) populations are given in Table 3.11. Only population figures for species/stocks which have been assessed in detail, such as the minke, fin, bowhead and gray whales, are provided. Estimates of whale populations are subject to considerable scientific uncertainty; some of the problems encountered in obtaining whale population estimates were discussed in the second edition of the 'UNEP Environmental Data Report' (UNEP, 1989).

The numbers of large whales caught for commercial, scientific and subsistence purposes during the period 1980–1991 can be seen in Table 3.12. Some incidental catches are also included. Figure 3.11 shows the proportion of the total catch during this period attributed to individual countries. Japan, the former USSR and Norway took the largest numbers. Data compiled by the IWC on catches of selected small cetaceans are presented as Table 3.13. A large proportion of the takes are indirect catches, such as those occurring through entanglement in tuna purse seines and gillnets. The incidentally-caught mammals are generally thrown back into the sea or retained for scientific analysis. In less prosperous nations the catch is usually marketed locally for human consumption. In most cases the impacts of incidental catches on the populations of small cetacean species are unknown. However, there is little doubt that some of the dolphin populations have declined as a result of the tuna fishery in the eastern tropical Pacific (Perrin, 1989).

In recent years steps have been taken to reduce some of the mortality of small cetaceans due to incidental catches. In December 1991 the UN General Assembly adopted, by consensus, Resolution 46/215 which called upon all nations engaged in large-scale pelagic high seas drift net fisheries to reduce the fishing effort by 50 per cent by 30 June 1992, and to implement a global moratorium on all large-scale drift net fisheries by 31 December 1992 (WCMC, 1992).

References

FAO 1990 *Review of the State of World Fishery Resources*, Food and Agricultural Organization of the United Nations, Rome.

FAO 1992 World Fisheries Situation. Paper presented at the International Conference on Responsible Fishing, 6–8 May 1992, Cancun, Mexico. Unpublished paper.

Perrin, W. F. (Ed.) 1989 *Dolphins, Porpoises, and Whales: An Action Plan for the Conservation of Biological Diversity: 1989–1992*, International Union for the Conservation of Nature and Natural Resources, Gland.

IUCN/UNEP 1989 Dolphins, porpoises and whales of the world: the story of the Cetacean Red Data Book, *The Pilot*, September Issue, 12.

Iwama, G. K. 1991 Interactions between aquaculture and the environment, *Critical Reviews in Environmental Control*, **21**(2), 177–216.

UNEP 1989 *United Nations Environment Programme Environmental Data Report 1989/90*, Basil Blackwell, Oxford.

UNEP 1991 *United Nations Environment Programme Environmental Data Report 1991/92*, Basil Blackwell, Oxford.

World Conservation Monitoring Centre 1992 *Global Biodiversity: Status of the Earth's Living Resources*, Chapman and Hall, London.

World Resources Institute 1992 *World Resources 1992–93*, Oxford University Press, New York.

Wildlife Resources

The earth's biological diversity, or biodiversity as it has come to be known, is perhaps the most fundamental natural resource. Its maintenance, at least to a certain level, is required to provide the material basis for human life – at one level to maintain the biosphere as a functioning system and, at another, to provide the basic materials for agriculture, forestry, fisheries and other utilitarian needs. Global biodiversity is, however, being steadily eroded by a variety of human-induced threats, such as pollution, over-exploitation and development. Its conservation has thus come to be viewed as one of the key environmental challenges of the 1990s.

Simply speaking, biological diversity is the total variety of life on Earth. On the more technical level biodiversity can be divided into a number of separate elements – genetic diversity, species diversity and ecosystem diversity. Each element describes quite different aspects of living systems:

○ Genetic diversity represents the heritable variation within and between populations of organisms.
○ Species diversity is generally viewed as the variety of species within a taxonomic group, e.g., the number of mammalian species within an area.
○ Ecosystem diversity – the most difficult to define – encompasses the variety of ecological processes (such as successional development phases and regeneration gaps), communities and habitats within a region.

Traditionally the discussion of global biodiversity has focused on species diversity, this being, practically and conceptually, the easiest element of biodiversity to measure. The material presented in this section reflects this perception of biodiversity, in that it is centred on the measurement and assessment of the rate of loss of individual species and populations. In view of the space restrictions in a report of this nature, each successive edition of the 'UNEP Environmental Data Report' focuses on one or two particular species; in this edition marine turtles have been selected for detailed discussion. Evidence for the loss of various types of natural habitats, a known cause of species and population losses, is also briefly considered. Coral reefs are covered in the present edition; the loss of wetlands were considered in the third edition of this report (UNEP, 1991). In recognition of their potential role in wildlife conservation strategies, this section concludes with a brief overview of the current status of the protected area network.

Many of the data presented here are drawn from the data bases maintained by the WCMC in Cambridge, UK. This institute was established in 1988 by the IUCN, the World Wide Fund for Nature (WWF) and UNEP specifically to provide information on the status, security, management and utilization of the world's biological diversity in order to support conservation programmes and international conservation conventions. The WCMC has recently published a comprehensive guide to the status of global biodiversity, based largely on its data holdings (WCMC, 1992). In addition to the WCMC, a number of other international organizations, including UNEP, UNESCO, FAO, United Nations Development Programme (UNDP), the World Bank, IUCN and the WWF are involved in the provision of information of global biodiversity and conservation issues.

Global Species Diversity

Estimates of the total number of species existing on earth vary from 5×10^6 to 100×10^6 species, with a best estimate of around 12.5×10^6. To date only 1.7×10^6 species have actually been described (WCMC, 1992). Of these known species, 2.7 per cent are classified as vertebrates (Figure 3.12). When expressed as the percentage of the possible global total number of species, the proportion of vertebrates is lower, only 0.4 per cent (Figure 3.13). The number of described species within taxonomic groups, as assessed by WCMC in 1992, for individual countries are listed in Table 3.14.

There is considerable uncertainty about the number of described species, and even greater uncertainty about the true

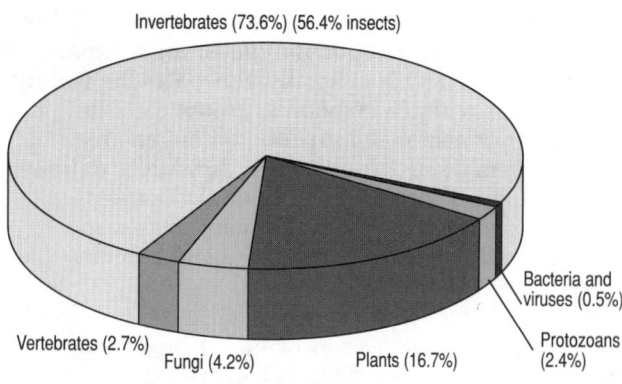

FIGURE 3.12 Major groups of organisms – described species (the number of species is expressed as a proportion of the known global total, assumed to be approximately 1.7×10^6)
Source: After WCMC, 1992

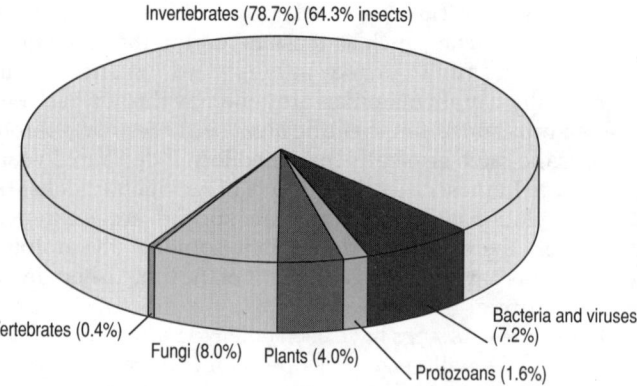

FIGURE 3.13 Major groups of organisms – possibly existing (the number of species is expressed as a proportion of the possible global total, assumed to be around 12.5×10^6)
Source: After WCMC, 1992

global total. This level of uncertainty stems from the relatively poor knowledge of the current status of global species' richness; even for the best known groups, for example, birds, catalogues and counts are very incomplete. The picture is further complicated by the fact that the term 'species' is in fact far from standard; the way in which species are defined varies considerably between groups and between taxonomists. In recognition of these shortcomings, work which will provide a more reliable basis for estimating the global number of species has been initiated. This involves intensive sampling of species-rich groups (such as insects) in especially species-rich areas, such as tropical moist forests (WCMC, 1992).

The Loss of Species Diversity

The loss of biological diversity can take many forms. At its most fundamental it involves the extinction of species. It is this aspect of biodiversity loss which will be covered here. However, it is important to note that the extinction of individual populations, as opposed to whole species, may also represent an important aspect of biodiversity loss (Ehrlich and Daily, 1993). This aspect will be considered in subsequent editions of this report. Losses of bird populations were briefly considered in the previous edition of this report (UNEP, 1991).

Extinctions of Species Despite the difficulties in quantifying the rates of species extinction, both at present and historically, it is beyond doubt that species extinctions caused directly or indirectly by human activities are presently occurring at a rate that far exceeds any reasonable estimates of natural, background extinction rates (WCMC, 1992). Direct causes of species extinctions include over-harvesting and inappropriate introduction of foreign plants and animals. Indirect causes comprise habitat destruction and environmental pollution.

On the basis of available information, WCMC has attempted to catalogue recent animal and plant extinctions (i.e., those that have occurred since 1600) and to attribute these extinctions to their most probable cause (WCMC, 1992). Although certain biases in the information base make analysis of extinction patterns problematic, some general trends are

nevertheless discernible. The most significant is the higher incidence of extinctions on islands compared with those on continental areas. About 75 per cent of the mammal and bird extinctions that have occurred since 1600 have been among island-dwelling species which, due to a certain degree of isolation from mainland species, are especially vulnerable to introduced species and other human impacts.

Figure 3.14 shows time series of known animal extinctions on both islands and continents. Two further trends are apparent: first, that documented island extinctions began almost two centuries earlier than continental extinctions; second, that both island and continental extinctions have increased rapidly from the early or mid-19th century to the mid-20th century. Of the documented animal extinctions that were assigned a cause, 39 per cent were attributed to the impact of introduced species and 36 per cent to habitat destruction. A further 23 per cent of extinctions were attributed to hunting and deliberate extermination. However, for the majority of animals the cause of extinctions is not known (WCMC, 1992).

Owing to the unreliability of historical records of animal and plant extinctions, current and future estimates of extinction rates are more usually derived from extrapolations of measured (or predicted) rates of habitat loss and estimates of species' richness in different habitats. Available estimates of current and projected extinction rates based on this type of analysis have been reviewed by WCMC (1992). On the assumption that the majority of terrestrial species are to be found in the tropical moist forests, most global analyses have been based on estimates of species'-richness in tropical forests and rates of deforestation. It has been estimated that at least 27,000 species per year are currently being lost in tropical forests. The reduction of biodiversity in other areas, notably coral reefs, wetlands, islands and montane environments, increases the total current extinction rate to some 30,000 species per year (Myers, 1993).

Extinction rates, although highly uncertain, tend to predict an accelerated and unprecedented loss of species diversity if current rates of tropical deforestation continue into the future (Bird, 1991, WCMC, 1992). Clearance of tropical rainforests also poses serious threats to genetic resources of many

a) Birds

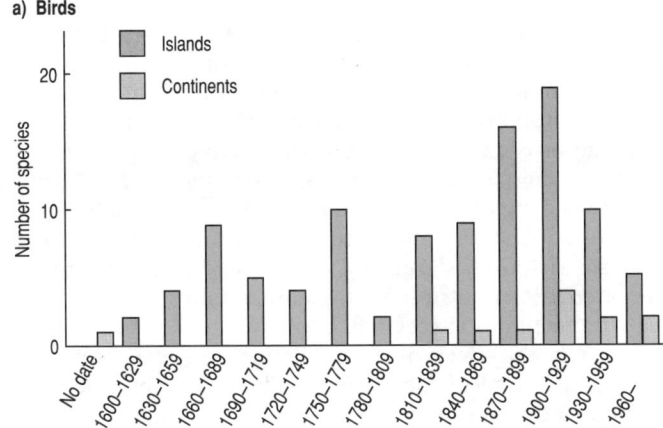

b) Mammals

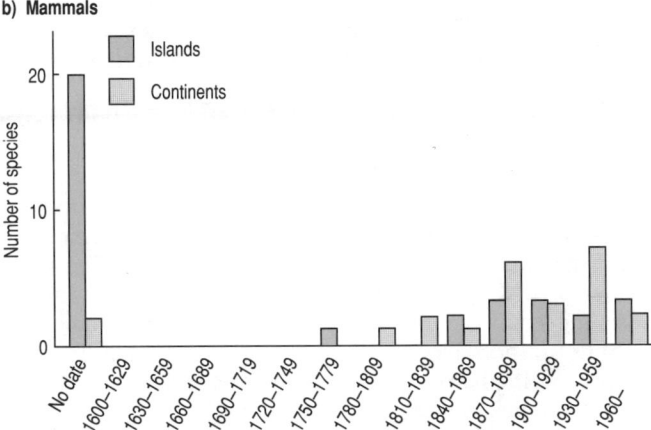

FIGURE 3.14 Time series of a) bird and b) mammal extinctions on islands and continents, 1600–1960
Source: After WCMC, 1992

economic plants and potential crops. Furthermore, the loss of tribal cultures is resulting in the disappearances of unique varieties of many of these crops, as well as special and potentially valuable knowledge about them (Smith and Schultes, 1990). Reid (1992) estimates that current rates of deforestation will lead to the loss of between 2 and 8 per cent of the planet's species in the next 25 years.

Threatened Species Broadly speaking, a threatened species is one that is believed to be at significant risk of extinction in the foreseeable future. During the 1960s, Sir Peter Scott devised the Red Data Book concept which attempted to categorize species at significant risk of extinction, according to the degree of threat. The early Red Data Books, which focused mainly on terrestrial vertebrates and included a wide range of information on each species, were compiled on a global basis by IUCN. Over time, changes have been made to the Red Data Books, including the shortening of range of data included to a direct listing of IUCN recognized threatened species plus the inclusion of data on threatened invertebrates and plants. Since its inception the Red Data Book concept has also been widely adopted at the national level.

The 'IUCN Red List of Threatened Animals' has been published by WCMC since 1986 and is updated every two years. It is based on information collated by the IUCN Species Survival Commission Specialist Groups (IUCN SSC). The International Council for Bird Preservation (ICBP) currently categorizes threatened bird species.

Species covered by the global Red Lists are assigned one of the following IUCN threat categories: Extinct (Ex), Endangered (E), Vulnerable (V), Rare (R), Indeterminate (I), or Insufficiently Known (K). Detailed definitions of these categories are given in the footnotes to Table 3.14 which lists the number of threatened species for each country, as assessed in 1990. In all, some 4,452 known animal species are listed as threatened in the 1990 Red List (IUCN, 1990), the majority of which are tropical birds and mammals. However, it should be noted that this figure is likely to be an underestimate as the IUCN Red List only includes species known to be threatened. Many more species may be under threat; for example, those that are as yet undescribed, and known species whose status have not been reviewed.

Although the causal factors currently putting selected species at the threat of extinction are broadly known, (e.g., habitation loss or modification, over-exploitation, incidental loss and introductions) it is often difficult, or even impossible, to attribute population declines to specific causes. Indeed, in many cases more than one class of threat is responsible. Part of the problem stems from the rather poor quality of available wildlife statistics; with a few notable exceptions many wildlife data have been gathered in an unsystematic manner, often based on irregularly repeated surveys. These problems aside, the assessment of population trends using ecological survey data is beset by statistical problems which hitherto have been largely underestimated. Inappropriate statistical analysis techniques, which fail adequately to allow for the effects of huge stochastic variations, over- and under-sampling and missing data – features which are typical of wildlife data sets – may bias results and lead to erroneous conclusions (Strien, 1993).

Some understanding of the relative impacts of different human activities on species' survival may be gained from an analysis of threats facing mammals (excluding marine mammals) of Australasia and the Americas, the results of which are summarized in Figure 3.15. It is concluded that of the 119 threatened species considered in the study, 94 were threatened by more than one factor, and 27 species faced four or more threats. By far the most common threat was habitat loss and modification, affecting over three-quarters of species. Over-exploitation affected half of the species, with hunting for meat dominating this threat category. Selected aspects of the live trade of animals, in the context of the Convention on International Trade in Endangered Species of Wild Fauna and Flora (CITES) have been covered in previous editions of this report (UNEP 1989, 1991).

Marine turtles are just one example of threatened species affected by a range of different threats. The majority are classified as threatened in both the 'European Red List of Globally Threatened Species' and the 'IUCN Red List of Threatened Animals.' Box 3.3 details some of the threats marine turtles face. Other threatened species including, for example, the African elephant have been described in earlier editions of this report (UNEP, 1991).

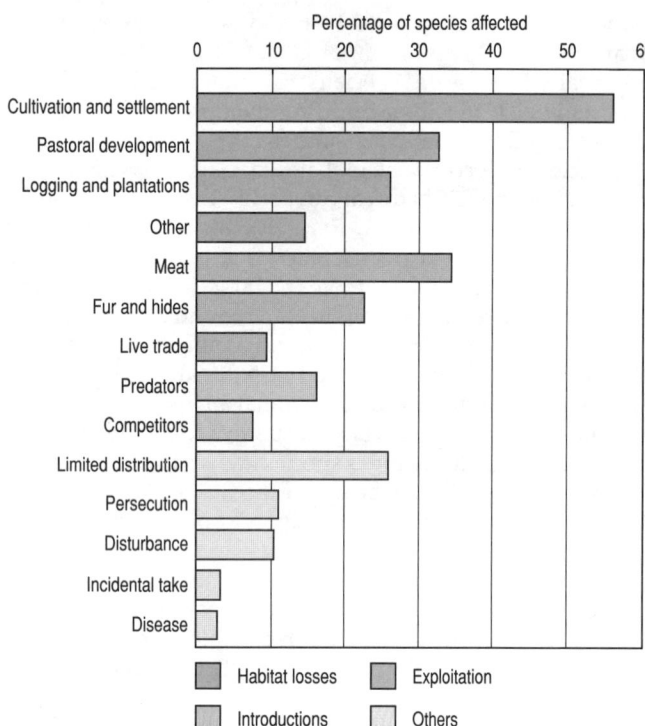

Percentage of species affected

FIGURE 3.15 Relative importance of different threats to mammals (based on an analysis of threats facing animals in Australasia and the Americas) Source: After WCMC, 1992

Loss of Natural Habitats

As noted above, habitat loss or modification is one, if not the, most important cause of loss of species and populations. Loss of natural habitats such as tropical rainforests, coral reefs, wetlands and mangroves are of particular concern in view of their high species diversity. Losses of coral reefs form the focus of discussion in this edition of the 'UNEP Environmental Data Report'; wetlands were covered in the previous edition (UNEP, 1991). Available information on the loss of tropical forests is reviewed in an earlier sub-section within this chapter.

Coral reefs are tropical shallow water ecosystems largely restricted to the seas between the latitudes 30° N and 30° S. The extent of coral reefs in the world is estimated at more than 10^4 km of reefs lining the coasts of 109 countries, or 0.17 per cent of the ocean floor. Reefs are second only to tropical rainforests in terms of biological diversity, and form what are believed to be the most species-rich ecosystems in the oceans. Like rainforests, coral reefs hold considerable untapped potential to science, e.g., reef organisms produce chemicals which have been found to be useful in medical research. Reefs also act as a natural sea defence by protecting coastal lands from erosion and are often desirable tourist destinations by virtue of their sandy beaches and clear lagoons. Locally, reefs are often an important source of fish.

At the present time, comprehensive assessments of the global extent of coral reef degradation and loss are not available. However, a considerable amount of qualitative information about coral reefs has been compiled by UNEP

and IUCN (UNEP/IUCN, 1988a,b,c). Based on this information the main uses of coral reef resources and disturbances to them due to human activities in individual countries have been documented and summarized here in tabular form (Table 3.15). Nutrient enrichment (which may occur as a result of an influx of nutrients from sewage and/or the leaching of soils), mining, fishing, siltation and sedimentation (e.g., due to the washing of unprotected soils into coastal waters from deforested land) are most frequently identified as causes of reef stress. In all, significant amounts of reef damage are reported in 93 countries. As an example, a survey of the Fasht-Al-Adham platform reef in Bahrain showed it to be suffering badly from the effects of sedimentation. At some points well over 50 per cent of the coral was dead. This high level of sedimentation is due to the frequent dredging and infilling along the coast and off-shore (UNEP, 1988). In recent years, rising sea-surface temperatures as a result of global warming have been identified as a further source of reef stress. The evidence for coral reef bleaching in response to temperature changes is briefly reviewed in Part 2: Climate.

It is now generally accepted that the conservation of species' biodiversity is fundamental to the success of the development process. Thus the Convention on Biological Diversity, negotiated at UNCED, includes provision for measures designed to protect animal and plant species and their habitats from extinction and destruction. The main contents of this convention and funding mechanisms for assisting countries to implement programmes which will help to preserve biodiversity (through the Global Environment Facility) are reviewed in Part 10: International Co-operation.

Protected Areas

Protected areas – legally established sites with conservation objectives in mind – are widely perceived as an essential means of conserving biodiversity (WRI/IUCN/UNEP, 1992). Protected areas may be established as part of a national programme, or may be designated as a result of a country's participation in an international programme or convention.

Nationally Protected Areas There are very considerable differences between countries in the mechanisms used to create and maintain systems of protected areas. The rationale behind designating selected sites as protected areas also varies widely across the globe. Box 3.4 contrasts the growth of protected area systems in Canada and China, which despite similar land sizes, have very different histories of protected area designation.

The IUCN, through its Commission on National Parks and Protected Areas (CNPPA), has developed a system of classification for different types of protected areas which enables a degree of inter-country comparison, in terms of the area protected, to be made. This system includes the following five classes of nationally protected areas which have conservation objectives:

I   – Scientific Reserve/Strict Nature Reserve

II   – National Park

III – Natural Monument/Natural Landmark

BOX 3.3 The Threats to Marine Turtles

Turtles the world over, particularly marine turtles, are disappearing rapidly for a number of reasons: These include: a continuing demand for turtle meat and eggs; a thriving market for tortoiseshell and leather products; accidental catching by shrimp trawlers; destruction and over-development of nesting grounds by humans; and environmental pollution (Lehrer, 1990).

All life stages of the sea turtle are vulnerable to the impacts of human activities. For example, the presence of people on the beach, either on foot or in vehicles, can adversely affect nesting, buried eggs and emerging hatchlings. The American Committee on Sea Turtle Conservation has concluded that for juveniles, subadults and breeders in USA coastal waters, the most important human-associated source of mortality is incidental capture in shrimp trawls. The Committee estimates that mortality from shrimping currently lies between 5,000–50,000 loggerheads and 500–5,000 Kemp's ridleys each year (National Research Council, 1990). In the Mediterranean, the incidental turtle catch from long-line fishing off the Balearics alone has been estimated to be as much as 20,000 turtles per year. Activities of other fishing fleets in the Mediterranean result in an annual mortality of between 3,000 and 12,000 maturing turtles (Corbett, 1989).

Many turtle species have been affected by overhunting including, for example, the Kemp's ridley turtle. The slaughter of adult turtles and egg collection at Tamaulipas, Mexico, has drastically reduced populations of this species at this site over the past 30 years. In the early 1960s, approximately 40,000 female turtles were nesting at Tamaulipas but, as egg collection increased, turtle numbers decreased to only 4,000 by the 1970s. Despite measures taken by the Mexican government to protect the turtles, only about 400 turtles were recorded in the late 1980s (Lehrer, 1990).

The world-wide trade in commercial marine turtle products has increased steadily during this century, the most affected species being the

hawksbill and the olive ridley and, to a lesser extent, the green turtle. Three of the nations most heavily involved in producing turtle goods are Japan, Mexico and Indonesia; Mexico and Ecuador alone harvested some 150,000 olive ridleys in 1979, primarily for their skins (Lehrer, 1990).

In the Mediterranean large-scale tourist development has been identified as the most direct and most acute threat to sea turtle nest sites (UN ECE, 1992). Problems include sand being removed for construction, bright lights disorientating nesting turtles, use of off-road vehicles over the nesting zones, and compaction of sand by the large number of tourists which makes it difficult for the hatchlings to emerge.

Both short-term and long-term pollution can have serious effects on turtle populations, especially in developed countries. Widespread oil pollution is another problem for turtles in the Mediterranean; not only can oil and tar produce lesions and sores on baby turtles' skins (which can lead to infection), but many die as a result of ingesting such substances. One fifth of loggerhead turtles examined off Malta were found to have oil contamination in the mouth and gut (Corbett, 1989). Other types of pollution which affect turtles include chemical contamination (e.g., heavy metals) and plastic debris. Turtles have been known to ingest plastic waste which can cause death by blocking their intestinal tract. Litter on the shore occasionally makes it difficult for adult females to negotiate the beach and especially hazardous for their hatchlings to reach the sea.

International trade in turtles of the family *Cheloniidae* – sea turtles – has been banned under Appendix 1 of CITES since 1981. They are also theoretically protected under the Berne Convention. In Europe, four of the six species of sea turtles have been classed as endangered or vulnerable in the 1991 'European Red List of Globally Threatened Animals and Plants' (UN ECE, 1991). *Dermochelys coriacea*, the single

turtle species in the leatherback *Dermochelyidae* family, is also listed as endangered. The marine turtle species (family *Cheloniidae* and family *Dermochelyidae*) are also listed in the '1990 IUCN Red List of Threatened Animals' (IUCN, 1990).

Although sea turtles are given full protection in Cyprus and in Turkey under their respective national fisheries regulations, the conservation status of marine turtles nesting in the Mediterranean is generally poor. The long-term survival prospects are not considered favourable without the rapid implementation of a rigorous protection programme. Such a programme would need to comprise: hatchery programmes, Turtle Exclusion Devices (TEDs) in trawl nets, protection of important nesting beaches, reduced pollution and increased monitoring (Corbett, 1989). Since its inception in 1988 the Mediterranean Association to Save the Sea Turtles (MEDASSET) has been active in co-ordinating and rationalizing efforts to protect marine turtles in the Mediterranean region (Venizelos, 1991).

References

Corbett, K. 1989 *Conservation of European Reptiles and Amphibians*, Christopher Helm (Publishing) Ltd., Bromley.

IUCN 1990 *1990 IUCN Red List of Threatened Animals*, The World Conservation Union, Gland.

Lehrer, J. 1990 *Turtles and Tortoises: A Photographic Survey*, Headline, London.

National Research Council 1990 *Decline of the Sea Turtles: Causes and Prevention*, National Academy Press, Washington DC.

UN ECE 1991 *European Red list of Globally Threatened Animals and Plants*, United Nations, New York.

UN ECE 1992 *The Environment in Europe and North America: Annotated Statistics 1992*, United Nations, New York.

Venizelos, L. E. 1991 Pressure on the endangered Mediterranean marine turtles is increasing: the role of MEDASSET, *Marine Pollution Bulletin*, **23**, 613–616.

IV – Managed Nature Reserve/Wildlife Sanctuary
V – Protected Landscape or Seascape

The number and area of nationally protected areas conforming to IUCN categories I–V inclusive are listed in Table 3.16. On the basis of these data, it may be said that protected areas now exist in 169 countries, and that there are currently 8,491 sites covering approximately 7,734,900 km², or 5.2 per cent of the earth's land area. The largest protected area is Greenland National Park, which covers 972,000 km². In 115 countries, 1,328 sites (covering 3,061,300 km²) have marine or

coastal elements within them. Of these, 94 sites include coral reefs. The largest single marine protected area is the Great Barrier Reef Marine Park in Australia, which covers 340,000 km² (WCMC, 1992). It should be noted that the data in Table 3.16 do not include sites under 10^3 ha (and therefore exclude numerous European small reserves) and what may be described as partially protected areas. These areas may make significant contributions to biodiversity conservation.

Owing to the diversity in national approaches to protected area management, it is difficult to make any statements about

BOX 3.4 The Growth of Protected Areas in Canada and China

The current proportion of protected land in Canada and China is not that dissimilar; however, the history of establishment and expansion of protected areas of various types in each country is quite different.

Canada's first national park was established as early as 1885 at Banff. The rationale for the initial efforts in establishing protected natural areas in Canada were twofold: to preserve outdoor scenic areas for outdoor recreation and tourism; and to protect wildlife habitats in order to ensure continued hunting opportunities. Since 1885 there has been a growth in protected areas (parks, reserves, protected landscapes, managed wildlife areas, conservation areas and sanctuaries) and currently about 7 per cent of the country has some degree of protection at federal, provincial, and territorial government level.

Some early protected areas were of immense proportions, so that the physical land area under protection accelerated rapidly during the 1920s. Areas designated in the subsequent decades were typically smaller, and thus the rate of expansion of protected areas, at least in terms of areal extent, declined. In the 1960s, a resurgence of interest in protected areas resulted in a renewed acceleration of the proportion of land protected in Canada. These trends are illustrated in the figure included here.

Although recreation, tourism, and consumptive use of resources are still common goals, the rationale for designating protected areas in Canada has broadened in the more recent decades. Today national parks, for example, are designated with a view to protecting species and ecosystems and to providing opportunities for education and enjoyment (Environment Canada, 1991).

China, on the other hand, did not adopt a natural protection policy until the 1950s and established her first nature reserve in 1956. In the following 10-year period 18 more reserves were

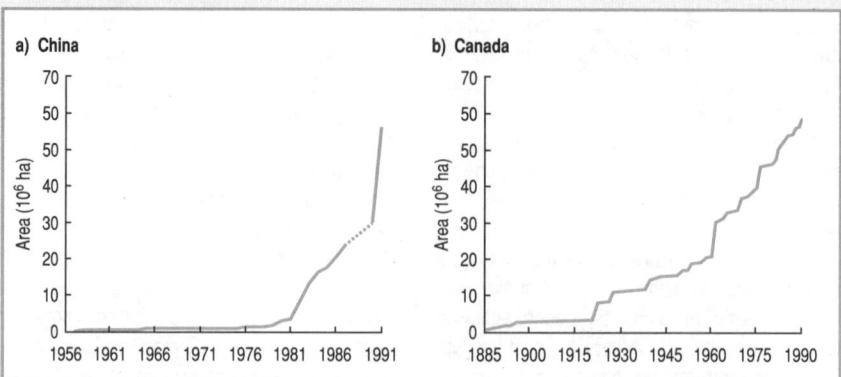

The growth of protected areas in a) Canada, 1885–1990 and b) China, 1956–1991
Sources: Data for Canada are extracted from the Protected Areas Database, Environment Canada. Data for China are taken from Anon., 1989 (data for 1987); Li and Zhao, 1989 (data for 1956–1986); NEPA, 1991 (data for 1991); and Zhu, 1991 (data for 1990)

set up, the majority of which were forest reserves. Between 1966 and 1976, however, a number of China's existing nature reserves were neglected and designation of new reserves was suspended (Li and Zhao, 1989).

Increased awareness of the need for conservation measures world-wide resulted in the declaration of a new set of laws and policies in China in 1976, giving nature conservation top priority. Since that date the number of nature reserves has steadily been increasing (see Figure). By 1991, the number of nature reserves had reached 708, covering a total area of 56×10^6 ha. This represents 5.6 per cent of the total land in China (NEPA, 1991).

China's existing reserves can be divided roughly into the following six categories, depending on the nature and goal of protection: conservation of an intact comprehensive ecosystem; conservation of the source of rare animals; protection of rare relict plants and special types of vegetation; conservation of natural landscape; preservation of special geological

sections and geomorphological features; and protection of the natural environment and natural resources of the coastline (Li and Zhao, 1989). Individual nature reserves are discussed by Zhao Ji et al. (1990) and National Environmental Protection Agency et al. (1989).

References

Anon. 1989 *People's Daily*, September 13, 1989.
Environment Canada 1991 *The State of Canada's Environment,* Government of Canada, Ottawa.
Li Wenhua and Zhao Xianying 1989 *China's Nature Reserves*, Foreign Language Press, Beijing.
National Environmental Protection Agency/Chinese Academy of Sciences/Changchun Institute of Geology 1989 *Atlas of Nature Conservation*, Science Publishing House, Beijing.
NEPA 1991 *Report on the State of the Environment in China 1991*, National Environmental Protection Agency, Beijing.
Zhao Ji, Zheng Guangmei, Wang Huadong, Xu Jialin 1990 *The Natural History of China*, Collins, London.
Zhu Xiaoy 1991 *Guide Journal of Environment*, **6**, 1.

recent trends in protected area establishment, other than the extent of protected areas is increasing in many world regions. In analysing protected areas in terms of types of biome covered, WCMC (1992) noted that regions such as mixed mountain systems or island systems (which typically are not intensively developed) are relatively well accounted for; temperate grasslands and lake systems, which in contrast are more likely to be intensively used by man, are not well represented. This finding suggests that socio-economic and political factors, rather than conservation ideals, may be playing a more

important role in determining the designation of protected areas, and thus brings into question the ability of the current protected area network to ensure adequate conservation of biodiversity (WCMC, 1992).

Internationally Protected Areas There are two international conventions and one international programme that include provision for designation of internationally important sites in any region of the world (within member countries): the Convention on Wetlands of International Importance Especially as Waterfowl Habitat (Ramsar site); the Convention Concerning the Protection of World Cultural and Natural Heritage (World Heritage Sites); and the UNESCO Man and Biosphere (MAB) Programme (Biosphere Resources). Table 3.16 lists the number and area of these three types of internationally protected areas; the footnotes to the table describe the international programme and conventions in more detail, while the discussion here focuses on the Biosphere Reserve concept.

The UNESCO Biosphere Conference in 1968 led to the launching of the MAB Programme in 1971. Out of MAB grew the Biosphere Reserve concept, which has since greatly influenced protected area management. Since 1976, 300 Biosphere Reserves have been established in 76 countries (Table 3.16). Biosphere Reserves are not exclusively designated to protect unique areas, but for a range of objectives which includes research, monitoring, training and demonstration, as well as conservation (WCMC, 1992).

Biosphere Reserves comprise of three zones: the core, which is legally protected and only the minimum disturbance occurs; the buffer zone, which is also legally protected but accommodates a larger variety of resources strategies; and the transition zone, in which reserve management co-operates with local socio-economic development. Thus the Biosphere Reserve concept, in theory, promotes sustainable development of the surrounding regions (Ishwaran, 1992).

In recent years extensions of the original Biosphere Reserve concept have been suggested. One option proposes the organization of Reserves into an integrated network according to a set of criteria based on ecosystem and landscape principles. The existing Biosphere Reserves would thus act as the nodes, or central places, and would be connected by corridors crossing through other types of ecosystems. These corridors can be used to investigate spatial and temporal variations that occur within the ecosystem and landscape of the nodes as a result of ecological and other processes (Dyer and Holland, 1991).

References

Bird, C. 1991 Medicines from the rainforest, *New Scientist*, **17**, 34–39.

Dyer, M. I. and Holland, M. M. 1991 The biosphere-reserve concept: needs for a network design, *BioScience*, **41**(5), 319–325.

Ehrlich, P. R. and Daily, G. C. 1993 Population extinction and saving biodiversity, *Ambio*, **22**(2–3), 64–68.

Ishwaran, N. 1992 Biodiversity, protected areas and sustainable development, *Nature & Resources*, **28**(1), 18–25.

IUCN 1990 *1990 IUCN Red List of Threatened Animals*, International Union for Conservation of Nature, Gland and Cambridge.

Myers, N. 1993 Biodiversity and the precautionary principle, *Ambio*, **22**(2–3), 74–79.

Reid, N. V. 1992 How many species will there be? In: *Tropical Deforestation and Species Extinction*, T. C. Whitmore and J. A. Sayer (Eds), Chapman and Hall, London, 55–73.

Smith, N. J. H. and Schultes, R. E. 1990 Deforestation and shrinking gene-pools in Amazonia, *Environmental Conservation*, **17**(3), 227–234.

Strien, A. J. van 1993 Statistical problems with assessing trends in ecological monitoring data, Background paper presented at the Joint UN ECE/EUROSTAT Work Session on Specific Methodological Issues in Environment Statistics, 20–23 September 1993, Bratislava, Slovakia. Unpublished paper.

UNEP 1988 *Bahrain National State of the Environment* Report 3, United Nations Environment Programme, Nairobi.

UNEP 1989 *United Nations Environment Programme Environmental Data Report 1989/90*, Basil Blackwell, Oxford.

UNEP 1991 *United Nations Environment Programme Environmental Data Report 1991/92*, Basil Blackwell, Oxford.

UNEP/IUCN 1988a *Coral Reefs of the World – Volume I: Atlantic and Eastern Pacific*, International Union for Conservation of Nature and Natural Resources, Gland, and United Nations Environment Programme. Nairobi.

UNEP/IUCN 1988b *Coral Reefs of the World – Volume II: Indian Ocean, Red Sea and Gulf*, International Union for Conservation of Nature and Natural Resources, Gland, and United Nations Environment Programme, Nairobi.

UNEP/IUCN 1988c *Coral Reefs of the World – Volume III: Central and Western Pacific*, International Union for Conservation of Nature and Natural Resources, Gland, and United Nations Environment Programme, Nairobi.

World Conservation Monitoring Centre 1992 *Global Biodiversity: Status of the Earth's Living Resources*, Chapman and Hall, London.

WRI/IUCN/UNEP 1992 *Global Biodiversity Strategy*, World Resources Institute, Washington DC/The World Conservation Union, Gland/United Nations Environment Programme, Nairobi.

TABLE 3.1 Land distribution in countries and changing usage, 1968–1970 and 1988–1990 (mean annual values)

| Region/country | Land area[a] (10⁵ ha) | Distribution (percentage of total land area) | | | | | | | | Major changes in distribution (3 percentage points change or more) 1968–70 to 1988–90 | | | |
| | | Cropland | | Permanent meadows and pastures | | Forest and woodland | | Other land | | Cropland | Pastures | Forest | Other |
		1968–70	1988–90	1968–70	1988–90	1968–70	1988–90	1968–70	1988–90				
WORLD	**130,792**	**10**	**11**	**25**	**26**	**32**	**31**	**32**	**32**				
AFRICA	**29,642**	**5**	**6**	**30**	**30**	**25**	**23**	**39**	**40**				
Algeria	2,381.7	3	3	16	13	2	2	80	82				
Angola	1,246.7	3	3	23	23	44	42	30	32				
Benin	110.6	14	17	4	4	42	32	40	47			−10.1	7.5
Botswana	566.7	2	2	58	58	20	19	20	20				
Burkina	273.8	8	13	37	37	29	24	27	26	4.9		−4.4	
Burundi	25.7	45	52	25	36	2	3	27	10	6.6	10.2		−17.4
Cameroon	465.4	13	15	19	18	58	53	11	14			−4.7	3.3
Cape Verde	4.0	10	10	6	6	–	–	84	84				
Cent. African Rep.	623.0	3	3	5	5	58	57	34	34				
Chad	1,259.2	2	3	36	36	12	10	50	52				
Comoros	2.2	40	45	7	7	16	16	37	33	4.3			−4.3
Congo	341.5	–	–	29	29	63	62	7	8				
Côte d'Ivoire	318.0	9	12	41	41	40	24	11	24			−15.7	12.8
Djibouti	23.2	0	0	9	9	–	–	91	91				
Egypt	995.5	3	3	0	0	–	–	97	97				
Equatorial Guinea	28.1	8	8	4	4	46	46	42	42				
Ethiopia	1,101.0	12	13	42	41	27	25	20	22				
Gabon	257.7	1	2	19	18	78	78	2	2				
Gambia	10.0	13	18	9	9	28	16	50	57	4.8		−11.8	7.0
Ghana	230.0	13	12	22	22	41	35	24	31			−6.1	6.9
Guinea	245.9	3	3	25	25	64	60	8	12			−4.9	4.7
Guinea-Bissau	28.1	10	12	38	38	38	38	13	12				
Kenya	569.7	4	4	67	67	5	4	25	25				
Lesotho	30.4	12	11	70	66	0	0	18	24		−3.6		5.1
Liberia	96.8	4	4	59	59	23	18	14	19			−5.2	5.1
Libya	1,759.5	1	1	6	8	–	–	92	91				
Madagascar	581.5	4	5	58	58	32	27	5	9			−5.2	3.9

Continued

TABLE 3.1 Continued

Region/country	Land area[a] (10⁵ ha)	Distribution (percentage of total land area)								Major changes in distribution (3 percentage points change or more) 1968–70 to 1988–90			
		Cropland		Permanent meadows and pastures		Forest and woodland		Other land		Cropland	Pastures	Forest	Other
		1968–70	1988–90	1968–70	1988–90	1968–70	1988–90	1968–70	1988–90				
Malawi	94.1	22	26	20	20	53	40	6	15	3.7		−12.9	9.2
Mali	1,220.2	1	2	25	25	6	6	68	68				
Mauritania	1,025.2	–	–	38	38	5	4	57	57				
Mauritius	1.9	56	57	4	4	32	31	8	8				
Morocco	446.3	17	20	40	47	17	20	26	13	3.6	6.5		−12.8
Mozambique	784.1	4	4	56	56	21	18	19	22			−3.0	
Namibia	823.3	1	1	46	46	23	22	30	31				
Niger	1,266.7	2	3	8	7	3	2	87	89				
Nigeria	910.8	33	35	44	44	20	13	4	7			−6.6	4.0
Réunion	2.5	22	21	3	4	39	35	36	39			−4.0	3.3
Rwanda	24.7	28	47	33	19	25	23	13	12	18.6	−14.2		
São Tomé & Príncipe	1.0	36	39	1	1	0	0	63	60				
Senegal	192.5	12	12	16	16	59	55	12	17			−4.3	4.3
Seychelles	0.3	19	22	0	0	19	19	63	59	3.7			−3.7
Sierra Leone	71.6	7	9	31	31	30	29	32	31				
Somalia	627.3	2	2	69	69	15	14	15	15				
South Africa	1,221.0	11	11	68	67	3	4	18	19				
St Helena	0.3	6	6	6	6	3	6	84	81			3.2	−3.2
Sudan	2,376.0	5	5	41	46	22	19	32	30		4.8		
Swaziland	17.2	9	12	78	69	8	6	6	14		−9.4		8.2
Tanzania	886.0	4	4	40	40	49	46	8	10				
Togo	54.4	11	12	33	33	33	30	23	25			−3.7	
Tunisia	155.4	28	30	21	22	3	4	47	45				
Uganda	199.6	25	34	9	9	32	28	35	29	8.8		−3.5	−5.2
Western Sahara	266.0	–	–	19	19	0	0	81	81				
Zaire	2,267.6	3	3	7	7	80	77	10	13				
Zambia	743.4	7	7	40	40	41	39	12	14				
Zimbabwe	386.7	6	7	13	13	54	50	28	30			−4.1	

Continued

TABLE 3.1 Continued

| Region/country | Land area[a] (10^5 ha) | Distribution (percentage of total land area) | | | | | | | | Major changes in distribution (3 percentage points change or more) 1968–70 to 1988–90 | | | |
| | | Cropland | | Permanent meadows and pastures | | Forest and woodland | | Other land | | | | | |
		1968–70	1988–90	1968–70	1988–90	1968–70	1988–90	1968–70	1988–90	Cropland	Pastures	Forest	Other
NORTH AMERICA	**21,377**	**12**	**13**	**17**	**17**	**33**	**33**	**37**	**37**				
Antigua & Barbuda	0.4	17	18	7	9	15	11	61	61			−3.8	
Aruba	0.2	11	11	0	0	0	0	89	89				
Bahamas	10.0	1	1	–	–	32	32	67	66				
Barbados	0.4	77	77	9	9	0	0	14	14				
Belize	22.8	2	2	2	2	46	44	50	51				
Bermuda	0.1	0	0	0	0	20	20	80	80				
Br. Virgin Is	0.2	20	27	33	33	7	7	40	33	6.7			−6.7
Canada	9,221.0	5	5	2	3	35	39	58	53			3.8	−4.7
Cayman Is	0.3	0	0	8	8	23	23	69	69				
Costa Rica	51.1	10	10	25	45	52	32	13	12		20.0	−19.7	
Cuba	109.8	22	30	22	27	21	25	35	17	8.8	4.7	4.3	−17.8
Dominica	0.8	23	23	3	3	47	41	28	33			−5.3	5.3
Dominican Rep.	48.4	23	30	43	43	14	13	20	14	6.7			−5.9
El Salvador	20.7	30	35	29	29	9	5	32	30	5.3		−3.7	
Greenland	341.7	0	0	1	1	–	–	99	99				
Grenada	0.3	47	38	3	3	12	9	38	50	−8.8			11.8
Guadelupe	1.7	26	18	10	15	36	40	28	28	−8.7	4.9	3.7	
Guatemala	108.4	14	17	11	13	47	35	27	35			−12.0	7.2
Haiti	27.6	29	33	23	18	3	1	46	48	3.9	−4.8		
Honduras	111.9	14	16	19	23	44	30	23	31		3.4	−14.5	8.7
Jamaica	10.8	23	25	22	18	19	17	36	40		−4.3		4.1
Martinique	1.1	24	19	12	18	42	42	23	22	−5.0	6.0		
Mexico	1,908.7	12	13	39	39	28	23	20	25			−5.9	5.1
Montserrat	0.1	20	20	10	10	10	40	60	30			30.0	−30.0
Neth. Antilles	0.8	9	10	0	0	0	0	91	90				
Nicaragua	118.8	10	11	36	45	48	29	5	15		8.8	−18.9	9.5
Panama	76.0	7	8	15	20	59	44	19	27		5.4	−15.2	8.6
Puerto Rico	8.9	26	14	37	38	15	20	22	28	−11.4		5.1	5.4
St Kitts & Nevis	0.6	26	30	5	5	20	13	49	52	3.3			3.3
St Lucia	0.4	41	39	8	3	19	17	31	42		−5.6	−6.6	10.2

Continued

TABLE 3.1 Continued

Region/country	Land area[a] (10⁵ ha)	Distribution (percentage of total land area)								Major changes in distribution (3 percentage points change or more) 1968–70 to 1988–90			
		Cropland		Permanent meadows and pastures		Forest and woodland		Other land					
		1968–70	1988–90	1968–70	1988–90	1968–70	1988–90	1968–70	1988–90	Cropland	Pastures	Forest	Other
St Pierre & Miquelon	0.2	13	13	0	0	4	4	83	83				
St Vincent & Grenadines	0.4	26	28	3	5	36	36	36	31				−5.1
Trinidad & Tobago	5.1	19	23	2	2	47	43	32	31	4.1		−3.9	
Turks & Caicos Is	0.4	2	2	0	0	0	0	98	98				
USA	9,166.6	21	21	27	26	33	32	19	21				
US Virgin Is	0.3	18	21	26	26	6	6	50	47				
SOUTH AMERICA	**17,529**	**5**	**6**	**25**	**28**	**52**	**48**	**18**	**18**			**−4.8**	
Argentina	2,736.7	9	10	53	52	22	22	15	16				
Bolivia	1,084.4	2	2	26	25	54	51	19	22				
Brazil	8,456.5	4	7	18	22	64	59	14	13		3.7	−5.7	
Chile	748.8	5	6	14	18	12	12	69	64		3.5		−4.3
Colombia	1,038.7	5	5	34	39	54	49	7	7		4.9	−5.8	
Ecuador	276.8	9	10	8	18	59	40	23	31		10.2	−19.0	8.1
Falkland Is	12.2	0	0	99	98	0	0	1	2				
French Guiana	88.2	–	–	–	–	92	83	8	17			−9.1	8.9
Guyana	196.9	2	3	5	6	92	83	1	8			−9.3	7.5
Paraguay	397.3	2	6	36	52	53	36	8	6	3.3	16.1	−17.1	
Peru	1,280.0	2	3	21	21	58	54	19	22			−3.9	3.1
Suriname	156.0	–	–	–	–	96	95	4	4				
Uruguay	174.8	8	7	78	77	3	4	10	11				
Venezuela	882.1	4	4	19	20	41	35	36	41			−6.6	4.6
ASIA	**26,790**	**16**	**17**	**24**	**28**	**22**	**20**	**37**	**35**		**4.2**		
Afghanistan	652.1	12	12	46	46	3	3	39	39				
Bahrain	0.7	3	3	6	6	0	0	91	91				
Bangladesh	130.2	70	71	5	5	17	15	9	10				
Bhutan	47.0	2	3	5	6	52	55	40	36			3.5	−4.4
Brunei	5.3	3	1	1	1	87	45	8	53			−42.7	44.8
Cambodia	176.5	17	17	3	3	76	76	4	4				
China	9,326.4	11	10	29	43	16	14	44	33		13.9		−10.9

Continued

TABLE 3.1 Continued

Region/country	Land area[a] (10^5 ha)	Cropland 1968–70	Cropland 1988–90	Permanent meadows and pastures 1968–70	Permanent meadows and pastures 1988–90	Forest and woodland 1968–70	Forest and woodland 1988–90	Other land 1968–70	Other land 1988–90	Major changes Cropland	Major changes Pastures	Major changes Forest	Major changes Other
Cyprus	9.2	17	17	1	1	13	13	69	69				
East Timor	14.9	5	5	10	10	67	74	17	11			6.7	−6.7
Hong Kong	1.0	13	7	0	1	11	12	76	79	−5.3			
India	2,973.2	55	57	4	4	22	22	18	17				
Indonesia	1,811.6	10	12	7	7	68	63	15	19			−5.3	3.5
Iran	1,636.0	10	9	27	27	11	11	53	53				
Iraq	437.4	11	12	9	9	4	4	75	74				
Israel	20.3	20	21	6	7	5	5	69	66				−3.0
Japan	376.5	15	12	1	2	67	67	18	19				
Jordan	88.9	3	4	9	9	1	1	87	86				
Korea	98.7	23	22	–	1	67	66	9	12				
Korea, Dem.	120.4	15	17	–	–	74	74	10	9				
Kuwait	17.8	–	–	8	8	–	–	92	92				
Laos	230.8	4	4	3	3	64	55	29	37			−8.7	8.4
Lebanon	10.2	31	29	1	1	9	8	58	62				3.4
Macau	–	0	0	0	0	0	0	100	100				
Malaysia	328.6	13	15	–	–	73	60	14	25			−12.5	11.0
Maldives	0.3	10	10	3	3	3	3	83	83				
Mongolia	1,566.5	–	1	89	79	10	9	1	11		−10.1		10.4
Myanmar	657.5	16	15	1	1	49	49	35	35				
Nepal	136.8	14	19	12	15	18	18	56	48	5.4			−7.6
Oman	212.5	–	–	5	5	0	0	95	95				
Pakistan	770.9	25	27	6	6	3	5	66	62				−3.6
Philippines	298.2	24	27	3	4	54	35	19	34	3.2		−19.0	14.5
Qatar	11.0	–	–	5	5	0	0	95	95				
Saudi Arabia	2,149.7	1	1	40	40	1	1	59	59				
Singapore	0.6	20	2	0	0	5	5	75	93	−17.5			17.5
Sri Lanka	64.6	29	29	7	7	28	32	36	32			3.4	−3.5
Syria	183.9	32	30	41	44	3	4	25	23				
Thailand	510.9	26	43	1	2	45	28	28	27	16.8		−17.2	

Continued

TABLE 3.1 Continued

Region/country	Land area[a] (10^5 ha)	Cropland 1968–70	Cropland 1988–90	Permanent meadows and pastures 1968–70	Permanent meadows and pastures 1988–90	Forest and woodland 1968–70	Forest and woodland 1988–90	Other land 1968–70	Other land 1988–90	Cropland	Pastures	Forest	Other
		Distribution (percentage of total land area)								Major changes in distribution (3 percentage points change or more) 1968–70 to 1988–90			
Turkey	769.6	35	36	14	11	26	26	24	26		−3.0		
United Arab Em.	83.6	–	–	2	2	0	–	97	97				
Viet Nam	325.5	19	20	1	1	43	30	38	49			−12.4	
Yemen[b]	528.0	3	3	30	30	8	8	59	59				10.5
EUROPE	**4,727**	**31**	**29**	**19**	**18**	**32**	**33**	**18**	**20**				
Albania	27.4	21	26	24	15	45	38	10	21	4.5	−9.0		11.3
Austria	82.7	20	18	27	24	39	39	14	19				4.7
Belgium[c]	32.8	28	25	25	21	21	21	26	33	−3.5	−4.1		7.2
Bulgaria	110.6	41	38	13	18	33	35	13	9	−3.6	5.3		−3.5
Czechoslovakia	125.4	43	41	14	13	35	37	8	9				
Denmark	42.4	64	61	7	5	11	12	18	23	−3.0			4.5
Faeroe Is	1.4	2	2	0	0	0	0	98	98				
Finland	304.6	9	8	–	–	73	76	18	15			3.3	
France	550.1	35	35	25	21	25	27	15	17		−3.9		
German Dem. Rep.	105.2	46	47	14	12	28	28	13	13				
Germany, Fed. Rep.	244.1	31	31	22	18	30	30	17	21		−3.5		3.6
Greece	128.9	30	31	41	41	20	20	9	8				
Hungary	92.3	61	57	14	13	16	18	10	12	−3.4			
Iceland	100.3	–	–	23	23	1	1	76	76				
Ireland	68.9	20	14	62	68	3	5	14	13	−6.3	5.8		
Italy	294.1	51	41	18	17	21	23	10	20	−10.3			9.4
Liechtenstein	0.2	25	25	31	38	19	19	25	19		6.3		−6.3
Malta	0.3	44	41	0	0	0	0	56	59	−3.1			3.1
Netherlands	33.9	26	27	39	33	9	9	25	31		−6.5		5.8
Norway	306.8	3	3	–	–	25	27	72	70				
Poland	304.4	50	48	14	13	28	29	8	10				
Portugal	92.0	34	34	9	9	32	32	26	24				
Romania	230.3	46	45	19	20	27	28	8	8				
Spain	499.4	41	41	23	20	28	31	7	8			3.3	
Sweden	411.6	7	7	2	1	67	68	23	24			3.3	

Continued

TABLE 3.1 Continued

Region/country	Land area[a] (10^5 ha)	Distribution (percentage of total land area)								Major changes in distribution (3 percentage points change or more) 1968–70 to 1988–90			
		Cropland		Permanent meadows and pastures		Forest and woodland		Other land		Cropland	Pastures	Forest	Other
		1968–70	1988–90	1968–70	1988–90	1968–70	1988–90	1968–70	1988–90				
Switzerland	39.8	10	10	45	40	25	26	21	23		−4.3		4.0
JK	241.6	30	28	49	46	8	10	12	16		−3.2		
Yugoslavia	255.4	32	30	25	25	35	37	8	8				
FORMER USSR	**22,273**	**10**	**10**	**17**	**17**	**41**	**42**	**32**	**31**				
Armenia	29.8[d]						14[e]						
Azerbaijan	86.6[d]						12[e]						
Belarus	207.6[d]						35[e]						
Estonia	45.1[d]						42[e]						
Georgia	69.7[d]						41[e]						
Kazakhstan	2,717.3[d]						6[e]						
Kyrgyzstan	198.5[d]						5[e]						
Latvia	64.6[d]						43[e]						
Lithuania	65.2[d]						30[e]						
Moldova	33.7[d]						11[e]						
Russia	17,075.4[d]						52[e]						
Tajikistan	143.1[d]						5[e]						
Turkmenistan	488.1[d]						17[e]						
Ukraine	603.7[d]						15[e]						
Uzbekistan	447.4[d]						8[e]						
OCEANIA	**8,453**	**5**	**6**	**53**	**51**	**22**	**19**	**19**	**24**			**−3.8**	**4.9**
American Samoa	0.2	20	20	0	0	70	70	10	10				
Australia	7,644.4	5	6	0	55	18	14	19	25		55.0	−4.1	5.5
Cantor Is	0.1	0	0	0	0	0	0	100	100				
Christmas Is.	0.1	0	0	0	0	0	0	100	100				
Cocos Is	–	0	0	0	0	0	0	100	100				
Cook Is.	0.2	26	26	0	0	0	0	74	74				
Fiji	18.3	12	13	4	3	65	65	19	19				
French Polynesia	3.7	7	7	5	5	31	31	56	56				
Guam	0.6	22	22	15	15	18	18	45	45				

Continued

TABLE 3.1 Continued

Region/country	Land area[a] (10⁵ ha)	Distribution (percentage of total land area)								Major changes in distribution (3 percentage points change or more) 1968–70 to 1988–90			
		Cropland		Permanent meadows and pastures		Forest and woodland		Other land					
		1968–70	1988–90	1968–70	1988–90	1968–70	1988–90	1968–70	1988–90	Cropland	Pastures	Forest	Other
Kiribati	0.7	51	52	0	0	3	3	46	45				
Nauru	–	0	0	0	0	0	0	100	100				
New Caledonia	18.3	1	1	14	15	39	39	47	45		3.4		–3.2
New Zealand	268.0	2	2	48	51	27	27	23	20				
Niue	0.3	27	27	3	4	19	19	51	50				
Norfolk Is.	–	0	0	25	25	0	0	75	75				
Pacific Is. Tr. Tr.	1.8	31	33	10	13	22	22	37	31		3.4		–5.8
Papua New Guinea	452.9	1	1	–	–	85	84	14	15				
Samoa	2.8	40	43	–	–	51	47	9	9	3.1		–3.5	
Solomon Is	28.0	2	2	1	1	91	91	5	5				
Tokelau	–	0	0	0	0	0	0	100	100				
Tonga	0.7	61	67	3	6	13	11	24	17	6.0			–7.4
Tuvalu	–	0	0	0	0	0	0	100	100				
Vanuatu	12.2	7	12	2	2	75	75	16	11	4.4			–4.4

a Land area data refer to 1990. 10⁵ha = 10³km².

b Data refer to the unified Republic of Yemen.

c Includes Luxembourg.

d Includes inland water bodies.

e Data are for 1988 and refer to "Total forest area" which includes all stocked and unstocked forest land. "Stocked forest land" is defined as land that supports trees which have a basal area of at least 30 per cent of the "normal" basal area for the respective type of stand, or as land that supports a stand of trees for which the crown closure is at least 30 per cent of the "normal" for the respective type of stand. "Unstocked forest land" is forest land that does not meet the minimum criteria of stocked forest land and includes areas destroyed by fire or cut over with no regeneration or replanting.

Data on land area and land use are provided to the FAO by national governments in response to annual questionnaires. When official estimates are lacking, FAO prepares its own estimates based on national agricultural censuses or unofficial data.

Definitions of land area and land use may vary considerably between countries. Land area data typically exclude the area under major inland water bodies, national claims to the continental shelf and maritime exclusive economic zones. For some countries, national land area may include overseas territories. The world land area total excludes Antarctica. In general, "Cropland" refers to arable land and land under temporary or permanent cultivation; "Permanent meadows and pasture" includes land and pastures used for five years or more for natural or cultivated forage crops; "Forest and woodland" refers to natural or planted stands of trees and includes cleared forest land that is to be reforested; "Other land" includes uncultivated land, wetlands, barren or wasteland, parks, built-on areas and roads.

Land area and land-use data are revised periodically as new information becomes available and therefore values may change significantly from year to year. Moreover, the definitions of land-use categories are sometimes adjusted and the data revised accordingly. Consequently, apparent changes in land use should be viewed with caution. For more detailed notes on definitions and individual countries, please refer to the latest edition of the 'FAO Production Yearbook'.

Sources:
Data extracted from the *AGROSTAT PC* data base of the Food and Agriculture Organization of the United Nations, Rome (1992). Europa Publications Limited 1990 *The Europa World Yearbook 1990, Volume II – Kenya–Zimbabwe*, Europa Publications Limited, London (land area estimates for the former USSR). State Committee for Forestry 1990 *Statistics Concerning the National Forest Inventory, Volume I*, State Committee for Forestry, Moscow (forest areas estimates for the former USSR; in Russian).

TABLE 3.2a   Estimates of the land area affected by all types of human-induced soil degradation, 1980s

Region	Degraded land Area (10⁶ ha)	% of "total"	Degree of degradation (10⁶ ha) Light	Moderate	Strong	Extreme
WORLD	**1,964.4**	**17**	**749.0**	**910.5**	**295.7**	**9.3**
AFRICA	494.2	22	173.6	191.8	123.6	5.2
NORTH AMERICA	158.1	8	18.9	112.5	26.7	0.0
SOUTH AMERICA	243.4	14	104.8	113.5	25.0	0.0
ASIA	748.0	20	294.5	344.3	107.7	0.5
EUROPE	218.9	23	60.6	144.4	10.7	3.1
OCEANIA	102.9	13	96.6	3.9	1.9	0.4

Data in the above table are derived from a world map of human-induced soil degradation prepared by the International Soil Reference and Information Centre (ISRIC) as part of the UNEP-sponsored Global Assessment of Soil Degradation (GLASOD). See Figure 3.3. The map represents a compilation of information submitted to ISRIC by soil science experts from around the world; investigators were requested to assess the status of human-induced soil degradation in the recent past (i.e., averaged over the last 5–10 years) in specific map units. For the purposes of this assessment, human-induced soil degradation is defined as the lowering of the current and/or future capacity of soils to produce goods or services. Two main types of human soil degradation are identified; the first deals with degradation by displacement of soil material, principally by water erosion and wind erosion. The second refers to internal soil deterioration by chemical and physical processes. The degree to which soil is degraded is estimated in relation to changes in agricultural suitability, in relation to declined productivity and in some cases, in relation to biotic functions. Thus the following degrees of soil degradation are defined:

Light: Light degradation is said to have occurred where there has only been a small decline in agricultural productivity and where soils can be fully restored given a change in current land use practices. The original biotic functions are still largely intact.

Moderate: Moderate degradation is said to have occurred where agricultural productivity is greatly reduced and restoration is only possible with major improvements in land management systems. The original biotic functions are partly destroyed.

Strong: Strong degradation is said to have occurred where agricultural land use is no longer possible under local land use management systems. Major engineering works are required for terrain restoration. The original biotic functions are largely destroyed.

Extreme: Extreme degradation is said to have occurred where the terrain is no longer suitable for agriculture and is beyond restoration. The original biotic functions are completely destroyed.

Information regarding the probable cause of human-induced soil degradation was also collated as part of the GLASOD study. Investigators compiling the data base identified one or two of the following five soil degradation causes for each land unit:

Vegetation removal: "Vegetation removal" refers to the complete removal of natural vegetative cover (usually forest or woodland) as a result of agricultural clearing, logging or urban and industrial development.

Overexploitation: "Overexploitation" encompasses the removal of natural vegetation for domestic purposes such as fuelwood, fencing and construction. Overexploitation does not usually lead to the complete removal of vegetation but can lead to degraded vegetation cover and consequently degraded soil.

Overgrazing: "Overgrazing" includes both the actual overgrazing by livestock (reducing vegetation cover) and other effects of livestock such as trampling (which can cause compaction).

Agricultural activities: The category "agricultural activities" encompasses all forms of improper land management practices that can lead to soil degradation including insufficient or excessive use of manure and fertilizers; cultivation on steep slopes or in arid areas without proper anti-erosion measures; improper irrigation; and use of heavy machinery on soils with weak structural stability.

Bio-industrial activities: "Bio-industrial activities" are those activities which give rise to soils contaminated with pollutants, for example, through waste discharge, overuse of pesticides and excessive fertilization.

The GLASOD map has been subsequently digitized to allow calculation of the area of terrain affected by type, degree and cause of soil degradation in any given world region. These data are presented in the above table for all types of soil degradation. Table 3.2b gives regional estimates of land areas affected by soil degradation broken down by cause. The following tables in this series (Tables 3.2c–3.2f) refer to specfic types of soil degradation (i.e., by water and wind erosion and by physical and chemical degradation).

The GLASOD map covers the land surface of the globe between 72°N and 57°S; all data therefore relate to that portion of the earth's surface (13,013 x 10⁶ ha). The "total" land area in the above table refers to the sum of the terrain that is impacted by soil degradation plus that which is classified by GLASOD as "wasteland" (i.e., where few human activities occur), as "stable" (i.e., either naturally stable or stabilized by human activities) and as "other terrain" (i.e., non-degraded by human activities). In all tables in this series the former USSR west of the Ural Mountains is included under Europe and the eastern part is included under Asia.

Source:
Oldeman, L. R., Hakkeling, R. T. A. and Sombroek, W. G. 1991 *World Map of the Status of Human-induced Soil Degradation: An Explanatory Note*, International Soil Reference and Information Centre, Wageningen, and United Nations Environment Programme, Nairobi (2nd Edition).

TABLE 3.2b Causes of human-induced soil degradation, 1980s

Region	Vegetation removal		Overexploitation		Overgrazing		Agricultural activities		Bio-industrial activities	
	Total area (10^6 ha)	% of all degraded land	Total area (10^6 ha)	% of all degraded land	Total area (10^6 ha)	% of all degraded land	Total area (10^6 ha)	% of all degraded land	Total area (10^6 ha)	% of all degraded land
WORLD	578.6	30	132.7	7	678.7	35	551.6	28	22.8	1
AFRICA	66.8	14	62.8	13	243.0	49	121.4	24	–	–
NORTH AMERICA	17.9	11	11.4	7	37.9	24	90.5	57	–	–
SOUTH AMERICA	100.1	41	12.0	5	67.9	28	63.5	26	0.0	0
ASIA	297.8	40	46.1	6	197.3	26	204.4	27	1.4	–
EUROPE	83.8	38	0.5	–	50.0	23	63.9	29	20.6	9
OCEANIA	12.3	12	0.0	0	82.5	80	7.9	8	–	–

In the above table estimates of the land area affected by soil degradation due to human activities are given in terms of the area affected in hectares and as a percentage of the total land area affected by all soil degradation processes.

See also notes to Table 3.2a.

Source: see Table 3.2a.

TABLE 3.2c Estimates of the land area affected by human-induced soil degradation due to water erosion, 1980s

Region/degradation type	Degraded land		Degree of degradation (10^6 ha)			
	Area (10^6 ha)	% of all degraded land	Light	Moderate	Strong	Extreme
WORLD	1,093.7	56	343.2	526.7	217.2	6.6
Topsoil loss	920.3	32	301.2	454.5	161.2	3.8
Terrain deformation	173.3	9	42.0	72.2	56.0	2.8
AFRICA	227.4	46	57.5	67.4	98.3	4.2
Topsoil loss	204.9	41	53.9	60.5	86.6	3.8
Terrain deformation	22.5	5	3.6	6.9	11.7	0.4
NORTH AMERICA	106.1	67	14.5	68.2	23.4	0.0
Topsoil loss	80.9	51	14.2	60.1	6.5	0.0
Terrain deformation	25.2	16	0.2	8.1	16.9	0.0
SOUTH AMERICA	123.2	51	45.9	65.1	12.1	0.0
Topsoil loss	95.1	39	34.9	51.9	8.3	0.0
Terrain deformation	28.1	12	11.0	13.2	3.8	0.0
ASIA	440.6	59	124.5	241.7	73.4	0.0
Topsoil loss	365.2	49	99.8	215.0	50.5	0.0
Terrain deformation	74.4	10	24.7	26.7	22.9	0.0
EUROPE	114.5	52	21.4	81.0	9.8	2.4
Topsoil loss	92.8	42	18.9	64.7	9.2	0.0
Terrain deformation	21.8	10	2.5	16.3	0.6	2.4
OCEANIA	82.8	81	79.4	3.2	0.2	0.0
Topsoil loss	81.7	79	79.4	2.2	0.1	0.0
Terrain deformation	1.1	1	0.0	1.0	0.1	0.0

In the above table estimates of soil degradation due to water erosion are given in terms of the area affected in hectares and as a percentage of the total area affected by all types of soil degradation processes.
Topsoil loss refers to the removal of topsoil by wind or water action.
Terrain deformation is defined as the uneven displacement of soil; it creates rills,

gullies and landslides when caused by water.
See also notes to Table 3.2a.

Source: see Table 3.2a.

TABLE 3.2d   Estimates of the land area affected by human-induced soil degradation due to wind erosion, 1980s

Region/degradation type	Degraded land		Degree of degradation (10^6 ha)			
	Area (10^6 ha)	% of all degraded land	Light	Moderate	Strong	Extreme
WORLD	**548.3**	**28**	**268.6**	**253.6**	**24.3**	**1.9**
Topsoil loss	454.2	23	230.5	213.5	9.4	0.9
Terrain deformation	82.5	4	38.1	30.0	14.4	0.0
Overblowing	11.6	1	0.0	10.1	0.5	1.0
AFRICA	**186.5**	**38**	**88.3**	**89.3**	**7.9**	**1.0**
Topsoil loss	170.7	35	79.1	84.2	7.4	0.0
Terrain deformation	14.3	3	9.2	5.1	0.0	0.0
Overblowing	1.5	–	0.0	0.0	0.5	1.0
NORTH AMERICA	**39.2**	**25**	**2.6**	**34.9**	**1.7**	**0.0**
Topsoil loss	37.5	24	2.5	33.3	1.7	0.0
Terrain deformation	1.7	1	0.1	1.6	0.0	0.0
Overblowing	0.0	0	0.0	0.0	0.0	0.0
SOUTH AMERICA	**41.9**	**17**	**25.8**	**16.1**	**0.0**	**0.0**
Topsoil loss	22.7	9	12.7	10.0	0.0	0.0
Terrain deformation	18.4	8	13.1	5.3	0.0	0.0
Overblowing	0.8	–	0.0	0.8	0.0	0.0
ASIA	**222.2**	**30**	**132.4**	**75.1**	**14.5**	**0.2**
Topsoil loss	165.8	22	116.7	48.9	0.0	0.2
Terrain deformation	47.5	6	15.7	17.3	14.5	0.0
Overblowing	8.9	1	0.0	8.9	0.0	0.0
EUROPE	**42.2**	**19**	**3.2**	**38.2**	**0.0**	**0.7**
Topsoil loss	42.2	19	3.2	38.2	0.0	0.7
Terrain deformation	0.0	0	0.0	0.0	0.0	0.0
Overblowing	0.0	0	0.0	0.0	0.0	0.0
OCEANIA	**16.4**	**16**	**16.3**	**0.0**	**0.1**	**0.0**
Topsoil loss	16.4	16	16.3	0.0	0.1	0.0
Terrain deformation	0.0	0	0.0	0.0	0.0	0.0
Overblowing	0.0	0	0.0	0.0	0.0	0.0

In the above table estimates of the land area affected by human-induced soil degradation due to wind erosion are given in terms of the area affected and as a percentage of the total area affected by all types of soil degradation processes.

Topsoil loss refers to the removal of topsoil by wind or water action.

Terrain deformation is defined as the uneven displacement of soil; it creates dunes and deflation hollows when caused by wind.

Overblowing is defined as the coverage of land by wind-carried particles and is an off-site effect of the wind erosion processes described above.

See also notes to Table 3.2a.

Source: see Table 3.2a.

TABLE 3.2e Estimates of the land area affected by human-induced soil degradation due to chemical deterioration, 1980s

Region/degradation type	Degraded land Area (10^6 ha)	% of all degraded land	Degree of degradation (10^6 ha) Light	Moderate	Strong	Extreme
WORLD	**239.1**	**12**	**93.0**	**103.3**	**41.9**	**0.8**
Nutrient loss	135.3	7	52.4	63.1	19.8	0.0
Salinization	76.3	4	34.8	20.4	20.3	0.8
Pollution	21.8	1	4.1	17.1	0.5	0.0
Acidification	5.7	–	1.7	2.7	1.3	0.0
AFRICA	**61.5**	**12**	**26.0**	**27.0**	**8.6**	**0.0**
Nutrient loss	45.1	9	20.4	18.8	6.2	0.0
Salinization	14.8	3	4.7	7.7	2.4	0.0
Pollution	0.2	–	0.0	0.2	0.0	0.0
Acidification	1.4	–	1.1	0.3	0.0	0.0
NORTH AMERICA	**7.0**	**4**	**0.5**	**5.7**	**0.8**	**0.0**
Nutrient loss	4.2	3	0.1	4.0	0.1	0.0
Salinization	2.3	1	0.3	1.5	0.5	0.0
Pollution	0.4	–	0.0	0.2	0.2	0.0
Acidification	0.1	–	0.1	0.0	0.0	0.0
SOUTH AMERICA	**70.3**	**29**	**26.3**	**31.4**	**12.6**	**0.0**
Nutrient loss	68.2	28	24.5	31.1	12.6	0.0
Salinization	2.1	1	1.8	0.3	0.0	0.0
Pollution	0.0	–	0.0	0.0	0.0	0.0
Acidification	0.0	–	0.0	0.0	0.0	0.0
ASIA	**73.2**	**10**	**31.8**	**21.5**	**19.5**	**0.4**
Nutrient loss	14.6	2	4.6	9.0	1.0	0.0
Salinization	52.7	7	26.8	8.5	17.0	0.4
Pollution	1.8	–	0.0	1.5	0.3	0.0
Acidification	4.1	1	0.4	2.5	1.2	0.0
EUROPE	**25.8**	**12**	**8.1**	**17.1**	**0.6**	**0.0**
Nutrient loss	3.2	1	2.9	0.3	0.0	0.0
Salinization	3.8	2	1.0	2.3	0.5	0.0
Pollution	18.6	8	4.1	14.3	0.1	0.0
Acidification	0.2	–	0.1	0.1	0.0	0.0
OCEANIA	**1.3**	**1**	**0.2**	**0.7**	**0.0**	**0.4**
Nutrient loss	0.4	–	0.2	0.2	0.0	0.0
Salinization	0.9	1	0.0	0.5	0.0	0.4
Pollution	0.0	–	0.0	0.0	0.0	0.0
Acidification	0.0	–	0.0	0.0	0.0	0.0

In the above table estimates of the land area affected by soil degradation due to chemical deterioration are given in terms of the area affected in hectares and as a percentage of the total land area affected by all types of soil degradation processes.

Loss of nutrients is caused by insufficient manuring or fertilizing of fields in poor or moderately fertile areas. Nutrient losses also occur when organic matter is lost following the clearing of vegetation. The loss of nutrients by erosion of fertile topsoil is considered to be a side-effect of erosion and is not distinguished separately.

Salinization is an increase in the salt content of soils. Human-induced soil salinization can occur as a result of poorly managed irrigation schemes, salt water intrusion into the ground water and/or the accumulation of salts from saline ground water or parent rock because of high moisture evaporation rates in intensively cultivated agricultural areas.

Soils degraded as a result of pollution are those which are contaminated by pesticides, urban and industrial wastes, acids, oil and other substances.

Acidification of soils is caused by the over-application of acidifying fertilizers or the drainage of pyrite-containing soils.

See also notes to Table 3.2a.

Source: see Table 3.2a.

TABLE 3.2f   Estimates of the land area affected by human-induced soil degradation due to physical deterioration, 1980s

Region/degradation type	Degraded land Area (10⁶ ha)	% of all degraded land	Degree of degradation (10⁶ ha) Light	Moderate	Strong	Extreme
WORLD	**83.3**	**4**	**44.2**	**26.8**	**12.3**	**0.0**
Compaction	68.2	3	34.8	22.1	11.3	0.0
Waterlogging	10.5	1	6.0	3.7	0.8	0.0
Subsidence of organic soils	4.6	–	3.4	1.0	0.2	0.0
AFRICA	**18.7**	**4**	**1.8**	**8.1**	**8.8**	**0.0**
Compaction	18.2	4	1.4	8.0	8.8	0.0
Waterlogging	0.5	–	0.4	0.1	0.0	0.0
Subsidence of organic soils	0.0	0	0.0	0.0	0.0	0.0
NORTH AMERICA	**5.9**	**4**	**1.3**	**3.8**	**0.8**	**0.0**
Compaction	1.0	1	0.5	0.5	0.0	0.0
Waterlogging	4.9	3	0.8	3.3	0.8	0.0
Subsidence of organic soils	0.0	0	0.0	0.0	0.0	0.0
SOUTH AMERICA	**7.9**	**3**	**6.8**	**0.8**	**0.3**	**0.0**
Compaction	4.0	2	2.9	0.8	0.3	0.0
Waterlogging	3.9	2	3.9	0.0	0.0	0.0
Subsidence of organic soils	0.0	0	0.0	0.0	0.0	0.0
ASIA	**12.1**	**2**	**5.7**	**6.0**	**0.4**	**0.0**
Compaction	9.8	1	4.6	5.0	0.2	0.0
Waterlogging	0.4	–	0.4	0.0	0.0	0.0
Subsidence of organic soils	1.9	–	0.7	1.0	0.2	0.0
EUROPE	**36.4**	**17**	**27.9**	**8.1**	**0.4**	**0.0**
Compaction	33.0	15	24.8	7.8	0.4	0.0
Waterlogging	0.8	–	0.5	0.3	0.0	0.0
Subsidence of organic soils	2.6	1	2.6	0.0	0.0	0.0
OCEANIA	**2.3**	**2**	**0.7**	**0.0**	**1.6**	**0.0**
Compaction	2.3	2	0.7	0.0	1.6	0.0
Waterlogging	0.0	0	0.0	0.0	0.0	0.0
Subsidence of organic soils	0.0	0	0.0	0.0	0.0	0.0

In the above table estimates of the land area affected by soil degradation due to physical deterioration are given in terms of the area affected in hectares and as a percentage of the total area affected by all types of soil degradation processes.

Compaction, sealing and crusting of soils occur when the soil structure deteriorates as a result of the pressure of heavy machinary or cattle trampling. Sealing and crusting of topsoils may occur if the soil cover does not provide sufficient protection against the impact of raindrops.

Waterlogging includes the flooding or inundation of soils by river water and submergence by rain water as a result of human interference with natural drainage systems.

Subsidence of organic soils occurs when the agricultural potential of the land is reduced by drainage and/or oxidation.

See also notes to Table 3.2a.

Source: see Table 3.2a.

TABLE 3.3   Indices of agricultural production and food production, 1961–1991 (relative to a base period, 1979–1981 in which annual average output = 100)

Year	Agricultural production						Food production					
	World		Developed countries		Developing countries		World		Developed countries		Developing countries	
	Total	Per capita	Total	Per capita	Total	Per capita	Total	Per capita	Total	Per capita	Total	Per capita
1961	61.9	89.5	68.4	81.8	55.8	87.3	61.0	88.3	67.3	80.5	55.0	85.9
1962	64.1	91.0	70.4	83.1	58.3	89.1	63.2	89.6	69.2	81.7	57.4	87.6
1963	65.9	91.6	71.5	83.3	60.7	90.6	64.9	90.1	70.1	81.7	59.8	89.2
1964	68.5	93.2	74.0	85.3	63.4	92.3	67.5	91.9	72.7	83.8	62.5	91.0
1965	69.7	92.9	74.1	84.4	65.7	93.2	68.5	91.3	72.8	83.0	64.3	91.3
1966	72.2	94.3	77.8	87.9	67.0	92.7	71.4	93.2	77.2	87.1	65.9	91.1
1967	74.8	95.7	80.5	90.0	69.5	93.7	74.1	94.8	80.1	89.6	68.3	92.1
1968	76.7	96.1	82.9	91.9	70.9	93.2	76.1	95.3	82.4	91.3	69.9	91.9
1969	76.9	94.4	81.2	89.1	72.9	93.5	76.2	93.5	80.6	88.4	71.9	92.2
1970	79.6	95.8	82.8	90.0	76.7	95.9	79.1	95.1	82.1	89.3	76.2	95.3
1971	82.0	96.6	86.2	92.8	78.0	95.2	81.4	96.0	85.7	92.3	77.3	94.2
1972	81.3	93.9	85.2	90.9	77.6	92.4	80.5	93.0	84.5	90.1	76.7	91.3
1973	85.7	97.1	90.7	95.8	81.2	94.3	85.3	96.5	90.3	95.5	80.4	93.4
1974	87.0	96.7	90.9	95.3	83.4	94.6	86.5	96.1	90.7	95.1	82.4	93.5
1975	89.1	97.1	91.6	95.3	86.7	96.2	88.9	97.0	91.5	95.1	86.4	95.9
1976	91.4	97.9	94.8	97.8	88.2	95.8	91.6	98.1	94.8	97.7	88.6	96.2
1977	93.4	98.3	95.8	98.1	91.1	96.9	93.2	98.1	95.5	97.8	91.0	96.7
1978	97.7	101.1	99.9	101.5	95.6	99.6	97.8	101.2	100.1	101.7	95.5	99.5
1979	98.4	100.1	100.0	100.7	96.9	99.0	98.4	100.2	100.0	100.8	96.9	99.0
1980	98.9	98.9	98.7	98.7	99.1	99.2	99.1	99.1	98.9	98.9	99.3	99.3
1981	102.7	101.0	101.3	100.6	104.0	101.9	102.5	100.7	101.1	100.4	103.8	101.7
1982	106.1	102.5	104.3	102.8	107.9	103.5	106.2	102.6	104.5	103.0	107.8	103.5
1983	105.9	100.6	99.6	97.5	111.9	105.1	106.1	100.8	100.0	97.9	112.0	105.3
1984	111.7	104.3	106.6	103.7	116.5	107.2	111.5	104.1	107.0	104.0	115.9	106.7
1985	114.5	105.0	108.1	104.4	120.5	108.6	114.2	104.7	108.2	104.5	120.0	108.1
1986	115.6	104.2	108.5	104.1	122.2	107.9	116.1	104.7	109.1	104.7	122.9	108.5
1987	116.7	103.3	107.8	102.7	124.9	107.9	116.7	103.3	108.1	102.9	125.0	108.0
1988	118.8	103.4	105.4	99.7	131.4	111.1	118.7	103.3	105.4	99.7	131.5	111.3
1989	122.6	104.9	109.6	103.0	134.8	111.7	122.9	105.1	110.1	103.5	135.3	112.1
1990	125.4	105.4	110.6	103.5	139.3	113.0	125.5	105.5	110.7	103.7	139.8	113.4
1991	125.2	103.4	106.5	99.2	142.5	113.2	124.7	103.0	106.4	99.0	142.5	113.1

Indices of agricultural and food production relate to the disposable output (net of animal feed and seed use) of a region's agricultural sector. The index numbers are calculated using Laspeyres formula and are relative to the base year period 1979–1981. The agricultural production index includes all crop and livestock products. The food production index includes all agricultural products which are considered edible and contain nutrients; coffee and tea, however, have virtually no nutritive value and are therefore excluded. For further details on the agricultural and food production indices calculation method, please refer to the latest edition of the 'FAO Production Yearbook'. "Developed countries" include all the European nations, Australia, Canada, Israel, Japan, New Zealand, South Africa, USA and the former USSR.

Source:
Data extracted from the *AGROSTAT PC* data base of the Food and Agriculture Organization of the United Nations, Rome (April 1992).

TABLE 3.4 Total and per capita production of selected agricultural commodities, 1970–1991

Year	World Total (10⁹ kg)	World Per capita (kg)	Developed countries Total (10⁹ kg)	Developed countries Per capita (kg)	Developing countries Total (10⁹ kg)	Developing countries Per capita (kg)	World Total (10⁹ kg)	World Per capita (kg)	Developed countries Total (10⁹ kg)	Developed countries Per capita (kg)	Developing countries Total (10⁹ kg)	Developing countries Per capita (kg)
				Cereals							*Root crops*	
1970	1,204	326	618	589	587	221	561	152	259	247	302	114
1971	1,311	348	710	671	602	222	531	141	238	225	292	108
1972	1,270	330	687	644	583	210	518	135	230	216	288	104
1973	1,377	351	753	700	624	219	585	149	261	242	324	114
1974	1,340	335	705	649	635	218	551	138	238	219	313	107
1975	1,372	336	690	630	683	229	548	134	223	204	325	109
1976	1,480	357	783	710	697	229	543	131	225	203	319	105
1977	1,471	348	773	695	698	224	564	134	227	204	337	108
1978	1,601	373	854	763	747	235	589	137	234	209	355	112
1979	1,553	355	805	713	748	231	578	132	241	214	337	104
1980	1,565	352	796	701	769	232	525	118	184	162	341	103
1981	1,646	364	836	731	810	240	543	120	210	184	333	99
1982	1,707	371	876	761	832	241	547	119	206	179	341	99
1983	1,641	350	749	646	892	253	552	118	204	176	348	99
1984	1,801	378	880	754	921	256	579	121	226	194	353	98
1985	1,840	379	915	779	925	252	565	116	216	184	349	95
1986	1,854	376	910	771	944	251	567	115	224	190	343	91
1987	1,788	356	856	721	932	243	575	114	211	178	363	95
1988	1,744	341	763	639	981	250	563	110	193	162	369	94
1989	1,884	362	879	732	1,005	251	580	111	201	167	379	95
1990	1,971	373	933	774	1,038	254	574	108	191	158	383	94
1991	1,884	350	842	694	1,042	250	575	107	181	149	394	94
				Pulses							*Oil crops*	
1970	44	12	12	12	32	12	35	10	15	14	21	8
1971	43	11	12	11	31	12	37	10	15	14	22	8
1972	42	11	11	11	30	11	36	9	15	14	21	8
1973	43	11	12	12	30	11	40	10	18	17	22	8
1974	43	11	13	12	30	10	39	10	16	14	24	8
1975	40	10	10	9	31	10	42	10	17	16	25	8
1976	45	11	12	11	33	11	40	10	15	14	25	8
1977	43	10	11	10	32	10	45	11	20	18	26	8
1978	45	10	12	11	33	10	48	11	21	18	27	8
1979	41	9	9	8	33	10	51	12	23	21	28	9
1980	41	9	11	10	30	9	50	11	20	18	30	9
1981	41	9	9	8	32	10	54	12	21	18	33	10
1982	46	10	12	10	34	10	57	12	23	20	34	10
1983	48	10	14	12	34	10	54	11	19	17	34	10
1984	50	10	15	13	34	10	60	13	22	19	37	10
1985	51	10	17	14	34	9	65	13	24	21	41	11
1986	53	11	16	14	37	10	65	13	24	20	41	11
1987	55	11	20	17	34	9	68	13	26	22	42	11
1988	56	11	20	17	36	9	69	13	23	20	45	12
1989	55	11	20	17	35	9	72	14	25	21	47	12
1990	59	11	21	18	37	9	75	14	25	21	50	12
1991	60	11	20	16	40	10	77	14	27	22	50	12
				Vegetables							*Fruits*	
1970	272	74	118	113	154	58	237	64	116	110	122	46
1971	283	75	121	114	162	60	239	63	113	106	126	47
1972	284	74	120	113	164	59	236	61	108	101	128	46

Continued

TABLE 3.4 Continued

	World		Developed countries		Developing countries		World		Developed countries		Developing countries	
Year	Total (10^9 kg)	Per capita (kg)	Total (10^9 kg)	Per capita (kg)	Total (10^9 kg)	Per capita (kg)	Total (10^9 kg)	Per capita (kg)	Total (10^9 kg)	Per capita (kg)	Total (10^9 kg)	Per capita (kg)
1973	295	75	129	119	166	58	258	66	128	119	130	46
1974	302	75	129	118	173	59	259	65	121	112	138	47
1975	313	77	130	118	183	61	264	65	123	113	140	47
1976	320	77	128	116	192	63	271	65	125	113	146	48
1977	332	79	135	122	196	63	266	63	115	104	150	48
1978	346	80	141	125	205	65	276	64	120	108	155	49
1979	354	81	142	126	212	65	295	68	135	120	160	49
1980	355	80	138	121	217	66	302	68	134	118	168	51
1981	364	80	142	124	222	66	295	65	124	108	170	50
1982	378	82	149	130	229	66	317	69	143	124	175	51
1983	380	81	146	126	234	66	315	67	138	119	177	50
1984	405	85	156	134	249	69	315	66	130	111	186	52
1985	415	86	151	129	264	72	315	65	123	104	193	52
1986	425	86	151	128	274	73	331	67	133	112	199	53
1987	435	87	152	128	283	74	335	67	126	106	209	55
1988	438	86	153	128	285	73	342	67	125	104	217	55
1989	447	86	158	131	289	72	346	66	127	106	219	55
1990	451	85	155	129	296	72	345	65	123	102	222	54
1991	452	84	154	127	298	71	348	65	118	97	230	55
			Meat						*Milk*			
1970	101	27	70	67	31	12	395	107	317	302	78	29
1971	105	28	73	69	31	12	398	105	317	300	81	30
1972	108	28	75	70	33	12	408	106	324	304	84	30
1973	108	28	75	69	34	12	415	106	330	306	85	30
1974	114	28	80	73	34	12	423	106	334	308	89	30
1975	116	28	80	73	36	12	427	105	335	305	93	31
1976	118	29	81	74	37	12	431	104	334	303	97	32
1977	123	29	84	76	38	12	443	105	345	310	99	32
1978	127	30	87	77	40	13	450	105	349	312	101	32
1979	132	30	89	78	43	13	455	104	351	311	104	32
1980	136	31	90	79	46	14	462	104	354	311	108	33
1981	138	31	90	79	48	14	465	103	353	308	113	33
1982	140	30	90	78	50	14	476	103	360	312	116	34
1983	144	31	93	80	51	15	495	106	374	323	120	34
1984	148	31	95	81	53	15	498	104	375	322	123	34
1985	153	32	96	82	57	15	508	105	378	322	130	35
1986	157	32	98	83	59	16	517	105	383	325	134	36
1987	163	32	101	85	62	16	516	103	378	319	138	36
1988	169	33	103	86	66	17	524	102	380	319	143	37
1989	172	33	103	86	69	17	530	102	382	318	149	37
1990	177	33	104	87	72	18	539	102	383	318	156	38
1991	179	33	104	85	75	18	528	98	370	305	159	38

Production data for agricultural commodities are compiled by the FAO on an annual basis from questionnaire responses and from decennial agricultural censuses.

Data for cereals refer to crops harvested for dry grain only; cereals harvested for hay, green feed or for grazing purposes are excluded. Data for pulses refer to crops harvested for dry grain only. Data for root crops exclude root crops grown for animal feed. Production data for oil crops are expressed in terms of oil equivalent. Data for vegetables relate to crops grown primarily for human consumption, and generally refer to crops grown in fields and market gardens for sale. Data for fruits refer to total production of fresh fruit irrespective of end use and generally refer to plantation or orchard crops. Production from scattered trees and wild plants used mainly for home consumption is not included. Data for meat production relate to both commercial and farm slaughter, and exclude offal and carcass fat. Data for milk refer to cow, buffalo, sheep and goat milk.

Per capita production values are based on UN population statistics. "Developed countries" include all the European nations, Australia, Canada, Israel, Japan, New Zealand, South Africa, USA and the former USSR.

Sources:
Data extracted from the *AGROSTAT PC* data base of the Food and Agriculture Organization of the United Nations, Rome (April 1992).
UN Population Division 1991 *World Population Prospects 1990*, United Nations, New York.

TABLE 3.5   Agricultural resources and inputs in selected countries, 1968–1970 and 1988–1990 (mean annual values)

Region/country	Fertilizer consumption				Arable land			Irrigated land			
	Total (10³ t a⁻¹)		Per cropping area (10³ kg ha⁻¹ a⁻¹)		Potential area (10³ ha)	Actual area, 1988–90		Area (10³ ha)		As % of arable land	
	1968–70	1988–90	1968–70	1988–90		(10³ ha)	(% of total)	1968–70	1988–90	1968–70	1988–90
WORLD	**64,200**	**142,219**	**46.8**	**98.6**		**1,442,292**		**164,687**	**234,003**	**12.0**	**16.2**
AFRICA	**1,478**	**3,647**	**9.1**	**20.2**		**180,951**		**8,715**	**11,188**	**5.4**	**6.2**
Algeria	83	138	12.2	18.2	7,700	7,606	99	237	336	3.5	4.4
Angola	10	16	2.9	4.8	77,300	3,400	4				
Benin	4	6	2.8	3.1	6,300	1,860	30	2	6	0.1	0.3
Botswana	2	1	1.4	0.7	1,700	1,380	81	1	2	0.1	0.1
Burkina	–	17	0.2	4.6	10,700	3,564	33	4	18	0.2	0.5
Burundi	–	3	0.3	2.3	1,000	1,336	134	24	71	2.1	5.3
Camercon	17	28	2.9	4.0	31,500	7,008	22	6	28	0.1	0.4
Cape Verde						39		2	2	5.0	5.1
Cent. African Rep.	2	1	1.1	0.4	35,800	2,006	6				
Chad	2	5	0.6	1.6	7,000	3,205	46	5	10	0.2	0.3
Congo	6	1	43.0	6.4	21,700	168	1	–	4	0.2	2.4
Côte d'voire	15	39	5.6	10.5	14,100	3,670	26	17	62	0.6	1.7
Egypt	347	990	122.9	381.9	2,900	2,591	89	2,826	2,591	100.0	100.0
Equatoʳial Guinea	2	0	7.6			230					
Ethiopia	5	96	0.4	6.9	25,000	13,930	56	154	162	1.2	1.2
Gabon	0	1		2.4	12,900	454	4				
Gambia	–	1	2.1	6.6	500	177	35	7	12	5.7	6.8
Ghana	2	11	0.7	4.2	11,000	2,710	25	5	8	0.2	0.3
Guinea	3	1	3.9	0.9	7,500	727	10	5	24	0.7	3.3
Guinea-Bissau	0	–		1.4		335					
Kenya	42	119	20.5	49.1	6,700	2,428	36	26	52	1.3	2.1
Lesotho	–	5	1.0	14.3	300	320	107				
Liberia	2	2	4.8	5.7	4,200	373	9	1	2	0.2	0.5
Libya	10	81	5.2	37.9	2,100	2,150	102	170	242	8.4	11.3
Madagascar	11	8	4.8	2.7	32,800	3,091	9	330	903	14.1	29.2
Malaw	8	51	4.1	21.3	4,100	2,405	59	4	19	0.2	0.8
Mali	3	15	2.0	7.4	16,800	2,093	12	75	205	4.4	9.8
Mauritania	–	2	0.8	12.1	1,400	204	15	8	12	2.9	5.9
Mauritius	22	29	209.5	277.0	100	106	106	15	17	14.4	16.0
Morocco	93	315	12.5	34.8	7,700	9,056	118	915	1,265	12.3	14.0

Continued

TABLE 3.5 Continued

Region/country	Fertilizer consumption Total (10³ t a⁻¹) 1968–70	1988–90	Per cropping area (10³ kg ha⁻¹ a⁻¹) 1968–70	1988–90	Arable land Potential area (10³ ha)	Actual area, 1988–90 (10³ ha)	(% of total)	Irrigated land Area (10³ ha) 1968–70	1988–90	As % of arable land 1968–70	1988–90
Mozambique	7	2	2.3	0.7	41,400	3,110	8	24	113	0.8	3.6
Namibia						662		4	4	0.6	0.6
Niger	–	2	0.1	0.5	11,800	3,605	31	17	38	0.6	1.1
Nigeria	9	363	0.3	11.3	47,900	32,057	67	801	865	2.7	2.7
Réunion	14	13	254.1	252.4		53		5	6	9.3	10.8
Rwanda	–	1	0.2	1.2	800	1,154	144	4	4	0.6	0.3
Senegal	9	17	3.9	7.1	9,700	2,350	24	107	178	4.6	7.6
Sierra Leone	2	1	3.8	1.8	2,600	643	25	5	33	1.0	5.2
Somalia	3	3	2.9	2.4	1,800	1,039	58	95	116	10.1	11.2
South Africa	524	797	39.8	60.5		13,174		977	1,128	7.4	8.6
Sudan	38	59	3.3	4.6	64,200	12,867	20	1,610	1,890	13.9	14.7
Swaziland	5	8	35.1	37.6	900	199	22	46	62	29.8	31.1
Tanzania	12	46	3.6	13.7	36,600	3,366	9	36	148	1.1	4.4
Togo	–	12	0.5	17.9	2,100	664	32	4	7	0.6	1.1
Tunisia	34	96	7.7	20.8	4,600	4,613	100	85	273	1.9	5.9
Uganda	5	–	1.1	–	10,700	6,707	63	4	9	0.1	0.1
Zaire	4	6	0.6	0.7	177,700	7,853	4	0	10	0.0	0.1
Zambia	24	74	4.8	14.2	51,100	5,258	10	7	31	0.1	0.6
Zimbabwe	95	162	40.6	57.5	15,900	2,811	18	46	219	2.0	7.8
NORTH AMERICA	**16,767**	**23,314**	**63.0**	**85.1**		**273,893**		**20,660**	**26,568**	**7.8**	**9.7**
Bahamas	1	–	118.5	36.7							
Barbados	6	3	166.8	88.9		33					
Belize	3	5	59.4	86.2		56		–	2	0.7	3.6
Bermuda	0	–									
Canada	751	2,124	17.2	46.2		45,963		410	840	0.9	1.8
Costa Rica	50	106	103.0	200.9	2,600	528	20	26	117	5.3	22.1
Cuba	437	613	184.5	184.1	5,400	3,332	62	427	889	18.0	26.7
Dominica	0	4		211.8							
Dominican Rep.	28	74	25.2	51.3	1,700	1,443	85	123	225	11.0	15.6
El Salvador	59	84	93.9	113.9	1,000	733	73	20	120	3.2	16.4
Guadeloupe	10	9	222.4	290.8		30		2	3	4.5	10.0
Guatemala	40	128	25.8	68.3	4,500	1,875	42	54	78	3.5	4.1
Haiti	–	2	0.3	2.5	900	904	100	55	73	6.9	8.1

Continued

TABLE 3.5 Continued

Region/country	Fertilizer consumption				Arable land			Irrigated land			
	Total (10³ t a⁻¹)		Per cropping area (10³ kg ha⁻¹ a⁻¹)		Potential area (10³ ha)	Actual area, 1988–90		Area (10³ ha)		As % of arable land	
	1968–70	1988–90	1968–70	1988–90		(10³ ha)	(% of total)	1968–70	1988–90	1968–70	1988–90
Honduras	22	41	14.0	22.7	3,900	1,805	46	67	89	4.4	4.9
Jamaica	23	26	91.7	98.5	400	269	67	24	35	9.7	12.9
Martinique	15	21	582.7	1,090.9		20		1	4	4.0	20.3
Mexico	532	1,681	22.9	68.0	45,400	24,710	54	3,494	5,143	15.1	20.8
Nicaragua	27	49	22.3	38.6	5,400	1,271	24	35	85	2.9	6.7
Panama	19	37	35.0	57.5	3,100	640	21	19	31	3.5	4.9
Puerto Rico						128		39	39	17.0	30.5
St Kitts & Nevis	2	2	130.7	116.8							
St Lucia	3	5	177.1	270.5		18		1	1	6.3	5.6
St Vincent & Grenadines	2	3	236.7	284.8		11		1	1	10.0	9.1
Trinidad & Tobago	8	6	84.0	46.5	200	120	60	15	22	15.2	18.3
USA	14,729[a]	18,290[a]	78.3[a]	96.3[a]		189,915		15,845	18,771	8.4	9.9
US Virgin Is	1	1	116.7	185.7							
SOUTH AMERICA	**1,341**	**5,336**	**16.4**	**47.5**		**112,235**		**5,431**	**8,606**	**6.7**	**7.7**
Argentina	78	162	3.0	6.0	85,800	27,200	32	1,247	1,667	4.8	6.1
Bolivia	2	9	1.3	4.1	44,400	2,296	5	80	165	4.8	7.2
Brazil	745	3,420	22.3	58.5	504,300	58,450	12	752	2,567	2.3	4.4
Chile	112	311	27.9	69.4	5,700	4,472	78	1,160	1,263	29.0	28.2
Colombia	144	538	28.7	99.9	37,900	5,382	14	247	515	4.9	9.6
Ecuador	46	75	18.2	27.6	11,900	2,720	23	465	550	18.3	20.2
French Guiana	0	1		87.5		10		1	2	60.0	19.4
Guyana	10	14	26.7	28.7	10,600	495	5	115	130	31.2	26.3
Paraguay	5	14	5.2	6.4	22,700	2,216	10	40	67	4.4	3.0
Peru	78	164	28.4	44.0	30,200	3,728	12	1,082	1,250	39.5	33.5
Suriname	2	2	51.3	23.0	11,300	68	1	25	58	67.0	85.8
Uruguay	66	71	46.2	54.1	10,600	1,304	12	45	110	3.2	8.4
Venezuela	53	555	15.2	142.7	39,700	3,893	10	172	263	4.9	6.8
ASIA	**10,534**	**53,828**	**24.0**	**118.2**		**455,305**		**107,094**	**147,456**	**24.4**	**32.4**
Afghanistan	17	52	2.1	6.4	8,200	8,054	98	2,320	2,750	29.0	34.1
Bahrain	0	–		242.7		2		1	1	50.0	50.0
Bangladesh	127	871	14.0	94.9	9,400	9,178	98	1,244	2,673	13.7	29.1

Continued

TABLE 3.5 Continued

Region/country	Fertilizer consumption Total (10³ t a⁻¹) 1968–70	1988–90	Per cropping area (10³ kg ha⁻¹ a⁻¹) 1968–70	1988–90	Arable land Potential area (10³ ha)	Actual area, 1988–90 (10³ ha)	(% of total)	Irrigated land Area (10³ ha) 1968–70	1988–90	As % of arable land 1968–70	1988–90
Bhutan	0	–		0.8		131		15	34	14.9	26.0
Brunei	0	2		252.4		7		0	1	0.0	14.3
Cambodia	4	4	1.4	1.3	8,000	3,056	38	93	92	3.0	3.0
China	3,526	25,942	34.3	268.7		96,544		37,232	45,998	36.2	47.6
Cyprus	26.0ᵇ	22.2ᵇ	163.6ᵇ	142.5ᵇ	400	156	39	30	34	18.9	22.0
Gaza Strip						24		10	11	47.6	45.8
Hong Kong						7		8	2	61.5	31.8
India	2,000	11,748	12.1	69.5	169,000	169,127	100	28,880	43,046	17.5	25.5
Indonesia	230	2,416	12.8	111.0	47,900	21,767	45	4,284	7,550	23.8	34.7
Iran	87	1,139	5.5	75.7	14,000	15,050	108	5,117	5,750	32.7	38.2
Iraq	14	211	2.8	38.8	7,800	5,450	70	1,453	2,546	29.1	46.7
Israel	52	106	127.8	244.4		435		167	213	40.9	48.9
Japan	2,072	1,907	372.2	411.1		4,638		3,420	2,868	61.5	61.8
Jordan	3	23	9.9	60.2	500	388	78	34	62	10.8	15.9
Korea	525	938	227.5	441.7	2,200	2,125	97	1,170	1,355	50.6	63.8
Korea, Dem.	270	820	152.1	410.5	2,300	1,997	87	500	1,393	28.1	69.8
Kuwait	0	1		214.6		4		1	2	66.7	50.0
Laos	1	1	1.1	0.8	3,700	904	24	17	121	2.0	13.3
Lebanon	34	25	105.3	83.7	300	301	100	68	86	21.2	28.6
Malaysia	156	839	35.5	171.8	10,000	4,880	49	237	341	5.4	7.0
Mongolia	1	17	0.9	12.5		1,376		8	77	1.0	5.6
Myanmar	27	87	2.6	8.6	20,900	10,038	48	824	1,008	7.9	10.0
Nepal	4	66	2.3	24.9	3,800	2,631	69	114	948	6.0	36.0
Oman	0	7	250.0	114.9		58		28	55	90.2	94.9
Pakistan	279	1,841	14.5	88.9	23,100	20,713	90	12,930	16,133	67.1	77.9
Philippines	184	543	26.2	68.1	13,300	7,970	60	802	1,537	11.4	19.3
Qatar	0	1		208.7							
Saudi Arabia	4	490	2.9	209.9	1,300	2,335	180	363	850	26.3	36.4
Singapore	3	6	250.0	4,200.0		1					
Sri Lanka	108	197	57.3	103.7	2,700	1,900	70	427	535	22.6	28.2
Syria	33	278	5.7	49.9	6,000	5,563	93	491	671	8.3	12.1
Thailand	95	874	7.0	39.5	16,400	22,114	135	1,862	4,217	13.8	19.1
Turkey	419	1,766	15.4	63.4	28,000	27,853	99	1,700	2,340	6.2	8.4

Continued

TABLE 3.5 Continued

Region/country	Fertilizer consumption				Arable land			Irrigated land			
	Total (10³ t a⁻¹)		Per cropping area (10³ kg ha⁻¹ a⁻¹)		Potential area (10³ ha)	Actual area, 1988–90		Area (10³ ha)		As % of arable land	
	1968–70	1988–90	1968–70	1988–90		(10³ ha)	(% of total)	1968–70	1988–90	1968–70	1988–90
United Arab Em.	0	10		256.6		39		4	5	40.6	12.8
Viet Nam	233	558	38.6	84.6	10,100	6,598	65	980	1,830	16.2	27.7
West Bank	–					206		7	10	3.6	4.8
Yemen[c]		19	0.2	12.0	3,000	1,596	53	255	310	18.1	19.4
EUROPE	**23,435**	**29,957**	**160.2**	**215.2**		**139,205**		**10,648**	**17,008**	**7.3**	**12.2**
Albania	39	102	66.8	143.3		709		263	422	45.0	59.4
Austria	398	310	237.0	203.6		1,523		4	4	0.2	0.3
Belgium[d]	510	408	546.9	498.6		819		1	1	0.1	0.1
Bulgaria	724	814	159.3	196.4		4,147		978	1,253	21.5	30.2
Czechoslovakia	1,197	1,504	224.0	294.5		5,107		124	291	2.3	5.7
Denmark	577	633	214.2	246.6		2,565		85	428	3.2	16.7
Finland	457	477	171.1	195.3		2,443		13	63	0.5	2.6
France	4,300	5,929	223.7	309.9		19,133		740	1,159	3.8	6.1
German Dem. Rep.	1,507	1,785[e]	311.4	362.8[e]		4,920		135	150	2.8	3.0
Germany, Fed. Rep.	3,023	2,972[e]	396.8	397.4[e]		7,479		280	330	3.7	4.4
Greece	325	674	83.1	171.3		3,933		718	1,190	18.4	30.3
Hungary	722	1,103	128.8	208.6		5,287		232	185	4.1	3.5
Iceland	23	22	3,488.5	2,687.8		8					
Ireland	389	696	280.3	730.2		953					
Italy	1,243	1,897	82.6	157.8		12,023		2,537	3,100	16.9	25.8
Malta	1	1	37.7	47.7		13		1	1	7.1	7.7
Netherlands	611	598	686.6	642.8		930		370	550	41.6	59.1
Norway	192	211	232.3	241.9		873		30	95	3.6	10.9
Poland	2,376	2,800	154.5	189.7		14,757		238	100	1.5	0.7
Portugal	171	282	55.4	89.0		3,170		621	631	20.1	19.9
Romania	539	1,303	51.1	126.9		10,267		630	3,228	6.0	31.4
Spain	1,132	2,041	55.0	100.3		20,345		2,376	3,358	11.6	16.5
Sweden	477	357	156.7	125.4		2,850		29	112	1.0	3.9
Switzerland	143	174	369.5	422.8		412		25	25	6.4	6.1
UK	1,771	2,467	243.2	364.3		6,772		92	156	1.3	2.3
Yugoslavia	589	893	71.5	115.2		7,755		126	176	1.5	2.3
USSR	**9,146**	**24,432**	**39.2**	**105.8**		**230,830**		**10,605**	**21,020**	**4.5**	**9.1**

Continued

TABLE 3.5 Continued

Region/country	Fertilizer consumption				Arable land			Irrigated land			
	Total (10³ t a⁻¹)		Per cropping area (10³ kg ha⁻¹ a⁻¹)		Potential area (10³ ha)	Actual area, 1988–90		Area (10³ ha)		As % of arable land	
	1968–70	1988–90	1968–70	1988–90		(10³ ha)	(% of total)	1968–70	1988–90	1968–70	1988–90
OCEANIA	**1,498**	**1,706**	**34.6**	**34.2**		**49,873**		**1,534**	**2,156**	**3.5**	**4.3**
Australia	1,054	1,315	25.3	27.2		48,275		1,435	1,877	3.4	3.9
Fiji	6	24	26.1	99.0		240		—	1	0.4	0.4
French Polynesia	0	1	—	33.3							
New Caledonia	0	1	—	70.0							
New Zealand	437	351	784.4	838.3		419		98	278	17.6	66.4
Papua New Guinea	2	14	4.7	35.7		390					

a   Includes Puerto Rico.
b   From 1974 onwards data refer to part of the country only.
c   Data refer to the unified Yemen Arab Republic.
d   Includes Luxembourg.
e   Average for two years only, 1988–1989.

With respect to the fertilizer data, a zero in the above table signifies zero magnitude or data not available. A dash signifies negligible quantity, generally less than half the unit specified. The mean annual consumption of commercial fertilizers represents the quantity of nutrients applied to cropland in the form of nitrogen (N), phosphate (P_2O_5) and potash (K_2O). The fertilizer year is from 1 July to 30 June; thus for most countries data refer to the year beginning in July.

Data on fertilizer production, trade and consumption are compiled by the FAO on an annual basis from questionnaire responses and from decennial agricultural censuses. For further information relating to these data, please refer to the most recent edition of the 'FAO Fertilizer Yearbook'.

Arable land is defined as the physical land area used for growing crops, both annual and perennial. Data listed represent an annual average over the three-year period 1988–90. Regional totals for arable land areas may include countries not listed.

"Potential" arable land refers to the land area which is at present arable or is potentially arable, i.e., suitable for crops when developed. Estimates of potential arable land as presented above have been derived as part of the FAO Agro-Ecological Zone Study. In this study data from the FAO/UNESCO Soil Map of the World were combined with climatic data in order to estimate the extent of lands that were suited to the production of various major crops. Some adjustments for factors that were not taken into account in the original FAO study have since been made to the data. Countries reporting actual arable land areas that are greater than their potential arable land (i.e., arable land as a percentage of potential is greater than 100 per cent) have expanded their cropland to marginal areas.

Irrigated land refers to areas purposefully provided with water, either several times or only once during the year. Irrigated land includes areas flooded by river water for crop production or pasture improvement.

Data on arable land area and irrigated land area are compiled by the FAO on an annual basis from questionnaire responses and from decennial agricultural censuses.

Sources:
Data extracted from the *AGROSTAT PC* data base of the Food and Agriculture Organization of the United Nations, Rome (1992). Food and Agriculture Organization of the United Nations 1988 *World Agriculture: Toward 2000: An FAO Study*, (N. Alexandratos Ed.), Belhaven Press, London (estimates of potential arable land areas).

TABLE 3.6   Forest resources in temperate zones as assessed by the UN ECE/FAO, 1981–1990

| Region/country | Year(s) covered | Forest and other wooded land | | | | Exploitable forest | | | Net change in forest and other wooded land, 1981–90 (10³ ha) |
		Total area (10³ ha)	Area per capita (ha)	Forest area (10³ ha)	Other wooded land area (10³ ha)	Area (10³ ha)	As % of total forest and other wooded land	Per capita (ha)	
TOTAL		2,063,575	1.62	1,432,467	631,109	897,540[a]	43	0.70	
NORTH AMERICA		749,289	2.71	456,737	292,552[b]	307,673	41	1.11	
Canada	1986	453,300	17.09	247,164	206,136[b]	112,077	25	4.23	[c]
USA	1987	295,989	1.18	209,573	86,416[b]	195,596	66	0.78	–3,165
EUROPE[d]		194,953	0.35	149,305	45,649	132,958[a]	68	0.24	1,909[e]
Albania	1990	1,449	0.45	1,046	403[f]	910	62	0.28	1
Austria	1986–90	3,877	0.50	3,877	[g]	3,330	86	0.43	142
Belgium	1980	620	0.06	620	–	620	100	0.06	19
Bulgaria	1990	3,683	0.41	3,386	298	3,222	87	0.36	78
Cyprus	1990	280	0.40	140	140[h]	88	31	0.13	2[b]
Czechoslovakia	1988	4,491	0.29	4,491	–	4,491	100	0.29	20
Denmark	1979	466	0.09	466	–	466	100	0.09	10
Finland	1980–89	23,373	4.68	20,112	3,261	19,511	83	3.91	55
France	1976–88	14,155[i]	0.25	13,110	1,044	12,460	88	0.22	80
Germany	1987–89	10,735[b]	0.13	10,490[b]	245[b]	9,852[b]	92[b]	0.12[b]	
German Dem. Rep.	1988–89	2,981	0.18	2,938	43	2,476	83	0.15	33
Germany, Fed. Rep.	1987–89	7,754	0.12	7,552	202	7,376	95	0.12	436
Greece	1964	6,032	0.60	2,512	3,520	2,289	38	0.23	9
Hungary	1990	1,675	0.16	1,675	–	1,324	79	0.13	82
Iceland	1970–85	134	0.54	11[j]	123				
Ireland	1989	429	0.12	396	33	394	92	0.11	48
Israel	1989–90	124	0.03	102	22	80	65	0.02	
Italy	1988	8,550	0.15	6,750	1,800[h]	4,387[h]	51	0.08	1[h]
Luxembourg	1989	87	0.24	85	3	82[h]	94	0.22	
Netherlands	1982–85	334	0.02	334	–	331	99	0.02	10[k]
Norway	1980–86	9,565	2.26	8,697	868	6,638	69	1.57	
Poland	1989	8,672	0.23	8,672	–	8,460	98	0.22	50
Portugal	1980–86	3,102	0.29	2,755	347	2,346	76	0.22	138
Romania	1990	6,265	0.27	6,190	75	5,413	86	0.23	2
Spain	1990	25,622	0.66	8,388	17,234	6,506	25	0.17	9

Continued

TABLE 3.6 Continued

Region/country	Year(s) covered	Forest and other wooded land				Exploitable forest			Net change in forest and other wooded land, 1981–90 (10³ ha)
		Total area (10³ ha)	Area per capita (ha)	Forest area (10³ ha)	Other wooded land area (10³ ha)	Area (10³ ha)	As % of total forest and other wooded land	Per capita (ha)	
Sweden	1985–89	28,015	3.27	24,437	3,578	22,048	79	2.58	0[h]
Switzerland	1983–85	1,186	0.18	1,130	56	1,093	92	0.16	66
Turkey	1990	20,199	0.34	8,856	11,343	6,642	33	0.11	31
UK	1988–89	2,380	0.04	2,207	173	2,207	93	0.04	242
Yugoslavia	1987–88	9,453	0.40	8,370	1,083	7,768	82	0.33	345
USSR	**1988**	**941,530**	**3.26**	**754,958**	**186,572**	**414,015**	**44**	**1.43**	**22,600**
Byelorussian SSR	1988	6,256	0.61	6,016	240	5,392	86	0.53	273
Ukranian SSR	1988	9,239	0.18	9,213	26	5,820	63	0.11	240
ASIA AND OCEANIA		**177,803**[b]	**1.23**	**71,467**	**106,336**	**42,894**[b]	**24**	**0.30**	**-42**
Japan	1985	24,718	0.20	24,158	560[l]	23,829[b]	96[b]	0.19	-48
Australia	1990	145,613[b]	8.52	39,837	105,776[b]	17,005	12	1.00	6
New Zealand	1987–89	7,472	2.23[b]	7,472	–	2,060[b]	28[b]	0.61	

a Excluding Iceland.
b Secretariat estimates.
c According to literary sources, during the period 1977–1986 the stocked productive non-reserved forest land declined at a rate of 474,000 hectares per year
d Including Cyprus, Israel and Turkey.
e Excluding Iceland, Israel and Turkey.
f Data refer to pastures.
g No survey of wooded lands other than forests conducted.
h Estimates by national correspondents.
i The value given for "Forest and other wooded land" should be considered as a rough estimate only. A range of estimates were given by French experts in response to UN ECE/FAO's enquiry: these ranged from 13,504,000 to 15,250,000 hectares.
j Data refer to natural forest area.
k No detailed information available. Results from the National Forest Survey show no significant change in area of forest and other wooded land.
l The figure given in response to the UN ECE/FAO questionnaire for "Other wooded land area" was inconsistent with data provided by Japan for total forest and other wooded land. The value given here is thus an UN ECE/FAO estimate.

Data presented in the above table represent the results of the joint UN ECE/FAO assessment of forest resources in temperate zone developed countries. The survey covers all forests in Europe, Cyprus, Israel, Turkey, the former USSR, Canada, USA, Japan, Australia and New Zealand. Data were obtained primarily from official sources in response to questionnaires and have been augmented by other sources as necessary.

"Forest" is defined as an area with tree crowns covering more than 20 per cent of the land area and is used primarily for forestry.

"Other wooded land" includes forests that are not used for agriculture and land areas having 5–20 per cent of their areas covered by tree crowns or stunted trees covering more than 20 per cent of their area. A dash in this column signifies that the category, "Other wooded land", does not exist in the national forest inventory and thus no data are available.

"Exploitable forest" is defined as forest on which there are no legal, economic or technical restrictions on wood production. It may include areas where, although there are no such restrictions at the present time, harvesting is not currently taking place. Examples in the latter category would include areas subject to long-term national utilization plans.

Regional totals only include listed countries reporting data. Regional totals may not tally due to independent rounding.

Source:
United Nations Economic Commission for Europe and Food and Agriculture Organization of the United Nations 1992 *The Forest Resources of the Temperate Zones: Main findings of the UN ECE/FAO 1990 Forest Resource Assessment*, United Nations, New York.

TABLE 3.7 Preliminary FAO estimates of tropical forest area and rate of deforestation in 87 tropical countries, 1981–1990

Region/subregion	Number of countries studied	Total land area (10^3 ha)	Forest area, 1980 (10^3 ha)	Forest area, 1990 (10^3 ha)	Annual average rate of deforestation, 1981–90	
					(10^3 ha a^{-1})	(% per year)
TOTAL	87	4,815,700	1,884,100	1,714,800	16,900	−0.9
AFRICA	40	2,243,400	650,300	600,100	5,000	−0.8
West Sahelian Africa	8	528,000	41,900	38,000	400	−0.9
East Sahelian Africa	6	489,600	92,300	85,300	700	−0.8
West Africa	8	203,200	55,200	43,400	1,200	−2.1
Central Africa	7	406,400	230,100	215,400	1,500	−0.6
Tropical Southern Africa	10	557,900	217,700	206,300	1,100	−0.5
Insular Africa	1	58,200	13,200	11,700	200	−1.2
LATIN AMERICA	32	1,675,700	923,000	839,900	8,300	−0.9
Central America and Mexico	7	245,300	77,000	63,500	1,400	−1.8
Caribbean Subregion	18	69,500	48,800	47,100	200	−0.4
Tropical South America	7	1,360,800	797,100	729,300	6,800	−0.8
ASIA	15	896,600	310,800	274,900	3,600	−1.2
South Asia	6	445,600	70,600	66,200	400	−0.6
Continental Southeast Asia	5	192,900	83,200	69,700	1,300	−1.6
Insular Southeast Asia	4	258,100	157,000	138,900	1,800	−1.2

Data represent preliminary results of the first phase (i.e., the statistical phase) of the second decadal FAO Tropical Forest Resources Assessment Project. The 1990 forest assessment, like its 1980 predecessor, defines forests as land covered by a minimum of 10 per cent crown cover of trees and/or bamboos that is not subject to agricultural practices. The same definition of the term deforestation is also applied, i.e., changes associated with the transfer from forest to non-forest land-use (e.g., agriculture, pasture or shifting cultivation) or a reduction in tree crown cover to less than 10 per cent. Note that changes within the forest class that negatively affect the stand, in particular those that lower the production capacity, are not reflected in the estimate of deforestation as defined here.

The statistical phase of the 1990 forest assessment broadly corresponds to the work carried out as part of the 1980 assessment; estimates of forest area and rates of deforestation for the period 1981–1990 are thus based on a combination of national reports and a model of forest exploitation. Totals may not tally due to independent rounding.

West Sahelian Africa includes Burkina, Chad, Gambia, Guinea–Bissau, Mali, Mauritania, Niger and Senegal.
East Sahelian Africa includes Djibouti, Ethiopia, Kenya, Somalia, Sudan and Uganda.
West Africa includes Benin, Côte d'Ivoire, Ghana, Guinea, Liberia, Nigeria, Sierra Leone and Togo.
Central Africa includes Cameroon, Cent. African Rep., Congo, Equatorial Guinea, Gabon, São Tomé & Príncipe and Zaire.

Tropical Southern Africa includes Angola, Botswana, Burundi, Malawi, Mozambique, Namibia, Rwanda, Tanzania, Zambia and Zimbabwe.
Insular Africa refers to Madagascar.

Central America and Mexico includes Costa Rica, El Salvador, Guatemala, Honduras, Mexico, Nicaragua and Panama.
Caribbean Subregion includes Antigua & Barbuda, Bahamas, Belize, Bermuda, Cayman Is, Cuba, Dominica, Dominican Rep., French Guiana, Grenada, Guadeloupe, Guyana, Haiti, Jamaica, Puerto Rico, Suriname, St Lucia and Trinidad & Tobago.
Tropical South America includes Bolivia, Brazil, Colombia, Ecuador, Paraguay, Peru and Venezuela.

South Asia includes Bangladesh, Bhutan, India, Nepal, Pakistan and Sri Lanka.
Continental Southeast Asia includes Cambodia, Laos, Myanmar, Thailand and Viet Nam.
Insular Southeast Asia includes Indonesia, Malaysia, the Philippines and Singapore.

Source:
Food and Agriculture Organization of the United Nations 1991 *Forest Resources Assessment Project: Second Interim Report on the State of Tropical Forests.* Paper presented at the 10th World Forestry Congress, September 1991, Paris, (revised 15 October 1991).

TABLE 3.8   IUCN estimates of the original and current extent of closed canopy tropical moist forests in selected countries

AFRICA	Original extent (10³ ha)	Current extent (10³ ha)	% of original	Date of estimate
Benin	1,680	42	3	1979,1989–90
Burundi	1,060	41	4	1984
Cent. African Rep.	32,450	5,224	16	1985
Cameroon	37,690	15,533	41	1985
Côte d'Ivoire	22,940	2,746	12	1989–90
Equatorial Guinea	2,600	1,700	65	1960
Gambia	410	50	12	1985
Ghana	14,500	1,584	11	1989–90
Guinea	18,580	766	4	1989
Liberia	9,600	4,124	43	1989–90
Madagascar	27,509	4,172	15	1985
Malawi	1,070	32	3	
Nigeria	42,100	3,862	9	1989–90
Rwanda	940	155	17	
São Tomé & Príncipe	96	30	31	1985
Senegal	2,770	205	7	1985
Sierra Leone	7,170	506	7	1989–90
Tanzania	17,620	1,613	9	1985
Togo	1,800	136	8	1989–90
Uganda	10,340	740	7	
Zaire	178,400	119,074	67	1990
Zimbabwe	770	8	1	

ASIA AND OCEANIA	Original extent (10³ ha)	Current extent (10³ ha)	% of original	Date of estimate
Australia	1,100	1,052	96	1988
Bangladesh	13,000	973	7	1981–86
Brunei	500	469	94	1988
Cambodia	16,000	11,325	71	1971
China[a]	34,000	2,586	8	1979
Fiji	1,800	697	39	1980s
India	91,000	22,833	25	1986
Indonesia	170,000	117,914	69	1985–89
Laos	22,500	12,460	55	1987
Malaysia	32,000	20,045	63	
Peninsular	13,000	6,978	54	1986
Sabah	7,000	3,600	51	1984
Sarawak	12,000	9,467	79	1979
Myanmar	60,000	31,185	52	1987
Papua New Guinea	45,000	36,675	82	1975
Philippines	29,500	6,602	22	1988
Singapore	50	2	4	1980s
Solomon Is.	2,850	2,559	90	1980s
Sri Lanka	2,600	1,276	47	1988
Thailand	25,000	10,690	43	1985
Viet Nam	28,000	5,668	20	1987

[a]   Southern China and Taiwan only.

Data refer to closed canopy tropical moist forests, i.e., tropical rain forests and tropical monsoon (or seasonal) forests. Open canopy woodlands (i.e., woodlands having less than 40 per cent tree canopy cover), forest plantations and areas of shifting cultivation are excluded.

Estimates of original and remaining tropical moist forest areas are based on a compilation of available maps and related information that has been generated as part of the IUCN's Tropical Forest Conservation Programme. These maps, which are derived from both published and unpublished information sources, represent a compilation of what IUCN consider to be the best information available at the end of 1989. The maps have been digitized using GIS technology to provide the statistical estimates of the tropical moist forest areas given in the above table.

Estimates of the 'original' extent of forests refer to pre-industrial times and are based on estimates of potential vegetation.

For further details and information on an individual country basis, please refer to the source documents.

Sources:
The World Conservation Union/World Conservation Monitoring Centre 1991 *The Conservation Atlas of Tropical Forests: Asia and the Pacific*, N. M. Collins, J. A. Sayer and T. C. Whitmore (Eds), Macmillan Press Ltd, London.
The World Conservation Union/World Conservation Monitoring Centre 1992 *The Conservation Atlas of Tropical Forests: Africa*, N. M. Collins, J. A. Sayer and T. C. Whitmore (Eds), Macmillan Press Ltd, London.

TABLE 3.9   Trends in total and per capita water withdrawals in OECD countries, 1960–1980s

Country	Total water withdrawal (10^6 m^3 a^{-1})						Per capita water withdrawal (m^3 a^{-1})					
	1960	1970	1975	1980	1985	Late 1980s	1960	1970	1975	1980	1985	Late 1980s
Australia[a]			17,500						1,280			
Austria[a]			2,620	2,190	2,120	2,120	981[b]		346	290	280	279
Belgium	8,981[b]	9,481		9,030			902	982		917		
Canada	16,152	24,057[c]	28,128	37,864[d]	41,470	43,888		1,130[c]	1,239	1,575[d]	1,635	1,684
Denmark		720	1,205[e]			1,170[f]		146	238[e]			228[f]
Finland	4,000[g]	3,300	3,550	3,700	4,000	3,001	906[g]	716	754	774	816	605
France[h]		23,500	27,000	37,000[d]	43,172	43,673[f]		463	512	698[d]	783	783[f]
Germany, Fed. Rep.[i]	18,712	29,488	33,544	42,206[l]	41,216[k]	44,582[l]	338	486	543	686[l]	675[k]	729[l]
Greece[m]		4,254	5,847	6,945				484	646	720		
Ireland				793[g,j]						235[g,j]		
Italy[l,n]		41,900		56,200	52,000	56,200[l]		781		996	910	984[l]
Japan			87,600	88,200	89,200	89,290[l]			786	755	739	733[l]
Luxembourg[o]					67	59					159	
Netherlands[o]		12,130[c]	13,360[p]	14,794[d]	14,471[q]			930[c]	1,000[p]	1,046[d]	993[q]	
New Zealand[r,s]		990	1,045	1,200	1,900			351	339	382	585	
Norway[t]			2,380		2,025				594		490	
Portugal[u]				1,476	1,271	1,290				158	131	125
Spain[i,v]	15,620[w]	24,600	36,080	39,920	45,250	45,845[q]	513[w]	731	1,016	1,068	1,175	1,184[q]
Sweden[i,x]			3,979		2,888[k]	2,996			486		346[k]	356
Switzerland		1,140	1,129	1,103	1,143	1,166		182	176	173	175	170
Turkey[y]		11,760	16,041	16,200	19,400	23,750		330	398	362	389	434
UK[z]		15,583[aa]	13,085	15,547	13,998	14,502		319[aa]	265	276	247	253
USA[ab]	293,935	440,100	470,000	523,000	467,000			2,146	2,180	2,305	1,952	
Yugoslavia			7,370[g]	8,767[g]			1,627		345[g]	393[g]		

a   Surface-water withdrawal excludes withdrawal for agricultural purposes (including irrigation), and for industrial processes other than cooling. Ground-water withdrawal excludes withdrawal for industry cooling and cooling of electrical power plants.
b   Data refer to 1965.
c   Data refer to 1972.
d   Data refer to 1981.
e   Data refer to 1977.
f   Data refer to 1988.
g   OECD Secretariat estimate.
h   Withdrawal data for 1975 are estimates based on four basins.
i   Excludes withdrawal for agricultural purposes other than irrigation.
j   Data refer to 1979.
k   Data refer to 1983.
l   Data refer to 1987.
m   Excludes withdrawal for cooling of electrical power plants.
n   Data for 1970 and 1980 include an estimate for withdrawal for industry cooling which refers to 1973.

o   Excludes withdrawal for agricultural purposes (including irrigation).
p   Data refer to 1976.
q   Data refer to 1986.
r   Excludes withdrawal for irrigation, industry and electrical power plant cooling.
s   Data for 1980 are estimates based on withdrawal for agricultural purposes other than irrigation in 1975, for industrial purposes other than cooling in 1975 plus withdrawal for public water supply in 1980.
t   Data for 1985 refer to 1983 but include 1978 industry data.
u   Data for 1980 refer to withdrawal for public water supply and for cooling of electrical power plants only. Data for 1985 and the late 1980s refer to cooling of electrical power plants only.
v   Ground-water withdrawal excludes that for industrial purposes.
w   Data represent composite totals.
x   Data for 1975 exclude withdrawal for agricultural purposes and industry cooling.
y   Data for 1970, 1975 and 1980 exclude withdrawal for agricultural purposes other than irrigation and for cooling of electrical

power plants. Data for 1985 exclude withdrawal for cooling of electrical power plants.
z   Data for 1970 and 1975 refer to England and Wales only.
aa   Data refer to 1971.
ab   Data for 1960 exclude withdrawal for industry cooling.

Water withdrawal refers to the quantity of fresh water that is taken from ground or surface water sources and conveyed to a place of use; withdrawals are counted each time the same water is withdrawn.

Sources:
OECD 1987 OECD Environmental Data Compendium 1987, Organisation for Economic Co-operation and Development, Paris.
OECD 1991 OECD Environmental Data Compendium 1991, Organisation for Economic Co-operation and Development, Paris.
United Nations Population Division 1991 World Population Prospects 1990, United Nations, New York.

TABLE 3.10 Marine and fresh-water fish catches by selected countries, 1978–1980 and 1988–1990 (mean annual values)

Region/country	Marine catch (10^3 t a^{-1})			Fresh-water catch (10^3 t a^{-1})			Total catch (10^3 t a^{-1})		
	1978–80	1988–90	% change	1978–80	1988–90	% change	1978–80	1988–90	% change
WORLD	63,756	84,966	33	7,254	13,915	92	71,010	98,880	39
AFRICA	2,738	3,357	23	1,324	1,872	41	4,062	5,228	29
Algeria	40	99	145	0	–		40	99	146
Angola	101	99	2	8	8	0	109	107	2
Benin	4	9	128	33	31	4	36	40	10
Burundi	0	0	0	14	14	2	14	14	2
Cameroon	62	59	4	20	20	0	82	79	3
Cent. African Rep.	0	0	0	13	13	0	13	13	0
Chad	0	0	0	113	112	1	113	112	1
Congo	19	22	16	8	23	192	27	45	68
Côte d'Ivoire	71	69	4	12	31	154	84	100	20
Egypt	30	73	141	96	224	134	126	297	136
Gabon	18	21	16	2	2	7	20	22	15
Gambia	10	12	19	3	3	0	13	15	15
Ghana	209	314	50	39	58	50	247	372	50
Guinea	15	31	102	1	3	174	16	33	106
Kenya	5	8	65	44	134	207	49	142	193
Liberia	8	13	56	4	4	4	12	16	36
Madagascar	17	68	296	40	34	15	58	102	78
Malawi	0	0	0	65	84	31	65	84	31
Mali	0	0	0	86	64	25	86	64	25
Mauritania	26	88	231	6	6	0	32	94	189
Mauritius	7	16	144	–	–	350	7	16	144
Morocco	302	544	80	1	2	158	303	546	80
Mozambique	26	34	29	5	–	95	31	34	9
Namibia	26	114	331	–	–	146	27	114	331
Nigeria	149	190	28	116	108	7	265	298	13
Senegal	211	266	26	15	17	13	226	283	25
Sierra Leone	43	36	14	10	16	63	52	52	0
Somalia	11	18	56	0	–		11	18	60
South Africa	954	903	5	1	2	56	955	905	5
Sudan	1	1	47	26	32	21	27	33	22
Tanzania	40	49	22	167	334	100	207	382	85
Togo	10	15	51	1	1	21	11	16	46
Tunisia	58	97	66	0	0	0	58	97	66
Uganda	0	0	0	190	224	18	190	224	18
Zaire	1	2	122	108	161	50	109	163	50
Zambia	0	0	0	49	64	30	49	64	30
Zimbabwe	0	0	0	10	24	130	10	24	130
NORTH AMERICA	6,327	8,995	42	155	535	245	6,482	9,531	47
Canada	1,328	1,552	17	50	50	–	1,378	1,602	16
Costa Rica	21	20	6	–	1	204	22	21	5
Cuba	179	185	3	5	19	283	184	204	11
Dominican Rep.	7	17	153	1	2	22	8	18	131

Continued

TABLE 3.10 Continued

Region/country	Marine catch (10^3 t a^{-1})			Fresh-water catch (10^3 t a^{-1})			Total catch (10^3 t a^{-1})		
	1978–80	1988–90	% change	1978–80	1988–90	% change	1978–80	1988–90	% change
El Salvador	10	10	3	2	3	63	12	13	6
Greenland	87	140	61	0	0	0	87	140	61
Honduras	7	17	162	–	–	127	7	17	161
Jamaica	9	7	24	–	3	6,320	9	10	8
Mexico	979	1,233	26	8	182	2,318	987	1,414	43
Panama	175	157	10	0	–		175	158	10
St Pierre & Miquelon	11	13	26	0	0	0	11	13	26
USA	3,449	5,578	62	88	274	211	3,537	5,852	65
SOUTH AMERICA	**7,752**	**14,659**	**89**	**258**	**343**	**33**	**8,010**	**15,003**	**87**
Argentina	477	501	5	13	11	18	491	512	4
Brazil	620	618	–	162	208	29	781	826	6
Chile	2,458	5,618	129	–	1	3,406	2,458	5,620	129
Colombia	22	53	138	51	43	16	73	96	31
Ecuador	619	667	8	–	2	1,355	619	669	8
Guyana	34	35	5	1	1	54	34	36	6
Paraguay	0	0	0	3	11	249	3	11	249
Peru	3,254	6,755	108	14	36	161	3,268	6,790	108
Uruguay	100	106	6	–	–	50	101	107	6
Venezuela	162	292	80	11	24	113	173	316	82
ASIA	**25,924**	**35,173**	**36**	**4,387**	**9,668**	**120**	**30,312**	**44,840**	**48**
Bangladesh	119	250	109	528	591	12	647	840	30
Cambodia	7	29	335	24	59	151	30	88	191
China	3,090	6,342	105	1,138	4,882	329	4,228	11,225	165
Hong Kong	176	232	32	6	6	4	182	238	31
India	1,512	2,124	40	854	1,396	63	2,366	3,520	49
Indonesia	1,307	2,183	67	435	757	74	1,742	2,939	69
Iran	53	200	275	5	49	796	59	249	324
Iraq	27	5	84	18	12	30	45	17	63
Israel	11	11	8	16	16	1	28	27	3
Japan	9,963	10,962	10	228	202	11	10,191	11,164	10
Korea	2,078	2,737	32	38	33	12	2,115	2,770	31
Korea, Dem.	1,265	1,613	28	65	103	58	1,330	1,717	29
Laos	0	0	0	20	20	0	20	20	0
Malaysia	703	592	16	3	16	443	706	607	14
Maldives	29	74	151	0	0	0	29	74	151
Myanmar	412	583	41	149	144	3	561	727	30
Nepal	0	0	0	3	13	275	3	13	275
Oman	76	135	77	0	0	0	76	135	77
Pakistan	250	352	41	41	105	157	291	457	57
Philippines	1,155	1,544	34	353	562	59	1,508	2,106	40
Saudi Arabia	26	48	82	0	1		26	49	86
Singapore	16	14	14	1	–	83	16	14	16
Sri Lanka	151	153	1	18	36	100	169	189	12
Thailand	1,809	2,487	37	140	204	46	1,949	2,691	38
Turkey	315	458	46	25	47	84	340	505	48

Continued

TABLE 3.10 Continued

Region/country	Marine catch (10^3 t a^{-1})			Fresh-water catch (10^3 t a^{-1})			Total catch (10^3 t a^{-1})		
	1978–80	1988–90	% change	1978–80	1988–90	% change	1978–80	1988–90	% change
United Arab Em.	64	92	43	0	0	0	64	92	43
Viet Nam	406	617	52	175	247	41	581	864	49
Yemen[a]	71	78	11	0	0	0	71	78	11
EUROPE	**12,207**	**11,861**	**3**	**334**	**477**	**43**	**12,541**	**12,338**	**2**
Albania	6	7	17	3	6	93	9	13	43
Belguim	48	40	15	0	1		48	41	14
Bulgaria	95	81	15	11	11	3	106	92	13
Czechoslovakia	0	0	0	17	22	31	17	22	31
Denmark	1,819	1,777	2	17	29	69	1,836	1,806	2
Faeroe Is	286	314	10	0	0	0	286	314	10
Finland	98	101	2	32	9	72	130	110	16
France	759	857	13	9	44	388	769	901	17
German Dem. Rep.	205	143	31	13	24	84	219	167	24
Germany, Fed. Rep.	342	203	41	16	28	74	358	231	36
Greece	95	125	32	9	10	7	104	135	30
Hungary	0	0	0	33	36	9	33	36	9
Iceland	1,575	1,590	1	1	1	22	1,576	1,591	1
Ireland	115	228	98	0	1		115	229	98
Italy	449	492	10	31	59	93	479	551	15
Netherlands	327	424	29	2	5	163	330	429	30
Norway	2,553	1,831	28	–	–	21	2,553	1,832	28
Poland	584	526	10	20	38	88	604	564	7
Portugal	255	331	30	0	2		255	333	31
Romania	114	143	25	49	64	30	163	207	26
Spain	1,314	1,507	15	26	30	15	1,340	1,537	15
Sweden	203	251	23	10	6	44	213	256	20
UK	927	845	9	3	17	571	929	863	7
Yugoslavia	35	44	26	24	25	6	59	70	18
FORMER USSR	**8,414**	**10,014**	**19**	**790**	**997**	**26**	**9,204**	**11,010**	**20**
OCEANIA	**393**	**907**	**131**	**6**	**23**	**260**	**400**	**930**	**133**
Australia	126	196	55	1	4	191	128	200	57
Fiji	17	29	70	1	4	529	18	33	88
Kiribati	11	29	165	0	–		11	29	165
New Zealand	151	560	270	–	–	44	152	561	270
Papua New Guinea	39	11	73	4	15	248	43	25	42
Solomon Is	33	56	71	0	0	0	33	56	71

[a] Data refer to the unified Republic of Yemen.

Data refer to the nominal catch of fish, crustaceans, molluscs and other aquatic products (excluding crocodiles and alligators, aquatic mammals and plants, and miscellaneous aquatic animal products such as sponges and corals) taken for all purposes except recreational. The nominal catch is defined as the live weight equivalent of the landed quantities, i.e., landings of each species adjusted for on-board processing such as gutting, filleting and drying. Country totals include production from aquaculture and quantities caught by vessels flying the national flag but landed in foreign ports. Only countries reporting on annual mean fish catch of 10,000 t in 1988–90 are listed in the above table. Regional and world totals do, however, include the catches for the unlisted countries.

Fisheries data are compiled by the Fishery Department of the FAO on an annual basis from questionnaire responses submitted by national fishery offices and regional fishery commissions. For further details regarding these data please refer to the latest edition of the FAO 'Yearbook of Fishery Statistics'.

Source:
Data extracted from the Database of the Fisheries Information, Data and Statistics Service, Fisheries Department, Food and Agriculture Organization of the United Nations, Rome, (June 1992).

TABLE 3.11 Estimates of selected whale populations by species and in different geographical stocks, 1980s

Region/species	Geographical stock	Assessment date	Best estimate	Range
NORTH ATLANTIC				
Minke	Northeastern	1987–89[a]	87,000	61,000–117,000
Minke	Central	1987–89[a]	28,000	21,600–31,400
Minke	West Greenland	1987–89[a]	3,270	1,790–5,950
Sei	Central	1989	10,300	6,100–17,700
Humpback	Western	1979–86[a]	5,500	8,120–2,890
Pilot	Central/Eastern	1989	780,000	440,000–1,370,000
Fin	Nova Scotia/Labrador/Newfoundland	1969	10,800	5,390–21,700
Fin	West Greenland	1987–89[a]	1,100	560–2,130
Fin	East Greenland/Iceland	1987–89	15,600	10,100–24,000
Fin	North Norway	1988–89[a]	1,480	850–2,580
Fin	West Norway	1989	340	80–1,350
Fin	Iberia	1987	17,300	10,400–28,900
Fin	Faeroes/Hebrides/Ireland	1987–89[a]	680	340–1,330
NORTH PACIFIC				
Minke	Northwest Pacific (north of 39° N, west of 170° E)	1989–90[a]	5,800	2,800–12,000
Minke	Okhotsk Sea	1989–90[a]	19,200	10,000–36,000
Gray	Eastern North Pacific	1987/88	21,000	19,800–22,500
SOUTHERN HEMISPHERE				
Minke	Area I (120° W – 60° W)	1982/83	73,000	45,000–120,000
Minke	Area II (60° W – 0°)	1986/87	122,000	84,000–177,000
Minke	Area III (0° – 70° E)	1987/88	89,000	52,000–150,000
Minke	Area IV (70° E – 130° E)	1988/89	75,000	45,000–123,000
Minke	Area V (130° E – 170° W)	1985/86	295,000	225,000–386,000
Minke	Area VI (170° W – 120° W)	1983/84	107,000	63,000–182,000

[a] Values shown represent an aggregate of estimates made during the period stated.

The range data represent the 95% confidence interval.

Source:
Data supplied by R. Gambell, International Whaling Commission, Cambridge, UK, (March 1993).

TABLE 3.12   Whale catches, 1980–1991

Species	1980	1981	1982	1983	1984	1985	1986	1987	1988	1989	1990	1991
TOTAL	**14,779**	**14,108**	**12,999**	**11,490**	**9,298**	**8,377**	**6,729**	**1,868**	**666**	**714**	**656**	**649**
Fin	471	410	356	278	281	219	85	89	77	82	19	16
Sperm	2,092	1,452	621	414	463	400	211	211	8			
Humpback	18	14	16	16	15	8	2	2	2	2	1	2
Sei	103	100	71	100	95	38	40	20	10	2		
Bryde	970	648	802	697	709	357	317	317				
Gray	181	136	168	171	169	170	171	158	151	180	163	170
Minke	10,910	11,320	10,946	9,796	7,541	7,168	5,875	1,040	389	422	429	414
Bowhead	34	28	19	18	25	17	28	31	29	26	44	47

Data refer to the number of whales caught by the following countries: Denmark, Iceland, Norway, Spain, Portugal, St Vincent & Grenadines, Japan, Korea, Taiwan, USA, USSR, Brazil, Chile, Peru, Philippines, Indonesia and Italy during the period 1980–1991.

Years refer to the calendar year except in the case of whales caught in the Antarctic where catches are given for the whaling season of December to March. Catches include commercial catches, accidental catches in driftnets, whales taken for scientific purposes and whales caught by subsistence whaling operations. Gray whales caught by Siberian Aleuts are included in the country totals of the former USSR; bowhead whales caught by Alaskan eskimos are included in the country totals of the USA; minke and fin whales caught by Greenlanders are included in the country totals for Denmark; and humpbacks taken by Bequia islanders are included in the country totals for St Vincent & Grenadines. Whales that have been struck and then lost by Alaskan subsistence whaling operations are included in the numbers of whales caught.

Source:
Data supplied by R. Gambell, International Whaling Commission, Cambridge, UK (March 1993).

TABLE 3.13   Catches of selected small cetaceans, 1981–1991

Species		1981	1982	1983	1984	1985	1986	1987	1988	1989	1990	1991
TOTAL[a]		**112,006**	**74,879**	**59,939**	**71,546**	**83,215**	**150,940**	**124,391**	**131,326**	**168,724**	**105,833**	**50,012**
Baird's beaked whale	D	31	60	37	40	40	40	40	57	54	54	54
False killer whale	D	356	1	290[b]	60	43	2	2	48	31	126	54
	I	21					1[b]		2		30	
Long-finned pilot whale	D	2,775	2,652	1,690	1,921	2,606	1,709	1,422	1,690	1,258	818	720
	I	3		2		20		3[b]	15[b]	4		28
Short-finned pilot whale	D	686	395	503	672	701	375[b]	386	569	250	167	355
	I	6	3[b]			2	3[b]	9[b]	23[b]	9		
Spotted dolphin	D	1,440	3,799[b]	2,945	743	863[b]	693		1,875	189	11	153
	I	22,589	18,779[b]	3,397[b]	18,682[b]	30,478[b]	67,245[b]	57,544[b]	38,329[b]	55,956[b]	33,851[b]	13,991[b]
Spinner dolphin	I	15,218	6,677[b]	4,142[b]	16,395[b]	17,340[b]	31,566[b]	22,135[b]	22,366[b]	23,547[b]	12,330[b]	8,854[b]
Striped dolphin	D	16,247	2,018[b]	2,219	3,737	4,217[b]	2,921	2,176	2,227	1,225	749	1,022
	I	518	594[b]	34[b]	17[b]	18[b]	86[b]	3,477[b]	264	306	369	6
Bottlenose dolphin	D	3,493	839[b]	743[b]	462	474[b]	230	1,813	823	403	1,363	499
	I	21	3[b]	24[b]	2[b]	3[b]	455[b]	44[b]	27[b]	48	67	39
Dall's porpoise	D	6,178	12,833	12,766	9,764	9,604	10,534	13,406	39,737	29,048	21,804	17,634
	I	9,158	4,961[b]	3,083	3,355	3,767[b]	248	816[b]	1,724	3,879	3,125	143

[a]   The "Total" includes catches of species not listed individually here and comprises direct, indirect and live catches of small cetaceans. The total given represents a minimum estimate in that where ranges for catches are available, the lower end of the range was used to compute the total.

[b]   Data refer to minimum estimates. These estimates exclude catch data for countries where information on catch levels is unavailable, or represent the lower bound of given ranges.

"D" refers to direct catches.
"I" refers to indirect/incident catches.
Baird's beaked whale *Beradius bairdii*; False killer whale *Pseudorca crassidens*; Long-finned pilot whale *Globicephala melaena*; Short-finned pilot whale *Globicephala macrorhynchus*; Bottlenose dolphin *Tursicops truncatus*; Spotted dolphin *Stenella attenuata*; Spinner dolphin *Stenella longirostris*; Striped dolphin *Stenella coeruleoalba*; Dall's porpoise *Phocoenoides dalli*.

Data refer to the number of small cetaceans (small whales, dolphins and porpoises) caught directly and indirectly by the following countries: Argentina, Australia, Brazil, Canada, Chile, Denmark, France, Iceland, Japan, the Netherlands, New Zealand, Norway, Seychelles, South Africa, Sri Lanka, Sweden, Taiwan, Turkey, UK, the USSR and the Eastern Tropical Countries (ETP). ETP countries include Colombia, Costa Rica, Ecuador, El Salvador, Mexico, Panama, Spain, USA, Vanuatu and Venezuela.

Source:
See Table 3.12.

TABLE 3.14 Diversity and current status of animal species and plant taxa

Region/country	Mammals		Birds		Reptiles		Amphibians		Plants	
	No. of species known	No. threatened	No. of species known	No. threatened	No. of species known	No. threatened	No. of species known	No. threatened	No. of taxa known	No. rare or threatened
AFRICA										
Algeria	92	12	192	15		0		8	3,139–3,150	145
Angola	276	14	872	12		2		0	5,000	19
Benin	188	11	630	1		2		0	2,000	3
Botswana	154	9	569	6	143	1	36	0	2,600–2,800	4
Burkina	147	10	497	1		2		0	1,096	0
Burundi	107	4	633	5		1		0	2,500	0
Cameroon	297	27	848	17		2		1	8,000	74
Cape Verde		0	36	3	12	1	0	0	659	1
Cent. African Rep.	209	12	668	2		2		0	3,600	0
Chad	134	18	496	4		2		0	1,600	14
Comoros	12	3	99	5	22	0		0	416	3
Congo	200	12	500	3		2		0	4,000	4
Côte d'Ivoire	230	18	683	9		1		1	3,660	70
Djibouti		6	311	3		0		0	534	3
Egypt	102	9	132	16	83	2	6	0	2,085	91
Equatorial Guinea	184	15	392	3		2		1	6,283	8
Ethiopia	255	25	836	14		1		0	8,000	44
Gabon	190	17	617	4		2		0	530	80
Gambia	108	7	489	1		2		0		0
Ghana	222	13	721	8		2		0	3,600	34
Guinea	190	17	529	6		1		1	1,000	36
Guinea-Bissau	108	5	376	2		2		0		0
Kenya	309	17	1,067	18	187	2	88	0	6,500	144
Lesotho	33	2	288	7		0		0	1,591	7
Liberia	193	18	590	10	62	2	38	0		1
Libya	76	12	80	9		1		0	1,600–1,800	58
Madagascar	105	50	250	28	252	10	144	0	10,000–12,000	194
Malawi	195	10	630	7	124	1	69	0	3,600	61
Mali	137	16	647	4	16	2		0	1,600	15
Mauritania	61	14	49	5		1		0	1,100	3
Mauritius		3	102	10		6	2	0	800–900	269

Continued

TABLE 3.14 Continued

Region/country	Mammals No. of species known	Mammals No. threatened	Birds No. of species known	Birds No. threatened	Reptiles No. of species known	Reptiles No. threatened	Amphibians No. of species known	Amphibians No. threatened	Plants No. of taxa known	Plants No. rare or threatened
Morocco	105	9	209	14		0		0	3,500–3,600	194
Mozambique	179	10	666	11		1	62	0	5,500	84
Namibia	154	11	640	7		2	32	0	3,159	17
Niger	131	15	473	1		1		0	1,178	1
Nigeria	274	25	831	10	100	2	60	0	4,614	9
Réunion	2	0	33	1		0		0	720	96
Rwanda	151	11	669	7		2		0	2,150	0
St Helena		0		1		0		0	320	
São Tomé & Príncipe	8	1	124	7	16	0	9	0		1[a]
Senegal	155	11	625	5		2		0	2,100	32
Seychelles	147	1	126	9	15	2	12	3	274	75
Sierra Leone	171	13	614	7		2		0	2,480	12
Somalia	171	17	639	7	193	1	27	0	3,000	52
South Africa	247	25	774	13	299	3	95	1	23,000	1,016
Sudan	267	17	938	8		1		0	3,200	9
Swaziland	47	0	381	5	106	1	39	0	2,715	25
Tanzania	306	30	1,016	26	245	3	121	0	10,000	158
Togo	196	9	630	1		2		0	2,302	0
Tunisia	78	6	173	14		1		0	2,120–2,200	26
Uganda	315	16	989	12	119	1	44	0	5,000	11
Western Sahara	15	5	60	5		0		0	300	0
Zaire	415	31	1,086	27		2		0	11,000	3
Zambia	229	10	732	10		2	83	0	4,600	1
Zimbabwe	196	9	635	6	153	1	120	0	5,428	96
NORTH AMERICA										
Anguilla	5	0		0		0				
Antigua & Barbuda	7	0		2	9	0	2	0	724	1
Bahamas	12	2	88	4	24	3	5	0	1,350	24
Barbados	6	1	24	1		0		0	700	1
Belize	125	8	528	4	107	3		0	3,240	36
Bermuda		0		2		0		0	165	11
Br. Virgin Is		0		3		1		0		1

Continued

188 Natural Resources

TABLE 3.14 Continued

Region/country	Mammals No. of species known	Mammals No. threatened	Birds No. of species known	Birds No. threatened	Reptiles No. of species known	Reptiles No. threatened	Amphibians No. of species known	Amphibians No. threatened	Plants No. of taxa known	Plants No. rare or threatened
Canada	139	5	426	6	41	0	40	0	3,220	12
Cayman Is	8	0	45	2		2		0		0
Costa Rica	205	10	848	14	214	2	162	0	8,000	419
Cuba	31	11	159	15	100	4	41	0	7,000	860
Dominica	12	0	59	3	13	0	2	0	1,600	62
Dominican Rep.	20	1	125	5	73	4	23	0	2,500	50
El Salvador	135	6	450	2		1		0		26
Greenland		2		1		0		0		0
Grenada	14	0	50	2	12	0	3	0		4
Guadeloupe	10	0		1		0		0	2,800	14
Guatemala	184	10	480	10	231	4	88	0	8,000	282
Haiti	20	1		4		4		0		13
Honduras	173	7	159	11	152	3	56	0	5,000	43
Jamaica	22	5	53	2		3		0	3,582	10
Martinique	9	0		3		0		0		12
Mexico	439	25	961	35	717	16	284	4	20,000	883
Montserrat	8	0	43	1		0		0		1
Neth. Antilles				3		2		0		0
Nicaragua		8		7	161	2	59	0	5,000	68
Panama	218[b]	13	922[b]	14	226[b]	2	164	0	8,000–9,000	549
Puerto Rico	13	2	94	4	46	5	22	1	3,000	84
St Kitts & Nevis	7	0	40	1	9	0	3	0		0
St Lucia	8	0	51	5	15	0	4	0		3
St Vincent & Grenadines	9	0	108	3	16	0	4	0		
Trinidad & Tobago	100	1	258	3		0		0	2,281	5
Turks & Caicos Is		0	184	0		1		0		1
USA	346	27	650	43		25		22	20,000	2,262
US Virgin Is		0		3		1		0		10
SOUTH AMERICA										
Argentina	258	23	1,257	53	250	4	123	1	9,000	159
Bolivia	280	21	1,573	34		4	110	0	15,000–18,000	29
Brazil	394	40		123	468	11	502	0	55,000	218

Continued

TABLE 3.14 Continued

Region/country	Mammals No. of species known	Mammals No. threatened	Birds No. of species known	Birds No. threatened	Reptiles No. of species known	Reptiles No. threatened	Amphibians No. of species known	Amphibians No. threatened	Plants No. of taxa known	Plants No. rare or threatened
Chile	91	9	432	18	78	0	39	0	5,500	284
Colombia	359	25	1,721	69	383	10	407	0	45,000	327
Ecuador	271	21	1,435	64	337	8	343	0	10,000–20,000	256
French Guiana	152	10		5		2		0	6,000–8,000	47
Guyana	193	12		9		3		0	6,000–8,000	68
Paraguay	156	14	650[b]	34	120	4	85	0	7,000–8,000	15
Peru	344	29	1,705	75	298	6	241	1	20,000	360
Suriname	187	11		6		1		0	4,500	68
Uruguay	81	5		11		2		0		14
Venezuela	288	19	1,308	34		3		0	15,000–20,000	106
ASIA										
Afghanistan	123	13	456	13	103	1	6	1	3,000	4
Bahrain		1		4	25	0		0		0
Bangladesh	109	15	354	27	119	14	19	0	5,000	33
Bhutan	109	15	448	10	19	1	24	0	5,000	15
Br. Ind. Oc. Tr.		0		0		0		0		0
Brunei	155	9	359	10	44	3	76	0		40
Cambodia	117	21	305	13	82	6	28	0		11
China	394	40	1,100	83	282	7	190	1	30,000	350
China, Taiwan	62	4	160	16	67	0	26	0	4,300	95
Cyprus	21	1	80	17	23	1	4	0	2,000	43
Hong Kong	38	1	107	9	61	2	23	0		5
India	317	39	969	72	389	17	206	3	15,000	1,336
Indonesia	515	49	1,519	135	511	13	270	0		70
Iran	140	15		20	164	4	11	0	7,000	301
Iraq	81	9	145	17	81	0	6	0	2,937	1
Israel		8	169	15		1		1	2,317	3
Japan	90	5	250	31	63	0	52	1	4,022	41
Jordan		5	132	11		0		0	2,200	752
Korea, Dem.		5		25	19	0	13	0		0
Korea	49	6		22	18	0	13	0	2,838	33
Kuwait		5	27	7	29	0	2	0	350	1

Continued

TABLE 3.14 Continued

Region/country	Mammals No. of species known	Mammals No. threatened	Birds No. of species known	Birds No. threatened	Reptiles No. of species known	Reptiles No. threatened	Amphibians No. of species known	Amphibians No. threatened	Plants No. of taxa known	Plants No. rare or threatened
Laos	173	23	481	18	66	5	37	0		3
Lebanon	52	4	124	15		1		0		5
Malaysia	264	23	501	35	268	12	158	0	3,000	522
Maldives		1	24	1		0		0		0
Mongolia		9		13		0		0		0
Myanmar	300	23	867[b]	42	203	10	75	0	7,000	33
Nepal	167	22	629	20	80	9	36	0	6,500	2
Oman	46	6		8	64	0		0	1,100	
Pakistan	151	15	476	25	143	6	17	0	5,500–6,000	14
Philippines	166	12	395	39	193	6	63	0	8,900	159
Qatar		0		3	17	0		0		0
Saudi Arabia	57	9	59	12	84	0		0	3,500	2
Singapore		4	118	5		1		0	2,030	19
Sri Lanka	86	7	221	8	144	3	39	0	3,700	220
Syria		4	165	15		1		0	3,000	11
Thailand	251	26	616	34	298	9	107	0	12,000	68
Turkey	116	5	284	18	102	5	18	1	10,150	1,944
United Arab Em.		4		7	37	0		0		0
Viet Nam[c]	273	28	638	34	180	8	80	1	8,000	338
Yemen[c]		6		9	77	0		0		134
EUROPE										
Albania	68	2	215	14	31	1	13	0	3,100–3,300	76
Andorra		0	104	1		0		0		0
Austria	83	2	227	13	14	0	20	0	2,900–3,100	25
Belgium	58	2	180	13	8	0	17	0	1,600–1,800	9
Bulgaria	81	3	242	15	33	1	17	0	3,500–3,650	88
Czechoslovakia	81	2	227	18	12	0	19	0	2,600–2,750	29
Denmark	43	1	185	16	5	0	14	0	1,000	7
Faeroe Is		0	75	2	0	0	0	0		0
Finland	60	3	230	12	5	0	5	0	1,150–1,450	11
France	93	6	267	21	32	2	32	1	300–4,450	143

Continued

TABLE 3.14 Continued

Region/country	Mammals No. of species known	Mammals No. threatened	Birds No. of species known	Birds No. threatened	Reptiles No. of species known	Reptiles No. threatened	Amphibians No. of species known	Amphibians No. threatened	Plants No. of taxa known	Plants No. rare or threatened
Germany	76	2	237	17	12	0	20	0	5,000	526
Greece	95	4	244	19	51	3	15	0	2,400	21
Hungary	72	2	203	16	15	0	17	0		
Iceland	11	1	80	2	0	0			470	2
Ireland	25	0	141	10	1	0	3	0	1,000–1,150	4
Italy	90	3	254	19	40	2	34	7	4,750–4,900	210
Liechtenstein	64	0	134	3	7	0	10	0		0
Luxembourg	55	1	130	8	7	0	14	0	1,000	1
Malta	22	0	28	13	8	0	1	0	900	4
Monaco		0		0	6	0	3	0		0
Netherlands	55	2	187	13	7	0	16	0	1,400	7
Norway	54	3	235	8	5	0	5	0	1,600–1,800	13
Poland	85	4	224	16	9	0	18	0	2,250–2,450	16
Portugal	63	6	214	18	29	0	17	1	2,400–2,600	240
Romania	84	2	249	18	25	1	19	0	3,300–3,400	67
San Marino		0		0		0		0		0
Spain	82	6	275	23	53	5	25	3	4,750–4,900	936
Sweden	60	1	249	14	6	0	13	0	1,600–1,800	10
Switzerland	75	2	201	15	14	0	18	1	2,600–2,750	18
UK	50	3	219	22	8	0	7	0	1,700–1,850	24[d]
Yugoslavia	95	3	245	17	41	1	23	2	4,750–4,900	190
FORMER USSR	**276**	**20**		**38**	**168**	**3**	**37**	**0**	**21,000**	**531**
OCEANIA										
American Samoa	3	1	38	1	11	0	0	0		
Australia	282	38	571	39	700	9	180	3	18,000	2,024
Cook Is.		0	28	1		0	0	0		0
Fiji	4	1	87	5	25	4	2	1	1,500	25[e]
French Polynesia	0	0	67	20		0	0	0		65
Guam		2	23	4	10	0	0	0	331	12
Kiribati		0	15	2		0	0	0		0
Marshall Is		0	18	1	7	0	0	0		0
Micronesia		5	47	3		1	0	0		0

Continued

TABLE 3.14 Continued

Region/country	Mammals		Birds		Reptiles		Amphibians		Plants	
	No. of species known	No. threatened	No. of species known	No. threatened	No. of species known	No. threatened	No. of species known	No. threatened	No. of taxa known	No. rare or threatened
Nauru		0	9	2		0	0	0		0
New Caledonia	7	1	116	5	32	0	0	0	3,250	168
New Zealand		1	285	26	40	1	3	3	2,000	232
Niue	1	0	16	0	4	0	0	0		0
Northern Marianas Is		1	31	2		0	0	0		8
Papua New Guinea	242	5	578	25	249	1	183	0	2,150	88
Solomon Is	47	2	163	20	57	3	15	0		28
Tokelau	0	0	5	0	7	0	0	0		0
Tonga	1	0	39	2	6	1	0	0		0
Tuvalu		0	9	1		0	0	0		0
Vanuatu	12	1	84	3	22	1	0	0		8
Wallis Is		0	14	0		0	0	0		
Western Samoa	3	1	44	2	8	0	0	0		12

a Total for São Tomé only.

b Where data refer to mammals or reptiles, the number of species known may include marine species. Where data refer to birds, the number of species known may include non-breeding species.

c Data refer to the unified Republic of Yemen.

d Includes Gibraltar (UK = 23; Gibraltar = 1).

e Includes Rotuna.

The World Conservation Union (IUCN) has defined the following group of threat categories which are used to derive the IUCN Red List of Threatened Animals, upon which the above table is based:

Extinct (Ex): Species not definitely located in the wild during the past 50 years.

Endangered (E): Taxa in danger of extinction and whose survival is unlikely if the cause of the threat continues to operate.

Vulnerable (V): Taxa which could fall into the Endangered category in the future if the threat continues.

Rare (R): Taxa with small global populations which are at risk.

Indeterminate (I): Taxa known to be Endangered, Vulnerable or Rare but cannot be categorized due to a lack of information.

Insufficiently Known (K): Taxa which are suspected to be under threat but are not definitely known to belong to any particular group because of insufficient information.

The number of threatened animal species includes those species that are classified as Endangered, Vulnerable, Rare, Indeterminate or Insufficiently Known. Species known to be extinct and introduced species are excluded. Only full species are accounted for. Mammal data exclude marine mammals unless otherwise stated. Threatened migratory bird species can be included in the total for countries were they breed and also in the total for countries where they over-winter.

The IUCN Red List of Threatened Species is compiled every two years by the WCMC in collaboration with the IUCN Species Survival Commission Network of Specialist Groups. Information on threatened birds is routinely collected by the International Council for Bird Preservation (ICBP). Although the IUCN Red List represents a comprehensive global compendium of animal species known to be threatened, many more species than those listed are likely to be threatened. These include species that are as yet undescribed, and species which have been described but whose status has not been reviewed. Only 50 per cent of mammal species, and probably less than 20 per cent of reptiles, 10 per cent of amphibians and 5 per cent of fish are estimated to have been reviewed. The ICBP review of birds is considered to be more comprehensive.

Data on threatened plants are compiled by WCMC from a variety of sources, including national Red Data Books. The same IUCN threat categories have been applied to plants. Thus, the number of rare and threatened plant taxa includes plants classified as Endangered, Vulnerable, Rare or Indeterminate. Numbers may include many taxa below species level.

Sources:
World Conservation Monitoring Centre 1992 *Global Biodiversity: Status of the Earth's Living Resources*, Chapman & Hall, London.
World Resources Institute 1992 *World Resources 1992–93*, Oxford University Press, New York.

TABLE 3.15 Coral reef resources and disturbances

Region/country	Reef uses						Reef disturbances								
	Fishing	Tourism	Recreation	Coral collection	Minerals	Education	Pollution	Construction	Nat. disasters	Siltation	Extraction	Fishing	Over-collection	Tourism	Recreation
AFRICA															
Comoros	●	●								●	●	●			
Djibouti		●								●		●	●		
Egypt	●	●					●					●	●		
Ethiopia	●	●										●	●		●
Kenya	●	●	●				●			●			●		
Madagascar	●	●		●					●	●		●			
Mauritius	●	●	●				●		●	●	●		●		
Mozambique	●						●			●		●			
Réunion	●	●					●			●		●	●		
Seychelles	●							●			●				
Somalia	●									●					
South Africa	●		●	●											
Sudan	●	●					●							●	
Tanzania	●	●								●		●			
NORTH AMERICA															
Angiulla	●	●							●			●			
Antigua & Barbuda	●	●										●			
Bahamas	●	●					●				●	●			
Barbados	●	●		●			●	●	●					●	
Belize	●	●	●				●		●						
Bermuda	●	●									●	●			
Br. Virgin Is	●	●	●				●			●	●	●		●	
Cayman Is	●	●	●				●		●		●	●	●	●	
Costa Rica	●	●					●			●		●	●		
Cuba	●	●	●				●								
Dominica	●	●		●					●						
Dominican Rep.	●	●							●	●		●			
Grenada	●	●					●	●	●		●				
Guadeloupe	●	●					●		●			●	●		
Haiti	●						●			●			●		
Honduras	●	●		●						●					
Jamaica	●	●					●		●			●	●		
Martinique	●	●							●	●		●			
Mexico	●	●					●					●		●	
Montserrat	●		●								●				
Neth. Antilles	●	●	●	●			●					●	●		
Panama	●								●						
Puerto Rico	●	●					●		●	●					
St Kitts & Nevis	●	●										●			
St Lucia	●	●		●	●					●	●	●	●		●

Continued

TABLE 3.15 Continued

	Reef uses						Reef disturbances								
Region/country	Fishing	Tourism	Recreation	Coral collection	Minerals	Education	Pollution	Construction	Nat. disasters	Siltation	Extraction	Fishing	Over-collection	Tourism	Recreation
St Vincent & Grenadines	●	●					●				●	●	●		
Trinidad & Tobago			●								●				
Turks & Caicos Is	●	●										●			
USA	●	●	●	●		●	●		●		●	●			
US Virgin Is	●	●	●	●			●		●		●	●	●	●	
SOUTH AMERICA															
Brazil	●	●								●	●				
Chile	●												●		
Colombia	●	●					●	●	●			●			
Ecuador	●	●							●				●		
Venezuela	●	●					●			●					
ASIA															
Bahrain	●		●					●							
China	●								●	●	●		●		
China, Taiwan	●	●					●			●	●	●	●	●	
Hong Kong	●		●				●	●		●	●	●	●		●
India	●			●			●			●	●				
Indonesia	●	●		●	●				●	●	●	●	●		
Iran							●								
Israel		●					●								
Japan	●		●				●	●			●		●		
Jordan			●				●					●			
Kuwait	●		●				●								
Malaysia	●	●		●			●			●	●	●			
Maldives	●	●		●	●		●				●				●
Myanmar	●	●													
Oman	●		●										●		
Philippines	●	●		●	●				●	●		●	●		
Qatar	●						●								
Saudi Arabia	●		●	●								●			
Singapore		●	●	●			●			●			●		
Sri Lanka		●		●	●		●				●	●	●		
Thailand	●	●		●						●		●			
United Arab Em.			●												
Yemen	●														
OCEANIA															
American Samoa	●	●					●				●	●	●		
Australia	●	●												●	
Cook Is.	●	●					●	●	●	●					
Fiji	●	●					●		●	●	●	●	●		

Continued

TABLE 3.15 Continued

Region/country	Reef uses						Reef disturbances								
	Fishing	Tourism	Recreation	Coral collection	Minerals	Education	Pollution	Construction	Nat. disasters	Siltation	Extraction	Fishing	Over-collection	Tourism	Recreation
French Polynesia	●	●		●			●	●	●		●			●	
Guam	●		●	●			●		●			●			
Kiribati	●	●										●			
Marshall Is	●	●							●		●				
Micronesia	●	●					●	●			●	●			
Nauru	●						●								
New Caledonia	●	●		●			●								
New Zealand	●		●												
Niue	●	●		●											
Northern Marianas Is	●	●										●			
Papua New Guinea	●	●		●			●					●			
Solomon Is	●	●	●	●	●		●					●			
Tokelau	●									●					
Tonga	●	●		●			●	●		●		●	●		
Tuvalu	●			●						●					
Vanuatu	●	●								●			●		
Wallis Is	●														
Western Samoa	●	●									●	●			

● Indicates that the specified use or disturbance is at a significant level or is at an increasing level.

Reef uses refer to present use.
"Fishing" refers to fishing for fish, turtles, crabs, clams, lobsters and other seafood.
"Tourism" refers to those activities mainly participated in by tourists.
"Recreation" comprises diving, speed boating and other local activities.
"Coral collection" refers to the collection of both corals (live and dead) and shells.
"Minerals" refers to the extraction of limestone (from the coral), sand and gravel.
"Education" includes research.

Reef disturbances refer to present disturbances.
"Pollution" includes contamination by agricultural run-off, sewage, oil, thermal and chemical effluents.
"Construction" refers to coastline construction and development as well as landfill and land reclamation.

"Natural disasters" includes severe weather conditions.
"Siltation" includes sedimentation.
"Extraction" comprises the extraction of minerals through blasting, mining and dredging.
"Fishing" refers to the fishing industry and includes over-fishing and exploitation, dynamite fishing and damage caused by boats (especially by the anchors).
"Over-collection" refers to the over-collection of corals and shells.
"Tourism" refers to the tourist industry and tourist developments.
"Recreation" includes diving and speed boats.

Source:
World Conservation Monitoring Centre 1992 *Global Biodiversity: Status of the Earth's Living Resources*, Chapman & Hall, London.

TABLE 3.16 Countries reporting nationally and internationally designated protected areas, status at the beginning of 1992

Region/country	Nationally protected areas			World Heritage Sites	Biosphere Reserves		Ramsar Wetlands	
	Number	Area (ha)	% of total land area	Number	Number	Area (ha)	Number	Area (ha)
WORLD	8,491	773,490,101	5.2	95	300	161,944,969	538	32,336,169
AFRICA	698	134,409,606	4.5	28	43	20,229,937	43	4,005,302
Algeria	18	12,695,295	5.3	1	2	7,276,438	2	4,900
Angola	6	2,641,200	2.1					
Benin	2	843,500	7.5	0	1	880,000		
Botswana	9	10,025,000	17.4					
Burkina	11	2,642,700	9.6	0	1	16,300	3	296,300
Burundi	3	86,735	3.1	0				
Cameroon	13	2,034,425	4.3	1	3	850,000		
Cent. African Rep.	12	5,856,000	9.4	1	2	1,640,200		
Chad	2	414,000	0.3				1	195,000
Congo	10	1,333,100	3.9	0	2	246,000		
Côte d'Ivoire	12	19,992,850	6.2	3	2	1,500,000		
Djibouti	1	10,000	0.4					
Egypt	13	800,400	0.8	0	1	1,000	2	105,700
Ethiopia	11	2,534,100	2.5	1				
Gabon	6	1,045,000	3.9	0	1	15,000	3	1,080,000
Gambia	3	18,440	1.7	0				
Ghana	8	1,074,637	4.5	0	1	7,770	1	7,260
Guinea	3	167,370	0.7	1	2	133,300		
Guinea-Bissau	0	0	0.0				1	39,098
Kenya	36	3,470,226	6.0		4	851,359	1	18,800
Lesotho	1	6,805	0.2					
Liberia	1	130,747	1.2					
Libya	3	155,000	0.1	0				
Madagascar	37	1,115,299	1.9	1	1	140,000		
Malawi	9	1,057,600	11.2	1				
Mali	11	4,011,989	3.2	0	1	771,000	3	162,000
Mauritania	4	1,746,000	1.7	1			1	1,173,000
Mauritius	3	4,023	2.2		1	3,954		
Morocco	10	362,120	0.8	0			4	10,580
Mozambique	1	2,000	–	0				
Namibia	11	10,370,602	12.6					
Niger	6	9,696,740	8.2	1			1	220,000
Nigeria	21	2,872,665	3.1	0	1	460		
Réunion	2	5,942	2.4	0			0	0
Rwanda	2	327,000	12.4		1	15,065		
St Helena	2	17,600	7.3	0			0	0
Senegal	10	2,180,709	11.1	2	3	1,093,756	4	99,720
Seychelles	4	38,568	95.5	2				
Sierra Leone	2	82,103	1.1					
Somalia	1	180,000	0.3					
South Africa	229	7,389,517	6.2				12	232,344
Sudan	14	9,357,500	3.7	0	2	1,900,970		

Continued

TABLE 3.16 Continued

Region/country	Nationally protected areas			World Heritage Sites	Biosphere Reserves		Ramsar Wetlands	
	Number	Area (ha)	% of total land area	Number	Number	Area (ha)	Number	Area (ha)
Swaziland	4	45,920	2.6					
Tanzania	28	12,999,975	13.8	4	2	2,337,600		
Togo	11	646,906	11.4					
Tunisia	7	44,867	0.3	1	4	32,425	1	12,600
Uganda	32	1,870,798	7.9	0	1	220,000	1	15,000
Zaire	8	8,577,000	3.7	4	3	297,700		
Zambia	20	6,360,900	8.5	1			2	333,000
Zimbabwe	25	3,067,823	7.9	2				
NORTH AMERICA	**1,703**	**262,138,537**	**10.8**	**22**	**69**	**94,624,670**	**63**	**15,368,626**
Antigua & Barbuda	1	4,128	9.3	0				
Bahamas	5	123,389	8.9					
Belize	7	117,990	5.1					
Bermuda	1	12,000	a	0			0	0
Br. Virgin Is	3	673	4.4	0			0	0
Canada	426	49,452,283	5.0	6	6	1,049,978	30	12,937,549
Cayman Is	2	5,041	19.5[b]	0			0	0
Costa Rica	31	623,048	12.2	1	2	728,955	2	29,769
Cuba	32	686,000	6.0	0	4	323,600		
Dominica	1	6,872	9.2					
Dominican Rep.	17	964,159	19.9	0				
El Salvador	9	26,152	1.2					
Greenland	2	98,250,000	45.2		1	70,000,000	11	1,044,500
Guadeloupe	2	21,000	11.8	0			0	0
Guatemala	17	834,966	7.7	1	1	1,000,000	1	48,372
Haiti	3	9,700	0.4	0				
Honduras	35	717,869	6.4	1	1	500,000		
Jamaica	2	37,953	3.3	0				
Martinique	1	70,150	65.0[b]	0			0	0
Mexico	63	10,073,115	5.1	1	6	1,288,454	1	47,480
Neth. Antilles	2	7,760	9.7	0			5	2,010
Nicaragua	11	362,738	2.5	0				
Panama	16	1,326,140	16.9[b]	2	1	597,000	1	80,765
Puerto Rico	14	28,548	3.2		2	15,346		
St Lucia	1	1,494	2.4					
St Vincent & Grenadines	2	8,284	21.3[b]					
Trinidad & Tobago	7	15,528	3.0					
Turks & Caicos Is	14	97,532	a	0			1	37,270
USA	975	98,239,946	10.5	10	44	19,115,210	10	1,140,841
US Virgin Is	1	14,079	40.1	0	1	6,127	0	0
SOUTH AMERICA	**599**	**107,326,936**	**6.0**	**9**	**26**	**13,781,071**	**7**	**322,085**
Argentina	115	9,395,408	3.4	2	5	2,409,980		
Bolivia	27	9,860,765	9.0	0	3	435,000	1	5,240
Brazil	172	21,566,704	2.5	1	2	1,862,100		
Chile	66	13,722,125	18.3	0	7	2,406,633	1	4,877

Continued

TABLE 3.16 Continued

Region/country	Nationally protected areas			World Heritage Sites	Biosphere Reserves		Ramsar Wetlands	
	Number	Area (ha)	% of total land area	Number	Number	Area (ha)	Number	Area (ha)
Colombia	41	9,048,185	8.0	0	3	2,514,375		
Ecuador	18	10,748,387	6.0	2	2	1,446,244	2	90,000
Guyana	1	11,655	0.1	0				
Paraguay	14	1,204,231	3.0	0				
Peru	20	2,687,846	2.1	4	3	2,506,739		
Suriname	13	735,970	4.5				1	12,000
Uruguay	8	32,086	0.2	0	1	200,000	1	200,000
Venezuela	104	28,313,574	31.0				1	9,968
ASIA	**2,180**	**118,201,269**	**5.2**	**13**	**38**	**12,885,459**	**40**	**1,354,493**
Afghanistan	5	183,438	0.3	0				
Bangladesh	8	96,790	0.7	0				
Bhutan	5	906,138	19.4					
Brunei	5	77,742	13.5					
China	396	28,357,804	3.0	1	8	1,966,722		
China,Taiwan	5	288,577	8.0					
Cyprus	1	2,000	0.2	0				
Hong Kong	12	37,821	35.6	0			0	0
India	362	13,770,557	4.4	5			6	192,973
Indonesia	194	19,230,879	10.0	2	6	1,482,400		
Iran	60	7,528,976	4.6	0	9	2,609,731	18	1,087,550
Israel	21	206,745	10.0					
Japan	684	4,663,543	12.7		4	116,000	3	9,892
Jordan	8	100,400	1.0	0			1	7,372
Korea, Dem.	2	57,890	0.5		1	132,000		
Korea	26	756,833	7.7	0	1	37,430		
Kuwait	1	30,000	1.2					
Lebanon	1	3,500	0.3	0				
Malaysia	51	1,488,047	4.5	0				
Mongolia	15	6,167,840	4.0	0	1	5,300,000		
Myanmar	2	173,271	0.3					
Nepal	13	1,126,000	8.0	2			1	17,500
Oman	2	54,000	0.2	0				
Pakistan	53	3,654,969	4.6	0	1	31,355	9	20,990
Philippines	27	572,866	1.9	0	2	1,174,345		
Saudi Arabia	10	21,197,560	8.8	0				
Singapore	1	2,715	4.4					
Sri Lanka	43	783,708	11.9	1	2	9,376	1	6,216
Thailand	90	5,513,986	10.8	1	3	26,100		
Turkey	18	269,176	0.3	1				
Viet Nam	59	897,498	2.7	0			1	12,000
EUROPE	**2,150**	**42,426,943**	**8.0**	**11**	**91**	**4,787,243**	**328**	**3,782,517**
Albania	13	44,500	1.6	0				
Austria	178	2,090,796	25.0		4	27,600	5	102,369
Belgium	2	71,829	2.4				6	9,607

Continued

TABLE 3.16 Continued

Region/country	Nationally protected areas			World Heritage Sites	Biosphere Reserves		Ramsar Wetlands	
	Number	Area (ha)	% of total land area	Number	Number	Area (ha)	Number	Area (ha)
Bulgaria	50	261,417	2.4	2	17	39,922	4	2,097
Czechoslovakia	65	2,058,780	16.1		6	364,170	8	16,958
Denmark	66	409,768	9.5	0			27	734,468
Finland	35	807,250	2.4	0			11	101,343
France	81	5,357,561	9.9	1	6	575,583	8	422,585
Germany	440	5,859,453	16.4	0	9	701,849	29	360,894
Greece	21	104,553	0.6	0	2	8,840	11	107,400
Hungary	54	576,966	6.2	0	5	128,884	13	110,389
Iceland	22	915,924	8.9				2	57,500
Ireland	6	26,810	0.4		2	8,808	21	12,562
Italy	144	2,008,617	6.7	0	3	3,798	46	56,950
Liechtenstein	1	6,000	38.0				1	90
Malta	0	0	0.0	0			1	11
Netherlands	67	352,589	8.6		1	260,000	11	306,348
Norway	88	4,982,974	12.9	0	1	1,555,000	14	16,256
Poland	80	2,241,701	7.2	1	4	25,836	5	7,090
Portugal	25	556,135	6.1	0	1	395	2	30,563
Romania	40	1,088,638	4.6	1	3	41,213	1	647,000
Spain	163	3,504,990	6.9	1	10	537,717	17	98,887
Sweden	195	2,918,552	6.5	0	1	96,500	30	382,750
Switzerland	112	752,892	18.2	0	1	16,870	8	7,049
UK	140	4,639,706	19.0	2	13	44,258	45	173,257
Yugoslavia	62	788,542	3.1	3	2	350,000	2	18,094
FORMER USSR	**213**	**24,374,326**	**1.1**	**0**	**20**	**10,891,366**	**12**	**2,987,185**
OCEANIA	**936**	**84,360,135**	**9.9**	**12**	**13**	**4,745,223**	**45**	**4,515,961**
American Samoa	1	3,725	18.9[b]	0			0	0
Australia	748	81,309,448	6.1	9	12	4,743,223	40	4,477,862
Cook Is.	1	160	0.7	0			0	0
Fiji	2	5,342	0.3					
French Polynesia	6	12,747	3.2		1	2,000		
Kiribati	3	26,630	38.9[b]					
New Caledonia	14	61,676	3.2	0			0	0
New Zealand	152	2,909,062	11.0	2			5	38,099
Northern Marianas Is	3	1,129	2.4					
Palau	1	1,200	3.3					
Papua New Guinea	5	29,016	0.1					
Pitcairn Is	0	0	0.0	1			0	0
ANTARCTICA	**12**	**252,349**	**–**	**0**	**0**	**0**	**0**	**0**
Antarctica	11	215,649	–					
French Southern Tr.	1	36,700	5.1	0			0	0

Continued

TABLE 3.16   Continued

[a] The percentage protected is omitted as the nationally designated protected areas are mainly marine.

[b] The percentage protected is inflated as the nationally designated protected area includes marine areas while the land area is used for percentage calculation.

An additional 39 countries have reported that they have no nationally protected areas; of these countries, 19 have specified that they have no World Heritage Sites and 6 countries have also reported no Ramsar Wetlands. These countries have been excluded from the table presented here but are included in the listings given in the source document.

A system of classification for different types of nationally designated protected areas, based on land management objectives, has been developed by the IUCN. This system forms the basis for the UN List of National Parks and Protected Areas upon which the above table is based. Here nationally designated protected areas are defined as an area over 10^3 ha that falls within IUCN Management Categories I to V, inclusive. Areas within these categories include Scientific Reserves/Strict Nature Reserves, National Parks, Natural Monuments/Natural Landmarks, Managed Nature Reserves/Wildlife Sanctuaries and Protected Landscapes and Seascapes.

Two international conventions include provision for the designation of internationally protected areas. The Convention Concerning the Protection of the World Cultural & National Heritage (adopted in Paris in 1972, entered into force 1975) provides for the designation of areas of 'outstanding universal value' as World Heritage Sites. Only numbers of natural sites have been listed here; Article 2 of the World Heritage Convention considers as natural heritage: natural features consisting of physical and biological formations or groups of such formations which are of outstanding universal value from an aesthetic or scientific point of view; geological or physiographical formations and precisely delineated areas which constitute the habitat of threatened species of animals and plants of outstanding universal value from the point of view of science or conservation; and natural sites or precisely delineated areas of outstanding universal value from the point of view of science, conservation or natural beauty. The Convention on Wetlands of International Importance Especially as Waterfowl Habitat (adopted in Ramsar in 1971, entered into force in 1975) provides a framework for the conservation of wetland habitats with special obligations pertaining to those wetlands which have been designated to the List of Wetlands of International Importance', i.e., Ramsar Sites.

Biosphere Reserves established as part of UNESCO's Man and Biosphere Programme are designated to serve a range of objectives; these include research, monitoring, training, demonstration as well as conservation.

Source:
World Conservation Monitoring Centre 1992 *Global Biodiversity: Status of the Earth's Living Resources*, Chapman & Hall, London.

Part 4

Population and Development

P opulation and levels of economic activity have increased more rapidly in the last four decades than at any time in human history. Since 1950 the world's population has grown from 2.5 billion to 5.3 billion; much of this growth has taken place in the developing world. The rapid pace of demographic change and the growth in economic activity are major driving forces for environmental change. Important trends in these factors highlighted in this chapter include:

○ Global average total fertility rates have fallen from 5.0 to 3.4 over the past three decades as more countries complete their transition to lower fertility rates. However, relatively high levels of fertility in some developing countries, particularly those in Africa, coupled with the momentum of a young age structure, will maintain world population growth well into the next century.

○ Over half the world's population currently resides in rural areas; in 35 years' time over 60 per cent will be inhabitants of urban agglomerations.

○ Although per capita incomes have increased in the developing world as a whole, population growth and increasing inequalities in the distribution of wealth over the last decade have led to an increase in the number of people living in poverty.

○ Over the past three decades progress has certainly been made in key aspects of human development. Nevertheless, population growth has outpaced some of the gains made in some sectors, for example, the provision of drinking water and sanitation services.

It was only with the publication, in 1987, of the Brundtland report, 'Our Common Future' (WCED, 1987), that the significance of the links between population, environment, resources and development were fully realized by the international community. Since that time, growing interest and appreciation of issues involved have led to their being closely considered during the deliberations leading to the United Nations Conference of Environment and Development (UNCED) in June 1992. In adopting the action plan commonly known as Agenda 21, the Rio conference formally recognized the integrative nature of environment and development and called for the integration of development and environment at levels of political and economic decision-making. It also stressed the need to formulate integrated national policies for environment and development which take into account demographic trends and factors (UN, 1992).

In line with this perception that population, environment and development are one and the same issue, this chapter of the 'UNEP Environmental Data Report' has been expanded to include key socio-economic data which reflect patterns of human and economic development. Moreover the nature of the interrelationships between population, development, resources and environment are briefly explored. Concepts such as sustainable development and integrated environmental economic accounting are also introduced.

Population

The third quarter of the 20th century has witnessed an acceleration in the growth-rate of world population which has been accompanied by a parallel increase in resource consumption. In effect, more people are now consuming more resources than at any time in history; this raises questions about the capacity of the global environment to sustain present rates of population growth.

Growing populations and material expectations inevitably place increasing demands on the environment, not only in terms of space and as a source of resources, but also as a sink for wastes. However, to assume a straightforward correlation between the increase in the number of consumers on the one hand and the disruption of the environment and resource depletion on the other, is to oversimplify the issue. That population growth is a driving force for environment change is not in dispute; what is controversial is the relative importance of population among the other factors generating environmental pressures (such as inappropriate policies regulating land tenure, trade and agriculture, poverty, and careless use of high-waste generating technologies). Also controversial is the conceivably positive role of population growth in resource use. Some demographers maintain that there are positive aspects to population growth; by stimulating adaptive and technological change, population growth may actually increase the prospects for development and improve the condition of human life. What is apparent from this debate is the fact that population growth and environmental degradation are engaged in a complex multi-linkage relationship, where one often serves to exacerbate the adverse impacts of the other.

Attempts have been made to quantify the relative impact of population growth as an environmental stress factor through statistical analyses. Commoner (1989), for example, concluded that in a group of 65 developing countries for impacts arising from motor vehicles, the use of commercial energy and the application of nitrogen-based fertilizers, the nature of the technologies of production were more important determinants of environmental degradation than population growth. The influence of population growth is, however, found to be much more significant in other spheres of change, especially those in which technology change is slow; in the developing countries as a whole population growth was considered responsible for about 79 per cent of the recorded deforestation, 72 per cent of arable land expansion and 69 per cent of the growth in livestock numbers (UNFPA, 1992).

In order further to unravel and understand the consequences of population growth on the environment and vice versa, detailed information on population size, growth and distribution will continue to be of fundamental importance. Population data are assembled on a systematic basis by a number of international organizations and United Nations (UN) departments, including the UN Statistical Office (UNSTAT), the UN Population Fund (UNFPA), the Population Division of the UN Department for Economic and Social Information and Policy Analysis (UNDESIPA) (formerly the UN Department of Economic and Social Development (UNDESD)) as well as the World Bank.

Statistics on population size and other demographic variables presented in UNSTAT's 'Demographic Yearbook' represent official country estimates, supplied to the UN by national statistical services. These data are forwarded by means of a questionnaire, and supplemented where necessary by data taken from official national publications or by direct correspondence with appropriate government offices. The Population Division of the UNDESIPA, on the other hand, prepares official UN population estimates and projections for all countries; these data are calculated by the UN using a combination of demographic models and available data, adjusted where necessary. These estimates and projections provide a standard, consistent set of population figures which are widely used throughout the UN system. The estimates and projections are periodically revised, generally on a biennial basis. The latest revisions were completed in 1992, the results of which have since been published as a series of publications (UNDESD, 1993; UNDESIPA, 1993a,b). Data are also available in machine-readable format.

The 1992 revisions comprise UN estimates of population size, growth rate, and age and sex distributions for 1950–1990, plus four variants of projections for 1990–2025 (based on the assumption of low, medium, high and constant rates of fertility change) for individual countries and world regions, and form the basis of the demographic data tables included here (Tables 4.1–4.3). The UN data also include estimates of urban, rural and city populations (see Urbanization) as well as other demographic variables, such as birth and death rates, fertility rates, life expectancy and infant mortality (see also Part 5: Human Health).

The 1992 revisions have had to accommodate a number of extraordinary world events, including the unification of Germany and Yemen and the separation of the former USSR into 15 individual republics. Population estimates and projections have been prepared for these new entities. The rise in refugee numbers in Africa, plus new migratory movements within and to Europe, have meant that the levels of international migration have had to be revised for an unusually large number of countries. Finally, the new projections have taken into account the potential demographic impact of the AIDS pandemic in the highest prevalence countries of Africa.

Population Size and Growth

According to the 1992 revisions of the official UN population estimates and projections, the world population has more than doubled since 1950 and is currently (mid-1992) estimated to be 5.5 billion persons (Table 4.1). It is currently growing at a rate of 1.7 per cent per year (UNDESIPA, 1993a).

The growth in the world population since 1950 has followed a number of distinct phases. Phase I corresponds to a period of rapid population growth during the 1950s and 1960s, peaking at a rate of 2.1 per cent per year during 1965–70 (Table 4.2). This phase of rapid population growth is attributed to the combination of a rapid decrease in crude death rates and a smaller decline in crude birth rates (Table 4.3) brought about largely through improvements in health care in the developing world. Between 1970 and 1975 a fall in crude birth rates has led to a relatively steep decline in the population growth rates – Phase II. This fall in crude birth rates was largely fertility-driven, i.e., a result of declines in the total fertility rate (TFR) or the average number of children born to women of child-bearing age. Since 1975, however, a slow-down in the rate of decline of fertility and thus crude birth rates has contributed to a stagnation in the decline of world population growth rates (Phase III). According to the medium-variant projections, the growth rate of the world population is expected to decline again after 1990. This corresponds to Phase IV in world population growth (Table 4.2).

The total population is projected to reach 6.2 billion by the end of the century and 8.5 billion by 2025 (Table 4.1 and Figure 4.1). Long-range projections of population size, which were revised by the UN Population Division in 1991, suggest that world population will double its 1990 level (5.3 billion) by the middle of the 21st century (i.e., around 2050) and will stabilize at around 11.6 billion shortly after 2200 (UNDESD, 1991).

Examination of regional and national data reveals increasingly divergent trends in population size and growth rates among the major areas and regions of the world (Tables 4.1–4.3). In effect these variations are a reflection of which stage of population growth an individual region or country is at. The developed countries reached the fourth phase of population growth around 1960 when, after a period of constant growth, population growth rates have declined from an average rate of 1.2 per cent per year during 1960–1965 to 0.6 per cent per year in 1985–1990 (Table 4.2). In addition to low rates of population growth, the developed countries are characterized by low TFRs, stable crude death rates (CDRs), relatively low and slowly declining crude birth rates (CBRs) (Table 4.3), and aging populations (i.e., an increase in the proportion of the elderly).

In contrast the developing countries are for the most part less far along the demographic transition pathway. By the late

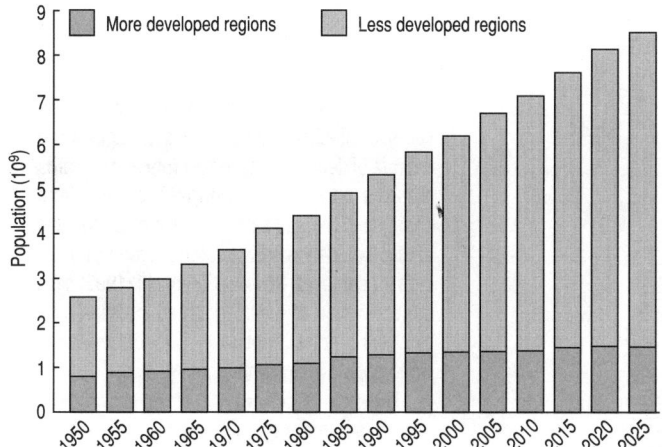

FIGURE 4.1 Population growth in developing and developed countries, and in the world, 1950–2025
Source: Based on UNDESIPA, 1993a

1980s many developing countries had reached phase III of population growth and thus have been characterized by fairly stable (albeit at relatively high) levels of population growth since 1970.

Aggregated data for the developing world tend to mask significant intra-regional differences in demographic trends. Within the less developed countries two distinct groups of countries, distinguishable by the timing of fertility declines, are beginning to emerge (Horiuchi, 1992). These comprise what can be referred to as the pre-transition group of countries, which have not yet started or are at early stages of fertility declines (i.e., Phase II), and the late-transition countries which began significant fertility declines after 1950. The group of early-transition countries, i.e., those which experienced significant fertility declines before 1950, roughly equates to the more developed countries.

The pre-transition countries include most of the countries in sub-Saharan Africa and a number of countries in South Asia (excluding India). These countries are characterized by rapidly growing populations (annual average growth rates of 2–3 per cent), relatively high TFRs (around 6–7 births per woman), stable or only slowing declining CBRs and a fast decline in CDRs throughout the 1950–1990 period. In sub-Saharan Africa recent evidence suggests that fertility could even be rising in some countries (Blacker, 1993).

Elsewhere in the developing world, in particular in East and South-east Asia and Latin America, population growth rates rose to levels of around 2.5–2.8 per cent per year in the 1960s but thereafter have declined, reaching rates of around 2 per cent per year in Latin America and South-eastern Asia and 1.4 in Eastern Asia in 1985–1990 (Table 4.2). The decline in population growth rate in these world regions is linked to dramatic decreases in fertility starting in the late 1960s. For example, in Latin America TFRs have fallen by about a third, from around 5.5 in the late 1960s to around 3.5 in 1985–1990. In Asia the TFR decrease is even more pronounced, falling from about 6 to 3.4 over the same time period. Trends in Asia are dominated by those in China where the fertility decline

from a TFR of 6.0 to 2.4 over the past 25 years is quite remarkable. Demographic trends in China are discussed in more detail in Box 4.1.

It is sometimes asserted that the place that developing countries occupy on the pathway to demographic transition is linked to their level of development. Certainly the experience of some countries would appear to suggest that a certain economic threshold is needed before mortality rates decline and couples start wanting to have fewer children. However, in general the relationship between population growth and economic development is not clear cut and has been the subject of much controversy for more than two decades.

Statistical analyses of past trends in rates of population and economic growth have been reviewed by UNFPA (1992). It is reported that the majority of studies have found no significant correlations between income growth and population growth in either the developed or developing countries up until the mid-1970s. However, over the past decade and a half there has been a tendency for income growth to be slower in countries having high rates of population growth. Conversely, countries with slower population growth rates have tended to fare better economically in that per capita incomes have risen more quickly than in these countries with rapid population growth (UNFPA, 1992).

High population growth rates, both current and past, have tremendous implications for future population size. Countries currently growing at a rate of 3 per cent per year can expect a doubling of their population in about 23 years. According to the 1992 revisions, 53 countries – which currently account for around 10 per cent of the world population – have annual growth rates of 3 per cent or more (UNDESIPA, 1993a). Thus although fertility has started to decline in several of the less developed countries, the rate of population growth will continue at a relatively high level for several decades to come, particularly in Africa. Projections indicate that over the next 35 years 94 per cent of the world population increase will take place in the less developed world and only 6 per cent in the more developed.

Age Structure

Declining fertility rates, coupled with increases in life expectancy have the effect of decreasing the number of young while increasing the numbers of the elderly. The result is a population with a higher proportion of the old. This so-called aging process tends to occur in all countries as they move along the path of demographic transition.

In the developed countries the aging process is already well under way. In 1950, the median age of population of the developed countries was 28.2 years and the proportion of the population aged 65 or over was 7.6 per cent. The developed countries currently have a median age of about 33 years and about 12 per cent of people are aged 65 and over. It is projected that by 2025 the median age may exceed 40 years and the proportion aged 65 and over may be almost 20 per cent (El-Badry, 1992). The progressive aging of the developed world population can be clearly seen in Figure 4.2.

The population structure of the developing countries, on the other hand, is dominated by large numbers of the young

BOX 4.1 Demographic Trends and Population Policy in China

China is the world's most populous country; in 1992 China's population was estimated by the UN to be 1,188 million persons or about one fifth of the world's total population. China is currently in what is widely referred to as phase III of population growth, i.e., a period of relatively constant population growth. Since 1975 the growth rate has remained around 1.4 per cent per year and is expected to maintain this level until 1990–95. Thereafter, annual population growth rates are projected to decline to about 0.5 per cent by 2020–25, with the total population topping the 1.5 billion mark by 2020 (medium variant projections) (UNDESIPA, 1993).

Population growth rates peaked in China at around 2.1 per cent per year during the 1965–70 period. In the early to mid-1970s growth rates decreased sharply (phase II) as a result of an extraordinarily pronounced transition to lower fertility rates; TFRs fell from a 1968 value of 6.4 to 2.2 in 1980. Since then the decline in fertility has stagnated, fluctuating around 2.5 throughout the 1980s (Horiuchi, 1992), despite the introduction of a "one-child" population policy in 1979.

The substantial decline in fertility in the 1970s is attributed to a number of factors including improvement of maternal and child health-care, wider provision of family planning services and a trend towards later marriages and spaced births. Detailed statistical analyses of Chinese fertility surveys suggest that much of the decline in fertility that occurred during the 1970s was in the numbers of births of orders of three or over. A noticeable reduction in the proportion of women having a second child was only apparent in the early 1980s, i.e., as the one-child policy came into operation. This trend was reversed, however, in 1985 as a result of a relaxation in the one-child policy at this time (Blacker, 1993).

The net effect of China's population policies and fertility declines over the past 20 years has been a reduction of about 200 million potential births; in other words, had China's birth rate remained at pre-1970 levels throughout the 1970s

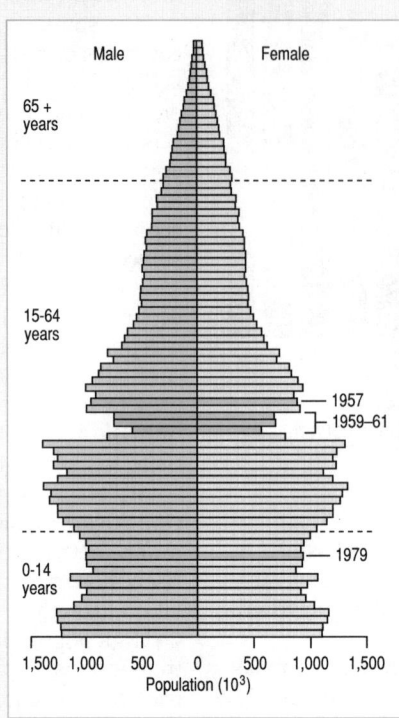

Age structure of the Chinese population, 1990 (populations are plotted at yearly intervals)
Source: State Statistical Bureau of the People's Republic of China, 1991

and 1980s, the population of mainland China would now be in excess of 1.3 billion (Hua, 1991).

Despite the rapid decline in fertility and the introduction of population control policies, the population of China is still increasing by around 17 million persons per year (or 1.4 per cent). This is largely a consequence of the age structure of the Chinese population, coupled with ineffective family planning programmes in some rural areas. The age structure of the Chinese population, based on 1990 national census data, is shown here. A rapid rise in the number of people born in the 1960s has provided a wide population base and has sustained continuing increases in the population, as

women born at this time reach their child-bearing years. In effect, the combination of a stagnation in the rate of decline of the fertility rate and upwards pressure on the crude birth rate resulting from changes in the age structure have acted to keep annual population growth rates at around 1.4 per cent for some 15–20 years. Horiuchi (1992) cites the stagnation of fertility declines in China as a significant factor in influencing recent trends in world population growth rates.

The population age structure shown here reflects a number of other key historical events which have influenced Chinese demographic trends, the most significant of which occurred around 1960. The years 1959 to 1961 were known as "the terrible years" in which natural disasters and famine cut birth rates and raised death rates to produce a near stable, if not actually decreasing, population (Stein, 1990). This is represented in the 1990 population structure as a dramatically smaller number of people aged between 29 years and 31 years of age. A dip in the birth rate during the mid-1980s is linked to the small birth cohort of the famine years.

References

Blacker, J. G. C. 1993 Trends in demographic change, *Transactions of the Royal Society of Tropical Medicine and Hygiene*, **87**(Supplement 1), 3–8.
Horiuchi, S. 1992 Stagnation in the decline of the world population growth rate during the 1980s, *Science*, **257**, 761–765.
Hua, Xin 1991 China's population tops 1.16 billion, *China Today*, **40**(2), 23–24.
State Statistical Bureau of the People's Republic of China 1991 *China Statistical Yearbook 1991*, China Statistical Information and Consultancy Service Centre, Beijing.
Stein, D. 1990 Family planning with Chinese characteristics, *China Now*, **133**, 15.
UNDESIPA 1993 *World Population Prospects: The 1992 Revision*, Department for Economic and Social Information and Policy Analysis (Population Division), United Nations, New York.

(Figure 4.2). Fertility declines have, however, begun to reduce the proportion of children and to initiate the aging process in some countries, particularly those in Latin America and parts of Asia. In the developing countries as a whole the proportion of the population aged over 65 years has remained more or less constant up to 1975 (around 4 per cent) but is projected to reach about 5 per cent by the year 2000, and to increase more

rapidly during the next century (El-Badry, 1992). The aging process has important implications; in the developed countries the welfare of increasing numbers of elderly people will depend on the productivity of a shrinking labour force. In contrast, in the developing world the labour force will grow rapidly. These trends have potentially profound implications for future economic performance.

a) More developed countries

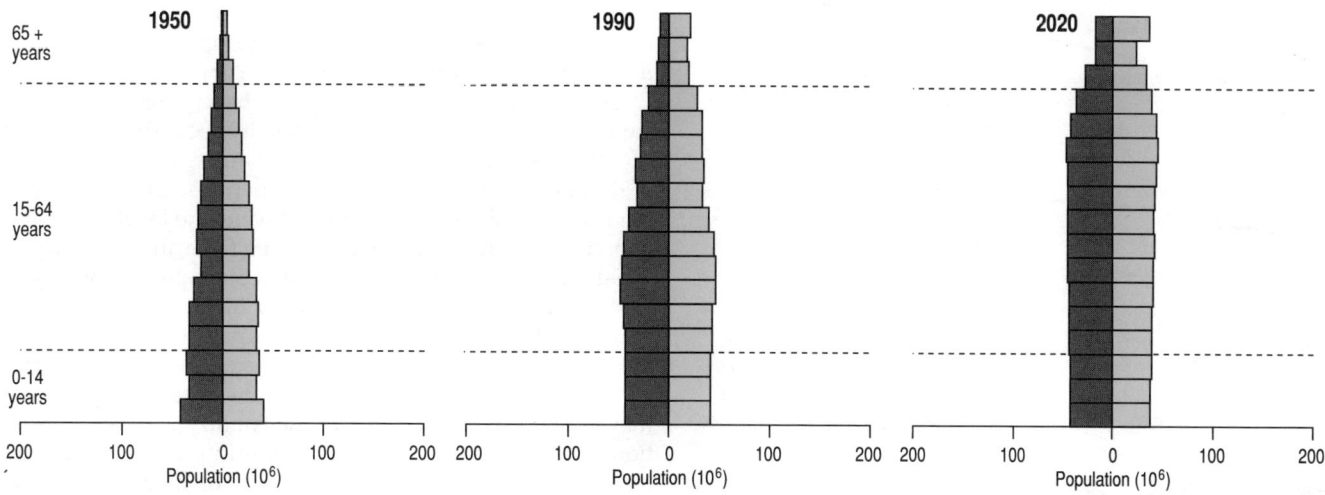

b) Less developed countries

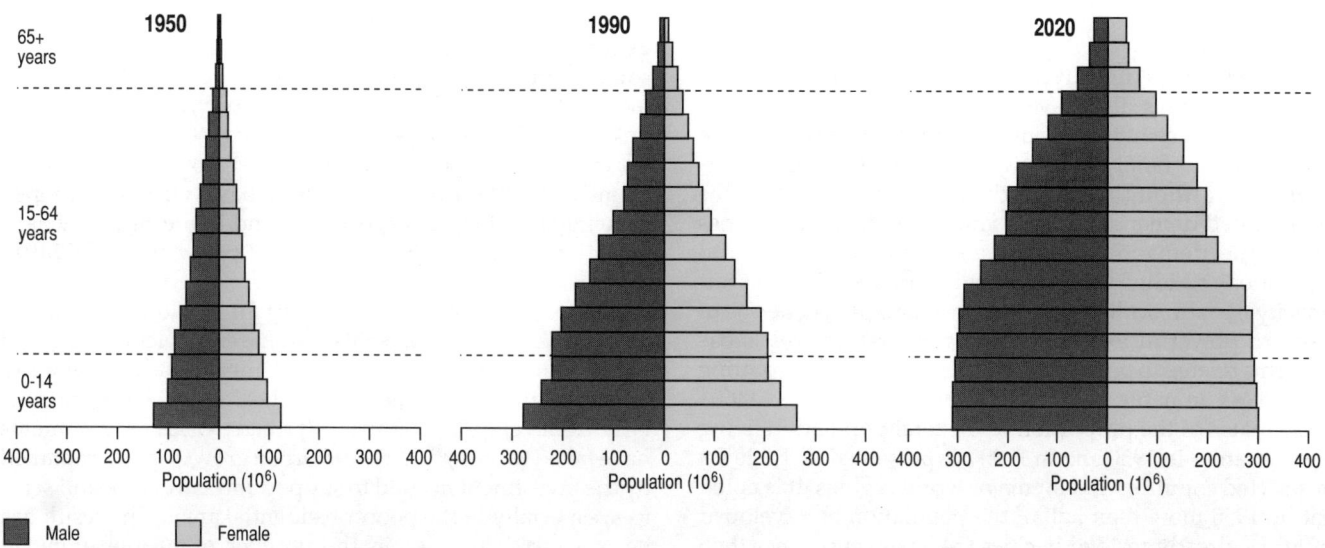

■ Male   ☐ Female

FIGURE 4.2 Population pyramids for developed and developing countries: 1950, 1990 and 2025
Source: Based on data from UNDIESA, 1991

Urbanization

In addition to aging, this century has witnessed another notable demographic development, namely a significant increase in the number and proportion of persons living in urban areas. Currently more than two of every five persons (or 43 per cent of the world's population) are urban dwellers; this compares with only one in ten at the turn of the century. The combination of population growth and a transition from an agriculture-based to a more industrialized economy are the main factors responsible for the rural-urban migration of the population in the majority of the world's nations.

The distribution of people between rural and urban areas has important implications for the type of stress placed on the environment. Urban areas and cities concentrate human activity and thereby create relatively high demands for natural resources (e.g., energy, fresh water and land), basic services and infrastructure (e.g., sanitation and waste disposal services, education and health care, roads and public transport), and employment. Furthermore, cities typically represent a disproportionate source of national pollutant emissions and wastes (liquid and solid) and are thus often associated with high levels of air pollution and other forms of environmental contamination.

On the positive side, however, urban areas are an essential part of economic development and can bring major benefits to human well-being by providing health, educational and social services to their residents (the concentration of

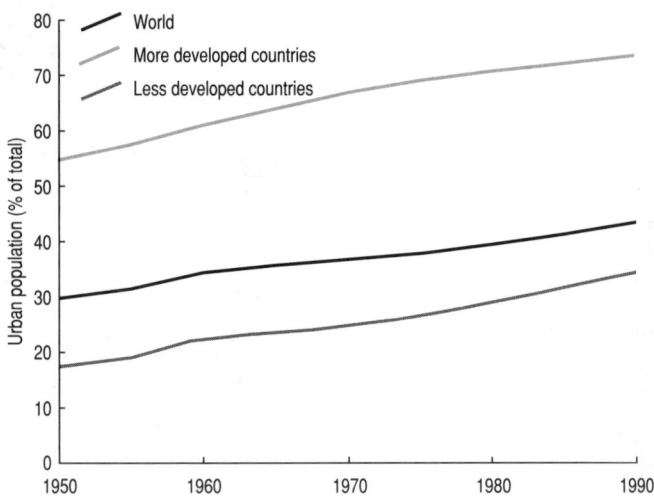

FIGURE 4.3 Trends in the percentage of the population residing in urban areas – world, more developed regions and less developed regions, 1950–1990
Source: Based on UNDESIPA, 1993b

population lowers the unit cost for the supply of many services). Urban areas also have far-reaching implications for population policy in that residents in urban areas often have better access to family planning means and education. Furthermore, although rural-urban migration generates concentrated demands for land and other resources, it does relieve the demand for the same needs amongst rural populations residing in areas already strained beyond their capacity. Paradoxically therefore urbanization may be one of the most powerful long-term means of dealing with those problems related to population growth and which are putting great stress on natural resources (UNCHS (Habitat), 1992).

Estimates of the proportion of the total population living in urban areas in 1950 and in 1990 are presented in Table 4.4 for selected countries and for major world regions. It is noted that by 1950 more than half of the population of developed countries already resided in cities (54.3 per cent); since then this percentage has grown steadily to reach 72.7 per cent in 1990 (Figure 4.3). By the year 2025, it is projected that at least 8 out of 10 residents will live in cities in the developed world (UNDESIPA, 1993b).

As a whole, developing countries have a lower level of urbanization; only 34 per cent of the developing world population currently reside in urban areas (Table 4.4). However, the trend towards urbanization has been proceeding at a much faster rate than in the more developed nations (Figure 4.3); in 1985–90 the annual rate of growth in the urban population of the less developed regions averaged 3.8 per cent compared with only 1.0 per cent in the more developed regions (Table 4.4). By the year 2025, it is estimated that the urban proportion of the developing world population will rise to 56.7 per cent (UNDESIPA, 1993b).

Although the degree of urbanization is higher in the developed world, the much higher absolute population size in the less developed regions means that there are more urban

dwellers in the developing world; UN projections indicate that in the next 35 years (i.e., by 2025) 2,904 million persons will be added to the world's urban population, 2,609 million of whom will be added to the urban population of the less developed countries (UNDESIPA, 1993b).

On regional and national scales, urbanization is characterized by a high degree of heterogeneity. Less than one third of Africans and Asians live in urban areas. In contrast 7 out of 10 persons in Europe, Northern America and Latin America are urban residents. On a national scale, the level of urbanization is even more disparate, ranging from almost complete urbanization (such as in Monaco and Singapore) to very low levels of urbanization (i.e., less than 20 per cent).

A significant proportion of the urban population of the world resides in large urban agglomerations. In some countries a single "megacity" dominates the urban structure and may contain more than half of the country's urban population. In 1990, about one third of the world's urban population resided in agglomerations of 1 million or more inhabitants and 13 per cent resided in agglomerations of 5 million or more. Since 1950 the number of urban agglomerations having more than 10 million inhabitants has risen from 1 to 13 (Table 4.5).

The size and number of large urban agglomerations are generally growing at a faster pace in the developing countries; some of the fastest growing cities in the 1985–1990 period include Brasilia in Brazil (7.8 per cent per year), Peshawar in Pakistan (7.8 per cent per year) and Kuala Lumpur in Malaysia (5.5 per cent per year). In contrast, some large cities in countries of the developed world have virtually stopped growing; London, Chicago, Osaka and Rome currently have growth rates of less than 0.4 per cent per year (UNDESIPA, 1993b).

The growth of very large urban agglomerations, or "megacities" often represents extreme examples of rapid and massive urban growth and development and has given rise to significant environmental and health problems, particularly in developing countries. In many cases, governments have failed to ensure that rapid urban growth is accompanied by the investment needed to support infrastructure and services, especially in the poorer residential areas. The result has been a rapid increase in the number of people living in overcrowded conditions and in illegal or informal settlements (sometimes referred to as squatter camps) without access to adequate basic services. Urban air problems in megacities are reviewed in Part 1: Environmental Pollution, Atmosphere.

It has been difficult to estimate with any precision what proportion of urban (and rural) dwellers live in inadequate housing without proper access to water, sanitation and other services. Case studies of cities in Africa, Asia and Latin America suggest that it is common for between 30 and 60 per cent of the population to live in either illegal settlements with little or no infrastructure or services, or in overcrowded and often deteriorating tenements and cheap houses. This extrapolates to some 600 million urban dwellers world-wide who currently reside in health-threatening houses and under conditions characterized by overcrowding and lack of proper services (WHO, 1992a).

Many of the available data describing conditions in human settlements, such as those compiled by the UN Centre for

Human Settlements (UNCHS Habitat), the UNFPA, the UN Development Programme (UNDP) and the World Health Organization (WHO), relate to city-wide averages. Yet considerable diversity exists in urban settlements. These large intra-urban differences in housing quality, access to basic services and in health status, have important implications for municipal governance and development action. In order to improve their usefulness for policy making, settlement statistics need to incorporate this diversity, i.e., be more disaggregated. To date, some progress has already been made in developing an improved methodology for monitoring the environmental conditions of different neighbourhoods within cities; case studies have been conducted in a number of cities including Jakarta, Accra and São Paulo (Goldstein, 1993). This work forms part of the wider UNDP/World Bank/UNCHS (Habitat) Urban Management Programme.

Strategies for improvement of conditions in human settlements have been proposed by a number of international organizations, including UNCHS (Habitat) and the WHO Commission on Health and Environment (WHO, 1992) and have since been affirmed in Agenda 21. Moreover, several programmes are now under way which embody new integrative approaches to urban management. The WHO's Health City Project, is one such example; through networking and awareness building the project aims to generate the opportunities for municipal governments, individuals and community groups to address collectively their environmental and health problems.

References

Blacker, J. G. C. 1993 Trends in demographic change, *Transactions of the Royal School of Tropical Medicine and Hygiene*, **87**(1), 3–8.

Commoner, B. 1989 Rapid population growth and environmental stress. In: *Consequences of Rapid Population Growth in Developing Countries*, Proceedings of a UN Expert Group Meeting, 1988, New York, USA, ESA/P/WP.110, United Nations, New York, 231–263.

El-Badry, M. A. 1992 World population change: a long-range perspective, *Ambio*, **21**(1), 18–23.

Goldstein, G. 1993 Personal communication, World Health Organization, Geneva, Switzerland.

Horiuchi, S. 1992 Stagnation in the decline of the world population growth rate during the 1980s, *Science*, **257**, 761-765.

UN 1992 *Report of the United Nations Conference on Environment and Development (Rio de Janeiro, 3–14 June 1992)*, UN document A/CONF.151/26, 12 August 1992 (Vol I, II, II and IV), United Nations Conference for Environment and Development, Geneva.

UNDESD 1991 *Long-Range World Population Projections* Department for Economic and Social Development (Population Division), United Nations, New York.

UNDESD 1993 *The Sex and Age Distribution of the World's Population: The 1992 Revision*, Department for Economic and Social Development (Population Division), United Nations, New York.

UNDESIPA 1993a *World Population Prospects: The 1992 Revision*, Department for Economic and Social Information and Policy Analysis (Population Division), United Nations, New York.

UNDESIPA 1993b *World Urbanization Prospects: The 1992 Revision*, Department for Economic and Social Information and Policy Analysis (Population Division), United Nations, New York. In press.

UNCHS (Habitat) 1992 *Improving the Living Environment for a Sustainable Future*, United Nations Centre for Human Settlements (Habitat), Nairobi.

UNDIESA 1991 *The Sex and Age Distributions of Population*, Department of International, Economic and Social Affairs, United Nations, New York.

UNFPA 1992 *The State of the World Population 1992: A World in Balance*, United Nations Population Fund, New York.

WCED 1987 *Our Common Future: Report of the World Commission on Environment and Development*, Oxford University Press, Oxford.

WHO 1992 *Our Planet, Our Health: Report of the WHO Commission on Health and the Environment*, World Health Organization, Geneva.

Development

Meeting people's basic needs for food, water, shelter and employment, thereby improving human well-being, are the central goals of development. Although considerable progress has been made over the past few decades, it is nevertheless estimated that currently more than one billion people still live in acute poverty and suffer grossly inadequate access to the resources (education, health services, infrastructure, land and credit) required to give them a chance for a better life.

Over the past decade concerns about inequalities in the development process have been supplemented with new fears regarding its sustainability; namely, whether development will cause serious environmental damage through resource degradation and pollution, and/or whether environmental constraints will limit development. For instance, it is now widely acknowledged that environmental problems have the capacity to undermine economic and human development in a number of ways. Firstly, human health may be adversely affected as a result of exposure to toxic contaminants in the environment and the indirect effects of resource degradation. Secondly, the "amenity" value of the environment, i.e., pleasure derived from a "clean" natural environment, may be reduced. Thirdly, and perhaps most significantly, environmental degradation in the form of air and water pollution, soil degradation and depletion of natural resources can compromise future economic productivity. The finding that is increasingly emerging from analyses of environmental and economic issues is that continued development depends ever more on the maintenance of the natural resource base. Paradoxically, however, it appears that the misuse and overuse of the renewable (e.g., water, forests, fisheries), rather than the non-renewable (fossil fuels, metal ores and minerals) resources currently represent the more immediate threats to future economic security (see also Part 3: Natural Resources and Part 7: Industry and Transport).

The idea that there needs to be a balance between protection of the environment on the one hand, and the demands of human and economic development on the other, is one that has become embodied in the now popular term "sustainable development". The notion of sustainable development was brought into common use in 1987 by the work of the World Commission on Environment and Development which defined sustainable development as "meeting the needs of the present generation without compromising the needs of future generations" (WCED, 1987). Since then this concept has gained widespread acceptance and endorsement as a universal goal.

The adoption of the concept of sustainable development has given rise to considerable debate regarding the ability of traditional measures or indicators of economic performance and human welfare to provide an appropriate yardstick of national progress towards sustainable development. The main arguments in this ongoing debate are briefly outlined here.

Economic Growth

The System of National Accounts (SNA), first developed in the 1940s, provides a standard, internationally accepted framework for calculating national incomes and setting up national income accounts. The central concepts Gross National Product (GNP) and Gross Domestic Product (GDP), which represent the total national output of goods and services expressed in monetary terms, have become the principal measures by which economic progress is judged and, as such, are corner stones in the present-day policy-making process.

Estimates of GNP and GDP in absolute amounts (i.e., US$), on a per capita basis and broken down by economic sector as calculated by the World Bank for 1990, are presented for the majority of the world's market economies in Table 4.6. The annual average growth rate of GDP during 1980–1990 and estimates of real GDP per capita, a modification of GDP which takes into account the difference in the domestic purchasing powers of two currencies, are also provided. The footnotes to the table define GDP and GNP more precisely and summarize some of the assumptions made in their calculation.

Per capita national income statistics (e.g., GNP per capita or real GDP per capita) are the most frequently used bases for measuring economic progress and for comparing living standards between countries. They also serve to divide the world into "developed" and "developing" countries. Comparisons of per capita income measures between countries should, however, be viewed with caution. The methods used and the level of accuracy achieved in the calculation of national incomes can vary considerably between countries. Moreover, some, and in a number of countries significant, parts of the production process do not show up as market transactions; thus their value is not recorded by the SNA system and they therefore are not counted as contributing to GNP. This is especially the case in countries in which subsistence agriculture is important.

However, it is the limitations of the existing measures of economic activity with respect to environmental and natural resource considerations have been the focus of debate in more recent years. In this respect there are two main criticisms of the existing framework. Firstly, neither the SNA nor GDP take into account the depletion of natural resources. Whereas the depreciation of capital goods (i.e., equipment and buildings) is taken into account in national accounts (by a depreciation allowance which is subtracted from GDP), the investment in human capital and depreciation of environmental capital (e.g., natural resources such as forests) is not measured. Secondly, the deleterious impact of economic activity on the environment in the form of environmental quality degradation (e.g., air pollution) and concomitant costs to human health and welfare, are neglected. Thus the traditional economic measures, by failing to take account of environmental

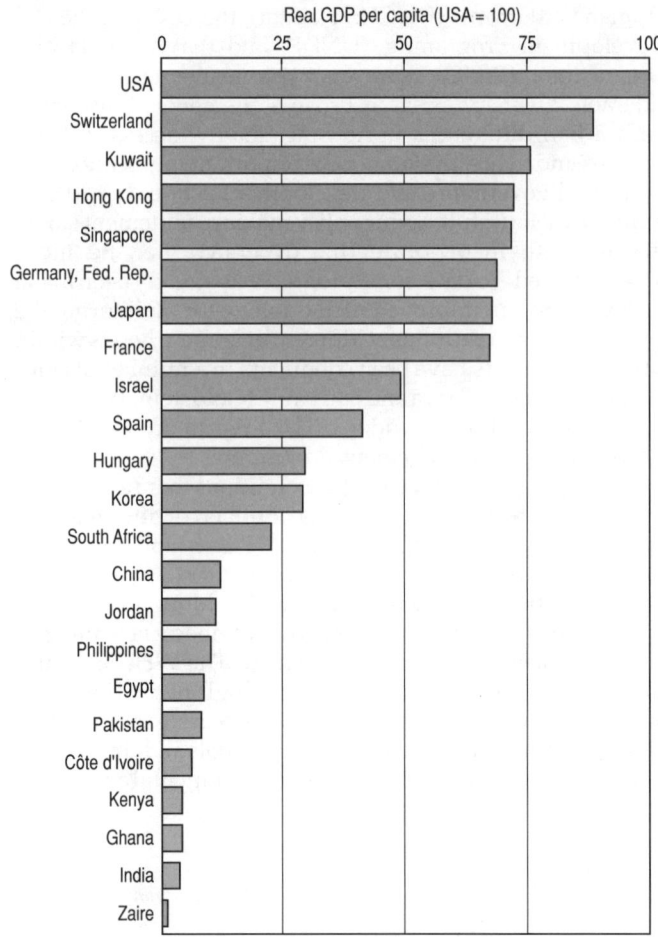

FIGURE 4.4 Per capita real GDP in selected countries, using the USA as a base (USA = 100), 1989
Source: After UNDP, 1992

resource degradation, are considered by many to overestimate economic growth and thereby provide misleading information for the policy maker. Alternative economic measures and modifications to the methods of calculation for GNP, which attempt to take account of the limitations outlined above, have been proposed in recent years. Some of these are briefly reviewed in Box 4.2.

Existing per capita income measures reveal huge disparities between individual countries; per capita real GDP, for example, ranges from over US$ 20,000 in the richest nations (e.g., USA) to less than US$ 500 in the poorest nations (e.g., Zaire). Figure 4.4 compares per capita real GDP in selected countries. These national per capita income statistics mask discrepancies in the patterns of income distribution within countries; although two countries may have the same income-per-head figures, the standards of living will be very different if in one the income is fairly evenly distributed, while in the other the income distribution is very unequal.

The past few decades have seen marked regional differences in the pace of economic growth. Trends in the high-income countries have been characterized by a moderation of the fairly high rates of growth in average per capita

BOX 4.2 Alternative Economic Indicators and "Green GDP"

In view of the now widely recognized limitations of the traditional measures of economic growth, various proposals have been made for adjusting existing economic accounting systems to take account more fully of the environmental impacts of economic activities (Abaza, 1992). For the purposes of this discussion, two main approaches to correcting the environmental blind spots of the existing accounting systems can be distinguished:

○ Adjustment of the core System of National Accounts (SNA) in order to introduce data on the environment;
○ Development of satellite accounts outside the core of the existing SNA, but based on national accounting principles, and/or SNA.

Many experts concur that the first approach, i.e., adjustment of the core SNA, represents the best way forward in the longer-term perspective. If environmental damage and net changes in natural capital (i.e., natural resources) could be appropriately incorporated into SNA, the resulting statistic (an Environmentally-adjusted Net Domestic Product or a so-called Green GDP) would conceivably indicate the maximum consumption path that would not impair the potential well-being of future generations; in effect an adjusted GDP would provide a good measure of sustainability (Pearce and Mäler, 1991). From the practical point of view, however, calculation of a "green GDP" at the present point in time is fraught with difficulties. Aside from the difficult questions of how to treat environmental data in economic and accounting terms, and in particular how to assign a monetary value to environmental services and natural resources, current weaknesses in the environment statistics are a major handicap for the development of a "green GDP" (OECD, 1991; Pearce and Mäler, 1991).

It is, however, generally considered that some partial modifications to the present SNA could be made which would better reflect environmental concerns. These could comprise a breakdown of environmental expenditures and inclusion of depreciation in the natural resource assets that can be estimated in moneterized form. Repetto and his associates at the World Resources Institute (WRI) have convincingly shown that even partial

reforms of accounting systems have the capacity to change our ideas about whether an economy has been growing sustainably or not (Repetto et al., 1989). For example, in an experimental study which incorporated depreciation of three marketed natural resources into the national accounts for Indonesia (oil, timber and topsoil) they revised Indonesia's official annual average growth rate in GNP between 1971 and 1984 from about 7 per cent, downwards to around 4 per cent. A more recent and sophisticated analysis, which also looked at the depreciation in three forms of natural capital – but this time in Costa Rica – similarly indicated a rapid depletion of natural capital. It was estimated that in two decades, Costa Rica's forests, soils and fisheries depreciated by an amount which exceeded the average value of one year's GDP during this time (Repetto et al., 1991).

The UN Statistical Office (UNSTAT) is currently engaged in a revision of the SNA, details of which are due to be published at the end of 1993 (UN, 1993a). The proposed methodologies for a revised SNA make some provision for the inclusion of the balance sheets for those natural assets which provide direct economic benefits. It is hoped that these proposed revisions will lead to further elaboration of changes in natural assets within the framework of internationally-approved SNA.

Until such time that methodologies which fully integrate environmental concerns into national economic accounting systems can be developed and gain international currency, several groups including the Organisation for Economic Co-operation and Development (OECD) and the UNSTAT have given priority to the second approach, that is, the establishment of a system of satellite accounts to accompany the core SNA accounts (Bartelmus, et al.,1991; OECD, 1991). The main objective of satellite accounts is to reflect expenditure for environmental protection and enhancement, the market value of natural assets and services and the environmental costs of economic activities within an accounting framework which maintains as far as possible SNA concepts and principles. These methodologies build on the integrated natural resource accounting systems, in physical terms, pioneered by Norway and France.

To this end UNSTAT, working in conjunction with other agencies, has devised a "Satellite System for Integrated Environmental and Economic Accounting" (SEEA) as a satellite system to the core SNA. The basic principles and structure of the SEEA have been outlined by Bartelmus and Tardos (1993). The SEEA methodologies have been tested and refined through a number of pilot country studies (e.g., Mexico, Papua New Guinea and Thailand) and have been shown to be not only feasible but to offer, albeit in a tentative form, an invaluable information base for integrated development planning and policy formulation (Bartelmus and Tardos, 1993). A handbook outlining the satellite approach to environmental accounting is due to be published in 1993 (UN, 1993b).

References

Abaza, H. (Ed.) 1992 *The Present State of Environmental and Resource Accounting and its Potential Application in Developing Countries*, Environmental Economics Series, Paper No. 1, United Nations Environment Programme, Nairobi.

Bartelmus, P., Stahmer, C. and von Tongeren, J. 1991 Integrated environmental and economic accounting: framework for a SNA Satellite System, *Review of Income and Wealth*, 37(2), 111–148.

Bartelmus, P. and Tardos, A. 1993 Integrated environmental and economic accounting – methods and applications, *Journal of Official Statistics*, 9(1), 179–188.

OECD (Organisation for Economic Co-operation and Development) 1991 Environmental Indicators: Progress Report. Background paper presented at the Environment Committee meeting at Ministerial Level, 30–31 January 1991, Paris, France. Unpublished paper.

Pearce, D. and Mäler, K. 1991 Environmental economics and the developing world, *Ambio*, 20(2), 52–54.

Repetto, R. et al. 1991 *Accounts Overdue: Natural Resource Depreciation in Costa Rica*, World Resources Institute, Washington DC.

Repetto, R., Magrath, W., Wells, H., Beer, C. and Rossini, F. 1989 *Wasting Assets: Natural Resources in the National Income Accounts*, World Resources Institute, Washington DC.

UN 1993a *1993 System of National Accounts*, United Nations, New York. In press.

UN 1993b *Handbook of National Accounting – Integrated Environmental and Economic Accounting*, United Nations, New York, In press.

incomes attained during the 1960s to more modest levels of just over 2 per cent per year in more recent decades. Within the developing world, only Asia has achieved a continuous rate of growth in per capita incomes throughout the 1960–1990 period. In contrast, the economies of many countries in Latin America and Africa, which sustained only modest or low rates of the growth in the 1960s and 1970s, have since stagnated or even declined (World Bank, 1992). The combination of deteriorating economic infrastructures, unfavourable terms of trade, low prices for agricultural commodities and debt problems are generally considered to be responsible for the decline in per capita incomes in Africa and Latin America over the past decade. Regional trends in per capita incomes are reflected in the table below (values represent average annual percentage changes in real per capita incomes):

Country group	1960–70	1970–80	1980–90	1990
High-income countries	4.1	2.4	2.4	2.1
Developing countries	3.3	3.0	1.2	–0.2
Sub-Saharan Africa	0.6	0.9	–0.9	–2.0
Asia and Pacific	2.5	3.1	5.1	3.9
Middle East/N. Africa	6.0	3.1	–2.5	–1.9
Latin America/Caribbean	2.5	3.1	–0.5	–2.4
Europe	4.9	4.4	1.2	–3.8

Source: World Bank, 1992

This sustained growth in per capita incomes in the high-income countries over the past 30 years, albeit at lower rates of growth, coupled with the decline in per capita incomes in the developing world as a whole, has meant that income disparities between rich and poor nations have actually increased over the last 30 years. In 1960, countries with the richest 20 per cent of the world's population represented 70.2 per cent of global GNP and had average per capita incomes 30 times those of the countries with the poorest 20 per cent. By 1989, the richest 20 per cent of the population increased their share of global GNP to 82.7 per cent and had incomes 60 times those of the poorest 20 per cent (UNDP, 1992).

In addition to widening disparities between rich and poor nations, the past decade is also noted for its lack of progress towards alleviating poverty. Indeed, the 1980s have often been described as "the lost decade" for development. According to an analysis by the World Bank, which uses a threshold per capita income (i.e., real GDP per capita) to define a poverty line, the proportion of poor in the world population has remained more or less constant during the second half of the 1980s. The absolute number of poor, has, however, increased in the developing countries from just over 1 billion to more than 1.1 billion persons during 1985–1990, i.e., at almost the same rate as population growth over the same period (World Bank, 1992).

These aggregated data mask marked regional differences in the incidence of poverty. Whereas some progress has been made in raising per capita incomes in some Asian countries, many poverty measures, including per capita incomes, indicate an increase in the prevalence of poverty in other developing world regions, particularly sub-Saharan Africa. These trends are reflected in the summary statistics presented below:

Region	Population below poverty line (%)		Number of poor (10⁶)	
	1985	1990	1985	1990
All developing countries	30.5	29.7	1051	1133
South Asia	51.8	49.0	532	562
East Asia	13.2	11.3	182	169
Sub-Saharan Africa	47.6	47.8	184	216
Middle East and North Africa	30.6	33.1	60	73
Eastern Europeᵃ	7.1	7.1	5	5
Latin America and the Caribbean	22.4	25.5	87	108

ᵃ Excludes the former USSR.
The poverty line used here – US$ 370 annual income per capita in 1985 purchasing power parity dollars – is based on estimates of poverty lines from a number of countries with low average incomes. In 1990 prices the poverty line would be approximately US$ 420 annual income per capita. The estimates for 1985 have been updated to incorporate new data and to ensure comparability across years.
Source: World Bank, 1992

Poverty is widely perceived as both a cause and consequence of environmental degradation in that the poor are both victims and agents of environmental damage. Over half of the world's poor live in ecologically fragile or marginal areas where overuse of land to support demands for fuelwood and for agricultural production can lead to rapid environmental degradation (through soil erosion, nutrient losses, etc.) and, in turn, dwindling crop yields. In these situations poor families are forced to "mine" environmental capital to meet short-term needs, but typically lack the resources to protect their environment – for example, they lack access to technology (e.g., fertilizers) and to credit and insurance markets. They also often have poorly defined property rights. Other factors such as droughts and rapid population growth may exacerbate this mutually reinforcing effect of poverty and environmental damage. The current situation in sub-Saharan Africa is a particularly clear example of this nexus.

At the other end of the scale, poverty increases vulnerability to environmental risks. Poor people tend to live in crowded squatter settlements and have limited access to safe water and sanitation services, health care and education (see below). They also tend to be exposed to greater environmental health risks arising from natural disasters, industrial accidents and air pollution. Poverty is therefore viewed as one of the main threats to the physical environment and to the sustainability of human well-being. Its alleviation thus represents one of the central components of sustainable development.

Human Development

The traditional concept of development has been one centred on economic growth – on the assumption that growth will ultimately benefit everyone. In more recent years recognition of the fact that growth in national income does not necessarily increase the well-being of the whole population has led to a broader notion of development, one of human development, which has been defined by the UNDP as the "process of enlarging people's choices". This concept widens the development issue from a discussion of merely GNP growth to one which encompasses a wide range of human needs and aspirations – access to basic services and a sound physical environment coupled with economic and political freedom.

In this section of the 'UNEP Environmental Data Report' selected social measures of human development are briefly reviewed. Access to safe drinking water and sanitation services, provision of health care and adult literacy rates have been singled out in the present edition. The status of the provision of municipal waste services and the implications for human health are mentioned in Part 8: Wastes and Waste Management.

Assessed on a national-average basis, available data and analyses (including those summarized below) indicate that some progress has been made in recent decades with respect to the advancement of human well-being. For example, the differences in basic human survival (i.e., life expectancy, infant and child mortality, rudimentary education and access to safe water) between the more and less developed nations have narrowed considerably over the past three decades. However, the disparities in technology and information are widening; this is reflected by the increasing gaps between developed and developing countries in the numbers of scientific and technical personnel, expenditure on research and development, the level of tertiary education enrolment and in the numbers of radios and telephones (UNDP, 1992). These widening gaps in technology have far-reaching implications; technology innovation is the engine of economic progress and an important factor in determining the sustainability of the development process.

In an attempt to ensure that future development planning is directed towards improving people's well-being rather than the pursuit of economic growth alone, the UNDP has proposed a new policy-making tool, the Human Development Index (HDI). The HDI is a composite measure of human progress in that it combines indicators of national income with those for life expectancy and educational attainment. Its derivation is described in more detail in Box 4.3.

The introduction of the HDI has not only renewed the debate on the relevance of socio-economic measures of development but has also highlighted the weaknesses in the existing statistical data bases upon which it relies. In many countries social data are either incomplete, outdated or entirely lacking. The UNDP, along with other UN agencies, is thus helping many countries to improve their data collection systems.

It is envisaged that the HDI will be further refined and developed. One of the proposed refinements is to devise methods of disaggregating the HDI according to population groups (e.g., by gender or income groups) and region. This is to take account of the fact that national statistics disguise many important disparities, not only between urban and rural areas, but also between male and female, between rich and poor, and between different ethnic groups (UNDP, 1992). Recognition of these disparities has important implications for national and local government and policy action. Methodology development aside, one of the main impediments to the calculation of disaggregated HDI (and indeed other socio-economic measures) at the present time, is the general lack of socio-economic statistics in a disaggregated format. Thus there is an urgent need to collect such data on a more disaggregated basis in the future. Some efforts to develop disaggregated data bases are already under way; the Global City Data Programme of UNCHS (Habitat), for example, is promoting the collection of gender-disaggregated data.

Education In today's technological world, investment in "human capital" in terms of skills and education is assuming ever increasing importance. Analyses conducted by the World Bank in recent years have demonstrated a statistical correlation between economic performance and the level of education. In the developing world, where an average adult had more than 3.5 years of schooling, the country's GDP grew at an average rate of 5.5 per cent per year between 1965 and 1987. Where the average length of schooling was less than that, GDP grew at only 3.8 per cent per year (World Bank, 1991). Education – especially for women – is also associated with lower fertility and smaller families; it thus provides the important link between economic and population growth. Where women's status and education are good, the use of family planning services is higher and fertility is lower. The converse is also true; studies in both developed and developing countries have shown that children from small families tend to have more years of school and tend to fare better than those from larger families (UNFPA, 1992; World Bank, 1992).

Available data suggest that the level of rudimentary education is improving. Adult literacy rates in all regions of the world are increasing, as is the number of children enrolled in primary schools (UNFPA, 1992). Although the proportion of illiterates has dropped, the absolute number of people who cannot read or write has in fact increased because of the effect of population growth. Thus during the past 15 years of "development" a further 65 million illiterates have been added to the previous total of 842 million (UNFPA, 1992).

Data presented in Table 4.7, i.e., adult literacy rates in selected developing countries, reflect the significant disparities that still exist in the level of educational attainment between men and women in some world regions. These disparities are particularly evident in the African region and in parts of Asia. The prevalence of illiteracy, especially amongst women, thus continues to represent a major challenge for the Third World.

Access to Safe Drinking Water and Sanitation Services The provision of safe water and the management of waste waters play a central role in reducing the incidence of life- and health-threatening water-borne or water-related communicable diseases and, as such, can be considered as important determinants of human welfare. Indeed, estimates of the numbers of people lacking access to water and sanitation services provide the best means to date of assessing the numbers of people at risk from water-related diseases (WHO, 1992). Statistics on access to water and sanitation services, which are compiled by WHO, are summarized in Table 4.7 for 1990. For data on the incidence of selected water-borne communicable diseases please refer to Part 5: Human Health.

Some analysts believe that, in some cases, official figures of service coverage misrepresent the actual situation. For instance, data may indicate that people with water taps in their settlements are adequately served, but in practice the

BOX 4.3   The Human Development Index

It is becoming increasingly accepted that economic measures alone, i.e., GNP and GDP, do not necessarily provide a reliable measure of national well-being or human welfare. In recognition of this fact several attempts have been made to devise alternative yardsticks of human development.

The Human Development Index (HDI), first proposed by the United Nations Development Programme (UNDP) in 1990, is one such effort. The HDI, measured on a scale of 0 to 1, is an aggregated measure combining three key components: longevity, knowledge and national income. Longevity is represented by life expectancy at birth (in years), knowledge or educational attainment by a weighted (2:1) combination of two variables, adult literacy and mean years of schooling, and national income by an adjusted value of real GDP per capita.

The component indicators of the HDI are expressed in different units and therefore a conversion to some common denominator is required in order to combine the components into a single number. This is achieved by expressing the difference between the maximum reported value of a parameter and that of a given country as a percentage of the difference between the maximum and minimum value of that parameter. For example, maximum life expectancy is attained by Japan (78.6 years), the minimum by Sierra Leone (42.0 years). Life expectancy in, for example, Singapore is 74.0 years. The Singapore life expectancy deprivation is thus;

$$(78.6 - 74.0)/(78.6 - 42.0) = 0.126$$

Similar calculations are performed for the other two components. The life expectancy,

Human Development Index 1990

Rank	Country	HDI value
1	Canada	0.982
2	Japan	0.981
3	Norway	0.978
4	Switzerland	0.977
5	Sweden	0.976
6	USA	0.976
7	Australia	0.971
8	France	0.969
9	Netherlands	0.968
10	UK	0.962
11	Iceland	0.958
12	Germany [a]	0.955
13	Denmark	0.953
14	Finland	0.953
15	Austria	0.950
......	....	......
146	Mozambique	0.153
147	Bhutan	0.146
148	Mauritania	0.141
149	Benin	0.111
150	Chad	0.088
151	Somalia	0.088
152	Guinea-Bissau	0.088
153	Djibouti	0.084
154	Gambia	0.083
155	Mali	0.081
156	Niger	0.078
157	Burkina	0.074
158	Afghanistan	0.065
159	Sierra Leone	0.062
160	Guinea	0.052

[a] Data refer to the unified Federal Republic of Germany
Source: UNDP, 1992

educational attainment and GDP deprivations are then averaged to give an average deprivation, the value of which is subtracted from unity to generate the HDI (UNDP, 1992). In effect therefore, the HDI does not measure human development in absolute terms, but ranks countries in relation to each other, according to how far they have come from the lowest levels of achievement. Values of the index, based on 1990 data, are given in the accompanying table for the top 15 and bottom 15 countries.

Efforts to develop methods of disaggregating the HDI according to population groups (e.g., gender or income groups) and regions are ongoing. A gender-sensitive HDI has, for example, been calculated for 33 countries for which suitable data exist (UNDP, 1992). This preliminary work suggests that many countries fall in their ranking when gender sensitivity is introduced. Canada for example no longer occupies the top slot but falls to eighth position, largely due to disparities in income levels between men and women. Similarly, Japan drops from second to eighteenth. In contrast Sweden, by virtue of its greater male–female parity, rises to number one.

The formulation of the HDI is outlined in detail in the 1990 'Human Development Report'; revisions to the original methodology and updated rankings are published in subsequent editions of the report.

Reference

United Nations Development Programme 1992 *Human Development Report 1992*, Oxford University Press, New York.

low number of communal taps means that people often have to wait in long queues for water. Furthermore, piped water systems in some tropical countries may only function intermittently for a few hours a day (WHO, 1992). Both of these factors tend to reduce per capita water consumption, sometimes to below that which is required to maintain good health.

In 1980, the UN General Assembly proclaimed the period 1981–1990 as the International Drinking Water Supply and Sanitation Decade (IDWSSD). At that time the primary goal envisaged was the attainment of full access to water supply and to sanitation by all inhabitants in the developing countries by the year 1990 (WHO/UNICEF, 1992). Significant progress in the provision of services was indeed achieved during the 1980s, with an additional 1,347 million people having access to safe water supplies and an extra 748 million persons having sanitation facilities by the end of the decade. This progress was achieved through community participation,

especially via the involvement of women in the decision-making, planning and management phases and through the adoption of a range of traditional, low-cost water and sanitation technologies (WHO/UNICEF, 1992). Figure 4.5 shows the percentage change in access to safe drinking water and sanitation services between 1975 and 1990 for selected countries.

However, despite these major accomplishments continuing population growth has meant that only relatively modest gains have been made in the level of service coverage, especially in urban areas. For example, the levels of coverage for water supply in developing countries rose from 76 per cent to 88 per cent in urban populations; in rural areas service coverage rose from 29 per cent to 68 per cent. The WHO estimates that over 1,200 million people (30 per cent of the developing world's population) still lacked access to safe water and 1,700 million (40 per cent) were without adequate sanitation at the end of the decade (WHO/UNICEF, 1992). With this in mind, the developing countries and the External Support Agencies (ESAs) reached a broad consensus to continue the existing thrust of the IDWSSD beyond 1990 and to harness that drive to coincide with the goal of "Health for All by the Year 2000" (WHO/UNICEF, 1992).

Availability of Health Care Although considerable progress has been made in recent decades in terms of public health care provision, it is estimated that in 1990 approximately 1.5 billion persons still lacked access to primary health care and thus to the low cost-effective preventive and curative interventions that such a service can provide (WHO, 1992; UNFPA, 1992). Service coverage is particularly low in some African countries; coverage levels are well below 50 per cent in some cases (Table 4.7). Data presented in Table 4.7 also reflect the pronounced urban and rural differences in access to local health care services; not surprisingly urban residents tend to be better served.

Levels of health care service are reflected by a number of other statistical measures which include expenditure on health care (as a proportion of GNP), density of trained personnel (i.e., numbers of doctors, nurses, midwifes and dentists per 10 thousand persons or population per doctor/nurse) and hospital beds (i.e., number of hospital beds per 10 thousand persons). Data on health care personnel, which are compiled routinely by the WHO, have been included in previous editions of this report (UNEP, 1991). On the assumption that 10 doctors/physicians per 10 thousand population ensures an adequate level of service (Southwick, 1985), many countries (particularly those in Africa and Asia) are still below this threshold level.

Expenditure on health care in absolute terms in developing countries is low compared with that in the industrialized world. In the early 1980s, annual per capita expenditure on health care averaged US$ 4 in 92 developing countries compared with an average of US$ 320 in 32 rich countries. Since then, health spending per person has declined in almost all countries (WHO, 1992).

Other measures or indicators are more specific. Examples include immunization coverage and availability and use of oral rehydration therapy (ORT). Again, both of these topics have been covered in previous editions of this report (UNEP 1989, 1991). Available data indicate significant progress in the level of immunization for the six common preventable childhood diseases (poliomyelitis, tetanus, measles, whooping cough, diphtheria and tuberculosis) in recent decades. Immunization coverage which has reached levels in excess of 70 per cent in most world regions (the exception is Africa), has been attained largely through the efforts of the WHO's Expanded Programme on Immunization (EPI). Nevertheless, in spite of these achievements, many hundreds of millions of children remain unimmunized. In Africa two fifths of children do not receive immunization (WHO, 1992). Similarly, despite rapid growth in the availability of ORT in latter years, in 1988 only one third of diarrhoeal episodes in children under five were treated (WHO, 1992).

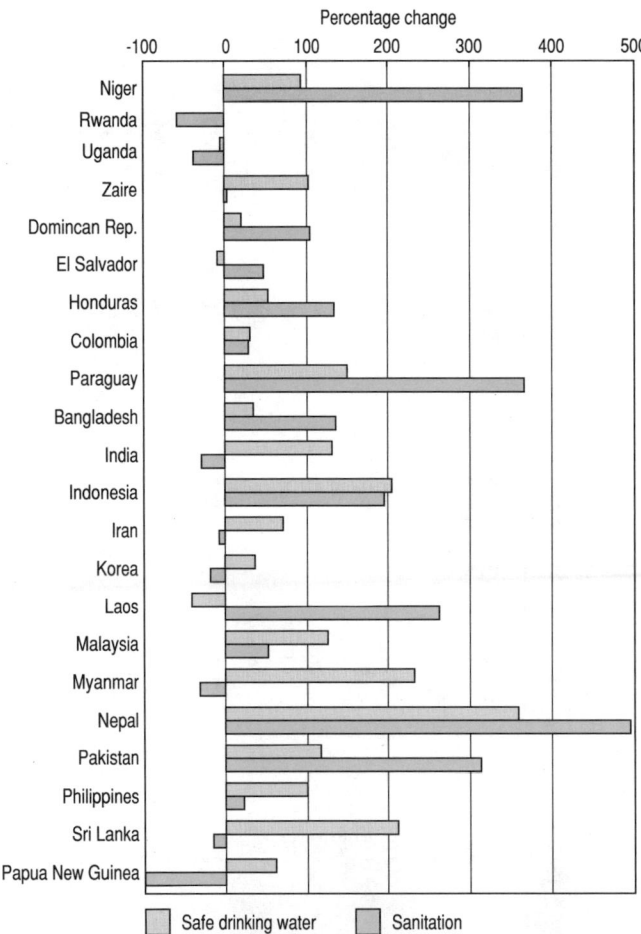

FIGURE 4.5 The percentage change in access to safe drinking water and sanitation services between 1975 and 1990 in selected developing countries
Source: Based on WHO data

References

Southwick, C. H. 1985 The human condition, economics and health. In: *Global Ecology*, C. H. Southwick (Ed.), Sinaver Associates Inc., Sunderlund, Massachussetts, 243–252.

UNEP 1989 *United Nations Environment Programme Environmental Data Report 1989/90*, Basil Blackwell, Oxford.

UNEP 1991 *United Nations Environment Programme Environmental Data Report 1991/92*, Basil Blackwell, Oxford.

UNFPA 1992 *The State of World Population 1992: A World in Balance*, United Nations Population Fund, New York.

United Nations Development Programme 1992 *Human Development Report 1992*, Oxford University Press, New York.

WCED 1987 *Our Common Future: Report of the World Commission on Environment and Development*, Oxford University Press, Oxford.

WHO 1992 *Our Planet, Our Health: Report of the WHO Commission on Health and Environment*, World Health Organization, Geneva.

WHO/UNICEF 1992 *Water Supply and Sanitation Services Monitoring Report 1990*, World Health Organization, Geneva.

World Bank 1991 *World Development Report 1991*, Oxford University Press, New York.

World Bank 1992 *World Development Report 1992: Development and the Environment*, Oxford University Press, New York.

TABLE 4.1 Trends in world population size, 1950–2025 (million persons)

Region	1950	1960	1970	1980	1990	2000	2010	2025
WORLD	**2,516**	**3,019**	**3,697**	**4,447**	**5,295**	**6,228**	**7,149**	**8,472**
More developed regions[a]	832	945	1,048	1,136	1,211	1,278	1,340	1,403
Less developed regions[b]	1,684	2,074	2,648	3,310	4,084	4,950	5,809	7,069
AFRICA	**222**	**248**	**363**	**479**	**642**	**856**	**1,116**	**1,583**
Eastern Africa	65	73	109	145	195	264	350	516
Middle Africa	26	29	40	52	70	95	127	190
Northern Africa	51	58	83	107	140	179	221	280
Southern Africa	15	17	25	33	43	54	67	85
Western Africa	63	71	105	141	193	262	350	511
NORTHERN AMERICA	**166**	**182**	**226**	**252**	**277**	**306**	**330**	**360**
LATIN AMERICA	**165**	**189**	**283**	**359**	**441**	**523**	**600**	**701**
Caribbean	17	18	24	29	33	38	43	50
Central America	36	42	67	89	113	140	165	199
South America	111	128	191	240	294	344	391	452
ASIA	**1,377**	**1,513**	**2,101**	**2,583**	**3,118**	**3,691**	**4,213**	**4,900**
Eastern Asia	671	732	986	1,176	1,350	1,520	1,629	1,762
South-eastern Asia	182	200	286	360	444	531	611	715
Southern Asia	481	531	754	948	1,191	1,469	1,757	2,136
Western Asia	42	48	73	98	131	171	215	286
EUROPE	**398**	**414**	**466**	**492**	**509**	**523**	**536**	**541**
Eastern Europe	70	75	85	92	96	99	103	107
Northern Europe	78	79	87	89	92	94	69	97
Southern Europe	109	113	128	138	144	146	149	148
Western Europe	141	145	165	170	175	183	187	188
(FORMER) USSR[c]	**174**	**190**	**235**	**258**	**281**	**297**	**317**	**344**
OCEANIA	**12**	**14**	**19**	**22**	**26**	**31**	**35**	**41**
Australia – New Zealand	10	11	15	17	20	23	26	29
Melanesia	2.1	2.5	3.3	4.2	5.2	6.4	7.9	10
Micronesia	0.15	0.17	0.24	0.31	0.42	0.54	0.67	0.88
Polynesia	0.23	0.27	0.40	0.47	0.54	0.62	0.70	0.79

a "More developed regions" comprise countries in Northern America (i.e., Bermuda, Canada, Greenland, St Pierre & Miquelon and the USA), Europe as well as the former USSR, Australia, New Zealand and Japan.

b "Less developed regions" comprise countries in Africa, Latin America, Asia (excluding Japan) and Oceania (excluding Australia and New Zealand).

c Includes 12 of the 15 republics of the former USSR; Estonia, Latvia and Lithuania are included in Northern Europe.

Data for the years 1950 to 1990 are mid-year estimates.
Data for the years 2000 to 2025 are medium variant projections.

Source:
UN Department for Economic and Social Information and Policy Analysis (Population Division) 1993 *World Population Prospects: The 1992 Revision*, United Nations, New York.

TABLE 4.2 Annual average rate of change in the size of the world population, 1950–2025 (per cent per year)

Region	1950–55	1955–60	1960–65	1965–70	1970–75	1975–80	1980–85	1985–90	1990–95	95–2000	2000–05	2005–10	2010–15	2015–20	2020–25
WORLD	**1.79**	**1.85**	**1.99**	**2.06**	**1.96**	**1.73**	**1.75**	**1.74**	**1.68**	**1.57**	**1.42**	**1.33**	**1.25**	**1.13**	**1.02**
More developed regions[a]	1.28	1.25	1.19	0.90	0.86	0.74	0.63	0.64	0.54	0.54	0.50	0.45	0.38	0.30	0.23
Less developed regions[b]	2.04	2.13	2.35	2.54	2.38	2.08	2.13	2.08	2.01	1.84	1.66	1.54	1.44	1.30	1.18
AFRICA	**2.22**	**2.39**	**2.55**	**2.65**	**2.66**	**2.88**	**2.91**	**2.95**	**2.93**	**2.81**	**2.70**	**2.61**	**2.50**	**2.33**	**2.15**
Eastern Africa	2.30	2.48	2.69	2.78	2.64	3.00	2.83	3.04	3.12	2.96	2.84	2.80	2.75	2.62	2.39
Middle Africa	1.80	2.00	2.22	2.40	2.47	2.82	2.97	3.04	3.12	2.98	2.88	2.84	2.80	2.69	2.51
Northern Africa	2.26	2.32	2.37	2.52	2.41	2.68	2.80	2.60	2.47	2.37	2.22	2.03	1.78	1.51	1.45
Southern Africa	2.29	2.41	2.59	2.47	2.80	2.69	2.63	2.48	2.43	2.34	2.15	1.98	1.79	1.59	1.37
Western Africa	2.27	2.50	2.67	2.76	2.92	2.98	3.12	3.19	3.11	3.00	2.92	2.83	2.71	2.53	2.30
NORTHERN AMERICA	**1.80**	**1.78**	**1.49**	**1.13**	**1.06**	**1.07**	**0.92**	**0.96**	**1.06**	**0.94**	**0.80**	**0.74**	**0.66**	**0.60**	**0.49**
LATIN AMERICA	**2.70**	**2.73**	**2.78**	**2.58**	**2.45**	**2.28**	**2.16**	**1.96**	**1.79**	**1.61**	**1.45**	**1.31**	**1.18**	**1.04**	**0.90**
Caribbean	1.77	1.86	2.08	1.84	1.81	1.35	1.43	1.45	1.36	1.26	1.19	1.15	1.11	1.04	0.96
Central America	2.79	3.06	3.18	3.16	3.06	2.63	2.43	2.32	2.24	2.01	1.77	1.57	1.39	1.23	1.08
South America	2.81	2.74	2.74	2.48	2.32	2.27	2.15	1.88	1.67	1.49	1.35	1.22	1.09	0.95	0.81
ASIA	**1.89**	**1.95**	**2.18**	**2.44**	**2.27**	**1.86**	**1.91**	**1.85**	**1.78**	**1.60**	**1.39**	**1.26**	**1.14**	**1.00**	**0.88**
Eastern Asia	1.75	1.54	1.98	2.43	2.11	1.40	1.37	1.39	1.31	1.05	0.74	0.65	0.61	0.55	0.41
South-eastern Asia	1.92	2.28	2.37	2.52	2.44	2.14	2.15	2.02	1.88	1.70	1.49	1.31	1.18	1.04	0.94
Southern Asia	1.99	2.29	2.33	2.38	2.35	2.23	2.34	2.22	2.15	2.04	1.88	1.70	1.50	1.25	1.15
Western Asia	2.70	2.79	2.76	2.77	2.92	2.83	3.11	2.78	2.69	2.55	2.39	2.20	2.03	1.92	1.75
EUROPE	**0.79**	**0.81**	**0.92**	**0.67**	**0.59**	**0.45**	**0.31**	**0.38**	**0.27**	**0.30**	**0.27**	**0.20**	**0.14**	**0.07**	**0.00**
Eastern Europe	1.40	1.11	0.88	0.69	0.77	0.73	0.51	0.32	0.18	0.36	0.41	0.39	0.32	0.24	0.18
Northern Europe	0.39	0.53	0.76	0.56	0.38	0.20	0.22	0.32	0.23	0.26	0.22	0.17	0.13	0.09	0.03
Southern Europe	0.84	0.78	0.88	0.76	0.77	0.80	0.46	0.30	0.16	0.19	0.19	0.13	0.04	-0.03	-0.13
Western Europe	0.66	0.82	1.06	0.63	0.47	0.15	0.14	0.50	0.43	0.36	0.27	0.18	0.12	0.04	-0.01
(FORMER) USSR[c]	**1.74**	**1.79**	**1.50**	**1.00**	**0.94**	**0.86**	**0.89**	**0.84**	**0.51**	**0.57**	**0.65**	**0.68**	**0.65**	**0.52**	**0.47**
OCEANIA	**2.26**	**2.20**	**2.09**	**1.98**	**1.81**	**1.49**	**1.51**	**1.64**	**1.52**	**1.46**	**1.37**	**1.29**	**1.16**	**1.05**	**0.91**
Australia – New Zealand	2.33	2.18	1.99	1.85	1.67	1.27	1.30	1.49	1.33	1.25	1.14	1.06	0.94	0.84	0.72
Melanesia	1.87	2.20	2.48	2.49	2.43	2.40	2.31	2.10	2.15	2.15	2.10	1.99	1.82	1.65	1.43
Micronesia	2.08	2.43	2.40	2.57	2.49	2.13	2.62	3.62	2.53	2.40	2.27	2.13	1.98	1.81	1.59
Polynesia	3.04	2.70	2.75	2.44	1.61	1.47	1.48	1.40	1.31	1.31	1.28	1.18	0.96	0.73	0.57

a "More developed regions" comprise countries in Northern America (i.e., Bermuda, Canada, Greenland, St Pierre & Miquelon and the USA), Europe as well as the former USSR, Australia, New Zealand and Japan.

b "Less developed regions" comprise countries in Africa, Latin America, Asia (excluding Japan) and Oceania (excluding Australia and New Zealand).

c Includes 12 of the 15 republics of the former USSR; Estonia, Latvia and Lithuania are included in Northern Europe.

Data for the periods 1990–95 onwards are medium variant projections.

Source:
See Table 4.1.

TABLE 4.3 Demographic data for selected countries and major world regions, around 1950 and 1990

	Population (10³)		Growth rate (% per year)		Total fertility rate (per woman)		Crude birth rate (per 10³)		Crude death rate (per 10³)	
	1950	1992	1950–55	1985–90	1950–55	1985–90	1950–55	1985–90	1950–55	1985–90
WORLD	2,516,190	5,479,046	1.8	1.7	5.0	3.4	38	27	20	10
More developed regions[a]	832,425	1,224,744	1.3	0.6	2.8	1.9	23	15	10	9
Less developed regions[b]	1,683,765	4,254,301	2.0	2.1	6.2	3.9	45	31	24	10
AFRICA	222,462	681,685	2.2	3.0	6.7	6.2	49	45	27	15
Algeria	8,753	26,346	2.1	2.7	7.3	5.4	51	36	24	8
Angola a	4,131	9,888	1.4	2.8	6.4	7.2	50	51	35	21
Benin	2,046	4,918	0.6	3.0	6.8	7.1	43	49	37	19
Botswana	389	1,313	2.1	3.0	6.5	5.5	49	40	23	11
Burkina	3,654	9,513	1.9	2.7	6.3	6.5	49	47	31	19
Burundi	2,456	5,823	1.8	2.9	6.8	6.8	48	47	25	17
Cameroon	4,466	12,198	1.6	2.9	5.7	6.1	43	42	27	14
Cent African Rep.	1,314	3,173	1.5	2.7	5.5	6.2	44	45	29	18
Chad	2,658	5,846	1.3	2.0	5.8	5.9	45	44	32	20
Comoros	173	585	2.3	3.6	6.3	7.1	47	49	24	13
Congo	808	2,368	1.9	3.0	5.7	6.3	44	44	25	15
Côte d'Ivoire	2,776	12,910	3.0	3.7	6.9	7.4	53	50	28	15
Egypt	20,330	54,842	2.5	2.4	6.6	4.5	49	35	24	11
Ethiopia	19,573	52,981	2.0	2.9	6.7	7.0	52	50	32	20
Gabon	469	1,237	0.3	3.3	4.1	5.0	30	39	27	17
Gambia	294	908	1.3	2.9	6.1	6.5	47	47	34	21
Ghana	4,900	15,959	3.2	3.1	6.9	6.4	48	44	22	13
Guinea	2,550	6,116	2.1	2.9	7.0	7.0	55	51	34	22
Guinea-Bissau	505	1,006	0.7	2.0	5.1	5.8	41	43	30	23
Kenya	6,265	25,230	2.8	3.4	7.5	6.8	53	46	25	11
Lesotho	734	1,836	1.6	2.5	5.8	5.0	42	36	27	11
Liberia	824	2,751	2.1	3.2	6.3	6.8	48	47	27	16
Libya	1,029	4,875	1.8	3.7	6.9	6.9	48	44	23	9
Madagascar	4,229	12,827	2.2	3.2	6.6	6.6	48	46	26	14
Malawi	2,881	10,356	1.9	5.3	6.8	7.6	52	56	31	21
Mali	3,520	9,818	2.1	3.0	7.1	7.1	53	51	32	21
Mauritania	825	2,143	1.8	2.7	6.5	6.5	48	46	31	19
Mauritius[c]	493	1,098	2.9	1.0	6.3	2.1	47	20	16	7
Morocco	8,953	26,318	2.5	2.6	7.2	4.8	50	36	26	10
Mozambique	6,198	14,872	1.7	0.9	6.2	6.5	46	46	30	19

Continued

TABLE 4.3 Continued

	Population (10³)		Growth rate (% per year)		Total fertility rate (per woman)		Crude birth rate (per 10³)		Crude death rate (per 10³)	
	1950	1992	1950–55	1985–90	1950–55	1985–90	1950–55	1985–90	1950–55	1985–90
Namibia	511	1,534	2.0	3.1	6.0	6.0	46	43	25	12
Niger	2,400	8,252	2.3	3.1	7.1	7.1	54	52	32	20
Nigeria	32,935	115,664	2.4	3.3	6.8	6.9	51	49	27	16
Réunion	242	624	3.3	1.9	5.7	2.5	39	23	13	6
Rwanda	2,120	7,526	2.4	3.3	7.1	8.5	47	52	23	18
Senegal	2,500	7,736	2.4	2.8	6.7	6.5	49	46	28	18
Sierra Leone	1,944	4,376	1.4	2.5	6.1	6.5	48	48	34	23
Somalia	3,072	9,204	2.0	2.0	7.0	7.0	52	50	32	20
South Africa	13,683	39,818	2.3	2.4	6.5	4.4	43	33	20	10
Sudan	9,190	26,656	2.0	2.9	6.7	6.4	47	45	27	16
Swaziland	264	792	2.0	2.6	6.5	5.3	50	38	28	12
Tanzania	7,886	27,829	2.5	3.4	6.7	6.8	51	48	27	14
Togo	1,329	3,763	1.2	3.1	6.6	6.6	47	45	29	14
Tunisia	3,530	8,401	1.8	2.1	6.9	4.0	46	29	23	7
Uganda	4,762	18,674	3.1	3.1	6.9	7.3	51	51	25	20
Zaire	12,184	39,882	2.2	3.3	6.0	6.7	48	48	26	15
Zambia	2,440	8,638	2.4	3.4	6.6	6.8	50	49	26	16
Zimbabwe	2,730	10,583	3.5	3.5	7.2	5.8	52	43	23	11
NORTH AMERICA	**220,360**	**435,857**								
Canada	13,737	27,367	2.7	1.1	3.7	1.7	28	15	9	7
Costa Rica	862	3,192	3.5	2.8	6.7	3.4	47	29	13	4
Cuba	5,850	10,811	1.9	1.0	4.1	1.8	30	18	11	7
Dominican Rep.	2,353	7,471	3.0	2.2	7.4	3.8	51	31	20	7
El Salvador	1,940	5,396	2.6	1.8	6.5	4.5	48	35	20	9
Guatemala	2,969	9,745	2.9	2.9	7.1	5.8	51	41	22	9
Haiti	3,261	6,755	1.5	2.0	6.3	5.0	43	36	28	13
Honduras	1,401	5,462	3.2	3.2	7.1	5.6	51	40	22	8
Jamaica	1,403	2,469	1.9	0.9	4.2	2.7	35	24	12	7
Mexico	27,297	88,153	2.8	2.2	6.8	3.6	46	30	17	6
Nicaragua	1,109	3,955	3.0	2.6	7.4	5.6	54	44	23	9
Panama	893	2,515	2.5	2.1	5.7	3.1	40	27	13	5
Puerto Rico	2,219	3,594	0.3	0.9	5.0	2.2	37	19	9	7
Trinidad & Tobago	636	1,265	2.5	1.3	5.3	3.0	38	26	11	7
USA	152,271	255,159	1.7	1.0	3.5	1.9	24	16	10	9

Continued

TABLE 4.3 Continued

	Population (10³)		Growth rate (% per year)		Total fertility rate (per woman)		Crude birth rate (per 10³)		Crude death rate (per 10³)	
	1950	1992	1950–55	1985–90	1950–55	1985–90	1950–55	1985–90	1950–55	1985–90
SOUTH AMERICA	**111,594**	**304,454**	**2.8**	**1.9**	**5.7**	**3.3**	**42**	**27**	**15**	**8**
Argentina	17,150	33,100	2.0	1.3	3.2	3.0	25	21	9	9
Bolivia	2,766	7,524	2.1	2.5	6.8	5.0	47	37	24	11
Brazil	53,444	154,113	3.2	1.9	6.2	3.2	45	27	15	8
Chile	6,082	13,600	2.2	1.7	5.1	2.7	37	24	14	6
Colombia	11,946	33,424	2.8	1.8	6.8	2.9	47	26	17	6
Ecuador	3,310	11,055	2.8	2.5	6.9	4.1	47	32	19	7
Guyana	423	808	2.8	0.2	6.7	2.8	43	27	18	8
Paraguay	1,351	4,519	2.8	2.9	6.8	4.6	47	35	9	7
Peru	7,632	22,451	2.6	2.1	6.9	4.0	47	31	22	9
Uruguay	2,239	3,130	1.2	0.6	2.7	2.4	21	18	11	10
Venezuela	5,009	20,186	4.1	2.4	6.5	3.5	47	29	12	5
ASIA	**1,377,262**	**323,968**	**1.9**	**1.9**	**5.9**	**3.5**	**43**	**28**	**24**	**9**
Afghanistan	8,958	19,062	1.7	2.6	6.7	6.9	48	49	32	23
Bahrain	116	533	2.9	3.2	7.0	4.1	45	28	16	4
Bangladesh	41,783	119,288	1.7	2.4	6.7	5.1	47	39	24	15
Bhutan	734	1,612	1.5	2.2	6.0	5.9	43	40	28	18
Cambodia	4,346	8,774	2.2	2.6	6.3	4.6	45	42	24	16
China	554,760	1,187,997	1.9	1.5	6.2	2.4	44	22	25	7
Cyprus	494	716	1.4	1.1	3.7	2.4	27	19	11	9
East Timor	433	791	1.3	2.8	6.4	5.4	47	44	35	22
Hong Kong	1,974	5,800	4.6	0.9	4.4	1.4	38	13	9	6
India	357,561	879,548	2.0	2.0	6.0	4.2	44	31	25	11
Indonesia	79,538	191,170	1.7	1.9	5.5	3.5	43	29	26	9
Iran	16,914	61,565	2.4	3.5	7.1	6.5	48	43	25	8
Iraq	5,158	19,290	2.7	3.3	7.2	6.2	49	40	22	7
Israel	1,258	5,131	6.6	1.9	4.2	3.1	33	23	7	7
Japan	83,625	124,491	1.4	0.4	2.8	1.7	24	11	9	6
Jordan	1,237	4,291	3.1	3.2	7.4	6.2	47	39	26	6
Korea	20,357	44,163	1.0	1.2	5.2	1.7	37	17	32	6
Korea, Dem.	9,726	22,618	-1.4	1.8	5.2	2.5	37	24	32	5
Kuwait	152	1,970	5.4	4.4	7.2	3.9.	45	28	11	2
Laos	1,755	4,469	2.1	3.1	6.2	6.7	46	45	25	17
Lebanon	1,443	2,838	2.2	0.5	5.7	3.4	41	28	19	8
Malaysia	6,110	18,792	2.7	2.6	6.8	4.0	45	32	20	6

Continued

TABLE 4.3 Continued

	Population (10³)		Growth rate (% per year)		Total fertility rate (per woman)		Crude birth rate (per 10³)		Crude death rate (per 10³)	
	1950	1992	1950–55	1985–90	1950–55	1985–90	1950–55	1985–90	1950–55	1985–90
Mongolia	761	2,310	2.2	2.7	6.0	5.0	44	36	22	9
Myanmar	17,832	43,668	1.9	2.2	6.0	4.5	45	34	27	13
Nepal	8,182	20,577	1.2	2.7	5.6	6.0	46	42	27	15
Oman	413	1,637	1.9	3.8	7.2	7.2	51	43	32	6
Pakistan	39,513	124,773	2.2	3.2	6.5	6.8	50	44	29	12
Philippines	20,988	65,186	2.6	2.4	7.3	4.3	49	33	20	7
Saudi Arabia	3,201	15,922	2.3	3.7	7.2	6.8	49	37	26	6
Singapore	1,022	2,769	4.9	1.1	6.4	1.7	44	17	11	5
Sri Lanka	7,678	17,666	2.6	1.3	5.7	2.7	39	23	12	6
Syria	3,495	13,276	2.5	3.6	7.1	6.7	47	44	21	7
Thailand	20,010	56,129	2.6	1.3	6.6	2.6	47	23	19	6
Turkey	20,809	58,362	2.7	2.1	6.9	3.8	48	30	24	8
United Arab Em.	70	1,670	2.5	3.3	7.0	4.8	48	23	23	4
Viet Nam[d]	29,954	69,485	1.3	2.1	6.1	4.2	42	32	29	10
Yemen[d]	4316	12,535	1.9	3.6	7.3	7.7	52	52	32	16
EUROPE	**398,140**	**512,023**	**0.8**	**0.4**	**2.6**	**1.7**	**20**	**13**	**11**	**10**
Albania	1,230	3,315	2.4	1.9	5.6	3.1	38	25	14	6
Austria	6,935	7,776	0.0	0.4	2.1	1.5	15	12	12	11
Belgium	8,639	9,998	0.5	0.2	2.3	1.6	17	12	12	11
Bulgaria	7,251	8,952	0.7	0.1	2.5	1.9	21	13	10	12
Czechoslovakia	12,389	15,731	1.1	0.2	2.9	2.0	22	14	11	12
Denmark	4,271	5,158	0.8	0.1	2.5	1.5	18	11	9	11
Estonia	1,101	1,582	1.0	0.6	2.1	2.2	17	16	13	12
Finland	4,009	5,008	1.1	0.3	3.0	1.7	23	13	10	10
France	41,829	57,182	0.8	0.6	2.7	1.8	20	14	13	10
Germany[e]	68,376	80,253	0.6	0.5	2.2	1.4	16	11	11	12
Greece	7,566	10,182	1.0	0.4	2.3	1.5	19	11	7	9
Hungary	9,338	10,512	1.0	-0.2	2.7	1.8	21	12	11	14
Ireland	2,969	3,486	-0.3	-0.3	3.4	2.3	21	16	13	9
Italy	47,104	57,782	0.6	0.2	2.3	1.3	18	10	10	10
Latvia	1,949	2,679	0.7	0.6	2.0	2.1	16	15	12	12
Lithuania	2,567	3,755	0.5	0.8	2.7	2.1	22	16	11	10
Netherlands	10,114	15,158	1.2	0.6	3.1	1.6	22	13	8	9
Norway	3,265	4,288	1.0	0.5	2.6	1.8	19	13	8	11
Poland	24,824	38,417	1.9	0.5	3.6	2.2	30	16	11	10
Portugal	8,405	9,866	0.5	-0.1	3.1	1.6	24	12	12	10

Continued

TABLE 4.3 Continued

	Population (10³)		Growth rate (% per year)		Total fertility rate (per woman)		Crude birth rate (per 10³)		Crude death rate (per 10³)	
	1950	1992	1950–55	1985–90	1950–55	1985–90	1950–55	1985–90	1950–55	1985–90
Romania	16,311	23,327	1.4	0.4	2.9	2.3	25	16	12	11
Spain	28,009	39,092	0.8	0.3	2.6	1.5	20	11	10	8
Sweden	7,014	8,652	0.7	0.5	2.2	1.9	16	13	10	11
Switzerland	4,694	6,813	1.2	0.7	2.3	1.6	17	12	10	9
UK	50,616	57,696	0.2	0.3	2.2	1.8	16	14	12	12
Yugoslavia	16,346	23,949	1.4	0.6	3.7	2.0	29	15	12	9
(FORMER) USSR[f]	**174,459**	**284,528**	**1.7**	**0.8**	**2.8**	**2.4**	**26**	**19**	**9**	**10**
OCEANIA	**12,616**	**27,529**	**2.3**	**1.6**	**3.8**	**2.5**	**28**	**20**	**12**	**8**
Australia[g]	8,219	17,596	2.3	1.6	3.2	1.9	23	15	9	7
Fiji	289	739	3.0	0.8	6.6	3.2	46	26	13	5
New Zealand	1,908	3,455	2.3	0.9	3.5	2.0	26	17	9	8
Papua New Guinea	1,613	4,056	1.6	2.3	6.2	5.3	44	34	29	12

a   "More developed regions" comprise countries in Northern America (i.e., Bermuda, Canada, Greenland, St Pierre & Miquelon and the USA), Europe as well as the former USSR, Australia, New Zealand and Japan.

b   "Less developed regions" comprise countries in Africa, Latin America, Asia (excluding Japan) and Oceania (excluding Australia and New Zealand).

c   Includes Agalesa, Rodrigues and St Brandon.

d   Data refer to the unified Republic of Yemen.

e   Data refer to the unified Federal Republic of Germany.

f   Includes 12 of the 15 republics of the former USSR; Estonia, Latvia and Lithuania are included in Europe.

g   Data include Christmas Island, Cocos (Keeling) Islands, and Norfolk Island.

Populations are mid-year populations.
"Growth rate" is the average annual increase of the population, averaged over the period stated. "Total fertility rate" is an estimate of the average number of live births that are typically born to a woman of childbearing age. "Crude birth rate" is the number of live births in a given year divided by the mid-year population, multiplied by 1,000. "Crude death rate" is the number of deaths in a given year divided by the mid-year population, multiplied by 1,000. Fertility rates, and crude birth and death rates, are annual averages for the periods stated.

Aggregated data are not given for the North American region due to differences in regional classification. See source document for further details of regional classifications. Countries with populations of less than 500,000 in 1992 are excluded from the country listings: however, regional and world totals do include those countries not listed here.

Source:
UN Department for Economic and Social Information and Policy Analysis (Population Division) 1993 *World Population Prospects: The 1992 Revision*, United Nations, New York.

TABLE 4.4 Trends in urban population size and average annual urban growth rate, 1950–1990

	Urban population (% of total)		Average annual urban growth rate (%)			Urban population (% of total)		Average annual urban growth rate (%)	
	1950	1990	1955–60	1985–90		1950	1990	1955–60	1985–90
WORLD	29	43	3.6	2.7	Senegal	31	40	3.0	3.8
					Seychelles	27	59	1.4	3.7
More developed regions[a]	54	73	2.4	1.0	Sierra Leone	9	32	4.9	5.1
Less developed regions[b]	17	34	5.2	3.8	Somalia	13	24	5.2	2.8
					South Africa	43	49	3.2	2.8
AFRICA	15	32	4.7	4.5					
					Sudan	6	23	6.8	4.3
Algeria	22	52	5.2	4.4	Swaziland	1	26	12.6	6.4
Angola	8	28	4.8	5.7	Tanzania	4	21	4.9	6.8
Benin	4	38	8.5	4.8	Togo	7	29	4.4	4.5
Botswana	–	25	18.3	7.9	Tunisia	31	56	3.6	3.2
Burkina	4	15	4.1	8.4					
					Uganda	3	11	8.3	5.5
Burundi	2	5	3.4	5.1	Western Sahara	68	57	6.5	5.7
Cameroon	10	40	5.2	5.3	Zaire	19	28	3.9	3.4
Cape Verde	8	29	10.0	4.4	Zambia	9	42	9.1	3.9
Cent. African Rep.	16	47	5.1	4.6	Zimbabwe	11	29	4.8	5.9
Chad	4	32	7.2	6.0					
					NORTH AMERICA				
Comoros	3	28	12.8	5.4					
Congo	31	41	2.4	4.3	Antigua & Barbuda	46	32	–0.4	1.4
Côte d'Ivoire	13	40	7.1	5.2	Bahamas	62	64	3.9	2.5
Djibouti	41	81	4.9	3.7	Barbados	34	45	0.7	1.5
Egypt	32	44	4.1	2.4	Belize	57	51	2.5	3.0
					Bermuda	100	100	1.3	1.7
Equatorial Guinea	16	29	5.7	3.1					
Ethiopia	5	12	5.5	4.2	Canada	61	77	3.8	1.3
Gabon	11	46	4.5	5.5	Cayman Is	100	100	3.9	4.7
Gambia	11	23	4.0	5.2	Costa Rica	34	47	4.6	3.7
Ghana	15	34	7.8	4.2	Cuba	49	74	2.7	1.7
					Dominican Rep.	24	60	5.7	3.9
Guinea	6	26	5.4	5.8					
Guinea-Bissau	10	20	3.8	3.9	El Salvador	37	44	3.5	2.5
Kenya	6	24	5.7	7.1	Greenland	76	78	3.8	1.4
Lesotho	1	19	14.5	6.3	Guadeloupe	42	49	2.3	3.0
Liberia	13	45	6.1	5.7	Guatemala	30	39	3.8	3.6
					Haiti	12	29	4.1	3.9
Libya	19	82	5.6	5.0					
Madagascar	8	24	5.4	5.8	Honduras	18	44	5.8	5.1
Malawi	4	12	4.3	7.9	Jamaica	27	52	3.4	2.1
Mali	9	24	4.9	5.5	Martinique	28	75	6.3	2.2
Mauritania	2	47	10.9	6.8	Mexico	43	73	4.7	3.1
					Montserrat	22	12	–4.3	0.9
Mauritius[c]	29	41	4.3	0.7					
Morocco	26	46	3.9	3.6	Nicaragua	35	60	4.3	3.7
Mozambique	2	27	6.4	7.3	Panama	36	53	3.9	2.7
Namibia	9	28	6.8	5.1	Puerto Rico	41	74	1.9	1.8
Niger	5	20	4.1	6.9	St Kitts & Nevis	22	49	2.7	1.1
					St Lucia	38	44	1.2	2.1
Nigeria	10	35	6.1	5.8					
Réunion	24	64	6.5	3.3	St Pierre & Miquelon	80	91	0.6	1.0
Rwanda	2	6	5.6	4.9	St Vincent & Grenadines	13	20	2.8	2.3
St Helena[d]	30	17	–0.2	3.3	Trinidad & Tobago	64	65	3.2	1.7
São Tomé & Príncipe	13	42	2.7	4.9	Turks & Caicos Is	41	51	2.5	5.6

Continued

TABLE 4.4 Continued

	Urban population (% of total)		Average annual urban growth rate (%)			Urban population (% of total)		Average annual urban growth rate (%)	
	1950	1990	1955–60	1985–90		1950	1990	1955–60	1985–90
USA	64	75	2.5	1.1	Maldives	11	29	2.4	5.8
US Virgin Is	45	47	3.3	0.7	Mongolia	19	58	8.4	3.8
					Myanmar	16	25	3.9	2.8
SOUTH AMERICA	**43**	**75**	**4.5**	**2.8**	Nepal	2	11	4.7	7.6
					Oman	2	11	5.9	7.4
Argentina	65	86	2.8	1.6					
Bolivia	38	51	2.6	3.8	Pakistan	18	32	4.7	4.6
Brazil	36	75	5.1	3.1	Philippines	27	43	3.9	3.7
Chile	58	85	3.7	2.0	Qatar	63	90	6.3	4.0
Colombia	37	70	5.4	2.7	Saudi Arabia	16	77	8.5	4.8
					Singapore	100	100	4.5	1.2
Ecuador	28	56	4.9	4.2					
French Guiana	54	75	4.0	3.9	Sri Lanka	14	21	4.7	1.6
Guyana	28	33	3.5	1.2	Syria	31	50	4.6	4.3
Paraguay	35	48	3.0	4.3	Thailand	11	22	4.7	4.0
Peru	36	70	5.2	2.8	Turkey	21	61	5.8	5.1
					United Arab Em.	25	81	7.1	4.3
Suriname	47	48	3.1	2.8					
Uruguay	78	89	1.6	1.0	Viet Nam	12	20	4.0	2.5
Venezuela	53	91	6.0	3.1	Yemen[e]	6	29	6.5	7.0
ASIA	**16**	**31**	**5.1**	**3.6**	**EUROPE**	**57**	**73**	**1.6**	**0.8**
Afghanistan	6	18	5.2	4.2	Albania	20	36	6.9	2.4
Bahrain	64	83	4.9	3.5	Andorra	79	63	5.3	0.3
Bangladesh	4	16	4.4	6.1	Austria	49	58	0.5	1.1
Bhutan	2	5	3.4	5.5	Belgium	92	96	0.7	0.3
Brunei	27	58	9.6	2.6	Bulgaria	26	68	4.9	1.0
Cambodia	10	12	2.4	4.1	Channel Is	43	31	−0.3	1.4
China	11	26	8.3	4.5	Czechoslovakia	37	77	3.0	1.4
Cyprus	30	53	3.3	2.4	Denmark	68	85	1.8	0.2
East Timor	10	13	1.8	5.0	Estonia	49	72	2.4	0.9
Gaza Strip	51	94	4.9	4.1	Faeroe Is	18	31	3.9	1.7
Hong Kong	83	94	4.5	1.2	Finland	32	60	2.6	0.3
India	17	26	2.7	3.0	France	56	73	2.0	0.4
Indonesia	12	29	3.7	4.5	Germany[f]	72	85	1.2	0.8
Iran	27	57	4.8	4.8	Gibraltar	100	100	0.0	1.6
Iraq	35	72	5.6	4.2	Greece	37	63	2.3	1.2
Israel	65	92	5.4	2.2	Hungary	39	64	1.2	1.0
Japan	50	77	3.3	0.6	Iceland	74	91	2.9	1.3
Jordan	35	68	5.2	4.4	Ireland	41	57	0.3	0.0
Korea	21	72	5.6	3.3	Isle of Man	53	74	−1.2	1.9
Korea, Dem.	31	60	6.0	2.2	Italy	54	69	1.5	0.6
Kuwait	59	96	8.5	4.8	Latvia	51	71	2.1	1.0
Laos	7	19	3.2	6.3	Liechtenstein	20	20	1.5	1.3
Lebanon	23	84	8.0	1.7	Lithuania	31	69	3.6	1.9
Macau	97	99	−1.5	3.4	Luxembourg	59	84	1.1	1.0
Malaysia	20	43	4.5	4.7	Malta	61	87	1.1	0.9

Continued

TABLE 4.4 Continued

	Urban population (% of total)		Average annual urban growth rate (%)	
	1950	1990	1955–60	1985–90
Monaco	100	100	1.8	0.4
Netherlands	83	89	1.6	0.7
Norway	50	75	0.8	1.0
Poland	39	62	3.7	1.1
Portugal	19	34	1.9	1.4
Romania	26	54	3.3	1.3
San Marino	18	92	8.0	1.6
Spain	52	78	1.7	0.9
Sweden	66	84	1.5	0.6
Switzerland	44	62	2.8	1.5
UK	84	89	0.6	0.3
Yugoslavia	22	56	3.3	2.5
(FORMER) USSR[g]	39	66	3.4	1.4

	Urban population (% of total)		Average annual urban growth rate (%)	
	1950	1990	1955–60	1985–90
OCEANIA	**61**	**71**	**2.9**	**1.6**
American Samoa	38	47	1.2	4.4
Australia[h]	75	85	2.9	1.6
Cook Is	32	25	1.9	–0.4
Fiji	24	39	5.1	1.2
French Polynesia	28	65	6.4	3.3
Guam	17	53	3.4	4.6
Kiribati	10	36	7.7	3.4
New Caledonia	49	60	3.6	2.3
New Zealand	73	84	2.7	0.9
Niue	24	23	0.1	–3.6
Papua New Guinea	1	16	15.4	4.3
Samoa	13	22	4.8	0.8
Solomon Is	8	15	3.3	6.6
Tonga	13	35	6.1	4.3
Vanuatu	6	19	6.2	2.8

[a] "More developed regions" comprise countries in Northern America (i.e., Bermuda, Canada, Greenland, St Pierre & Miquelon and the USA), Europe as well as the former USSR, Australia, New Zealand and Japan.

[b] "Less developed regions" comprise countries in Africa, Latin America, Asia (excluding Japan) and Oceania (excluding Australia and New Zealand).

[c] Includes Agalesa, Rodrigues and St Brandon.

[d] Includes Ascension and Tristan da Cunha.

[e] Data refer to the unified Republic of Yemen.

[f] Data refer to the unified Federal Republic of Germany.

[g] Includes 12 of the 15 republics of the former USSR; Estonia, Latvia and Lithuania are included under Northern Europe.

[h] Includes Christmas Island, Cocos (Keeling) Islands, and Norfolk Island.

The definition of "urban" may vary between countries. Please refer to the source document for a list of individual country definitions.

Aggregated data for the North American region are not given here due to differences in regional classifications. See source document for further details of regional classifications.

Source:
UN Department of Economic and Social Development (Population Division) 1993 *World Urbanization Prospects: The 1992 Revision*, United Nations, New York. In press.

TABLE 4.5   The world's 30 largest urban agglomerations ranked according to population size in 1950 and in 1990

1950				1990			
Rank	Agglomeration	Country	Population (10^6)	Rank	Agglomeration	Country	Population (10^6)
1	New York	USA	12.3	1	Tokyo	Japan	25.0
2	London	UK	8.7	2	São Paulo	Brazil	18.1
3	Tokyo	Japan	6.9	3	New York	USA	16.1
4	Paris	France	5.4	4	Mexico City	Mexico	15.1
5	Moscow	USSR	5.4	5	Shanghai	China	13.4
6	Shanghai	China	5.3	6	Bombay	India	12.2
7	Essen	Germany, Fed. Rep.	5.3	7	Los Angeles	USA	11.5
8	Buenos Aires	Argentina	5.0	8	Buenos Aires	Argentina	11.4
9	Chicago	USA	4.9	9	Seoul	Korea	11.0
10	Calcutta	India	4.4	10	Rio de Janeiro	Brazil	10.9
11	Osaka	Japan	4.1	11	Beijing	China	10.9
12	Los Angeles	USA	4.0	12	Calcutta	India	10.7
13	Beijing	China	3.9	13	Osaka	Japan	10.5
14	Milan	Italy	3.6	14	Paris	France	9.3
15	Berlin	Germany	3.3	15	Tianjin	China	9.2
16	Mexico City	Mexico	3.1	16	Jakarta	Indonesia	9.2
17	Philadelphia	USA	2.9	17	Moscow	USSR	9.0
18	Leningrad[a]	USSR	2.9	18	Metro. Manila	Philippines	8.9
19	Bombay	India	2.9	19	Cairo	Egypt	8.6
20	Rio de Janeiro	Brazil	2.9	20	Delhi	India	8.2
21	Detroit	USA	2.8	21	Karachi	Pakistan	7.9
22	Naples	Italy	2.8	22	Lagos	Nigeria	7.7
23	Manchester	UK	2.5	23	London	UK	7.3
24	São Paulo	Brazil	2.4	24	Bangkok	Thailand	7.1
25	Cairo	Egypt	2.4	25	Chicago	USA	6.8
26	Tianjin	China	2.4	26	Tehran	Iran	6.7
27	Birmingham	UK	2.3	27	Dhaka	Bangladesh	6.6
28	Frankfurt	Germany, Fed. Rep.	2.3	28	Istanbul	Turkey	6.5
29	Boston	USA	2.2	29	Lima	Peru	6.5
30	Hamburg	Germany, Fed. Rep.	2.2	30	Essen	Germany, Fed. Rep.	6.4

[a]   Leningrad is now known as St Petersburg.

The definitions of an "urban agglomeration" may vary between countries and over time.

Source:
UN Department of Economic and Social Development (Population Division) 1993 *World Urbanization Prospects: The 1992 Revision*, United Nations, New York. In press.

TABLE 4.6 Selected measures of economic development, 1990 (unless otherwise stated)

	Total GDP (10^6 US$)	Distribution of GDP			GDP growth rate, 1980–90 (% per year)	GNP per capita, (US$)	Real GDP per capita, 1989 (PPP)
		Agriculture (%)	Industry (%)	Services[a] (%)			
WORLD	22,298,850				3.2	4,200	
AFRICA							
Algeria[b]	42,150	13	47	41	3.1	2,060	3,088
Angola	7,700	13	44	43			1,225
Benin[b]	1,810	37	15	48	2.8	360	1,030
Botswana[b]	2,700	3	57	40	11.3	2,040	3,180
Burkina	3,060	32	24	44	4.3	330	617
Burundi	1,000	56	15	29	3.9	210	611
Cameroon[b]	11,130	27	28	46	2.3	960	1,699
Cent. African Rep.	1,220	42	17	41	1.5	390	770
Chad[b]	1,100	38	17	45	5.9	190	582
Congo[b]	2,870	13	39	48	3.6	1,010	2,382
Côte d'Ivoire	7,610	47	27	26	0.5	750	1,381
Egypt	33,210	17	29	53	5.0	600	1,934
Gabon[b]	4,720	9	49	42	2.3	3,330	4,735
Ghana[b]	6,270	48	16	37	3.0	390	1,005
Guinea[b]	2,820	28	33	39		440	602
Kenya	7,540	28	21	51	4.2	370	1,023
Lesotho	340	24	30	46	3.1	530	1,646
Madagascar[b]	2,750	33	13	54	1.1	230	690
Malawi	1,660	33	20	46	2.9	200	620
Mali[b]	2,450	46	13	41	4.0	270	576
Mauritius	2,090	12	33	55	6.0	2,250	5,375
Morocco[b]	25,220	16	33	51	4.0	950	2,298
Mozambique	1,320	65	15	21	−0.7	80	1,060
Niger[b]	2,520	36	13	51	−1.3	310	634
Nigeria	34,760	36	38	25	1.4	290	1,160
Rwanda[b]	2,130	38	22	40	1.0	310	680
Senegal[b]	5,840	21	18	61	3.0	710	1,208
Sierra Leone	840	32	13	55	1.5	240	1,061
Somalia	890	65	9	26	2.4	120	861
South Africa	90,720	5	44	51	1.3	2,530	4,852
Tanzania	2,060	59	12	29	2.8	110	557
Togo[b]	1,620	33	22	46	1.6	410	752
Tunisia	11,080	16	32	52	3.6	1,440	3,329
Uganda	2,820	67	7	26	2.8	220	499
Zaire[b]	7,540	30	33	36	1.8	220	380
Zambia[b]	3,120	17	55	29	0.8	420	767
Zimbabwe	5,310	13	40	47	2.9	640	1,469
NORTH AMERICA							
Canada	570,150				3.4	20,470	18,635
Costa Rica[b]	5,700	16	26	58	3.0	1,900	4,413
Dominican Rep.[b]	7,310	17	27	56	2.1	830	2,537

Continued

TABLE 4.6 Continued

	Total GDP (10^6 US$)	Distribution of GDP			GDP growth rate, 1980–90 (% per year)	GNP per capita, (US$)	Real GDP per capita, 1989 (PPP)
		Agriculture (%)	Industry (%)	Services[a] (%)			
El Salvador[b]	5,400	11	21	67	0.9	1,110	1,897
Guatemala[b]	7,630	26	19	55	0.8	900	2,531
Haiti[b]	2,760				−0.6	370	962
Honduras	2,360	23	24	53	2.3	590	1,504
Jamaica[b]	3,970	5	46	49	1.6	1,500	2,787
Mexico[b]	237,750	9	30	61	1.0	2,490	5,691
Panama[b]	4,750	10	9	80	0.2	1,830	3,231
Trinidad & Tobago	4,750	3	48	49	−4.7	3,610	6,266
USA[b]	5,392,200				3.4	21,790	20,998
SOUTH AMERICA							
Argentina[b]	93,260	13	41	45	−0.4	2,370	4,310
Bolivia[b]	4,480	24	32	44	−0.1	630	1,531
Brazil	414,060	10	39	51	2.7	2,680	4,951
Chile[b]	27,790				3.2	1,940	4,987
Colombia	41,120	17	32	51	3.7	1,260	4,068
Ecuador[b]	10,880	13	42	45	2.0	980	3,012
Paraguay[b]	5,260	28	23	49	2.5	1,110	2,742
Peru[b]	36,550	7	37	57	−0.3	1,160	2,731
Uruguay	8,220	11	34	55	0.3	2,560	5,805
Venezuela[b]	48,270	6	50	45	1.0	2,560	5,908
ASIA							
Bangladesh	22,880	38	15	46	4.3	210	820
Bhutan	280	43	27	29	7.5	190	750
China[b]	364,900	27	42	31	9.5	370	2,656
Hong Kong	59,670	0	26	73	7.1	11,490	15,180
India	254,540	31	29	40	5.3	350	910
Indonesia[b]	107,290	22	40	38	5.5	570	2,034
Iran	116,040	21	21	58	2.5	2,490	3,120
Israel[b]	53,200				3.2	10,920	10,448
Japan[b]	2,942,890	3	42	56	4.1	25,430	14,311
Jordan[c]	3,330	8	26	66		1,240	2,415
Korea[b]	236,400	9	45	46	9.9	5,400	6,117
Kuwait[b]	23,540	1	56	43	0.7		15,984
Malaysia[b]	42,400				5.2	2,320	5,649
Nepal	2,890	60	14	26	4.6	170	896
Oman[b]	7,700	3	80	18	12.8		10,573
Pakistan	35,500	26	25	49	6.3	380	1,789
Philippines[b]	43,860	22	35	43	0.9	730	2,269
Saudi Arabia[b]	80,890				−1.8	7,050	10,330
Singapore[b]	34,600	0	37	63	6.4	11,160	15,108
Sri Lanka	7,250	26	26	48	4.0	470	2,253
Syria[b]	14,730	28	22	50	2.1	1,000	4,348
Thailand[b]	80,170	12	39	48	7.6	1,420	3,569
Turkey	96,500	18	33	49	5.1	1,630	4,002

Continued

TABLE 4.6 Continued

	Total GDP (10^6 US$)	Distribution of GDP			GDP growth rate, 1980–90 (% per year)	GNP per capita, (US$)	Real GDP per capita, 1989 (PPP)
		Agriculture (%)	Industry (%)	Services[a] (%)			
United Arab Em.	28,270	2	55	43	−4.5	19,860	23,798
Yemen[b]	6,690	20	28	47			1,560
EUROPE							
Austria[b]	157,380	3	37	60	2.1	19,060	13,063
Belgium[b]	192,390	2	31	67	2.0	15,540	13,313
Bulgaria	19,910	18	52	31	2.6	2,250	5,064
Czechoslovakia[b]	44,450	8	56	36	1.4	3,140	7,420
Denmark	130,960	5	28	67	2.4	22,080	13,751
Finland	137,250	6	36	58	3.4	26,040	14,598
France[b]	1,190,780	4	29	67	2.2	19,490	14,164
Germany, Fed. Rep.[b]	1,488,210	2	39	59	2.1	22,320	14,507
Greece	57,900	17	27	56	1.8	5,990	6,764
Hungary[b]	32,920	12	32	56	1.3	2,780	6,245
Ireland	42,500				3.1	9,550	7,481
Italy[b]	1,090,750	4	33	63	2.4	16,830	13,608
Netherlands[b]	279,150	4	31	65	1.9	17,320	13,351
Norway	105,830				2.9	31,120	16,838
Poland[b]	63,590	14	36	50	1.8	1,690	4,770
Portugal[b]	56,820				2.7	4,900	6,259
Romania	34,730	18	48	34	1.2	1,640	3,000
Spain[b]	491,240				3.1	11,020	8,723
Sweden	228,110	3	35	62	2.2	23,660	14,817
Switzerland[b]	224,850				2.2	32,680	18,590
UK	975,150				3.1	16,100	13,732
Yugoslavia	82,310	12	48	40	0.8	3,060	5,095
OCEANIA							
Australia[b]	296,300	4	31	64	3.4	17,000	15,266
New Zealand[b]	42,760	9	27	65	1.9	12,680	11,155
Papua New Guinea[b]	3,270	29	31	40	1.9	860	1,834

[a] "Services" includes unallocated items.
[b] GDP and its components are at purchaser values.
[c] With the exception of "Real GDP per capita", data for Jordan refer to the East Bank only.

Gross Domestic Product (GDP) provides a measure of the total output of goods and services for final use that is produced by an economy; it includes net exports of goods and non-factor services. Most countries estimate GDP by the production method which sums the final outputs of the various sectors of economy (e.g., agriculture, manufacturing and services) from which the value of the inputs (e.g., investment capital, labour and raw materials) is subtracted. GDP is given in US$, converted from domestic currencies using single-year official exchange rates. "Agriculture" covers forestry, hunting and fishing as well as agriculture. "Industry" comprises value added in mining; manufacturing; construction; and electricity, water and gas. Value added in all other branches of economic activity, including imputed bank service charges, import duties, and any statistical discrepancies noted by national compilers, are categorized as "Services".

Gross National Product (GNP) is the sum of two components: GDP and net factor income from abroad. Net factor income from abroad is the income in the form of overseas workers' remittances, interest on loans and other factor payments, less payments made to non-residents for factor services (e.g., labour and capital). Thus GNP provides a measure of the domestic and foreign value added to the economy. For most countries GNP is calculated by the Atlas method, full details of which are given in the 'World Development Report'. Total GNP is divided by the mid-year population to give GNP per capita.

The use of official exchange rates to convert national currency figures to US$ does not reflect the relative domestic purchasing power of currencies. The International Comparisons Program (ICP) has thus developed an alternative economic measure, real GDP per capita, which uses purchasing power parities (PPPs) as conversion factors for national currencies instead of exchange rates. Real GDP per capita is expressed in international dollars, and is considered to have greater international comparability than GDP. The values presented here are preliminary and are subject to further revision as the methodology is improved.

Sources:
World Bank 1992 *World Development Report: Development and the Environment*, Oxford University Press, New York.
United Nations Development Programme 1992 *Human Development Report 1992*, Oxford University Press, New York.

TABLE 4.7   Selected measures of human development in selected developing countries, 1990

Region/country	Adult literacy (rate)		Access to safe drinking water (% of population)		Access to sanitation services (% of population)		Access to local health care[a] (% of population)	
	Female	Male	Urban	Rural	Urban	Rural	Urban	Rural
AFRICA								
Angola	29	56	73	20	25	20		
Benin	16	32	73	43	60	35		
Botswana	65	84	100	88	100	85	100	85
Burkina	9	28		70				
Burundi	40	61	92	43	64	16		
Cameroon	43	66	42	45			44	39
Cent. African Rep.	25	52	19	26	45	46		
Côte d'Ivoire	40	67	57	80	81	100	61	11
Egypt	34	63	95	86	80	26		
Equatorial Guinea	37	64	65	18	54	24		
Gambia	16	39	100	48	100	27		
Ghana	51	70	63		63	60	92	45
Guinea	13	35	100	37		0		
Mali	24	41	41	4	81	10		
Mauritius			100	100	100	100	100	100
Morocco	38	61	100	18	100		100	50
Namibia			90	37	24	11		
Niger	17	40	98	45	71	4	99	30
Nigeria	40	62	100	22	80	11	75	30
Rwanda	37	64	84	67	88	17	60	25
Senegal	25	52	65	26	57	38		
Sierra Leone	11	31	80	20	55	31		
Uganda	35	62	60	30	32	60	90	57
Zaire	61	84	68	24	46	11	40	17
Zimbabwe	60	74	95	80	95	22	100	62
NORTH AMERICA								
Bahamas			98	75	98	2		
Barbados			100	100	100	100	100	100
Belize			95	53	76	22	100	50
Br. Virgin Is			100	100	100	100		
Cuba			100	91	100	68		
Dominican Rep.	82	85	82	45	95	75		
El Salvador	70	76	87	15	85	38	80	40
Guatemala	47	63	92	43	72	52	47	25
Haiti	47	59	56	35	44	17	80	70
Honduras	71	76	85	48	89	42	85	65
Mexico	85	90	94		85			
Trinidad & Tobago			100	88	100	92		
SOUTH AMERICA								
Bolivia	71	85	76	30	38	14	90	36
Brazil	80	83	95	61	84	32		
Colombia	86	88	87	82	84	18		

Continued

TABLE 4.7 Continued

Region/country	Adult literacy (rate)		Access to safe drinking water (% of population)		Access to sanitation services (% of population)		Access to local health care[a] (% of population)	
	Female	Male	Urban	Rural	Urban	Rural	Urban	Rural
Ecuador	84	88	63	44	56	38	90	30
Guyana	95	98	100	71	97	81		
Paraguay	88	92	61	9	31	60	90	38
Peru	79	92	68	24	76	20		
Uruguay	96	97	100					
Venezuela	90	87		36		72		
ASIA								
Afghanistan	14	44	40	19	13		80	17
Bahrain	69	82	100	0	100	0		
Bangladesh	22	47	39	89	40	4		
Bhutan	25	51	60	30	80	3		
China	62	84	87	68	100	81		
Cyprus			100	100	96	100	100	100
Hong Kong			100	96	90	50		
India	34	62	86	69	44	3		
Indonesia	62	84	35	33	79	30		
Iran	43	65	100	75	100	35	95	60
Iraq	46	70	93	41	96		97	70
Jordan	70	89	100	97	100	100	98	95
Korea	94	99	100	76	67	12		
Laos			47	25	30	8		
Malaysia	70	87	96	66	94	94		
Maldives			77	68	95	4	100	95
Mongolia			100	58	100	47		
Myanmar	72	89	79	72	50	13	100	
Nepal	13	38	66	34	34	3		
Pakistan	21	47	82	42	53	12	99	35
Philippines	90	90	93	72	79	63		
Qatar			100		100	85		
Singapore			100		99		100	
Sri Lanka	84	93	80	55	68	45		
Thailand	90	96		85		86		
Viet Nam	84	92	47	33	23	10	100	75
OCEANIA								
Cook Is			100	100	100	100		
Fiji			96	69	91	65		
French Polynesia			100	18	98	95		
Kiribati			91	63	91	49		
Marshall Is			100	45	100	45		
Micronesia				38	99	46		
Niue			0	100	0	100		
Northern Marianas Is			100	0	100	71		
Palau			100	97	95	100		
Papua New Guinea	38	65	94	20	57			

Continued

TABLE 4.7   Continued

Region/country	Adult literacy (rate)		Access to safe drinking water (% of population)		Access to sanitation services (% of population)		Access to local health care[a] (% of population)	
	Female	Male	Urban	Rural	Urban	Rural	Urban	Rural
Samoa			100	77	100	92	100	
Solomon Is			82	58	73	2		
Tonga			92	98	88	78		

[a]   Data refer to 1987–90.

Adult literacy rates refer to the percentage of people over the age of 15 who can read and write.

Safe drinking water includes treated surface water and untreated water from protected springs, boreholes and wells. The WHO defines access to safe drinking water in urban areas as piped water to housing units or to public stand pipes (within 200 metres). In rural areas reasonable access implies that fetching water does not take up a disproportionate part of the day. Urban and rural populations with access to sanitation services were defined as those served by connections to public sewers or household disposal systems such as pit privies, pour flush latrines, septic tanks and communal toilets.

Access to local health care is defined as the proportion of the population having treatment for common diseases and injuries and a regular supply of at least 20 essential drugs available within one hour's walk or travel.

Sources:
UNESCO 1990 *Compendium of Statistics on Illiteracy – 1990 Edition*, United Nations Educational, Scientific and Cultural Organization, Paris (adult literacy data).
United Nations Development Programme 1992 *Human Development Report 1992*, Oxford University Press, New York (health care data).
WHO 1992 *The International Drinking Water Supply and Sanitation Decade*, World Health Organization, Geneva (sanitation and safe drinking water data).

Part 5
Human Health

The state of human health is in many ways a reflection of the environment. Health is the result of interactions between humans and the full range of factors in their physical, socio-economic, cultural and political environments. The physical environment consists of the natural and the built environment, comprising physical, chemical and biological conditions within the home, neighbourhood and workplace. The social environment includes such aspects as access to health care and the level of development, which in turn has a bearing on access to safe water and sanitation.

○ It is increasingly recognized that the health of populations is at risk from a variety of environmental and pollution hazards although the interactions are still not well understood and direct links are difficult to establish.

○ Infection and parasitic diseases account for 45 per cent of all deaths on a global scale while cardiovascular disease accounts for approximately 25 per cent. Within the next decade deaths from cardiovascular diseases are expected, in some developing countries, to exceed deaths from infectious diseases.

○ The high prevalence of infectious diseases in developing countries is linked to malnutrition, inadequate water supply and sanitation, poor hygiene practices and overcrowded living conditions, as well as poor immunization coverage and resistance of disease vectors to pesticides.

○ Waterborne diseases are a major contributor to infant mortality in developing countries, responsible for about 4 million deaths a year.

○ Whereas efforts to eradicate some diseases, such as dracunculiasis, are proving successful, new strains of other diseases, such as cholera, are causing serious epidemics.

Although the relationships between health, development and the environment are complex, health can be directly related to wealth. The wealthiest nations, and especially their most favoured citizens, generally enjoy the best health and well-being. The wealthiest people have greater access to medical attention, eat more nutritiously, smoke less, and have the opportunity to live far removed from the influence of industry and its effluents. Conversely, poverty can be associated with ill health and premature death. Malnutrition,

poor hygiene and sanitation, lack of safe water, inadequate shelter and housing, and illiteracy, contribute towards disease and ill health (WHO, 1992a). One estimate suggests that about 2.2 billion people live in poverty in the less developed nations and, of these, about 700 million live in extreme poverty; 9 out of 10 have no access to safe drinking water and most have no health care services (WHO, 1992b). The deteriorating urban environment in many developing countries, including the increase in shanty towns and squatters, detailed in Part 4: Population and Development, can create serious threats to human health. Countries rapidly undergoing industrialization, and many developing countries, are ill-prepared to deal with most environmental health problems (Schaefer, 1991).

Poor air quality, insufficient and unsafe water, inadequate solid waste disposal, absence of treatment of excreta, and risks from hazardous chemicals are typical areas where pollution impacts on human health (WHO 1985, 1988). However, direct links between global environmental contamination and human health in the general population are hard to verify. Nevertheless, some risks associated with environmental exposure have been established, such as lead contamination causing neurological and psychological problems in children and asbestos exposure causing asbestosis or mesothelioma in exposed workers (Brunekreef, 1986; WHO, 1986). Risks associated with the ever growing number of synthetic chemicals used in industry today are difficult to assess because only about 20 per cent have ever been tested for their carcinogenicity, neurotoxic, immunotoxic or other toxic potential (Landrigan, 1992).

This edition of the 'UNEP Environmental Data Report' places greater emphasis on environment-related health problems, particularly those associated with water. General health status indicators are being retained, as in previous editions (UNEP 1987, 1989, 1991), to enable the environmentally-related diseases to be placed within the wider health context.

The World Health Organization (WHO) is the specialized agency of the United Nations (UN) with primary responsibilities for international health matters and public health. As such, it is involved in collecting and and disseminating health statistics, including injuries, causes of death and data based on the International Classification of Diseases. Environment and health is specifically considered in a broad-based programme concerned with chemical, physical and biological health hazards, paying special attention to problems related to water and air quality, food, the use of chemicals and housing (see Part 1: Environmental Pollution). These programmes provide a basis for developing effective chemical pollution control programmes and emerging response systems which are detailed in the various sections of this volume.

The WHO Commission on Health and the Environment (WHO, 1992a) has highlighted that health and environment are issues that need to be considered in a broad development context, requiring the participation of many disciplines and programmes. The work of the WHO Commission was supported by four panels of experts in areas of special interest, i.e., energy, industry, urbanization and food and agriculture (WHO, 1992b,c,d,e). The reports from these expert panels draw attention to the numerous environment-related conditions that impair human health and to how human activities can create or enhance these threats to health. Agenda 21, an action plan adopted by more than 150 member states at the United Nations Conference on Environment and Development (UNCED) (UN, 1992a,b,c), provides a sound and comprehensive basis for developing and implementing a global strategy for health and the environment. The section on Protection and Promotion of Human Health contains many of the recommendations of the WHO Commission, although they are presented within a different framework because not all health-related issues are concerned with environmental health.

Co-operation between all of the UN organizations is an essential component of the future strategy for health and environment. The United Nations Food and Agriculture Organization (FAO) and WHO, through the Joint Meeting on Pesticide Residues (JMPR) and the Joint Expert Committee on Food Additives (JECFA), are examining the effects of food and agriculture on health, particularly indiscriminate use of pesticides and food contamination. Health effects of priority chemicals are evaluated by the International Programme on Chemical Safety (IPCS) and by the United Nations Environment Programme's (UNEP) International Register of Potentially Toxic Chemicals (IRPTC) in co-operation with the International Labour Organization (ILO). The WHO (European Region) Healthy Cities project, which involves health services, health promotion and environmental health has been found to be an effective way for supporting community action for health (see Part 4: Populations and Development). The International Committee of the Red Cross and the International Federation of Red Cross and Red Crescent Societies, the Save the Children Fund and other non-governmental organizations (NGOs) are also involved in improving human health around the world.

References

Brunekreef, B. 1986 *Childhood Exposure to Environmental Lead*, MARC Report No. 34, Monitoring and Assessment Research Centre, King's College London, University of London.

Landrigan P. J. 1992 Commentary: Environmental disease – a preventable epidemic, *American Journal of Public Health*, **82**(7), 941–943.

Schaefer, M. 1991 *Combating Environmental Pollution: National Capabilities for Health Protection*, Report WHO/PEP/91.14, World Health Organization, Geneva.

UN 1992a *Report of the United Nations Conference on Environment and Development* (Rio de Janeiro, 3–14 June 1992) A/CONF.151/26 (Vol. I), United Nations, New York.

UN 1992b *Report of the United Nations Conference on Environment and Development* (Rio de Janeiro, 3–14 June 1992) A/CONF.151/26 (Vol. II), United Nations, New York.

UN 1992c *Report of the United Nations Conference on Environment and Development* (Rio de Janeiro, 3–14 June 1992) A/CONF.151/26 (Vol. III), United Nations, New York.

UNEP 1987 *United Nations Environment Programme Environmental Data Report 1987/88*, Basil Blackwell, Oxford.

UNEP 1989 *United Nations Environment Programme Environmental Data Report 1989/90*, Basil Blackwell, Oxford.

UNEP 1991 *United Nations Environment Programme Environmental Data Report 1991/92*, Basil Blackwell, Oxford.

WHO 1985 *Preliminary Assessment of National Programmes for Health Protection Against Environmental Hazards*, Report WHO/PEP/85.5, World Health Organization, Geneva.

WHO 1986 *Asbestos and Other Natural Mineral Fibres*, Environmental Health Criteria 53, World Health Organization, Geneva.

WHO 1988 *National Capabilities and Needs in Aspects of Environmental Health in Rural and Urban Development and Housing*, Report WHO/EHE/RUD/88.1, World Health Organization, Geneva.

WHO 1992a *Our Planet, Our Health*, Report of the WHO Commission on Health and Environment, World Health Organization, Geneva.

WHO 1992b Report of the Panel on Urbanization, WHO Commission on Health and Environment, WHO/EHE 92.5, World Health Organization, Geneva.

WHO 1992c Report of the Panel on Industry, WHO Commission on Health and Environment, WHO/EHE 92.4, World Health Organization, Geneva.

WHO 1992d Report of the Panel on Energy, WHO Commission on Health and Environment, WHO/EHE 92.3, World Health Organization, Geneva.

WHO 1992e Report of the Panel on Food and Agriculture, WHO Commission on Health and Environment, WHO/EHE 92.2, World Health Organization, Geneva.

General Health Status Indicators

Global indicators of health were agreed upon at the World Health Assembly in 1981. Twelve global indicators were originally chosen for monitoring progress towards health. These have recently been revised, reformulated and finally adopted in 1990. The indicators range from economic (e.g., percentage of gross national product spent on health) to health-specific (e.g., the percentage of the population having access to safe water, or receiving immunization) to population-related (e.g., infant mortality rate and life expectancies) (WHO, 1993). This edition of the 'UNEP Environmental Data Report' concentrates on life expectancies, infant and child mortality, causes of death and maternal mortality. Previous editions have also concentrated on nutritional status (UNEP 1987, 1989), availability of health care (UNEP 1987, 1989, 1991) and immunization status (UNEP 1989, 1991).

Health statistics are collected in each country throughout the world and form the basis for the WHO information system in Member States. International agreement on classification systems for diseases/disability and causes of death has been reached, although it is continually being refined. For the purposes of health monitoring, it is more informative to use incidence rates (i.e., cases per 100,000 inhabitants) rather than the total number of cases. For the elucidation of trends, age-specific rates are also more appropriate. In general, data on causes of death are available in most developed countries and are the only health statistics for which comparatively long term series are available. Variations in diagnostic practice and coding procedures will, however, affect the comparability of data between countries. Systems for registering cases of important communicable diseases exist in most countries. The WHO, the United Nations Children's Fund (UNICEF) and the International Agency for Research on Cancer (IARC) are major organizations involved in health data collection and collation.

In rapidly urbanizing areas, the lack of data about vital events (births and deaths), the incidence of disease and the size and characterization of the population precludes statistical assessment of the effects of the urban environment on health. Although such data usually show the level of urban health to be higher than that for rural areas in the same nation, much ill-health among the poor and among the "illegals" or squatters in the informal settlements is under-reported (UNCHS (Habitat), 1992). Informal settlements may account for up to 70 per cent of city populations (UNCHS (Habitat), 1991). Since available mortality and morbidity data are seldom disaggregated by socio-economic status or neighbourhood, the full extent of the health problems and important differences between groups within city populations often remain conjectural (WHO, 1991a). With well over 1,000 million people living in poverty, mainly in developing countries, and around one-third of urban dwellers living in urban slums, reliable health information is often lacking (WHO, 1991a).

Women and children are often very vulnerable to the health risks associated with environmental degradation, inadequate standards of living, poverty, poor education, discriminating economic and social practices, early marriage, and early and multiple pregnancies (Royston and Armstrong, 1989). In many low-income settlements there is a significantly higher incidence of particular diseases among women. Respiratory problems associated with smoky living conditions (see Part 1: Environmental Pollution, Human Exposure) and diseases associated with contaminated and insufficient water and lack of sanitation are major health risks (WHO, 1992a). The specific effects and impacts of tropical diseases on women of all ages have been identified as an area where greater attention should be paid, including future research programmes (Wijeyaratne, Rathgeber and St-Onge, 1992).

Life Expectancy, Infant and Child Mortality

Life expectancy at birth provides a simple measure of the state of health in a country. The main factors responsible for the trends and current differences in life expectancy are changes in social and physical environment, personal behaviour and medical care. The state of the economy of a country has a strong influence on its health level (WHO, 1992a). Data from a cross-section of countries illustrate that life expectancy is positively associated with a country's level of economic development as measured by gross domestic product (GDP) per capita (Figure 5.1). In high-income countries of the Organisation for Economic Co-operation and Development (OECD), life expectancy is approximately 76 years compared with less than 60 years in the low-income countries. In general, expenditure on health rises in line with increasing GDP per capita. This association between health and economic development can also be seen over time in individual countries, although variations in health levels do occur among countries with similar incomes (WHO, 1992b).

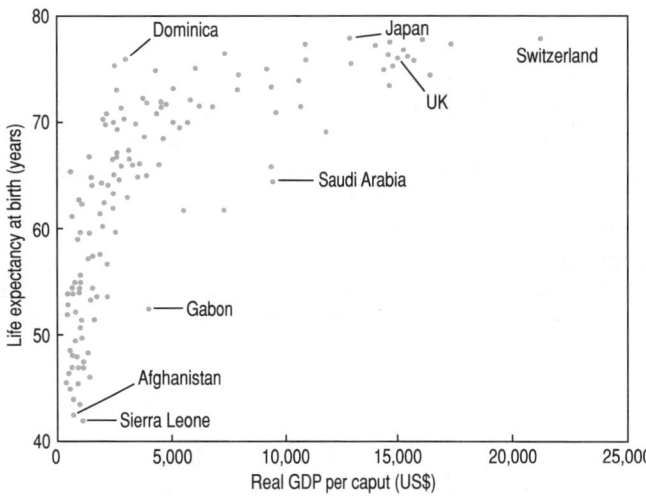

FIGURE 5.1 The relationship between life expectancy and level of economic development in terms of GDP per capita
Source: Adapted from WHO, 1992a

Real GDP in the OECD countries grew five-fold between 1960 and 1980. In the same period, life expectancy at birth increased by 2.7 years for males and 4.4 years for females.

Life expectancy at birth for the period 1985–1990 has been estimated at 61.5 years for the world as a whole, 73.4 years for more developed regions and 59.7 years for less developed regions (WHO, 1990). These data suggest an improvement since 1960–1965 when the annual average was approximately 50 years. Ninety-four countries, representing 66 per cent of the world population, have achieved a life expectancy of 60 years or more. There are, however, substantial regional differences as shown in Table 5.1. For many countries in Africa, Southeast Asia and the Eastern Mediterranean life expectancy is below 60 years. Trends in life expectancy at birth can be analysed separately for urban and rural populations as illustrated by the data from the former USSR (Table 5.2). These data show that the increase in life expectancy between 1958 and 1989 has been small (Virganskaya and Dmitriev, 1992).

The infant mortality rate (defined as the death rate of infants less than one year of age per thousand live births) and the child mortality rate (defined as the death rate of children between one year and five years of age per thousand one-year-old children) are also used as indicators of the general health status of countries (Table 5.1). Factors which influence the death rates of children under five years of age include drinking water quality and sanitation, nutritional status, housing conditions and maternal health.

Globally, the infant and child mortality rates have declined since the 1950s, although there are marked variations between countries and geographical regions. Despite an 8 per cent increase in the number of births in the developing countries as a whole between 1985 and 1990, the estimated number of child deaths fell from 13.5 million in 1985 to 12.9 million in 1990 (WHO, 1992b). The global improvement in child survival more than compensated for the increase in the population of young children, with the net result that the total number of child deaths declined by about 4–5 per cent over the period.

Causes of Death

Over the period 1985–1990, an estimated 50 million deaths occurred in the world each year (WHO, 1990). An overview of the data in Table 5.3 shows that in developing countries most of the deaths occur in the under-60 age group, compared with only 30 per cent in developed countries. Differences between the sexes for the various age groups are also shown, together with the proportion of deaths due to infectious and parasitic diseases compared with neoplasms (cancer). Vital statistics on causes of death in developing countries are, however, often not available or may be unreliable, at least at the national level (WHO, 1990). Deaths from infectious and parasitic diseases in developing countries are higher than those for developed countries and are particularly high in the younger age groups. By contrast, malignant neoplasms are higher in the older age groups. Cardiovascular diseases account for the highest proportion of all causes of death. In developing countries they account for about 16 per cent compared with 54 per cent in developed countries (WHO, 1992a,b). By the year 2000, cardiovascular diseases are projected to claim about 1.5 times as many deaths as infectious and parasitic diseases in Asia (WHO, 1992b). The projected increase in cardiovascular diseases in developing countries is also related to the progressive ageing of the population.

Age-specific death rate and death rate by cause can also be analysed for rural and urban populations as has been done for the former USSR (Tables 5.4 and 5.5) Age-specific mortality rates of the rural population in the age group less than 50 years are higher than those of the urban population. The difference between urban and rural death rates increased between the 1960s and the 1980s for most age groups of males and females below 50 years. Conversely, in the over-50 age group death rates are greater for urban populations and differences between urban and rural populations had decreased from the 1960s to the 1980s, although less for women than men (Table 5.4). A specific feature of these changes was the fact that they occurred against the background of a general cut in the life expectancy rate of the population (Table 5.2). Adverse trends in the development of mortality rates were more pronounced in the countryside, and the substantial reduction in the life expectancy rate was due to this.

Age-mortality differences between the former USSR as a whole and selected individual states have also been examined and grouped according to causes of death (Table 5.5). Differences can be seen in the cause-specific mortality prevailing in the selected states. Environmental, climatic and topographic features, as well as social and economic development and lifestyle, including nutritional habits, are major factors influencing health and cause of death. Large urban and rural differences are also noted which may be accounted for by the restricted access to, and low quality of, health care in rural areas (Virganskaya and Dmitriev, 1992).

Maternal Mortality

The need for improvement in the social status of women, and access to the same care and attention as men is seen to be a major requirement for development world-wide. The resolution passed by WHO's World Health Assembly in 1992, calling upon the Director-General to set up a "global commission on women's health" represents a major attempt to make policy makers more aware of women's health issues. That women have a vital role in environmental management and development was further stressed by the adoption of Principal 20 of the Rio Declaration on Environment and Development at UNCED (UN, 1992a) which stated that the full participation of women was essential to achieve sustainable development.

Maternal mortality rate has been proposed as a more indicative measure of regional disparities in socio-economic and environmental conditions, simply because the needs of a pregnant mother are more diverse than those of an individual. In many developing countries pregnancy and childbirth account for more than a quarter of all deaths of women of child-bearing age. About half a million women, 99 per cent of them in the developing world, die in childbirth each year (WHO, 1992a,b). The risk of women dying from causes related to pregnancy and childbirth is 100–200 times greater among the poor of the less developed countries than on average in industrialized countries (WHO, 1991b). The risks associated with becoming pregnant are highest in Eastern, Middle and Western Africa, all of which have shown increases in the number of maternal deaths per 100 thousand live births (WHO, 1991b). A woman in this part of the world is 75 times more likely to die as a result of becoming pregnant than is a woman in Western Europe. Conversely, maternal mortality is considerably lower in Northern and Southern Africa. Recent information has shown that mortality risks for women in Asia and parts of Latin America have decreased (Figure 5.2). The number of maternal deaths in the region as a whole appears to have declined by about one quarter. There are also signs of improvement in Asia, where decreases in the risks of pregnancy and the number of deaths are apparent in all subregions except Eastern Asia (although data may be affected by improved reporting in China).

References

Royston, E. and Armstrong, S. 1989 *Preventing Maternal Deaths*, World Health Organization, Geneva.

UN 1992a *Report of the United Nations Conference on Environment and Development* (Rio de Janeiro, 3–14 June 1992) A/CONF.151/26 (Vol. I), United Nations, New York.

UNCHS (Habitat) 1991 Environmental health and human settlements, *Habitat News* **13**, 24–26.

UNCHS (Habitat) 1992 *The Role of Human Settlements in Improving Community Health in Africa*, Presented at International Conference on Community Health in Africa, 4–6 September 1992, Brazzaville, Congo.

UNEP 1987 *United Nations Environment Programme Environmental Data Report 1987/88*, Basil Blackwell, Oxford.

UNEP 1989 *United Nations Environment Programme Environmental Data Report 1989/90*, Basil Blackwell, Oxford.

UNEP 1991 *United Nations Environment Programme Environmental Data Report 1991/92*, Basil Blackwell, Oxford.

Virganskaya, I. M. and Dmitriev, V. I. 1992 Some problems of the medicodemographic development in the former USSR, *World Health Statistics Quarterly* **45**, 4–14, World Health Organization, Geneva.

WHO 1990 *Global Estimates for Health Situation Assessments and Projections*, WHO/HST/90.2, World Health Organization, Geneva.

WHO 1991a *Environmental Health in Urban Development*, WHO Technical Report 807, World Health Organization, Geneva.

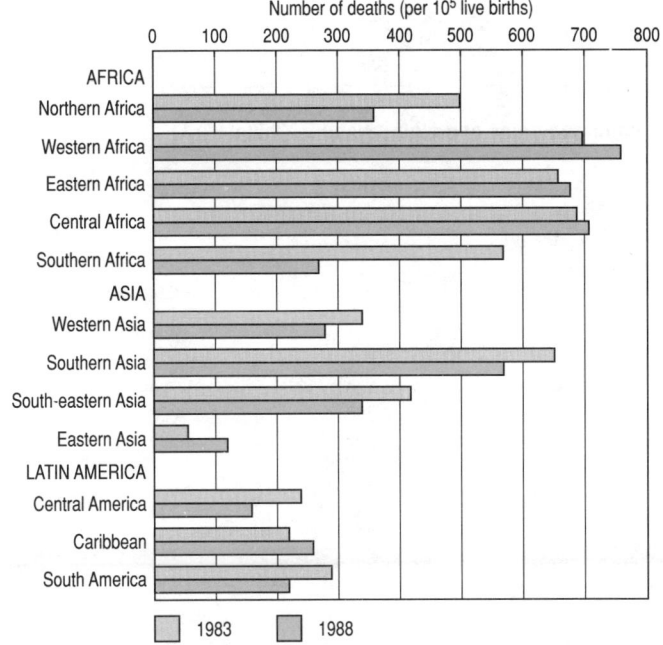

Number of deaths (per 10^5 live births)

FIGURE 5.2 Maternal mortality risks in regions of Africa, Asia and Latin America, 1983 and 1988
Source: WHO, 1991b

WHO 1991b New estimates of maternal mortality. *Weekly Epidemiological Record*, **47**(66), 345–352.

WHO 1992a *Our Planet, Our Health*, Report of the WHO Commission on Health and Environment, World Health Organization, Geneva.

WHO 1992b *World Health Statistics Annual 1991*, World Health Organization, Geneva.

WHO 1993 *World Health Statistics Annual 1992*, World Health Organization, Geneva.

Wijeyaratne, P., Rathgeber, E. M. and St-Onge, E. (Eds) 1992 *Women and Tropical Diseases*, International Development Research Centre, Ottawa.

Environmentally-related Diseases

The greatest causes of morbidity and mortality on a global scale are infectious diseases. The high prevalence of infectious diseases in developing countries accounts for 72 per cent of deaths of children under five years and 45 per cent over all. This contrasts with developed countries where cardiovascular diseases and neoplasms are major causes of death accounting for 54 per cent and 21 per cent respectively, but only for 17 per cent and 7 per cent respectively in developing countries (WHO, 1992a). The high prevalence of infectious diseases in developing countries is linked to malnutrition, inadequate water supply and sanitation, poor hygiene practices and overcrowded living conditions. In addition, poor immunization coverage and resistance of disease vectors to pesticides are contributing factors.

Communicable diseases are usually classified according to the nature of the pathogen (e.g., the classification used as part of the United Nations Development Programme (UNDP)/ World Bank/WHO Special Programme for Research and Training in Tropical Diseases (WHO, 1990a). Many of the communicable diseases are associated with water and can be classified according to the various aspects of the environment that human intervention can alter (WHO, 1992a):

○ *Waterborne diseases* arise from the presence in water of human or animal faeces or urine infected by pathogenic viruses or bacteria and which are transmitted when the water is used for drinking or in the preparation of food (e.g., cholera and typhoid).

○ *Water-washed diseases* occur where scarcity of water make cleanliness difficult. This category includes the waterborne diseases as well as infestation with lice or mites, which are vectors of various forms of typhus.

○ *Water-based diseases* are those where water provides the habitat for the intermediate host organisms in which some parasites pass part of their life cycle (e.g., schistosomiasis and dracunculiasis). The host organism may be ingested when drinking water or eating fish or other freshwater foods.

○ *Water-related diseases* are those in which insect vectors of parasitic diseases rely on water as a habitat but transmission is not due to direct human contact with the water. Diseases in this category include malaria, filariasis, onchocerciasis, dengue, yellow fever and Japanese encephalitis.

○ *Water-dispersed infections* are those where the infectious agents (e.g., the bacterium *Legionella*) proliferates in fresh water and is inhaled into the respiratory tract with minute water droplets. These infections are primarily associated with developed countries.

Access to safe water, sufficient water supplies and adequate sanitation are useful indicators of the number of people at risk from waterborne, water-washed and water-related diseases. Provision of basic water and sanitation facilities cannot keep pace with the rate of urbanization in many areas. The WHO has estimated that 170 million urban and 770 million rural inhabitants lack access to safe water and over 1,700 million people lack adequate sanitation; 330 million urban and 1,390 million rural inhabitants. Most urban centres in Africa and Asia have no sewage systems at all, including many cities with 1 million or more inhabitants (WHO, 1992a). An example of a national assessment of risk from waterborne diseases is given in Box 5.1.

Environmental factors are most prominent for the tropical diseases which, in the majority of cases, are caused by infections from parasites requiring one or more intermediate hosts and vectors for their development (WHO, 1992a). Table 5.6 lists the water-related diseases and estimates their impact on mortality and morbidity and the size of the population at risk (WHO, 1992a).

Waterborne diseases are the largest single category of communicable diseases contributing to infant mortality in developing countries, i.e., greater than 1,500 million episodes of diarrhoea and some 4 million deaths per year. Few fatal cases of waterborne diseases are now recorded in developed countries and outbreaks are infrequent. Diarrhoea remains

Box 5.1 Drinking Water Quality and Waterborne Diseases in China

In 1983 in response to the International Drinking Water Supply and Sanitation Decade (1981-1990), the Chinese National Committee of Patriotic Health Campaign and the Ministry of Public Health jointly devised a programme for a 'Nationwide Survey on Drinking Water Quality and Waterborne Diseases'. Each of 28,800 sampling sites was tested for major water quality indices over a three-year period. The data were presented in the form of maps in the 'Drinking Water Atlas of China' (Li, 1990). The Atlas provides information on 13 water quality indices (colour, turbidity, pH, total hardness, iron, manganese, fluoride, arsenic, nitrate-nitrogen, chloride, oxygen consumed, sulphate, total coliforms). Maps are provided showing per capita consumption of drinking water as well as waterborne diseases and endemic diseases.

The survey classifies drinking water into three grades according to standards set by WHO, national conditions, and in reference to the 'Sanitary Standard for Drinking Water' of China:

○ Grade I which is very satisfactory, or sanitary safe water;
○ Grade II which is satisfactory, or sanitary allowable water;
○ Grade III which is unsatisfactory, or sanitary unsafe water.

If one or more of the 13 water quality indicators does not meet the standards given in the table for each grade, the water is classified as the next (less satisfactory) grade.

The drinking water is also categorized as community or non-community supplies. Community supplies are divided according to the levels of water treatment applied. Non-community supplies are divided into three groups: motor-pumped well water, hand-pumped well water, and that collected by hand from sources such as rivers, lakes or springs.

Grade III water is the most widely used on a national scale (70 per cent). Non-community water supplies are generally of a lower quality than community water sources. Between 54 and 65 per cent of the population using well-water supplies and 85 per cent of the population using hand drawn water supplies have Grade III water. This compares with only 21 per cent of the population with completely treated tap water and 47 per cent with partially treated tap water. A high proportion of the population with non-community supplies has water which does not meet the standard for coliforms as shown by the analysis of proportion of population by province in the mid-1980s.

Standards for each indicator and grade of drinking water in China

Indicator (unit)	Grade I	Grade II	Grade III
Colour (degree)	15	50	>50
Turbidity (degree)	5	25	>25
pH	6.5–8.5	6.0–9.0	<6.0 or >9.0
Iron (mg l^{-1})	0.3	1.0	>1.0
Manganese (mg l^{-1})	0.1	0.5	>0.5
Oxygen consumed (mg l^{-1})	3	6	>6
Hardness (CaCO$_3$, mg l^{-1})	450	700	>700
Chloride (mg l^{-1})	250	600	>600
Sulphate (mg l^{-1})	250	400	>400
Fluoride (mg l^{-1})	1.0	1.0	>1.0
Arsenic (mg l^{-1})	0.05	0.1	>0.1
Nitrate-nitrogen (N, mg l^{-1})	20	23	>23
Coliforms (MPN l^{-1})	<3	60	>60

Source: Li, 1990

Proportion of population supplied with Grade III drinking water in respect to total coliforms, mid-1980s and occurrences of waterborne diseases

Provinces	Community (%)	Non-community (%)	Waterborne Bacillary Dysentry 1959–1983 Outbreaks	Cases	Waterborne Typhoid 1958–1984 Outbreaks	Cases
All China	20–25	60–65	157	50,934	353	45,535
Beijing	5–10	30–35	16	2,481	13	504
Gansu	10–15	55–60			14	8,132
Guizhou	45–50	85–90	4	10,443	30	6,965
Henan	20–25	55–60	1	624		
Hubei	40–45	70–75	6	7,403	5	438
Hunan	35–40	85–90	25	3,925	42	3,636
Jiangsu	5–10	55–60	16	1,660	46	2,370
Jiangxi	20–25	60–65	3	338	4	323
Liaoning	0–5	15–20	17	8,625	22	2,863
Nei Mongol	5–10	40–45	22	4,777	50	4,681
Shaanxi	25–30	60–65			3	398
Shanghai	0	65–70	21	1,127	13	355
Shanxi	10–15	55–60	5	1,271	9	1,093
Sichuan	30–35	70–75	4	849	4	491
Yunnan	45–50	90–95	3	12,310	80	11,815
Zhejiang	25–30	70–75	15	1,019	16	851

Source: Li, 1990

Contamination with coliforms is believed to be the most serious health problem with respect to drinking water supplies in China (Li, 1990).

Analysis of the outbreaks of waterborne diseases by province between 1958 and 1984 shows that these outbreaks frequently occurred in provinces which have a large proportion of the population served by water with unsatisfactory coliform concentrations. The outbreaks of waterborne diseases are related to the overall sanitary conditions of the populations of the provinces, especially the drinking water sources which are polluted with human excreta and domestic refuse (Li, 1990).

Reference

Li, Z. (Ed.) 1990 *Drinking Water Atlas of China*, China Cartographic Publishing House, Beijing.

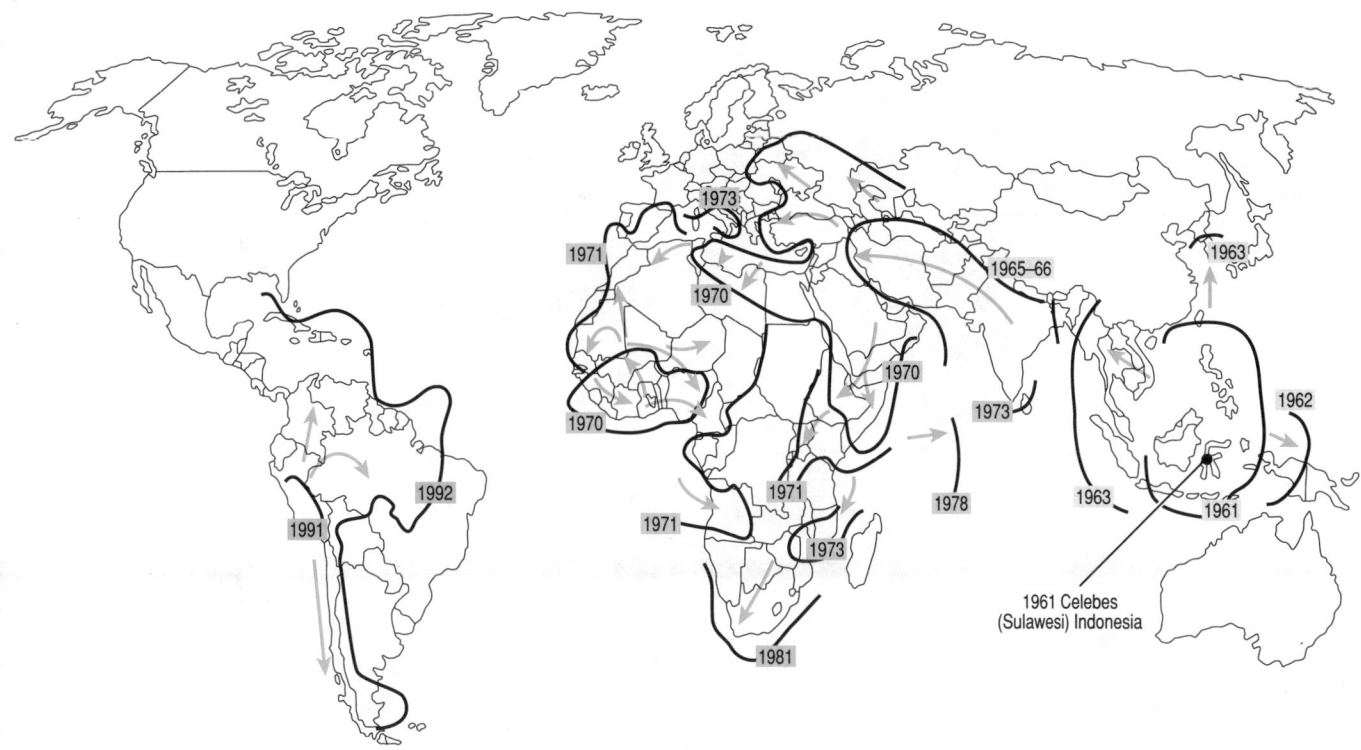

FIGURE 5.3 The progressive spread of cholera from Indonesia in 1961 to the known global distribution in 1992
Source: Based on WHO 1992a,b, 1993a

one of the most pressing health problems and is usually caused by one of a number of waterborne pathogens including *Giardia*, vibrio and rotavirus, although it can also result from non-enteric infection (see the previous edition of this report (UNEP, 1991) for detailed coverage of diarrhoea).

Efforts to alleviate problems associated with waterborne diseases in urban areas by the health sector are often limited because it is difficult to address all the factors affecting people's health. They are also hindered by inadequate health data and a poor understanding of the factors involved (WHO, 1991a). Estimates of the likely health outcomes of improving the water supply and of developing sanitary disposal of excreta for a range of diseases are given below (WHO, 1992a):

Disease	Projected reduction in morbidity (%)
Cholera, typhoid, leptospirosis, scabies, guinea-worm infection	80–100
Trachoma, conjunctivitis, yaws, schistosomiasis	60–70
Tularaemia, paratyphoid, bacillary dysentery, amoebic dysentery, gastroenteritis, louse-borne diseases, diarrhoeal disease, ascariasis, skin infections	40–50

Source: WHO, 1992a

This edition of the 'UNEP Environmental Data Report' focuses on cholera within the waterborne and water-washed category; dracunculiasis within the water-based category; and trypanosomes and malaria within the water-related insect vector category. Additional information on these and other communicable diseases including onchocerciasis, schistosomiasis, yellow fever and plague, as well as diarrhoeal diseases, is also available in earlier editions (UNEP 1987, 1989, 1991). Environmental aspects of neoplastic diseases are also discussed in this edition.

Cholera

Cholera originated in Asia and is one of the oldest diseases of humans. During the 19th century it reached Europe for the first time and caused six major pandemics, earning its reputation as a killer disease (WHO, 1991b). After the sixth pandemic, cholera returned to Asia. The seventh pandemic began in 1961 in Indonesia and the subsequent global spread of cholera from 1961 to 1992 is shown in Figure 5.3. The disease ultimately became endemic in many areas, particularly in coastal areas, where the temperature, humidity, rainfall and population density are conducive to its persistence.

The most recent estimates of the incidence of cholera, detailed by WHO region for 1989 to 1992, are reported in Table 5.7. The rise in the number of cases in Africa and South America during 1991 is particularly striking. More countries reported cholera in 1991 than at similar times in previous years. The African countries reported nine times as many cases in 1991 than in any other year of this pandemic. The overall case-fatality rate in 1991 (10.2 per cent) has only been exceeded twice since cholera struck Africa, and was 10 times higher than that reported in the Americas that year (WHO, 1991c).

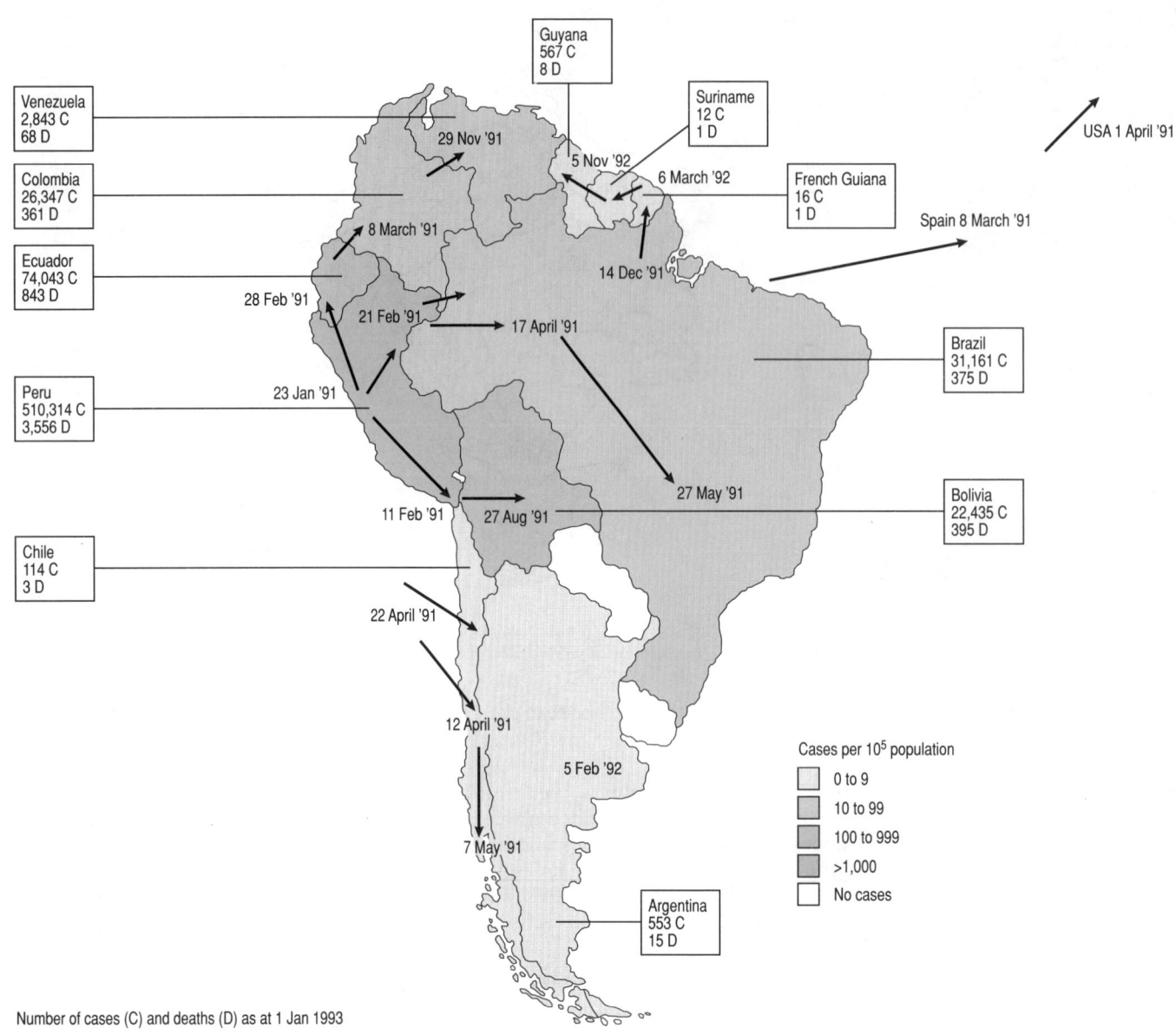

Venezuela
2,843 C
68 D

Colombia
26,347 C
361 D

Ecuador
74,043 C
843 D

Peru
510,314 C
3,556 D

Chile
114 C
3 D

Guyana
567 C
8 D

Suriname
12 C
1 D

French Guiana
16 C
1 D

Brazil
31,161 C
375 D

Bolivia
22,435 C
395 D

Argentina
553 C
15 D

USA 1 April '91

Spain 8 March '91

29 Nov '91
5 Nov '92
6 March '92
8 March '91
28 Feb '91
21 Feb '91
17 April '91
14 Dec '91
23 Jan '91
11 Feb '91
27 Aug '91
27 May '91
22 April '91
12 April '91
5 Feb '92
7 May '91

Cases per 10^5 population

0 to 9
10 to 99
100 to 999
>1,000
No cases

Number of cases (C) and deaths (D) as at 1 Jan 1993

FIGURE 5.4　The spread of cholera in South America, 1991–1992.
Source: Adapted from WHO, 1993a

The situation improved substantially in 1992 (Table 5.7). Since 1990, cholera has been occurring more frequently in peri-urban and urban areas, resulting in larger epidemics which have been overwhelming the available health care facilities (WHO, 1993a). The recurrent unplanned movements of populations, often resulting in crowded refugee camps, are contributing to the spread of the disease in Africa. A single case in November 1992 in a refugee camp in Zimbabwe (the first case since 1986) resulted in 2,039 cases and 105 deaths amongst refugees and indigenous populations by the end of the year (WHO, 1993a).

Towards the end of January 1991 a cholera epidemic broke out on the coast of Peru (WHO, 1992c). Its location was close to the seaports on the Pacific Coast with suggestions that it could have been introduced by maritime traffic from the Pacific. The disease struck Peru with an intensity unprecedented during the present pandemic, causing on average over 1,700 cases per day during the first four months and up to 20,000 cases per week during the seventh, eighth and ninth weeks of the epidemic. The scarcity of safe drinking water, the inadequate sanitation, the close relationship between the inhabitants, and polluted water have continued to promote the spread and persistence of the epidemic. With over 300,000 cases in 1991 and over 200,000 cases in 1992, Peru has reported the largest total number of cholera cases in the world to WHO. However, the national case-fatality rate for the disease has remained very low (0.95 per cent).

The epidemic has spread rapidly across to all but two

countries in Central and South America (Figure 5.4), although it has been decreasing in intensity as it proceeds further away from its Peruvian focus. It has now been introduced into Ecuador, Colombia, Brazil, Chile and other countries, although the rate of spread and intensity of the epidemic has varied, showing clear seasonal peaks in some countries (WHO, 1993a). The strain of cholera responsible for this epidemic (*Vibrio cholerae* 01 biotype El Tor) has been shown to survive in the aquatic environment for long periods of time, thereby increasing the likelihood that it will become established as an endemic disease in Latin America (WHO, 1993a). Chemoprophylaxis has been practised in the various countries at different times as summarized in WHO (1992c).

In December 1992, a new epidemic of a cholera-like disease affecting mostly adults broke out in Bangladesh (Cholera Working Group, 1993). By March 1993 it had accounted for 1,473 deaths. This new epidemic is thought to be of a new strain of *Vibrio cholerae* (known as *V. cholerae* 0139 Bengal), which explains the apparent lack of previous immunological experience in the population. Recent studies isolating *V. cholerae* from water samples in Bangladesh suggest that this new strain is hardier than the usual *V. cholerae* 01 strain and that its apparent transmission through water contributes to its pandemic potential (Islam et al., 1993).

Dracunculiasis

The presence in water of organisms that are vectors of pathogens gives rise to risks of a different nature where surface waters are used directly for drinking. Dracunculiasis (guinea worm disease) is caused by the ingestion of a copepod crustacean (*Cyclops*) present in surface waters. The copepod contains the larvae of a worm, *Dracunculus medinensis*, which invade human tissues and develop into worms up to 1 m long giving rise to a debilitating chronic disease. The total cases reported from endemic countries (Table 5.8) showed another year of overall decline in 1992. However, incidence figures published by countries are considered to be gross underestimates (WHO, 1990b). Other estimates suggest that 19 million people in 17 African countries and parts of India and Pakistan are affected each year (WHO, 1991d,e).

A strategic plan to eradicate dracunculiasis based on filtration of unsafe waters, chemical control of the crustacean host and education of villagers, was one of the goals for the 1990s specifically endorsed by the heads of states of 71 countries at the World Summit for Children (WHO, 1991d). Following the development of national guinea worm eradication programmes, numbers of cases of dracunculiasis have declined substantially. The latest information available from WHO reports that in 1992 the total number of cases of dracunculiasis was estimated at less than 3 million, which is a substantial reduction from earlier estimates of 10 million or more (WHO, 1993b). Indeed, most of the endemic countries have as a target the complete eradication of the disease by the end of 1995 (WHO, 1992d).

The efforts of national eradication programmes in Pakistan and India involving hygiene education, water filtration and pesticide application to reduce populations of the copepod hosts, show that effective intervention in endemic areas is

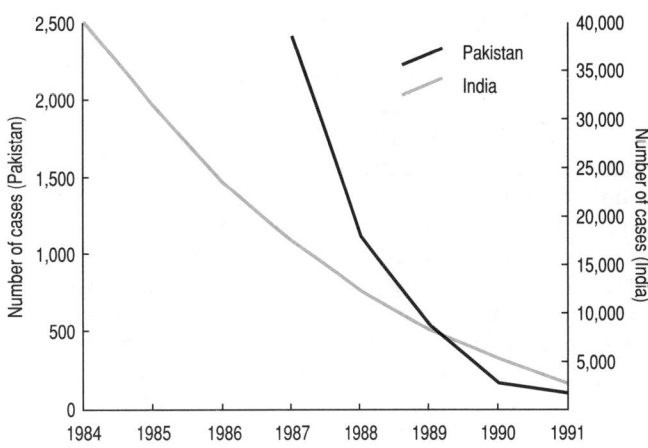

FIGURE 5.5   Numbers of new cases of dracunculiasis in India and Pakistan, 1984–1991
Source: Data from Table 5.8

worthwhile (WHO, 1991e). The decline in reported cases, especially since the late 1980s, is evident in Table 5.8 and Figure 5.5. As only seven endemic villages remained in Pakistan in 1992, the disease will soon be certified by WHO as eradicated from Pakistan (WHO, 1993c). In India the number of cases declined by 45 per cent and the number of endemic villages by 57 per cent in 1992, compared with 1991 (WHO, 1993c). Twelve districts have now been deleted from the endemic register after three years of surveillance during which no cases were found. It has therefore been concluded that, with the remarkable decrease in the effective and potential risks of dracunculiasis to the population, the goal of eradication is not very far off (WHO, 1992d).

Chagas Disease

Chagas disease, also known as South American trypanosomiasis, is caused by the protozoan parasite *Trypanosoma cruzi* and is an example of a disease where the vector relies on water as a habitat. Transmission is by the blood-sucking insect hosts of the sub-family Triatominae. These insects colonize poor housing, living in the roofs, crevices in walls and the floor. Infection may also occur through blood transfusions from infected donors. In endemic areas of Argentina and Brazil between 6 and 20 per cent of blood donors are sero-positive for *T. cruzi* and up to 63 per cent are sero-positive in Bolivia. Approximately 90 million people in Central and South America are at risk of the disease with around 16–18 million cases being reported (WHO, 1990a,b). Thus one-quarter of the total population of Latin America are at risk of infection. In Brazil alone, the endemic area covers 44.5 per cent of a vast area of 3×10^6 km^2 (WHO, 1992d). Prevalence of Chagas disease in Latin America is shown in Table 5.9.

About 10 per cent of infected people develop chronic Chagas cardiopathy, which is responsible in some areas for up to 10 per cent of deaths among adults. However, there are no accurate figures, because the disease mostly affects poor rural areas that are often without access to health services, and where there are many deficiencies and inaccuracies in the reporting of causes of deaths (Vlassoff and Tanner, 1992). New intense foci of transmission have appeared in the peri-urban slums of various Latin American cities where poor housing, overcrowding and inadequate health services contribute to transmission. Thus the migration of many rural people infected with Chagas disease to urban areas has made it a significant cause of morbidity and mortality (WHO, 1992a). In the Federal District of Brazil, for example, which is considered to be free of the insect vector of Chagas disease, it was found to be responsible for about 10 per cent of deaths among people between 25 and 64 years of age (i.e., roughly equivalent to cancer, heart disease and stroke in its contribution to causes of death).

Malaria

Malaria remains one of the major health problems of the world, being present in more than 100 countries throughout the tropics, but also occurring in temperate regions. The causative agents are single-celled protozoan parasites of the genus *Plasmodium*. Four species infect people: *P. falciparum* throughout tropical Africa, Asia and Latin America; *P. vivax* world-wide in tropical and some temperate zones; *P. ovale* mainly in tropical West Africa; and *P. malariae* world-wide but with patchy distribution (WHO, 1990a,b). Transmission is through the blood-sucking mosquito host, which relies on water for part of its life cycle.

The spread of malaria has been facilitated by the extension of perennial irrigation agriculture. The presence of large bodies of standing water in the vicinity of human dwellings, providing ideal breeding grounds for the mosquito host, increases the likelihood of infection, especially in Asia and Latin America. In sub-Saharan Africa the infection rate is already so high that it does not appear to be possible for the situation to worsen (WHO, 1992a).

Over two billion people, or 40 per cent of the population of the world, are at risk of the disease. It has been estimated that 270 million people are infected, with 110 million clinical cases a year. More than a million deaths a year are considered as a realistic estimate, three-quarters of which occur in children aged under five years (WHO, 1992a). Malaria is therefore one of the main killers of children in that age group.

Numbers of malaria cases reported to WHO are listed in Table 5.10. Reporting is fragmentary and irregular in many countries, particularly in Africa, but an upward trend in the number of malaria cases has been reported in the Americas and some Asian countries (WHO, 1990b). About 83 per cent of the total number of cases reported annually to WHO (excluding the Africa Region) are concentrated in nine countries: Afghanistan, Brazil, China, India, Mexico, the Philippines, Sri Lanka, Thailand and Viet Nam. In Africa, south of the Sahara, between 2 and 7 million cases are reported each year, but estimates suggest 90 million clinical malaria cases may occur every year, and that prevalence of infection may be in the order of 250 million parasite carriers (WHO, 1990b).

Malaria has reached epidemic proportions in a number of countries. The incidence of malaria can only be drastically reduced by effective and large-scale action, including selective indoor spraying of insecticide and insect-repelling products to control the host mosquito, the use of personal protective measures (by children in particular), early treatment and better management of water resources. Resistance of the parasite to drugs, resistance of the vectors to insecticides, climatic changes and the rudimentary nature of health services in some areas have all combined to increase the malaria problem. Civil wars and mass movements of refugees also add to the problem.

Neoplastic Diseases

Neoplastic diseases (cancers) are a multi-causal, multi-stage group of diseases, the mechanisms of which are still only partially known. Early examples of increased cancer risk in humans tended to arise from high exposure situations, particularly occupational exposure involving chemical, physical and ionizing substances. Attempts to control exposure to carcinogens is becoming increasingly complex as more information is gathered on the carcinogenic risk of chemicals and other factors including multiple and low-level exposures and environmental factors. Within the framework of WHO, IARC conducts major programmes of research and data collection concentrating particularly on the epidemiology of cancer and the study of potential carcinogens in the human environment (Vainio et al., 1992).

Although specific incidents where contamination levels are exceptionally high (for examples see Part 9: Environmental Disasters) suggest direct links between environmental contamination and human health, it is difficult to establish direct links between general levels of environmental contamination and population health as a whole. The links are especially difficult to ascertain because individuals and populations are potentially exposed to multiple contaminants (sometimes at barely detectable concentrations) and the possible interactions with their social, occupational, nutritional and behavioural patterns are difficult to determine. Extrapolation from industrial exposures indicates that urban exposure to smoke (in conjunction with other compounds) may have contributed to a 10 per cent increase in lung cancers in cities (Doll, 1992).

Although it is commonly claimed that increases in cancer and congenital abnormalities are due to industrial and agricultural development, there is little evidence to support this for the general population (Doll, 1992). The lack of direct relationships should, however, not deter policy makers from considering the real and potential risks associated with toxic chemicals in the environment, especially as many of the risks, particularly long-term risks, are still to be investigated. Additional risks may already exist from the synthetic chemicals with unknown carcinogenic potential which have already been released into the environment (Landrigan, 1992). Evidence suggesting that atmospheric pollution and ozone-layer depletion (see Part 1: Environmental Pollution, Atmosphere) may lead to increased risks of mortality from skin cancer is discussed in Box 5.2.

Attempts have been made to ascertain the background cancer risk from global pollutants (particularly polychlorinated biphenyls (PCBs), dioxins and some pesticides) as well as other

BOX 5.2   Malignant Melanoma of the Skin

In view of the ozone layer depletion arising from the impact of chlorofluorocarbons in the atmosphere (see Part 1: Environmental Pollution, Atmosphere) concern has been expressed about the likely health impacts from increased ultraviolet (UV) exposure (IARC, 1992). Models based on melanoma mortality in white-skinned populations in the USA over a 30-year period together with calculated UV flux data from satellite measurements, have projected a 2–7 per cent increase in melanoma mortality, depending on latitude (Pitcher and Longstreth, 1991). The data produced on melanoma mortality are comparable with earlier estimates by Elwood (1989) which were also based on a projected 10 per cent increase in solar radiation.

Descriptive studies, of white-skinned populations in North America, Australia and several other countries, show a positive correlation between incidence of, and mortality from, melanoma of the skin and residence at lower latitudes (IARC, 1992). The neoplasms occur predominantly on the skin of the face and neck, which is most commonly exposed to sunlight although differences occur between males and females.

Early studies, discussed in the second edition of the this report (UNEP, 1989), indicated that malignant melanoma of the skin is one of the most common concerns in white-skinned populations. Indeed, skin melanoma has been second only to lung cancer in its rate of increase in incidence over the past 40 years in the USA. Rapid increases in both incidence and mortality have been observed in people of each sex almost

everywhere in the world. The largest average annual increases are about 5 per cent in USA, 6 per cent in the Nordic countries, 7 per cent in New Zealand and as much as 11 per cent in the Jewish population of Israel (IARC, 1990).

Rates of melanoma in dark-skinned populations are low. In the black population of the USA there has been almost no change in incidence or mortality rates (IARC, 1990). The age-standardized incidence ratio in India is 0.2 per 100,000 individuals compared with about 30 in Queensland, Australia. Rates are lower in Hispanics than in other whites illustrating that ethnic origin, skin pigmentation, socio-economic status and recreational outdoor exposure are relevant in the determination of risk.

Large, carefully conducted population-based studies have been carried out in Western Australia, Queensland, western Canada and Denmark. Generally, the results are consistent with positive associations with residence in sunny environments throughout life, in early life and even for short periods in early adult life, although the assessment of exposure is difficult and complex as is mentioned in Part 1: Environmental Pollution, Human Exposure.

Data on incidence and mortality rates of melanoma in migrants in Australia, New Zealand, the USA, Israel and other countries with high exposure to the sun, provide useful results. For example, most migrants to Australia come from higher latitudes which have lower levels of exposure to the sun than resident Australians. Residents of Australia, mostly of British origin, experienced incidence and mortality rates of

melanoma approximately twice those of British immigrants. This increased risk is related to the duration of residence in the new country and to age at immigration. Detailed case-control studies within countries have been evaluated by IARC (1992). Recent evaluations conclude that there is sufficient evidence in humans for carcinogenicity of solar radiation (UVR) and that ultraviolet A radiation (UVA) and ultraviolet B radiation (UVB) are probably carcinogenic (IARC, 1992). The potential for reducing mortality from malignant melanoma of the skin has also been considered by IARC (1990). It was concluded that surveillance of populations at high risk may prove to be cost effective.

References

Elwood, J. M. 1989 Epidemiology of melanoma: its relationship to ultraviolet radiation and ozone depletion. In: *Ozone Depletion: Health and Environmental Consequences*, R. R. Jones and T. Wiley (Eds), John Wigley, New York, 169–189.
IARC 1990 *Cancer: Causes, Occurrence and Control*, L. Tomatis (Ed.), IARC Scientific Publication 100, International Agency for Research on Cancer, Lyon.
IARC 1992 *IARC Monographs on the Evaluation of Carcinogenic Risks to Humans; Solar and Ultraviolet Radiation, Vol 55*, International Agency for Research on Cancer, Lyon.
Pitcher, H. M. and Longstreth, J. D. 1991 Melanoma mortality and exposure to ultraviolet radiation: an empirical relationship, *Environment International* **17**, 17–21.
UNEP 1989 *United Nations Environment Programme Environmental Data Report 1989/90*, Basil Blackwell, Oxford.

chemicals to which populations are commonly exposed during their daily lives (e.g., volatile organics). Travis and Hester (1991) used USA Environmental Protection Agency (EPA) methods to ascertain background cancer risk to 11 contaminants commonly found in indoor air, water and the food chain. They concluded that measured exposures suggested the background cancer risk accounted for 1–2 per cent of actual annual cancer deaths. New approaches, including at the molecular level, are continually being sought which can link environmental exposure and cancer risk. For example, exposure to environmental pollutants, such as polycyclic-aromatic-hydrocarbons (PAH), may cause modification of deoxyribose nucleic acid (DNA) and chromosomal mutations which can be linked to cancer risk (Perera et al., 1992). Such molecular connections may lead to the development of biological markers for early detection of the risk and onset of the disease.

A selection of recent data on cancer incidence is shown in Table 5.11. Cancer is the second most frequent cause of death in developed countries; accounting for 21 per cent of all

deaths. In developing countries it accounts for only 7 per cent of all deaths (WHO, 1992a). However, as the incidence of cancer is strongly dependent on age, the different age structures of the population in the two groups of countries may be largely responsible for these differences in mortality. Differences of lifestyle and environmental exposure in the two groups may also give rise to different types of risks. The rates of new occurrences of major neoplasms in developed and developing countries are shown in Figure 5.6. A substantial fraction of the excess mortality from cancer of the colon, rectum and breast in developed countries is probably related to dietary factors, and the very high mortality from lung cancer reflects the high frequency of smoking (IARC, 1990). In developing countries the high exposure to hepatitis B and to aflatoxins in foodstuffs contributes to the high incidence of liver cancer.

Geographical analyses of incidence rates of cancer have been published for a number of countries. An atlas of cancer mortality using data from 22 cancer sites from 170 areas in the

Number of new cases per year (10³)

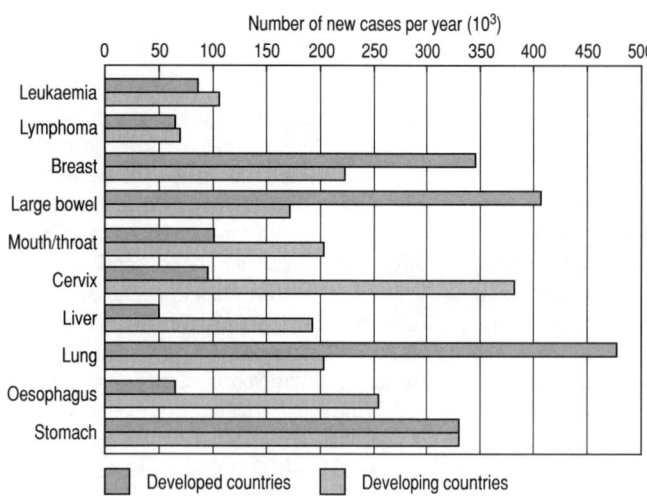

FIGURE 5.6 The rate of new occurrences of major neoplasms in developed and developing countries
Source: Anon, 1991

former USSR will be available shortly (Dowd, 1992). Recent results for Canada reveal higher rates for cancers of the tongue, mouth, pharynx, lung and bladder in Quebec (probably related to higher levels of tobacco use in that province), a relatively high rate of stomach cancer in Newfoundland (consistent with greater use of salted and smoked foods), and higher rates of melanoma of the skin in Ontario and British Columbia (possibly related to variations in exposure to sunlight) (National Cancer Institute of Canada, 1992; Berkel et al., 1992).

References

Anon 1991 Cancer will overwhelm the third world, *New Scientist* **1799**(132), 9.

Berkel, J. et al. 1992 *Atlas of Cancer Incidences in Western Canada, 1984–1988*, Alberta Cancer Board, Health and Welfare Canada, Alberta.

Cholera Working Group 1993 Large epidemic of cholera-like disease in Bangladesh caused by *Vibrio cholerae* 0139 synonym Bengal, *The Lancet*, **343**, 387–390.

Doll, R. 1992 Health and the environment in the 1990s. *American Journal of Public Health*, **82**(7), 933–941.

Dowd, J. E. 1992 An overview of relevant data sources in the former USSR for studies in demographic trends, aging and non-communicable disease problems, *World Health Statistics Quarterly* **45**, 68–74.

IARC 1990 *Patterns of Cancer in Five Continents*, IARC Scientific Publications 102, International Agency for Research on Cancer, Lyon.

Islam, M. S., Hasan, M. K., Miah, M. A., Qadri, F., Yunus, M., Sack, R. B. and Albert, M. J. 1993 Isolation of *Vibrio cholerae* 0139 Bengal from water in Bangladesh, *The Lancet*, **342**, 430.

Landrigan P. J. 1992 Commentary: Environmental disease – a preventable epidemic, *American Journal of Public Health*, **82**(7), 941–943.

National Cancer Institute of Canada 1992 *Canadian Cancer Statistics 1992*, Toronto.

Perera, F. P., Hemminki, K., Gryzbowska, E., Motykiewicz, G., Michalska, J., Santella, R. M., Young, T–L., Dickey, C., Brandt-Rauf, P., DeVivi, I., Blaners, W., Tsai, W–Y. and Chorazy, M. 1992 Molecular and genetic damage in humans from environmental pollution in Poland, *Nature*, **360**, 256–258.

Travis, C. C. and Hester, S. T. 1991 Global chemical pollution, *Environmental Science and Technology*, **25**(5), 815–819.

UNEP 1987 *United Nations Environment Programme Environmental Data Report 1987/88*, Basil Blackwell, Oxford.

UNEP 1989 *United Nations Environment Programme Environmental Data Report 1989/90*, Basil Blackwell, Oxford.

UNEP 1991 *United Nations Environment Programme Environmental Data Report 1991/92*, Basil Blackwell, Oxford.

Vainio, H., Magee, P., McGregor, D. and McMichael, A. J. 1992 *Mechanisms of Carcinogenesis in Risk Identification*, IARC Scientific Publications 116, International Agency for Research on Cancer, Lyon.

Vlassoff, C. and Tanner, M. 1992 The relevance of rapid assessment to health research and interventions, *Health Policy and Planning* **7**, 1–9.

WHO 1990a *Tropical Diseases 1990*, World Health Organization, Geneva.

WHO 1990b *Global Estimates for Health Situation Assessments and Projections 1990*, WHO/HST/90.2, World Health Organization, Geneva.

WHO 1991a *Environmental Health – Urban Development*, WHO Technical Report Series 807, World Health Organization, Geneva.

WHO 1991b Cholera, *Weekly Epidemiological Record* **9**, 61–63.

WHO 1991c Cholera in Africa, *Weekly Epidemiological Record* **42**, 305–311.

WHO 1991d Dracunculiasis, *Weekly Epidemiological Record* **66**, 225–232.

WHO 1991e Dracunculiasis eradication, update 1990, *Weekly Epidemiological Record* **66**, 41–43.

WHO 1992a *Our Planet, Our Health*, Report of the WHO Commission on Health and Environment, World Health Organization, Geneva.

WHO 1992b *Choléra 1991 – Vieil ennemi, nouveau visage*, *World Health Statistics Quarterly* **45**, 208–219.

WHO 1992c Cholera in the Americas, *Weekly Epidemiological Record* **67**, 33–40.

WHO 1992d *World Health Statistics Annual 1991*, World Health Organization, Geneva.

WHO 1993a Cholera in 1992, *Weekly Epidemiological Record* **21**, 149–155.

WHO 1993b Global health situation IV. Selected infections and parasitic diseases due to identified organisms, *Weekly Epidemiological Record* **68**, 43–44.

WHO 1993c Dracunculiasis global surveillance summary 1992, *Weekly Epidemiological Record*, **68**(18), 125–132.

TABLE 5.1 Trends in life expectancy and infant mortality, 1980–1990

Region/country	Infant mortality rate[a]				Life expectancy at birth (years)			
	1980–82	1983–85	1986–87	1990	1980–82	1983–85	1986–87	1990
WORLD	**73.7**	**68.5**	**61.9**	**63**	**59.9**	**61.0**	**61.9**	**63**
More developed regions[b]	14.8	13.3	12.5	12	73.2	73.8	74.3	75
Less developed regions[c]	91.1	84.7	79.8	77	55.9	57.2	58.2	60
AFRICA								
Algeria	95.2	83.8	76.8	74	59.6	61.6	63.3	64
Angola	152.1	145.0	139.3	137	41.4	42.8	44.0	46
Benin	102.4	97.0	92.0	90	43.4	44.6	45.6	48
Botswana	77.8	73.3	68.8	67	55.3	56.8	58.1	60
Burkina	153.1	145.9	140.0	138	44.6	45.8	46.8	49
Burundi	131.7	126.2	121.1	119	45.0	46.1	47.1	51
Cameroon	105.6	100.4	95.8	94	49.2	50.7	52.0	52
Cape Verde	57.7	49.9	45.7	44	62.9	64.7	65.7	63
Cent. African Rep.	116.4	111.0	106.0	104	45.9	47.1	48.1	47
Chad	146.2	139.6	134.2	132	42.4	43.8	45.0	47
Comoros	112.8	106.3	101.1	99	51.4	52.6	53.6	54
Congo	82.2	78.6	74.6	73	50.1	51.3	52.3	50
Côte d'Ivoire	108.5	102.4	97.9	96	49.7	51.0	52.0	54
Djibouti	135.2	128.9	123.7	122	44.4	45.6	46.6	–
Egypt	119.8	100.0	75.0	65	55.8	57.3	58.6	63
Equatorial Guinea	140.7	134.1	128.7	127	43.4	44.6	45.6	48
Ethiopia	156.0	152.5	141.6	137	40.6	41.2	43.2	43
Gabon	114.8	109.2	104.8	103	48.4	49.8	51.0	53
Gambia	158.0	150.8	145.1	143	40.4	41.6	42.6	45
Ghana	99.6	95.7	91.3	90	51.7	52.6	53.6	56
Guinea	160.1	153.5	147.6	145	40.0	41.1	42.1	44
Guinea-Bissau	167.2	159.8	153.8	151	38.9	40.1	41.1	47
Kenya	82.4	77.6	73.6	72	55.2	56.7	57.9	60
Lesotho	114.5	107.7	102.2	100	52.7	54.2	55.5	58
Liberia	156.9	149.6	144.4	142	50.9	51.9	52.7	56
Libya	99.6	92.2	85.0	82	57.5	58.9	60.1	63
Madagascar	136.0	127.0	122.0	120	50.9	52.1	53.1	55
Malawi	167.2	159.2	153.0	150	44.4	45.6	46.6	48
Mali	183.3	176.7	171.2	169	41.4	42.6	43.6	46
Mauritania	140.7	134.1	128.7	127	43.4	44.6	45.6	48
Mauritius	31.4	26.8	24.1	23	66.2	67.4	68.5	70
Morocco	100.5	92.2	85.0	82	57.4	59.0	60.2	63
Mozambique	155.4	149.9	143.9	141	44.2	45.1	46.1	48
Namibia	119.0	113.0	108.0	106	53.0	54.5	55.7	58
Niger	149.2	142.4	136.8	135	41.9	43.1	44.1	46
Nigeria	117.2	111.4	106.8	105	47.9	49.1	50.1	52
Rwanda	134.4	128.9	123.7	122	46.0	47.1	48.1	50
Senegal	101.5	94.0	89.0	87	44.5	45.9	46.9	47
Sierra Leone	170.4	162.8	156.7	154	38.4	39.6	40.6	43
Somalia	144.6	139.6	134.2	132	42.7	43.6	44.6	46

Continued

TABLE 5.1 Continued

Region/country	Infant mortality rate[a]				Life expectancy at birth (years)			
	1980–82	1983–85	1986–87	1990	1980–82	1983–85	1986–87	1990
South Africa	86.7	80.1	74.7	72	57.3	58.7	59.9	62
Sudan	121.6	114.7	110.0	108	47.0	48.4	49.4	51
Swaziland	132.3	125.7	120.2	118	52.1	53.8	55.0	57
Tanzania	118.0	112.2	107.5	106	50.4	51.6	52.6	53
Togo	108.5	101.6	96.0	94	49.7	51.2	52.5	55
Tunisia	76.0	65.2	55.8	52	62.2	63.8	65.1	67
Uganda	112.6	109.2	104.6	103	48.7	49.6	50.6	53
Zaire	95.3	89.3	84.8	83	49.4	50.6	51.6	54
Zambia	89.8	85.5	81.4	81	50.8	52.0	53.0	55
Zimbabwe	79.0	73.2	68.1	66	55.2	56.6	57.8	60
NORTH AMERICA								
Barbados	18.6	13.8	11.8	11	72.6	73.6	74.3	74
Canada	9.8	8.3	7.6	7	75.4	76.2	76.6	77
Costa Rica	23.0	19.4	18.4	18	72.7	73.9	74.4	75
Cuba	18.8	16.4	15.4	15	73.8	74.4	75.0	74
Dominican Rep.	77.7	72.0	67.0	65	63.4	64.6	65.5	67
El Salvador	81.6	73.8	66.8	64	57.0	58.5	61.1	66
Guatemala	73.6	66.7	61.2	14	58.2	59.8	61.3	64
Haiti	111.9	104.7	99.2	97	52.2	53.4	54.4	56
Honduras	85.9	78.1	71.6	69	60.6	62.5	63.5	65
Jamaica	20.4	17.7	17.2	17	71.0	71.7	72.3	74
Mexico	52.7	47.9	44.4	43	66.8	67.8	68.6	70
Nicaragua	81.1	71.8	64.8	62	58.8	60.9	62.6	66
Panama	27.4	24.8	23.5	23	70.4	71.3	71.8	72
Trinidad & Tobago	20.6	17.5	16.3	16	69.0	70.0	70.7	71
USA	12.2	10.9	10.1	10	74.1	74.8	75.3	76
SOUTH AMERICA								
Argentina	37.5	34.8	32.8	32	69.4	69.9	70.4	71
Bolivia	128.2	119.8	112.8	110	50.1	51.5	52.7	–
Brazil	73.4	68.6	64.6	63	63.0	63.9	64.6	66
Chile	29.9	22.1	20.6	20	69.9	71.2	71.4	71
Colombia	46.4	40.7	40.2	40	66.2	67.5	68.1	65
Ecuador	73.4	67.7	64.6	63	63.4	64.6	65.5	66
Guyana	64.2	60.9	57.4	56	61.0	61.8	62.8	70
Paraguay	46.2	44.1	42.6	42	70.4	71.3	71.8	67
Peru	100.5	95.5	90.3	88	58.1	59.4	60.8	64
Suriname	40.8	37.6	34.4	33	66.6	67.7	68.5	70
Uruguay	36.2	30.8	26.2	24	70.6	71.3	71.8	71
Venezuela	40.2	38.1	36.6	36	68.5	69.1	69.5	70
ASIA								
Afghanistan	183.0	179.7	174.3	172	40.3	40.8	41.3	43
Bahrain	26.8	20.2	17.2	16	68.4	69.3	70.1	71
Bangladesh	130.7	125.5	121.0	119	48.0	49.2	50.3	52
Bhutan	141.0	135.5	130.3	128	45.3	46.5	47.3	49
Cambodia	190.9	150.8	135.7	130	39.8	45.0	47.5	50

Continued

TABLE 5.1 Continued

Region/country	Infant mortality rate[a]				Life expectancy at birth (years)			
	1980–82	1983–85	1986–87	1990	1980–82	1983–85	1986–87	1990
China	40.0	37.2	33.8	32	67.2	68.3	69.1	70
Cyprus	17.3	14.9	12.8	12	74.6	75.2	75.6	76
India	115.0	106.8	101.1	99	54.6	56.1	57.4	60
Indonesia	94.5	85.5	78.0	75	55.2	57.4	59.4	58
Iran	84.6	70.2	57.2	52	60.0	62.0	64.3	67
Iraq	78.8	74.5	70.3	69	62.1	62.9	63.6	66
Israel	15.3	13.5	12.4	12	74.0	74.7	75.2	76
Japan	7.2	6.0	5.3	5	76.5	77.3	78.0	78
Jordan	57.4	51.2	46.3	44	62.9	64.3	65.4	67
Korea	31.3	28.2	25.8	25	67.0	68.2	69.0	70
Korea, Dem.	31.3	29.1	28.2	28	67.0	68.3	69.3	70
Kuwait	26.3	21.4	18.8	18	71.0	72.0	72.6	73
Laos	125.9	118.4	112.4	110	45.2	46.7	48.0	50
Lebanon	48.0	48.1	48.4	48	65.0	65.0	65.0	68
Malaysia	29.7	26.8	24.9	24	67.2	68.4	69.2	70
Mongolia	80.8	75.0	70.3	68	58.0	59.5	60.8	65
Myanmar	83.0	77.1	72.2	70	56.7	58.3	59.5	62
Nepal	141.0	135.5	130.3	128	47.5	49.0	50.3	53
Oman	73.5	54.0	44.0	40	57.2	60.5	62.9	58
Pakistan	123.0	116.7	111.1	109	53.2	54.7	56.0	59
Philippines	51.6	48.9	46.2	45	51.6	48.9	46.2	–
Qatar	40.4	35.8	32.6	31	66.5	67.4	68.2	70
Saudi Arabia	89.5	80.8	73.7	71	59.9	61.5	62.8	65
Singapore	10.8	9.6	8.7	8	71.5	72.3	73.1	73
Sri Lanka	37.7	32.9	29.5	28	68.2	69.3	70.0	71
Syria	62.2	55.7	50.6	48	61.8	63.3	64.5	67
Thailand	42.7	34.3	29.8	28	62.3	63.4	64.6	67
Turkey	107.4	94.1	80.9	76	61.2	62.3	63.6	66
United Arab Em.	33.5	30.0	27.3	26	68.0	68.9	69.6	71
Viet Nam	80.1	72.3	66.6	64	58.0	59.7	61.0	63
Yemen	138.9	130.1	123.0	120	46.8	48.3	49.5	53
EUROPE								
Albania	46.4	43.0	40.0	39	70.0	70.8	71.4	72
Austria	13.5	11.8	11.1	11	72.8	73.5	74.1	75
Belgium	11.9	10.7	9.9	10	73.3	74.1	74.6	75
Bulgaria	18.9	17.1	16.5	16	71.3	71.6	71.9	73
Czechoslovakia	17.1	15.9	15.5	15	70.6	70.8	71.1	72
Denmark	8.3	7.7	7.3	7	74.4	74.7	75.2	–
Finland	7.1	6.2	6.0	6	73.4	74.2	74.8	75
France	9.8	8.8	8.2	8	74.4	75.1	75.7	76
German Dem. Rep.	11.8	10.5	9.6	9	72.0	72.6	73.3	74
Germany, Fed. Rep.	12.1	10.2	9.6	9	73.4	74.2	74.8	75
Greece	18.2	15.7	16.3	17	74.4	75.0	75.5	76
Hungary	22.3	20.2	19.8	20	69.7	69.8	70.1	71
Iceland	7.1	6.1	5.6	5	76.6	77.0	77.4	78
Ireland	10.8	9.1	9.0	9	72.8	73.4	73.9	76
Italy	14.2	12.2	11.6	11	74.3	74.9	75.4	76

Continued

TABLE 5.1 Continued

Region/country	Infant mortality rate[a]				Life expectancy at birth (years)			
	1980–82	1983–85	1986–87	1990	1980–82	1983–85	1986–87	1990
Luxembourg	10.3	9.5	9.9	10	72.8	73.6	74.2	75
Malta	13.6	12.1	10.8	10	71.4	72.0	72.6	
Netherlands	8.7	8.1	7.7	8	75.8	76.3	76.7	77
Norway	8.4	8.1	7.7	7	75.8	76.3	76.6	77
Poland	20.8	19.2	18.2	18	70.9	71.1	71.4	71
Portugal	23.3	18.9	16.5	15	71.6	72.6	73.2	74
Romania	27.6	24.9	23.2	22	69.6	69.8	70.1	71
Spain	12.3	10.5	10.1	10	75.3	76.0	76.5	77
Sweden	7.1	6.6	6.2	6	76.0	76.5	76.9	77
Switzerland	8.1	7.4	7.3	7	75.9	76.5	76.9	77
UK	11.5	10.1	9.4	9	73.7	74.4	75.0	76
Yugoslavia	31.7	28.6	25.7	25	70.6	71.2	71.8	73
FORMER USSR	**26.6**	**25.5**	**24.6**	**24**	**67.9**	**68.5**	**69.5**	**70**
Armenia			25.6[d]	23[e]				71[e]
Azerbaijan			27.4[d]	28[e]				70[e]
Belarus			13.2[d]	13[e]				72[e]
Estonia			12.6[d]					
Georgia			22.2[d]	22[e]				72[e]
Kazakhstan			29.7[d]	22[e]				69[e]
Kyrgyzstan			37.7[d]	36[e]				68[e]
Latvia			11.0[d]					
Lithuania			11.3[d]					
Moldova			23.2[d]	24[e]				68[e]
Russia			18.8[d]	19[e]				70[e]
Tadjikistan			50.5[d]	46[e]				70[e]
Turkmenistan			55.8[d]	54[e]				65[e]
Ukraine			14.2[d]	14[e]				71[e]
Uzbekistan			44.3[d]	43[e]				69[e]
OCEANIA								
Australia	10.7	9.5	8.6	8	74.7	75.5	75.9	76
Fiji	32.6	29.7	27.8	27	61.4	62.5	63.4	71
New Zealand	12.5	11.6	11.2	11	73.3	74.0	74.6	75
Papua New Guinea	77.3	69.5	61.9	59	51.3	52.5	53.5	56

[a] Number of deaths of infants less than one year of age per 10^3 live births.

[b] "More developed regions" includes Canada, USA, Japan, Europe, Australia, New Zealand and the USSR.

[c] "Less developed regions" includes Africa, Americas (excluding Canada and the USA) South East Asia and Western Pacific (excluding Australia and New Zealand).

[d] Data are for 1988 and are taken from: Ermakov, S. P. and Kiselev, A. A. 1992 Economic aspects of health, *World Health Statistics Quarterly*, **45**, 50–60.

[e] Data are taken from: UNDESD Population Division 1992 *World Population 1992*, United Nations, New York.

Revised estimates of infant mortality rate and life expectancy at birth for the periods 1950–55 to 2020–25 are now available in UNDESD 1993 *World Population Prospects: The 1992 Revision (Annex Tables)*, Population Division, UN Department of Economic and Social Development, New York (advance summary report).

Sources:

Sadik, N. 1990 *The State of the World Population*, United Nations Population Fund, United Nations, New York (life expectancy for 1990).

UNDESD Statistical Office 1992 *Statistical Yearbook 1988/89, Thirty-seventh issue*. United Nations, New York (IMR for 1990).

WHO 1992 *World Health Statistics Annual 1991*, World Health Organization, Geneva.

TABLE 5.2   Trends in life expectancy at birth in the former USSR, 1958–1989

	Year	Life expectancy at birth (years)			
		Both sexes	Male	Female	Difference[a]
Total population	1958–1959	68.6	64.4	71.1	6.7
	1964–1965	70.4	66.1	73.8	7.7
	1984–1985	68.1	62.9	72.7	9.8
	1985–1986	69.0	64.2	73.3	9.1
	1986–1987	69.8	65.0	73.8	8.8
	1988	69.5	64.8	73.6	8.8
	1989	69.5	64.6	74.0	9.4
Urban population	1958–1959	68.1	63.7	71.4	7.7
	1964–1965	69.9	65.5	73.3	7.8
	1984–1985	68.8	63.8	73.3	9.5
	1985–1986	69.8	65.0	73.8	8.8
	1986–1987	70.4	65.8	74.3	8.5
	1988	70.1	65.6	73.9	8.3
	1989	70.1	65.2	74.4	9.2
Rural population	1958–1959	68.9	64.9	71.8	6.9
	1964–1965	70.8	66.3	74.1	7.8
	1984–1985	66.5	61.0	71.6	10.6
	1986–1987	68.5	63.5	72.8	9.3
	1988	68.3	63.2	72.8	9.6
	1989	68.5	63.5	73.2	9.7

[a]   Difference in life expectancy between males and females (females–males).

Source:
Virganskaya, I. M. and Dmitriev, V. I. 1992 Some problems of the medicodemographic development in the former USSR, *World Health Statistics Quarterly*, **45**, 4–14.

TABLE 5.3 Mortality in selected countries by cause, age group and sex

Region/country	Year	Age group (years)	Total no. of deaths Male	Total no. of deaths Female	Proportional mortality[a] Infectious and parasitic diseases Male	Infectious and parasitic diseases Female	Malignant neoplasms Male	Malignant neoplasms Female	Diseases of the circulatory system Male	Diseases of the circulatory system Female	Respiratory diseases Male	Respiratory diseases Female	Injuries and poisonings Male	Injuries and poisonings Female
AMERICAS														
Canada	1989	All ages	104,106	85,905	0.5	0.7	27.0	26.4	39.5	43.4	9.1	7.6	5.9	3.5
		0–14	2,426	1,721	1.1	1.2	5.3	5.2	1.6	2.0	3.0	3.3	17.2	12.7
		15–44	9,228	3,994	0.5	1.1	11.2	32.0	10.9	10.8	1.7	3.1	35.1	19.5
		45–64	22,224	12,542	0.5	0.7	36.2	51.5	37.4	24.1	4.5	4.6	4.8	3.4
		65+	70,209	67,636	0.5	0.6	26.9	21.9	45.2	50.0	11.7	8.5	2.0	2.3
Cuba[b]	1988	All ages	38,384	29,560	1.4	1.3	19.9	18.0	42.1	45.4	8.5	8.9	14.5	9.2
		0–14	2,032	1,440	8.6	10.9	3.5	4.1	1.4	2.2	8.2	7.8	21.3	14.3
		15–44	4,850	3,143	1.2	1.5	8.9	17.2	13.8	14.5	2.8	5.1	59.8	40.9
		45–64	7,602	5,514	1.0	0.8	24.3	31.2	41.9	39.0	5.3	6.2	14.3	6.2
		65+	23,874	19,459	0.9	0.8	22.2	15.5	51.4	55.5	10.8	10.3	4.8	4.5
Ecuador	1988	All ages	29,253	23,479	11.4	12.4	7.9	11.2	15.0	18.8	10.5	11.4	13.2	4.7
		0–14	8,677	7,153	24.0	26.4	1.0	0.9	1.1	0.8	17.8	19.7	8.1	5.4
		15–44	5,610	3,179	8.0	12.4	4.8	11.7	8.9	13.1	3.6	4.6	33.5	9.7
		45–64	4,890	3,322	7.1	6.5	12.7	26.0	20.2	19.6	5.1	5.2	15.1	5.3
		65+	9,936	9,699	4.5	4.1	13.4	13.6	28.1	30.9	10.5	7.7	5.1	2.4
Puerto Rico	1989	All ages	15,242	10,743	2.1	2.2	15.7	14.8	31.9	39.6	9.6	12.2	6.3	2.6
		0–14	691	530	1.7	2.3	2.9	3.6	3.0	3.4	9.0	8.1	8.4	3.8
		15–44	2,662	794	2.5	2.3	4.8	18.0	8.1	13.1	4.1	8.4	19.0	10.7
		45–64	3,167	1,715	1.6	2.2	18.3	25.3	30.5	32.4	5.8	9.1	6.0	2.6
		65+	8,659	7,692	2.2	2.1	19.3	12.9	42.1	46.4	11.0	13.5	2.3	1.7
Trinidad & Tobago	1988	All ages	4,311	3,725	2.6	2.0	12.0	14.2	36.4	39.6	7.7	5.8	0.1	0.2
		0–14	315	225	5.4	4.0	1.9	4.0	2.9	1.8	12.7	8.9	1.0	1.8
		15–44	653	371	4.0	2.7	6.0	17.5	14.4	14.6	5.1	5.4	0.2	0.5
		45–64	1,102	886	1.9	2.6	13.2	19.8	43.7	36.1	4.2	3.3	0.1	0.0
		65+	2,240	2,239	2.2	1.5	14.6	12.4	44.0	49.0	9.6	6.5	0.0	0.0
Uruguay	1989	All ages	16,207	13,414	1.9	1.9	24.5	21.5	35.0	43.7	8.5	6.1	6.3	3.5
		0–14	863	611	6.5	7.5	2.3	2.9	2.7	1.1	6.0	9.2	14.6	12.4
		15–44	1,189	652	2.6	2.1	13.8	27.1	11.9	19.0	3.4	4.9	36.4	17.5
		45–64	4,241	2,066	1.8	1.7	32.0	39.7	31.0	28.9	6.2	3.9	6.3	3.9
		65+	9,821	10,027	1.4	1.5	24.6	18.6	42.4	50.9	8.7	6.3	1.9	1.9

Continued

TABLE 5.3 Continued

Region/country	Year	Age group (years)	Total no. of deaths		Proportional mortality[a]									
					Infectious and parasitic diseases		Malignant neoplasms		Diseases of the circulatory system		Respiratory diseases		Injuries and poisonings	
			Male	Female	Male	Female	Male	Female	Male	Female	Male	Female	Male	Female
USA	1988	All ages	1,125,540	1,042,459	1.3	1.6	22.9	21.8	42.0	48.0	8.9	8.3	5.8	3.0
		0–14	31,709	23,555	1.9	2.2	3.1	3.2	3.3	3.7	2.9	3.5	15.9	12.6
		15–44	123,950	50,808	1.6	2.1	8.5	23.8	12.1	13.1	2.3	3.8	29.2	20.9
		45–64	239,283	147,938	1.3	1.5	29.6	42.4	40.3	30.6	5.5	6.1	4.5	3.0
		65+	730,181	819,988	1.3	1.5	23.7	18.5	49.4	54.6	11.3	9.2	1.9	1.6
EUROPE														
Austria	1990	All ages	38,386	44,566	0.5	0.3	25.2	21.7	46.0	56.0	5.7	4.5	5.8	3.0
		0–14	573	440	1.6	2.0	2.8	5.0	1.4	0.9	3.7	3.6	14.1	11.4
		15–44	2,798	1,114	0.5	0.6	11.3	31.9	12.7	14.7	1.9	2.9	35.8	16.5
		45–64	8,503	4,231	0.8	0.6	32.7	47.6	35.3	26.8	3.6	2.6	6.1	3.2
		65+	26,536	38,781	0.4	0.2	24.7	18.8	53.9	61.0	6.8	4.8	2.4	2.5
Bulgaria	1990	All ages	59,780	48,828	0.6	0.4	14.9	12.6	58.0	65.9	6.6	5.1	5.1	1.7
		0–14	1,451	1,010	4.1	4.1	4.5	3.6	5.0	5.6	20.6	23.0	15.9	13.2
		15–44	4,358	1,778	1.3	1.6	14.1	32.0	25.7	22.5	3.7	4.0	30.2	14.8
		45–64	16,036	7,653	0.7	0.6	23.3	29.9	49.5	49.1	3.6	4.2	5.8	2.3
		65+	37,935	38,387	0.3	0.2	11.8	8.5	67.3	72.8	5.2	4.9	1.6	0.6
Byelorussian SSR	1990	All ages	53,453	56,129	0.9	0.5	18.8	13.3	44.4	57.0	7.8	5.9	10.4	2.6
		0–14	1,676	1,136	4.4	7.3	5.5	7.1	1.1	1.8	11.8	11.2	20.5	15.8
		15–44	7,327	2,117	1.4	1.2	9.0	28.3	19.0	15.4	1.6	2.9	41.1	19.7
		45–64	19,429	9,551	0.3	0.6	28.2	30.0	42.9	46.2	6.6	4.4	9.2	5.2
		65+	25,008	43,312	0.4	0.2	15.3	9.1	56.0	62.9	10.3	6.3	1.7	0.9
Czechoslovakia[b]	1990	All ages	96,731	87,054	0.3	0.3	22.8	18.8	50.8	60.2	5.7	4.5	5.8	4.2
		0–14	1,965	1,374	0.7	1.1	6.4	6.6	0.9	1.2	7.6	9.1	13.8	10.6
		15–44	7,798	2,839	0.4	0.5	15.3	35.4	19.6	15.8	3.6	3.3	28.1	13.4
		45–64	27,434	12,065	0.4	0.4	30.3	38.0	43.3	39.2	4.1	3.2	5.5	2.7
		65+	59,507	70,776	0.2	0.3	20.9	15.2	60.1	66.6	6.6	4.6	2.7	4.0
Denmark	1990	All ages	30,944	29,645	0.9	0.5	7.0	3.8	43.5	46.3	8.0	7.3	7.1	5.4
		0–14	406	276	3.4	3.3	0.0	0.0	0.2	0.7	3.0	1.8	13.1	12.0
		15–44	1,742	840	6.2	1.9	1.7	3.3	10.3	10.1	2.2	2.5	46.5	30.6
		45–64	5,736	3,900	1.0	0.7	10.1	9.4	34.9	20.5	4.3	6.8	9.0	7.6
		65+	23,058	24,628	0.4	0.4	6.8	2.9	48.9	52.2	9.5	7.5	3.5	4.1

Continued

TABLE 5.3 Continued

Region/country	Year	Age group (years)	Total no. of deaths		Proportional mortality[a]									
					Infectious and parasitic diseases		Malignant neoplasms		Diseases of the circulatory system		Respiratory diseases		Injuries and poisonings	
			Male	Female	Male	Female	Male	Female	Male	Female	Male	Female	Male	Female
Finland	1989	All ages	24,529	24,600	0.6	0.7	20.5	19.3	46.6	52.4	7.7	6.6	7.9	3.8
		0–14	337	242	2.4	0.4	3.6	6.2	1.5	2.5	3.0	2.5	16.3	12.4
		15–44	2,553	821	0.6	0.7	7.3	26.3	14.7	12.7	1.9	3.3	30.3	17.4
		45–64	6,325	2,539	0.4	1.1	23.0	40.5	46.0	31.5	4.3	3.1	10.0	7.2
		65+	15,314	20,998	0.6	0.7	22.0	16.6	53.2	57.1	10.2	7.2	3.1	2.8
France	1989	All ages	274,263	255,020	1.2	1.4	30.7	21.3	29.6	38.3	6.9	6.4	6.9	5.9
		0–14	4,959	3,573	1.7	1.8	4.7	5.5	2.2	2.5	2.6	2.5	15.3	13.8
		15–44	24,011	9,356	0.9	1.0	13.6	26.1	9.0	8.8	2.0	2.4	30.6	20.0
		45–64	60,862	24,728	0.9	1.1	43.9	47.9	16.0	16.3	3.5	3.1	6.2	5.3
		65+	184,431	217,363	1.3	1.4	29.3	18.3	35.9	42.7	8.7	7.1	3.8	5.2
German Dem. Rep.	1989	All ages	91,091	114,621	0.4	0.2	18.8	15.4	51.4	62.3	6.9	4.2	4.5	2.8
		0–14	1,415	921	1.9	3.3	4.4	5.9	2.5	3.1	4.3	5.1	16.8	12.7
		15–44	6,187	2,611	0.6	0.5	12.7	30.6	14.7	13.3	2.3	3.9	27.0	13.7
		45–64	23,400	12,973	0.5	0.4	28.2	37.2	36.7	30.2	5.3	3.7	4.7	3.2
		65+	60,089	95,107	0.3	0.2	16.0	12.6	62.0	70.6	8.0	4.5	1.7	2.5
Germany, Fed. Rep.	1989	All ages	326,008	371,722	0.8	0.7	26.2	22.9	44.8	52.9	7.2	4.8	3.4	2.4
		0–14	4,179	2,974	1.7	2.1	4.5	4.6	1.8	2.7	2.2	2.4	13.1	9.6
		15–44	18,667	8,700	0.9	1.1	14.7	32.9	13.5	12.2	1.5	2.3	24.3	12.5
		45–64	76,743	38,326	0.8	0.8	34.4	48.0	35.2	24.2	3.8	2.9	3.3	2.1
		65+	226,409	321,722	0.7	0.7	24.8	19.8	51.5	57.9	8.8	5.2	1.5	2.1
Greece	1989	All ages	48,635	44,085	0.7	0.7	24.1	16.9	47.3	56.5	5.4	5.0	6.2	3.2
		0–14	839	625	1.8	2.4	5.0	5.8	1.1	1.4	4.5	4.0	18.4	13.3
		15–44	2,811	1,160	0.2	0.5	15.2	34.3	18.1	43.9	4.2	3.7	44.6	24.5
		45–64	9,621	4,916	0.3	0.6	36.5	46.0	40.8	79.8	3.1	3.5	7.7	5.1
		65+	35,363	37,384	0.7	0.6	22.0	12.8	52.5	49.7	6.2	5.2	2.5	2.1
Hungary	1990	All ages	76,521	68,174	1.0	0.6	22.4	19.5	47.5	57.7	5.6	3.8	6.2	4.9
		0–14	1,499	1,146	1.8	1.9	4.9	4.8	1.3	0.9	7.2	5.2	10.3	9.7
		15–44	7,221	2,886	1.4	0.9	12.5	26.9	20.6	20.1	2.9	3.5	22.2	11.1
		45–64	23,913	11,682	1.2	0.7	19.5	33.4	40.2	37.7	4.2	3.1	5.9	3.8
		65+	43,888	52,460	0.8	0.5	21.8	16.4	57.6	65.4	6.8	4.0	3.6	4.8

Continued

TABLE 5.3 Continued

| Region/country | Year | Age group (years) | Total no. of deaths | | Proportional mortality[a] | | | | | | | | |
| | | | Male | Female | Infectious and parasitic diseases | | Malignant neoplasms | | Diseases of the circulatory system | | Respiratory diseases | | Injuries and poisonings | |
					Male	Female	Male	Female	Male	Female	Male	Female	Male	Female
Iceland	1990	All ages	910	794	0.5	0.8	26.8	26.4	43.4	44.3	9.6	14.6	6.7	3.0
		0–14	26	13	0.0	7.7	3.8	7.7	0.0	0.0	0.0	0.0	30.8	7.7
		15–44	83	30	0.0	0.0	13.3	30.0	7.2	23.3	1.2	0.0	36.1	20.0
		45–64	150	118	0.0	0.8	38.7	63.6	38.7	19.5	5.3	5.1	8.7	3.4
		65+	651	633	0.8	0.6	26.7	19.7	50.8	50.9	12.1	17.2	1.5	2.1
Ireland	1989	All ages	17,058	15,053	0.6	0.6	22.6	23.1	46.1	47.5	14.0	13.2	4.3	2.5
		0–14	410	266	2.9	2.6	5.6	4.1	0.7	1.1	4.6	3.8	15.9	9.4
		15–44	958	416	0.9	1.4	15.1	42.1	15.7	14.4	2.7	2.9	35.7	15.9
		45–64	3,027	1,827	0.7	1.1	30.4	47.9	47.9	29.8	6.9	7.0	4.3	2.8
		65+	12,663	12,544	0.5	0.5	21.9	19.3	49.4	52.1	16.8	14.7	1.6	1.8
Israel	1990	All ages	15,417	13,759	1.6	1.9	18.2	19.6	41.3	42.8	6.9	6.4	5.5	3.9
		0–14	767	650	2.0	3.2	3.7	3.7	0.7	0.8	3.3	3.8	14.5	10.2
		15–44	1,102	540	1.0	2.4	15.9	36.9	12.8	10.7	2.6	3.1	25.9	13.1
		45–64	2,662	1,851	1.3	1.4	25.1	39.1	38.9	30.6	4.3	3.8	5.3	3.9
		65+	10,886	10,718	1.7	1.8	17.8	16.4	47.6	49.2	8.0	7.0	2.9	3.1
Italy	1988	All ages	281,149	258,277	0.5	0.4	29.9	22.7	39.3	48.8	7.6	5.1	4.8	3.6
		0–14	4,260	3,142	1.2	1.4	6.0	6.1	2.0	2.8	2.9	3.1	10.9	7.4
		15–44	15,487	6,800	0.7	0.9	16.5	39.5	14.1	13.2	1.6	3.0	32.8	15.2
		45–64	63,636	31,133	0.5	0.5	43.5	50.1	29.7	23.9	3.7	2.5	4.8	3.1
		65+	197,766	217,202	0.4	0.3	27.1	18.5	45.2	54.2	9.4	5.6	2.4	3.2
Malta	1990	All ages	1,391	1,318	1.0	1.0	21.7	18.0	45.8	55.2	10.4	5.2	3.1	2.8
		0–14	42	28	9.5	3.6	2.4	7.1	2.4	7.1	7.1	3.6	4.8	3.6
		15–44	56	33	1.8	0.0	23.2	57.6	12.5	15.2	1.8	0.0	21.4	12.1
		45–64	275	186	0.0	0.0	32.4	38.2	41.5	44.1	4.7	0.5	1.5	3.2
		65+	1,018	1,069	0.9	1.1	19.5	13.6	50.6	59.6	12.3	6.2	2.5	2.4
Netherlands	1989	All ages	67,089	61,816	0.5	0.7	30.0	24.7	39.0	41.4	9.4	6.7	3.1	2.8
		0–14	1,090	814	2.2	3.8	4.4	4.9	1.3	1.7	3.3	1.5	12.7	9.5
		15–44	3,616	2,115	0.4	1.2	17.7	37.1	16.5	11.1	1.3	2.0	20.7	10.5
		45–64	13,006	7,097	0.4	0.6	38.5	51.9	37.5	22.7	3.5	3.0	2.4	2.1
		65+	49,377	51,790	0.5	0.6	29.2	20.8	41.9	45.8	11.7	7.5	1.8	2.5

Continued

TABLE 5.3 Continued

Region/country	Year	Age group (years)	Total no. of deaths		Proportional mortality[a]									
					Infectious and parasitic diseases		Malignant neoplasms		Diseases of the circulatory system		Respiratory diseases		Injuries and poisonings	
			Male	Female	Male	Female	Male	Female	Male	Female	Male	Female	Male	Female
Norway	1989	All ages	23,652	21,589	0.6	0.7	22.5	20.8	46.3	47.4	9.0	10.7	5.2	3.8
		0–14	384	263	1.0	4.6	3.9	2.3	0.8	0.8	2.1	2.3	14.3	7.6
		15–44	1,288	557	0.8	0.7	12.0	36.3	14.6	11.0	1.2	1.3	29.9	16.2
		45–64	3,753	1,896	0.5	0.9	28.3	51.8	43.1	23.9	3.6	4.0	6.5	3.4
		65+	18,227	18,873	0.6	0.7	22.4	17.5	50.2	51.5	10.7	11.8	2.9	3.5
Poland	1990	All ages	209,333	179,107	1.0	0.6	20.1	17.2	48.0	57.6	4.8	3.1	7.7	3.2
		0–14	6,992	5,020	5.3	5.4	4.0	5.6	1.8	2.1	4.6	5.1	14.3	10.1
		15–44	24,660	7,804	0.9	1.4	10.2	32.4	22.0	21.3	1.6	2.7	33.0	14.2
		45–64	65,208	28,874	1.1	0.7	28.3	63.9	43.2	39.8	3.7	2.7	6.8	3.0
		65+	112,473	137,409	0.6	0.3	18.5	15.2	59.3	65.4	6.2	2.1	2.3	2.4
Portugal	1990	All ages	53,439	49,676	1.0	0.6	19.2	15.9	39.2	49.5	8.1	6.3	6.3	2.4
		0–14	1,290	903	4.0	2.9	4.2	6.5	1.4	2.2	5.4	5.8	16.7	14.2
		15–44	4,812	1,814	1.5	2.1	10.4	27.3	9.7	10.7	3.0	3.9	33.6	15.3
		45–64	11,264	5,810	1.5	1.2	28.1	37.6	29.2	28.7	5.8	4.0	6.6	3.6
		65+	36,073	41,149	0.6	0.3	18.1	12.6	47.6	55.1	9.6	6.7	2.1	1.4
Romania	1988	All ages	130,178	116,492	1.1	0.6	13.5	11.3	51.8	65.1	11.6	9.1	9.4	3.6
		0–14	7,973	5,830	6.3	6.4	3.0	4.0	1.0	0.9	31.6	37.1	17.3	12.7
		15–44	11,400	5,381	3.0	2.3	12.0	25.3	18.4	18.3	6.5	7.1	44.6	22.1
		45–64	38,186	21,040	1.3	0.6	22.7	41.3	41.7	46.4	9.3	5.4	10.9	5.3
		65+	72,619	84,241	0.2	0.1	10.0	8.6	68.0	77.2	11.4	8.2	2.2	1.3
Sweden[b,c]	1988	All ages	50,427	46,329	0.6	0.8	21.1	20.7	51.7	52.5	8.1	8.0	3.6	2.6
		0–14	547	383	1.1	2.1	4.4	4.4	1.5	3.1	2.9	2.6	12.4	8.4
		15–44	2,437	1,202	0.5	1.7	11.4	33.2	13.5	10.0	2.6	3.1	24.3	11.5
		45–64	7,464	4,098	0.5	0.7	27.7	49.3	44.5	24.9	3.6	4.9	4.4	2.8
		65+	39,979	40,646	0.7	0.8	20.7	17.6	56.0	57.0	9.3	8.6	2.0	2.3
Switzerland	1990	All ages	32,492	31,247	1.6	0.9	28.4	24.1	40.2	47.8	8.5	7.8	10.1	6.5
		0–14	470	363	4.3	4.7	5.5	5.2	1.9	2.5	3.4	3.6	17.0	10.5
		15–44	2,512	986	10.0	6.6	11.0	30.7	9.6	9.3	1.2	2.4	56.5	36.6
		45–64	5,689	2,980	1.6	0.9	37.9	53.0	30.8	18.8	4.6	3.8	11.9	9.4
		65+	23,821	26,918	0.7	0.6	28.3	32.0	46.4	53.1	10.3	8.4	4.6	5.0

Continued

TABLE 5.3 Continued

| Region/country | Year | Age group (years) | Total no. of deaths | | Proportional mortality[a] | | | | | | | | | |
| | | | Male | Female | Infectious and parasitic diseases | | Malignant neoplasms | | Diseases of the circulatory system | | Respiratory diseases | | Injuries and poisonings | |
					Male	Female	Male	Female	Male	Female	Male	Female	Male	Female
Ukranian SSR	1990	All ages	297,584	332,018	1.5	0.4	19.1	13.4	44.9	60.0	7.6	8.7	2.8	0.4
		0–14	8,723	5,823	4.7	5.3	6.2	6.7	1.6	1.7	8.9	34.1	11.4	3.1
		15–44	36,315	11,619	3.5	1.9	9.8	28.8	16.5	14.6	1.6	120.3	6.1	3.6
		45–64	107,664	53,731	2.1	0.8	28.3	31.8	39.5	44.6	7.0	19.2	2.4	0.9
		65+	144,335	260,656	0.5	0.2	15.4	9.1	58.8	66.6	9.3	0.9	1.6	0.1
USSR	1990	All ages	1,462,389	1,522,769	2.2	1.0	18.3	14.0	43.4	62.0	8.0	5.0	10.9	3.2
		0–14	98,245	70,094	11.5	13.0	2.9	3.0	0.9	1.0	25.5	28.3	16.5	12.2
		15–44	228,301	72,676	3.3	3.1	8.1	23.8	16.5	15.5	2.2	3.7	36.9	18.7
		45–64	536,529	265,248	2.0	0.8	27.2	30.0	41.6	44.6	6.5	4.2	9.0	5.2
		65+	596,991	1,113,799	0.5	0.2	16.7	10.3	62.4	73.0	8.6	4.5	1.7	1.2
UK	1990	All ages	314,601	327,198	0.5	0.4	26.7	23.6	45.4	46.8	11.3	10.9	2.7	1.7
		0–14	5,070	3,669	2.9	2.8	4.5	5.2	2.0	2.3	5.3	5.3	9.8	7.4
		15–44	15,029	7,664	1.2	1.3	15.1	40.5	16.8	12.0	3.5	4.1	26.1	12.3
		45–64	57,490	36,068	0.5	0.6	34.1	49.2	45.9	29.0	5.5	6.4	2.8	1.8
		65+	237,012	279,797	0.3	0.3	26.2	20.0	48.0	50.6	13.3	11.7	1.0	1.3
Yugoslavia	1989	All ages	113,819	101,664	1.5	1.2	18.7	15.3	46.7	57.7	5.2	4.1	5.8	2.6
		0–14	5,558	4,629	10.2	12.4	2.7	2.6	2.9	3.3	10.8	12.0	7.7	5.1
		15–44	9,259	4,131	1.7	2.1	14.7	29.3	17.2	17.8	2.0	2.7	29.8	13.5
		45–64	35,709	18,731	1.3	0.9	28.2	31.5	38.2	41.1	3.6	2.9	6.0	3.0
		65+	63,160	74,114	0.8	0.5	15.3	11.2	59.7	67.5	6.1	4.0	2.0	1.8
EASTERN MEDITERRANEAN														
Bahrain	1988	All ages	890	633	0.6	0.9	10.4	9.6	34.3	30.3	4.4	4.6	4.7	1.3
		0–14	181	166	0.6	0.0	0.0	0.6	0.0	0.0	2.8	2.4	7.7	2.4
		15–44	156	48	0.0	2.1	6.4	31.3	37.2	27.1	1.3	6.3	11.5	2.1
		45–64	227	130	0.4	0.8	16.7	19.2	46.3	30.0	3.1	5.4	3.1	1.5
		65+	326	289	0.9	1.4	13.8	6.9	43.6	48.4	7.7	5.2	0.9	0.3
WESTERN PACIFIC														
Australia	1988	All ages	65,082	54,784	0.6	0.6	25.9	23.2	42.7	49.8	8.6	6.2	6.0	3.3
		0–14	1,792	1,302	1.9	2.1	4.7	3.7	0.9	1.5	3.0	4.0	17.0	13.7
		15–44	6,160	2,584	1.2	0.7	11.7	30.0	12.3	10.4	2.2	3.6	35.9	22.2
		45–64	13,910	7,421	0.5	0.6	35.2	49.0	39.7	26.9	5.7	6.3	4.0	3.4
		65+	43,210	43,475	0.5	0.5	25.8	19.0	49.6	57.5	10.7	6.4	1.7	1.9

Continued

TABLE 5.3 Continued

Region/country	Year	Age group (years)	Total no. of deaths		Proportional mortality[a]									
					Infectious and parasitic diseases		Malignant neoplasms		Diseases of the circulatory system		Respiratory diseases		Injuries and poisonings	
			Male	Female	Male	Female	Male	Female	Male	Female	Male	Female	Male	Female
China[b,d]	1989	All ages	392,900	327,757	4.4	3.1	21.4	15.4	31.6	36.4	37.1	41.1	9.9	8.6
		0–14	30,292	26,975	5.3	5.0	2.5	2.1	0.6	0.6	40.7	42.0	23.6	18.3
		15–44	43,307	30,812	7.4	6.7	21.2	18.3	9.8	12.3	7.0	9.1	40.4	37.5
		45–64	104,116	66,253	5.6	4.2	34.2	28.8	28.8	34.3	25.3	26.1	7.5	6.7
		65+	215,185	206,717	3.1	2.0	17.9	12.1	41.7	44.8	48.4	50.0	3.1	3.5
Hong Kong	1989	All ages	16,131	12,345	3.4	3.7	33.3	26.0	25.3	31.6	18.0	16.7	3.7	1.9
		0–14	407	318	1.5	3.5	6.6	7.5	2.2	1.9	6.1	9.4	11.1	7.9
		15–44	1,446	747	1.9	3.2	30.7	38.7	11.5	9.8	7.5	7.2	20.1	7.0
		45–64	4,815	2,101	3.1	2.0	46.6	46.1	22.6	23.9	11.2	7.3	2.9	2.3
		65+	9,451	9,179	3.9	4.1	28.2	21.0	29.9	36.1	23.6	19.9	1.3	1.1
Japan	1990	All ages	427,114	361,480	1.6	1.2	29.8	23.6	33.4	42.0	12.1	10.0	5.1	2.5
		0–14	6,186	4,709	2.7	2.5	6.9	7.5	6.0	5.4	5.8	5.6	21.5	16.0
		15–44	28,169	14,474	1.2	1.6	18.5	37.6	18.4	15.5	3.1	4.0	26.5	10.2
		45–64	107,652	52,554	1.6	1.5	40.9	46.4	26.5	25.5	4.5	4.2	6.0	3.4
		65+	284,721	289,686	1.5	1.1	27.2	19.1	37.0	46.9	16.0	11.4	2.3	1.8
Singapore	1989	All ages	7,952	6,122	3.1	3.1	24.8	21.4	34.5	38.3	16.0	14.7	4.0	1.2
		0–14	264	220	1.5	2.3	5.7	8.2	3.4	3.2	6.8	10.5	7.6	3.6
		15–44	990	554	3.1	3.6	15.8	33.0	16.9	12.1	7.2	7.6	19.3	4.2
		45–64	2,363	1,256	2.2	2.2	32.7	33.8	40.2	37.6	8.5	7.6	2.5	0.7
		65+	4,284	4,065	3.7	3.3	24.0	16.8	37.6	44.2	22.5	18.1	1.0	0.7

a Percentage contribution to all deaths.
b Deaths caused by violence and suicide are included in the "Injuries and poisonings" category.
c Data are from the reporting areas within the country only.
d Deaths classified according to the 8th revision of the International Classification of Diseases.

Countries included are those reporting statistics for 1988 or more recent years.

All deaths are classified according to the 9th revision of the International Classification of Diseases except those marked d. The category "Injuries and poisonings" does not include death by violence or suicide unless marked b. Where the numbers of deaths in the specified age groups do not sum to the total, the difference represents deaths for which age or cause was not known.

The proportional mortalities do not add up to 100 per cent because only five causes of death are listed; these five are the leading causes of death.

Source:
WHO 1992 World Health Statistics Annual 1991, World Health Organization, Geneva (and earlier editions).

TABLE 5.4 Correlation of age-specific death rates in rural and urban populations in the former USSR, 1964–1989

Age group (years)	Males			Females		
	1964–1965	1986–1987	1989	1964–1965	1986–1987	1989
<5	20.0	80.0	71.0	30.0	104.0	92.0
5–9	30.0	50.0	33.0	60.0	50.0	25.0
10–14	0.0	20.0	40.0	20.0	30.0	33.0
15–19	10.0	10.0	25.0	20.0	80.0	60.0
20–24	30.0	50.0	45.0	50.0	100.0	50.0
25–29	40.0	90.0	27.0	60.0	120.0	57.0
30–34	20.0	40.0	24.0	40.0	60.0	44.0
35–39	10.0	40.0	22.0	20.0	50.0	36.0
40–44	10.0	30.0	21.0	10.0	30.0	29.0
45–49	10.0	20.0	14.0	−6.0	20.0	20.0
50–54	−1.0	4.0	8.0	−7.0	10.0	14.0
55–59	−10.0	−3.0	−4.0	−14.0	−1.0	0.0
60–64	−11.0	−5.0	−7.0	−17.0	−8.0	−9.0
65–69	−23.0	−8.0	−8.0	−28.0	−13.0	−11.0
70+	−25.0	−5.0		−17.0	−10.0	
All ages	20.0	30.0	16.0	20.0	30.0	24.0

The death rates of the urban population are assumed to be 100 per cent.
Positive values indicate that the death rate in the rural population exceeds that of the urban population. Conversely, negative values indicate that the death rate in the urban population exceeds that of the rural population.
The larger the value the greater the difference between urban and rural death rates.

Source:
Virganskaya, I. M. and Dmitriev, V. I. 1992 Some problems of the medicodemographic development in the former USSR, *World Health Statistics Quarterly*, **45**, 4–14.

TABLE 5.5 Mean age at death by cause in rural and urban populations in selected Republics of the former USSR, 1989

Republic	Causes of death											
	Infectious and parasitic diseases		Neoplasms		Cardio-vascular diseases		Respiratory system diseases		Digestive system diseases		Injuries and poisonings	
	Urban	Rural	Urban	Rural	Urban	Rural	Urban	Rural	Urban	Rural	Urban	Rural
					Males							
FORMER USSR	44.3	33.7	65.6	63.2	72.7	73.3	65.3	60.4	64.1	60.7	44.9	41.7
Russia	47.2	43.3	65.5	62.9	72.2	72.5	67.2	65.3	64.4	61.1	44.3	40.8
Ukraine	49.2	46.0	65.4	61.9	73.5	73.5	69.5	69.8	64.2	59.5	45.7	41.5
Georgia	42.5	38.8	65.6	63.4	74.3	75.3	55.7	58.4	64.5	62.7	46.6	42.9
Moldova	43.6	37.4	65.4	60.6	74.5	74.9	65.1	61.7	63.5	61.3	47.6	40.2
Tajikistan	31.9	19.5	66.2	67.2	74.3	77.5	57.8	60.0	61.5	63.3	45.5	41.1
					Females							
FORMER USSR	37.2	23.5	67.8	64.2	80.2	80.2	67.6	63.0	70.4	65.3	55.6	46.5
Russia	43.2	32.1	68.0	64.4	80.2	80.5	71.2	71.4	71.1	68.6	55.4	47.5
Ukraine	48.4	40.8	67.0	62.5	80.1	80.2	73.3	75.3	69.9	64.5	55.2	45.1
Georgia	42.0	20.1	66.8	62.8	80.5	80.7	53.7	60.8	69.7	69.2	58.6	48.9
Moldova	37.3	28.8	66.8	61.0	79.6	78.0	69.2	65.4	67.9	61.6	57.6	47.0
Tajikistan	24.1	20.3	68.7	65.6	80.4	79.9	60.5	62.9	66.6	65.4	52.8	37.8

Source: As Table 5.4

TABLE 5.6 Global estimates of morbidity, mortality and populations at risk from selected water–related infectious diseases, late 1980s

Disease	Causative agent	Vector	Present distribution	Morbidity	Mortality[a]	Population at risk
WATERBORNE AND WATER–WASHED						
Diarrhoeal diseases[b]	Various	Various	Cosmopolitan	>1.5 billion[c]	4 million[c]	>2 billion
Enteric fevers[d]	Various	Various		500,000	25,000	
Poliomyelitis	Viral	Direct contact	Cosmopolitan	200,000	25,000	
Ascariasis (roundworm)	*Ascaris lumbricoides*	Contaminated water, food or hands		800–1,000 million	20,000	
WATER–WASHED						
Skin and eye infections						
Trachoma	*Chlamydia trachomatis*	Contact with trachomatous material	Asia/Africa	6–9 million blind		500 million
Leishmariasis (oriental sore)	*Leishmania* spp. (protozoa)	Sandflies	Asia/S. Europe/ Africa/S.America	12 million[e]		350 million
WATER–BASED						
Penetrating skin						
Schistosomiasis (bilharzia)	*Schistosoma* spp. (trematodes)	Water snails	Tropics/Subtropics	200 million	200,000	500–600 million
Ingested						
Dracunculiasis (Guinea worm)	*Dracunculus medinensis* (nematodes)	Copepods	Africa/India/Pakistan	>10 million		>100 million
WATER–RELATED INSECT VECTORS						
Biting near water						
African trypanosomiasis (sleeping sickness)	*Trypanosoma* spp. (protozoa)	Tsetse flies	Tropical Africa	20,000[f]		50 million
Breeding in water						
Lymphatic filiariasis (elephantiasis)	*Wuchereria bancrofti Brugia* spp. (nematodes)	Various mosquito species	Tropics/Subtropics	90 million		900 million
Malaria	*Plasmodium* spp. (protozoa)	Various mosquito species	Tropics/Subtropics	267 million[g]	1–2 million[h]	2,100 million

Continued

TABLE 5.6 Continued

Disease	Causative agent	Vector	Present distribution	Morbidity	Mortality[a]	Population at risk
Onchocerciasis (river blindness)	Onchocerca volvulus (nematodes)	Simulium blackflies	Africa/Latin America	18 million[i]	20,000–50,000	85–90 million
Yellow fever	Yellow fever virus	Various mosquito species	Africa/The Americas	10,000–20,000		
Dengue fever (breakbone fever)	Dengue virus	Aedes mosquitoes	Tropics/Subtropics	30–60 million		
Japanese encephalitis	Japanese encephalitis virus	Various mosquito species	Asia/Pacific Islands			

a   Number of deaths per year unless otherwise stated.
b   Includes salmonellosis, shigellosis, rotavirus, amoebiasis, giardiasis and diseases associated with Campylobacter and E.coli.
c   In children under five years.
d   Paratyphoid and typhoid.
e   400,000 new cases each year.
f   Estimate refers to the number of new cases per year. It is thought to be an under-estimate.

g   107 million clinical cases.
h   Three-quarters of the deaths are in children under five.
i   Over 300,000 blind.

Sources:
WHO 1992 Our Planet, Our Health: Report of the WHO Commission on Health and Environment, World Health Organization, Geneva.

Weihe, W. H. and Mertens, R. 1991 Human well-being, diseases and climate. In: Climate Change: Science, Impacts and Policy, Proceedings of the Second World Climate Conference, J. Jäger and H. L. Ferguson (Eds), Cambridge University Press, Cambridge, UK, 345–359.

UNEP 1992 The World Environment 1972–1992: Two Decades of Challenge, M. K. Tolba and O. A. El-Kholy (Eds), Chapman and Hall, London.

TABLE 5.7   Incidence of cholera in selected countries as reported to WHO, 1989–1992

Region/country	1989 Cases	1990 Cases	1991 Cases	1991 Deaths	1992 Cases	1992 Deaths
WORLD	53,970 (205)	70,084 (104)	594,694 (121)	19,295	461,783	8,072
AFRICA	35,951	38,683	153,367	13,998	91,081	5,291
Algeria	393	1,293			69	
Angola	17,601	9,527	8,590	582	3,608	184
Benin			7,474	259	413	17
Burkina			537	61		
Burundi	94[a]	82[b]	3		479	29
Cameroon	4	16[b]	4,026	491	1,268	66
Chad			13,915	1,344		
Côte d'Ivoire			604	116	37	7
Ghana		2,937	13,172	409	228	23
Kenya	918				3,388	80
Liberia	28		132	40		
Malawi	8,351	13,457	8,088	245	298	8
Mauritania	700					
Mozambique	371	4,152	7,847	328	30,802	726
Niger	166		3,238	367		
Nigeria	1,078		59,478	7,654	7,671	686
Rwanda	1		679	35	530	32
São Tomé & Príncipe	3,953	804	3	1		
South Africa			10		11	
Tanzania	2,150	2,230	5,676	572	18,526	2,173
Togo			2,396	28	753	49
Uganda			279	28	5,072	104
Zaire	99	468	4,066	294	1,949	59
Zambia	44	3,717	13,154	1,091	11,659	913
AMERICAS	(1)	7(5)	391,220 (23)	4,002	354,089	2,401
Argentina					553	15
Belize					159	4
Bolivia			206	12	22,260	383
Brazil			1,567	26	30,309	363
Canada	(1)		(2)		(4)	
Chile			41	2	73 (5)	1
Colombia			11,979	207	15,129	158
Costa Rica					12 (8)	
Ecuador			46,320	697	31,870	208
El Salvador			947	34	8,106	45
French Guiana			(1)		16 (6)	1
Guatemala			3,674	50	15,395	207
Honduras			11		384	17
Mexico			2,690	34	8,162	99
Nicaragua			1		3,067	46
Panama			1,178	29	2,416	49
Peru			322,562	2,909	212,642	727
Suriname					12	1

Continued

TABLE 5.7 Continued

Region/country	1989 Cases	1990 Cases	1991 Cases	1991 Deaths	1992 Cases	1992 Deaths
USA		7 (5)	26 (9)		102 (98)	1
Venezuela			15 (11)	2	2,842 (5)	68
ASIA	**18,007**	**30,979 (65)**	**49,791 (73)**	**1,286**	**16,299**	**372**
Bhutan					494	6
Cambodia			770	97	1,229	120
China	6,158	639	205		580	1
Hong Kong	29 (23)	5 (3)	5 (2)		3 (1)	
India	5,026	3,583	6,993	149	6,911	55
Indonesia	67	155[b]	6,202	55	25	
Iran	5,222	178	1,880	32	97	4
Iraq			877	6	97	
Japan	99 (37)	73 (62)	90 (65)	1	(46)	
Korea, Dem.			113	4		
Kuwait	(133)					
Macao	(3)	1				
Malaysia	350	2,071	506	6	474	8
Myanmar	597		924	39	826	50
Nepal	141	23,888[c]	30,648[d]	873[d]	764	15
Philippines					345	
Singapore	39	26	34 (6)	2		
Sri Lanka			70	2	121	3
Viet Nam	143	358	52	1	4,260	110
Yemen					(4)	
EUROPE	**11 (8)**	**349 (30)**	**316**	**9**	**18**	
Austria		(2)				
Belgium					(1)	
Denmark		(1)				
France	(1)	(6)	(7)			
Germany	(1)[e]	(1)			(1)	
Netherlands		(3)				
Norway	(1)					
Romania		270	226	9	3	
Spain	3 (2)	(11)	(1)			
Switzerland					(1)	
UK	(1)	(6)	(4)			
Yugoslavia	4(2)					
FORMER USSR[f]		**49**				
Russia			3(2)		(6)	
Ukraine			75			
OCEANIA		**66(4)**			**296**	**8**
Australia		(2)			(3)	
Micronesia		34[b]				
New Zealand		3 (2)				
Tuvalu		27[b]			293	8

Continued

TABLE 5.7   Continued

[a]   Incomplete figures.
[b]   Provisional figures.
[c]   Case definition based mainly on physicians' diagnosis – out of a sample of 566 specimens, 243 (43%) were positive for *Vibrio cholerae* 01, El Tor.
[d]   Estimated cases.
[e]   Data for 1989 are for the Federal Republic of Germany only.
[f]   Data for the whole of the former USSR were not reported in 1991 or 1992.

Numbers in parentheses refer to the number of imported cases.

In this table a blank space represents no reported incidences of cholera.

Sources:
WHO 1990 *Weekly Epidemiological Record*, **65**(19), 141–148.
WHO 1991 *Weekly Epidemiological Record*, **66**(19), 133–140.
WHO 1992 *Weekly Epidemiological Record*, **67**(34), 253–260.
WHO 1993 *Weekly Epidemiological Record*, **68**(21), 149–156.

TABLE 5.8   Number of cases of dracunculiasis in selected countries as reported to WHO, 1984–1992

Region/country	1984	1985	1986	1987	1988	1989	1990	1991	1992
AFRICA									
Benin				400	33,962	7,172	37,414[a]	4,006	4,315
Burkina	1,739	458	2,558	1,957	1,266	45,004[a]	42,187[a]		11,784
Cameroon	0	168	86		752[a]	871[a]	742[a]	393[b]	127[b]
Cent. African Rep.		31	0	1,322			10		
Chad	1,472	9	314						156[a]
Côte d'Ivoire	2,573	1,889	1,177	1,272	1,370	1,555	1,360	12,690	
Ethiopia	2,882	1,467	3,385	2,302	1,487	3,565	2,333		303[a]
Gambia	0	0	0	0			0		
Ghana	4,244	4,501	4,717	18,398	71,767	179,556[a]	123,793[a]	66,697[b]	33,464[b]
Guinea	0	0	0	0		1	0		
Kenya						5[a]	6[a]		
Mali	5,008	4,072	5,640	435	564	1,111	884	16,024[a]	
Mauritania	1,241	1,291		227	608	447	8,301[a]		1,557
Niger		1,373		699		288		32,829[a]	500
Nigeria	8,777	5,234	2,821	216,484	653,492[a]	640,078[a]	394,082[a]	281,937[b]	183,169[b]
Senegal		62	128	132	138		38[a]	1,341[a]	728
Sudan			822	399	542				2,447[a]
Togo	1,839	1,456	1,325		178	2,749	3,042[a]	5,118	8,179[b]
Uganda	6,230	4,070			1,960	1,309	4,704	120,259[a]	126,369[a]
ASIA									
India[a]	39,792	30,950	23,070	17,031	12,023	7,881	4,798	2,185	1,081
Pakistan				2,400	1,110[a]	534[a]	160[a]	106[b]	23[b]

[a]   National survey.
[b]   Village-based reporting.

Sources:
WHO 1990 Dracunculiasis global surveillance summary, 1989, *Weekly Epidemiological Record*, **65**(30), 292–236.

WHO 1991 Dracunculiasis global surveillance summary, 1990, *Weekly Epidemiological Record*, **66**(31), 225–232.
WHO 1992 Dracunculiasis global surveillance summary, 1991, *Weekly Epidemiological Record*, **67**(17), 121–128.
WHO 1993 Dracunculiasis global surveillance summary, 1992, *Weekly Epidemiological Record*, **68**(18), 125–132.

TABLE 5.9 Prevalence of human *Trypanosoma cruzi* infection (Chagas disease) in Latin America, 1980–1985

Country	Population at risk		Population infected	
	No. of persons (10^3)	% of total population	No. of persons (10^3)	% of total population at risk
TOTAL	88,987	25	15,849	18
GROUP I				
Argentina	6,900	33	2,640	38
Brazil	41,054	32	6,340	26
Chile	1,800	15	1,460	81
Ecuador	3,823	41	30	1
Honduras	1,824	42	300	16
Paraguay	1,475	45	397	27
Peru	6,676	34	643	10
Uruguay	975	33	37	4
Venezuela	11,392	68	1,200	11
GROUP II				
Bolivia	1,800	30	500	28
Colombia	3,000	10	900	30
Costa Rica	1,112	45	130	12
GROUP III				
El Salvador	2,146	43	322	15
Guatemala	4,022	52	730	18
Panama	898	42	220	24

Countries listed are those in which human *T. cruzi* infection is endemic. However, a small number of cases have been reported in the following countries: Belize, Mexico, Nicaragua, Trinidad & Tobago and the USA.

GROUP I countries include those in which the high prevalence of human *T. cruzi* infection and high triatomine house infestation rates have induced the national health authorities to establish control activities and primary health care.
GROUP II countries include those in which there is evidence of intradomiciliary transmission with a clear association with *T. cruzi* infection and electrocardiographic alterations as well as other pathologies associated with Chagas disease. No formal control programme has been introduced.

GROUP III countries are those in which there is evidence of domiciliary transmission but more accurate epidemiological data are needed to support the evidence of a clear correlation between human *T. cruzi* infections and clinical pictures. The acute phase of Chagas disease is frequently observed.

Sources:
WHO 1990 Chagas disease – frequency and geographical distribution, *Weekly Epidemiological Record*, **65**(34), 257–264.
Moncayo A. 1992 Chagas disease: epidemiology and prospects for interruption of transmission in the Americas, *World Health Statistics Quarterly*, **45**(2/3), 276–279.

TABLE 5.10 Number of malaria cases in selected countries as reported to WHO, 1987–1990

Region/country	Number of reported cases				Number of reported cases per 10^3 population[a]			
	1987	1988	1989	1990	1987	1988	1989	1990
AFRICA								
Algeria	64	189			0.0	0.0		
Egypt	33	225			0.0	0.0		
Libya	75	76			0.1	0.1		
Morocco	1,287	550			0.0	0.0		
NORTH AMERICA								
Belize	3,258	2,725	3,285	3,033	19.2	16.0	18.3	16.1
Costa Rica	883	1,016	699	1,151	1.1	1.3	0.9	1.4
Dominican Rep.	1,206	1,072	1,275	356	0.2	0.2	0.2	0.0
El Salvador	12,834	9,095	9,605	9,269	2.8	2.0	2.0	2.0
Guatemala	57,662	52,561	42,453	41,711	16.8	15.8	11.7	11.1
Haiti	12,134	12,306	23,231	4,806	2.5	2.8	4.6	0.8
Honduras	19,095	29,737	45,922	53,095	4.2	6.7	9.9	11.7
Mexico	102,984	116,238	101,241	44,513	2.3	2.7	2.3	1.0
Nicaragua	17,011	33,047	45,982	35,785	4.9	9.4	12.3	9.2
Panama	1,195	1,000	427	381	0.5	0.5	0.2	0.2
SOUTH AMERICA								
Argentina	1,521	666	1,620	1,660	0.4	0.2	0.4	0.4
Bolivia	24,891	22,258	25,367	19,680	9.5	9.1	8.9	6.8
Brazil	508,864	559,535	577,520	560,396	8.4	8.9	9.0	8.5
Colombia	90,014	100,850	100,286	99,489	4.6	5.2	4.9	4.5
Ecuador	63,503	53,607	23,274	71,670	11.1	9.4	3.8	11.5
French Guiana	3,318	3,188	6,284	5,909	24.7	35.4	69.8	59.7
Guyana	34,142	35,470	20,822	22,681[b]	34.5	35.8	20.4	28.5[b]
Paraguay	3,741	2,884	5,247	1,660	1.1	0.9	1.5	0.4
Peru	36,136[b]	32,359[b]	32,114	28,882[b]	5.7[b]	4.7	4.5	3.9[b]
Suriname	2,044	2,691	1,704	1,608	7.1	9.4	5.8	5.3
Venezuela	17,988	45,827	43,374	46,910	0.9	3.2	2.9	1.7
ASIA								
Afghanistan	428,128[b]	378,896	257,282	317,479	42.7[b]	49.9	28.4	34.4
Bangladesh	35,848	33,824	50,738	53,875	0.3	0.3	0.5	0.5
Bhutan	13,134	11,314	19,162		61.9	60.7	100.4	
Cambodia				123,796				73.5
China[c]	210,614	93,170	88,569	86,628	0.1	0.1	0.1	0.1
India	1,611,189	1,854,830	2,017,823	1,777,253	2.1	2.4	2.6	2.2
Indonesia[d]	19,309	32,471	22,802	26,018	0.1	0.2	0.1	0.2
Iran		53,319	59,175	77,470	1.3	1.3	1.4	1.8
Iraq	3,742	6,883	4,963	1,832	0.4	0.4	0.3	0.1
Laos	34,960	37,724	34,637	21,499	23.3	12.2	10.9	6.5
Malaysia	36,657	50,743	65,363	50,500	2.2	3.0	3.9	2.8
Myanmar	61,650	94,736	135,194	23,167[b]	1.7	2.6	3.6	0.6[b]
Nepal	26,690	23,751	22,333	20,338	2.0	2.1	2.2	1.7
Oman	15,514	24,619	17,867	32,720	11.7	17.8	12.6	21.8
Pakistan	64,342	57,811	107,739	79,689	0.6	0.5	1.0	0.7

Continued

TABLE 5.10 Continued

Region/country	Number of reported cases				Number of reported cases per 10^3 population[a]			
	1987	1988	1989	1990	1987	1988	1989	1990
Philippines	154,091	154,943	115,542	86,172	9.0	8.8	6.4	5.4
Saudi Arabia	17,650	9,797	6,475	15,666	3.3	2.1	1.3	3.2
Sri Lanka	676,569	383,294	258,727	287,384	31.0	31.0	20.9	23.0
Syria	150	105	83	107	0.0	0.0	0.0	0.0
Thailand	321,510	349,291	299,137	272,634	6.9	7.5	6.4	5.6
Turkey	20,134	16,245	12,112	8,680	0.4	0.3	0.2	0.1
United Arab Em.		3,056	2,823	3,514		2.0	1.8	2.2
Viet Nam	130,690	151,520	142,818	139,588	2.9	3.3	3.1	3.0
Yemen	2,551	11,291	11,411	11,783	0.8	1.9	1.9	1.9
OCEANIA								
Papua New Guinea	164,228	83,944	121,796	104,939	47.2	23.6	33.9	28.4
Solomon Is	72,000[b]	63,893	65,241	116,449	248.0[b]	213.0	209.0	372.0
Vanuatu	26,631	26,894	25,422	28,558	190.2	179.3	169.5	194.3

[a] These figures are with respect to the population in originally malarious areas not the national population.
[b] Provisional figures.
[c] Microscopically confirmed cases only.
[d] Java and Bali only.

Sources:
WHO 1990 World malaria situation 1988, *World Health Statistics Quarterly*, **43**(2), 68–79.
WHO 1992 World malaria situation 1990, *World Health Statistics Quarterly*, **45**(2/3), 257–266.
WHO 1992 World malaria situation in 1990 Part I, *Weekly Epidemiological Record*, **67**(22), 161–167.
WHO 1992 World malaria situation in 1990 Part II, *Weekly Epidemiological Record*, **67**(23),169–176.

TABLE 5.11 Cancer incidence in selected countries and sub-populations by site, around 1980 (percentage of all cancers)

Region/country	Gender	Stomach	Lung	Prostate	Colon	Bladder	Pancreas	Liver	Kidney	Rectum	Oesophagus	Nasopharynx	Hypopharynx	Larynx	Mouth	Tongue	Lip	N.H.L.	Gall-bladder	Thyroid	Nervous system	Myeloma	Melanoma	Breast	Ovary	Cervix uteri	Corpus uteri	Other
AMERICAS																												
Brazil																												
Fortaleza	M	20.9	10.5	14.5	3.3	2.7	2.6				3.4			3.6	2.5			3.6										32.4
	F	7.8	2.2		3.0													2.3						23.3	3.6	3.6	3.3	27.5
Recife	M	12.7	9.4	18.7		4.2		3.9			5.3			3.4		3.0		3.2	2.8	2.6								36.2
	F	3.2			2.3			3.1																18.1	2.7	21.6	2.5	34.6
Porto Alegre	M	9.4	20.9	10.7	4.0	3.6	2.8			2.3	8.1			4.4														30.6
	F	5.0	4.8		5.2					2.9	3.1								2.9					24.5	3.9	9.6	4.5	33.2
São Paulo	M	16.8	11.5	10.4	4.1	4.9				2.8	5.1			5.6	2.5	2.3												34.0
	F	9.1	2.8		4.3					3.3								2.1						23.6	3.7	12.7	5.0	23.4
Canada	M	4.6	21.2	15.1	8.3	7.0	3.0			5.3				2.4				3.3					2.5					27.1
	F	2.5	7.1		9.5					4.1								3.0						28.1	4.9	4.4	7.1	26.8
Alberta	M	4.7	18.4	18.0	7.4	6.8	3.4		2.7	5.7							3.6	3.7					2.7					25.6
	F	2.5	6.8		8.3		2.6			4.7								3.6						28.6	5.3	4.2	7.4	25.8
British Columbia	M	4.5	20.3	18.7	8.2	4.2	3.3		2.9	5.8								3.5					2.5					26.1
	F		8.8		9.0		2.5			4.2								3.1					3.6	29.6	4.7	4.0	6.5	24.0
Manitoba	M	4.8	21.1	15.2	9.2	6.1	3.0		3.2	5.4							3.7	3.6					3.6					24.7
	F		6.9		9.6		2.6			3.5								3.3						28.7	4.9	6.6	7.6	23.8
Maritime Provinces	M	5.6	23.1	13.9	9.6	7.0	3.0		2.8	5.6				2.6				3.2					2.5					23.6
	F	2.8	7.5		11.0					4.7								2.7						29.2	4.6	5.5	6.2	23.3
New Brunswick	M	6.2	22.9	14.2	9.4	6.9	3.9		2.9	5.4				2.3				3.4					2.5					23.1
	F	3.0	7.2		11.1					4.8								2.8						27.8	4.0	5.8	6.1	24.9
Nova Scotia	M	5.1	24.2	13.5	10.0	7.1	2.6			5.6				3.0				3.0					2.6					23.3
	F	2.7	8.1		10.9					4.9								2.7						30.3	5.1	5.1	6.5	21.1
New Foundland	M	9.5	20.2	10.3	9.1	6.9	2.8		3.3	6.2				2.3			5.7						2.6					23.7
	F	5.7	4.3		12.5				2.4	6.1										2.1				24.9	4.7	6.6	5.2	25.5
Northwest Territories & Yukon	M	2.2	29.8	11.5	5.9	4.2			6.2	8.9	2.9							3.6			1.7							24.8
	F		21.7		10.7				3.0	3.8														24.4	3.4	8.4		21.2
Ontario	M	3.9	20.8	14.1	9.4	7.5	2.9		2.5	4.7								3.5										28.1
	F		7.5		10.1	2.5				3.7								3.2						26.8	5.0	4.1	7.2	26.7
Quebec	M	5.0	23.1	13.7	7.9	7.5	2.7		2.7	5.3				3.4				2.9					3.2					25.8
	F	2.8	6.0		9.5	2.6				4.4								2.8						28.8	4.8	5.0	7.2	26.1
Saskatchewan	M	4.1	17.3	20.4	7.2	6.3	3.4		3.1	5.9							4.4	3.7					2.7					24.2
	F		6.9		9.1		2.6			4.0								3.3						29.8	5.4	3.2	7.2	25.8
Colombia																												
Cali	M	23.6	12.1	14.6	2.5	4.6	2.5			1.6				2.6				2.6			2.2							31.1
	F	11.9	4.4		2.8					1.7									4.2	2.4				15.8	3.5	21.8	2.7	28.8

Continued

TAELE 5.11 Continued

Region/country	Gender	Stomach	Lung	Prostate	Colon	Bladder	Pancreas	Liver	Kidney	Rectum	Oesophagus	Nasopharynx	Hypopharynx	Larynx	Mouth	Tongue	Lip	N.H.L.	Gall-bladder	Thyroid	Nervous system	Myeloma	Melanoma	Breast	Ovary	Cervix uteri	Corpus uteri	Other
Costa Rica	M	29.6	9.0	13.2	2.7	3.9	3.1	2.6		2.5				2.5				2.6										
	F	13.9	3.7		2.9					2.7									3.2	2.5				16.9	3.0	20.4	3.1	27.7
Hawaii																												
White	M	3.4	19.1	16.8	8.1	7.2			3.0	4.0				2.9				3.1					6.6					25.8
	F		11.3		7.4		2.8			2.6								2.3					6.5	29.2	3.8	2.8	8.1	23.2
Japanese	M	12.3	15.6	13.5	14.8	4.1	3.2	2.7	3.0	7.7								2.9		3.3								20.2
	F	7.8	5.7		12.2					5.0								3.0						27.8	4.4	3.6	8.6	18.6
Hawaiian	M	10.0	26.6	13.1	6.4	3.2	2.9		2.6	5.2								3.2										23.3
	F	5.0	13.3		4.0		2.6			2.3	3.5									3.5				31.6	4.7	4.1	8.5	20.4
Filipino	M	3.9	15.4	17.5	3.9	2.7	3.3	3.5		8.5				15.4				4.2		3.5								26.5
	F	2.9	10.6		7.3	2.5				2.3								3.5		11.8				20.7	5.7	4.6	7.1	23.3
Chinese	M	5.5	17.9	12.3	11.0	6.6		3.8		5.9								4.1		4.3								23.2
	F	3.6	9.4							2.9										2.9			5.4	30.7	4.2	3.4	10.0	18.2
Martinique	M	12.3	5.4	24.4	3.0						7.3			4.4		4.3						2.9						32.4
	F	7.2	3.2		4.9						2.8							3.6			3.2	3.7		20.4	2.9	19.2	5.0	28.1
Neth. Antilles[a]	M	10.7	19.1	15.0	4.8	3.2	4.9			2.8	7.4							3.2										28.1
	F	5.2	3.2		5.8		2.6			2.6	3.9													26.1	6.1	13.3	3.9	27.3
USA																												
Puerto Rico	M	9.3	9.5	16.2	5.4	5.1				3.9	6.3			3.7	3.4	3.0		2.7										34.2
	F	5.8	4.3		6.5					3.6	2.4											2.2		23.4	4.0	10.4	5.4	31.5
California, Alameda County: White	M	3.4	22.0	15.9	9.8	6.7	2.7		2.7	5.2								3.6					3.1					24.9
	F		12.2		8.1	2.5				3.9								2.8					3.3	28.6	4.4	2.8	9.1	22.3
California, Alameda County: Black	M	4.2	25.7	22.1	8.3	2.9	4.1			3.6	3.1			2.4														21.4
	F	2.8	12.3		11.2		3.7			3.4								3.6				2.2		26.9	4.1	5.5	4.2	24.0
San Francisco Bay: White	M	3.3	21.1	16.0	9.8	7.0	2.9		1.9	4.9													3.3					25.4
	F		11.3		8.0		2.3		2.7	3.7								2.9					3.1	29.5	4.4	3.0	8.7	23.1
San Francisco Bay: Black	M	4.9	26.0	21.2	7.4	3.1	4.0			3.2				2.5				2.0				2.0						22.3
	F		12.0		11.1		3.3			3.2								3.5						26.6	3.6	5.7	4.6	25.5
San Francisco: Chinese	M	4.0	24.1	6.6	11.0	4.5	3.1	8.4		5.5		6.6						3.8										22.7
	F		12.1		9.1					5.1		4.0												22.1	4.3	6.1	7.5	22.8
San Francisco: Filipino	M	2.5	19.8	17.9	8.4		4.4	5.6		3.9								4.4		3.1		2.7						27.3
	F		5.6		5.6		1.8			3.0								3.2		8.8				26.9	6.2	5.5	5.3	25.8
San Francisco Bay: Japanese	M	12.9	17.5	8.7	15.8	4.7	2.7	3.4	3.4	7.2								5.6			2.6							16.8
	F	6.1	6.9		11.8		2.7			7.0														27.7	5.0	3.4	11.1	15.7

Continued

TABLE 5.11 Continued

Region/country	Gender	Stomach	Lung	Prostate	Colon	Bladder	Pancreas	Liver	Kidney	Rectum	Oesophagus	Nasopharynx	Hypopharynx	Larynx	Mouth	Tongue	Lip	N.H.L.	Gall-bladder	Thyroid	Nervous system	Myeloma	Melanoma	Breast	Ovary	Cervix uteri	Corpus uteri	Other
ASIA																												
China																												
Shanghai	M	23.8	22.3		3.5	2.9	2.2	14.1		3.8	8.5	1.8									1.5							15.6
	F	15.9	12.0		4.9		2.4	7.5		4.6	5.8													12.4	3.3	5.5		25.7
Tianjin	M	20.4	24.1		2.2	3.5	2.5	10.2		3.6				1.9							2.3							16.0
	F	10.4	20.0		3.0			6.2		4.4														12.7	2.6	9.6		22.6
Hong Kong	M	6.8	20.6		5.6	5.6		11.4		4.4	6.6	10.6		3.3				2.3										22.7
	F	4.9	12.3		6.3			3.8		4.7		6.6												14.5	3.0	12.3	3.2	28.7
India																												
Bangalore	M	12.5	5.9	4.9		4.0	4.1	4.5	3.1	2.2	9.3		5.4	5.0	4.5	5.0												39.0
	F	5.2														11.4								14.5	4.3	29.3	1.5	23.4
Bombay	M	6.2	10.9	5.7			3.9	3.4		3.1	6.1		6.9	6.9	4.6	6.5												35.6
	F	4.8	2.8		2.3			2.0							4.0	2.7				2.1				19.3	5.8	16.6		31.4
Madras	M	15.0	6.3	3.4			4.0			3.3	8.3		4.6	5.8	8.8	4.8												39.5
	F	5.6									3.2				8.3	1.7				1.5				17.4	3.6	38.5		18.6
Nagpur	M	6.5	6.4	4.0	1.6					3.6	11.8		5.3	9.8	4.3	1.9												41.5
	F	6.3								1.8	8.5			2.1	4.3									17.9	3.8	27.3	1.6	24.5
Poona	M	7.9	8.3	3.6	1.5		2.7			3.6	11.0		3.5	9.7	6.3	3.6		2.2										40.3
	F	4.4	2.4							2.3	8.5				4.4									17.7	3.5	24.3	2.2	28.8
Israel																												
All Jews	M	7.5	12.9	8.7	7.5	9.4	4.1			7.7								4.6			4.2		3.4					30.3
	F	4.3	4.2		6.9					6.2								3.3			3.9			28.3	5.5		4.5	29.5
Jews born in Israel	M	5.8	11.4	11.5	6.1	11.3	3.9			6.8								4.6			3.9		4.2					30.5
	F	3.1	2.8		6.3					4.9								3.7			4.8			31.0	4.9			28.2
Jews born in Europe or America	M	7.4	12.0	9.1	8.5	8.4	4.0		3.1	9.1								4.2			4.4		6.4					29.8
	F	4.1	4.2		7.0					6.3								3.1			3.3			27.7	6.0		3.9	29.5
Jews born in Africa or Asia[b]	M	5.2	13.3	7.1	3.9	7.7		3.1		4.3	15.1										4.5		3.7					31.0
	F	2.7			5.0					3.3								4.8			5.9			23.6	2.4		4.9	26.8
Non-Jews	M	6.7	20.0	5.5	4.0	10.6	2.7			2.6				4.2				5.7	8.9		4.7							33.3
	F	6.1	4.4		5.6													5.3	4.3		4.1			17.5	4.8	3.8	2.3	40.3
Japan																												
Hiroshima City	M	34.3	12.7	2.9	4.9	4.5	3.2	8.9		4.9	2.6							2.3										18.8
	F	21.6	6.0		5.7		3.6	3.6		3.9										3.2				16.8	3.9	13.2	2.2	20.0
Miyagi Prefecture	M	37.0	13.8	2.9	4.5	3.0	4.2	5.2		4.6	6.2								2.7									15.9
	F	24.8	6.0		6.5		3.5	2.8		5.1									3.7					15.2	2.9	6.9		22.6

Continued

TABLE 5.11 Continued

Region/country	Gender	Stomach	Lung	Prostate	Colon	Bladder	Pancreas	Liver	Kidney	Rectum	Oesophagus	Nasopharynx	Hypopharynx	Larynx	Mouth	Tongue	Lip	N.H.L.	Gall-bladder	Thyroid	Nervous system	Myeloma	Melanoma	Breast	Ovary	Cervix uteri	Corpus uteri	Other
Nagasaki City	M	31.8	12.7	3.4	7.1	4.1	2.9	10.0		5.0	3.2							3.3										16.5
	F	21.8	7.1		5.8			4.7		4.6									4.1	3.6				12.2	4.2	9.8		22.1
Osaka Prefecture	M	31.7	15.0	2.1	5.0	3.2	3.2	13.2		4.4	3.3							2.1										16.8
	F	23.8	7.0		5.4			5.2		4.1									2.9					13.0	2.9	10.6		22.1
Kuwait																												
Kuwaitis	M	5.1	19.2	8.3		6.2		6.1		4.1	2.8	2.9	2.8	4.7				4.0										36.6
	F		8.8								2.8							2.5		8.8		2.5		22.0	4.6	5.4	2.5	37.3
Non-Kuwaitis	M	4.5	17.8	7.2		10.0		4.6		3.0	4.0		2.7	4.8				5.3										35.0
	F		7.9		3.7	6.1			4.1		3.0									7.2				18.9	5.9	5.3	4.2	34.8
Philippines																												
Rizal Province	M	6.7	26.5	8.0	3.4		2.4	12.6		3.6		3.4		2.5				3.3										27.6
	F	4.9	6.4		3.0			5.2							3.3					3.5				22.0	5.2	12.3		34.2
Singapore																												
Chinese	M	13.7	26.8	2.4	6.0	2.9		11.6		5.3	5.0	6.6		3.0														16.7
	F	8.8	13.0		9.1			4.1		6.0														15.5	4.9	9.7	2.9	21.5
Malay	M	7.9	22.6	6.4	4.4	3.3		13.3	2.5	4.2		3.4						4.6										27.4
	F	6.2	6.2		7.0			4.8		4.5										4.5				19.0	8.9	8.9	3.5	26.5
Indian	M	10.1	13.8	5.8	5.2	3.8		9.2		4.1	5.6			5.3	5.6			4.7										32.4
	F	9.7	3.9		11.7					5.4	2.3							2.0						16.1	2.5	16.4		24.4
EUROPE																												
Czechoslovakia																												
Slovakia	M	12.2	27.0	6.1	4.6	4.6	3.2		2.5	5.9				4.1		2.5												27.3
	F	8.8	4.1		5.6		2.7			5.3									3.8					18.9	5.6	9.1	8.3	27.8
Denmark	M	5.5	21.7	10.6	7.2	9.5	3.5		3.6	6.7										3.2							28.5
	F		6.9		7.9		2.9			4.6										3.2		3.5	26.2	6.0	7.7	6.3	24.8
Finland	M	9.6	29.0	13.4	3.9	5.0	3.9		3.3	3.9							3.0	3.0	2.5								22.5
	F	7.4	4.0		5.9					3.8														25.5	5.6	3.2	6.9	31.1
France																												
Bas-Rhin	M	4.7	18.2	8.3	7.0	6.6			2.3	5.4			3.6	3.7	4.1			2.7					2.3					33.4
	F	3.6			7.2					4.2														30.8	5.5	7.8	6.3	27.3
Calvados	M	6.0	16.4	10.1	4.8	4.1				5.4			6.2	3.9				2.0					2.8					26.5
	F	5.0	2.0		5.8					4.8														33.8	5.0	10.0	5.8	28.3

Continued

TABLE 5.11 Continued

Region/country	Gender	Stomach	Lung	Prostate	Colon	Bladder	Pancreas	Liver	Kidney	Rectum	Oesophagus	Nasopharynx	Hypopharynx	Larynx	Mouth	Tongue	Lip	N.H.L.	Gall-bladder	Thyroid	Nervous system	Myeloma	Melanoma	Breast	Ovary	Cervix uteri	Corpus uteri	Other
Doubs	M	5.4	18.0	8.8	6.5	6.2				6.2	4.3		3.1	4.5				2.1			2.7		2.9					37.0
	F	3.9			8.3					5.4															30.9	5.3	7.2	5.6	25.7
Isere	M	4.7	18.4	9.6	6.6	5.3			2.8	6.2	4.2			4.8					2.6			2.6							34.8
	F	3.2	2.3		7.7					4.9									2.5						34.3	4.7	7.6	5.7	24.5
German Dem. Rep.	M	11.4	26.7	9.0	5.3	5.3	3.2		3.9	6.3				2.4						3.9									26.5
	F	6.7	3.1		6.2					5.4															22.4	6.6	13.3	7.5	22.5
Germany, Fed. Rep.																												
Hamburg	M	9.6	26.3	10.7	6.3	6.0	2.9			5.9														2.2					27.4
	F	5.5	5.2		7.3		2.2			4.3															27.2	6.0	9.5	5.3	25.3
Saarland	M	8.0	24.5	9.7	7.1	7.0	2.3		3.5	7.3				2.5					2.2	2.5		2.6							25.9
	F	5.7	3.3		8.1					6.2															26.6	4.3	6.5	6.7	27.5
Hungary																												
County Szabolcs-Szatmar	M	17.1	28.1	6.7	4.1	4.0	2.8	1.8		5.7				5.0			4.4			5.5									20.3
	F	11.1	6.1		6.1		3.0			5.4															19.9	5.1	8.2	6.7	22.9
County Vas	M	14.3	25.5	7.7	5.3	3.5	3.8		2.4	9.1				4.1			2.8			4.2									21.5
	F	7.3	3.6		6.2		3.5			6.6															22.1	5.7	11.2	9.3	20.3
Iceland	M	14.0	11.0	16.1	6.4				5.4	3.5										2.5	3.8								26.3
	F	6.0	7.9		6.2				3.3											5.7	4.3			25.7	5.9	5.0	5.7	24.3
Ireland[c]	M	6.4	18.9	10.6	9.2	4.5	3.4			6.8	3.4						6.0					3.6		3.5					27.2
	F		6.5		9.5					4.7	2.6											2.9			32.9	5.0	3.5	3.0	25.9
Italy																												
Lombardy Region, Varese Province	M	11.6	24.1	6.1	6.2	8.2	2.5		2.8					4.8				2.9			2.2							25.6
	F	8.4	3.0		8.1					4.4								2.2			2.7			29.2	5.0	5.0	6.1	26.4
Parma Province	M	16.1	21.8	6.4	6.9	6.5	2.8	3.1		4.1				4.5														24.3
	F	9.9	3.7				2.3			4.7														29.1	5.8	6.6	6.6	24.2
Ragusa Province	M	10.2	17.0	9.8	6.3	6.1	2.3	3.1		3.6				5.6			5.7					3.0							28.2
	F	5.8	1.9		7.2			2.8														3.3			32.3	5.5	6.7	8.5	23.4
Netherlands																												
Eindhoven	M	7.1	32.5	9.7	7.1	6.8	2.4		3.1	5.2				2.7				2.6					2.9					20.8
	F	4.6	2.8		9.7				2.1	4.4															35.1	5.7	3.7	5.1	23.9

Continued

TABLE 5.11 Continued

Region/country	Gender	Stomach	Lung	Prostate	Colon	Bladder	Pancreas	Liver	Kidney	Rectum	Oesophagus	Nasopharynx	Hypopharynx	Larynx	Mouth	Tongue	Lip	N.H.L.	Gall-bladder	Thyroid	Nervous system	Myeloma	Melanoma	Breast	Ovary	Cervix uteri	Corpus uteri	Other
Norway	M	7.6	13.0	17.6	7.3	7.1	3.5		3.7	6.2												3.5		3.7					26.8
	F	4.3	3.4		8.3					4.8												3.6		5.0	24.4	7.2	7.3	5.7	26.0
Poland																												
Cracow	M	13.5	30.1	5.7	3.3	4.3	3.9		2.6	4.2				3.8								3.6							25.0
	F	7.4	6.7		3.4					3.4										4.4					21.7	6.6	11.1		27.1
Nowy Sacz Rural Areas	M	21.2	22.8	5.0		5.1	2.8	2.7		3.6				6.0			3.2					3.3							24.3
	F	13.7	3.8		3.9		3.1			3.3												5.0			14.8	9.6	9.9	3.5	29.4
Warsaw City	M	11.1	26.8	5.5	5.1	4.7			3.2	4.4				5.5								3.3							26.6
	F	5.7	7.8		5.6					3.6										5.4					20.6	6.7	9.3	5.9	26.7
Romania																												
County Cluj	M	19.6	19.9	5.6		5.4	3.7	4.6		4.9				4.6			4.8					2.5							24.4
	F	3.3	4.3		3.0			2.9		4.8															22.6	4.9	14.4	3.3	30.4
Spain																												
Catalonia, Tarragona	M	8.3	15.5	10.4	5.6	11.7		3.3		5.1	2.6			5.2								2.5							29.8
	F	5.1	2.1		5.6	2.5		2.5		5.4															29.6	4.3	5.1	7.4	30.4
Navarra	M	13.0	14.4	8.4	4.6	8.8		3.2		4.6				7.1			3.2					3.2							29.5
	F	8.8			5.3			3.1		4.9										2.9		3.7			25.2	4.2	3.7	8.1	30.1
Zaragoza	M	10.9	17.9	8.9	3.1	7.4		3.8		3.8				8.1			3.2					3.6							29.3
	F	8.1	2.8		3.7			4.3		4.2												3.6			27.3	4.2	4.8	5.9	31.1
Sweden	M	6.4	10.8	19.7	7.2	6.6	3.7		4.8	5.0								3.2			4.0		3.7					28.6
	F	3.3	3.4		7.1					3.5											4.2			27.1	6.8	4.4	5.9	30.6
Switzerland																												
Basel	M	6.2	19.8	15.9	7.3	8.9		2.7	3.0	5.0								3.5					3.1					27.7
	F	4.0	3.8		7.1				3.1	4.3														33.2	4.5	3.2	6.7	27.0
Geneva	M	4.2	22.8	12.4	7.6	6.2		3.2	3.0	4.0													2.8					31.0
	F	2.8	4.9		7.3		2.3			4.4	2.8													32.5	5.4	3.7	6.8	25.6
Neuchatel	M	6.4	20.6	11.7	6.7	7.0	3.7	2.7		4.8				3.6									3.0					30.4
	F	3.0	3.5		5.7	2.4				5.1	2.7													29.8	5.0	6.6	7.5	28.4
Vaud	M	5.4	21.1	12.6	6.9	4.8				5.2				2.9				4.1					3.9					30.9
	F	2.5	3.4		7.5		2.8			4.5	3.3							3.3						32.2	4.5	5.0	6.4	26.8
Zurich	M	5.5	18.6	17.1	6.9	7.3			3.3	5.4								3.3					2.9					29.7
	F	3.2	3.7		6.1					4.3								2.8					4.4	29.5	5.2	5.2	8.4	27.2
UK																												
England and Wales	M	7.8	30.2	8.7	7.0	7.1	3.3			5.7	2.5							2.4			2.2							23.1
	F	4.1	9.9		7.7	2.4	2.6			4.1														28.2	5.8	6.1	4.3	24.8

Continued

TABLE 5.11 Continued

Region/country	Gender	Stomach	Lung	Prostate	Colon	Bladder	Pancreas	Liver	Kidney	Rectum	Oesophagus	Nasopharynx	Hypopharynx	Larynx	Mouth	Tongue	Lip	N.H.L.	Gall-bladder	Thyroid	Nervous system	Myeloma	Melanoma	Breast	Ovary	Cervix uteri	Corpus uteri	Other
Scotland	M	7.2	32.1	8.2	7.2	7.0	2.9			4.7	2.8																	23.3
	F	4.3	11.8		8.4	2.9	2.6			3.7														26.7	5.1	5.4	3.0	26.1
Yugoslavia																												
Slovenia	M	14.8	24.5	8.0	3.7	4.0	2.9		2.0	6.1	3.0			3.7														27.3
	F	9.5	4.2		4.9	2.9				5.5									2.8					23.7	6.0	8.5	7.3	25.1
OCEANIA																												
Australia																												
New South Wales	M	4.7	19.6	12.4	8.7	6.3	2.8		2.6	5.6								3.4					6.3					27.6
	F	2.8	5.5		9.7					4.5								3.2					7.8	25.6	4.3	5.3	4.0	27.3
Queensland	M	4.3	18.8	12.7	8.2	5.3			2.8	4.7							3.5	3.3					10.1					26.3
	F		4.8		11.5					3.8								3.0					12.8	25.0	3.8	5.5	4.4	23.2
South Australia	M	4.7	18.3	15.1	8.1	6.3	2.8		2.8	5.7							3.3	3.9			2.2		4.9					26.9
	F	2.8	4.9		9.8					4.7								3.8					8.0	26.2	4.0	4.3	5.3	26.2
Tasmania	M	4.9	21.1	12.2	9.0	6.0	2.8		3.1	6.2								3.4					4.2					27.1
	F	2.3	6.3		9.9					5.5								3.2					6.7	25.1	4.3	6.0	3.9	26.8
Victoria	M	4.8	18.1	13.4	9.0	7.6	2.6		2.7	6.2								3.5					4.5					27.6
	F		5.5		9.0	2.8				5.3								3.6					6.8	27.0	4.3	4.5	4.1	25.5
Western Australia	M	5.7	19.5	12.0	8.0	6.8	2.2		2.5	5.5								3.6					7.9					26.3
	F		6.7		8.0	2.8				4.6								3.4					10.3	25.1	4.2	6.3	4.7	23.9
New Zealand																												
Maori	M	9.2	31.4	11.0	3.3		3.8	3.5		3.6								2.4			2.8							29.0
	F	6.6	22.9		3.9		1.7			1.6								1.8						20.0	3.7	9.7	5.2	22.9
Non-Maori	M	5.2	19.7	12.6	10.0	5.1	2.8		2.8	6.4								2.2			2.7		5.9					26.8
	F	2.6	5.9							5.1													9.3	25.1	4.6	5.1	4.3	23.4
Pacific Polynesian Islanders	M	8.8	24.0	11.6	2.9	5.2	3.0	6.9				3.7		2.5	2.7			1.9		5.1								28.7
	F	4.3	6.4		5.2																2.4	2.2		22.7	7.3	18.2	5.5	24.0

a Excludes Aruba.

b Under "Jews born in Africa or Asia" the following should be included for females: Other endocrines: 7.6% Lymphoid leukaemia: 11.5%

c Southern Ireland.

N.H.L = Non-Hodgkin lymphoma.

Source:
IARC 1990 Patterns of Cancer in Five Continents, IARC Scientific Publications 102, International Agency for Research on Cancer, Lyon.

Part 6

Energy

There is an adequacy of commercial energy resources in the world; world-wide proved reserves of oil and natural gas have continued to rise and are expected to be able to sustain current levels of production for 40 and 60 years, respectively. There are sufficient reserves of coal to sustain production for at least 200 years. Consequently, fossil fuels are likely to continue to dominate the energy sector for the foreseeable future. Other key trends in the energy sector of environmental significance are highlighted below:

○ Despite significant increases in commercial energy consumption in developing countries as a whole over the past decade, biomass fuels are still the major source of energy for nearly one half of the world's population.

○ Relative to the early 1970s energy intensity has fallen in many of the developed countries, but continues to rise in many of the rapidly industrializing nations. Trends in pollutant emissions from energy sources tend to follow the same pattern.

○ Although the technologies already exist for reducing most of the pollutant emissions from coal and fuel oil, they have yet to be widely adopted. Without the implementation of such technologies and given the probable continued reliance on fossil fuels, pollutant emissions are likely to increase significantly, especially in the developing countries.

○ Reductions in the level of radioactivity discharged in liquid effluents from nuclear power stations and reprocessing plants have occurred; however, problems relating to the disposal of high-level radioactive wastes continue to concern the nuclear industry.

The provision of energy is central to the development of national economies, both as an industrial sector in itself and as an essential factor input to almost all other economic activities. However, the production, transportation, conversion and use of energy – particularly of that derived from fossil fuels – is responsible for some of the most serious environmental problems facing the world today. Thus energy, through its effects on the environment, is widely perceived as one of the key variables which determine the sustainability of the development process.

At present the world's energy supply is not produced and used in a sustainable way. Moreover, continued population and economic growth are likely to increase demands for energy; current predictions suggest that total primary energy demand will increase by as much as 50 per cent from 1980 levels by the year 2000. The issue of unsustainability of energy use was raised at the recent United Nations Conference on Environment and Development (UNCED) and, in recognition of the current state of affairs, Agenda 21 has identified the following as priorities for action (UN, 1992):

○ Co-operation in identifying and developing economically viable and environmentally sound energy sources;
○ Promotion of the development of methodologies for making integrated energy and environment and economic policy decisions; and
○ Promotion of research, development, transfer and use of improved and energy-efficient technologies.

The aim of this particular section of the 'UNEP Environmental Data Report' is to provide a selective overview of environmentally significant trends in the energy sector. The analysis presented here is not intended to be a comprehensive treatment of the issue; space restrictions prohibit coverage of the complete range of the environmental impacts associated with the energy cycle for all fuel types. Thus this edition will focus on biomass fuels and nuclear power. Selected impacts of these fuel types are highlighted and proposed indicators which broadly reflect the potential impact of energy on the environment are introduced. Renewable forms of energy, other than biomass, were highlighted in the previous edition of this report (UNEP, 1991). Throughout the discussion brief mention will be made of some recent developments within the energy sector which seek to address issues raised by Agenda 21 mentioned above.

Energy Resources

The world relies on a wide variety of energy sources. Fossil fuels (coal, oil and natural gas) and other forms of commercial energy are the dominant sources of energy in the developed world; in the developing countries biomass fuels are the main source of energy, followed by fossil fuels. Although important in some countries, hydropower and nuclear energy are relatively minor sources of energy at the global level.

The World Energy Council (WEC) is widely recognized as one of the most authoritative sources of global data and related information pertaining to energy resources, both fossil fuels and the renewable resources. By means of a question-naire sent to all 90 member countries, the WEC periodically updates its detailed inventory of energy resources and reserves at the national level. These questionnaire data are supplemented with information obtained via various governmental and non-governmental organizations, specialist journals, international companies and other bodies working in the energy field and are published as the 'Survey of Energy Resources' (WEC, 1992). Considerable care is taken when compiling these data to use standardized definitions of resources and reserves. Resources are generally considered to be the occurrences of material in a recognizable form, i.e., the "amount in place"; the proportion of these resources which can be recovered for the benefit of people are referred to as "reserves". The latter depend on physical and economic factors and the degree of certainty regarding the amount in place. For historical reasons, countries differ in their precise nomenclature of resources and reserves; such discrepancies in the definitions thus limit the intercomparability of national data (WEC, 1992).

Commercial Energy Resources

Global, regional and national estimates of proved recoverable reserves of the principal commercial energy resources as assessed in 1990 by the WEC are given in Table 6.1 and Figure 6.1. These data represent a revision of previous WEC surveys of energy resources and thus supersede those published in previous editions of this report.

At the present time coal accounts for the largest proportion of proved recoverable reserves of fossil fuels when measured in terms of weight and in terms of tonnes of oil equivalent (toe). Following re-evaluation of what is considered "exploitable" under expected local economic conditions with existing available technology, China no longer reports the largest reserves of coal. These have been re-assessed at 115×10^9 t, i.e., considerably less than the 731×10^9 t reported earlier (see UNEP, 1991). The other major reserves of coal are to be found in the USA (241×10^9 t) and the former USSR (241×10^9 t).

In recent decades coal has further improved its long-term position as the world's most widely available fossil energy source. Since 1950 proved reserves of coal have risen by around 20 per cent, from 450×10^9 toe to 570×10^9 toe (World Bank, 1992). It is estimated that there are sufficient coal reserves to sustain the current level of production (around 4.7×10^9 t per year) for more than 200 years (Table 6.2).

Proved reserves of oil and natural gas have also continued to rise especially in recent years (Figure 6.2). Oil reserves rose by 10 per cent between 1987 and 1990; this increase is largely attributed to a re-evaluation of existing reserves in the Middle East (Saudi Arabian crude oil reserves alone account for 25.9 per cent of the global total). The oil reserves/production ratio for the world currently stands at more than 40 years; possible upgrading of reserves in Russia could enhance this position yet further (WEC, 1992). Present estimates of global reserves of natural gas represent more than 60 years of production at

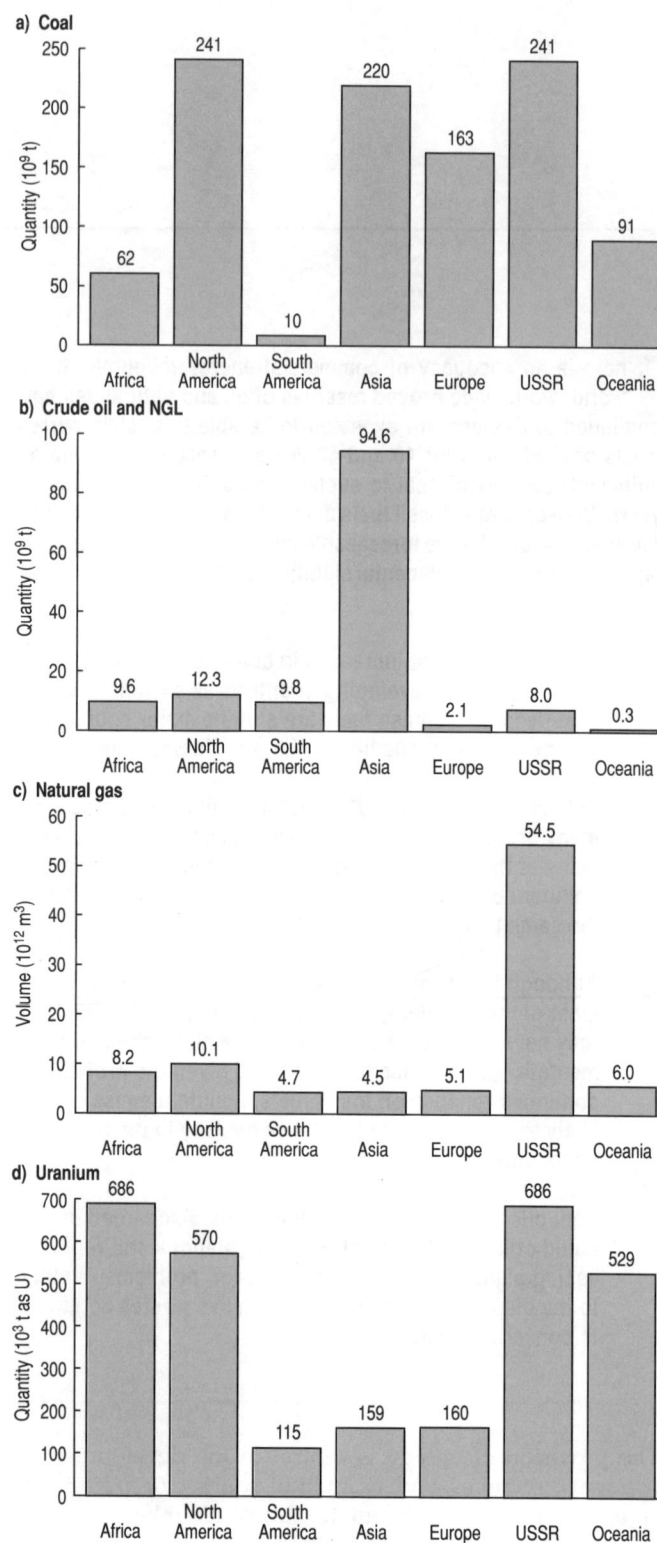

FIGURE 6.1 Regional distribution of proved recoverable reserves of a) coal, b) crude oil, c) natural gas and d) uranium, 1990
Source: Data from Table 6.1

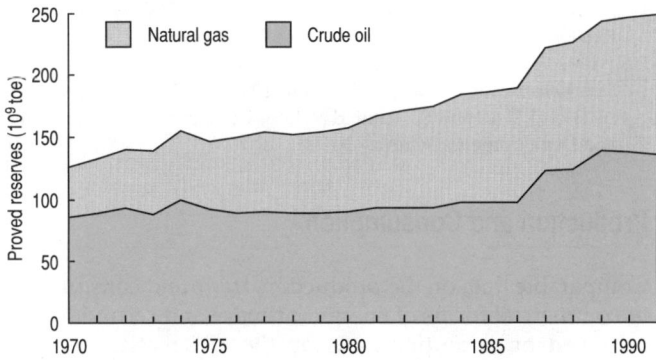

FIGURE 6.2 Variations in proved recoverable reserves of crude oil and natural gas (in 10⁹ toe equivalent), 1960–1990
Source: BP, 1992

the current rate (Table 6.2). The development of reserves in the former USSR is one of the major phenomena in the gas history in recent decades (Figure 6.1).

There is no physical resource restraint at the present time on further expansion of nuclear power, despite a downwards revision of the "known" quantity of uranium resources in the latest WEC survey. Similarly most renewable sources of energy, including biomass, hydro-power, solar and wind energy, are not constrained by physical resource limitations, but rather more by environmental, economic and technological factors. It is anticipated that, for the foreseeable future at least, renewable forms of energy are unlikely to offer economically viable alternatives to fossil fuels and thus will continue to have a limited impact on the world's overall energy balance. Nevertheless, a selective growth of renewables is considered probable (WEC, 1992).

Biomass Fuels

Biomass in all forms supplies about 14 per cent of the world's primary energy. As noted earlier, biomass fuels are particularly important in the developing world where they account for, on average, 35 per cent of primary energy. In some countries, and in rural areas in particular, this proportion rises to as much as 90 per cent (Hall and Rosillo-Calle, 1991).

Biomass fuels comprise processed biomass (e.g., charcoal, methane – from biogas plants and landfills – logging waste and sawdust, and alcohol produced from fermentation processes) and unprocessed biomass which is exactly what the term implies – wood, twigs, straw, dung, vegetable matter and agricultural wastes. Processed biomass produces more energy per unit of fuel than unprocessed or raw biomass.

The availability and use of biomass resources for energy generation is not well documented for the vast majority of countries. Selected data on production of wood and other forms of biomass for energy purposes have been compiled by the WEC (WEC, 1992) and are summarized here as Tables 6.3 and 6.4, respectively. These data are based on information provided by the Food and Agriculture Organization of the United Nations (FAO) (wood fuels) and the Biomass Users Network (Hall, 1991), supplemented with material from the scientific literature.

Data presented in Table 6.3 indicate that some 1.4 x 10⁹ t of wood, or roughly half the world's total wood consumption, are currently used for energy generation, either directly as fuel wood (90 per cent) or in the production of charcoal (10 per cent). This quantity corresponds to approximately 0.5 x 10⁹ toe (WEC, 1992). Of this, some 1.1 x 10⁹ t or 0.4 x 10⁹ toe is consumed by the developing countries; this equates to around 17 per cent of the total consumption of energy in these countries. However, in 40 of the world's poorest countries wood is the source of more than 70 per cent of national energy consumption. The contribution of wood fuel to national energy consumption in selected countries is shown graphically in Figure 6.3.

Estimates of available non-wood fuel biomass resources in selected countries are shown in Table 6.4. Listed here are the traditional biomass fuels which are widely used in developing countries; also included are data on peat reserves, sewage and municipal solid waste. Energy may be generated from sewage in a number of ways; in New Zealand and Japan, for example, biogas is produced by the anaerobic digestion of sewage. Sewage can also be used for electricity generation. Biogas extracted from landfills is utilized in a number of countries; in the Netherlands this biogas is injected into the national gas grid, whereas in Denmark biogas from landfill is used for electricity and/or heat production. Biogas production from a range of raw materials is a particularly important fuel source in Asia; around 4 x 10⁶ biogas plants are reported to be in operation in China and around 1 x 10⁶ in India (Ellegård, 1991).

Nearly half the world's population, some 2.5 billion people, rely mainly or exclusively on biomass fuels for their daily energy needs, i.e., for cooking, space heating and illumination (Hall and Rosillo-Calle, 1991; Smith, 1991). Of these some 800 million people rely on agricultural residues and dung as sources of fuel. These provide readily available and relatively cheap energy resources which can also offer environmentally acceptable means of disposing of otherwise unwanted or even polluting wastes. There are four main categories of agricultural residues which are used for fuel in different parts of the world: woody crop residues such as coconut shells and jute sticks; cereal residues such as rice, wheat straw and maize

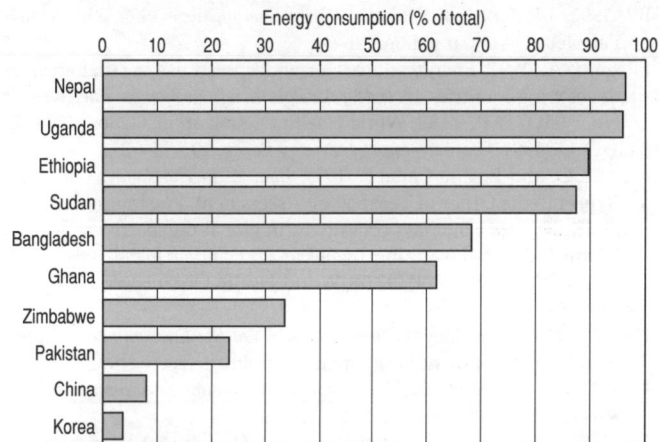

FIGURE 6.3 Wood fuel as a percentage of national energy consumption in selected countries
Source: UNEP, 1990

stalks; green crops such as groundnut plants; and crop processing residues such as rice husks (Table 6.4).

The use of biomass fuels, despite offering important benefits, nevertheless presents a number of environmental, agricultural and health problems. On the environmental side, the rate at which wood fuel is being used in many parts of the world is not only unsustainable, but excessive cutting and felling of trees for fuel has, in some areas, led to deforestation and concomitant soil erosion and flooding (see also Part 3: Natural Resources). Efforts are being made in many countries to remedy acute shortages of wood fuel by promoting community tree planting and extensive commercial fuelwood planting. These efforts have been particularly successful when tree planting meets a multiplicity of needs, for example, for fruits, fodder and timber as well as for fuelwood (WEC, 1992).

From the viewpoint of the global carbon cycle and climate change the combustion of biomass fuels offers the advantage of producing no net emissions of carbon dioxide (CO_2), but only when the biomass is regrown as fast as it is burnt. However, Smith (1991) has calculated that the use of biomass fuels contributes 1–5 per cent of all methane emissions, 6–14 per cent of all carbon monoxide emissions, 8–25 per cent of all non-methane hydrocarbon emissions, and thus 1–3 per cent of all human-induced global warming. The wisdom of using agricultural residues as fuel has also been the subject of much debate (Ellegård, 1991). When residues are returned to the soil, rather than being burnt, they not only provide nutrients and organic matter, but also improve the ability of the soil to retain moisture. Hence if residues are burnt, the soil is deprived of these important constituents; this may have detrimental effects on agriculture and food supply.

The combustion of forms of raw biomass, including wood fuels, produces a range of air pollutants including suspended particulate matter (smoke), oxides of carbon, nitrogen and sulphur, hydrocarbons, and aldehydes (WHO, 1992). Thus potentially serious health effects can arise from the domestic use of biomass fuels. This issue is discussed separately in Part 1: Environmental Pollution, Human Exposure.

References

BP 1992 *BP Statistical Review of World Energy*, June 1992, The British Petroleum Company, London.

Ellegård, A. 1991 Energy sources for widespread use in rural areas of developing countries. In: *Indoor Air Pollution from Biomass Fuel*, Report No. WHO/PEP/92.3B, World Health Organization, Geneva, 29–40.

Hall, D. O. 1991 Biomass energy, *Energy Policy*, **19**, 711–737.

Hall, D. O. and Rosillo-Calle, F. 1991 *Biomass in Developing Countries*, Report to the Office of Technology Assessment, Washington DC.

Smith, K. R. 1991 Biomass cookstoves in global perspectives: energy, health and global warming. In: *Indoor Air Pollution from Biomass Fuel*, Report No. WHO/PEP/92.3B, World Health Organization, Geneva, 165–184.

UNEP 1990 *Green Energy: Biomass Fuels and the Environment*, United Nations Environment Programme, Nairobi.

UNEP 1991 *United Nations Environment Programme Environmental Data Report 1991/92*, Basil Blackwell, Oxford.

WEC 1992 *1992 Survey of Energy Resources* (16th Edition), World Energy Council, London.

WHO 1992 *Report of the Panel on Energy*, WHO Commission on Health and Environment, WHO/EHE/92.3, World Health Organization, Geneva.

World Bank 1992 *World Development Report 1992: Development and the Environment*, Oxford University Press, New York.

UN 1992 *Report of the United Nations Conference on Environment and Development*, Rio de Janeiro, 3–14 June 1992, UN Report A/Conf.151/26, (Vols I, II, III and IV), United Nations Conference on Environment and Development, Geneva.

Production and Consumption

Comparable data on the production, trade and consumption of commercial forms of energy (primary and secondary) are compiled on a routine basis by the UN Statistical Office (UNSTAT) and published in its 'Energy Statistics Yearbook'. These data are compiled primarily by means of a questionnaire distributed to national statistical offices. Where official data are lacking or are of uncertain reliability, UNSTAT prepares its own estimates from governmental, professional or commercial data sources. Detailed energy statistics are also available from the International Energy Agency (IEA) of the Organisation for Economic Co-operation and Development (OECD), for both OECD and selected non-OECD countries. The IEA/OECD data bases comprise comprehensive energy supply and demand statistics, in both the original units and expressed in a common unit (toe), for coal, oil, natural gas and manufactured gases, electricity and heat. Energy balance sheets, i.e., energy statistics expressed in common units, are particularly useful policy-making tools, in that such data allow easy comparison of the relative contributions of fuel types to the economy. Both the UNSTAT and IEA/OECD data bases are available on computer tapes and diskette.

Global, regional and national data on production and consumption of commercial energy forms in 1990 as compiled by UNSTAT are presented in Table 6.5. Global and regional data for 1980 are included for comparison.

World primary energy production has increased by approximately 19 per cent between 1980 and 1990. Analysis of the relative contribution of different commercial energy supplies to primary production indicates that the use of fossil fuels continues to dominate the energy sector. Within the group of fossil fuels it is evident that during this time both natural gas and coal have increased their role in the world energy supply at the expense of oil (i.e., liquid fuels):

Fuel type	Contribution to global primary energy production (%)	
	1980	1990
Solids	28.6	30.5
Liquids	48.1	41.9
Natural gas	20.1	22.9
Electricity	3.3	4.7

"Electricity" refers to the generation of primary electricity from nuclear, hydro and geothermal sources.
Source: Data from Table 6.5

In the OECD countries alone, production of primary energy has increased by almost 40 per cent since 1970. This production increase has been more than sufficient to cover the rise in total primary energy requirements (i.e., the total energy needed for both final use and for energy transformation, e.g., from fossil fuels to electricity) of OECD countries over the same period. This growing reliance on indigenous production has increased the stress placed on the environment within the OECD region by the activities associated with the production and transportation of primary energy supplies. Areas in which production has increased over the last 20 years include a number of formerly unexploited locations, for example, the North Sea and Alaska (OECD, 1991).

In both the developed and developing world there has been a marked shift away from primary to secondary energy, (i.e., electricity production) as an end-use energy source for industry, services and households. In OECD countries the volume of electricity produced has approximately doubled over the past two decades; in the developing world as a whole, the rise has been even greater, averaging some 10 per cent per year over the same period. Fossil fuels provide 64 per cent of the global electricity production, with coal alone accounting for 40 per cent of the global total. The increasing share of electricity as an end-use energy form can have both desirable and less desirable implications for the environment; the net effect depends on the type of fuel used for electricity production, the extent of emissions controls and the potential savings in end-use energy achieved by switching to electricity (OECD, 1991).

The structure of electricity production varies markedly between countries and depends, to some extent, on the availability of national resources. Some countries, for example Finland, have a very diverse electrical power generating base involving coal, oil, hydroelectricity, natural gas, peat and nuclear fuels. In other countries, such as Poland, one fuel source dominates production; in this case coal. Variations in the structure of electricity production for the major fuel types within member countries of the UN Economic Commission for Europe (UN ECE) are illustrated in Figure 6.4.

Trends in final energy consumption broken down by broad categories of end-users (i.e., industry, transport and "other" sectors) are illustrated in Figure 6.5 for selected countries. Whereas countries of the industrialized world are typified by stationary, or only slowly increasing, trends in energy consumption, since 1970 trends in the developing world are predominantly upwards. However, despite the more rapid growth in the developing world in latter decades, there is still a huge disparity in commercial energy consumption between the developed and developing countries:

Year	Percentage share of world final energy consumption	
	Developed countries	Developing countries[a]
1973	90.53	9.47
1980	87.11	12.89
1985	83.64	16.36
1990	81.96	18.04

[a]   Excluding China; when China is included the share of the developing countries increases to 19 per cent in 1980, 23 per cent in 1985, and 27 per cent in 1990.
Sources: UNIDO, 1991; UNSTAT, 1993

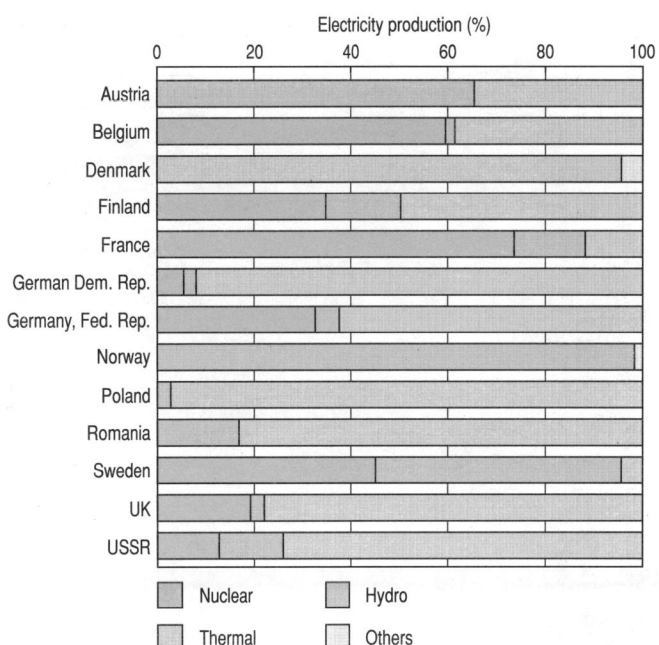

FIGURE 6.4 Production of electricity by fuel source in selected UN ECE countries, 1980 and 1990
Source: UN ECE, 1992

This dramatic disparity between developed and developing regions is more pronounced when energy consumption is measured in per capita terms and per unit of output; in 1985 OECD countries consumed on average nearly 10 times more energy per capita than developing countries but at the same time achieved a 15 times greater per capita gross domestic product (GDP). In effect, relative to the developed world, the developing countries are consuming more energy per unit of GDP and industrial output (UNIDO, 1991). Differences in energy consumption are also pronounced between countries of the developing world; data presented in Table 6.5 reflect these large variations in per capita energy consumption. Values range from less than 5×10^9 J in some African countries to over 500×10^9 J in the oil-exporting developing countries. To some extent these comparisons of energy consumption between developed and developing countries based on commercial energy forms are a distortion of the true picture; as noted above, traditional biomass fuels represent an important energy supply in many developing countries, the contribution of which is not included in commercial energy statistics (WEC, 1992).

The fact, noted above, that over the past 20 years in much of the industrialized world the growth in energy requirements and final consumption has been lower than that in economic growth is of particular significance. Since 1970, in the OECD countries as a whole, GDP has increased by 72 per cent while energy requirements have only grown by 30 per cent (OECD, 1991). That is to say, increased economic output has been sustained without a commensurate increase in energy consumption, or that there has been a "delinking" or "decoupling" of industrial production from energy use (OECD,

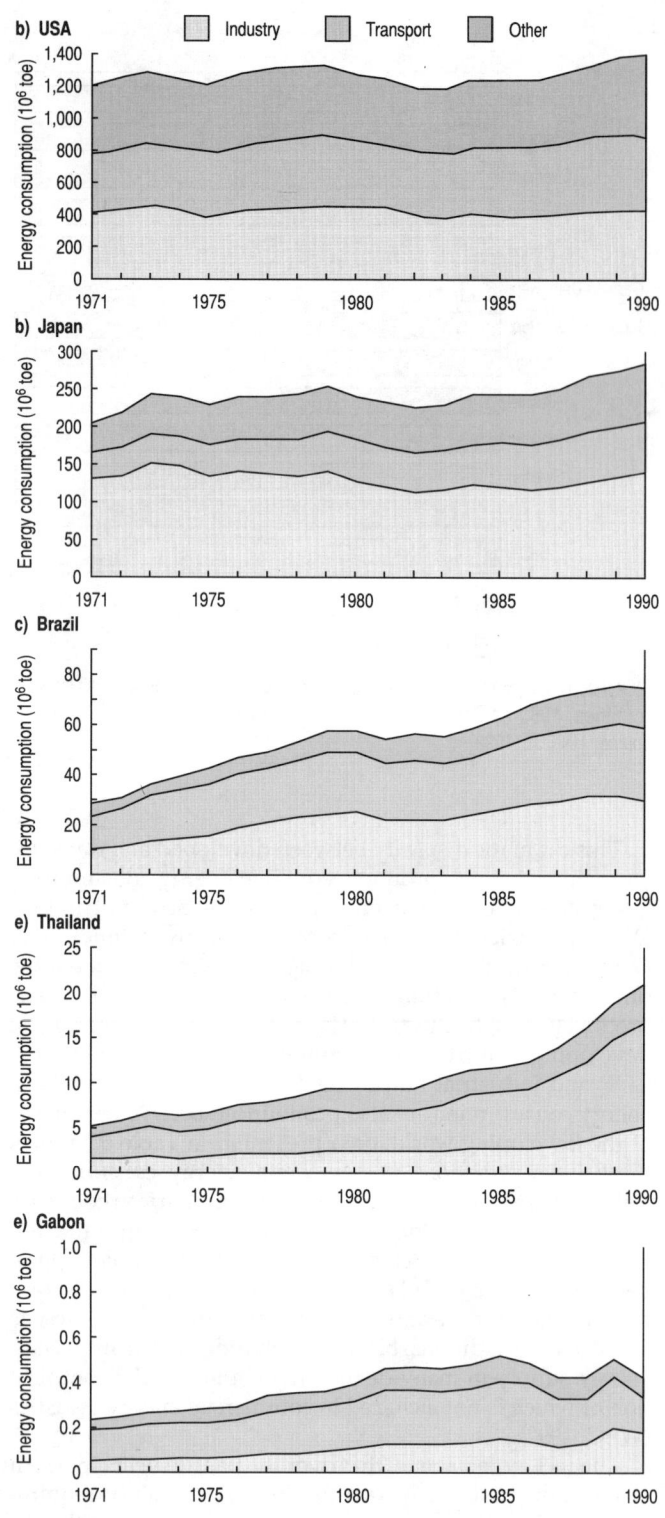

FIGURE 6.5 Trends in final energy consumption by end-use in selected countries, 1971–1990
Source: IEA/OECD, 1992

1991; UNIDO, 1991). This delinking has been made possible by a number of factors, namely a structural change (from energy-intensive materials-processing industries to less energy-intensive high-technology industries) and the wider adoption of energy conservation measures and more energy-efficient technologies. At the same time these changes have helped to reduce some of the environmental impacts associated with the energy sector; over the past 20 years emissions of sulphur dioxide (SO_2) and particulates have declined substantially in many OECD countries while growth in emissions of some other pollutants – CO_2 and oxides of nitrogen (NO_x) – may have been limited. These relationships between GDP, energy requirements and pollutant emissions are illustrated in Figure 6.6a.

The position of the developed nations is in sharp contrast with that of the developing nations where there is a strong positive relationship between economic growth and energy consumption (Figure 6.6b). This positive link is partly explained by the expansion of energy-intensive raw materials processing industries in the developing world (UNIDO, 1991) (see also Part 7: Industry and Transport). It is also evident that emissions of key air pollutants (SO_2, NO_x and CO_2) are increasing in many developing countries (see also Part 1: Environmental Pollution, Atmosphere). However, slowing down the rate of the industrialization process in the developing world would not necessarily relieve the pressure on the environment. Instead, the transfer of technology from the industrialized world is widely perceived as the best option for improving energy efficiency and thus mitigating the adverse effects of energy use on the environment (UNIDO, 1991). The ratio of primary energy requirements and GDP, otherwise known as energy intensity, and its application as a possible indicator of sustainability is discussed in Box 6.1.

It is highly probable that, for the forthcoming decades at least, the production and consumption of fossil fuels will continue to dominate the energy sector. However, this prospect raises a number of serious environmental concerns, most of which are centred on the potential impacts of further increases in emissions of the sulphur and nitrogen oxides and, more significantly, of CO_2. The World Bank (1992) estimates that, under an "unchanged practices" scenario, i.e., one in which control technologies are not widely implemented, emissions of pollutants from fossil fuel generation of electric power will increase fourfold in the next 20 years and tenfold in the next 40 years.

The technological achievements of the past decade or so mean that options are now available (or are emerging) for reducing all the significant pollutant emissions, with the exception of CO_2, from both coal and fuel oil to relatively low levels per unit of output. However, these modern emissions control technologies have yet to be widely deployed, especially in the developing countries. Some of the technological options for the control of SO_2 and NO_x and the extent of their application will be reviewed in a subsequent edition of this report.

It is feasible to reduce CO_2 emissions in the long term through the increased use of renewable sources of energy where appropriate and through greater efficiency in energy production and use. However, in order to achieve a

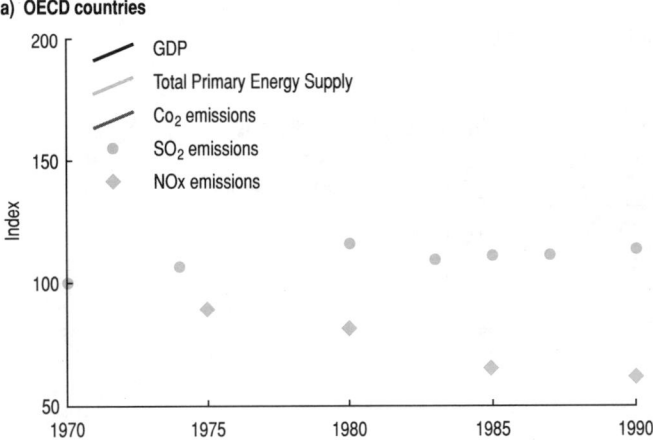

a) OECD countries

b) Developing countries

FIGURE 6.6 Trends in GDP, primary energy requirements and emissions of selected air pollutants in a) OECD countries and b) the developing world (data are expressed relative to a base year, 1970 = 100)
Sources: OECD, 1991; CDIAC,1992; IAEA, 1993

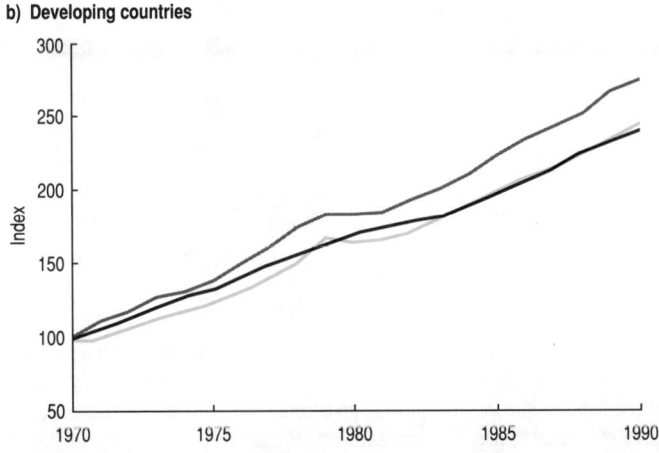

stabilization of national emissions of CO_2 at 1990 levels in the shorter term as recommended at UNCED, many countries will need to make changes in their current energy policies. The key elements of an interim response strategy have been recently summarized by the International Atomic Energy Agency (IAEA); these include upgrading the efficiency of existing fossil fuel plants, expanding the use of natural gas in new plants, deploying improved and advanced nuclear power plants, expanding the use of hydropower resources, promoting other renewable energy sources where economically viable, and accelerating investment in cost-effective measures for demand management and end-use efficiency improvement (IAEA, 1991a). Policy guidance on greenhouse gas abatement strategies is also provided by UNEP's Collaborating Centre on Energy and Environment, based in Risø, Denmark. Recent work involves the development of approaches for estimating the cost of reducing national greenhouse gas emissions.

Data indicate that emissions of CO_2 are beginning to stabilize

in a number of the more developed countries (Figure 6.6a; see also Part 1: Environmental Pollution, Atmosphere). Countries which have increased their reliance on natural gas and, in particular, on nuclear power have even managed to reduce emissions of CO_2. France is such a case; since the early 1970s total emissions of CO_2 have declined by some 20 per cent (OECD, 1991). These trends are, however, a reflection of structural changes in national economies and in energy supply, rather than an indication of pro-active national efforts to reduce CO_2 emissions as a result of concern over global warming.

There are, however, some signs that environmental concerns are beginning to influence energy policies. Environmental characteristics are increasingly determining the attractiveness of fuels; of the fossil fuels, natural gas is becoming the most attractive and is likely to increase further its growing share of the market. Oil's dominance in the transportation sector has yet to be really challenged, although regulations affecting fuel quality are becoming more widespread. Coal looks set to be affected adversely as sulphur restrictions tighten and concerns over CO_2 increase (WEC, 1992).

In order to advance further the adoption of environmental concerns in the policy-making process, it is evident that there is a clear need for improved comparative risk assessments of the different energy systems. This requirement was highlighted at a recent conference organized by the IAEA (IAEA, 1991a). Studies of this type have, however, been constrained by the lack of comprehensive, internationally co-ordinated data bases on the health and environmental impacts of the different energy sources. One such assessment of the health risks to both occupational workers and to the general public from a range of energy technologies has, nevertheless, been conducted by IAEA (IAEA, 1991b). In this particular study it was concluded that, under routine operating conditions, renewable and nuclear energy systems tend to be at the lower end of the spectrum of health risk. In contrast, the energy systems based on coal and oil are at the higher end of the health risk system.

Nuclear Power

According to the IAEA, at the end of 1991 there were 420 nuclear power reactors connected to electricity supply networks representing a total installed capacity of 326,600 MW(e) (IAEA, 1992a). A further 323 research reactors were in operation – only 7 of which were used to produce electricity (IAEA, 1991c) – and an additional 76 reactors were under construction.

Table 6.6 lists, by country, the number and capacity of nuclear power reactors in operation and under construction at the end of 1991. Whereas the USA currently produces almost a third of the world's nuclear electricity, France is the most heavily dependent on nuclear power. Belgium, Sweden, Korea Dem., India and several east European nations also rely on nuclear energy sources to provide half or more of their national needs. Figure 6.7 shows the types of reactors currently in operation and their contribution to power generation. Light water reactors, either as pressurized water reactors or boiling water reactors, predominate and have proved to be relatively economical, safe and reliable (IAEA, 1992a).

BOX 6.1 Energy Intensity: An Approximate Measure of Energy Efficiency

Traditionally, as economies have developed their demand for energy has been positively linked to economic output, as measured by the gross domestic product (GDP). In other words, increased economic output has demanded a greater energy consumption.

The ratio between total primary energy supply (TPES) and GDP is conventionally known as "energy intensity" and as such provides a measure of the amount of energy required to produce a given quantity of goods and services. As reductions in energy supply, through the more efficient use of energy, will tend to decrease this ratio, energy intensity can be viewed as an approximate measure of national energy efficiency. However, national energy intensity can be influenced by a number of other unrelated factors - such as changes in fuel type and shifts in the structure of the industrial sector - which will tend to reduce the direct relationship between the value of energy intensity and the level of energy efficiency.

Energy intensity is one of the 26 preliminary environmental indicators proposed by the Organisation for Economic Co-operation and Development (OECD). The OECD argues that reducing energy intensity through the adoption of energy conservation measures not only saves energy but has the added advantage of mitigating some of the adverse effects of increasing energy use on the environment. Energy intensity is thus perceived to be a potentially important indicator of sustainable development (OECD, 1991).

The data presented here illustrate the variation in energy intensity amongst the member countries of the OECD; 1990 values vary from 0.25 toe per US$ 10^3 in Japan up to 0.73 toe per US$ 10^3 in Turkey. Of particular significance is the marked decrease in energy intensity – in some cases by as much as 20 per cent – over the past 15–20 years that has occurred in a number of the OECD countries. This achievement can be attributed, but only in part, to the introduction of energy conservation measures and more energy-efficient technologies in use during this time. These were adopted in many OECD countries as a response to the high oil prices that dominated the 1970s.

The other factor that has had a profound impact on energy intensity in many OECD countries is the change in the structure of industrial production from energy-intensive, raw materials-processing industries to less energy-consuming high-technology and service industries; this pattern has characterized the economies of many

Trends in energy intensity in selected OECD countries, 1960–1990

Country	Energy intensity (toe per US$ 10^3)				
	1960	1970	1980	1985	1990
OECD	0.50	0.52	0.47	0.41	0.39
North America	0.57	0.59	0.53	0.46	0.42
Pacific	0.35	0.38	0.33	0.29	0.28
Europe	0.49	0.48	0.44	0.42	0.39
Australia	0.53	0.52	0.51	0.46	0.47
Belgium	0.67	0.72	0.60	0.55	0.51
Canada	0.63	0.69	0.64	0.56	0.52
Denmark	0.34	0.49	0.38	0.34	0.29
France	0.39	0.42	0.39	0.38	0.36
Germany, Fed. Rep.	0.49	0.52	0.47	0.44	0.39
Japan	0.31	0.36	0.31	0.27	0.25
Norway	0.34	0.47	0.38	0.35	0.34
Portugal	0.45	0.48	0.52	0.55	0.63
Turkey	0.33	0.48	0.76	0.73	0.73
UK	0.62	0.61	0.49	0.45	0.39
USA	0.56	0.59	0.52	0.47	0.41

Source: OECD, 1992

OECD countries over the last 10–15 years (OECD, 1991; UNIDO, 1991) (see also Part 7: Industry and Transport). Although the methodology of separating the impact of energy-saving technologies from the impact of structural change is difficult, it is estimated that technological innovation accounts for between 66 and 77 per cent of the observed drop in energy intensity (Beckel, 1992).

It is interesting to note that in the latter half of the 1980s the rate of improvement in energy intensity has fallen in a number of the OECD countries. In Canada, for example, the downwards trend in energy intensity was actually reversed in 1989 and 1990; this is attributed to strong economic growth and a decline in the relative cost of energy (Environment Canada, 1991).

Available data indicate that, in general, energy intensity is higher in the less developed countries owing to less efficient use of energy (UNIDO, 1991). Moreover, in a number of the more rapidly industrializing nations of South America and South-east Asia the rate of growth of energy requirements exceeds the rate of increase in economic output or GDP; consequently, energy intensity is at best stationary or is increasing in these countries (Beckel, 1992). In such

cases factors which are contributing to the surge in the demand for energy include rapid economic growth, an acceleration of the pace of urbanization (see Part 4: Population and Development), a tendency towards a dominance of the energy-intensive heavy industries (see Part 7: Industry and Transport) and a replacement of traditional (biomass fuels) by commercial forms of energy (UNIDO, 1991).

References

Beckel, J. 1992 Technical progress, competitiveness and sustainable development, *Industry and Environment*, **15**(1–2), 73–79.

Environment Canada 1991 *A Report on Canada's Progress Towards a National Set of Environmental Indicators*, SOE Report No. 91-1, Environment Canada, Ottawa.

OECD 1991 *Environmental Indicators: A Preliminary Set*, Organisation for Economic Co-operation and Development, Paris.

OECD 1992 *Energy Balances of OECD Countries: 1989-1990*, Organisation for Economic Co-operation and Development, Paris.

UNIDO 1991 *Industry and Development: Global Report 1991/92*, United Nations Industrial Development Organization, United Nations, New York.

BOX 6.2 Generation and Management of Radioactive Wastes Arising from the Nuclear Fuel Cycle

The nuclear power generating industry is currently the largest single source of radioactive wastes; other sources of wastes include medical, research, industrial and military installations where radioactive isotopes are used. Radioactive wastes are produced at each stage of the nuclear fuel cycle, i.e., uranium mining and enrichment, nuclear fuel fabrication, irradiation in the reactor, and spent fuel reprocessing. The wastes comprise atmospheric emissions (largely inert gases such as argon-41 and krypton-85 from nuclear power stations and reprocessing plants), liquid effluents and solid materials.

Radioactive wastes are typically divided into three categories:

○ Low-level wastes (LLWs), i.e., liquid and solid wastes lightly contaminated by short-lived radionuclides. Examples of LLWs include discarded protective clothing, contaminated building materials and uranium mine tailings.

○ Intermediate-level wastes (ILWs), i.e., waste materials contaminated by long-lived radionuclides such as plutonium and transuranic elements. Examples of ILWs include fuel cladding and liquids used to store spent fuel before reprocessing.

○ High-level wastes (HLWs), i.e., materials contaminated by highly active radionuclides with long half-lives which may cause significant increases in the temperature of the waste. HLWs chiefly comprise irradiated fuel discharged from reactors (if not reprocessed) and liquid wastes produced during reprocessing of spent fuel.

In most countries, LLWs from all sources constitute about 90 per cent (by volume) of annual radioactive waste arisings, but account for a relatively small proportion of the radioactivity in waste. In contrast, HLWs contain the largest share of radioactivity but are small in volume relative to the other two waste categories.

The United Nations Scientific Committee on the Effects of Atomic Radiation (UNSCEAR) periodically reviews and assesses data on the generation of radioactive materials (UNSCEAR, 1993). Available information suggests that in recent years there have been significant decreases in the level of radioactivity discharged as liquid effluents from both nuclear power stations and reprocessing plants (UNEP, 1991). In the UK, for example, reductions of both alpha and beta radiation in liquid discharges from the reprocessing plant at Sellafield have been achieved over the past 10 years. Discharges of beta radiation (excluding tritium) from this site have fallen from 4,306 TBq in 1980 to 71 TBq in 1990; the corresponding figures for alpha radiation are 39 TBq and 2 TBq (Brown, 1992). The reduced levels of radioactivity in Sellafield discharges is reflected in the analyses of sediments in the Irish Sea (see Part 1: Environmental Pollution, Soils and Sediments).

Disposal options for solid radioactive wastes largely revolve around containment followed by burial. Low-level wastes are currently disposed of at approved shallow waste disposal sites. In the past LLWs and selected ILWs, packaged in specially designed containers, have been dumped at sea at licensed sites in the North Atlantic and Pacific Oceans (UNEP, 1991). This practice has been suspended since 1983 following the adoption of a resolution by the contracting parties to the London Dumping Convention on the Prevention of Marine Pollution from the Dumping of Waste and Other Matters.

Various solutions to the key problem of the ultimate disposal of HLWs have been proposed. However, burial in suitable geological formations on land at minimum depths of several hundred metres is the only solution currently being seriously investigated. Facilities capable of receiving long-lived wastes, either as irradiated fuel or vitrified liquid wastes, are planned in France, Germany and the USA and are expected to become viable by the year 2010 (OECD, 1991). Until such time, HLWs and ILWs are generally stockpiled at nuclear establishments. By the beginning of 1989, 1,320 m³ of HLWs and a further 30,500 m³ of ILWs had accumulated at Sellafield, UK (Brown, 1992).

References

Brown, A. (Ed.) 1992 *The UK Environment*, HMSO, London.

OECD 1991 *The State of the Environment*, Organisation for Economic Co-operation and Development, Paris.

UNEP 1991 *United Nations Environment Programme Environmental Data Report 1991/92*, Basil Blackwell, Oxford.

UNSCEAR 1993 *Sources, Effects and Risks of Ionizing Radiation*, 1993 Report to the United Nations General Assembly, United Nations Scientific Committee on the Effects of Atomic Radiation , United Nations, New York. In press.

In 1991 nuclear power accounted for 17 per cent of total electricity generated world-wide. This represents a considerable increase with respect to the position in 1960; the table below shows the increase in nuclear electricity generation in terawatt hours (TWh⁻¹) and as a percentage of total electricity generation over the period 1960–1991:

Year	Nuclear electricity supplied	
	TW.h	% of total
1960	2.7	0.1
1965	24.7	0.7
1970	78.7	1.6
1975	351.7	5.5
1980	681.4	8.3
1985	1,448.5	15.3
1991	2,009.1	17.0

Source: IAEA, 1992c

The nuclear fuel cycle – from mining, conversion, preparation of reactor fuel, and effluent and spent fuel management – is associated with a range of environmental concerns. Public concerns centre particularly on the risks of major accidents (see Part 9: Environmental Disasters) and the problems associated with the disposal of radioactive wastes (Box 6.2). More recently operation of nuclear power plants, problems of decontamination and the high costs of decommissioning nuclear facilities, have also given rise to public concern.

On a global scale, nearly two-thirds of the nuclear power plants in operation at the end of 1991 are under 10 years old (Figure 6.8). In the USA, however, about two-thirds of reactors are over 10 years old. Similarly, in the UK, reactors dating from the mid- to late 1950s are still in operation. A number of other countries have plants that are 20 years old or more (IAEA, 1992a). By the year 2000, more than 60 nuclear power plants and 250 research reactors around the world may have reached the end of their useful life and there will be a need

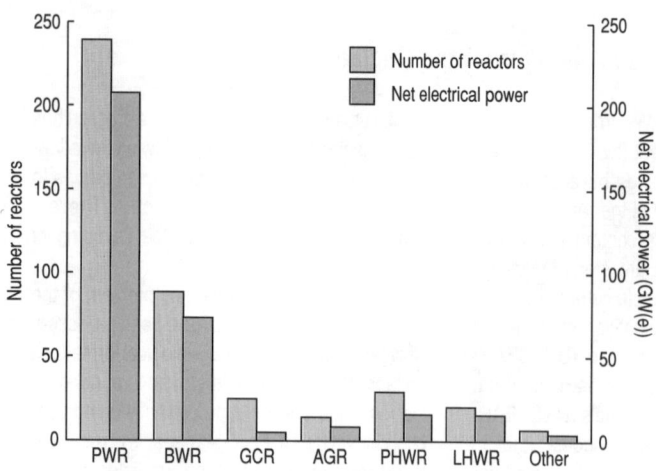

FIGURE 6.7 Type and capacity of nuclear power reactors in service at the end of 1991 (PWR = Pressurized water reactors (light water cooled and moderated); BWR = Boiling water reactors (light water cooled and moderated); GCR = Gas cooled reactors (carbon dioxide cooled and graphite moderated); AGR = Advanced gas cooled reactors (graphite moderated); PHWR = Pressurized heavy water reactors (heavy water cooled and moderated); LWGR = Light water cooled reactors (light water cooled and graphite moderated); Other = Miscellaneous types of reactors, including fast breeder reactors)
Source: IAEA, 1992b

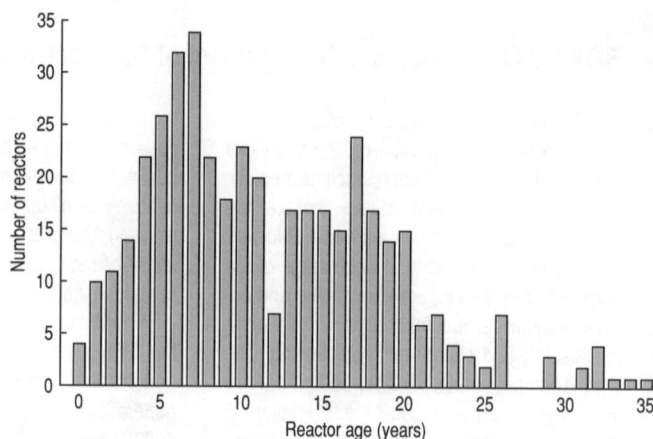

FIGURE 6.8 Age distribution of nuclear power reactors in operation at the end of 1991
Source: IAEA, 1992c

for extensive decommissioning. No large nuclear power plant has yet been completely dismantled, although experience has been gained in decommissioning many other types of nuclear facilities.

Possibly the most acute problem facing the nuclear power generating industry today is the lack of public confidence. Disposal of radioactive wastes, particularly high-level wastes, is widely perceived as an unsolved problem despite the range of satisfactory engineering options now available (see Box 6.2). The spectre of Chernobyl continues to haunt the industry and the relative age and unsafe conditions of selected power stations (particularly those in eastern Europe) heighten the fear of another disaster. On the positive side, however, the current pre-occupation with reducing emissions of CO_2 and other pollutants from fossil-fuel power stations may increase the acceptability of nuclear power (WEC, 1992).

References

CDIAC 1992 *Estimates of CO₂ Emissions from Fossil Fuel Burning and Cement Manufacturing, Based on the United Nations Energy Statistics and the US Bureau of Mines Cement Manufacturing Data*, Numeric Data Package-030/R4, Carbon Dioxide Information Analysis Center, Oak Ridge National Laboratory, Oak Ridge.

IAEA 1991a Senior Expert Symposium on Electricity and the Environ-ment: Executive Summary, Helsinki, Finland, 13–17 May 1991, International Atomic Energy Agency, Vienna.

IAEA 1991b Senior Expert Symposium on Electricity and the Environ-ment: Key Issues Papers, Helsinki, Finland, 13–17 May 1991, International Atomic Energy Agency, Vienna.

IAEA 1991c *Nuclear Research Reactors in the World*, Reference Data Series Number 3, International Atomic Energy Agency, Vienna.

IAEA 1992a *IAEA Yearbook 1992*, International Atomic Energy Agency, Vienna.

IAEA 1992b *Nuclear Power Reactors in the World*, Reference Data Series Number 2, International Atomic Energy Agency, Vienna.

IAEA 1992c *Nuclear Power. Nuclear Fuel Cycle and Waste Management: Status and Trends*. International Atomic Energy Agency, Vienna.

IAEA 1993 Personal communication, International Atomic Energy Agency, Vienna.

IEA/OECD 1992 *Energy Statistics and Balances, OECD (1960–1990), Non-OECD (1971–1990)*, Diskette Service, International Energy Agency, Paris, and Organisation for Economic Co-operation and Development, Paris.

OECD 1991 *The State of the Environment*, Organisation for Economic Co-operation and Development, Paris.

UN ECE 1992 *Annual Bulletin of Electric Energy Statistics for Europe*, (Volume 36), United Nations, New York.

UNIDO 1991 *Industry and Development: Global Report 1991/92*, United Nations Industrial Development Organization, United Nations, New York.

UNSTAT 1993 *Energy Statistics Yearbook 1991*, United Nations, New York.

WEC 1992 *1992 Survey of Energy Resources (16th edition)*, World Energy Council, London.

World Bank 1992 *World Development Report 1992: Development and the Environment*, Oxford University Press, New York.

TABLE 6.1   Proved recoverable reserves of commercial energy resources, 1990

Region/country	Bituminous coal/ anthracite (10^6 t)	Sub-bituminous coal/lignite (10^6 t)	Crude oil and NGL (10^6 t)	Natural gas (10^9 m^3)	Uranium (10^3 t)
WORLD	**460,600**	**516,319**	**136,754**	**128,584**	**2,084**
AFRICA	**60,811**	**1,267**	**9,605**	**8,179**	**686**
Algeria	43	–	2,400	3,300	26
Angola			250	51	
Benin			3		
Botswana	3,500	–			
Cameroon			55	110	
Cent. African Rep.	–	4			16
Congo			110	74	
Côte d'Ivoire			14	100	
Egypt	13	40	840	351	
Equatorial Guinea				24	
Ethiopia	–	–		25	
Gabon			100	14	16
Libya			3,150	1,218	
Madagascar				2	
Malawi	12				
Morocco	45	–	<1	2	
Mozambique	240	–		65	
Namibia				59	101
Niger	70	–			173
Nigeria	21	169	2,400	2,475	
Rwanda				57	
Somalia				6	7
South Africa	55,333	–	<1	51	344
Sudan			41	85	
Swaziland	–	999			
Tanzania	200	–		24	
Tunisia			230	85	
Zaire	600	–	12	1	2
Zambia	–	55			
Zimbabwe	734	–			2
NORTH AMERICA	**118,429**	**132,474**	**12,274**	**10,129**	**385**
Barbados			<1	<1	
Canada	4,509	4,114	927	3,116	
Cuba			15	3	
Greenland	–	183			27
Guatemala			24	<1	
Mexico	1,252	468	6,921	2,025	2
Trinidad & Tobago			80	252	
USA	112,668	127,892	4,307	4,732	356
SOUTH AMERICA	**5,648**	**4,062**	**9,829**	**4,696**	**113**
Argentina	–	130	215	579	11

Continued

TABLE 6.1 Continued

Region/country	Bituminous coal/anthracite (10^6 t)	Sub-bituminous coal/lignite (10^6 t)	Crude oil and NGL (10^6 t)	Natural gas (10^9 m^3)	Uranium (10^3 t)
Bolivia			25	118	
Brazil	–	2,359	400	115	100
Chile	31	1,150	41	131	<1
Colombia	4,240	299	280	108	
Ecuador	–	24	207	16	
Peru	960	100	52	200	2
Suriname			5		
Venezuela	417	–	8,604	3,429	
ASIA	**125,817**	**94,389**	**94,599**	**44,922**	**51**
Afghanistan	66	–		100	
Bahrain			11	177	
Bangladesh			<1	360	
Brunei			180	317	
China	62,200	52,300	3,264	1,127	51[a]
China, Taiwan	100	–	<1	17	
India	60,648	1,900	829	730	66[a]
Indonesia	962	31,101	726	1,817	4
Iran	193	–	12,700	17,000	
Iraq			13,600	2,690	
Israel			<1	<1	
Japan	827	17	7	36	7
Jordan			3	11	
Korea	203	–			31
Korea, Dem.	300	300			
Kuwait			12,785	1,360	
Malaysia	4	–	394	1,817	
Mongolia	–	–			
Myanmar	2	–	7	265	
Oman			580	280	
Pakistan	–	524	28	728	
Philippines	<1	262	19	17	
Qatar			507	4,589	
Saudi Arabia			35,070	5,214	
Syria			240	181	
Thailand	–	999	47	196	
Turkey	162	6,986	52	44	9
United Arab Em.			13,000	5,646	
Viet Nam	150	–		3	
Yemen[b]			550	200	
EUROPE	**61,337**	**102,704**	**2,173**	**5,137**	**133**
Albania	–	–	24	20	
Austria	–	59	16	19	
Belgium	410	–			
Bulgaria	30	3,700	2	7	
Czechoslovakia	1,870	3,500	2	14	

Continued

TABLE 6.1   Continued

Region/country	Bituminous coal/ anthracite (10^6 t)	Sub-bituminous coal/lignite (10^6 t)	Crude oil and NGL (10^6 t)	Natural gas (10^9 m^3)	Uranium (10^3 t)
Denmark			103	115	
Finland					2
France	178	32	23	53	40
Germany[c]	23,919	56,150	62	347	5
Greece	–	3,000	4	1	<1
Hungary	596	3,865	22	104	3
Ireland	14	–		19	
Italy	–	34	66	282	5
Netherlands	497	–	20	1,970	
Norway	–	13	1,100	1,227	
Poland	29,600	11,600	2	166	
Portugal	3	33			9
Romania	<1	3,117	157	150	26[a]
Spain	850	600	3	16	39
Sweden	–	1	<1		4
Switzerland				<1	
UK	3,300	500	535	545	
Yugoslavia	70	16,500	32	82	<1
USSR	**104,000**	**137,000**	**8,000**	**54,530**	**686[a]**
OCEANIA	**45,369**	**45,690**	**274**	**991**	**529**
Australia	45,340	45,600	225	493	529
New Caledonia	2	–			
New Zealand	27	90	22	98	
Papua New Guinea			27	400	

[a]   Reserve estimates are not consistent with standard WEC or NEA/IAEA resource terminology or definitions.

[b]   Data refer to the unified Republic of Yemen.

[c]   Data refer to the unfied Federal Republic of Germany.

In this table a dash signifies an unknown or zero quantity.

NGL = Natural gas liquids.

Data for coal, crude oil, NGL and natural gas refer to proved recoverable reserves, i.e., the tonnage (or volume) that can be recovered under present and expected local economic conditions with existing available technology. Data for uranium are proved reserves, i.e., recoverable uranium that occurs in known mineral deposits of such size, grade and configuration that could be recovered with currently proven mining and processing technology at a cost of less than US$ 130 per kg. Costs are expressed in terms of the US$ at 1 January 1991.

In the absence of internationally agreed standard classifications and definitions of energy reserves, comparisons of reserves of different countries must be made with caution. Although the WEC attempts to establish precisely worded standard definitions it is a matter of judgement for each country to determine the quantities which in its opinion, meet these definitions when reporting data to the WEC.

Regional and world totals do not take into account the reserves in countries for which data are not reported or are unavailable. In the case of uranium, regional and world totals exclude estimates for the following countries: Chile, China, Romania and the former USSR. Totals may not equal those reported in the source document due to differences in regional classification of countries.

Source:
WEC 1992 *1992 Survey of Energy Resources* (16th edition), World Energy Council, London.

TABLE 6.2   Fossil-fuel reserves to production ratios (R/P ratio) by world region, 1984, 1987 and 1990 (years)

Region	Fossil–fuel type	1984	1987	1990
AFRICA	Bituminous coal/anthracite	391	283	336
	Sub-bituminous coal/lignite	3	279	1,267
	Crude oil & NGL	31	34	32
	Natural gas	160	154	105
NORTH AMERICA	Bituminous coal/anthracite	173	190	175
	Sub-bituminous coal/lignite	1,636	372	394
	Crude oil & NGL	18	19	21
	Natural gas	20	17	16
SOUTH AMERICA	Bituminous coal/anthracite	96	465	235
	Sub-bituminous coal/lignite	91,150	365	580
	Crude oil & NGL	27	48	46
	Natural gas	71	90	72
ASIA	Bituminous coal/anthracite	172	562	93
	Sub-bituminous coal/lignite	272	2,129	1,716
	Crude oil & NGL	70	93	85
	Natural gas	239	213	174
EUROPE	Bituminous coal/anthracite	152	132	167
	Sub-bituminous coal/lignite	127	97	129
	Crude oil & NGL	10	13	10
	Natural gas	21	26	21
USSR	Bituminous coal/anthracite	225	189	192
	Sub-bituminous coal/lignite	731	652	729
	Crude oil & NGL	14	13	14
	Natural gas	67	57	66
OCEANIA	Bituminous coal/anthracite	257	252	277
	Sub-bituminous coal/lignite	1,099	972	914
	Crude oil & NGL	16	11	10
	Natural gas	71	69	38

NGL = Natural gas liquids.

The reserves to production ratio (R/P ratio) provides a measure of the remaining lifetime of reserves for a given annual rate of production. The R/P ratio in 1984 is thus the estimated length of time, in years, that reserves as assessed in 1984 would last at 1984 annual rates of production. Reserves are proved recoverable reserves as defined in the notes to Table 6.1.

Source:
WEC 1992 *1992 Survey of Energy Resources* (16th edition), World Energy Council, London.

TABLE 6.3 Estimates of forest area and fuelwood production in selected countries

Region/country	Total forest area (10^6 ha)	Productive forest area (10^6 ha)	Fuelwood production, 1990 (10^3 t)	Region/country	Total forest area (10^6 ha)	Productive forest area (10^6 ha)	Fuelwood production, 1990 (10^3 t)
WORLD	**3,878.0**	**2,542.0**	**1,415,168**	**NORTH AMERICA**	**819.8**	**483.2**	**172,994**
AFRICA	**703.5**	**363.7**	**340,658**	Canada	453.0	244.0	2,302
				Costa Rica	1.8	1.2	2,146
Algeria	2.2	1.2	1,378				
Angola	53.7	21.4	4,014	Cuba	1.6	1.1	1,828
Benin	3.9	1.1	3,460	Dominican Rep.	0.6	0.6	707
Botswana	32.6	0.2	944	El Salvador	–	–	3,203
Burundi	2.8	–	4,500	Guadeloupe	–	–	11
				Guatemala	4.6	3.0	5,586
Cameroon	25.6	19.7	7,928				
Cent. African Rep.	35.9	19.4	2,214	Haiti	–	–	4,059
Congo	21.4	13.7	1,507	Honduras	4.0	3.0	3,865
Côte d'Ivoire	9.9	5.1	7,066	Jamaica	–	–	9
Egypt	–	–	1,554	Martinique	–	–	7
				Mexico	49.0	27.0	16,700
Equatorial Guinea	1.3	1.0	324				
Ethiopia	27.2	4.5	29,560	Nicaragua	4.5	4.1	2,317
Gabon	20.6	19.9	1,860	Panama	4.2	2.9	1,238
Ghana	18.0	10.0	8,000	Puerto Rico	0.3	0.1	–
Guinea	10.6	4.3	2,502	Trinidad & Tobago	0.2	0.2	16
				USA	296.0	196.0	129,000
Kenya	2.5	1.2	24,503				
Liberia	2.0	1.9	3,543	**SOUTH AMERICA**	**830.9**	**630.2**	**175,490**
Libya	0.3	0.1	388				
Madagascar	13.5	7.2	5,282	Argentina	–	33.9	3,139
Malawi	4.3	0.6	5,662	Bolivia	66.8	37.3	972
				Brazil	518.0	423.0	135,131
Mali	7.2	1.3	3,791	Chile	8.4	7.5	7,700
Mauritania	–	–	5	Colombia	55.9	15.4	11,300
Mauritius	–	–	12				
Morocco	3.6	2.3	1,001	Ecuador	13.1	5.2	3,270
Mozambique	15.5	3.8	10,886	French Guiana	7.8	6.7	48
				Guyana	18.7	13.5	12
Namibia	18.4	2.0	–	Paraguay	3.0	1.5	6,229
Niger	2.6	0.3	3,370	Peru	70.7	43.6	5,116
Nigeria	14.9	4.5	72,365				
Réunion	–	–	22	Suriname	15.0	12.5	13
Rwanda	0.3	0.1	4,059	Uruguay	0.5	0.1	1,860
				Venezuela	53.0	30.0	700
Senegal	12.6	–	4,781				
Sierra Leone	2.1	0.4	2,136	**ASIA**	**518.0**	**346.8**	**605,521**
Somalia	9.1	0.1	5,103				
South Africa	1.3	1.1	5,129	Afghanistan	1.2	0.6	3,570
Sudan	47.8	31.5	14,987	Bangladesh	1.1	1.0	21,778
				Bhutan	2.1	1.8	2,135
Swaziland	0.2	0.1	406	Brunei	0.3	0.3	57
Tanzania	44.0	34.6	45,050	Cambodia	12.7	6.7	3,886
Togo	1.7	0.4	520				
Tunisia	0.4	0.4	2,234	China	128.0	119.0	134,357
Uganda	6.1	2.0	9,556	China, Taiwan	18.6	0.2	50
				Cyprus	–	–	15
Zaire	178.0	139.0	26,133	Hong Kong	–	–	140
Zambia	29.5	6.6	8,380	India	59.3	44.9	181,189
Zimbabwe	19.9	0.7	4,543	Indonesia	119.0	75.6	102,179

Continued

TABLE 6.3 Continued

Region/country	Total forest area (10^6 ha)	Productive forest area (10^6 ha)	Fuelwood production, 1990 (10^3 t)	Region/country	Total forest area (10^6 ha)	Productive forest area (10^6 ha)	Fuelwood production, 1990 (10^3 t)
Iran	3.8	1.6	1,778	Denmark	0.5	0.5	357
Iraq	1.2	0.9	76	Finland	20.1	18.8	4,600
Israel	–	–	328	France	15.2	13.3	11,250
Japan	25.0	22.5	1,500	Germany[b]	10.4	9.4	2,649
Korea	6.5	4.0	1,400	Greece	2.5	1.8	975
Korea, Dem.	4.8	3.0	2,965	Hungary	1.7	1.5	934
Laos	13.6	6.0	2,770	Ireland	0.2	0.2	33
Lebanon	–	–	328	Italy	8.6	7.7	10,000
Malaysia	21.0	15.6	6,318	Luxembourg	0.1	0.1	20
Mongolia	9.5	4.2	978	Netherlands	0.3	0.2	109
Myanmar	32.0	23.9	12,886	Norway	7.6	6.7	650
Nepal	2.1	1.3	12,795	Poland	8.7	7.7	1,800
Pakistan	2.6	1.4	17,362	Portugal	2.6	2.2	433
Philippines	15.9	1.0	24,322	Romania	6.3	5.2	2,494
Sri Lanka	2.4	1.2	10,000	Spain	22.8	8.8	2,107
Syria	0.2	0.1	10	Sweden	23.5	22.2	3,000
Thailand	23.2	–	25,064	Switzerland	0.9	0.7	636
Turkey	20.2	4.8	17,870	UK	2.0	2.0	185
Viet Nam	10.3	5.4	17,499	Yugoslavia	9.1	8.5	2,931
Yemen[a]	–	–	235	**USSR**	**740.0**	**535.0**	**58,478**
EUROPE	**156.5**	**128.4**	**51,075**	**OCEANIA**	**88.9**	**53.0**	**10,861**
Albania	1.2	0.9	1,165	Australia	41.7	36.7	6,300
Austria	3.7	3.2	1,942	Fiji	0.8	0.3	27
Belgium	0.7	0.6	414	New Caledonia	0.7	0.4	–
Bulgaria	3.4	2.5	1,100	New Zealand	7.5	1.5	525
Czechoslovakia	4.4	3.7	1,291	Papua New Guinea	38.2	14.1	4,009

[a] Data refer to the unified Republic of Yemen.
[b] Data refer to the unified Federal Republic of Germany.

In this table a dash signifies an unknown or zero quantity.

Total forest area refers to closed forests in temperate regions and to natural forest and plantations in tropical regions. Productive forest is that part of the forest which is accessible and able to sustain wood production. It excludes forest areas designated as reserved in national parks and nature reserves.

Fuelwood production is the tonnage of wood in the rough produced for direct use as a fuel or for the production of charcoal. It does not include wood residues recycled for energy use. In most cases, estimates of fuelwood production are based on FAO data for the volume of wood cut from forests for use as fuelwood. Volume of wood (in m^3) has been converted to a tonnage using the conversion factor $1\ t = 1.4\ m^3$.

Source:
WEC 1992 *1992 Survey of Energy Resources* (16th edition), World Energy Council, London.

TABLE 6.4 Estimates of available biomass energy resources other than fuelwood in selected countries, 1990 (10^6 t)

Region/country	Sugar cane	Crop wastes[a]	Animal dung	Biogas	MSW	Industrial/ wood wastes	Sewage	Various[b]	Peat reserves
WORLD	226	854	726		50				25,240
AFRICA									
Burundi				–					56
Ethiopia		–	–						
Senegal								<1	24
South Africa	–	3				–		<1	30
Tanzania		<1							
NORTH AMERICA									
Canada		–[c]			17				
Costa Rica		165							
Jamaica		1	1						233
Mexico		12							
USA	9[d]								13,000
SOUTH AMERICA									
Argentina	14	–							80
Brazil	159	56							
Colombia	26	7							
Ecuador		4		2					
Mexico								12	
Paraguay								–	
Peru		6	7						
Uruguay[e]		<1							
Venezuela		34	53				11		
ASIA									
Bangladesh		15	5						
China		237	13	–					328
China, Taiwan			500		40	50	300		
Hong Kong						<1			
India		41	73		23				
Indonesia		30							
Israel		1	<1		3			<1	386
Japan		19			34	29		2	
Jordan								24	
Korea		–			<1	–		–	
Malaysia		–		20				23	
Philippines			19			1		–	
Sri Lanka		<1							5
Syria			–						
Thailand		72							
Turkey								15	
EUROPE									
Austria								3	
Denmark				–					

Continued

TABLE 6.4 Continued

Region/country	Sugar cane	Crop wastes[a]	Animal dung	Biogas	MSW	Industrial/ wood wastes	Sewage	Various[b]	Peat reserves
Finland						11			340
France		27		–		8			
Germany[f]		29			–			2	
Ireland									114
Italy		20	22			3	12		
Netherlands				–					1,120
Norway		1	1		1			3	350
Poland									70
Portugal			13		2			–	70
Romania				–					13
Spain				94	11				70
Sweden		1				10			70
UK		7			30			4	
USSR		**60**	**20**		**50**	**10**	**6**		**8,881**
OCEANIA									
Australia	27	8							
New Zealand						–	–		

a   Crop wastes may include sugar cane husks, coffee husks, cocoa husks and straw.
b   Data included under the "Various" category generally refer to either a category of biomass not listed here, or to any combination of the other categories where estimates for individual resources are not available.
c 50×10^3 t of grain are available for ethanol production.
d   In 1989 9×10^6 t of corn were used for the production of ethanol.
e   Data refer to 1989.
f   Data refer to the unified Federal Republic of Germany.

MSW = Municipal solid waste.
In this table a dash signifies an unknown quantity.

Estimates of available biomass given in the above table refer to the quantity of raw materials available for the production of energy rather than the quantity of raw materials actually converted to energy. Biomass may be converted to liquid (i.e., alcohol) or gaseous fuels, or may be used as a direct energy source by combustion. Data under municipal solid waste refer to the quantity of waste made available for energy production, i.e., incinerated. (See also Table 8.5 in Part 8: Wastes and Waste Management).

Peat reserves are the tonnage of the proved amount in place that can be recovered under present and expected local economic and technological limits. Reserves may not necessarily be used for energy production.

Source:
WEC 1992 *1992 Survey of Energy Resources* (16th edition), World Energy Council, London.

TABLE 6.5 Production and consumption of commercial energy, by major world region in 1980 and 1990, and by country in 1990

Region/country	Year	Primary energy production (10^15 J)					Energy consumption (10^15 J and 10^9 J per capita)						P/C ratio[a]
		Total	Solids	Liquids	Gas	Electricity	Total	Solids	Liquids	Gas	Electricity	Per capita	
WORLD	1980	269,244	76,878	129,586	53,998	8,782	251,844	76,888	112,386	53,795	8,775	56	1.07
	1990	318,735	97,199	133,393	73,083	15,059	301,442	97,523	116,617	72,176	15,125	57	1.06
AFRICA	1980	16,921	2,879	12,790	1,032	219	5,591	2,175	2,493	704	219	12	3.03
	1990	20,730	4,085	13,679	2,772	195	8,050	2,956	3,359	1,551	184	13	2.58
Algeria		4,593	0	2,617	1,975	0	1,084	36	295	753	0	43	4.24
Angola		998		986	7	5	26	0	15	7	5	3	38.38
Benin		12		12			7		6		1	2	1.71
Burkina							8	0	8		0	1	
Cameroon		365	0	355		9	85	0	75		9	7	4.29
Congo		340	0	338	0	1	25	0	23	0	2	11	13.60
Côte d'Ivoire		19		13		6	67	0	61		6	6	0.28
Djibouti							5		5			12	
Egypt		2,331		2,047	255	29	1,131	30	817	255	29	22	2.06
Ethiopia		3				3	37	0	34		3	1	0.08
Gabon		571		565	4	3	25		18	4	3	21	22.84
Ghana		19		0		19	45		26		18	3	0.42
Guinea		1		0		1	14		14		1	2	0.07
Kenya		10				10	78	4	63		11	3	0.13
Liberia		1				1	7		6		1	3	0.14
Libya		3,239		2,868	371		629	0	306	323		138	5.15
Madagascar		1				1	14	0	13		1	1	0.07
Malawi		2				2	9	1	6		2	1	0.22
Mali		1				1	6		6		1	1	0.17
Mauritania		0				0	34	0	34		0	17	
Mauritius		0				0	16	2	13		0	15	
Morocco		22	15	1	2	4	270	52	211	2	4	11	0.08
Mozambique		1	1			0	14	2	11		1	1	0.07
Niger		5	5			1	13	5	8		1	2	0.38
Nigeria		3,779	3	3,623	145	8	598	2	444	145	8	6	6.32
Réunion		2				2	15		13		2	25	0.13
Rwanda		1			0	1	6		6	0	1	1	0.17
Senegal							30		30			4	
Sierra Leone							9	0	9			2	
Somalia							11		11			1	
South Africa[b]		3,919	3,902			17	3,114	2,653	454		7	76	1.26
Sudan		3				3	48	0	44		3	2	0.06

Continued

TABLE 6.5 Continued

Region/country	Year	Primary energy production (10¹⁵ J)					Energy consumption (10¹⁵ J and 10⁹ J per capita)						P/C ratio[a]
		Total	Solids	Liquids	Gas	Electricity	Total	Solids	Liquids	Gas	Electricity	Per capita	
Tanzania		2	0			2	29	0	27		2	1	0.07
Togo		0	0			0	6	0	5		1	2	
Tunisia		209		195	14	0	195	3	128	64	0	24	1.07
Uganda		2				2	13		12		2	1	0.15
Zaire		84	4	59		22	69	10	38		21	2	1.22
Zambia		37	9			28	50	9	18		23	6	0.74
Zimbabwe		158	145			13	185	145	24		16	19	0.85
NORTH AMERICA	**1980**	**74,201**	**19,398**	**28,511**	**23,230**	**3,063**	**81,767**	**16,814**	**38,555**	**23,336**	**3,063**	**219**	**0.91**
	1990	**80,457**	**24,872**	**28,138**	**22,763**	**4,683**	**86,853**	**21,549**	**38,484**	**22,133**	**4,686**	**203**	**0.93**
Aruba							7		7			117	
Bahamas							18	0	18			71	
Barbados		4		3	1		12	0	11	1		47	0.33
Bermuda							8	0	8			138	
Canada		10,684	1,547	3,814	3,992	1,331	7,971	987	3,236	2,419	1,329	301	1.34
Costa Rica		13				13	52		39		13	17	0.25
Cuba		37		36	1	0	449	8	440	1	0	42	0.08
Dominican Rep.		3				3	77	0	74		3	11	0.04
El Salvador		8				8	38		30		8	7	0.21
Greenland		0					8	0	8			143	
Guadeloupe							14		14			41	
Guatemala		15		7		8	51		43		8	6	0.29
Haiti		1				1	10		9		1	2	0.10
Honduras		3				3	26	0	22		4	5	0.12
Jamaica		0				0	60	0	59		0	24	
Martinique							18		18			53	
Mexico		7,530	209	6,191	1,011	118	4,551	220	3,190	1,027	113	51	1.65
Neth. Antilles							54		54			287	
Nicaragua		2				2	29		27		3	7	0.07
Panama		8				8	43	1	32	2	8	18	0.19
Puerto Rico		1				1	307	6	300		1	88	0.00
Trinidad & Tobago		545		330	215		257	0	42	215		201	2.12
USA		61,604	23,116	17,758	17,542	3,188	72,732	20,328	30,743	18,466	3,195	292	0.85
US Virgin Is							40		40			345	

Continued

TABLE 6.5 Continued

Region/country	Year	Primary energy production (10¹⁵ J)					Energy consumption (10¹⁵ J and 10⁹ J per capita)						
		Total	Solids	Liquids	Gas	Electricity	Total	Solids	Liquids	Gas	Electricity	Per capita	P/C ratio[a]
SOUTH AMERICA	**1980**	**9,889**	**265**	**7,635**	**1,285**	**704**	**7,241**	**443**	**4,807**	**1,287**	**703**	**30**	**1.37**
	1990	**13,292**	**735**	**9,254**	**2,097**	**1,206**	**8,720**	**677**	**4,659**	**2,092**	**1,292**	**29**	**1.52**
Argentina		1,980	7	1,088	793	91	1,729	34	722	878	95	53	1.15
Bolivia		164		46	113	5	76	0	44	28	5	10	2.16
Brazil		2,348	89	1,358	147	754	3,374	376	2,011	144	843	22	0.70
Chile		220	64	54	69	33	482	109	272	68	33	37	0.46
Colombia		1,713	517	935	164	98	745	144	339	164	99	23	2.30
Ecuador		653		630	5	18	225		203	5	18	21	2.90
French Guiana							7		7			71	
Guyana		0				0	9	0	9		0	11	
Paraguay		9				9	28		22		7	7	0.32
Peru		351	4	290	20	38	310	6	246	20	38	14	1.13
Suriname		16		12		4	22	0	18		4	52	0.73
Uruguay		23				23	68	0	49		18	22	0.34
Venezuela		5,815	55	4,840	786	134	1,643	7	716	786	134	83	3.54
ASIA	**1980**	**71,315**	**18,146**	**48,460**	**3,379**	**1,331**	**45,832**	**20,514**	**20,666**	**3,316**	**1,336**	**18**	**1.56**
	1990	**89,885**	**30,762**	**48,292**	**8,229**	**2,574**	**74,053**	**35,522**	**27,579**	**8,383**	**2,568**	**24**	**1.21**
Afghanistan		122	4	0	114	3	107	4	26	74	3	6	1.14
Bahrain		305		104	201		236	0	35	201		457	1.29
Bangladesh		162		5	154	3	248	12	79	154	3	2	0.65
Brunei		740		332	408		116	0	23	93		436	6.38
Cambodia		0				0	6	0	6		0	1	
China		29,360	22,586	5,782	594	398	27,027	22,256	3,774	594	403	24	1.09
Cyprus							52	3	49			74	
Hong Kong							304	200	110			52	
India		7,039	4,993	1,398	388	260	7,764	5,204	1,910	388	263	9	0.91
Indonesia[c]		4,774	215	2,947	1,579	33	1,624	104	1,052	435	33	9	2.94
Iran		7,667	38	6,672	934	24	2,648	51	1,690	882	24	48	2.90
Iraq		4,404		4,277	125	2	485	0	438	45	2	26	9.08
Israel		2		1			430	109	322	1		93	0.09
Japan		1,396	213	23	81	1,079	15,009	3,363	8,560	2,005	1,079	122	0.09
Jordan		1					127		127			32	0.01
Korea		538	324			213	3,102	1,037	1,726	127	213	72	0.17
Korea, Dem.		1,536	1,421			114	1,760	1,501	145		114	81	0.87
Kuwait		2,833		2,629	204		419		135	284		205	6.76
Lebanon		2				2	121	0	118		2	45	0.02

Continued

TABLE 6.5 Continued

Region/country	Year	Primary energy production (10^{15} J)					Energy consumption (10^{15} J and 10^9 J per capita)						P/C ratio[a]
		Total	Solids	Liquids	Gas	Electricity	Total	Solids	Liquids	Gas	Electricity	Per capita	
Macau							14	0	14		0	29	
Malaysia		1,801	4	1,274	498	26	790	56	576	133	25	44	2.28
Mongolia		90	90				115	84	30		1	53	0.78
Myanmar		86	2	38	42	4	74	3	24	42	4	2	1.16
Nepal		3				3	11	1	7		3	1	0.27
Oman		1,536		1,430	106		174		68	106		116	8.83
Pakistan		652	54	109	427	62	998	79	430	427	62	8	0.65
Philippines		75	23	10		42	550	62	447		42	9	0.14
Qatar		1,077		865	212		256		43	212		696	4.21
Saudi Arabia		15,293		14,191	1,101		2,648		1,547	1,101		187	5.78
Singapore							442	1	441		0	162	
Sri Lanka		11				11	65	0	54		11	4	0.17
Syria		989		960	11	17	391	0	363	11	17	31	2.53
Thailand		477	136	118	205	18	1,204	143	836	205	20	22	0.40
Turkey		712	500	121	7	84	1,701	654	846	120	81	30	0.42
United Arab Em.		5,283		4,495	788		940		277	663		592	5.62
Viet Nam		279	147	113	0	19	275	129	127	0	19	4	1.01
Yemen, Dem.		33		33			70		70			28	0.47
Yemen		355		355			46		46			5	7.72
EUROPE	**1980**	**37,125**	**19,156**	**5,974**	**9,524**	**2,472**	**64,148**	**21,382**	**28,689**	**11,547**	**2,529**	**132**	**0.58**
	1990	**38,704**	**16,472**	**9,050**	**8,526**	**4,655**	**62,832**	**19,852**	**24,952**	**13,250**	**4,778**	**126**	**0.62**
Albania		185	35	121	16	13	119	43	51	16	11	37	1.55
Austria		241	28	50	46	117	903	161	423	204	115	119	0.27
Belgium		218	63		0	155	1,707	473	707	385	141	173	0.13
Bulgaria		522	466	3	0	53	1,137	623	210	234	69	126	0.46
Czechoslovakia		1,739	1,607	5	23	103	2,569	1,610	356	487	115	164	0.68
Denmark		380	0	251	127	2	703	257	335	83	28	137	0.54
Faeroe Is		0				0	9	0	9		0	191	
Finland		165	56			109	865	244	368	105	147	174	0.19
France[d]		1,959	362	143	117	1,337	6,528	805	3,209	1,340	1,174	116	0.30
German Dem. Rep.		2,575	2,469	2	55	50	3,388	2,530	528	277	33	209	0.76
Germany, Fed. Rep.		4,264	2,985	150	522	607	10,014	3,028	4,139	2,240	608	163	0.43
Greece		329	283	35	4	7	896	321	562		10	89	0.37
Hungary		504	194	100	160	50	1,016	237	317	379	83	96	0.50
Iceland		17				17	46	3	27		17	182	0.37

Continued

TABLE 6.5 Continued

Region/country	Year	Primary energy production (10^15 J)					Energy consumption (10^15 J and 10^9 J per capita)						P/C ratio[a]
		Total	Solids	Liquids	Gas	Electricity	Total	Solids	Liquids	Gas	Electricity	Per capita	
Ireland		143	47		92	4	399	145	159	92	4	107	0.36
Italy[e]		944	12	196	599	138	6,150	565	3,779	1,543	263	108	0.15
Luxembourg		3				3	142	44	63	18	17	381	0.02
Malta							21	8	14			59	
Netherlands		2,721	0	167	2,541	13	3,159	447	1,232	1,434	46	211	0.86
Norway		5,019	9	3,444	1,131	436	854	41	335	99	379	203	5.88
Poland		4,117	3,999	9	98	12	4,187	3,301	506	373	8	109	0.98
Portugal		38	5			34	563	116	413		34	55	0.07
Romania		1,862	519	345	959	40	2,545	719	549	1,203	74	109	0.73
Spain		848	475	33	50	290	2,901	751	1,640	222	288	74	0.29
Sweden		509	0	0		509	1,172	99	552	20	502	139	0.43
Switzerland[f]		197				197	761	12	484	76	189	115	0.26
UK		8,167	2,154	3,848	1,904	262	8,397	2,481	3,425	2,186	305	146	0.97
Yugoslavia		1,037	706	149	82	100	1,681	791	560	230	98	71	0.62
USSR	1980	56,287	14,960	25,300	15,149	878	44,181	14,385	15,777	13,209	810	166	**1.27**
	1990	69,072	16,015	23,722	27,733	1,602	56,602	15,330	15,886	23,914	1,473	196	**1.22**
OCEANIA	1980	3,506	2,075	917	398	115	3,084	1,174	1,399	396	115	135	**1.14**
	1990	6,624	4,258	1,258	964	144	4,332	1,637	1,698	854	144	164	**1.53**
Australia[g]		6,219	4,202	1,181	784	53	3,724	1,585	1,412	674	53	182	1.67
Fiji		0					11	0	9		1	14	
French Polynesia		0				0	9		9		0	44	
Guam							21		21			178	
New Caledonia		2				2	22	5	15		2	132	0.09
New Zealand		400	57	78	180	86	497	47	185	180	86	142	0.80
Papua New Guinea		2				2	33	0	31		2	9	0.06

a Production/consumption ratio.
b South Africa Customs Union (South Africa, Botswana, Lesotho, Swaziland and Namibia)
c Includes East Timor
d Includes Monaco.
e Includes San Marino.
f Includes Liechtenstien.
g Includes Christmas Island.

Zero value equals less than 0.5 of the unit specified or nil. Solids include hard coal, lignite, peat and oil shale; liquids comprise crude petroleum and natural gas liquids; gas comprises natural gas; and electricity includes primary electricity generation from hydro, nuclear and geothermal sources.
Data on production refer to the first stage of production; for hard coal this is mine production; for crude petroleum and natural gas it is production at oil and gas wells, and for natural gas liquids it is production at wells and processing plants. Data on consumption refer to "apparent consumption" which is defined as production plus imports minus exports, minus bunkers and takes into account changes in stocks. Countries consuming less than 5 x 10^15 J are not listed separately but are included in the regional and world totals.

The P/C ratio given in the last column of the table provides an approximate measure of self sufficiency in commercial energy production. A value of unity implies self sufficiency; a value of less than one implies that the nation consumes more than it produces, i.e., it is a net importer of energy; and a value of greater than one implies that the nation produces more than it consumes, i.e., it is a net exporter of commercial energy.

Source:
UNSTAT 1992 1990 Energy Statistics Yearbook, United Nations, New York.

TABLE 6.6 Nuclear power reactors in operation and under construction, 1991

Region/country	Reactors in operation			Reactors under construction			Nuclear electricity supplied		Total operating experience to 31 December 1991 (years)
	No. of units	Capacity (MWe)[a]	% of world capacity	No. of units	Capacity (MWe)[a]	% of world capacity	(TWe.h)[b]	% of total	
WORLD	420	326,611	100	76	62,044	100	2,009.1		6,039
AFRICA									
South Africa	2	1,842	0.56				9.1	5.9	14
NORTH AMERICA									
Canada	20	13,993	4.28	2	1,762	2.84	80.1	16.4	263
Cuba				2	1,812	2.92			
Mexico	1	654	0.20	1	654	1.05	4.1	3.6	3
USA	111	99,757	30.54	3	3,480	5.61	612.6	21.7	1,592
SOUTH AMERICA									
Argentina	2	935	0.29	1	692	1.12	7.2	19.1	27
Brazil	1	626	0.19	1	1,245	2.01	1.3	0.6	10
ASIA									
China	1	288	0.09	2	816	1.32			<1
China, Taiwan	6	4,890	1.50				33.9	37.8	62
India	7	1,374	0.42	7	1,540	2.48	12.9	48.4	93
Iran				2	2,392	3.86			
Japan	42	32,044	9.81	10	9,192	14.82	209.5	23.8	514
Korea, Dem.	9	7,220	2.21	3	2,550	4.11	53.5	47.5	63
Pakistan	1	125	0.04				0.4	0.8[c]	20
EUROPE									
Belgium	7	5,484	1.68				40.4	59.3	108
Bulgaria	6	3,538	1.08				13.2	34.0	59
Czechoslovakia	8	3,264	1.00	6	3,336	5.38	22.2	28.6	68
Finland	4	2,310	0.71				18.4	33.3	51
France	56	56,873	17.41	5	7,005	11.29	314.9	72.7	655
Germany	21	22,390	6.86				140.0	27.6	426
Hungary	4	1,645	0.50				12.9	48.4	26
Italy	2[d]	1,120[d]	–						81
Netherlands	2	508	0.16				3.5	4.9	42
Romania				5	3,125	5.04			
Spain	9	7,067	2.16				53.2	35.9	111
Sweden	12	9,817	3.01				73.5	51.6	171
Switzerland	5	2,952	0.90				21.7	40.0	84
UK	37	11,710	3.59	1	1,188	1.91	62.0	20.6	926
Yugoslavia	1	632	0.19				4.7	6.3	10
USSR	45	34,673	10.62	25	21,255	34.26	212.1	12.6	559

[a] MWe = Megawatts (10^{12}) (electrical).
[b] TWe.h = 10^{12} watt-hours (We.h) = 10^6 megawatt-hours (mWe.h). For an average power plant, 1 TWe.h = 0.39 x 10^6 t of coal equivalent (input) or 0.23 x 10^6 toe of oil equivalent (input).
[c] Estimate.
[d] Data are for 1990.

Years of operating experience are rounded to the nearest whole year.

Sources:
IAEA 1992 *IAEA Yearbook 1992*, International Atomic Energy Agency, Vienna.
IAEA 1992 *Operating Experience with Nuclear Power Stations in a Number of States in 1991*, International Atomic Energy Agency, Vienna.

Part 7

Industry and Transport

Global industrial production grew rapidly between the 1950s and the early 1970s. Since then the rate of growth has slowed overall but is faster in the less developed countries. The relocation of some heavy manufacturing industry to less developed countries is part of a recent trend towards international specialization in industry. High technology industries concentrated in the more developed countries showed faster growth in the 1980s than other industrial sectors. Regional and national changes in emphasis of industrial production will lead to associated changes in regional and national environmental impacts. This chapter discusses the background to the trends highlighted below.

○ The environmental impact of industry is decreasing in terms of the quantitative amounts of energy and raw materials used and the quantity of wastes discharged or emitted.

○ World reserves of most metals have a lifetime in excess of 20 years. Nevertheless, greater awareness and concern for the environment are leading to an increase in the use of recycled metals; nearly a third of aluminium used and about a tenth of steel produced world-wide come from recycled material.

○ Production of hardwood has been rising steadily since the 1960s but growth in the production of softwood has declined since the late 1980s. Rapid growth in the production of wood products, particularly paper and paperboard and wood-based panels, has taken place in the less developed countries.

○ All transport, but particularly road transport, is a major contributor to air pollution and greenhouse gases. Road transport has been increasing world-wide. Nevertheless, major reductions in emissions from road transport are being achieved in some countries by strict controls on vehicle exhaust emissions.

Economic development and industrial growth are important goals for all countries. In the last two decades, however, there has been an increasing awareness that such development cannot be pursued without consideration of the possible adverse effects on the environment (see Part 4: Population and Development). Thus, whilst being crucial to the development of a nation, industry and transport are also responsible for

a) consuming energy and resources, b) contributing to the generation of waste, c) polluting air, water and soil, and d) generating substances that, in certain circumstances, can be a risk to human health.

Although many nations have given attention to defining environmental policies, these have rarely been linked to economic policies. Therefore, a crucial theme of the United Nations Conference on Environment and Development (UNCED), held in Rio de Janeiro in June 1992, was the integration of environmental and socio-economic factors in sustainable development. To this end, one of the objectives of the action plan for the 1990s and into the 21st century (commonly known as Agenda 21 – see Part 10: Environmental Co-operation) which was adopted by governments at UNCED, was that industry should aim to increase efficiency of resource use, including reuse and recycling of residues, and to reduce environmental impacts (UN, 1992). Issues relating to waste generation and management are discussed in detail in Part 8: Wastes and Waste Management.

Industry is now being included in the 'UNEP Environmental Data Report' on a regular basis, particularly with respect to the impacts of production and manufacturing on the environment. This, and future editions, of the 'UNEP Environmental Data Report' will focus on at least two industries which have major implications for the state of the environment. This edition concentrates on mining and processing of metals, and on the wood and wood product industries. This chapter also highlights recent trends in the transport sector, together with the potential environmental impact associated with those trends.

Industry

Industry is fundamental to the development of nations and provides an important source of national income through trade. The contribution of industry to the gross domestic product (GDP) of low-income countries increased from 28 per cent in 1965 to 37 per cent in 1989, whereas in industrial market economies the contribution fell from 42 per cent to 35 per cent over the same time period (see Part 4: Population and Development). The impact of industry on the environment is, however, more closely linked to the nature of the resources used, products manufactured, the structure and location of the industry, the techniques employed and the methods of treating residues, than to the general level of economic activity. The Organisation for Economic Co-operation and

Development (OECD) has identified those industrial sectors which are mainly responsible for raw material consumption and pollution of the environment (OECD, 1991). These are the agro-foodstuffs industry, metals extraction and processing, cement works, the pulp and paper industry, oil refining and the chemicals industry.

Various components of the United Nations (UN) collect and synthesize data and information on industry, some of which are referred to in this section. The United Nations Industrial Development Organization (UNIDO) specifically gathers information concerning industry, economic growth and development, and publishes detailed information in its 'Industry and Development' reports (UNIDO, 1992). A growing need for environmental information relevant to industrial decision-making has prompted UNIDO, through its Environment Programme to compile and publish 'Industry and Environment: A Guide to Sources of Environmental Information' (UNIDO, 1991). This guide lists a selection of sources of environmental information frequently used at UNIDO, concentrating on sources which have databases with on-line search facilities.

The United Nations Environment Programme (UNEP) established its Industry and Environment Programme Activity Centre (IE/PAC) in 1975. This centre is located in Paris and through its activities brings industry, governments and non-governmental organizations together with the common aim of working towards environmentally sound industrial development. The IE/PAC provides access to practical information and develops co-operative on-site activities which are backed up by regular assessment. Information is made available through technical reviews and guidelines, the quarterly 'Industry and Environment' review and a query-response service. Examples of recent activities include: the Awareness and Preparedness for Emergencies at Local Level (APELL) programme, which is supported by the chemical industry and governments and has led to the creation of a network of experts and centres aimed at accident prevention and emergency response (see also Part 9: Environmental Disasters); the OzonAction programme which includes an information component collecting and disseminating information on policies and technologies to reduce the use of ozone depleting substances; and the Cleaner Production programme which was established to raise awareness and help industry and governments develop cleaner products, technologies and production techniques. Greater emphasis is currently being placed on activities in the field of tourism and transport (IE/PAC, 1993a).

Global industrial production grew rapidly between the 1950s and the early 1970s but has since slowed down. Trends in industrial production or activity can be reflected by indices such as those constructed by the Statistical Office of the United Nations (UNSTAT) and published regularly in the 'Industrial Statistics Yearbook' series (UNSTAT, 1990). Trends in industrial production between 1973 and 1990 using the UNSTAT index are given in Table 7.1. Another index of industrial activity (specifically manufacturing) is manufacturing value added (MVA) which is: the gross output, less the cost of materials, supplies, fuel and electricity consumed, and services received. In recent years, world MVA has shown a positive but declining rate of growth (UNIDO, 1992).

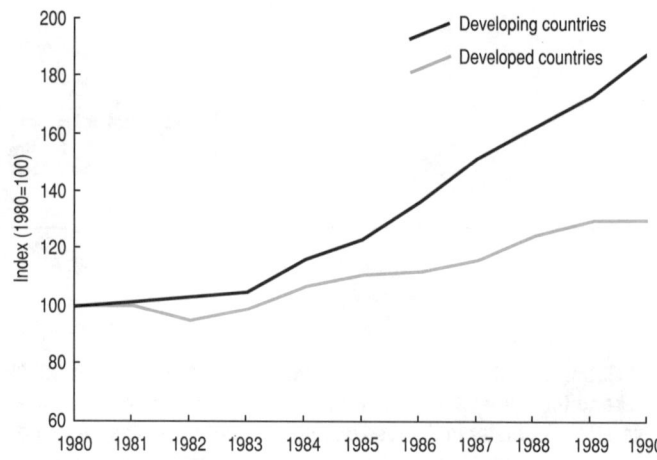

FIGURE 7.1 Growth in heavy manufacturing in developed and developing countries from 1980 to 1990 relative to the base year, 1980=100
Source: United Nations Conference on Trade and Development, 1992

Table 7.1 shows that industrial production has been increasing steadily and at a faster rate in developing countries than in developed countries. Developing countries are showing more rapid and greater growth in the manufacturing, and electricity, gas and water sectors. The rapid growth in the heavy manufacturing industry in developing countries is illustrated in Figure 7.1. This trend is partly associated with the relocation of heavy industry to those countries with low labour costs, or rich energy or raw material resources (OECD, 1991). This is part of an overall industrial trend towards international specialization. For example, aluminium manufacture is becoming concentrated in Australia, Brazil, Canada and Venezuela because electricity is cheaper in these countries, and the petrochemicals industry is concentrating in the oil-producing countries. In the developed countries, such as the members of the OECD, the decline in growth in heavy industry has been replaced by growth in the electronics and electrical industries, telecommunications, data processing and fine chemicals. In the OECD countries, growth in the high-technology sector between 1980 and 1989 averaged 8–10 per cent a year compared with only 2 per cent a year for the rest of industry (OECD, 1991).

The change in emphasis of industrial production in the developed and developing countries will lead to changes in the environmental impacts of industry in these countries. The concentration of the heavy manufacturing industries in the developing countries has potential implications for their environmental quality, because heavy industry is responsible for the major environmental emissions and is usually energy intensive (see Part 6: Energy). The restructuring of industry in the developed countries could reduce the relative pressure of industry on the local environment, but could also introduce a wider variety of environmental risks. The potential environmental risks of new technologies are mostly not well known, although those associated with new materials (such as polymers, alloys, ceramics and composites) are assumed to be limited (OECD, 1991). A

BOX 7.1 Environmental Audit and Life Cycle Analysis

In recent years industry has been under pressure to recognize that it has a responsibility to protect the environment and human health. This responsibility is now going beyond compliance with environmental legislation and is taking the form of the introduction of appropriate environmental policies and practices. The process of environmental audit is now widely accepted as the best way to determine the impact of industrial activities on the environment.

An environmental audit examines the activities of a company or organization through a systematic determination of the environmental impacts of all activities within the organization. These activities include impacts arising from emissions to land, water or air; the effects on the local communities, the landscape and the ecology; noise pollution; and an examination of the management and policies of the company with respect to its use of materials, products and processes. Special attention may be paid to the environmental implications of the raw materials used and the end products generated. Environmental performance is measured against predetermined targets (such as statutory conditions relating to discharges) and the results can be used to help ensure that statutory obligations are met and that environmental risks are minimized. The results can also be used to increase environmental awareness. The first internationally agreed definition of environmental auditing was published by the International Chamber of Commerce in 1989 and their detailed standard practical methodology was published in 1991 (ICC, 1991).

In an effort to satisfy public pressure many industries have been including environmental statements in their annual reports. However, many of these fall short of environmental audits, presenting only broad statements of environmental policy and general assessments of environmental performance. As a result there has been an increasing demand for quantified information, perhaps through the use of indicators of environmental performance. Indicators of impact would include such measurements as effluent loads or waste arisings. Indicators of performance for other aspects of a business, such as management systems or process technology, are still being devised.

Life cycle analysis is an assessment procedure which concentrates on the main impacts associated with the manufacture, use and disposal of a product. It consists principally of an inventory of materials consumed and releases to the environment. In collaboration with the Stockholm Environment Institute (SEI), UNEP has been developing an environmental data base containing emission factors for a wide range of energy systems (see Part 1: Environmental Pollution, Atmosphere). The main use of life cycle analyses has been in the marketing of products by manufacturers who wish to claim their products to be "environmentally friendly".

The problem of approaches to life-cycle analysis (particularly in the absence of standard methods and accreditation procedures) is highlighted by the different conclusions reached by different studies of the same products. For example, five European countries, Denmark, Germany, the Netherlands, Sweden and Switzerland, studied milk containers and came to different conclusions as to which container (refillable glass bottle, polycarbonate bottle or laminated paper carton) was the most environmentally benign. Some surprising results from life cycle analysis in the Federal Republic of Germany have also suggested that plastic bags are more environmentally benign than paper ones (The Warmer Campaign, 1993).

There are now moves to standardize environmental auditing at the international level with the new European Community (EC) regulations on eco-auditing which were approved by EC ministers in March 1993 (Doyle, 1993; EC Report, 1993). This voluntary scheme for participating companies includes a system of external verification of the company's published environmental statement and the accreditation of individuals responsible for environmental audits.

A new approach to environmental auditing now includes the use of ecological accounts, in which companies can use and perform calculations on environmental data in the same way as their existing financial data. Companies adopting this approach would take account of the environmental value of, for example, the resources they use in their production processes. Corporate accounting would therefore embrace environmental value as well as financial analysis. In its broadest sense this would include placing a value on the environmental damage caused by the production or extraction of the raw materials or products used in their own processes (see also Part 4: Population and Development where national accounting systems are discussed). Similarly the company's own products may also have adverse impacts on the environment during or after their consumption.

References

Doyle, P. 1993 Eco-management. *Waste Environment Today*, **6**(5), 1.
EC Report 1993 Eco-management and audit-scheme approved. *Waste Environment Today*, **6**(5), 19–21.
ICC 1991 *Guide to Effective Environmental Auditing*, International Chamber of Commerce, Paris.
The Warmer Campaign, 1993 *Life Cycle Analysis*, Warmer Information Sheet, The Warmer Campaign, Tunbridge Wells, UK.

---

major problem associated with the substitution of traditional materials for new ones, such as silicon and thermoplastics, is that the new products are not always easily biodegradable and present problems of waste disposal. However, some new technologies (such as high temperature electrical systems) eliminate the need for these new materials. Many of the new materials also present environmental and human health risks during the production processes (OECD, 1991).

Trade is an important aspect of industrial activity. The developing countries are net exporters of food, raw materials, minerals and fuels to industrial countries. Primary products usually dominate their total exports, accounting for as much as 98 per cent of total exports in some developing countries compared with only 24 per cent of total exports in the USA and 2 per cent in Japan (French, 1993). With increasing trends towards trade liberalization there is the possibility that some environmental problems could become worse in the absence of appropriate domestic controls. Developing countries are particularly at risk where the need for foreign income encourages the exploitation of natural resources and the rapid expansion of industry. Materials which are extracted in one country, exported to another for processing and then exported again for manufacture create environmental problems in all three countries, although the environmental costs are not distributed evenly between the countries. Full appreciation of the environmental issues associated with international trade has not yet been reached.

In addition to the consumption of raw materials and the associated environmental effects, industry is responsible for the generation of large quantities of waste in the form of emissions to air, waste-water discharges and solid and hazardous waste products. National statistics for several different categories of waste, including hazardous waste, are presented in Part 8: Wastes and Waste Management. Industry is responsible for a large proportion of the total emissions of some major pollutants. For example, industry in OECD countries was responsible for 25 per cent of NO_x emissions, 40–45 per cent of SO_x emissions and 50 per cent of total greenhouse gases in 1987 (OECD, 1991). On an individual country basis these proportions can be even greater (see also Part 1: Environmental Pollution, Atmosphere for national emission estimates for major pollutants).

Many industries are concerned about their public image in relation to their use of raw materials and their emissions and discharges to the environment. This concern has led to a recent interest in the application of environmental audits and the publication of company-based environmental statements (see Box 7.1). As a result of efforts to save energy and raw materials and to control pollution by those industrial sectors mainly responsible, the environmental impact of industry is decreasing when viewed in terms of the quantitative amounts of energy and raw materials used and the quantity of wastes discharged or emitted (OECD, 1991). Nevertheless, the major emissions to the environment associated with any particular industry may arise from "downstream" use and disposal of the manufactured product, thus making it difficult to establish clear links between industrial consumption and environmental effects. In response to public demands for more information concerning the nature of emissions from industry, new legislation is being introduced in many countries which requires manufacturing facilities to make information on their emissions to the environment more widely available (see Part 8: Wastes and Waste Management for an example in the USA).

Mining and Processing of Metals

Mineral resources play an important role in the world's industry, providing the raw materials for manufacturing, construction, energy and technological processes, as well as for agriculture. Minerals such as sand, gravel and lime (for cement) are used in the construction industry. Other non-metals such as phosphates and potash are used as fertilizers, and industrial minerals such as soda ash and salt are used in chemical processes. Minerals used for energy production are discussed in Part 6: Energy. This section concentrates on reserves, production and consumption of selected metals. Metals are the most important and valuable minerals extracted; world production in 1990 was estimated to be 552 x 10^6 t (Young, 1992). Data for production, trade and consumption are collected by a number of organizations, including the United States Bureau of Mines, the World Bureau of Metal Statistics, and specialized organizations such as the International Iron and Steel Institute.

Mining, mineral processing and the transport of minerals to the site of processing, contribute to environmental problems around the world, such as the destruction of ecosystems, land degradation, and air and water pollution. Despite regulations to minimize adverse effects and efforts to recycle metals, the processes of mining and extraction inevitably damage the environment (see Box 7.2). Technical guides to the environmental aspects and management of ore mining and production are produced by the UNEP IE/PAC. These guides emphasize the need for environmental control procedures at the company and national level (IE/PAC 1991, 1993b).

Reserves   When assessing the possible environmental impact of the exploitation of minerals, it is important to distinguish between metal resources and metal reserves. Metal resources consist of geological deposits which have been demonstrated to be present, for example, by surveys, but which are not yet economically viable. Metal reserves are the proportion of the resources which have been measured more precisely by detailed surveying and have been judged to be available for production over a specific time period and at a cost no higher than current market prices. For this reason metal reserves can apparently increase and decrease over periods of time. Table 7.2 details reserves of 15 metals as assessed in 1991 for individual countries, and Figure 7.2 shows the distribution of the reserves of the same metals by world region. The most extensive reserves for the 15 metals are concentrated in the former USSR, South Africa, the USA, China and Australia, which together hold over half of the world's metal reserves.

New exploration techniques are leading to the discovery

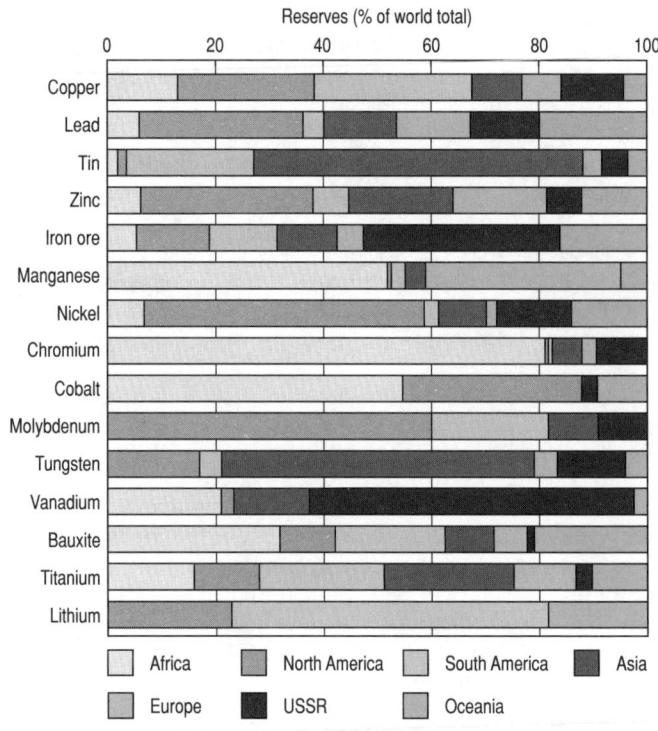

FIGURE 7.2 Distribution of major metal reserves by world region, 1991
Source: Data from Table 7.2

BOX 7.2 Environmental Impacts of Mining and Excavation

The environmental damage associated with production of any particular metal depends on the ecological character of the mining site, the quantity of material removed, the depth of the deposit, the chemical composition of the ore and surrounding rocks and soils, and the nature of the processes used to extract and purify the metal from the ore. The mining process can potentially cause: contamination of surface and ground waters with metals, acids and suspended solids; air pollution through the release of particles, gases and vapours; ecosystem damage; and aesthetic and socio-economic effects (IE/PAC, 1991). The potential impacts of mining can be summarized as follows (Balkau and Epps, 1993):

Environmental impacts

○ Destruction of natural habitat at the mining site and at waste disposal sites.
○ Destruction of adjacent habitats as a result of emissions and discharges.
○ Destruction of adjacent habitats arising from influx of settlers.
○ Changes in river regime and ecology due to siltation and flow modification.
○ Alteration in watertables.
○ Change in landform.
○ Land degradation due to inadequate rehabilitation after closure.
○ Land instability.
○ Danger from failure of structures and dams.
○ Abandoned equipment, plant and buildings.

Pollution impacts

○ Drainage from mining sites, including acid mine drainage and pumped mine water.
○ Sediment run-off from mining sites.
○ Pollution from mining operations in riverbeds.
○ Effluent from minerals processing operations.
○ Sewage effluent from the site.
○ Oil and fuel spills.
○ Soil contamination from treatment residues and spillage of chemicals.
○ Leaching of pollutants from tailings and disposal areas and contaminated soils.
○ Air emissions from mineral processing operations.
○ Dust emissions from sites close to living areas or habitats.
○ Release of methane from mines.

Occupational health impacts

○ Handling of chemicals, residues and products.
○ Dust inhalation.
○ Fugitive emissions within the plant.
○ Air emissions in confined spaces from transport, blasting, combustion.
○ Exposure to asbestos, cyanide, mercury or other toxic materials used on-site.
○ Exposure to heat, noise, vibration.
○ Physical risks at the plant or at the site.
○ Unsanitary living conditions.

Mining wastes represent one of the largest sources of solid waste in many countries (see Part 8: Wastes and Waste Management). Surface mining produces more waste than underground mining; in the USA in 1989, for example, surface mines produced eight times as much waste per tonne of ore as underground mines (Young, 1992). The grade of an ore (i.e., per cent metal content) is important in determining the impact of metal mining. As the metal content falls more ore needs to be processed to obtain the same amount of metal and this requires more energy. In addition, the amount of waste generated per tonne of metal (excluding overburden) for ores with low metal contents is higher, as indicated in the table on metal production and waste generation in 1991.

Up to 90 per cent of metal ore ends up as tailings (i.e., the finely ground residue from ore concentration) and these are commonly dumped in piles or ponds near the mine. Contaminants formerly bound up in the solid rock (e.g., arsenic, cadmium, copper, lead and zinc) become more accessible to water when the rock is in the form of fine tailings, especially in the presence of acid which is often associated with mine drainage. The sulphur commonly present in the metal-containing rocks combines with rainwater to form sulphuric acid. The acid from overburden can contaminate water and soil leading to the release from tailings of metals which are often acutely toxic to aquatic life (IE/PAC, 1991).

Reduction of environmental impacts of mining can be achieved through improved and appropriate technology in the production process, together with clear and practical environmental laws and standards, combined with effective enforcement (Balkau and Epps, 1993). Since a comprehensive system of environmental regulation and standards does not yet exist in many countries (Balkau and Epps, 1993), the mining companies themselves must ensure that all likely environmental issues are addressed, for example, through the use of environmental auditing techniques (see Box 7.1).

References

Balkau, F. and Epps, J. 1993 *Environmental management of mining activities*. Paper presented for the UN/CDG Workshop on Environmental Management Systems, September 1993, Namibia, United Nations Environment Programme Industry and Environment Programme Activity Centre, Paris.

IE/PAC 1991 *Environmental Aspects of Selected Non-Ferrous Metals (Cu, Ni, Pb, Zn, Au) Ore Mining: A Technical Guide*, Technical Report Series No. 5, United Nations Environment Programme Industry and Environment Programme Activity Centre, Paris.

Young, J. E. 1992 *Mining the Earth*, Worldwatch Paper 109, Worldwatch Institute, Washington DC.

World metal production and the associated waste generation in 1991

Metal	Production (10^6 t a^{-1})	Average grade of ore (%)	Waste generation (10^6 t a^{-1})
Copper	1,000	0.91	990
Gold	620	0.00033	620
Iron	906	40.0	540
Lead	135	2.5	130
Aluminium/Bauxite	109	23.0	84
Nickel	38	2.5	37
Tin	21	1.0	21
Manganese	22	30.0	16
Tungsten	15	0.25	15
Chromium/Chromite	13	30.0	9

Waste figures do not include overburden.
Source: Young, 1992

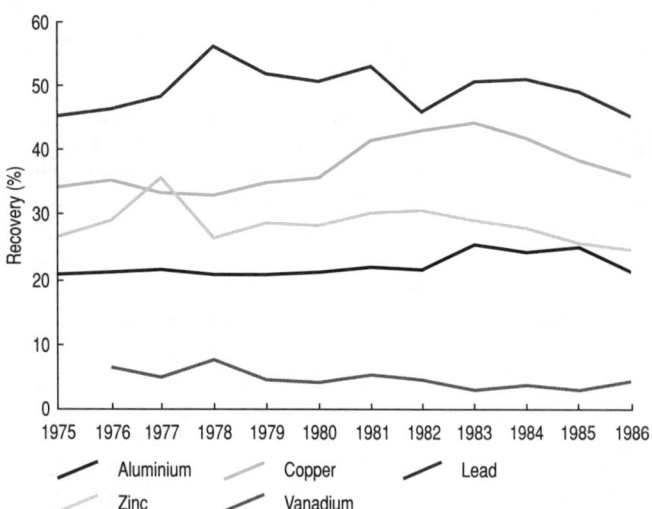

FIGURE 7.3 Percentage recovery of some non-precious metals in the European Community, 1975–1986
Source: EUROSTAT, 1991

of new deposits with better grades of ore (Crowson, 1992) and some of the best metal reserves are now in developing countries, although the quantity of the reserves is often small (Table 7.2). The search for new reserves is leading to potential conflicts over land use, particularly land which has been designated as national parks or globally significant, such as Antarctica (see Part 10: International Co-operation).

Apparent growth in metal reserves can be explained partly by improved technology and changing economic factors (Crowson, 1992). Production is unlikely therefore to be constrained by scarcity of deposits of the most important metals in the immediate future. The production-reserve ratios given in Table 7.3 show that world reserves for most metals have a lifetime of at least 20 years and, in the cases of bauxite and iron ore, lifetimes in excess of 150 years.

Production and Consumption Trends in production of selected metals for the world's 10 largest producers in 1990 are given in Table 7.3. Overall increases in world production between 1975 and 1990 were observed for bauxite, cadmium, copper, nickel, crude steel and zinc and decreases occurred for iron ore, lead, mercury and tin. The greatest increases in production over the 15-year period, 1975 to 1990, were generally observed in developing countries, where production of many metals was at least doubled. Many developing countries are now amongst the world's principal producers of important metals (McDivitt, 1993).

Consumption of the same metals over the same time period, 1975 to 1990, for the 10 largest consumers, together with their per capita consumption in 1990, is given in Table 7.4. These data represent consumption of raw and recovered metals. World consumption of all metals, apart from mercury, increased over the 15-year period. The developed countries were responsible for the largest proportion of the consumption, although some developing countries have substantially increased their consumption of two or more metals.

Nevertheless, the per capita consumption by these countries remains lower than for developed countries. The growth in consumption of metals by some developing countries can be attributed to their continuing expansion of their industrial sector whereas many developed nations are shifting towards services and high technology industries (Young, 1992), leading to a reduced requirement for metals in favour of new materials (see above).

The pattern of consumption of raw metals, particularly in the developed countries, has been influenced by growing concerns about depletion of resources and the environmental implications associated with the use of the metals. Extraction, processing, transformation and consumption of metals generate wastes and emissions which contaminate air, water and soil and lead to harmful effects in ecosystems, as well as presenting possible risks for human health (see Box 7.2). The process of smelting emits large quantities of contaminants to the air. Copper and other non-ferrous metals release about 6×10^6 t of sulphur dioxide to the atmosphere each year (Young, 1992). Non-ferrous smelters also emit arsenic, lead, cadmium and other heavy metals, and aluminium smelters emit fluoride when no pollution control measures are applied. In developed countries, smelters are now usually required to have pollution control equipment but in developing countries, and the formerly socialist countries, few smelters have such controls. Whereas emissions from smelters are being tightly controlled in most developed countries, control of the disposal of mining wastes is still weak (Young, 1992).

Potential environmental and health risks, together with concerns for the long-term future availability of primary sources of metals, have stimulated new emphasis on the use of recycled materials as a source of raw material in industry, together with the search for alternative materials. The most commonly recycled metals are those of high value such as the precious metals, gold, silver and platinum and those which are highly toxic such as lead. Trends in the recovery of non-precious metals are given in Figure 7.3. In the USA, 73 per cent of the lead consumed was supplied by recycling, mainly because lead is so toxic that its use is tightly regulated (Young, 1992).

Scrap metal is now becoming regarded by the metal industry as a secondary raw material rather than merely waste. The recycling of steel and aluminium cans is widespread in developed countries. In Europe as a whole, 21 per cent of used aluminium drinks' cans were recycled in 1991, equivalent to 32,350 t. The highest recycling rate was achieved by Sweden, where recycling began in 1984 and now accounts for 85 per cent of used cans (Anon, 1992). Nearly a third of aluminium used world-wide comes from recycled material and the remelting of scrap aluminium requires only 5 per cent of the energy needed to smelt new metal (Dermer, 1992). As a result of new legislation in some countries, which makes the manufacturer responsible for the ultimate disposal of their products, automobile manufacturers are beginning to build recyclability into their products (UNIDO, 1992). At least 10 per cent of world steel production now comes from scrap metal rather than fresh ore (Young, 1992). Estimated consumption, import and export of scrap metal for steel making in 1989 is given below for major world regions:

Region	Consumption (10³ t)	Imports (10³ t)	Exports (10³ t)
Eastern Europe (inc. USSR)	118,902	915	2,922
Western Europe	81,892	21,794	17,198
North America	74,110	2,471	12,039
Oceania	2,000	–	1,154
Africa	4,000	200	–
Western Asia	1,800	600	–
Asia	90,000	12,000	900
Latin America	15,500	1,000	–
TOTAL[a]	388,200	39,000	34,200

A dash indicates none or a negligible amount.

[a] Rounded figure.

Source: UNIDO, 1992

Recently there has been a trend towards the export of scrap metal for steel making to south and south-east Asia from developed regions (UNIDO, 1992). In the long term the increasing use of recycled metals will reduce the pressure on global reserves of ores, thereby reducing the environmental degradation associated with mining, as well as reducing the quantities of waste products, including solid wastes, released to the environment.

Wood and wood products

Forestry and related activities constitute a major industrial activity for some countries and in some world regions. In Malaysia, for example, logs and timber are responsible for about 11 per cent of the country's total export earnings (French, 1993). In addition, forest industries generate important employment opportunities in rural areas, and non-wood products collected and traded in local markets provide employment and additional income (FAO, 1992a). Forest resources therefore represent an important national asset to many countries and increasing attention is being paid to conservation and sustainable management of all aspects of these resources. Apart from over-exploitation for production of commercial timber or fuelwood, forest resources are under threat from atmospheric pollution (mainly in developed countries), forest fires, squatters and shifting agriculture (mainly developing countries) and inappropriate use for wildlife management. Forest resources and deforestation are discussed further in Part 3: Natural Resources.

Global data on production, trade and prices of wood and wood products are collected, assembled and published annually by the Food and Agriculture Organization of the United Nations (FAO). The annual average world total roundwood production, during the period 1988–1990, was $1,954 \times 10^6$ m³ a⁻¹ for hardwood and $1,357 \times 10^6$ m³ a⁻¹ for softwood (Tables 7.5 and 7.6). The developing countries are responsible for 79 per cent of the total hardwood production but only 19 per cent of the total softwood production, and 54 per cent of the total hardwood and softwood production combined. Production of hardwood has been rising steadily since 1961 but growth in production of softwood has ceased since the late 1980s (Figure 7.4).

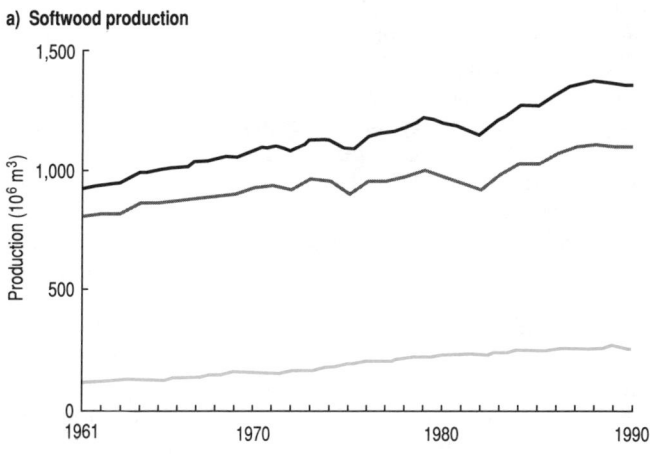

a) Softwood production

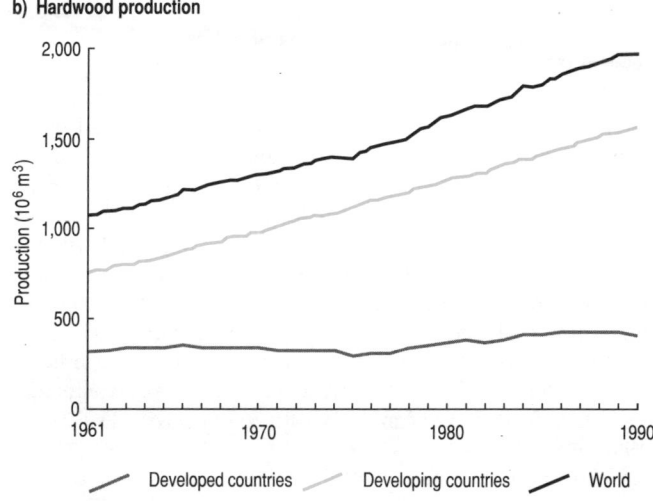

b) Hardwood production

FIGURE 7.4   Softwood and hardwood production in developed and developing countries, 1961–1990
Source: Data extracted from the FAO AGROSTAT data base

Whereas in developed countries forests provide mainly wood for industrial uses, in developing countries they are an important source of energy supply. Production of energy from wood in developed countries is only about 1 per cent compared with between 70 and 90 per cent in some of the poorer countries of Africa (Wardle, 1993 pers. comm.). Energy production in developing countries accounts for more than 80 per cent of total roundwood production (FAO, 1993a). Over the period 1968–1970 to 1988–1990, developing countries increased their production of fuelwood by 40 per cent for softwood and 57 per cent for hardwood (Tables 7.5 and 7.6). The use of biomass fuels, including fuelwood, is covered in more detail in Part 6: Energy.

Production of industrial roundwood (i.e., roundwood, other than fuelwood, which may be in the form of sawlogs, veneer logs, sleepers, pit props and pulpwood) has increased by 88 per cent for hardwood and 126 per cent for softwood in developing countries over the 20-year period 1968–1970 to 1988–1990, compared with only 19 per cent and 22 per cent respectively for developed countries. Most softwood is used as industrial roundwood (Table 7.6) but nearly three-quarters of hardwood is used as fuelwood (Table 7.5).

Production of wood products increased in developing countries over the period 1968–1970 to 1988–1990, particularly the production of paper and paperboard and wood-based panels (Table 7.7). The dramatic growth in production of wood products is clearly illustrated by Indonesia for wood-based panels and Korea for paper and paperboard. In 1991, the developing countries accounted for just over 20 per cent of world production of sawnwood and wood-based panels and about 16 per cent of the world production of paper and paperboard (FAO, 1993a). Production, particularly of mechanical wood products used by the construction industry, declined markedly in the developed countries in 1990 and 1991 as a result of economic trends reducing domestic demand in the building industry (FAO, 1993a).

The developed countries were responsible for over 80 per cent of world paper and paperboard production in 1990, particularly North America (36.7 per cent), Western Europe (25.8 per cent) and Japan (12.1 per cent) (UNIDO, 1992). The anticipated growth in wood pulp production, the main raw material of the paper industry, has been reduced by increased use of recycled fibre (FAO, 1992a). The pulp and paper industry is particularly associated with the discharge of chlorine, organochlorines (including dioxin) and organic matter in its effluents. As a result the industry, particularly in the developed countries, is under pressure to provide environmentally friendly products in the form of paper made from pulp which has not been bleached with any chlorine compounds, to use more recycled fibre and to reduce the effects of paper mills on the environment. The world-wide recovery rate of wastepaper increased from 31 per cent in 1985 to 36 per cent in 1991 (FAO, 1992b).

Pulp and paper mills produce effluents containing pulping and bleaching chemicals, and solids such as wood fibres and particulates. These effluents are usually discharged, before or after treatment, directly into rivers, lakes or marine waters where they have a variety of environmental impacts. Air emissions for pulp and paper processing are relatively less important. Considerable improvements in controlling effluents to water bodies from pulp and paper mills have been achieved in recent years as a result of scrutiny by environmentalists, regulators and the general public. Primary treatment of effluents can remove 80–90 per cent of settleable solids by screening and settling, and further secondary biological treatment methods can reduce the biological oxygen demand (BOD) (an indicator of pollutant load) by as much as 70–95 per cent as well as effectively reducing the presence of chlorinated compounds (Statistics Canada, 1993).

Consumption of wood products by developing countries is low, but growth in consumption has been more rapid than in developed countries. Consumption of paperboard in Association of South East Asian Nations (ASEAN) countries, for example, increased by 24 per cent from 1989 to 1990 (UNIDO, 1992). The fast growth in consumption of wood products has largely been met by growth in domestic production.

Wood and wood products are an important part of the world economy. World production of industrial roundwood was 1.607×10^6 t in 1992 (FAO, 1992b). Developing country exports accounted for 13.5 per cent of the value of total world exports of forest products in 1992 (FAO, 1992b). The export of wood products, e.g., wood-based panels and paper and paperboard, has grown faster than for industrial roundwood as shown in the following table of exports from developing countries:

Wood product	1970	1980	1990	1991
Industrial roundwood (10^6 m³)	37.9	42.2	35.6	35.7
Sawnwood and sleepers (10^6 m³)	5.9	10.9	11.7	11.5
Wood-based panels (10^6 m³)	2.7	4.7	13.0	13.4
Paper and paperboard (10^6 t)	0.3	0.9	3.7	4.3

Source: Data extracted from the *FAO AGROSTAT 1993* data base

Over 30 per cent of world plywood production was exported in 1991 (FAO, 1993b). The upward trend in exports of wood and wood products and the decline in exports of sawnwood have been aided by restrictions (such as export levies and quotas) placed on unprocessed roundwood exports by some developing countries (as well as some developed countries) in order to encourage the development of domestic wood processing industries and their exports (FAO, 1992b).

Forests are important for many other products besides timber and fuelwood. Most of these non-wood products are consumed or traded locally by rural people, although some are valuable for international trade, such as gums and resins, bamboos, various oils, rosin and turpentine, tanning materials, honey, seeds and spices, wildlife products, bark and tree leaves, and medicinal plants (FAO, 1992a). Products traded locally or regionally include insects, fruits, fungi, bushmeat, cola nuts and palm wines. Besides depending on some forest products for food, fuel, medicines, animal fodder, material for handicrafts and building for their own use, rural people obtain useful seasonal employment and income from harvesting, processing and marketing many of these products (FAO, 1992a). Reduction of forest resources therefore has widespread implications for the food security and income generation of local communities.

Countries belonging to the International Tropical Timber Organization (ITTO) have pledged to aim to provide all tropical timber exports (logs, sawnwood and panels) from sustainable sources by the year 2000. The ITTO originated from the Integrated Programme for Commodities of the United Nations Conference on Trade and Development (UNCTAD) and was created by the 1983 International Tropical Timber Agreement. This agreement was between 18 "producer" tropical countries and 22 "consumer" (mostly industrialized) countries and was mainly concerned with improving market conditions. However, it also included the objective of encouraging national policies aimed at maintaining ecological balance in the region concerned. The organization is currently involved in investigating multiple-use forestry practices, natural forest management programmes and plantation development (Collins et al.; 1991).

References

Anon 1992 1991 Aluminium can recycling statistics, *Aluminium Industry*, **11**(4), 23.

Collins, N. M., Sayer, J. A. and Whitmore, T. C. (Eds) 1991 *The Conservation*

Atlas of Tropical Forests: Asia and the Pacific, Macmillan Press Ltd, London.

Crowson, P. 1992 *The Infinitely Finite*, The International Council on Metals and the Environment, Ottawa.

Dermer, D. 1992 Aluminium recovery and recycling using eddy currents, *Aluminium Industry*, **11**(4), 20–23.

EUROSTAT 1991 *Raw Materials and Environment*, Rapid Reports No.1 Statistical Office of the European Community, Luxembourg.

FAO 1992a *The State of Food and Agriculture 1991*, Food and Agriculture Organization of the United Nations, Rome.

FAO 1992b *Commodity Review Outlook, 1992–93*, Food and Agriculture Organization of the United Nations, Rome.

FAO 1993a *State of Food and Agriculture 1992*, Food and Agriculture Organization of the United Nations, Rome.

FAO 1993b *Forest Products Yearbook 1991*, Food and Agriculture Organization of the United Nations, Rome.

French, H. E. 1993 *Costly Tradeoffs: Reconciling Trade and the Environment*, Worldwatch Paper 113, Worldwatch Institute, Washington DC.

IE/PAC 1991 *Environmental Aspects of Selected Non-Ferrous Metals (Cu, Ni, Pb, Zn, Au) Ore Mining: A Technical Guide*, Technical Report Series No. 5, United Nations Environment Programme Industry and Environment Programme Activity Centre, Paris.

IE/PAC 1993a *UNEP IE/PAC Activity Report 1992*, United Nations Environment Programme Industry and Environment Programme Activity Centre, Paris.

IE/PAC 1993b *Environmental Management of Nickel Production: A Technical Guide*, Technical Report Series No. 15, United Nations Environment Programme Industry and Environment Programme Activity Centre, Paris.

McDivitt, J. F. (Ed.) 1993 *International Mineral Development Sourcebook*, Forum for International Mineral Development, Golden, Colorado.

OECD 1991 *The State of the Environment*, Organisation for Economic Co-operation and Development, Paris.

Statistics Canada 1993 Compliance costs of regulation in the Canadian pulp and paper industry. Working Paper No. 15, Joint ECE/EUROSTAT Work Session on Specific Methodological Issues in Environmental Statistics, 20–23 September 1993, Bratislava, Slovakia.

UN 1992 *Report of the United Nations Conference on Environment and Development* (Rio de Janeiro, 3–14 June 1992), UN document A/CONF.151/26, 12 August 1992 (Vol I, II, III and IV). United Nations Conference for Environment and Development, Geneva.

UNIDO 1991 *Industry and Environment: A Guide to Sources of Information*, United Nations Industrial Development Organization, Vienna.

UNIDO 1992 *Industry and Development. Global Report 1992/93*, United Nations Industrial Development Organization, Vienna.

United Nations Conference on Trade and Development 1992 *Handbook of International Trade and Development Statistics*, United Nations, New York.

UNSTAT 1990 *Industrial Statistics Yearbook 1988*, Statistical Office of the United Nations, New York.

Wardle, P. 1993 Personal communication, Food and Agriculture Organization of the United Nations, Rome.

Young, J. E. 1992 *Mining the Earth*, Worldwatch Paper 109, Worldwatch Institute, Washington DC.

Transport

Transport systems are important to individuals and to the economic activity of nations. The transport industry provides an important source of employment within nations and the transport systems provide the means for national and international trade. The demand for transport mainly reflects the level of economic development of a country, with freight transport being particularly important to highly productive countries. There are few data available on the level of use of modes of transport in developing countries and it is, therefore, difficult to establish the relative importance of different types of transport to these countries.

Human- or animal-powered means of transport, such as bicycles and carts, have little or no impact on the environment during their use and are not discussed further here. Mechanized transport falls into four main categories: motor vehicles or road traffic; railways; marine and inland shipping; and air traffic. This edition of the 'UNEP Environmental Data Report' concentrates on passenger and freight transport, and roads and road traffic.

Transport statistics are collected by a number of international trade and commercial user organizations as well as by specialized UN organizations. The International Road Federation (IRF) publishes annual statistics on roads and road traffic, the Union Internationale des Chemins de Fer (UIC) compiles railway and rail traffic statistics, and the International Civil Aviation Organization (ICAO) and the International Air Transport Association (IATA) both compile information on air transport. The different criteria used by countries in their collection of statistics makes comparisons between countries difficult.

Transport activities have a wide variety of effects on the environment, such as air pollution and noise from road traffic, oil pollution from marine shipping, and consumption of energy. Raw materials, natural resources and land are consumed or used in the production of vehicles or construction of roads, railways and airports. Thus different forms of transport contribute in different ways to actual or potential environmental damage. The major environmental consequences of transport can be summarized as (OECD, 1991a):

○ Emissions of greenhouse gases, particulates, fuel and fuel additives;
○ Contamination of surface and ground water from surface run-off and spillages of petrol and oil and transported substances;
○ Modifications of hydrological regimes during construction of roads, ports, canals and airports;
○ Use and wastage of land and its associated ecosystems;
○ Excavation and use of minerals (e.g., gravels) for road construction; and
○ Generation of solid waste as vehicles are withdrawn from use.

Passenger and freight transport

The principal modes of transport vary within a country (particularly between rural and urban areas) and from country to country, depending on geographical features. National data for inland passenger and freight transport by different modes are given in Table 7.8 for 1981 and 1991. In the developed countries there was an overall increase in passenger transport by road and rail from 1981 to 1991, although increases in road transport were principally in the private sector.

Roads have become more important as the principal means

of passenger transport in most of the OECD area, except North America where passenger rail transport increased. The increase in road transport and the associated increases in motor vehicles have led to increased problems with air pollution from vehicle emissions (Box 7.3). Associated with this increase in road transport, there was a 4.1 per cent decrease in the total length of railways between 1970 and 1985 in European countries (OECD, 1991a).

In the European community the use of private cars increased by 56 per cent between 1970 and 1985 (CEC, 1992). In some countries, e.g., Denmark and Italy, passenger transport by road in the public sector showed a significant increase, but in others, e.g., USA and UK, it declined. In 1980, in all member countries of the United Nations Economic Commission for Europe (UN ECE), transport by road accounted for more that 80 per cent of inland passenger transport (UN ECE, 1992). In the OECD region as a whole there was a 77 per cent increase in passenger transport between 1970 and 1988 as indicated in the following table:

Region/country	Percentage change		
	By rail	By road	All modes
OECD	38.0	68.0	77.0
North America	329.3	–	–
USA	401.2	56.9	80.3
Japan	25.3	107.3	69.9
OECD – Europe	25.2	77.6	72.3
France	54.4	81.1	78.2
Germany, Fed. Rep.	8.3	52.3	48.4
Italy	33.5	121.6	111.2
UK	13.1	62.0	57.7

Data refer to surface transport of passengers only and are with respect to passenger-kilometres travelled.
Source: OECD, 1991b

Table 7.8 shows that for most developed countries goods transported by road increased between 1981 and 1991 whereas the use of railways and inland waterways declined for many countries. This trend is indicated more clearly by the following table of percentage change in volume of freight transport by road and rail in OECD countries between 1970 and 1988:

Region/country	Percentage change		
	By rail	By road	All modes
OECD	21.0	84.0	46.0
North America	27.0	72.0	–
USA	30.3	69.7	44.0
Japan	–62.7	79.4	37.2
OECD – Europe	–6.3	105.0	57.7
France	–22.6	68.5	13.9
Germany, Fed. Rep.	–17.0	94.1	27.4
Italy	8.3	179.4	124.5
UK	–25.7	46.8	37.0

Data refer to surface transport of freight only and are with respect to tonnes-kilometres travelled.
Source: OFCD, 1991b

Although there was an overall increase in freight transport by rail in the OECD region, the decline in some individual countries is evident from the negative percentage values. Reductions in the use of rail for freight have been compensated by large increases in the use of road transport. Changes in the relative importance of the individual modes of inland freight transport between 1980 and 1990 for countries submitting data to the UN ECE are given in Table 7.9. The decline in the use of railways and the increase in road transport is again evident. For most countries waterways and pipelines take only a small share of freight and their use has changed relatively little over the 10-year period. Freight data by mode of transport are also compiled for specific commodities, such as chemicals, on a national and regional basis. The UN ECE, for example, compiles such data according to quantities imported and exported within Europe (UN ECE, 1992). Data for accidents associated with the transport of hazardous substances are given in Part 9: Environmental Disasters.

Roads and Road Traffic

The discussion on passenger and freight transport above has highlighted the increase in the use of road transport over the last 20 years. In addition, road transport has been increasing its share of all transport in most countries. Changes in vehicle numbers and density, and traffic volume for individual countries, between 1981 and 1991 are given in Table 7.10. Some developing countries, such as China and India, show very large increases in the number of passenger cars in use, with three to four times as many cars in 1991 as in 1981. By comparison, most developed countries show relatively small increases over the same time period. Developing countries have increased their share of the global motor fleet from 20 per cent in 1970 to 32 per cent in 1989 (Faiz et al., 1992). The numbers of commercial vehicles in use reflect similar patterns of increase to passenger cars for most countries (Table 7.10). The relative increase of commercial vehicles in use in North America and of passenger cars in use in Europe are illustrated by the OECD statistics presented in Figure 7.5 for 1970 and 1988. The differences in the use of vehicles in developed and developing countries is more obvious when the vehicle density is considered in relation to numbers of persons. Vehicle density per thousand persons in Africa was less than 100 and often less than 10 (excluding South Africa) in 1991, compared with about 400 to 500 for most wealthy European countries and more than 600 in North America.

Data for traffic volume on an individual country basis (Table 7.10) are very incomplete but suggest large increases in commercial vehicles for some countries. The percentage change in traffic volume from 1970 to 1988 is illustrated in Figure 7.6 using OECD data. This figure highlights the large increase of 167 per cent in the use of passenger cars in Japan compared with only 56 per cent in North America. Similar changes were reflected in the trends in passenger transport (in passenger-kms travelled) given in the table above. The national trends for increase in the use of commercial vehicles were, however, reversed with North America showing 168 per cent increase compared with only 66 per cent in Japan. The

BOX 7.3 Motor Vehicle Emissions and Control Measures

The OECD countries have been calculated to be responsible for 73 per cent of the world motor vehicle emissions of carbon monoxide (CO), 75 per cent of oxides of nitrogen (NO_x) emissions and 73 per cent of hydrocarbon (HC) emissions (Faiz et al., 1992). The USA is the single largest contributor to automotive emissions of air pollutants and greenhouse gases.

The effect of vehicle emissions on air quality in urban areas is more pronounced than their share of emissions on a regional or global basis. In city centres traffic may account for 90–95 per cent of CO and lead, and 60–70 per cent of NO_x and HC (Faiz et al., 1992). Latin American countries have high levels of car ownership and as a consequence the contribution of motor vehicles to the urban levels of NO_x is significantly higher than for other developing regions. Ozone is also a major concern in urban areas word-wide (see also Part 1: Environmental Pollution, Atmosphere).

When volatile organic compounds (VOCs) and NO_x, both found in vehicle emissions, react in the presence of sunlight, ozone is formed. Ozone can cause respiratory distress symptoms in people, depending on the ozone concentration, duration of exposure, and health and activity of the individual. Ozone concentrations are sensitive to temperature, which leads to problems of high ozone concentrations in areas of high traffic density, such as cities in hot, sunny climates, e.g. some tropical developing countries.

Attempts to protect public health by setting standards for ozone concentrations in the USA have not been very successful in cities; the principal approach has been to control VOC emissions, including those from motor vehicles (Chang et al., 1992). However, recent progress in air-quality modelling suggests, that in many urban areas in the USA, the control of NO_x emissions will probably be necessary in addition to, or instead of, the control of VOCs (Committee on Tropospheric Ozone Formation and Measurement, 1991).

Diesel-powered vehicles emit less CO and VOC than petrol-driven cars without emission controls, but substantially more suspended particulate matter (SPM) as indicated in the table of vehicle emissions by engines, with and without emission controls. Such control mechanisms can substantially reduce the volume of pollutants emitted from diesel- and petrol-driven vehicles.

Catalytic converters on petrol driven cars are very effective at reducing CO, VOCs and NO_x. However, recent studies have found that catalytic converters lead to increased emissions of nitrous oxide (N_2O) compared with cars without catalysts.

Vehicle emissions by engines with and without emission controls (g kg^{-1} of fuel)

Vehicle type	CO	VOC	NO_x	SPM
Diesel engine: uncontrolled	20	5	70	3
Diesel engine: low emission	5	3	30	0.4
Petrol engine: uncontrolled	250	30	60	<0.3
Petrol engine: with catalyst	25	2	6	<0.1

Source: Faiz et al., 1992

Estimated emissions from vehicles with and without catalysts in the USA

Vehicle	Catalyst[a]	NO_x (10^9 kg)	N_2O (10^6 kg)
Cars	No catalyst	3.5	1
	Oxidation		26
	Three-way		74
Light-duty trucks	No catalyst	1.4	1.0
	Oxidation		14
	Three-way		4.6
Heavy-duty petrol	No catalyst	0.28	0.6
Heavy-duty diesel	No catalyst	1.9	3.8
TOTAL			125

[a] Oxidation catalysts increase the rate of reaction between oxygen in the exhaust and unburned HC and CO. In addition to this process, three-way catalysts promote the reduction of NO to nitrogen and oxygen.
Source: Dasch, 1992

Consumption of unleaded petrol (as % of total) in selected European countries

Country	1987	1988	1989	1990
Belgium		0.5	15.4	24.5
Denmark	0.3	0.3	40.1	56.6
France			2.4	14.5
Germany, Fed. Rep.	25.1	43.7	57.5	67.8
Ireland			6.4	18.8
Italy		0.7	2.1	5.2
Luxembourg		10.0	20.4	29.9
Netherlands	16.8	22.2	32.3	42.2
UK	0.1	1.1	19.4	34.0

Source: EUROSTAT, 1992

Emissions from non-catalyst cars in the USA averaged 3.6 mg N_2O per mile whereas emissions from vehicles with oxidation catalysts increased to an average of 29 mg N_2O per mile, vehicles with dual-bed catalysts increased to 61 mg N_2O per mile and vehicles with three-way catalysts increased to 45 mg N_2O per mile (Dasch, 1992). The table of estimated USA emissions from vehicles indicates that the USA is responsible for 125 x 10^6 kg of N_2O, which represents 3 per cent of world vehicular emissions.

Fuel impurities and additives such as lead also contribute to vehicular emissions. In 1983 emissions from motor vehicles in the OECD region were responsible for about 30 per cent of all vehicular emissions of lead (Faiz et al., 1992). Large decreases in total man-made emissions of lead in some OECD countries, such as the USA from 1980 to 1985, can be attributed to decreases from the transportation sector and specifically to the introduction of unleaded petrol. By contrast, lead

Continued overleaf

Continued

emissions in developing countries are still rising (Faiz et al., 1992). Public concern for the environment and more widespread publicity of the risks to human health associated with lead have contributed to the increase in use of unleaded petrol in many developed countries. Most of the countries listed in the table of consumption of unleaded petrol have achieved about a 25 per cent use of unleaded petrol, or better, since it became widely available in the mid-1980s. Growth in total consumption is largely a reflection of the growth in the number of retail outlets. In the Federal Republic of Germany, the Netherlands and Denmark, 100 per cent (90 per cent for Denmark) of retail outlets had unleaded petrol available by 1988 (CEC, 1992).

Alternative approaches to reducing vehicle emission are being researched, such as battery-powered cars, and some countries already use alternative fuels. For example, Brazil has been using a petrol/ethanol blend since 1975; Italy, Japan and other countries routinely use liquefied petroleum gas; and fuel oxygenate petrol additives are in widespread use in the USA (UNEP, 1993). These initiatives will help reduce emissions substantially.

References

CEC 1992 *The State of the Environment in the European Communities*, Commission of the European Communities (DG XI), Brussels.
Chang, T. Y., Chock, D. P., Hammerlle, R. H., Japar, S. M. and Salmeen, I. T. 1992 Urban and regional ozone air quality: issues relevant to the automobile industry.

Critical Reviews in Environmental Control, **22**(1/2), 27–66.
Committee on Tropospheric Ozone Formation and Measurement 1991 *Rethinking the Ozone Problem in Urban and Regional Air Pollution*, National Academy Press, Washington DC.
Dasch, J. M. 1992 Nitrous oxide emissions from vehicles. *Journal of the Air and Waste Management Association*, **42**(1), 63–67.
Faiz, A., Weaver, C., Sinha, K., Walsh, M., Carbajo, J. 1992 *Air Pollution from Motor Vehicles*, World Bank, Washington DC and United Nations Environment Programme Industry and Environment Office, Paris.
EUROSTAT 1992 *Environment Statistics 1991*, Statistical Office of the European Communities, Luxembourg and Brussels.
UNEP 1993 Transport and the environment: facts and figures. *Industry and Environment*, **16**(1–2), 4–6.

increases in freight transport (in tonnes-kms travelled) for the two countries over the same time period were roughly similar (see table above).

All forms of transportation contribute to some degree to air pollution, and motor vehicles are a major cause of air pollution in cities. In busy streets, concentrations of carbon monoxide (CO), oxides of nitrogen (NO_x) and hydrocarbons (HC) can reach levels which cause discomfort or even a health risk to some individuals. The transport sector is responsible for about 10–54 per cent of the world-wide anthropogenic emissions of CO, 29–32 per cent of NO_x, 47–49 per cent of HC, 10–20 per cent of particulates, 2–6 per cent of sulphur oxides (SO_x) and 14 per cent of carbon dioxide (CO_2) (Faiz et al., 1992). Transportation is therefore a major contributor to air pollution and greenhouse gases. In addition, in countries where leaded petrol is used, almost all lead in air emissions in cities come from vehicle exhausts (Faiz et al., 1992). Emissions from motor vehicles and associated control measures are discussed in Box 7.3.

National data for the percentage contribution of transport (i.e., road, rail and waterways) to the total emissions of major air pollutants have been assembled by the OECD and are given below:

a) Passenger cars

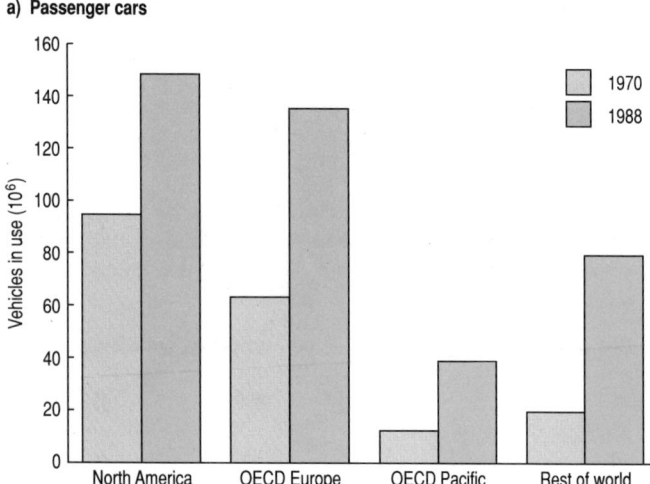

b) Commercial vehicles

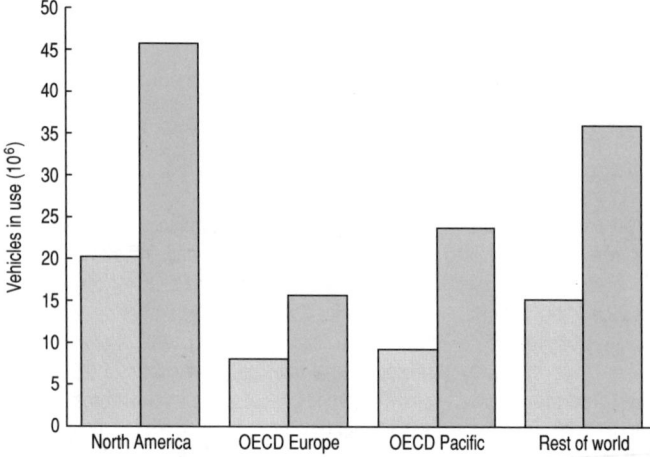

FIGURE 7.5 Passenger cars and commercial vehicles in use in OECD regions and the rest of the world, 1970 and 1988
Source: OECD, 1991a

Country/region	Percentage contribution			
	NO_x	CO	HC	SO_x
Canada	61	66	37	3
USA	41	67	33	4
France	76	71	60	10
Germany, Fed. Rep.	65	74	53	6
Italy	52	91	87	4
UK	49	86	32	2
Japan	44			18
OECD	49	71	39	4

Data refer to 1988 or the most recent year available.
Source: OECD, 1991a.

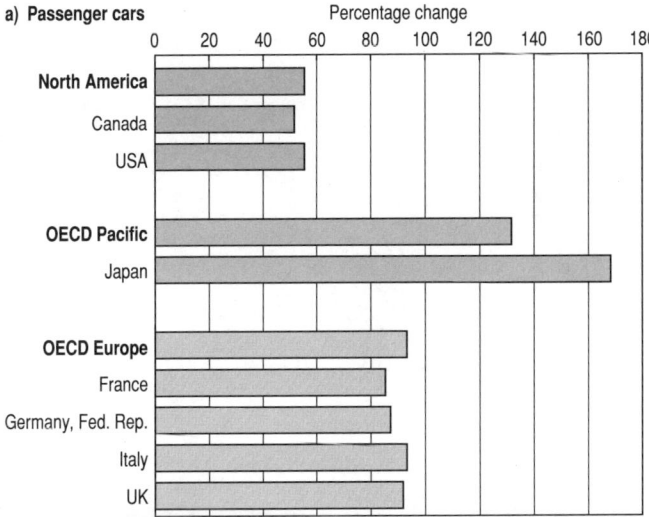

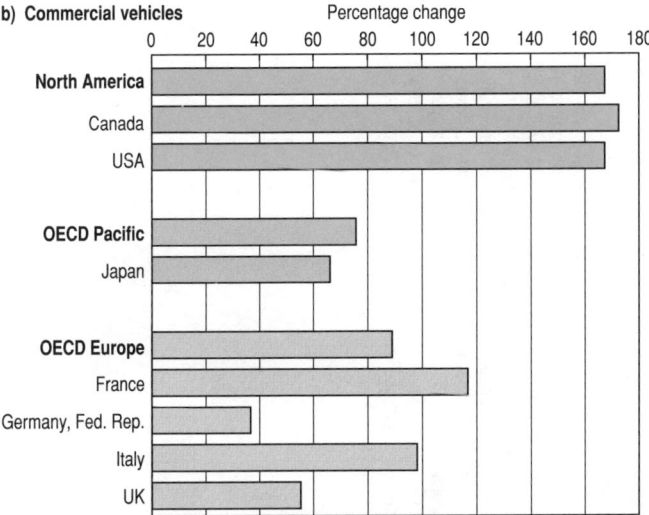

FIGURE 7.6 Percentage changes in traffic volume (vehicle-kms) in OECD regions between 1970 and 1988 for passenger cars and commercial vehicles
Source: OECD, 1991a

Standardized inventories of emissions due to transport and other sources have also been compiled on a national basis by the European Community Co-ordinated Information System on the State of the Environment and Natural Resources (CORINE) project (EUROSTAT, 1992).

Faiz et al. (1992) assembled emissions data for selected non-OECD countries from a variety of sources and found the relative contribution of transport to emissions of most air pollutants to be similar in developed and developing countries. The small differences occasionally noted were largely attributable to differences in the total number of vehicles. However, in OECD countries transport was responsible for the larger share of particulate matter emissions. This was thought to be due to the fact that in OECD countries emissions from stationary sources are subject to strict controls. Despite the large increase in commercial vehicle traffic in the USA (Figure 7.6), estimated emissions of particulates from mobile sources have increased only slightly over the period 1970 to 1988 (OECD, 1991b). The USA is considered to have the most stringent particulate emission standards in the world and other countries are basing their legislation on these standards (Faiz et al., 1992). Diesel-powered vehicles are the major source of particulates from motor traffic (see Box 7.3).

References

CEC 1992 *The State of the Environment in the European Communities*, Commission of the European Communities (DG XI), Brussels.

EUROSTAT 1992 *Environment Statistics 1991*, Statistical Office of the European Community, Luxembourg.

Faiz, A., Weaver, C., Sinha, K., Walsh, M., Carbajo, J. 1992 *Air Pollution from Motor Vehicles*, World Bank, Washington DC and United Nations Environment Programme Industry and Environment Office, Paris.

OECD 1991a *The State of the Environment*, Organisation for Economic Co-operation and Development, Paris.

OECD 1991b OECD *Environmental Data Compendium 1991*, Organisation for Economic Co-operation and Development, Paris.

UN ECE 1992 *Annual Bulletin of Transport Statistics for Europe*, United Nations Economic Commission for Europe, Geneva.

TABLE 7.1 Indices of industrial production – world, developed and developing countries, 1973–1990 (relative to a base year, 1980=100)

Year	All industry			Mining			Manufacturing			Electricity, gas and water		
	World	Developed countries	Developing countries	World	Developed countries	Developing countries	World	Developed countries	Developing countries	World	Developed countries	Developing countries
1973	82	88	84	94	82	108	81	89	68	74	78	54
1975	82	82	83	90	81	97	80	83	73	79	82	64
1979	101	101	104	106	97	113	100	101	96	97	98	93
1980	100	100	100	100	100	100	100	100	100	100	100	100
1981	99	100	94	92	103	84	101	100	102	102	101	104
1982	97	97	90	86	99	75	98	96	104	103	102	112
1983	99	99	91	85	99	73	102	99	106	108	105	121
1984	105	106	97	88	103	76	108	106	115	113	110	131
1985	108	109	99	87	106	72	112	109	121	118	114	140
1986	112	110	107	90	103	78	115	111	131	120	116	150
1987	116	114	113	91	104	79	120	114	141	125	120	163
1988	122	120	121	96	106	87	127	121	150	131	124	174
1989	127	125	130	100	106	95	132	126	159	135	128	186
1990	127	126	137	101	124	96	132	126	171	128	130	202

Index numbers of industrial production are listed here for all the major divisions of industry as defined by the International Standard Industrial Classification of all Economic Activities (ISIC codes), i.e., mining, manufacturing and electricity, gas and water. The series is compiled by the Statistical Office of the UN and is based on national replies to the Quarterly UN Index of Industrial Production Questionnaire

The indices provide a measure of the value added in constant US dollars. The 'value added' concept derives from the system of national accounts and is defined as the gross output less the cost of materials, supplies, fuel and electricity consumed and services received.

Time series of index numbers are calculated directly from national data using the Laspeyres formula (i.e., the series are base-weighted arithmetic means) for each ISIC industry category for each country and for the developing and developed market economies as a whole. In this instance developed countries comprise the European countries (excluding eastern Europe and the former USSR), Australia, Canada, Israel, Japan, New Zealand, South Africa and the USA. The developing countries comprise the remaining countries in Africa, the Americas, Asia (excluding China, Korea Dem., Mongolia and Viet Nam) and Oceania. Indices for eastern Europe and the former USSR are calculated separately according to a slightly different methodology and are combined with those for the developed and developing countries to form the time series for the world.

For a more detailed description of the industrial production index please refer to the latest edition of the 'UN Industrial Yearbook' (Volume I).

Source:
United Nations Conference on Trade and Development 1992 Handbook of International Trade and Development Statistics 1991, United Nations, New York.

TABLE 7.2 World reserves of major metals, as assessed in 1991

Region/country	Base metals (10^6 t, metal content)				Iron and ferro alloys (10^6 t, metal content)								Light metals (10^6 t)		
	Cu	Pb	Sn	Zn	Iron ore	Mn	Ni	Cr	Co	Mo	W	V	Bauxite[a]	Ti[b]	Li[c]
WORLD	321.0	70.4	5.9	150.0	64,648	812.8	47.7	418.9	3.3	5.5	2.4	4.3	21,559	288.6	2.2
AFRICA	42.0	4.0	0.1	9.0	3,454	422.8	3.1	341.3	1.8	0.0	–	0.9	6,874	45.0	–
Algeria	0.0	0.1[d]	0.0	0.1[e]	65	0.0	0.0	0.0	0.0	0.0	0.0	0.0	0	0.0	0.0
Angola	0.0	0.0	0.0	0.0	15	0.0	0.0	0.0	0.0	0.0	0.0	0.0	0	0.0	0.0
Botswana	0.4[e]	0.0	0.0	0.0		0.0	0.5	0.0	–[d]	0.0	0.0	0.0	0	0.0	0.0
Burkina	0.0	0.0	0.0	0.0			0.0	0.0	0.0	0.0	0.0	0.0	0	0.0	0.0
Cameroon	0.0	0.0		0.0	0	0.0	0.0	0.0	0.0	0.0	0.0	0.0	680	0.0	0.0
Congo	–[e]	–[e]	0.0	–[e]	0	0.0	0.0	0.0	0.0	0.0	0.0	0.0	0	0.0	0.0
Egypt	0.0	0.0	0.0	0.0	90	0.0	0.0		0.0	0.0	0.0	0.0	0	0.0	0.0
Ethiopia	0.0	0.0	0.0	0.0		0.0	0.0	0.0	0.0	0.0	0.0	0.0	0	0.0	0.0
Gabon	0.0	0.0	0.0	0.0	0	52.6	0.0	0.0	0.0	0.0	0.0	0.0	0	0.0	0.0
Ghana	0.0	0.0	0.0	0.0	0	0.9	0.0	0.0	0.0	0.0	0.0	0.0	450	0.0	0.0
Guinea	0.0	0.0	0.0	0.0	0	0.0	0.0	0.0	0.0	0.0	0.0	0.0	5,600	0.0	0.0
Kenya	0.0	–[e]	0.0	0.0	0	0.0	0.0	0.0	0.0	0.0	0.0	0.0	0	0.0	0.0
Liberia	0.0	0.0	0.0	0.0	500	0.0	0.0	0.0	0.0	0.0	0.0	0.0	0	0.0	0.0
Libya	0.0	0.0	0.1[d]	0.0		0.0	0.0	0.0	0.0	0.0	0.0	0.0	22	0.0	0.0
Madagascar	0.0	0.0	–[d]	0.0	0	0.0	0.0	2.1	0.0	0.0	0.0	0.0	0	0.0	0.0
Mauritania	0.0	0.0	0.0	0.0	200	0.0	0.0	0.0	0.0	0.0	0.0	0.0	0	0.0	0.0
Morocco	0.3[e]	1.0[d]	0.0	0.2[e]	30	0.0	0.0		0.0	0.0	0.0	0.0	0	0.0	0.0
Mozambique	0.0	0.0	0.0	0.0	0	0.0	0.0	0.0	0.0	0.0	0.0	0.0	2	2.4	0.0
Namibia	1.0	0.2[d]	0.1[d]	0.4[e]	0	0.0	0.0	0.0	0.0	0.0	–[f]	0.0	0	0.0	0.0
Niger	0.0	0.0	0.0	0.0	0	0.0	0.0	0.0	0.0	0.0	0.0	0.0	0	0.0	0.0
Nigeria	0.0	0.0[e]	–	0.0	0	0.0	0.0	0.0	0.0	0.0	0.0	0.0	0	0.0	0.0
Sierra Leone	0.0	0.0	0.0	0.0	11	0.0	0.0	0.0	0.0	0.0	0.0	0.0	140	3.0	0.0
South Africa	2.0	2.0	0.0[d]	3.0	2,500	369.2	2.5	295.2	0.1[d]	0.0	0.0	0.9	0	39.6	0.0
Sudan	0.0	0.0	0.0	0.0	0	0.0	0.0	0.5	0.0	0.0	0.0	0.0	0	0.0	0.0
Tanzania	0.0	0.0		0.0		0.0	0.0	0.0	0.0	0.0	0.0	0.0	0	0.0	0.0
Tunisia	0.0	0.6[d]	0.0	0.1[e]	13	0.0	0.0	0.0	0.0	0.0	0.0	0.0	0	0.0	0.0
Uganda	0.0	0.0		0.0	0	0.0	0.0	0.0	–[d]	0.0	–[e]	0.0	0	0.0	0.0
Zaire	26.0	0.0	–	5.0	0	0.0	0.0	0.0	1.4	0.0	–[f]	0.0	0	0.0	0.0
Zambia	12.0	0.2[d]	–[d]	0.2[e]		0.0	0.0	0.0	0.4	0.0	–[f]	0.0	0	0.0	0.0
Zimbabwe	0.3[e]	0.0		0.0	30	0.0	0.1	43.5	–[d]	0.0	0.0	0.0	2	0.0	–

Continued

TABLE 7.2 Continued

Region/country	Base metals (10⁶ t, metal content)				Iron and ferro alloys (10⁶ t, metal content)								Light metals (10⁶ t)		
	Cu	Pb	Sn	Zn	Iron ore	Mn	Ni	Cr	Co	Mo	W	V	Bauxite[a]	Ti[b]	Li[c]
NORTH AMERICA	**81.0**	**21.4**	**0.1**	**48.0**	**8,580**	**3.6**	**24.8**	**0.7**	**1.1**	**3.3**	**0.4**	**0.1**	**2,156**	**35.1**	**0.5**
Canada	12.0	7.0	0.1	22.0	4,600	0.0	6.2	0.0	0.0	0.5	0.3	0.0	0	27.0	0.2
Costa Rica	0.0	0.0	0.0	0.0	0	0.0	0.0	0.0	0.0	0.0	0.0	0.0	78	0.0	0.0
Cuba		0.0	0.0	0.0		0.0	18.1	0.7	1.0	0.0	0.0	0.0	0	0.0	0.0
Dominican Rep.	0.0	0.0	0.0	0.0	0	0.0	0.5	0.0	0.0	0.0	0.0	0.0	30	0.0	0.0
Greenland	0.0	0.3[e]	0.0	0.0		0.0	0.0	0.0	0.0	0.0	0.0	0.0	0	0.0	0.0
Haiti	0.0	0.0	0.0	0.0	0	0.0	0.0	0.0	0.0	0.0	0.0	0.0	10	0.0	0.0
Honduras	0.0	0.1[d]	0.0		0	0.0	0.0	0.0	0.0	0.0	0.0	0.0	0	0.0	0.0
Jamaica	0.0	0.0	0.0	0.0	0	0.0	0.0	0.0	0.0	0.0	0.0	0.0	2,000	0.0	0.0
Mexico	14.0	3.0	0.0	6.0	180	3.6	0.0	0.0	0.0	0.1	-[g]	0.0	0	0.0	0.0
Panama	0.0	0.0	0.0	0.0		0.0	0.0	0.0	0.0	0.0	0.0	0.0	0	0.0	0.0
USA	55.0	11.0	-	20.0	3,800	0.0	-	0.0	0.0	2.7	0.2	0.1	38	8.1	0.4
SOUTH AMERICA	**94.0**	**2.8**	**1.4**	**10.0**	**8,213**	**20.9**	**1.2**	**2.3**	**-**	**1.3**	**0.1**	**0.0**	**4,437**	**67.6**	**1.3**
Argentina		0.2[d]	-[d]	0.2[e]	60	0.0	0.0	0.0	0.0	0.0	-[g]	0.0	0	0.0	0.0
Bolivia		0.1[d]	0.1	0.6[e]	0	0.0	0.0	0.0	0.0	0.0	0.1	0.0	0	0.0	0.0
Brazil	1.0	0.5[d]	1.2	2.0	6,500	20.9	0.7	2.3	-[d]	0.0	-	0.0	2,800	67.6	1.3
Chile	85.0	-[e]	0.0	0.1[e]	220	0.0	0.0	0.0	0.0	1.1	0.0	0.0	0	0.0	0.0
Colombia	0.0	-[e]	0.0	-[e]	33	0.0	0.6	0.0	0.0	0.0	0.0	0.0	0	0.0	0.0
Ecuador	0.0	0.0	0.0	-[e]	0	0.0	0.0	0.0	0.0	0.0	0.0	0.0	0	0.0	0.0
French Guiana	0.0	0.0	0.0	0.0	0	0.0	0.0	0.0	0.0	0.0	0.0	0.0	42	0.0	0.0
Guyana	0.0	0.0	0.0	0.0	0	0.0	0.0	0.0	0.0	0.0	0.0[f]	0.0	700	0.0	0.0
Peru	8.0	2.0	-[d]	7.0	200	0.0	0.0	0.0	0.0	0.1	0.0	0.0	0	0.0	0.0
Suriname	0.0	0.0	0.0	0.0	0	0.0	0.0	0.0	0.0	0.0	0.0	0.0	575	0.0	0.0
Venezuela	0.0	0.0	0.0	0.0	1,200	0.0	0.0	0.0	0.0	0.0	0.0	0.0	320	0.0	0.0
ASIA	**30.0**	**9.6**	**3.7**	**29.0**	**7,207**	**30.8**	**4.3**	**23.8**	**-**	**0.6**	**1.4**	**0.6**	**1,960**	**69.8**	**0.0**
China	3.0	6.0	1.5	5.0	3,500	13.6	0.7	0.0	0.0	0.5	1.1	0.6	150	30.0	0.0
Cyprus	-[e]	0.0	0.0	0.0	0	0.0	0.0	0.0	0.0	0.0	0.0[f]	0.0	0	0.0	0.0
India	3.0	0.1[d]	0.0	11.0	3,300	17.2	0.0	18.1	-[d]	0.0	-[f]	0.0	1,000	35.4	0.0
Indonesia	3.0	0.0	0.7	0.0	16	0.0	3.2	0.2	-[d]	0.0	0.0	0.0	750	0.0	0.0
Iran	3.0	0.0	0.0	2.0	27	0.0	0.0	0.7	0.0	0.1	0.0	0.0	0	0.0	0.0
Iraq	0.0	0.0	0.0	0.0	13	0.0	0.0	0.0	0.0	0.0	0.0[f]	0.0	0	0.0	0.0
Japan	1.0	0.8[d]	0.1[d]	4.0	0	0.0	0.0	0.0	0.0	0.0	-[f]	0.0	0	0.0	0.0
Korea, Dem.	0.7[e]	2.0	0.0	4.0	140	0.0	0.0	0.0	0.0	0.0	0.1[g]	0.0	0	0.0	0.0

Continued

TABLE 7.2 Continued

Region/country	Base metals (10⁶ t, metal content)				Iron and ferro alloys (10⁶ t, metal content)								Light metals (10⁶ t)		
	Cu	Pb	Sn	Zn	Iron ore	Mn	Ni	Cr	Co	Mo	W	V	Bauxite[a]	Ti[b]	Li[c]
Korea	0.0	0.2[d]		0.7[e]	17	0.0	0.0	0.0	0.0		0.1	0.0	0	0.0	0.0
Laos	0.0	0.0		0.0		0.0	0.0	0.0	0.0	0.0	0.0	0.0	0	0.0	0.0
Malaysia	1.2[e]	0.0	1.1	0.0	34	0.0	0.0	0.0	0.0	0.0	–[g]	0.0	15		0.0
Mongolia	3.0	0.0	0.0	0.0	0	0.0	0.0	0.0	0.0		0.1[e]	0.0	0	0.0	0.0
Myanmar	0.2[e]	0.1[d]	–	0.1[e]	0	0.0	0.0	0.0	0.0	0.0	–	0.0	0	0.0	0.0
Oman	0.8[e]	0.0	0.0	0.0	0	0.0	0.0		0.0	0.0	0.0	0.0	0	0.0	0.0
Pakistan	0.0	0.0	0.0	0.0	0	0.0	0.0		0.0	0.0	0.0	0.0	20	0.0	0.0
Philippines	10.0	0.0	0.0	0.0	0	0.0	0.4	2.3	0.0	0.0	0.0	0.0	0	0.0	0.0
Saudi Arabia	–[e]	0.0	0.0	0.1[e]		0.0	0.0	0.0	0.0	0.0	0.0	0.0	0	0.0	0.0
Sri Lanka	0.0	0.0	0.0	0.0	0	0.0	0.0	0.0	0.0	0.0	0.0	0.0	0	4.4	0.0
Thailand	0.0	0.4[e]	0.3	1.0			0.0	0.0	0.0	0.0	–	0.0	0	0.0	0.0
Turkey	1.0	–[d]	0.0	1.0	160	0.0	0.0	2.5	0.0	0.0	0.1[g]	0.0	25	0.0	0.0
Viet Nam	0.0	0.0	0.0	0.2[e]		0.0	0.0	0.0	0.0	0.0	0.0	0.0	0	0.0	0.0
EUROPE	**23.0**	**9.7**	**0.2**	**26.0**	**3,214**	**0.0**	**0.9**	**11.2**	**–**	**–**	**0.1**	**0.0**	**1,342**	**33.4**	**0.0**
Albania	0.5	0.0	0.0	0.0	30	0.0	0.2	1.9	0.0	0.0	0.0	0.0	0	0.0	0.0
Austria	0.0	–[d]	0.0	0.2[e]	30	0.0	0.0	0.0	0.0	0.0	–	0.0	0	0.0	0.0
Bulgaria	1.5	2.0	0.0	0.8[e]	30	0.0		0.0	0.0	–	0.0	0.0	0	0.0	0.0
Czechoslovakia	0.0	–[e]	0.0	0.1[e]	30	0.0	0.0	0.0	0.0	0.0	–[e]	0.0	0	0.0	0.0
Finland	1.0	–[d]		1.0	27	0.0	0.1	8.9	–	0.0	0.0	0.0	0	1.4	0.0
France		0.1[d]	0.0	1.0	900	0.0	0.0	0.0	0.0	0.0	–	0.0	30	0.0	0.0
Germany, Fed. Rep.		0.1[d]	0.0	1.0	32	0.0	0.0	0.0	0.0	0.0	0.0	0.0	2	0.0	0.0
German Dem. Rep.		0.0	0.0	0.0	3	0.0		0.0	0.0	0.0	0.0	0.0	0	0.0	0.0
Greece	0.0	0.5[d]	0.0	1.0	12	0.0	0.5	0.4	0.0	0.0	0.0	0.0	600	0.0	0.0
Hungary	0.0	0.0	0.0	0.0	13	0.0	0.0	0.0	0.0	0.0	0.0	0.0	300	0.0	0.0
Ireland	0.0	0.7[d]	0.0	5.0	0	0.0	0.0	0.0	0.0	0.0	0.0	0.0	0	0.0	0.0
Italy	0.0	0.3[d]	0.0	2.0	3	0.0	0.0	0.0	0.0	0.0	0.0	0.0	5	0.0	0.0
Norway	1.0	–[d]	0.0	0.2[e]	200	0.0		0.0	0.0	0.0	0.0	0.0	0	32.0	0.0
Poland	10.0	0.6[e]	0.0	3.0	25	0.0	0.0	0.0	0.0	0.0	0.0	0.0	0	0.0	0.0
Portugal	3.0	0.0	0.1	2.0	0	0.0		0.0	0.0	0.0	–	0.0	0	0.0	
Romania	1.0	0.5[e]	0.0	0.6[e]	25	0.0	0.0	0.0	0.0	0.0	0.0[f]	0.0	50	0.0	0.0
Spain	1.0	1.7[d]		5.0	230	0.0	0.0	0.0	0.0	0.0	–[f]	0.0	5	0.0	0.0
Sweden		1.0[d]	0.0	1.0	1,600	0.0	0.0	0.0	0.0	0.0	–[f]	0.0	0	0.0	0.0
UK		0.1[d]	0.1	0.1[e]	14	0.0	0.0	0.0	0.0	0.0	–[f]	0.0	0	0.0	
Yugoslavia	4.0	2.0	0.0	2.0	40	0.0	0.2	0.0	0.0	0.0	0.0	0.0	350	0.0	0.0

Continued

TABLE 7.2 Continued

Region/country	Base metals (10⁶ t, metal content)				Iron and ferro alloys (10⁶ t, metal content)								Light metals (10⁶ t)		
	Cu	Pb	Sn	Zn	Iron ore	Mn	Ni	Cr	Co	Mo	W	V	Bauxite[a]	Ti[b]	Li[c]
USSR	37.0	9.0	0.3	10.0	23,500	294.8	6.6	39.6	0.1	0.5	0.3	2.6	300	8.4	0.4
OCEANIA	14.0	14.0	0.2	18.0	10,480	39.9	6.7	0.0	0.3	0.0	0.1	–	4,490	29.3	0.4
Australia	7.0	14.0	0.2	18.0	10,200	39.9	2.2	0.0	–	0.0	0.1	–	4,440	29.3	0.4
New Caledonia	0.0	0.0	0.0	0.0	0	0.0	4.5	0.0	0.2	0.0	0.0	0.0	0	0.0	0.0
New Zealand	0.0	0.0	0.0	0.0	280	0.0	0.0	0.0	0.0	0.0	–[e]	0.0	0	0.0	0.0
Papua New Guinea	7.0	0.0	0.0	0.0	0	0.0	0.0	0.0	0.0	0.0	0.0	0.0	0	0.0	0.0
Solomon Is	0.0	0.0	0.0	0.0	0	0.0	0.0	0.0	0.0	0.0	0.0	0.0	50	0.0	0.0

a   Dry weight of bauxite ore.
b   Tonnage is expressed in terms of titanium dioxide (TiO₂) content.
c   Tonnage is expressed in terms of metal content; the world total excludes reserves in Argentina, China, Namibia, Portugal and the former USSR because data are not available for these countries.
d   Estimate derived by the Minerals Availability Program of the US Bureau of Mines.
e   Estimate based on mine production data.
f   UNESCO estimate.
g   Estimate refers to 1985.

Cu = Copper; Pb = Lead; Sn = Tin; Zn = Zinc; Mn = Manganese; Ni = Nickel; Cr = Chromium; Co = Cobalt; Mo = Molybdenum; W = Tungsten; V = Vanadium; Ti = Titanium; Li = Lithium.

Estimates of mineral reserves given here are defined as those deposits whose magnitude and grade have been determined by sampling and measurement and can be profitably recovered at the time of the assessment. In contrast, the term mineral resources is used to describe those deposits the presence of which is confirmed by preliminary surveys and other geologic evidence, but for which extraction is not economically viable. Changes in geologic information, available technology, extraction and production costs and mined product prices can affect reserve estimates and move deposits from the resources to the reserves category. Reserves do not necessarily signify that extraction facilities are in place and operative.

Major national reserves of each metal are estimated by the US Bureau of Mines. These data are published regularly in their Mineral Commodity Summaries. Estimates for countries with minor reserves are taken from a variety of sources and are generally less reliable.

Reserves are generally expressed in terms of the metal content of the parent ore or mineral deposit, unless otherwise stated.

Sources:
Patterson, S. H., Kurtz, H. F., Olsen, J. C. and Neeley, C. L. 1986 World Bauxite Resources, Paper 1076-B, US Geological Survey, Washington DC.
UNESCO 1986 Geology of Tungsten, United Nations Educational, Scientific and Cultural Organization, Paris.
US BOM 1985 Mineral Facts and Problems 1985, US Bureau of Mines, Washington DC.
US BOM 1991 Minerals Availability Program, US Bureau of Mines, Washington DC (unpublished data).
US BOM 1992 Mineral Commodity Summaries 1992, US Bureau of Mines, Washington DC.
US BOM 1992 Minerals Yearbook 1990, US Bureau of Mines, Washington DC (and earlier editions).

TABLE 7.3 Production of selected metals, 1975–1990 (10^3 t a^{-1})

	1975	1980	1985	1990	R/P ratio (years)
Bauxite					
WORLD	**74,927**	**89,220**	**84,189**	**109,118**	**197.6**
TOTAL, 10 COUNTRIES	**58,817**	**75,588**	**73,974**	**98,987**	**180.0**
Australia	21,004	27,179	31,839	40,697	109.1
Guinea	8,406	11,862	11,790	16,500	339.4
Jamaica	11,571	12,054	6,239	10,921	183.1
Brazil	969	5,538	5,846	8,750	320.0
India	1,273	1,785	2,281	5,000	200.0
USSR	4,369	4,600	4,600	4,200	71.4
China	986	1,500	1,650	4,000	37.5
Suriname	4,928	4,646	3,738	3,267	176.0
Yugoslavia	2,306	3,138	3,538	2,952	118.6
Greece	3,005	3,286	2,453	2,700	222.2
Cadmium					
WORLD	**15.2**	**18.2**	**19.1**	**20.2**	[a]
TOTAL, 10 COUNTRIES	**12.1**	**13.3**	**14.1**	**14.5**	[a]
USSR	2.6	2.9	3.0	2.8	[a]
Japan	2.7	2.2	2.5	2.4	[a]
Belgium	0.9	1.5	1.3	1.8	[a]
USA	2.0	1.6	1.6	1.7	[a]
Canada	1.2	1.3	1.7	1.4	[a]
Germany[b]	1.0	1.2	1.1	1.2	[a]
Mexico	0.6	0.8	0.9	1.0	[a]
China	0.1	0.3	0.5	0.8	[a]
Italy	0.4	0.6	0.5	0.7	[a]
Australia	0.5	1.0	0.9	0.7	[a]
Copper					
WORLD	**6,739.0**	**7,204.0**	**7,870.0**	**8,814.0**	**36.4**
TOTAL, 10 COUNTRIES	**5,281.6**	**5,609.6**	**5,993.2**	**6,790.0**	**37.6**
Chile	831.0	1,063.0	1,359.8	1,603.2	53.0
USA	1,282.2	1,181.1	1,104.8	1,587.2	34.7
Canada	733.8	716.4	738.6	779.6	15.4
USSR	580.0	590.0	600.0	600.0	61.7
Zambia	676.9	595.8	452.6	445.0	27.0
Poland	230.4	343.0	431.3	380.0	26.3
China	99.8	115.0	185.0	375.0	8.0
Zaire	462.6	425.7	470.0	370.0	70.3
Peru	165.8	336.1	391.3	334.0	24.0
Australia	219.0	243.5	259.8	316.0	22.2
Iron Ore					
WORLD	**902,389**	**890,924**	**860,640**	**864,370**	**175**
TOTAL, 10 COUNTRIES	**721,683**	**754,039**	**744,495**	**832,077**	**171**
USSR	232,792	244,703	247,639	236,000	270
Brazil	89,890	114,727	128,251	150,000	74
China	65,024	68,072	80,000	118,000	76
Australia	97,647	95,529	97,447	110,000	146
USA	80,128	70,727	49,533	59,032	273
India	41,403	41,934	42,545	52,000	104
Canada	46,866	48,752	39,502	36,443	331
South Africa	12,298	26,310	24,414	30,347	133
Venezuela	24,771	16,102	14,710	20,365	100
Sweden	30,865	27,183	20,454	19,890	158

Continued

TABLE 7.3   Continued

	1975	1980	1985	1990	R/P ratio (years)
			Lead		
WORLD	**3,432.2**	**3,448.2**	**3,431.2**	**3,367.2**	**20.9**
TOTAL, 10 COUNTRIES	**2,579.7**	**2,477.8**	**2,539.9**	**2,711.3**	**21.0**
Australia	407.8	397.5	498.0	563.0	24.9
USA	563.8	550.4	424.4	495.2	22.2
USSR	480.8	420.0	440.0	450.0	20.0
China	99.8	160.0	200.0	315.0	19.0
Canada	349.1	296.6	268.3	236.2	29.6
Peru	184.5	189.1	201.5	189.0	10.6
Mexico	178.6	145.5	206.7	179.9	16.7
Korea, Dem.	117.9	125.0	110.0	120.0	16.7
Sweden	70.4	72.2	75.9	90.0	11.1
Yugoslavia	126.9	121.5	115.1	73.0	27.4
			Mercury		
WORLD	**8.7**	**6.8**	**6.1**	**5.8**	**22.5**
TOTAL, 10 COUNTRIES	**7.0**	**6.7**	**6.1**	**5.8**	**20.6**
USSR	1.9	2.1	2.2	2.1	4.8
Spain	1.5	1.5	0.9	1.5	50.7
China	0.9	0.7	0.7	0.8	13.3
Algeria	1.0	0.8	0.8	0.6	3.3
Mexico	0.5	0.1	0.4	0.3	14.5
Finland	0.0	0.1	0.1	0.2	[c]
Czechoslovakia	0.2	0.2	0.2	0.1	[c]
Turkey	0.2	0.2	0.2	0.1	23.1
Yugoslavia	0.6	0.0	0.1	0.1	142.9
USA	0.3	1.1	0.6	[d]	
			Nickel		
WORLD	**807.9**	**779.7**	**812.6**	**937.1**	**50.9**
TOTAL,10 COUNTRIES	**723.8**	**658.2**	**687.3**	**836.9**	**53.8**
USSR	152.4	154.2	185.1	259.0	25.6
Canada	242.2	184.8	170.0	201.9	30.6
New Caledonia	133.3	86.6	72.4	88.0	51.5
Australia	75.8	74.3	85.8	70.0	31.1
Indonesia	19.2	53.3	40.3	58.0	55.2
Cuba	36.6	36.6	32.1	41.0	442.5
South Africa	20.8	25.7	25.0	36.0	70.6
Dominican Rep.	26.9	16.3	25.4	33.0	13.7
Botswana	16.6	15.4	26.3	25.0	19.1
China		10.9	25.0	25.0	29.0
			Steel, crude		
WORLD	**644,208**	**713,813**	**718,131**	**771,979**	
TOTAL, 10 COUNTRIES	**495,632**	**533,918**	**527,431**	**570,148**	
USSR	141,327	147,944	154,670	154,000	
Japan	102,314	111,397	105,281	110,339	
USA	105,818	101,457	80,069	89,726	
China	25,402	37,121	46,721	66,000	
Germany[b]	46,889	51,147	48,350	44,022	
Italy	21,836	26,501	23,789	25,439	
Korea	2,009	85,559	13,539	23,125	
Brazil	8,308	15,339	20,456	20,572	
France	21,531	23,176	18,833	19,017	
UK	20,198	11,278	15,723	17,908	

Continued

TABLE 7.3   Continued

	1975	1980	1985	1990	R/P ratio (years)
			Tin		
WORLD	**222.3**	**247.3**	**180.7**	**219.3**	**27.0**
TOTAL, 10 COUNTRIES	**201.3**	**228.1**	**162.0**	**202.2**	**27.3**
China	22.0	14.6	15.0	40.0	37.5
Brazil	5.0	6.9	26.5	39.1	30.7
Indonesia	25.3	32.5	21.7	30.2	22.5
Malaysia	64.4	61.4	36.9	28.5	38.6
Bolivia	24.3	27.3	16.1	18.0	7.8
USSR	30.0	36.0	13.5	15.0	20.0
Thailand	16.4	33.7	16.9	14.6	18.4
Australia	9.6	11.6	6.4	7.4	27.1
Peru	0.2	1.1	3.8	5.1	7.8
UK	4.1	3.0	5.2	4.2	21.4
			Zinc		
WORLD	**5,849.7**	**5,961.6**	**6,758.3**	**7,325.0**	**20.5**
TOTAL, 10 COUNTRIES	**3,975.4**	**4,030.3**	**4,723.1**	**5,570.9**	**17.6**
Canada	1,229.5	1,059.0	1,172.2	1,177.0	18.7
Australia	500.9	495.3	759.1	937.0	19.2
USSR	689.5	785.0	810.0	750.0	13.3
China	99.8	160.0	300.0	619.0	8.1
Peru	384.8	487.6	523.4	576.8	12.1
USA	425.8	317.1	251.9	543.2	36.8
Mexico	288.9	235.8	275.4	322.5	18.6
Spain	85.3	183.1	234.7	258.0	19.4
Korea, Dem.	159.7	140.0	180.0	230.0	17.4
Sweden	111.3	167.4	216.4	157.4	6.4

a   A production-reserve ratio has not been calculated as production includes production of secondary raw materials and therefore would be misleading.
b   Data refer to the unified Federal Republic of Germany.
c   As national data for reserves are unavailable, it is not possible to calculate this ratio.
d   Data withheld.

Production data are expressed in terms of the metal content of the parent ore in the case of copper, lead, mercury, nickel, tin and zinc. Aluminium (i.e., bauxite) and iron ore production are given as gross weights of ore mined. Iron ore includes iron ore concentrates and iron ore agglomerates (sinter and pellets). Production of cadmium refers to production of the refined metal. The production of crude steel is defined here as the production of usable ingots, continuously cast semi-finished products plus liquid steel for castings.

The countries listed represent the ten largest producers in 1990 and are ranked in decending order of production.

The production-reserve ratio, given in the final column of the table represents an estimate of the remaining lifetime (in years) of the reserves given in Table 7.2 at 1990 levels of production from primary raw materials.

Sources:
US BOM 1991 *Minerals Yearbook 1989*, US Bureau of Mines, Washington DC (and earlier editions).
US BOM 1992 Unpublished data, US Bureau of Mines, Washington DC.

TABLE 7.4 Consumption of selected metals, 1975–1990 (10^3 t a^{-1})

	1975	1980	1985	1990	Per capita consumption, 1990 (kg per person)
Aluminium					
WORLD	**11,349.8**	**15,297.9**	**15,861.5**	**17,908.7**	**113.38**
TOTAL, 10 COUNTRIES	**8,739.7**	**11,778.7**	**11,796.8**	**13,157.4**	**4.52**
USA	3,265.0	4,453.5	4,282.0	4,352.3	17.46
Japan	1,170.8	1,639.0	1,694.8	2,414.3	19.56
USSR	1,580.0	1,850.0	1,750.0	1,700.0	5.89
Germany[a]	903.7	1,272.3	1,390.9	1,378.5	22.48
France	399.2	600.9	586.1	720.9	12.84
Italy	270.0	458.0	470.0	652.0	11.43
China	320.0	550.0	630.0	650.0	0.57
UK	392.7	409.3	350.4	453.7	7.93
India	145.0	233.8	297.6	420.0	0.49
Canada	293.3	311.9	345.0	415.7	15.67
Cadmium					
WORLD	**12.6**	**17.0**	**17.6**	**19.5**	**0.00**
TOTAL, 10 COUNTRIES	**10.2**	**14.5**	**15.4**	**17.8**	**0.01**
Japan	0.4	1.1	1.9	4.8	0.04
USA	3.0	3.9	3.7	3.1	0.01
Belgium	1.0	1.7	1.9	2.7	0.28
USSR	2.2	2.4	2.9	2.0	0.01
France	0.8	1.2	1.1	1.8	0.03
UK	1.0	1.3	1.4	0.9	0.02
Germany[a]	1.6	2.2	1.6	0.9	0.01
Mexico		0.3	0.2	0.5	0.01
China		0.3	0.4	0.4	0.00
Korea		0.2	0.3	0.4	0.01
Copper					
WORLD	**7,457.5**	**9,374.6**	**9,699.9**	**10,781.1**	**2.04**
TOTAL, 10 COUNTRIES	**5,816.9**	**7,201.3**	**7,418.6**	**8,242.1**	**3.95**
USA	1,396.5	1,867.7	1,958.0	2,142.5	8.60
Japan	827.4	1,158.3	1,226.3	1,576.5	12.77
Germany[a]	746.6	870.8	886.8	1,027.8	16.76
USSR	1,220.0	1,300.0	1,305.0	1,000.0	3.47
China	316.0	386.0	420.0	512.0	0.45
France	364.5	433.4	397.8	477.6	8.51
Italy	290.0	388.0	362.0	474.8	8.32
Belgium	177.4	303.9	309.6	389.5	39.56
Korea	28.0	84.0	206.6	324.2	7.58
UK	450.5	409.2	346.5	317.2	5.54
Iron ore					
WORLD	**902,389**	**890,924**	**860,640**	**924,869**[b]	**174.76**
TOTAL, 10 COUNTRIES	**669,381**	**658,846**	**659,475**	**731,768**	
USSR	189,177	197,840	203,760	199,700	691.96
China	66,436	120,394	140,354	183,963	161.50
Japan	132,689	108,693	102,215	107,395[b]	869.88
USA	127,531	90,832	64,679	73,002[b]	292.92
Germany, Fed. Rep.	48,193[c]	50,072	45,204	46,867[b]	604.19
Brazil	21,453	18,383	36,419	40,079	266.54
France	42,094[c]	37,875	26,606	25,750[b]	458.67
Korea	1,241	9,675	11,709	22,870	534.47
UK	23,696[c]	9,326	15,176	18,663	326.05
Belgium	16,871[c]	15,756	13,353	13,479[b]	

Continued

TABLE 7.4 Continued

	1975	1980	1985	1990	Per capita consumption, 1990 (kg per person)
			Lead		
WORLD	**4,526.2**	**5,348.3**	**5,440.7**	**5,531.3**	**1.05**
TOTAL, 10 COUNTRIES	**3,297.6**	**3,856.6**	**3,880.2**	**4,131.5**	**1.95**
USA	1,120.2	1,094.0	1,141.7	1,288.4	5.17
USSR	620.0	800.0	800.0	650.0	2.25
Germany[a]	373.5	433.1	440.0	447.5	7.30
Japan	189.4	392.5	394.9	416.9	3.38
UK	306.0	295.5	274.3	301.6	5.27
Italy	192.0	275.0	235.0	258.0	4.52
France	190.3	212.8	208.0	254.5	4.53
China	185.0	210.0	220.0	250.0	0.22
Korea	10.2	33.0	63.2	150.0	3.51
Spain	111.0	110.7	103.1	114.6	2.92
			Mercury		
WORLD	**7.3**	**7.7**[d]	**7.4**	**6.6**[e]	**0.001**[e]
TOTAL, 10 COUNTRIES	**4.6**	**5.4**	**4.2**	**4.3**	**0.002**
USA	1.8	2.0	1.7	1.2[b]	0.005[b]
Spain	0.2	0.2[d]	0.6	0.8[e]	0.021[e]
Algeria			0.2	0.7[e]	0.028[e]
UK	0.7	0.4[d]	0.3	0.4[e]	0.006[e]
China	0.5	0.5[d]	0.4	0.3[f]	<0.001[f]
Brazil			0.2	0.3[e]	0.002[e]
Germany, Fed. Rep.	0.4	0.5[g]	0.3	0.2[e]	0.004[e]
Mexico			0.2	0.2[e]	0.002[e]
Belgium	0.2	0.1[g]	0.3	0.1[e]	0.011[e]
USSR	0.9	1.8[d]			
			Nickel		
WORLD	**576.2**	**716.7**	**775.2**	**842.6**	**0.16**
TOTAL, 10 COUNTRIES	**485.6**	**593.7**	**625.7**	**666.3**	**0.32**
Japan	90.0	122.0	136.1	159.3	1.29
USA	132.9	143.1	143.1	124.6	0.50
USSR	115.0	132.0	138.0	115.0	0.40
Germany[a]	51.8	78.1	87.0	93.3	1.52
France	31.9	38.4	31.9	44.8	0.80
UK	20.8	22.8	24.8	32.6	0.57
China	18.0	18.0	21.0	27.5	0.02
Italy	17.0	27.1	29.0	27.3	0.48
Belgium	3.2	3.6	6.6	21.3	2.16
Spain	5.0	8.6	8.2	20.6	0.53
			Steel, crude		
WORLD	**644,153**	**718,921**	**720,568**	**794,470**[b]	**150.12**
TOTAL, 10 COUNTRIES	**463,827**	**511,379**	**524,306**	**577,016**[b]	
USSR	141,031	150,330	157,161	166,319[b]	576.30
USA	116,821	114,433	105,593	102,351[b]	410.69
Japan	68,080	79,007	73,377	93,278[b]	755.53
China	29,110	43,005	71,428	69,504[b]	61.02
Germany[a]	39,793	44,631	39,995	44,269[b]	570.70
Italy	17,778	26,764	21,880	27,994[b]	490.61
India	8,086	10,900	14,400	20,036[b]	23.49
Korea	2,964	6,100	11,310	18,300[b]	427.67
France	19,261	20,159	14,812	17,565[b]	312.88
UK	20,903	16,050	14,350	17,400[b]	303.98

Continued

TABLE 7.4 Continued

	1975	1980	1985	1990	Per capita consumption, 1990 (kg per person)
			Tin		
WORLD	**230.5**	**234.6**	**214.6**	**229.7**	**0.04**
TOTAL, 10 COUNTRIES	**168.7**	**175.6**	**158.2**	**170.8**	**0.08**
USA	55.8	56.4	37.8	36.8	0.15
Japan	28.1	30.9	31.6	34.8	0.28
Germany[a]	15.6	19.0	17.8	21.7	0.35
USSR	23.0	25.0	31.5	20.0	0.07
China	14.0	12.5	11.5	18.0	0.02
UK	14.4	9.9	9.4	10.4	0.18
France	10.0	10.1	6.9	8.3	0.15
Korea	0.6	1.8	2.6	7.8	0.18
Netherlands	3.9	5.0	4.5	6.9	0.46
Brazil	3.3	5.0	4.6	6.1	0.04
			Zinc		
WORLD	**5,062**	**6,283**	**6,552**	**6,965**	**1.32**
TOTAL, 10 COUNTRIES	**3,560**	**4,364**	**4,514**	**4,903**	**2.35**
USA	839	879	962	991	3.98
USSR	900	1,030	1,000	920	3.19
Japan	563	752	780	814	6.60
Germany[a]	360	474	480	530	8.63
China	180	259	349	500	0.44
France	223	330	247	284	5.06
Italy	150	236	218	270	4.73
Korea	35	68	120	227	5.31
UK	207	181	189	189	3.30
Belgium	103	155	169	178	18.04

[a] Data refer to the unified Federal Republic of Germany.
[b] 1989.
[c] 1976.
[d] 1978.
[e] 1987.
[f] 1986.
[g] 1979.

Consumption of metal (with the exception of iron ore) represents the domestic use of refined metals and includes metal refined from either primary (i.e., raw) sources or secondary (i.e., recovered) materials. Metal that is used in a product and then exported is considered to be consumed by the country of manufacture and not the importing country.

Consumption of iron ore refers to the quantity of iron ore and iron ore concentrates delivered to consuming industries. Data for Brazil, China, Korea and the USSR are apparent consumption, i.e., production plus imports minus exports, as opposed to "reported consumption" for which data are unavailable for those countries. Consumption data for iron ore are not strictly comparable between countries as different countries report different grades of iron ore. Furthermore "reported consumption" and "apparent consumption" can vary considerably for some countries, partly because changes in stocks are not accounted for in the calculation of apparent consumption. World production data have been used to represent world consumption; world-wide consumption of iron ore will be roughly equal to production assuming negligible changes in stocks.

Data on mercury consumption should be viewed with caution because they include estimates of consumption of secondary materials which are generally not well documented.

Crude steel consumption is apparent consumption; imports and exports have been converted to a crude steel equivalent to avoid distortion of the share of exports and imports relative to domestic production.

Per capita consumption data are based on 1990 population estimates compiled by the UN Population Division.

Sources:
EUROSTAT 1990 *Iron and Steel Statistical Yearbook 1990*, Statistical Office of the European Community, Luxembourg (and earlier editions).
ILZSG 1990 *Lead and Zinc Statistics 1960–88*, International Lead and Zinc Study Group, London.
Intergovernmental Group of Experts on Iron Ore 1991 *Iron Ore Statistics 1981–1990*, United Nations Conference on Trade and Development, Geneva.
International Iron and Steel Institute 1989 *Steel Statistical Yearbook 1989*, International Iron and Steel Institute, Brussels (and earlier editions).
OECD 1991 *The Iron and Steel Industry in 1989*, Organisation for Economic Co-operation and Development, Paris (and earlier editions).
Roskill 1980 *Statistical Supplement to the Economics of Mercury 1978* (4th edition), Roskill Information Services Limited, London.
Roskill 1984 *Roskill's Metals Databook 1984* (5th edition), Roskill Information Services Limited, London.
Roskill 1990 *The Economics of Mercury 1990* (7th edition), Roskill Information Services Limited, London.
USBOM 1989 *Mineral Industry Surveys: Mercury in 1989*, US Bureau of Mines, Washington DC.
USBOM 1992 Unpublished data, US Bureau of Mines, Washington DC.
WBMS 1992 *World Metal Statistics*, World Bureau of Metal Statistics, Ware (and earlier editions).

TABLE 7.5 Charcoal and hardwood production in selected countries, 1968–1970 and 1988–1990 (mean annual values)

Country	Total roundwood (10³ m³ a⁻¹)			Industrial roundwood (10³ m³ a⁻¹)			Fuelwood (10³ m³ a⁻¹)			Charcoal (10³ m³ a⁻¹)		
	1968–70	1988–90	% change	1968–70	1988–90	% change	1968–70	1988–90	% change	1968–70	1988–90	% change
WORLD	1,278,459	1,953,506	53	353,923	518,024	46	924,536	1,435,482	55	12,955	22,113	71
Developed countries[a]	328,472	417,920	27	212,989	253,034	19	115,483	164,886	43	306	961	214
Developing countries	949,987	1,535,587	62	140,934	264,990	88	809,053	1,270,596	57	12,649	21,152	67
TOTAL, 20 COUNTRIES	876,948	1,467,434	67	212,049	355,571	68	664,900	1,111,863	67	7,739	13,611	76
India	157,840	247,990	57	10,876	21,556	98	146,964	226,434	54	1,247	1,922	54
USA	84,960	181,083	113	69,761	97,873	40	15,198	83,209	447		500	
Brazil	96,497	175,306	82	12,088	42,146	249	84,410	133,161	58	3,681	5,807	58
Indonesia	98,993	167,563	69	9,555	29,936	213	89,439	137,627	54	82	126	54
China	90,142	145,544	61	15,609	35,557	128	74,532	109,987	48			
Nigeria	49,369	95,826	94	2,968	7,868	165	46,402	87,958	90	764	1,449	90
USSR	64,343	65,967	3	33,410	35,567	6	30,933	30,400	-2			
Malaysia	21,721	47,769	120	17,983	41,610	131	3,738	6,159	65	237	390	65
Philippines	32,230	38,187	18	12,702	5,572	-56	19,528	32,615	67		11	
Ethiopia	23,233	37,318	61	1,012	1,727	71	22,221	35,591	60	146	234	61
Zaire	19,663	34,963	78	1,840	2,764	50	17,823	32,199	81	391	609	56
Thailand	24,198	34,169	41	4,633	3,723	-20	19,565	30,446	56		488	
Tanzania	13,051	31,702	143	965	1,590	65	12,086	30,112	149	77	155	102
Bangladesh	17,995	30,145	68	1,112	873	-22	16,883	29,272	73			
Viet Nam	17,072	28,178	65	1,930	4,540	135	15,143	23,638	56			
Pakistan	12,161	23,478	93	296	1,449	390	11,865	22,029	86			
Myanmar	13,830	21,866	58	2,526	4,449	76	11,304	17,417	54			
Kenya	10,220	21,859	114	441	1,026	133	9,779	20,833	113	1,002	1,786	78
France	18,660	21,005	13	11,793	15,205	29	6,867	5,800	-16	95	106	12
Nepal	10,769	17,515	63	549	540	-2	10,220	16,975	66	16	28	69

a Developed countries include all the European nations, including Eastern Europe and the former USSR, Israel, South Africa, Canada, USA, Japan, Australia and New Zealand.

The 20 countries listed above are ranked in descending order of total roundwood production (hardwoods) during the period 1988–1990 and account for 77 per cent of the global production of hardwoods.

"Total roundwood" production refers to the volume of all rough wood felled or otherwise harvested from forests for either industrial or fuelwood purposes. "Industrial roundwood" production refers to all roundwood products, other than fuelwood, and includes sawlogs, veneer logs, sleepers, pitprops and pulpwood. "Fuelwood" and "Charcoal" production comprise all rough wood used for cooking, heating and power generating purposes.

Data on raw wood production are compiled by FAO on an annual basis from questionnaire responses submitted by national governments. Where data are lacking, data from other sources are used and, in some cases, FAO makes its own estimates. The latter applies to the case of fuelwood and charcoal production in particular as official data are lacking for many countries. FAO derives its own estimates of fuelwood and charcoal production from country-specific per capita consumption figures and population data.

For the most recent annual data and further information relating to the data in this table, please refer to the latest edition of the FAO 'Forest Products Yearbook'.

Source:
Data extracted from the *AGROSTAT PC* data base of the Food and Agriculture Organization of the United Nations, Rome, Italy in April 1992.

TABLE 7.6 Softwood production in selected countries, 1968–1970 and 1988–1990 (mean annual values)

Country	Total roundwood (10³ m³ a⁻¹)			Industrial roundwood (10³ m³ a⁻¹)			Fuelwood (10³ m³ a⁻¹)		
	1968–70	1988–90	% change	1968–70	1988–90	% change	1968–70	1988–90	% change
WORLD	**1,055,872**	**1,357,449**	**29**	**885,777**	**1,139,023**	**29**	**170,095**	**218,426**	**28**
Developed countries[a]	907,785	1,101,212	21	829,219	1,011,345	22	78,566	89,868	14
Developing countries	148,087	256,237	73	56,558	127,679	126	91,529	128,558	40
TOTAL, 20 COUNTRIES	**987,672**	**1,261,516**	**28**	**839,826**	**1,070,579**	**27**	**147,846**	**190,937**	**29**
USA	244,921	337,238	38	241,098	317,302	32	3,823	19,936	421
USSR	315,523	314,167	–	258,157	262,867	2	57,367	51,300	–11
Canada	108,000	152,587	41	106,707	150,830	41	1,294	1,757	36
China	75,328	133,919	78	25,328	60,099	137	50,000	73,820	48
Sweden	47,710	46,773	–2	46,310	45,101	–3	1,400	1,672	19
Brazil	19,553	45,301	132	10,173	30,503	200	9,381	14,798	58
Germany, Fed. Rep.	18,258	37,635	106	17,592	36,290	106	667	1,345	102
Finland	30,633	35,838	17	29,126	34,888	20	1,507	950	–37
France	16,731	22,516	35	12,731	18,516	45	4,000	4,000	0
Japan	28,346	19,931	–30	28,298	19,927	–30	48	4	–92
Poland	14,985	16,492	10	14,077	14,937	6	908	1,555	71
Czechoslovakia	10,676	14,152	33	9,671	13,122	36	1,005	1,029	2
Austria	9,973	13,717	38	9,519	12,398	30	454	1,319	190
Chile	3,788	10,666	182	3,161	9,790	210	626	876	40
Mexico	6,974	10,583	52	4,502	6,255	39	2,472	4,328	75
New Zealand	7,884	10,394	32	7,431	10,369	40	453	25	–94
Norway	7,233	10,360	43	6,963	10,093	45	270	267	–1
Spain	5,162	10,138	96	3,987	9,707	143	1,175	431	–63
India	5,791	9,787	69	1,234	2,766	124	4,557	7,021	54
Turkey	10,202	9,322	–9	3,763	4,817	28	6,440	4,505	–30

a   Developed countries include all the European nations, including Eastern Europe and the former USSR, Israel, South Africa, Canada, USA, Japan, Australia and New Zealand.

The 20 countries listed above are ranked in descending order of total roundwood production (softwoods) during the period 1988–1990 and account for 93 per cent of the global production of softwoods.

"Total roundwood" production refers to the volume of all rough wood felled or otherwise harvested from forests for either industrial or fuelwood purposes. "Industrial roundwood" production refers to all roundwood products, other than fuelwood, and includes sawlogs, veneer logs, sleepers, pitprops and pulpwood. "Fuelwood" production comprises all rough wood used for cooking, heating and power generating purposes.

Data on raw wood production are compiled by FAO on an annual basis from questionnaire responses submitted by national governments. Where data are lacking, data from other sources are used and, in some cases, FAO makes its own estimates. The latter applies to the case of fuelwood and charcoal production in particular as official data are lacking for many countries. FAO derives its own estimates of fuelwood and charcoal production from country-specific per capita consumption figures and population data.

For the most recent annual data and further information relating to the data in this table, please refer to the latest edition of the FAO 'Forest Products Yearbook'.

Source:
Data extracted from the *AGROSTAT PC* data base of the Food and Agriculture Organization of the United Nations, Rome, Italy in April 1992.

TABLE 7.7 Production of wood products – world and top 20 producers, 1968–1970 and 1988–1990 (mean annual values)

Sawnwood and sleepers (10^3 m^3 a^{-1})

Country	1968–70	1988–90	% change
WORLD	410,416.7	498,028.7	21.3
Developed countries[a]	358,838.7	384,524.3	7.2
Developing countries	51,578.0	113,504.3	120.1
TOTAL, 20 COUNTRIES	366,355.6	443,791.3	21.1
USA	86,610.7	104,826.7	21.0
USSR	117,178.3	98,333.3	-16.1
Canada	26,971.0	57,520.7	113.3
Japan	41,846.0	30,174.0	-27.9
China	14,138.3	24,759.0	75.1
Brazil	7,489.0	18,179.0	142.7
India	3,994.0	17,460.0	337.2
Sweden	11,522.0	11,528.0	0.1
Germany, Fed. Rep.	9,514.3	11,328.7	19.1
France	9,462.0	10,519.3	11.2
Indonesia	1,694.3	9,978.0	488.9
Malaysia	2,822.7	7,737.3	174.1
Finland	6,853.3	7,696.3	12.3
Austria	5,158.0	6,910.0	34.0
Poland	7,411.0	5,028.3	-32.2
Turkey	2,220.7	4,923.0	121.7
Czechoslovakia	3,717.3	4,896.7	31.7
Yugoslavia	3,065.7	4,526.3	47.6
Korea	1,091.3	4,057.0	271.8
Australia	3,595.7	3,409.7	-5.2

Paper and paperboard (10^3 t a^{-1})

Country	1968–70	1988–90	% change
WORLD	121,038.3	231,852.7	91.6
Developed countries[a]	113,032.0	195,239.3	72.7
Developing countries	8,006.3	36,613.3	357.3
TOTAL, 20 COUNTRIES	110,381.8	211,122.7	91.3
USA	45,578.7	70,355.3	54.4
Japan	11,413.7	26,507.3	132.2
Canada	11,001.7	16,553.3	50.5
China	2,805.7	15,209.7	442.1
Germany, Fed. Rep.	5173.3	11,236.0	117.2
USSR	6,236.3	10,710.7	71.7
Finland	3,986.7	8,670.7	117.5
Sweden	4,053.7	8,316.3	105.2
France	3,914.3	6,691.0	70.9
Italy	3,354.0	5,578.0	66.3
Brazil	966.0	4,778.3	394.6
UK	4,833.3	4,583.3	-5.2
Korea	256.3	4,067.0	1,486.8
Spain	1,176.3	3,433.3	191.9
Mexico	817.7	3,207.7	292.3
Austria	933.7	2,758.7	195.5
Netherlands	1,569.7	2,600.7	65.7
India	798.7	2,091.7	161.9
Australia	931.0	1,906.0	104.7
South Africa	581.0	1,867.7	221.5

Wood–based panels (10^3 m^3 a^{-1})

Country	1968–70	1988–90	% change
WORLD	65,328.3	127,351.7	94.9
Developed countries[a]	59,921.7	102,251.7	70.6
Developing countries	5,406.3	25,099.7	364.3
TOTAL, 20 COUNTRIES	56,485.7	110,237.8	95.2
USA	23,102.3	33,142.7	43.5
USSR	5,457.7	13,745.7	151.9
Japan	6,987.7	9,039.7	29.4
Indonesia	6.7	8,945.0	133,407.5
Germany, Fed. Rep.	4645.0	8256.3	77.7
Canada	3,365.0	6,469.0	92.2
Italy	1,486.7	4,288.0	188.4
China	852.7	3,625.7	325.2
France	2,107.0	2,992.3	42.0
Brazil	700.3	2,877.0	310.8
Belgium[b]	1,263.3	2,256.0	78.6
Spain	723.0	2,213.3	206.1
Poland	981.0	1,822.3	85.8
UK	411.3	1,680.7	308.6
Austria	543.0	1,618.0	198.0
Malaysia	261.7	1,571.7	500.6
Romania	797.3	1,438.0	80.4
Finland	1,373.0	1,434.7	4.5
Czechoslovakia	596.3	1,411.0	136.6
Korea	824.7	1,410.7	71.1

a Developed countries include all the European nations, including Eastern Europe, the former USSR, Israel, South Africa, Canada, USA, Japan, Australia and New Zealand.

b Includes Luxembourg.

The groups of 20 countries listed above are ranked in descending order of production of sawnwood and sleepers, paper and paperboard and wood-based panels during the period 1988–1990, respectively. The countries listed account for 89 per cent of the global production of sawnwood and sleepers, 98 per cent of the global production of paper and paperboard and 87 per cent of the global production of wood-based panels in 1988–1990.

Production of "Sawnwood and sleepers" refers to wood that has been sawn, planed or shaped in some way to produce wood products such as planks, beams, boards, rafters and railway ties. Wood flooring is, however, excluded from sawnwood production. "Wood-based panels" include all wood-based panel commodities such as veneer sheets, plywood, particleboard and fibreboard. "Paper and paperboard" production refers to the production of all newsprint, printing and writing paper and other paper and paperboard products.

Data on the production of wood products are compiled by the FAO on an annual basis from questionnaire responses submitted by national governments. Where data are lacking, data from other sources are used and, in some cases, FAO prepares its own estimates.

For the most recent annual data and further information relating to the data in this table, please refer to the latest edition of the FAO 'Forest Products Yearbook'.

Source:
Data extracted from the *AGROSTAT PC* data base of the Food and Agriculture Organization of the United Nations, Rome, Italy in April 1992.

TABLE 7.8 Inland passenger and freight transport by mode of transport in selected countries, 1981 and 1991

Region/country	Inland surface goods transport (10⁶ tonne-km)						Inland surface passenger transport (10⁶ passenger-km)					
	By road		By rail		By inland waterway		By road — Public transport		By road — Private transport		By rail	
	1981	1991	1980	1991	1980	1991	1980	1991	1980	1991	1980	1991
AFRICA												
Algeria		14,000		2,600								
Cent. African Rep.												
Ghana				200[a]		133		11,649[a]				663[a]
Kenya				2[a]		110[a]						
Madagascar			201	210[b]								196[b]
Niger		1,524[b]						3,203[b]		649[b]		580[b]
South Africa		1,538[b]	99,173	93,980[b]				171[b]				
Sri Lanka		5,285[b]		202[b]				16,745[b]		18,454[b]	1,011	2,130[b]
Tunisia	816	990[a]	1,720	2,049[a]			1,810	2,010[c]				1,039[a]
NORTH AMERICA												
USA	790,020	1,182,615[b]	1,351,810[b]	1,723,329[b]	599,830	743,358[b]	39,794	37,007[b]	3,831,300	4,888,734[b]	18,968[d]	21,400[b]
SOUTH AMERICA												
Chile			1,900	2.8[c]		3[a]	405	37[c]			1,700	1[c]
Colombia	16,227	22[a]	622	–		–	91,420					–
Costa Rica		2,022[a]		80[c]				4,433[a]		4,294[a]		72[c]
Ecuador		259[a]		8,180[a]				1[a]		1[a]		63[a]
Mexico		108,884		34,000				271[b]				5,405
ASIA												
China, Taiwan		11,543[b]		1,877[b]				7,955[b]		8,423[b]	1,724	8,322[b]
Hong Kong			67	65			18,005	21,515	5,073[e]	7,302		8,806
Israel				1,091								186
Japan	181,309	274,444	34,088	27,196		244,546	89,470	92,980	347,609	760,080	316,204	387,478
Korea	4,868	8,645[c]	10,815	13,784[c]		16,617[c]	67,315	85,325[c]	5,076	22,498[c]	22,786	25,978[c]
Laos		53		–	–	16		530	f	155		–
Turkey	39,233	61,969	5,943	7,521[a]			77,531	81,956		33,580	6,105	146[a]

Continued

TABLE 7.8 Continued

Region/country	Inland surface goods transport (10⁶ tonne-km)						Inland surface passenger transport (10⁶ passenger-km)					
	By road		By rail		By inland waterway		By road				By rail	
							Public transport		Private transport			
	1981	1991	1980	1991	1980	1991	1980	1991	1980	1991	1980	1991
EUROPE												
Albania		1,195		584		35		1,280				779
Austria	8,398		10,318	12,864	6,530	1,830		191[b]		500	7,375	173
Belgium	17,056	25,979[b]	7,528	8,354[b]	5,442	5,448[b]	5,487				7,829	6,400[c]
Bulgaria	11,226	11,090	18,052	7,661	65,943	<1[a]	22,365	22,696			6,962	4,866
Croatia		1,399		3,943[b]		118		4,194				1,503
Czechoslovakia		8,131		49,933		3,891		43,073				19,263
Denmark	8,100	10,400	1,000	1,100	1,800	1,600	7,600	10,400	37,400	55,000	4,500	4,900
Finland	17,700	3,560	8,391	7,634	4,900	3,560	8,500	8,100	34,600	46,400	3,274	3,230
France	111,000	147,700	57,000	49,400	7,600	6,800	57,000[g]	43,000	460,000[g]	599,000	55,806[h]	72,000
Germany, Fed. Rep.	1,217,000	169,800[b]	62,200	61,800[b]	50,000	45,800[b]	75,700	61,500[c]	444,000	555,600[c]	42,300	41,800[c]
Greece		12,347	813[i]	658[c]			595	5,295			1,463	2,172
Hungary	11,759	5,939[b]	24,342	16,781[b]	385	14,456[b]	21,788	22,723[b]			14,487	11,403[b]
Ireland			691	589[b]				3[a]			995	1,226[b]
Italy	129,136	167,228[a]	17,783	20,639[a]	195	129[a]	41,997	160,037[a]	387,153	528,092[a]	43,846	47,197[a]
Luxembourg	263		586	709[b]							310	261[b]
Netherlands	17,849	23,300	1,053	1,088	7,081	6,476	12,440	13,500	122,070	152,000	9,230	12,026
Norway		7,692[b]	1,529[g]	1,632[b]	9,765	8,394[b]	3,811[g]	3,956[b]	28,470	40,119[b]	2,530[g]	2,430[b]
Poland	36,847	49,000	109,835	65,146	1,914	737	48,514	41,720			48,238	41,912
Portugal	7,220[g]	10,923	994	1,748			8,650	10,500	42,500	67,000	5,856	5,688
Romania		28,994		37,853		2,030		20,639				25,429
Spain	95,400	150,000	10,783	14,466	30,225		33,500	38,640	139,380	145,370	14,753	15,022
Sweden	20,950	29,000	14,555	18,000	9,000[g]	7,800		9,000[b]		92,000	6,851	5,600
UK	44,800	136,200	17,500	15,800	44,800	52,500	42,000	41,000	384,000	568,000	35,000	41,000
Yugoslavia	16,673	21,796[b]	25,720	25,921[b]	4,201	5,007[b]	30,259	26,235[b]	f		10,510	11,653[b]

Continued

TABLE 7.8 Continued

| Region/country | Inland surface goods transport (10⁶ tonne-km) | | | | | | Inland surface passenger transport (10⁶ passenger-km) | | | | | |
| | By road | | By rail | | By inland waterway | | By road — Public transport | | By road — Private transport | | By rail | |
	1981	1991	1980	1991	1980	1991	1980	1991	1980	1991	1980	1991
USSR												
Armenia		1,140		4,177				3,398				320
Estonia		3,811		6,545		1		3,833		6		1,273
Latvia		4,857		16,739		344		5,331		3,577		3,929
Lithuania		7,019		17,748		141		6,498				934
Moldova		7,229		11,883		238		4,500				1,280
Ukraine		78,519		405,525		10,436		82,542				70,968
OCEANIA												
Australia		85,529ᶜ	1,123	81,200ᶜ		–		16,300ᶜ		199,200ᶜ	334	9,300ᶜ

a 1989.
b 1990.
c 1988.
d Not including mass urban transport.
e Including private cars, passenger vans, motor cycles and coaches.
f Included in previous column.
g 1983.
h National rail system only.
i 1982.

A dash in the above data table signifies a nil value.

Freight transport is expressed as tonne-kilometres; the tonne-kilometre is a unit of measurement representing the movement of one tonne in a road goods vehicle (when performing services for which it is primarily intended) over one kilometre. Passenger transport is expressed as passenger-kilometres; the passenger-kilometre is a unit of measurement representing the movement of one seat or authorized standing place available in a road vehicle (when performing the purpose for which it is primarily intended) over one kilometre.

Statistics on inland freight and passenger transport are compiled by the IRF on an annual basis from questionnaire responses submitted by national statistical offices.

Sources:
IRF 1986 *World Road Statistics 1981–1985: Edition 1986*, International Road Federation, Geneva and Washington DC.
IRF 1992 *World Road Statistics 1987–1991: Edition 1992*, International Road Federation, Geneva and Washington DC.

TABLE 7.9 Relative importance of the principal modes of inland freight transport, 1980 and 1990

Region/country	Percentage contribution of							
	Rail		Road		Waterways		Pipelines	
	1980	1990	1980	1990	1980	1990	1980	1990
NORTH AMERICA								
USA	47.6	39.1[a]	25.0	27.6[a]	10.6	10.5[a]	16.8	22.8[a]
EUROPE								
Bulgaria		42.0		51.2		4.8		2.0
Czechoslovakia	67.8	65.7[a]	19.9	21.8[a]	3.2	4.4[a]	9.1	8.1[a]
Denmark	17.8	12.2	81.9	75.2			0.3	12.7
Finland	30.9	23.8	68.3	74.9	0.8	1.3		
German Dem. Rep.		71.5[a]		20.5[a]		2.8[a]		5.3[a]
Germany, Fed. Rep.	25.8	24.2	48.9	54.5	20.2	17.6	5.1	3.8
Hungary	64.4	51.1[a]	25.0	34.6[a]	4.1	5.5[a]	6.5	8.7[a]
Ireland	11.3	9.2[a]	88.7	90.8[a]				
Italy		15.3[a]		77.6[a]		0.1[a]		7.0[a]
Netherlands	5.2	3.9	37.4	44.7	49.9	45.2	7.5	6.2
Norway	15.2	11.3[a]	34.8	39.4[a]			50.0	49.3[a]
Poland	67.6	65.0	22.6	23.4	1.2	0.8	8.6	10.8
Spain	10.5	7.3[a]	86.6	89.8[a]			2.9	2.9[a]
UK	14.3	8.9	77.7	84.0	0.4	0.1	7.6	7.0
Yugoslavia		35.6[a]		47.7[a]		12.1[a]		4.6[a]
USSR	**64.5**	**64.9**	**8.1**	**8.9**	**4.6**	**3.9**	**22.8**	**22.2**

[a] 1989.

Percentages in this table have been derived from data for tonne–kms submitted to the UN ECE. Owing to differences in the methods of compilation of the basic statistics the percentages given in this table must be considered only as orders of magnitude.

Data for waterways do not include national sea transport.

Source:
UN ECE 1992 *Annual Bulletin of Transport Statistics for Europe*, United Nations Economic Commission for Europe, Geneva.

TABLE 7.10 Vehicle numbers, vehicle density and traffic volume in selected countries, 1981 and 1991

Region/country	Vehicles in use — Passenger cars 1981	Passenger cars 1991	Commercial vehicles 1981	Commercial vehicles 1991	Vehicle density — Per 10^3 persons 1981	Per 10^3 persons 1991	Per km of road 1981	Per km of road 1991	Traffic volume (10^6 vehicle-km travelled) — Passenger cars 1981	Passenger cars 1991	Commercial vehicles 1981	Commercial vehicles 1991
AFRICA												
Cent. African Rep.	23,750	8,221	3,165	8,541	11.2	—	1.2	—	—	—	—	—
Congo		27,000		15,430		89.0		3.3				
Ethiopia	41,227	37,799[a]	13,327	20,939	1.7	1.1	1.5	3.1				
Kenya	114,197	157,166[a]	40,002	133,968[a]	8.8	12.0[a]	2.6	4.7[a]	58	1,082[a]	136	4,088[a]
Madagascar	25,208	41,900[a]	81	29,100[a]	3.0	—	3.0	—	<1	23,563[a]	<1	17,937[a]
Mauritius	31,709	50,016	8,043	16,849	39.6	62.0	21.4	37.0			54	
Niger	18,226	31,427[a]	7,962	8,768[a]	6.2	5.3[a]	2.2	3.5[a]	28	117[a]	87	265[a]
Rwanda	6,187	7,868[a]	9,056	2,048[a]	2.6	1.4[a]	2.0	0.7[a]				
South Africa	2,448,968	3,403,605[a]	1,349,454	1,497,607[a]	136.0	159.1[a]	18.9	26.4[a]	37,058		19,475	
Swaziland		26,415[a]		24,296[a]		64.8[b]		16.9[b]	169	295[a]		
Togo	2,687	4,920[a]	34	254[a]	1.5	1.5[a]	1.6	1.8[a]				
NORTH AMERICA												
Canada	10,199,388	11,722,506[b]	3,192,197		552.8	617.0[b]	14.4	30.2[a]	201,789	2,438,230[a]	632,466	1,001,697[a]
USA	123,461,507	143,549,627[a]	34,995,004	45,105,835	681	752[a]	25.6		1,792,957			
SOUTH AMERICA												
Chile	505,000	761,148[b,c]	230,900		65.2	75.1[b]	9.4		3,535		4,572	
Colombia	672,385	1,098,895[b]	168,096		30.1	41.0[b]					2,527	
Costa Rica	48,188	168,814[a]	27,143	108,256[a]		91.0[a]		7.8[a]		3,239[a]		
Honduras	58,920	88,982[a]	24,385	18,049[a]		43.0[a]		9.0[a]		1,468[a]		1,055[a]
Mexico	4,808,014	5,336,228[b]	1,777,425			104[b]						
ASIA												
Bahrain		107,657		24,523		255.0		49.5		17,300		91,000
Brunei		115,377		13,019		128.0		96.0		87		10
China, Taiwan	717,595	2,328,439[a]	85,352	653,869[a]		146.5[a]		150.1[a]				
Cyprus	96,097	189,701	27,939	84,024	239.0	470.9	11.5	26.8				
Hong Kong	227,658	260,196[a]	75,205	139,670	57.2	69.4	254.5	261.5	3,368	4,781[a]	1,500	4,016
India	898,143	2,056,696[a]	1,108,476	1,608,030[a]	2.3	4.1[b]	1.1	1.7[d]	7,180	13,574	3,341	5,916
Israel	459,178	857,381	105,079	174,085	140.0	204.2		77.6				

Continued

TABLE 7.10 Continued

Region/country	Vehicles in use				Vehicle density				Traffic volume (10⁶ vehicle-km travelled)			
	Passenger cars		Commercial vehicles		Per 10³ persons		Per km of road		Passenger cars		Commercial vehicles	
	1981	1991	1981	1991	1981	1991	1981	1991	1981	1991	1981	1991
Japan	24,612,277	37,076,065	15,020,133	22,936,389[e]	335.5	483.4	35.4	53.8	247,156	365,597	147,501	262,984
Jordan	104,078	172,075	43,058	35,854	62.9	77.1[f]	28.8	39.7[f]	528	–	213	–
Laos		21,269		14,790		9.0		2.6				
Saudi Arabia	757,395	2,664,028[a]	661,290	2,272,794[a]	189.2	422.0[a]	25.6	34.1[a]				
Singapore	165,198	285,298	638,554	126,389		147.3[a]		145.0				
Thailand	451,001	735,326	535,735	1,903,220	20.6	46.0	13.2	50.0	5,547	17,237	13,345	34,860
Turkey	626,105	2,143,680	827,048	954,376	23.6	52.9	4.6	8.4	8,044	13,432	7,281	12,624
Yemen[g]	72,698	165,438	105,388	237,957	20.8	30.5	10.8	7.8	396	4,352	1,042	5,193
EUROPE												
Austria	2,312,932	3,100,014	217,437	268,577	335	428	23.7	30.3[a]	21,000		9,927	
Belgium	3,206,472	3,928,906	303,555	379,111	353	430	28	31	41,594		4,735	
Bulgaria	836,397	1,316,644		133,375		162		39		9,875	707	2,946[a]
Croatia		735,650		43,891		163		29				
Czechoslovakia	2,475,774	3,341,774	524,056	341,809		236		5				1,807
Denmark	1,447,319	1,649,074[b,h]	316,627	251,334[i]	316	369	23.0	26.7	21,500	30,700	5,120	6,800
Finland	1,279,192	1,909,787	164,055	270,552	299	434	19.1	28.5	22,600	33,130	4,650	6,040
France	19,750,000	23,810,000	2,716,000	4,833,000	413	501	27.8	35.4	282,700	325,000	63,000	109,000
Germany, Fed. Rep.	23,680,911	31,309,165	1,548,397	1,569,458	406	511[a]	51.6	63.0[a]		406,000	34,500	48,000
Greece	912,016	1,777,484	471,583	802,947	143.0	251.4	37.3		4,360		1,704	
Iceland	90,258	120,062	10,678	16,012	435	527	8.6	12.1	1,014		219	
Ireland	774,594	828,225	69,858	154,511	245	279	9.2	10.6			114	
Luxembourg	133,315	191,588[a]	12,488	12,838	391	532	28	40	1,417		188	
Netherlands	4,594,000	5,569,000[a]	350,000	565,900[a]	345	405	45.0	58.3[a]	61,000	77,790	9,465	13,525
Norway	1,278,817	1,614,623	171,676	334,351	353	456	18	21[a]				
Poland	2,634,338	6,112,171	708,371	1,329,650	92.7	194.3	11.2	20.4	18,440	61,100	21,323	30,200
Portugal	1,345,988	1,800,000[a,j]	527,407	648,200	154.0	248.5	27.6	30.0[a]	18,040		2,419	
Romania		1,397,118		332,273		72[a]		23[a]				
Slovenia		578,268[a]		33,844[a]		306[a]		42[a]				
Spain	7,943,325	12,537,099	1,466,257	2,541,830	248	387	125	45.9	53,874	76,009	18,076	26,180
Sweden	2,893,242	3,621,114	502,658	324,345	372	457	24	29	35,000			
Switzerland	2,394,455	3,085,372[b]	278,950	291,457	408	500	40	47.5		86,200[a]		9,000[k]

Continued

TABLE 7.10 Continued

Region/country	Vehicles in use				Vehicle density				Traffic volume (10^6 vehicle-km travelled)			
	Passenger cars		Commercial vehicles		Per 10^3 persons		Per km of road		Passenger cars		Commercial vehicles	
	1981	1991	1981	1991	1981	1991	1981	1991	1981	1991	1981	1991
UK	14,943,000	19,742,000[a]	2,146,000	2,765,000[a]	306	405[a]	50	63[a]	201,390	329,700[a]	43,800	289,300[a]
Yugoslavia	2,567,961	3,323,940[a]	227,985	247,046[a]	124	151[a]	24	140[a]				
USSR												
Armenia		2,782[a]		12,034		4.1		1.9				
Estonia		261,086		85,685		222		23.4[a]				
Latvia		328,436		76,990		153		6.3		1,638		1,990
Lithuania		512,362		14,518		141		11.8				787
Moldova		210,385		14,254		–		–				568[a]
Ukraine		3,570		–		–		–				–
OCEANIA												
Australia	6,977,000	7,734,100[a]	554,300	1,915,400[f]	519	552	10.6	12.1	96,109	116,640	20,321	3,500
New Zealand	1,331,765	1,539,809	273,181	309,543[b]	502	608[f]	17.2	21.3[f]	16,653			

Data in the above table refer to four-wheeled vehicles only. Two and three-wheelers, caravans, trailers, semi-trailers and tractors are excluded, unless otherwise stated.
"Passenger cars" generally comprise cars that are licensed to carry up to nine persons including the driver. This category therefore includes taxis, minibuses and station wagons.
"Commercial vehicles" include buses, coaches and goods vehicles (i.e., trucks and vans).
Traffic volume is expressed as vehicle-kilometres travelled; the vehicle-kilometre is a unit of measurement representing the movement of a road vehicle over one kilometre.

Data generally refer to the situation at the year end, i.e., 31 December.
Road statistics are compiled by the IRF on an annual basis from questionnaire responses submitted by national statistical offices.

Sources:
IRF 1986 *World Road Statistics 1981–1985: Edition 1986*, International Road Federation, Geneva and Washington DC.
IRF 1992 *World Road Statistics 1987–1991: Edition 1992*, International Road Federation, Geneva and Washington DC.

a 1990.
b 1989.
c Includes light trucks.
d 1988.
e Includes three-wheeled vehicles.
f 1987.
g Data refer to the former Yemen Arab Republic.
h Includes vans under two tonnes.
i Includes vans over two tonnes.
j Includes minibuses.
k Incomplete total.

Part 8

Wastes and Waste Management

The lack of reliable statistics describing the generation, treatment and disposal of waste arisings continues to impair the formulation of appropriate waste management strategies and the assessment of their effectiveness. New initiatives reviewed here are, however, beginning to remedy this situation. Other emerging trends in the area of waste management which are highlighted in this report include:

○ Municipal waste generation continues to increase world-wide, in both absolute terms and on a per capita basis;

○ Partly due to definitional problems, the amount of hazardous wastes generated world-wide is still highly uncertain; rough estimates lie in the range 350–500 x 10^6 t a^{-1}. In OECD Europe alone the quantity of hazardous wastes generated is estimated to be 24 x 10^6 t a^{-1}; the estimate for the USA, some 275 x 10^6 t a^{-1}, includes liquid discharges that are classified as hazardous;

○ As pressures on landfill increase and sea disposal is subjected to increasingly strict regulation, incineration is emerging as a key component of future waste management strategies in some countries.

○ Recovery, recycling and adoption of cleaner, low-waste technologies are the means by which environmental damage per unit output can be reduced; although world-wide this philosophy is gaining increasing acceptance as an alternative to the traditional end-of-pipe technologies, in practice the introduction of cleaner production has been slow.

The quantity of waste arisings – solid, liquid and gaseous – are generally considered to be growing across the globe as a result of increases in the world's population, increasing industrialization, increasing urbanization and rising standards of living. Moreover, major advances in the development of new materials and chemicals have increased the diversity and complexity of the waste streams.

During the past two decades the imposition of progressively tighter regulatory controls aimed at mitigating the adverse environmental and human health impacts arising from the generation, treatment and disposal of wastes has, predictably, increased the costs of waste disposal.

Consequently, wastes are taking on a new economic importance, not only in terms of revenues generated by the waste treatment and disposal industry, but also because wastes may have a residual value as a secondary raw material which can be recovered or reused.

Wastes are increasingly becoming an international issue. Exporting wastes to countries with less stringent controls and a lower public awareness of the issues has been used to counter the rising costs of disposal and the growth of the Not In My Back Yard (NIMBY) syndrome in many industrialized countries. Most significantly, the recipients of some of these exports have been the less developed countries and the countries of Eastern Europe where the technology for their safe disposal is often not readily available.

During the past decade or so the area of waste management has been further characterized by a significant change in perception. In place of a preoccupation with the "safe management" of wastes and investment in pollution control equipment (i.e., end-of-pipe technologies), waste managers are increasingly turning to a more preventative or integrated approach to waste management, i.e., one which involves changing processes and products in order to minimize the total volume of wastes generated by a manufacturing process. This "cleaner production" concept was endorsed at the United Nations Conference for Environment and Development (UNCED) in June 1992, in the action plan commonly known as Agenda 21, as a means of achieving a more sustainable development.

The generation, movement, treatment and final disposal of solid waste arisings are covered by the material presented in this section of the 'UNEP Environmental Data Report'. Data and related information pertaining to gaseous discharges are included in Part 1: Environmental Pollution, Atmosphere. Some of the environmental and health implications of recent changes in waste management practices are discussed and the present legal instruments for the control of hazardous waste movements are reviewed. Also included in this section of the report is a brief overview of waste recovery and minimization practices. Since 1989 the United Nations Environment Programme (UNEP) has endeavoured to increase awareness of the cleaner production concept through the establishment of its Cleaner Production Programme, currently operated by the Paris-based Industry and Environment Programme Activity Centre (IE/PAC). The activities of this programme are also briefly described.

Traditionally, the collection of statistics on wastes has had a relatively low priority in many countries. Consequently the statistical data base for solid wastes is poor, especially in the historical context. This paucity in the availability of reliable statistics on both waste generation and disposal is seen by many as an impediment to identifying the best or most practical waste management options, establishing priorities and to assessing the effectiveness of policy action.

The poor statistical data base for wastes can be attributed to a number of factors. Firstly, responsibility for regulating and managing wastes is diverse, often resting with local levels of administration where financial constraints and other difficulties impair the establishment of nationally coherent infrastructure and procedures for data collection. Moreover, local authorities in many countries weigh relatively little of the waste volumes they handle. A review of procedures in England and Wales, for example, revealed that over 60 per cent of the local authorities weighed less than half of the wastes they handled; a further 24 per cent weighed none at all (ENDS, 1992). The complexity of many waste streams in terms of composition – i.e., many waste streams comprise mixtures of chemicals – exacerbates the problems of waste characterization.

The lack of standard classifications and definitions of waste types acts as a further impediment to the development of statistical descriptions of wastes. Moreover, inter-country differences in waste classifications limit the comparability of national estimates of waste arisings. In recent years the lack of a universal definition of a "hazardous" waste is an issue which has been at the forefront of concern, partly because of legal implications. Latterly some progress has been made in establishing standard classifications and definitions for wastes and, in particular, for hazardous wastes; the work of the Organisation for Economic Co-operation and Development (OECD), the United Nations Economic Commission for Europe (UN ECE) and the Statistical Office of the European Communities (EUROSTAT) are particularly significant in this regard, as is that of UNEP within the framework of the Basel Convention (see below).

Various efforts are also under way to improve the availability of policy-relevant information on waste generation and disposal practices around the world. For example, the International Maritime Organization (IMO) in 1991 initiated its Global Waste Survey (GWS) as a means of developing a plan of action for co-ordinating and assisting the efforts of international agencies and countries around the world to identify and implement environmentally-sound waste management options to minimize, and where possible avoid, the generation of hazardous wastes, and to eliminate the dumping of industrial wastes at sea. Although the IMO's project arose out of concern for protection of the marine environment (more specifically from the concern that not all countries would necessarily have the capabilities to meet decisions of the London Dumping Convention (LDC) regarding the practices of at-sea incineration of noxious liquid wastes and ocean dumping of industrial wastes), it was conceived as having a wider significance for the global management of industrial wastes (IMO, 1992a).

The GWS has been designed in co-operation with 10 other organizations (including UNEP) and comprises five principal tasks:

○ Task 1: Development of a Global Waste Inventory (GWI);
○ Task 2: More detailed evaluation of existing waste management practices in selected countries, e.g., development of "National Waste Management Profiles";
○ Task 3: Identification of clean technologies and processes for waste reduction;
○ Task 4: Implementation of case studies; and
○ Task 5: Development of strategy and action plans.

To date, Task 1 has been completed (IMO, 1992b) and Task 2 is under way: the entire project is scheduled for completion by the end of 1994. Completion of Task 1, the development of a GWI, has generated a computerized data base of qualitative (and some quantitative) information on waste generation and disposal options in over 100 countries world-wide; analyses of information contained in this data base are used throughout this chapter where appropriate.

For the purposes of this report, statistics on the quantities of solid wastes generated have been compiled largely from national "State of the Environment"–type reports and as such represent official government estimates. When interpreting the information presented it should be borne in mind that, as noted above, the methods of estimation and definitions employed for each category of waste in each country are far from standard. These comments also apply to the information included here on waste treatment and disposal.

Waste Generation

Generally speaking, wastes are produced in three ways: through the production and consumption of goods and services; through the processing of wastes from these activities; and through end-of-pipe control or treatment of emissions and effluents. Waste arisings are generally reported according to their source, that is within the following broad categories: mining and construction wastes, energy production wastes, agricultural wastes, municipal or household wastes, industrial wastes and sludges. Environmental and human health implications of these waste streams vary markedly.

Table 8.1 shows the quantities of solid waste generated in selected countries across a broad spectrum of consumption and production activities and in the case of sewage sludge, wastes generated by a pollution abatement process. Although the categories included have been expanded over those in the last edition of the 'UNEP Environmental Data Report', they are still not complete. This fact, combined with the lack of systematic surveys in many countries, means that the figures provide an incomplete picture of the total waste burden. Trends in the arisings of certain waste streams in selected countries are illustrated in Figure 8.1.

Owing to the paucity and limitations of the data on waste arisings, especially in the developing world regions, generalized statements regarding the generation of wastes are difficult to make – other than the fact that waste volumes are growing everywhere. These difficulties notwithstanding, the OECD has estimated that in 1990 its member countries

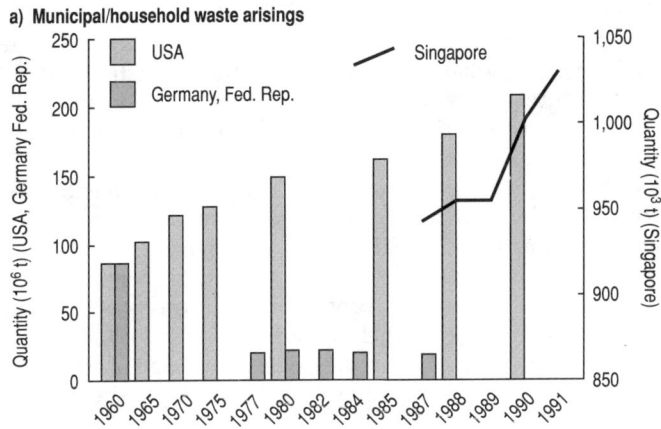

a) Municipal/household waste arisings

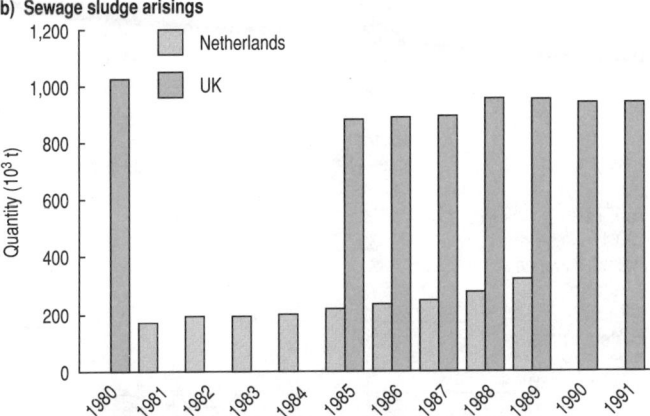

b) Sewage sludge arisings

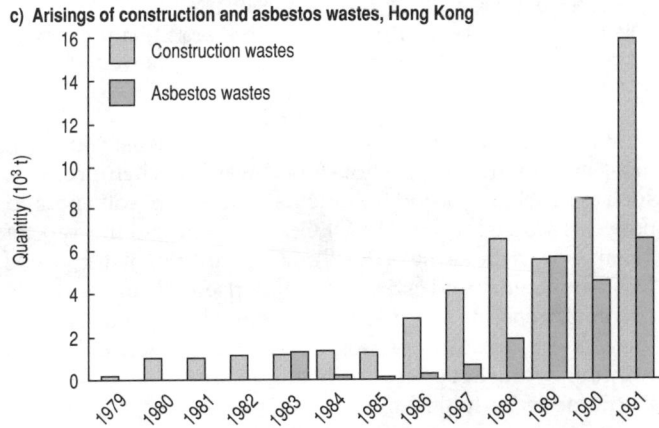

c) Arisings of construction and asbestos wastes, Hong Kong

FIGURE 8.1 Trends in arisings of certain waste streams in selected countries
Sources: a) Federal Environmental Agency, 1991 (Germany); Ministry of the Environment, 1992 (Singapore); Kaldijan, 1990 (USA); b) Department of the Environment, 1992 (UK); Central Bureau of Statistics, 1992 (Netherlands); c) Environmental Protection Department, 1991 (Hong Kong)

absolute tonnage generated, the agricultural and mining/quarrying sectors are often the largest waste producers. Wastes generated as a result of mining for metals are discussed in Part 7: Industry and Transport. Agricultural wastes will be considered in more detail in a subsequent edition of this report.

In the absence of detailed quantitative information, alternative methods of estimation of waste arisings have been explored. "Rapid assessment" methods involve using socio-economic data (e.g., population, Gross Domestic Product (GDP) and number of employees in different industrial sectors) to predict waste generation. At the national level such techniques are generally found to offer a simple, cost-effective initial assessment of the likely types and quantities of various waste streams in a particular country. However, rapid assessment methods cannot be considered to be substitutes for specific in-country survey data (IMO, 1992b).

Municipal Wastes

Relative to other waste streams, municipal wastes are quite well documented. Table 8.2 presents national data on annual municipal waste arisings, waste arisings per capita and the materials composition of the waste stream, generally for the most recent year available. Municipal wastes typically include household waste, certain white goods and bulky consumer wastes, as well as similar wastes from small commercial and industrial firms, institutions and markets, which are collected and disposed of by, or for, local authorities. There are nevertheless considerable variations in the exact definition of "municipal" wastes between countries.

In most countries the quantity of municipal waste is rising (see Figure 8.1). The small decline in arisings in Germany can be explained by the increase in the separate collection of recyclable materials such as glass and paper (see also Waste Recovery). Municipal waste generation is also growing on a per capita basis; in the OECD as a whole per capita arisings have increased since the mid-1970s from 407 kg a^{-1} (or 1.1 kg day^{-1}) to 513 kg a^{-1} (or 1.4 kg day^{-1}) by the end of the 1980s (OECD, 1991). Limited data for other world regions suggest similar levels of per capita municipal waste generation in the rapidly industrializing Asian nations, and lower rates of generation in the African countries (Table 8.2).

Levels of municipal waste collection services vary markedly; in most OECD member countries services have expanded to the point whereby over 90 per cent of the population (and 100 per cent of the urban population) have access to municipal waste collection services (UNEP, 1991). In contrast many developing countries have limited services. The failure to provide adequate collection services poses a potentially serious threat to human health in many developing countries. Refuse left uncollected accumulates on the streets and is a major factor in the spread of gastro-intestinal and parasitic diseases primarily as a result of the proliferation of insect and rodent vectors. Uncontrolled dumping of household and industrial wastes at the same sites in peri-urban areas and near squatter settlements increases the risk of poisoning, injury and exposure to other health hazards, particularly where scavenging on the sites is seen as an income opportunity (WHO, 1992).

produced some 9×10^9 t of solid wastes; this quantity included 420×10^6 t of municipal wastes and nearly $1,500 \times 10^6$ t of industrial wastes, of which at least 300×10^6 t are described as hazardous wastes. The remainder, some 7×10^9 t comprised more inert wastes such as fly ash (from energy production), agricultural wastes, mining spoils, demolition debris, dredge spoils and sewage sludge (OECD, 1991). Thus, in terms of the

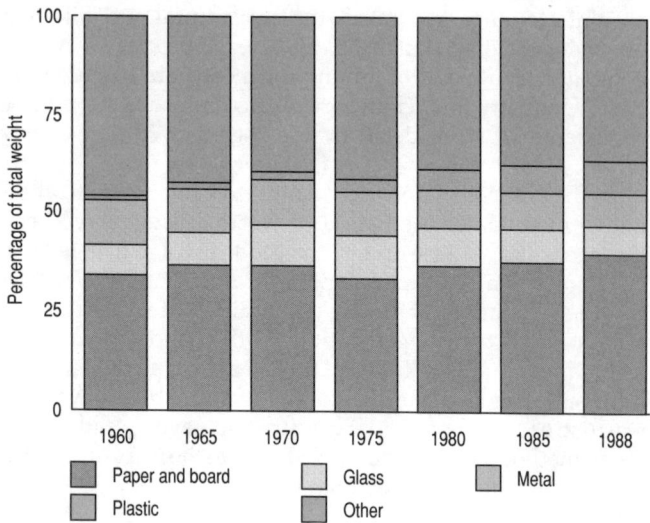

FIGURE 8.2 Changes in the composition of municipal solid waste in the USA, 1960–1988
Source: Kaldijan, 1990

Although the composition of municipal waste varies widely from country to country (Table 8.2), some general trends in composition can be distinguished. For example, organic waste continues to account for the largest part of municipal wastes in all countries for which such information is available. In the industrialized countries there is evidence of a decline in the proportion of paper and board, and an increase in the contribution of plastics. The proportions of glass and metals, however, have tended to remain relatively stable in most countries, with the exception of Japan where there has been a marked drop in the quantities of glass and metal in municipal wastes. Figure 8.2 shows that in the USA the proportion of glass and metals has varied little since 1960, remaining at between 5–10 per cent. This is despite the fact that these materials are tending to be replaced by plastics and an increasing proportion of them are removed from the waste stream through separate collection initiatives.

Industrial and Hazardous Wastes

Industrial wastes encompass a wide range of materials of varying environmental toxicity, but typically include general rubbish, packaging, food wastes, acids and alkalis, oils, solvents, resins, paints and sludges. In many countries the exact arisings of industrial wastes are largely unknown; some available estimates are shown in Table 8.1. In the OECD area alone, 1.5×10^9 t a^{-1} (1990 estimate) of industrial wastes are generated, a value which represents a 50 per cent increase over 1980 estimates (OECD, 1991).

A proportion of the wastes generated by industry are deemed to be "hazardous wastes" because they contain substances that are toxic to humans, plants or animals, are flammable, corrosive, or explosive, or have high chemical reactivity. As there are few comprehensive or consistent statistics characterizing hazardous waste streams, the assessment of the global rate of hazardous waste generation is difficult. In the first instance, many countries do not undertake routine and systematic surveys of hazardous wastes.

Moreover, compilation of such information is hampered by the fact that until very recently there has been no universal agreement on what constitutes a hazardous waste. Furthermore, there has been no scientific agreement on how we determine if, or which, chemicals are hazardous, or on how they should be reported.

Qualitative information on the types of hazardous wastes generated by individual nations world-wide has been obtained by the IMO, as part of its GWI. Figure 8.3 shows the percentage of respondents in each of the major regions reporting arisings of various categories of hazardous wastes. With the exception of the "polychlorinated biphenyls (PCBs)" and "contaminated soils" categories, over 50 per cent of countries surveyed indicated that the specified wastes were being generated (IMO, 1992b).

The OECD has compiled rough estimates of the volumes of industrial and hazardous wastes generated in OECD countries, Eastern Europe and the rest of the world. These estimates are summarized below (data refer to the late 1980s):

Region	Industrial wastes (10^6 t a^{-1})	Hazardous and special wastes (10^6 t a^{-1})
WORLD [a]	2,100	338
OECD	1,430	303
North America	821	278[b]
Europe	272	24
Pacific	333	
Eastern Europe	520	19
Rest of World	180	16

[a] Alternative estimates of the quantity of hazardous wastes generated world-wide put the figure somewhat higher, at 500×10^6 t a^{-1} or more (ISBC, 1993).
[b] The value for USA (275×10^6 t a^{-1}) used to derive the regional total for North America includes liquid wastes that are classified as hazardous.
Source: OECD, 1991

Hazardous wastes can contaminate other waste streams, most notably municipal/household wastes, when products such as pesticides, wood treatments, paints, oils, solvents and batteries are discarded. The OECD has identified this type of hazardous waste as the SQHW (small quantities of hazardous wastes) category and estimates that households in the OECD currently generate about 350×10^3 t of SQHW annually. Small businesses, including photographic laboratories, paint and printing workshops, educational establishments, research laboratories, hospitals and the agricultural sector also produce significant quantities of SQHW (OECD, 1991).

Table 8.3 provides national estimates of hazardous waste arisings taken selectively from publications and reworked. Owing to the varying detail of descriptions of the wastes in the source documents, data have been presented under broad categories of the substances which make up or contaminate the waste stream. It should not be assumed that the figures presented under each waste type represent all the arisings of that material. Totals for national estimates of hazardous waste arisings have not been given, except where that was the only piece of information available for a country.

It is anticipated that improved figures for industrial and hazardous wastes will become available in the future. Increasingly companies and manufacturers are monitoring and

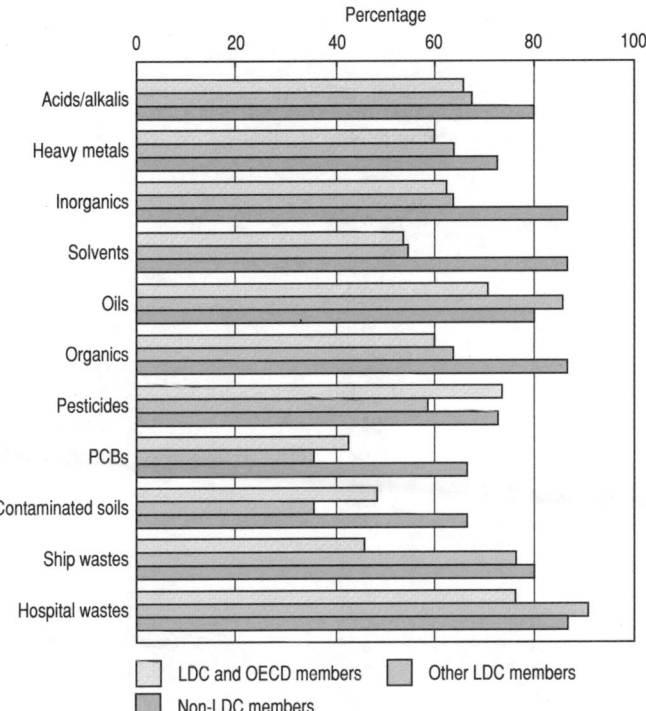

FIGURE 8.3 Percentage of GWI respondents reporting arisings of certain types of hazardous wastes (LDC = London Dumping Convention)
Source: IMO, 1992b

reporting their level of waste generation and discharges to the environment; some 20–25 chemical companies, for example, are among a growing number of large multinational companies who have started to publish detailed accounts of their gaseous and liquid discharges and solid waste arisings. The practice of conducting environmental audits and/or life-cycle analyses (LCA), which include a breakdown of waste arisings, is also on the increase (see Part 7: Industry and Transport).

At the national level several countries including Canada and Australia, are considering developing detailed inventories of releases of potentially toxic chemicals from major manufacturing installations, such as that already set up by the US Environmental Protection Agency (EPA) (Sasnet, 1993). The US EPA's Toxic Release Inventory (TRI) is described in more detail in Box 8.1. In order to extend this type of approach yet further the International Programme for Chemical Safety (IPCS) has recently taken a lead role in the co-ordination of national efforts to develop pollutant release and transfer registers (PRTRs). On the international level, qualitative and quantitative information on hazardous waste arisings will be collected as part of the provisions of the Basel Convention (see below). Furthermore, the reliability and comparability of waste statistics should improve as the standard definitions and catalogues developed as part of the Basel Convention and by the OECD, EUROSTAT and the UN ECE gain currency.

References

Central Bureau of Statistics 1992 Sludge from public sewage treatment plants 1981–1989, *Quarterly Bulletin of Environment Statistics*, **9**(1), 21–26 (in Dutch with English titles).

Department of the Environment 1992 *Digest of Environmental Protection and Water Statistics No. 14 1991*, HMSO, London.
ENDS 1992 DOE aims to improve waste statistics, *ENDS Report*, **208**, 15–16.
Environmental Protection Department 1991 *Environmental Hong Kong 1991, A Review of 1990*, Environmental Protection Department, Hong Kong.
Federal Environmental Agency 1991 *Facts and Figures on the Environment of Germany 1988/89*, Federal Environmental Agency, Berlin.
IMO 1992a *Global Waste Survey: Project Outline*, LCD.2/Circ.298 Annex 2, International Maritime Organization, London (revised 18 May 1992).
IMO 1992b *Final Report: Global Waste Survey: Task 1 Global Waste Inventory*, LCD.2/Circ.305, International Maritime Organization, London.
ISBC 1993 The Basel Convention: a global approach for the management of hazardous waste, Background paper prepared by the Interim Secretariat of the Basel Convention, Geneva.
Kaldijan, P. 1990 *Characterisation of Municipal Solid Waste in the United States: 1990 Update*, US Environmental Protection Agency, Washington DC.
Ministry of the Environment 1992 *Annual Report 1991*, Public Affairs Department, Ministry of the Environment, Singapore.
OECD 1991 *The State of the Environment*, Organisation for Economic Co-operation and Development, Paris.
Sasnet, S. 1993 Personal communication, Office of Toxic Substances, US Environmental Protection Agency, Washington DC.
UNEP 1991 *United Nations Environment Programme Environmental Data Report 1991/92*, Basil Blackwell, Oxford.
WHO 1992 *Our Planet, Our Health, Report of the WHO Commission on Health and the Environment*, World Health Organization, Geneva.

Movements of Hazardous Wastes

During the 1980s it became apparent that hazardous wastes were increasingly being exported and imported. Such movements can be justified if there is a need to find a site where the waste can be disposed of safely. However, trade in wastes took place not just between industrialized countries; arrangements and proposals to send hazardous wastes from industrialized countries to developing countries were widely reported in the world press. There have also been large-scale movements from Western to Eastern European countries (OECD, 1991; Yakowitz, 1993a). It is impossible to justify sending hazardous wastes from the more developed to the less developed regions since most developing countries have neither clear waste management policies nor the facilities for treating hazardous wastes. The reasons are instead largely economic; the current situation of widely differing prices for the disposal of similar classes of hazardous wastes in different countries promotes movements of wastes in search of the lowest cost of legal, or even illegal, disposal.

Information on the exact quantities of hazardous wastes traded is hard to obtain. It is estimated that in OECD Europe alone over 100,000 trans-frontier movements of hazardous wastes, representing as much as 2×10^6 t, are made per year; some of these shipments are destined for East European and developing countries. Of the 2×10^6 t, approximately half is destined for recycling and the other half for final disposal in a country other than the country of origin. These quantities of hazardous wastes, however, represent a relatively small proportion (less than 1 per cent) of the total tonnage of 'wastes' that

BOX 8.1 The US EPA's Toxic Release Inventory (TRI)

In the latter half of the 1980s, the USA government introduced legislation which requires industrial manufacturers to prepare annual reports itemizing their releases to the environment of a given list of chemical substances. Estimates of quantities of chemicals injected into underground wells, stored on site and/or transferred to off-site treatment or disposal facilities are also required. Since 1987, the first year of reporting, these data have been collated·by the US EPA and published on an annual basis as the Toxics Release Inventory (TRI). The TRI data have been widely disseminated and are available to all interested parties through a publicly accessible data base as well as through a variety of published reports.

The 1990 Reporting Year
In 1990 industrial manufacturing facilities with 10 or more employees which manufactured or processed more than 25,000 pounds (11.3 t), or used more than 10,000 pounds (4.5 t), of any of the 329 reportable chemicals were required to file toxic release reports. Based on reports filed from 23,638 facilities which meet these criteria, the US EPA estimates that a total of 4.83×10^9 pounds (2.2×10^6 t) of toxic chemicals were generated by USA manufacturing industries in 1990. Of this total, 2.20×10^9 pounds (1.0×10^6 t) were emitted to the air, 197×10^6 pounds (0.09×10^6 t) were released to water bodies, 441×10^6 pounds (0.20×10^6 t) released to land and 725×10^6 pounds (0.33×10^6 t) were injected underground. The balance, 1.26×10^9 pounds (0.57×10^6 t), was transferred for treatment and disposal elsewhere, either to municipal waste-water treatment plants (448×10^6 pounds (0.20×10^6 t)) or to other off-site facilities for treatment or disposal (815×10^6 pounds or 0.97×10^6 t).

The accompanying table shows that the chemical industry and the primary metals processing industries are the largest generators of toxic chemicals. Toluene, ammonia, methanol, acetone and trichloroethane are among the chemicals having the largest releases to air; phosphoric acid, ammonia and sulphuric acid together account for 76 per cent of the toxic chemicals discharged to surface waters; while metallic compounds, in particular those containing zinc, manganese and copper, dominate the chemical releases to land (US EPA, 1992).

Limitations of the TRI
The TRI data base provides a relatively detailed account of the quantities of toxic chemicals generated by industry; it also describes their disposal

TRI releases to the environment and transfer to off-site facilities by industry, 1990 (10^3 t)

Industry	Releases	Transfers	Total
TOTAL	1,619.35	571.54	2,190.89
Chemicals	719.41	236.37	955.78
Metals (primary)	257.92	138.98	396.90
Paper	131.09	31.98	163.07
Transportation	88.77	22.32	111.09
Plastics	87.64	14.20	101.83
Metals (manufactured)	59.47	34.16	93.62
Petroleum	53.21	7.71	60.92
Electrical	38.56	21.59	60.15
Food	18.60	23.32	41.91
Machinery	24.09	7.39	31.48
Furniture	27.26	2.13	29.39
Printing	23.59	2.13	25.72
Stone/Clay	14.47	6.26	20.73
Textiles	15.70	5.04	20.73
Lumber	17.24	3.36	20.59
Measure/Photo	14.88	5.26	20.14
Miscellaneous	11.29	3.22	14.52
Leather	5.53	5.53	11.07
Non-SIC code	7.53	1.95	9.48
Tobacco	1.13	<0.05	1.13
Apparel	0.73	0.14	0.86

Industries are grouped according to the Standard Industrial Classification (SIC codes 20–39 inclusive) and are ranked according to the total quantities of wastes generated, largest first. Releases and transfers reported here in units of 10^3 t have been converted from the original reporting units of the TRI, i.e., pounds, by using the conversion factor 1 pound = 0.4536 kg.
Source: US EPA, 1992

pathways. Nevertheless, in its present form, the TRI does not represent a complete picture of toxic chemical releases or hazardous waste generation, nor does it facilitate more detailed analysis of waste generation by industry with a view to pinpointing and assessing waste prevention opportunities (see also Waste Prevention).

In the first instance the inventory only covers manufacturing industries and within those it is limited to the larger facilities. Thus not all sources of toxic chemical releases are covered by the TRI. Secondly, the TRI suffers from problems of incomplete reporting by facilities which are required to submit reports. Moreover, the accuracy of data will vary between facilities and from year to year. This is because, in many cases, the releases given are based on estimates not direct measurements (US EPA, 1992).

A third limitation of the present TRI is that it excludes some toxic chemicals. Furthermore, the TRI list of chemicals is constantly changing as chemicals are selectively removed or added to the list. This limits its inter-annual comparability. Comparability across the years of reporting is

also impaired by the fact that the manufacturing and processing reporting threshold has dropped progressively from 75,000 pounds (34.0 t) in 1987 to the present 25,000 pounds (11.3 t).

Future Directions
Aware of the inherent limitations of the present TRI, the US EPA is currently evaluating various options for the improvement of the current system of reporting. For example, proposals to include an additional 313 chemicals in the existing list of reportable chemicals are currently under review. The US EPA is also considering incorporating other sources of toxic chemicals, such as waste management companies, chemical warehouses and commercial launderers into the inventory process. Furthermore, from 1991 facilities where required to report on waste prevention and recycling activities (US EPA, 1992).

Reference
US EPA 1992 *1990 Toxics Release Inventory: Public Data Release,*Office of Toxic Substances, US Environmental Protection Agency, Washington DC.

cross the borders of OECD member countries each year (Yakowitz, 1993b). Available figures for North America suggest that annually some 9,000 frontier crossings of hazardous wastes were made in the latter half of the 1980s (OECD, 1991). Quantities of hazardous wastes actually shipped to developing countries from the industrialized world are even harder to obtain; based on existing written and oral reports the OECD suggests a ball-park figure of several hundred thousand tonnes (Yakowitz, 1993a). In more recent years there have been verbal reports of movements of wastes between developing countries and from developing countries to Europe.

Selected data on quantities of hazardous wastes imported and exported by various countries are presented in Table 8.4. These data represent known or reported movements; it is unlikely that they reflect the true scale of trade in hazardous wastes. Further information has been compiled by the Non-Governmental Organization (NGO) Greenpeace (Vallette and Spalding, 1990).

Since the mid-1980s the transboundary movement of hazardous wastes has become the subject of increasingly stringent regulation. A brief synopsis of the development of international controls for the hazardous waste trade, initiated by OECD member countries and culminating in the ratification and entry into force of the Basel Convention in May 1992, is given in Box 8.2.

Increasing regulation should improve the availability and reliability of data on quantities of hazardous wastes traded. For instance, since 1988 most OECD member countries have been collecting data on movements of waste across their national boundaries for collation by the OECD. Article 13 of the Basel Convention commits signatory nations to annual reporting of the amount and characteristics of hazardous and other wastes imported and exported. These data will be held by the Secretariat of the Basel Convention, currently located in Geneva. However, the adoption of the principle of prior informed consent (see Box 8.2) may generate new data problems; the system of prior notification tends to encourage exporters to overestimate the amount of hazardous waste due to be shipped. Thus it may be difficult to reconcile figures provided by the importers and exporters.

References

OECD 1991 *The State of the Environment*, Organisation for Economic Co-operation and Development, Paris.

Vallette, J. and Spalding, H. (Eds) 1990 *The International Trade in Wastes: A Greenpeace Inventory*, 5th Edition, Greenpeace USA, Washington DC.

Yakowitz, H. 1993a Waste management: what now? what next? an overview of policies and practices in the OECD area, *Resources, Conservation and Recycling*, **8**, 131–178.

Yakowitz, H. 1993b Personal communication, Organisation for Economic Co-operation and Development, Paris.

Waste Treatment and Disposal

Waste management practices in terms of treatment and disposal vary enormously between countries. Nevertheless, some general trends are becoming apparent.

Waste treatment and disposal are subject to increasing regulation; in Europe for example, governments are prescribing the disposal option for a given waste stream. Growing regulatory control has resulted in a sharp increase in the cost of legal waste disposal. For example, in the USA prices of commercial landfill for hazardous wastes increased by almost a third between 1985 and 1987 and currently lie between US$ 250–750 per tonne (Yakowitz, 1993a). Furthermore, increasingly stringent legislation governing the siting of new waste management facilities has meant that operators are tending to expand existing facilities rather than build new ones.

At the present time the most widely used waste treatment and disposal routes include incineration, physical/chemical treatment, composting, landfill, sea dumping and recovery/recycling. Landfill remains the most commonly adopted approach for the majority of waste streams (see below). Incineration and recovery are, however, on the increase; it is anticipated that in Europe incineration capacity is likely to expand in the future and will play a key role in the management of municipal and hazardous wastes (Yakowitz, 1993a). In contrast, there is a general trend towards the reduction of sea disposal options, particularly for hazardous wastes (see Sea Disposal).

The above methods of waste treatment and disposal can be effective in minimizing many of the harmful impacts of waste generation on the environment and on human health when used judiciously and where plant design and management are of a high standard. Incineration, for example, effectively reduces the volume of material for ultimate disposal to landfill. Nevertheless, waste treatments and disposal, like all other human activity, inevitably give rise to some form of pollution. Incineration of waste, for example, when inadequately controlled can cause nuisance and atmospheric pollution as a result of the emission of unburnt waste material, acidic gases, heavy metals and trace quantities of organic compounds (see Box 8.3). Adverse environmental impacts can also arise from improperly managed landfill operations. These include leachate leakage (which may contaminate surface and ground-water resources) and emission of methane and carbon dioxide (both greenhouse gases). Methane emissions from landfill are briefly discussed in Part 1: Environmental Pollution, Atmosphere. Both issues will be reviewed in more detail in subsequent editions of this report.

The availability of various treatment and disposal facilities for waste streams was surveyed by the IMO as part of its GWI. Figure 8.4, which plots the percentage of respondents reporting given facilities, indicates, not surprisingly, that OECD countries rely more heavily on engineered technologies. The differences are most apparent in the case of land-based incineration, solidification and port reception facilities for treatment of ship-generated wastes (IMO, 1992).

Municipal Wastes

Table 8.5 gives figures on the disposal routes for municipal wastes. Burial in controlled landfills continues to be the most common means of disposing of municipal waste in the OECD region; about 70 per cent of municipal waste is disposed of

BOX 8.2 International Controls on the Trade in Hazardous Wastes

Moves to control the international trade in hazardous wastes were first made by OECD member countries in the mid-1980s. The OECD Council meetings held in 1984 and 1985 laid the foundations for future control of transfrontier movements with the adoption of the following guiding principles (OECD, 1991):

○ The principle of non-discrimination: OECD members will apply the same controls on transfrontier movements of hazardous waste involving non-member states as for movements between member states;

○ The principle of prior consent: movements of waste to non-member countries will not be allowed to take place without the consent of appropriate authorities in the importing country; and

○ The principle of adequacy of disposal facilities: movements of waste will not be permitted unless the wastes are directed to adequate disposal facilities in the importing country.

These OECD initiatives have since provided the basis for both European Community and global legislation in this area. European Community Directive 84/631/EEC adopted in December 1984 in effect implements the principles set out by the OECD Council. This directive has since been amended (Directive 86/279/EEC); these amendments and revisions include a clear definition of the terms "wastes", a list of potentially hazardous wastes and a stipulation that the wastes being transported must be identified and classified in accordance with the OECD's International Waste Identification Code

(IWIC) described by Yakowitz (1993).

The United Nations Environment Programme has been involved in the development of legal instruments for the control of transboundary movements of hazardous wastes since the early 1980s. Following recommendations made by the 1981 Montevideo Meeting of Senior Government Experts in Environmental Law, UNEP developed the Cairo Guidelines for the Environmentally Sound Management of Hazardous Wastes. The Cairo Guidelines, which included various principles for regulating the transboundary movements of hazardous wastes, were adopted by UNEP's Governing Council in 1987. In the same year, UNEP was requested to prepare a global legal instrument to control transboundary movements of hazardous wastes. This action led to the adoption, in March 1989, of the Basel Convention on the Control of Transboundary Movements of Hazardous Wastes and their Disposal. The Convention entered into force on 5 May 1992, and as of May 1993, 45 countries have ratified or acceded to the Convention. Parties to the convention include 5 countries in Africa, 11 in Asia, 12 in Western Europe, 7 in Eastern Europe and 10 in Latin America (see also Part 10: International Co-operation).

Although the Basel Convention entered into force in May 1992, its full effectiveness is somewhat inhibited by the fact that some of the main generators of hazardous wastes, including several European Community member countries and the USA, have not yet ratified the Convention. These countries have, however, confirmed their resolve to become parties as soon as their internal regulations permit (ISBC, 1993).

The provisions of the Convention are described in its 29 Articles and 6 Annexes. The Articles set out the scope of the Convention and the obligations of the Parties and of the Basel Convention Secretariat. Although broadly based on the principles established at 1984 and 1985 OECD Council meetings, the Basel Convention goes much further, incorporating requirements to control the generation of hazardous wastes and requiring such wastes to be managed "at home" unless there is no capacity to do so properly. The Convention also describes illegal traffic as criminal.

The Annexes define the wastes covered by the Convention. Included is a list of 47 categories of wastes to be controlled and a list of hazardous waste characteristics which, in effect, provides a global, universally-accepted definition and classification system for hazardous wastes. A Secretariat has now been set up in Geneva to process and disseminate information provided by the Parties, and to provide assistance to countries implementing the Convention.

References

ISBC 1993 The Basel Convention: Global Approach for the Management of Hazardous Wastes. Background paper prepared by the Interim Secretariat at the Basel Convention, Geneva.

OECD 1991 *The State of the Environment*, Organisation for Economic Co-operation and Development, Paris.

Yakowitz, H. 1993 Waste management: what now? what next? an overview of policies and practices in the OECD area, *Resources, Conservation and Recycling*, 8, 131–178.

---

in this way by the USA and by most European countries. However, in Belgium, Luxembourg, France, Italy, Switzerland and in Japan more than half of all municipal waste goes to some destination other than landfill.

Incineration is the next most common method of disposal for municipal wastes. Data in Table 8.5 indicate that in Japan and Switzerland more than 75 per cent of municipal waste is incinerated. In many instances energy can be recovered from waste incineration; this practice has a number of benefits in that it helps to conserve energy resources, to reduce emissions of carbon dioxide, and to reduce demand for landfill space. Figure 8.5 shows the increase in incineration of municipal waste in the USA since 1960 and the increasing use of incineration with energy recovery. Similar trends are evident in some other countries.

In many developing countries the prevalent method of municipal waste disposal is uncontrolled dumping or burning on open waste ground, or where no collection systems exist, in the city streets. Such landfills as do exist, are often poorly designed and controlled due to insufficient resources and a lack of trained staff. Scavenging of reusable materials on these uncontrolled sites is widespread in some countries.

Many developing countries are making attempts to apply proper treatment and disposal methods, including opening of "sanitary" landfills, expanding capabilities for recycling (in a controlled manner), composting and incineration. In the more developed countries, further advances in landfill technologies have been made as a result of the introduction of stricter regulation. Substantial progress has been made not only in preventing leachate problems by leakproofing, but also in reducing methane emissions to the atmosphere through the installation of degasification plants (OECD, 1991).

BOX 8.3 Environmental and Health Impacts of Incineration

Incineration of municipal wastes and the utilization of energy this produces, is increasingly chosen by a number of countries as a method of reducing the bulk of solid wastes. Sweden, Denmark and France, for example, incinerate more than 80, 60 and 40 per cent of their municipal wastes respectively, usually with energy recovery (The Warmer Campaign, 1990).

During the last decade, however, the general public has been increasingly concerned about the possible health implications of emissions of hazardous compounds from municipal waste incineration plants. Of the possible emissions from incinerators (i.e., particulate matter, heavy metals and trace organics) public attention has focused on the releases of polychlorinated dibenzo-*p*-dioxins (PCDDs) and polychlorinated dibenzo-*p*-furans (PCDFs), and in particular on 2,3,7,8-tetrachlorodibenzo-*p*-dioxin (TCDD). These compounds are formed as a result of the incomplete combustion of chlorinated hydrocarbons present in the waste stream as vegetable matter and plastics. Emissions of metals such as lead, mercury and cadmium (which are found in batteries) have also been a matter of some concern. In a number of countries incinerators now account for significant proportions of the national emissions of these pollutants; in the UK, for example, it is estimated that 23 per cent of the total cadmium, 13 per cent of the total mercury and 25 per cent of the total dioxin and furan emissions originate from municipal incinerators (WSL, 1989).

Owing to the complexity of many waste streams, especially in terms of their toxicity, interpretation of available data on the possible human health effects of various waste disposal options, including incineration, is extremely difficult. The small number of people exposed generates difficulties in establishing dose-response relationships (National Research Council, 1991). In addition there may well be heightened perception of the hazards involved in those persons living in the vicinity of incinerators (Neutra et al., 1991). Thus many studies which have attempted to assess the health impact of incinerators have, to date, failed to provide conclusive evidence of a causative relationship with health outcomes such as low birth weight, birth defects, cancer incidence and elevated mortality rates (Levine and Chitwood, 1985).

Considerable progress in incinerator technology has been made in recent years. Use of incinerators with improved combustion techniques and more effective control of particulate and gaseous emissions, coupled with the introduction of screening of waste prior to incineration (to remove critical components such as batteries) means that much of the unacceptable incinerator emission can be prevented. Indeed, under appropriate conditions, incineration of single waste streams, organics for example, can be carried out without major toxic emissions or residues. Further reductions in the potential impacts of emissions of hazardous substances can be achieved by fitting pollution control technology to the older, more-polluting incinerators. In Europe municipal solid waste incineration plants are now required to meet the terms of the EEC Municipal Waste Incineration Directives; these set emissions standards for new plants and a timetable for existing plants to be upgraded.

All signs point to an increased emphasis on incineration as a key waste management process in the future, especially as a means of reducing the dependence on land-fill. However, in the longer term, the adoption of waste minimization approaches will go further towards reducing the potential environmental and human health impacts of waste disposal options.

References

Levine, R. and Chitwood, D. D. 1985 Public health investigations of hazardous organic chemical waste disposal in the United States, *Environmental Health Perspectives*, **62**, 415–422.

National Research Council 1991 *Environmental Epidemiology, Public Health and Hazardous Wastes*, National Academy Press, Washington DC.

Neutra, R., Lipscomb, J., Kenneth, S. and Dennis, S. 1991 Hypothesis to explain the higher symptom rates observed around hazardous waste sites, *Environmental Health Perspectives*, **94**, 31–38.

The Warmer Campaign 1990 *Waste Incineration*, Warmer Factsheet January 1990, Tunbridge Wells.

WSL 1989 *Review of Municipal Solid Waste Incineration in the UK*, Report Number LR776, Warren Spring Laboratory, Stevenage.

The practice of burying municipal waste in landfills faces an uncertain future in many countries, particularly those in the industrialized world. In some cases existing sites are close to saturation; the pressure of space and opposition of residents to the creation of new sites limit the expansion of landfill disposal options. On the other hand, the imminent banning of sea disposal (see below) will inevitably increase demand for more disposal facilities on land, including that for landfill and incineration (OECD, 1991; Yakowitz, 1993a).

Industrial and Hazardous Wastes

As in the case of municipal wastes, landfill is currently the most widely used disposal option for industrial and hazardous waste (Table 8.6). Of the estimated 24 x 10⁶ t of hazardous wastes generated by the OECD member countries of Europe, approximately three quarters is disposed of onto land or by burying underground. A rough breakdown of disposal options for hazardous wastes in the European countries of the OECD is given below:

Disposal route	Quantity (10⁶ t a⁻¹)	As percentage of total (%)
Landfill	14–18	70–75
Ocean dumping	0.7	2–5
Incineration	1.5	5–8
Physico-chemico treatment	1.0	4
Recovery	1.5–3.0	5–12

Source: Yakowitz, 1993a

Up until the late 1970s little heed was paid to potential consequences of the landfill disposal of industrial or hazardous wastes. However, the realization that old landfills containing hazardous industrial wastes could entail serious risks to the environment and human health has led to increasingly stringent regulation of the management of these wastes in landfill. Moreover, the discovery of significant numbers of contaminated sites has necessitated costly remedial action programmes. Estimates of the numbers of contaminated sites are summarized below for selected countries:

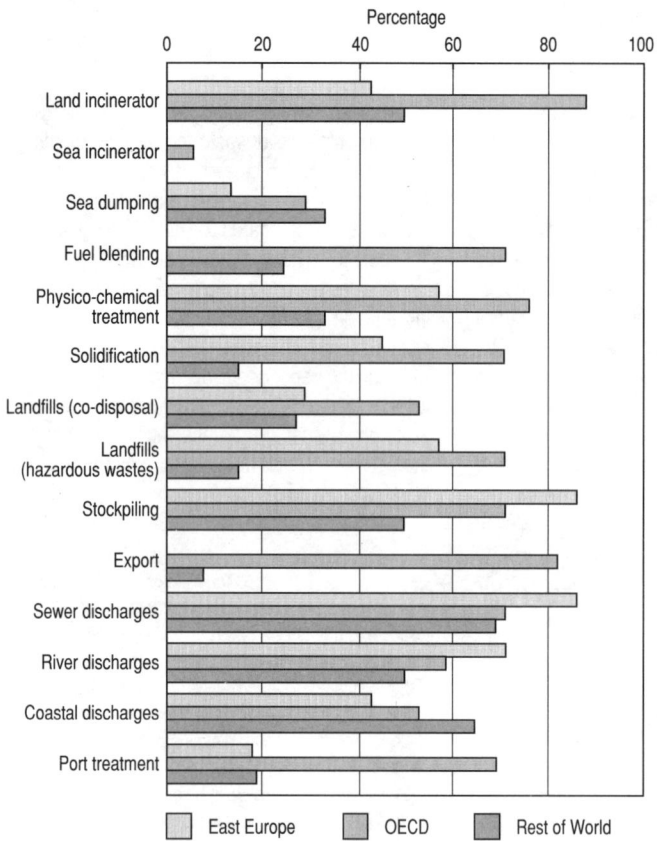

FIGURE 8.4 Percentage of GWI respondents reporting availability of different types of waste treatment and disposal facilities
Source: IMO, 1992

Region/country	No. of contaminated sites	
	Identified	Requiring immediate attention
NORTH AMERICA		
USA	2,000–10,000[a]	1,200[a]
EUROPE		
Denmark	3,115	
France	450	80
Germany	50,000	5,000
Netherlands	4,000	1,000

[a] It is estimated that between 2,000 and 10,000 contaminated sites would qualify for remedial action as part of the US EPA's Superfund Program; to date more than 1,200 sites have been identified and added to a priority list of sites for immediate funding.
Sources: OECD, 1991; Russell et al., 1992

Initial estimates of the cost of "clean-up" programmes in both Europe and North America have since been revised upwards, as more information about the extent of the contamination problem became available. Current estimates of the clean-up costs for hazardous waste sites in OECD Europe are of the order of 1.0–1.5 x 10^9 ECUs (Yakowitz, 1993b).

Within the industrialized world attitudes towards the use of landfill for hazardous waste vary widely. Some countries are continuing to promote controlled landfill, as opposed to uncontrolled dumping, as a disposal option for hazardous wastes.

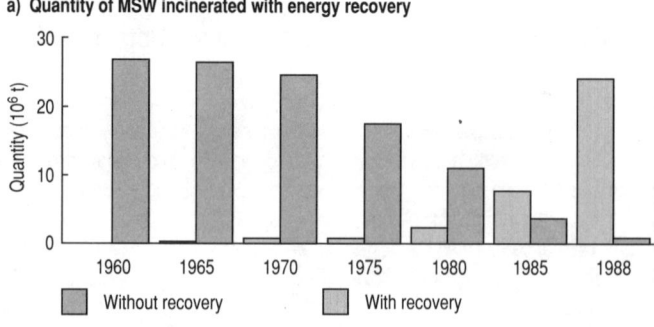

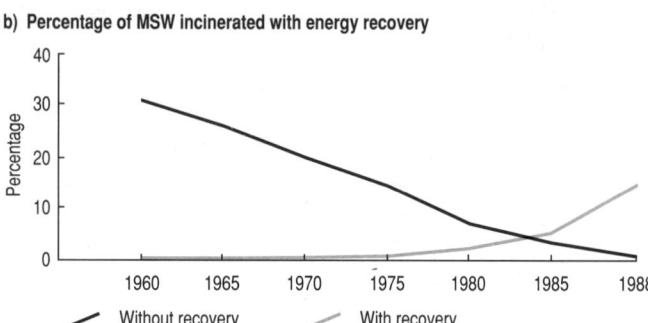

FIGURE 8.5 Trends in the amount of municipal waste incinerated and use of energy recovery in the USA, 1960–1988
Source: Kaldijan, 1990

Other countries are actively seeking to limit the scope of landfill for hazardous wastes. An increasing number of countries are imposing regulations requiring that hazardous wastes be subjected to some form of pre-treatment prior to landfilling. According to the IMO survey, over 50 per cent of OECD member countries have facilities for co-disposal (i.e., disposal of certain types of hazardous wastes with, or into, municipal wastes). Approximately three-quarters have dedicated hazardous waste landfill sites (Figure 8.4). In the developing world controlled landfilling is widely seen as an interim solution, i.e., the first step towards development of overall, longer-term hazardous waste management strategies (Wilson, 1992).

Although it is the second most common disposal option for hazardous wastes, incineration currently handles less than 10 per cent of the hazardous waste stream in Western Europe. It is generally the preferred disposal route for hazardous organic chemicals that cannot be recycled. Some countries now stipulate that certain organic liquid wastes must be incinerated (Yakowitz, 1993a).

Over the past two decades international organizations, including UNEP, have devoted considerable resources to developing and disseminating guidance on policy and strategies for hazardous waste management. Particular emphasis has been placed on the needs of developing countries; UNEP's IE/PAC has been especially active in this area. In addition to organizing workshops and conferences and other forms of training relating to hazardous waste management, IE/PAC is responsible for the preparation of a series of technical documents describing specific waste management technologies and their application. A recent initiative includes the preparation of a training manual on hazardous waste management policies and strategies (IE/PAC, 1992). A second handbook on landfilling of industrial and hazardous wastes is in preparation (IE/PAC, 1994). Technical guidelines for the environmentally sound management of wastes subject to the Basel Convention are also under preparation by an appointed Technical Working Group.

Guidance (developed through a 1985 Expert Meeting) and information on chemicals is provided by UNEP's International Register of Potentially Toxic Chemicals (IRPTC). In addition to its central computerized data bank of detailed information on individual toxic chemicals (see Part 1: Environmental Pollution), the IRPTC has developed Waste Management Data Files on treatment and disposal options for specific waste chemicals. The WHO, in collaboration with UNEP, the World Bank, the International Solid Waste Association (ISWA) and other organizations, is currently developing a series of specific guidance documents which address management options for commonly occurring waste streams in developing countries.

Sewage Sludge

The treatment of waste water, whether from domestic or industrial sources, generates sludges which need disposal. Sewage sludge is a slurry of organic-rich particles which has a highly variable chemical composition depending on the original source of the effluent, and type and efficiency of the treatment processes. Sewage sludge tends to concentrate heavy metals and water-soluble organic compounds, and may also contain oils, greases and bacteria.

The amounts of sewage sludge generated and disposed of through various disposal pathways in selected OECD countries are listed in Table 8.7. Sludge can be usefully disposed of on land by using it as a fertilizer (i.e., for agricultural purposes) or by processing it into compost or other products. However, contamination of sludges, particularly by heavy metals, can limit the amount of sludge that can be used in this way. Landfill, incineration or sea dumping may then be the only option.

Steady progress in the volume and proportion of both domestic and industrial waste waters receiving some form of treatment prior to discharges means that the volume of sludge residues requiring disposal is increasing in both the developed and developing world (UN ECE, 1992). For example, in the UK implementation of the EEC Urban Waste Water Treatment Directive is expected to contribute to a doubling in the amount of sewage sludge arisings from 1.1×10^6 t a^{-1} dry weight in 1991 to 2.2×10^6 t a^{-1} in 2006 (DoE, 1993). Likewise, in Egypt it is estimated that the proposed domestic waste-water treatment plant for Cairo and Alexandria will generate approximately $0.7-1.0 \times 10^6$ tons a^{-1} of sewage sludge. Wider adoption of waste-water treatment schemes throughout Egypt could increase the amount of sewage sludge for disposal tenfold, i.e., to 10×10^6 tons a^{-1} (Government of the Arab Republic of Egypt, 1992). Ending of sea disposal by 1998, by signatories of the Oslo Convention, will increase the pressure to dispose of sludges on land.

Sea Disposal

The practice of dumping wastes at sea, although not a major disposal route for most waste streams, has in recent years been the subject of strict regulation. Ocean disposal is controlled by the "International Convention for the Prevention of Marine Pollution by Dumping of Wastes and Other Matter", more commonly known as the London Dumping Convention (LDC). The IMO serves as its Secretariat. The 1972 LDC and its annexes stipulate the licensing of waste dumping at sea and ban the sea disposal of certain hazardous wastes. More recent resolutions refer to the incineration at sea of noxious liquid waste, a practice scheduled to be banned by 31 December 1994 and to the ocean dumping of industrial wastes, which is due to be phased out by 31 December 1995.

The sea disposal of selected waste streams, including dredged spoils and sludges is still widely practised throughout the world. In Hong Kong, for example, the quantity of dredged marine mud disposed of at sea has increased from around 20–30 x 10^6 t a^{-1} in the late 1980s to over 60 x 10^6 t a^{-1} in 1991 (Environmental Protection Department, 1992). The number of GWI respondents reporting the use of ocean disposal, according to individual world regions is given below:

Region	Number of countries
TOTAL	23
Sub-Saharan Africa	4
Other Africa	3
Latin America/Carribean	3
Asia & Pacific[a]	5
OECD	5
Eastern Europe	1

[a] Includes one country which plans to use sea disposal in the future.
Source: IMO, 1992

As current legislation requires the licensing of ocean disposal, the quantities of wastes dumped at sea, at least legally, are relatively well known. Previous editions of the 'UNEP Environmental Data Report' have included information on the disposal of industrial waste, dredged spoils, and sewage sludge at sea by contracting Parties to the Oslo Convention (UNEP, 1991). Data to update these tables were not available at the time of writing.

Radioactive Waste in the Marine Environment Detailed data on the disposal at sea of packaged low-level radioactive wastes have been compiled by the International Atomic Energy Agency (IAEA) and were included in the third edition of this report (UNEP, 1991). Two other reports are in preparation by the IAEA which will further contribute to an understanding of inputs of anthropogenic radionuclides to the marine environment. These additional studies will address the discharge of low-level radioactive effluents and inputs from marine accidents involving radioactive materials. Some data relating to the latter issue are included here in Part 9: Environmental Disasters.

Previously unavailable information on the dumping of radioactive material to sea by the former USSR has recently been released by the Russian Federation. A special Russian Commission has estimated that, since 1957, some 92×10^{15} Bq have been dumped into the Arctic and Pacific oceans. The table below summarizes the nature and quantities of radioactive material dumped at sea by the former USSR:

| Ocean region | Liquid wastes (TBq) | Solid wastes (TBq) | Nuclear reactors | | | |
| | | | With fuel | | Without fuel | |
			No.	Activity (kCi)	No.	Activity (kCi)
Barents Sea	564					
Kara Sea	315	574	7	2,300	10	Not known
Far Eastern Seas	456	225				

Unknown quantities of radioactive liquid and solid wastes have also been discharged into the Japanese and Bering Seas.
Source: IMO, 1993

Litter in the Marine Environment Litter constitutes only a very small part of the total quantity or volume of wastes, but the effects, particularly of plastic litter, on the marine environment are of widespread concern. Litter originates from both land and sea sources and typically comprises three types: fishing gear (such as nylon lines, buoys and nets); packaging bands, straps and synthetic ropes; and general litter (such as bags, bottles and plastic sheeting). Much of this material is washed up on beaches by winds and tides.

World-wide the annual loss of fishing gear is estimated at more than 150,000 t. Although the full extent of the problem regarding plastic waste is not known, it has been estimated that in 1985 over 450,000 plastic containers were dumped from the world's shipping fleet. Plastics accounted for up to 70 per cent of the debris collected in the Mediterranean and more than 80 per cent in the Pacific.

Few countries have produced national or even local data on the extent or types of marine litter. Table 8.8 provides a selection information on marine/beach litter extracted from a variety of published sources. Owing to a lack of standardization in the way the data are collected or described, the information is necessarily presented in very broad terms. Figure 8.6 shows the percentage of the UK shoreline surveyed on which certain categories of litter were found.

References

DoE 1993 *UK Sewage Sludge Survey*, Department of the Environment, London.

Environmental Protection Department 1992 *Environment Hong Kong: A Review of 1991*, Environmental Protection Department, Hong Kong.

Government of the Arab Republic of Egypt 1992 *Environmental Action Plan*, Government of the Arab Republic of Egypt, Cairo.

IE/PAC 1992 *Hazardous Waste: Policies and Strategies Training Manual*, Technical Report No.10, United Nations Environment Programme, Industry and Environment Programme Activity Centre, Paris.

IE/PAC 1994 *A Training Manual on Landfilling of Industrial and Hazardous Wastes*, United Nations Environment Programme, Industry and Environment Programme Activity Centre, Paris. In press.

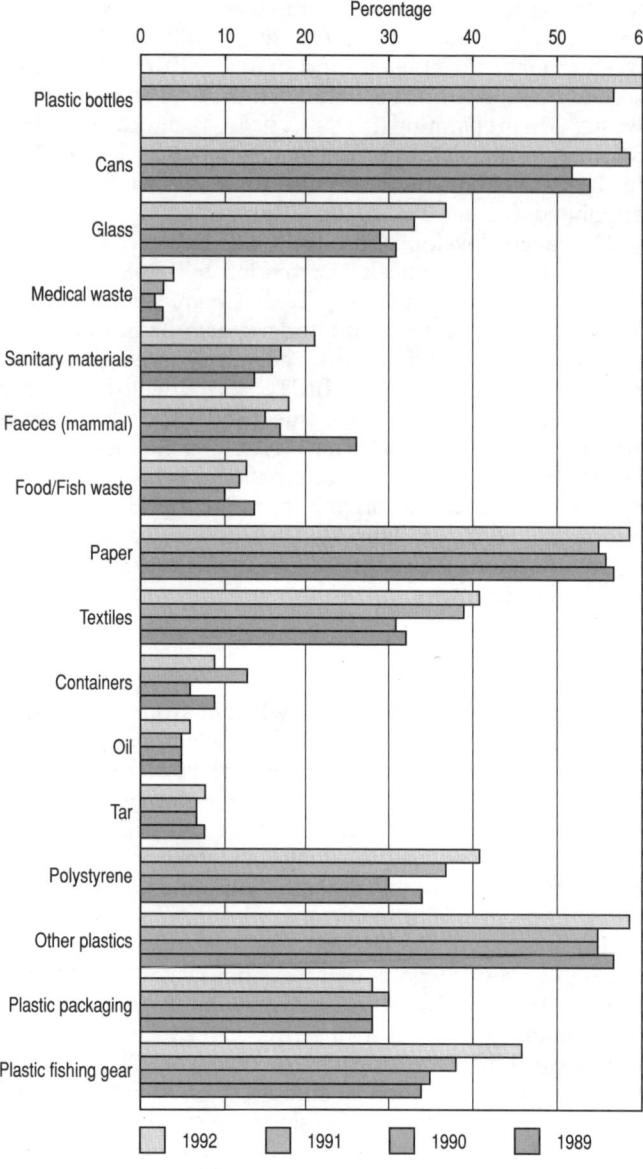

FIGURE 8.6 Categories of litter found on the upper shore of the UK coastline during the Coastwatch surveys, 1989, 1990, 1991 and 1992
Source: Data supplied by K. Pond, Coastwatch UK, Farnborough, UK

IMO 1992 *Final Report: Global Waste Survey: Task 1 Global Waste Inventory*, LCD.2/Circ.305, International Maritime Organization, London.

IMO 1993 *Nature and Quantities of Radioactive Materials Dumped at Sea by the Former USSR*. Report submitted by the Russian Federation to the meeting of the Inter-Governmental Panel of Experts on Radioactive Waste Disposal at Sea, 12–16 July, 1993, London, UK. Unpublished report.

Kaldijan, P. 1990 *Characterisation of Municipal Solid Waste in the United States 1990 Update*, US Environmental Protection Agency, Washington DC.

OECD 1991 *The State of the Environment*, Organisation for Economic Co-operation and Development, Paris.

Russell, M., Colazier, F. W. and Tonn, B. E. 1992 The US hazardous waste legacy, *Environment*, **34**(6), 12–15 and 34–39.

UN ECE 1992 *The Environment in Europe and North America: Annotated Statistics 1992*, United Nations, New York.

UNEP 1991 *United Nations Environment Programme Environmental Data Report 1991/92*, Basil Blackwell, Oxford.

Wilson, D. C. 1992 Hazardous wastes in developing countries, In: *WHO Commission on Health and the Environment: Report of the Panel on Industry*, World Health Organization, Geneva, 119–144.

Yakowitz, H. 1993a Waste management: what now? what next? an overview of policies and practices in the OECD area, *Resources, Conservation and Recycling*, **8**, 131–178.

Yakowitz, H. 1993b Personal communication, Organisation for Economic Co-operation and Development, Paris.

Waste Recovery, Recycling and Prevention

The benefits of recycling and, in particular, of waste prevention at source have long been recognized. These waste management strategies not only reduce natural resource consumption, but also off-set the cost and inadequacies of treatment, storage and disposal options for wastes. Despite clear advantages both these approaches do, nevertheless, have some drawbacks. Recycling, for example, may involve high initial or capital costs; it also necessitates the identification of markets for recycled material and requires educating the general public to accept the need to separate materials at source. Reducing waste arisings at source may require the regulation of industry, the development of economic instruments to encourage plant modification or redesign, and the education of consumers as to the benefits of "environment-friendly" products.

Recent years have witnessed a surge of interest in both waste recovery and prevention technologies. Indeed, there is now no shortage of examples of the successful adoption of recycling and waste prevention initiatives and programmes, a selection of which are briefly reviewed below. The very success of such programmes has, however, raised questions of how to measure national and/or industry's progress with respect to pollution and waste prevention and, in particular, how to assess the effectiveness of so-called "cleaner technologies" on waste prevention. Although large quantities of data have been collected over the last decade by national and local governments, as well as by private industry, the existing data bases on material throughputs, waste generation and discharges to environmental media (air, water and soils) are generally inadequate to support the assessment of pollution prevention progress. Even relatively sophisticated data systems, such as the US EPA's TRI (Box 8.1) are considered unable to provide a measure of waste prevention in their present form (Freeman et al., 1992). In order to generate the appropriate data, an expansion of data systems such as the TRI into materials accounting surveys, which incorporate data on material inputs and outputs, waste generation, discharges to air water and soil and recycled materials, is advocated by some researchers (Freeman et al., 1992).

Recovery and Recycling

If they can be used profitably, some waste products can accrue real economic value, for example:

○   By reusing a recovered product, as it is, for the same purpose it was originally used for (e.g., glass milk bottles);

○   By using a recovered product in a production cycle other than that from which it came; or

○   By putting a recovered product back into the production cycle it came from (e.g., aluminium beverage cans).

Recovery and recycling schemes tend to be most advanced for commodities such as paper, glass, aluminium and some other metals. National data reflecting progress in recycling of paper, glass and aluminium over the last 20 years or so have been included in previous editions of this report (UNEP, 1991a). It should be noted for several commodities national recycling figures sometimes include the recycling of production off-cuts within the factory; in such cases the level of post-consumer recycling is thus somewhat lower.

Table 8.9 shows the level of recovery of waste paper and glass (in tonnes and as a percentage of total consumption) achieved by selected countries in the most recent year for which data are available. The re-use of fly ash generated by coal-fired power stations (and incinerators) to make bricks and other construction materials is a long-standing practice in many countries. Estimates of fly ash re-use, expressed as a percentage of total production have thus also been summarized in Table 8.9. Re-utilization rates vary markedly between countries, and range from less than 10 per cent to as much as 90 per cent.

Recycling data for selected metals (both ferrous and non-ferrous) indicate that recycling, as a source of secondary raw materials, is generally growing in importance. This trend can be attributed, at least in part, to concerns about depletion of mineral resources and environmental quality. According to the results of the IMO's GWI, over 80 per cent of OECD countries surveyed had facilities for non-ferrous metal recycling; facilities are also available in other regions, but to a lesser extent (IMO, 1992). Within the European Community recovery rates (i.e., domestic recovery expressed as a percentage of total consumption) have reached levels of around 20 per cent for aluminium, and over 40 per cent for lead (see Figure 7.3). Metals recycling is discussed in greater detail in Part 7: Industry and Transport.

Whilst information on the total amount of the above materials collected and reused is available at national level, it is often difficult to distinguish the extent of materials recycling from the industrial and municipal waste streams. The OECD reports that the average recovery rate of paper and board from municipal wastes in the OECD region as a whole increased from 27 per cent in 1975 to 34 per cent in the late 1980s; for glass, the average recovery rate rose from 22 per cent in 1980 to 32 per cent in the late 1980s (OECD, 1991). Figure 8.7 shows the growth in the USA in the level of recovery of various materials from the municipal waste stream.

The range of commodities being considered for recycling is increasing. For instance, the selective collection and recovery of plastics, solvents, textiles and batteries is being encouraged in many countries, particularly those in the industrialized world. However, few data on the level of recycling are available for these materials. As indicated by the summary data presented below, recovery facilities in these higher-technology

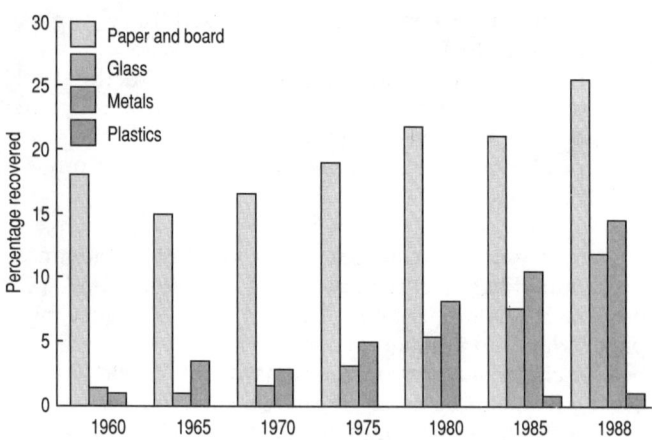

FIGURE 8.7 Recovery of selected materials from municipal solid waste in the USA, 1960–1988
Source: Kaldijan, 1990

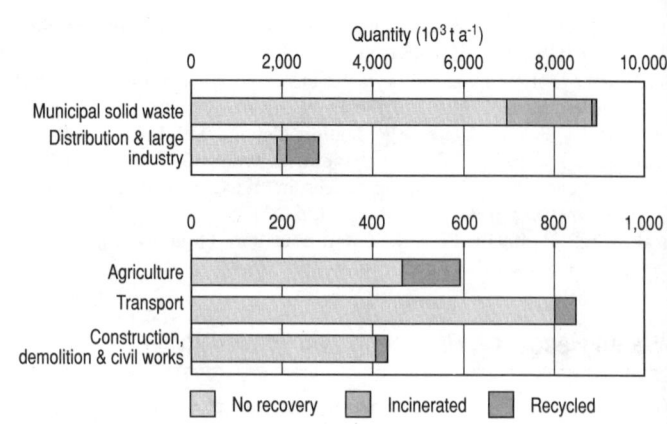

FIGURE 8.8 Recovery of post-user plastics by industry sector in the European Community, 1990
Source: After PWMI, 1992

areas are more developed in the OECD countries compared with those in Eastern Europe and the rest of the world:

Region	Respondents reporting facilities (%)		
	Battery recycling	Oil recycling	Solvent recycling
OECD	63	82	75
East Europe	0	34	17
Rest of world	22	38	12

Source: IMO, 1992

Qualitative information on the availability of recycling facilities for batteries (lead-acid), solvents and oil are reported by the IMO's GWI.

The level of plastics recycling in the European region has been assessed by the Brussels-based European Centre for Plastics in the Environment (PWMI). Based in a compilation of data provided by the European plastics industry, the PWMI has estimated that of the total plastics waste arisings generated by Western European countries in 1990 (13.6×10^6 t), almost 1×10^6 t (or 7 per cent) is reclaimed and re-used through mechanical recycling processes and a further 2.1×10^6 t (or 16 per cent) through thermal (i.e., energy) recovery (PWMI, 1992). Analysis of plastics recovery by sector indicates that the large industry/distribution and the agricultural sectors are the most developed in this respect, achieving recovery rates (through mechanical means) of around 20 per cent. Mechanical recovery rates for plastics from the municipal waste stream are much lower, typically around 1 per cent (Figure 8.8). These differences in recovery rates are largely related to the types of plastics used in the different sectors and ease of separation (PWMI, 1992).

Waste Exchange

An interesting idea to assist in the recovery of specific waste streams from industry is the use of "waste exchanges". Waste exchange schemes offer a service to industry which enables a variety of "wastes" (including process residues) to be used by another industry as a secondary raw material. They are particularly useful in informing potential users of the availability of "wastes" where no well-developed commercial recycling infrastructure exists (Tron, 1990).

According to the results of the IMO GWI, some 65 per cent of OECD member countries who responded to the survey reported ongoing waste exchange initiatives. In Eastern Europe less than 20 per cent of respondents claimed use of waste exchange services. In the rest of the world this proportion was even lower, only 12 per cent (IMO, 1992). Individual companies engaged in waste exchanges have reported marked increases in the number of exchanges handled in recent years; for example, a UK-based company has experienced a fourfold increase in the tonnage of wastes exchanged between 1991 and 1992; preliminary data for 1993 suggest a continuing upwards trend (Trayer, 1993).

Waste Prevention

Waste prevention at source is universally acknowledged as the best way to manage waste. Although the desirability of this approach to waste management has been widely accepted since the mid-1970s, the move from theory to practice has been slow. In more recent years there have, however, been encouraging signs of a switch in emphasis from control to prevention as a waste management strategy, both internationally within multi-national organizations and within individual countries and industries. The driving force behind this change in emphasis is the realization that technological transformation, i.e., more efficient production, is the primary strategy for reconciling the environment and economic development.

In recent years the term "cleaner production" has come to embody the principles of integrated preventative waste management strategies. The cleaner production philosophy has been adopted as a key component of environmental policy in most developed countries and, increasingly so, in the developing world. For industrial production this means adapting processes to:

O Conserve energy and raw materials;
O Reduce the use of toxic or environmentally harmful substances; and
O Reduce the quantity and toxicity of all wastes and pollutant discharges.

For individual products the cleaner production strategy involves reducing impacts along the entire life cycle of a product, from raw material extraction to ultimate disposal of the product. Life cycle analyses (LCAs) which have been widely developed as cleaner production assessment tools are briefly reviewed in Part 7: Industry and Transport.

During the last few years there has been a proliferation of programmes and industry initiatives based on the cleaner production approach. In some cases, government has provided the impetus for adopting cleaner production technologies by offering incentives and financial backing for the development of low-waste processes. Some countries, for example, Denmark, Finland, France and the Netherlands, have embraced the cleaner production philosophy to the extent that it is now regarded as a means of making their businesses more competitive, while at the same time reducing waste production (OECD, 1991). Efforts to implement cleaner production are not confined to the industrialized world; cleaner production technologies are increasingly finding applications in developing countries. Over 50 per cent of IMO's waste survey respondents, including 20 in the developing world, reported ongoing waste reduction initiatives in their countries (IMO, 1992).

One of the examples of the successful adoption of cleaner production given in Box 8.4 is the phasing out of ozone-depleting substances in favour of more environmentally-friendly technologies. According to UNEP's Technology and Economic Assessment Panel it is technically feasible, with existing technologies, to phase out virtually all consumption of the Montreal Protocol-controlled CFCs and halons by 1995–1997, at least in the developed countries. The barriers to elimination of controlled substances are considered to be not so much technical, or even economic, as administrative and informational (UNEP, 1991b). The dissemination of information and the transfer of technology, particularly to developing countries, is thus identified as an important area for action if Montreal Protocol phase-out dates are to be met.

The above example serves to illustrate the fact that the transition towards more efficient and sustainable technologies is not necessarily impeded by a lack of technical know-how. Indeed, studies of major sectors of the economy have demonstrated that technologies which have the capacity to reduce environmental problems while maintaining or even raising economic productivity already exist (Heaton et al., 1991). However, the present mix of regulation and economic market forces has done little to encourage the adoption of new, cleaner production technologies. The continued reliance on best available technology, for example, has tended to entrench existing control technologies, and has not provided the incentive to do better than standards dictate. The challenge

is thus to find the correct balance of regulation, economic incentives and support to industry which will further encourage the increased use of cleaner production. As noted above, the dissemination of information and technology transfer also have key roles to play.

In response to the need for information on cleaner production technologies, including those for CFC and halon replacements, the last decade has seen a growth in the number of data bases on these issues, at both the national and international level. As part of its Cleaner Production Programme (established in 1989), UNEP's IE/PAC currently operates an "International Cleaner Production Information Clearing House (ICPIC)", a computerized information system which comprises:

O Descriptions of almost 650 technology and programme case studies, each highlighting the wastes involved;
O Detailed descriptions of country and corporate cleaner production programmes; and
O A bibliographic data base containing over 1,500 references to publications dealing with cleaner production.

In order further to foster the adoption of cleaner technology world-wide, the Cleaner Production Programme also provides training and technical assistance in the form of workshops and seminars for government, industry and academia and, more recently, has set up a number of National Cleaner Production Centres (a joint programme with the United Nations Industrial Development Organization (UNIDO)). A variety of technical publications on various aspects of cleaner production are also produced.

More recently, UNEP's IE/PAC has developed an information clearing house specifically to provide a mechanism for the transfer of information on policy and technical options for the phasing-out of controlled ozone-depleting substances (ODS). The Ozone Action Information Clearing House (OAIC) is structured in the same way as the ICPIC, in that it contains information on alternative technologies, national and corporate programme summaries, a data base of ODS reduction products and services, a directory of experts and a bibliographic data base.

References

Freeman, H., Harten T., Springer, J., Randall, P., Curran, H. A. and Stone, K. 1992 Industrial pollution prevention: a critical review, *Journal of the Air and Waste Management Association*, **42**(5), 618–656.

Heaton, G., Repetto, R. and Sorbin, R. 1991 *Transforming Technology: An Agenda for Environmentally Sustainable Growth in the 21st Century*, World Resources Institute, Washington DC.

IMO 1992 *Final Report: Global Waste Survey: Task 1 Global Waste Inventory*, LCD.2/Circ.305, International Maritime Organization, London.

Kaldijan, P. 1990 *Characterisation of Municipal Solid Waste in the United States: 1990 Update*, US Environmental Protection Agency, Washington DC.

OECD 1991 *The State of the Environment*, Organisation for Economic Co-operation and Development, Paris.

PWMI 1992 *Overview of Plastics Recycling in Europe*, European Centre for Plastics in the Environment, Brussels.

BOX 8.4 Approaches to Cleaner Production – Selected Examples

The scientific literature is full of examples of successfully applied cleaner production technologies. The approaches adopted are wide-ranging, encompassing:

○ The development of new and more efficient low-waste processes;
○ Changing production processes to optimize performance and reduce waste generation;
○ The use of alternative less toxic chemicals in production processes, and;
○ The development of more environmentally-friendly products.

The diversity in approaches to cleaner production is illustrated by specific examples listed in the accompanying table.

Nowhere has industry demonstrated its ability to react to environmental concerns more than in the case of the development of technologies leading to the elimination of ozone-depleting substances. In recent years considerable progress has been made in developing substitute chemicals and alternative technologies in order to meet the terms of the Montreal Protocol (see Part 10: International Co-operation). As a result of these efforts a significant reduction in the world-wide production and consumption of selected CFCs and halons has been achieved; CFC consumption is currently 50 per cent below that in 1986, the baseline year for reduction targets (see also Part 1: Environmental Pollution, Atmosphere).

These reductions in CFC consumption have been made by using hydrocarbons as aerosol propellants and foam blowing agents; by switching to aqueous and semi-aqueous systems, alcohols and other solvents for cleaning and degreasing applications; by developing "no-clean" technologies and by making use of ammonia and HCFCs (CFC replacement chemicals with lower ozone-depleting potentials) for refrigeration and air conditioning devices. Recovery and recycling of CFCs in refrigeration units is also actively being pursued. Since 1991 reductions in halon consumption have also been achieved, largely by developing alternative methods of fire protection in new facilities and through improved regulation of releases made during training and equipment testing (UNEP, 1991).

References

IE/PAC 1993 *Cleaner Production Worldwide*, United Nations Environment Programme, Industry and Environment Programme Activity Centre, Paris.
UNEP 1991 *Synthesis of the Report of Technology and Economic Assessment Panel*, Report prepared by the Assessment Chairs for the Parties to the Montreal Protocol, United Nations Environment Programme, Nairobi.

Selected Examples of the Application of Cleaner Production Technologies

Country	Industry/process	Cleaner production	Advantages
France	Metal processing – steel galvanizing	Development of a new process for zinc coating of steel	Total suppression of conventional plating waste; lower operating costs and improved quality control
India	Textile – sulphur black colour dyeing using sodium sulphide to convert sulphur dyes	Substitution of sodium sulphide, with hydrol a by-product of the maize starch industry	Reduction of sulphide in the effluent; less corrosion in the treatment plant and reduction of odour problems
Indonesia	Cement production	Optimization of kiln operation and improved process control through the installation of expert control systems	Some reduction in NO_x and SO_2 emissions, energy savings and reduction in, the quantities of off-specification material
Poland	Automobile component manufacture – Cu/Ni/Cr and Zn electro-plating of aluminium alloy and steel components	Modification of the rinsing systems and addition of facilities which permit water recycling and raw materials recovery	A decrease in water and raw materials consumption; reductions in contaminant levels in waste-water streams of between 80–98%
Sweden	Metal fabrication – de-greasing of metal sections using trichloroethylene	A switch to an alkaline de-greasing procedure which utilizes biodegradable cutting oils	Zero emission of trichloroethylene and reduction in the quantity of hazardous waste sludge produced
UK	Adhesives	Development of a new range of water-based adhesives in place of solvent-based ones	Compared to solvent-based adhesives, water-based ones are less toxic and require no special handling yet still offer a wide range of applications

Source: IE/PAC, 1993

Trayer, D. 1993 Personal Communication, Waste Exchange Service Limited, Stockton-on-Tees, UK.
Tron, R. 1990 *The Role of 'Waste Exchange' in Waste Utilization*, Report Number LR 742 (MR), Warren Spring Laboratory, Stevenage, UK.

UNEP 1991a *United Nations Environment Programme Environmental Data Report 1991/92*, Basil Blackwell, Oxford.
UNEP 1991b *Synthesis of the Report of Technology and Economic Assessment Panel*, Report prepared by the Assessment Chairs for the Parties to the Montreal Protocol, United Nations Environment Programme, Nairobi.

TABLE 8.1   Quantities of solid waste generated in selected countries by source category, most recent available year (10^3 t a^{-1})

Region/country	Year of data	Municipal	Industrial	Agricultural	Mining/ quarrying	Sewage sludge	Energy production	Demolition/ construction	Source(s)
AFRICA									
Côte d'Ivoire	1991		100[a]						1
Egypt	1990	5,600	200–1,000[b]	20,600[c]					2
Namibia	1990	44.4[d]							3
Tunisia	1991	1,200[e]	100[f]						4
NORTH AMERICA									
Canada	1989	16,400			10,529	500			5
Dominica	1990		2[g]						6
USA	1986	208,760	760,000[h]	150,566	14,000	10,400	99,247	31,500	7
ASIA									
China	1990		577,970						8
China, Taiwan	1990	0.96[i]	30,000[j]						9
Hong Kong[k]	1990	5,851.2[l]	1,879[m]					8,442.9	10
Japan	1988	48,283	60						5
Korea[n]	1988	78							11
Pakistan	1978	1.05[o]							12
Singapore	1991	1,247	903			150			13
Turkey	1989	19,500							5
EUROPE									
Austria	1988	2,700							5
Belgium	1988	3,470[p]	26,700[q]			687[r]	1,069[r]	680[r]	5,14
Bulgaria	1987							669.5	7
Czechoslovakia	1987	2,600	43,800	663.9	787,616			23,600	15
Denmark	1985	2,400	2,300		507,600[s]	1,263	1,532	1,203	16
Finland	1989	3,100	11,500	23,000	21,600	1,000		5,000–10,000	17
France	1989	17,000	50,000	400,000	10,000[t]	620			5,14
Germany, Fed. Rep.[u]	1987	21,000[v]	72,378		9,733	1,750	13,026	110,887	5,14,18
Greece	1989	3,147	4,304	90	3,900		7,680		5,14
Italy	1989	17,300	39,978			3,500		34,374	5,14
Luxembourg	1990	170	1,300			15		4,000	5,14
Netherlands	1988	6,900	6,687	86,000[w]		326	1,482	7,700	5,14
Norway	1989			18,000	9,000	100		2,000	5,7,19
Poland	1989	2,070[x]	1,637,900	98,700	150,418				7

Continued

TABLE 8.1 Continued

Region/country	Year of data	Municipal	Industrial	Agricultural	Mining/quarrying	Sewage sludge	Energy production	Demolition/construction	Source(s)
Portugal	1987	662		202					5,14
Spain	1988	12,546	5,108	45,000					5,14
Switzerland	1989	2,850					164[y]	3,000	7
UK	1990	20,000[z]	56,000	80,000[aa]	107,000[ab]	260	13,000	32,000[ad]	20
USSR	**1987**	**48,340**		**2,155.9**	**51,079.7**	**1,000[ac]**		**34,355.7**	**21**

a Extrapolated from known arisings of 70,000 t in Abidjan.
b Excludes bulk amounts of non-hazardous wastes from cement, steel and mining industries.
c Data refer to crop residues only.
d Data comprise 37,000 t of municipal solid waste from Windhoek and about 7,500 t from 15 small towns.
e Data refer to the quantity of municipal solid waste collected from the urban population.
f Data refer to quantities of industrial waste generated in Tunis only.
g Data refer to wastes generated from the processing of grapefruits and limes only.
h Includes wastewaters which meet the US definition of solid waste.
i Data are in units of kilograms per capita per day.
j 1985.
k Figures in tonnes per day.
l Data refer to household, public cleansing and commercial wastes.
m Estimate comprises 1,605 t day^{-1} collected as part of municipal wastes and a further 274 t day^{-1} of chemical wastes (the latter is a daily average assuming an annual generation of 100,000 t).
n Data refer to the average daily quantity of wastes generated.
o Wastes generated in Karachi only in grams per capita per day.
p Data include 1982 figures for the Brussels region, 1987 figures for the Flanders region and 1989 figures for the Walloon region.
q Data include 1986 figures for the Brussels region, 1987 figures for the Flanders region and 1989 figures for the Walloon region.
r Wastes generated in the Flanders region only.
s Wastes arising from the mining of brown coal only.
t Includes demolition waste.
u Provisional figures.
v Data include household wastes, commercial wastes similar to household wastes and bulky wastes collected by public refuse collection services.
w Data refer to the amount dumped at sea and include manure generated by housed livestock in intensive husbandry.
x Includes some production waste, car wrecks and bulky wastes. 1989.
y
z Data refer to household waste only and represent typical arisings for any 12-month period in recent years.
aa Data refer to the production of manure from housed livestock only (wet weight).
ab Excludes waste from open cast coal mining.
ac Dry weight.
ad Includes road planings.

Data on solid wastes presented here represent official country statistics and have been extracted from State of the Environment-type reports. Definitions of solid waste categories vary between countries; inter-country comparisons should therefore be made with caution.

Sources:
1. Ministry of the Environment, Construction and Urbanization 1992 National Report on the State of the Environment in Côte d'Ivoire, Ministry of the Environment, Construction and Urbanization, Côte d'Ivoire (in French).
2. Government of the Arab Republic of Egypt 1992 Environmental Action Plan, Egyptian Environmental Affairs Agency, Cairo.
3. Ministry of Wildlife, Conservation and Tourism 1992 Namibia's Green Plan (Draft), Ministry of Wildlife, Conservation and Tourism, Windhoek.
4. National Environmental Protection Agency 1992 Tunisia National Report to the United Nations Conference on Environment and Development, National Environmental Protection Agency, Tunis.
5. OECD 1991 OECD Environmental Data Compendium 1991, Organisation for Economic Co-operation and Development, Paris.
6. The Government of the Commonwealth of Dominica 1991 Dominica Country Environmental Profile, The Caribbean Conservation Association, St Michael.
7. UN ECE 1992, The Environment in Europe and North America: Annotated Statistics 1992, United Nations, New York.
8. State Statistical Bureau of the People's Republic of China 1991 China Statistical Yearbook 1991, China Statistical Information and Consultancy Service Centre, Beijing.
9. Environmental Protection Administration 1991 Current Measures and Strategies of Environmental Protection in Taiwan, the Republic of China, The Environmental Protection Administration, Taipei.
10. Environmental Protection Department 1991 Environment Hong Kong 1991: A Review of 1990, Environmental Protection Department, Hong Kong.
11. Ministry of Environment 1991 National Report of the Republic of Korea to UNCED 1992, International Affairs Office, Ministry of Environment, Korea.
12. Federal Bureau of Statistics 1987 Environment Statistics of Pakistan 1986, Manager of Publications, Karachi.
13. Ministry of the Environment 1992 Annual Report 1991, Public Affairs Department, Ministry of the Environment, Singapore.
14. EUROSTAT 1992 Environment Statistics 1991, Statistical Office of the European Communities, Luxembourg.
15. Ministry of the Environment of the Czech Republic and Czechoslovak Academy of Sciences 1990 Environment of the Czech Republic Part II, Czech Ministry of Environment, Prague.
16. Denmark's Statistics 1990 Figures on Nature and the Environment, Denmark's Statistics, Copenhagen (in Danish with English titles).
17. Ministry for Foreign Affairs 1991 Finland National Report to UNCED 1992, Ministry of Environment, Helsinki.
18. Federal Environment Agency 1991 Facts and Figures on the Environment of Germany 1988/89, Federal Environmental Agency, Berlin.
19. Central Bureau of Statistics 1992 Natural Resources and the Environment 1991, Central Bureau of Statistics, Oslo (in Norwegian with English titles).
20. Department of the Environment 1992 Digest of Environmental Protection and Water Statistics No. 14 1991, HMSO, London.
21. Data supplied by the Institute of Global Climate and Ecology, Russian Academy of Sciences, Moscow.

TABLE 8.2 Generation and composition of municipal/household wastes, most recent year available

Region/country	Annual arisings			Composition (% of total weight)							
	Year of data	Total (10^3 t)	Per capita (kg day^{-1})	Year of data	Paper and cardboard	Plastics	Glass	Metals	Other	Organic	Source(s)
AFRICA											
Egypt[a]	1990	5,600	0.3–0.8	1980	10	1	2		87		1,2
Namibia[b]	1990	44,484		1991	23	11	24	13		29	3
Nigeria[c]			0.05	1980[d]							2,4
Tunisia[e]		1,200	0.48		15			5	80		5
NORTH AMERICA											
Canada	1989	16,000	1.71	1989	36.5	4.7	6.6	6.6	45.7		6,7
Mexico[f]		40.8									8
USA	1990	293,613	3.23	1988	40	8	7	8.5	36.5		7,9
ASIA											
China	1987		1[g]						60[h]		8
Hong Kong	1990	15.9[i]	0.9[j]								10
Israel[k]	1986		1.5	1986	16–21	10–12	3–5	3–5		50–54	11
Japan	1988	48,283	1.08	1988[l]	45.5	8.3	1[m]	1.3	43.9	77.2[n]	6,7
Korea	1989	78[f]	2.22	1989	12.3			4.8	82.9		12
Pakistan	1978		1.05	1978	15	2			74		13
Singapore	1991	1,247	1.1[c]				4	5			14
Thailand				1985	18.4	14.5[o]	6[p]		60.1		15
Turkey	1989	19,500	0.97								6,7
EUROPE											
Austria	1988	2,700	0.97	1986[q]	33.6	7.0	10.4	3.7	45.3	60.5[r]	6,7
Belgium	1989	3,470	0.96	1989	28.3	7.7	7.6	3.7	52.7	47.6	6,7
Finland	1989	2,500	1.4	1985[j]	40	8	4	3	45	85	6,7
France	1989	17,000	0.83	1989[j]	27.5	4.5	7.5	6.5	54.0	59	6,7
Germany, Fed. Rep.	1987	19,483	0.87	1985[j]	17.9	5.4	9.2	3.2	64.3	63.4	6,7
Greece	1989	3,147	0.71	1989	20	7	3	4	66	57	6,7
Italy	1989	17,300	0.82	1986	22.3	7.2	6.2[s]	3.1	61.2[s]	64.4	6,7
Luxembourg	1990	170	1.27	1985	17.2	6.4	7.2	2.6	66.6	44	6,7
Netherlands	1988	6,900	1.27	1989[j]	24.2	7.1	7.2	3.2	58.3	88.3	6,7
Norway	1989	2,000	1.29	1988	30	5	3	7	55	77	6,7

Continued

TABLE 8.2 Continued

Region/country	Annual arisings			Composition (% of total weight)							
	Year of data	Total (10³ t)	Per capita (kg day⁻¹)	Year of data	Paper and cardboard	Plastics	Glass	Metals	Other	Organic	Source(s)
Spain	1988	12,546	0.88	1988	20	7	6	4	63	49	6,7
Switzerland	1989	2,850	0.87	1989	32	13	7	6	42	70	6,7
UK	1990	20,000		1990	35	7	12	10	36	63	7,16 ✳
USSR	1990	48,340	0.21	1990ᵗ	32.9	3.3	5.1	3.9	54.8		17

a Annual arisings refer to urban wastes only. Compositional data are for Cairo only where arisings were estimated to be 0.6 kg per person per day.
b Annual arisings from urban areas only, comprising 37,000 t from Windhoek and the rest from 15 small towns. Compositional data refer to Windhoek only.
c Data refer to domestic solid waste.
d Compositional data refer to Onitsha only where arisings per person per day in 1980 were estimated to be 0.7 kg.
e Data refer to urban areas only. The per capita figure is an average for Tunisia, but daily arisings range from 0.5 kg in Tunis to 2.5 kg in Sidi Bin Said.
f Estimate is in tonnes per day.
g Data refer to domestic waste from urban dwellers.
h It is estimated that 60 per cent of domestic waste consists of coal cinders.
i Arisings refer to the average daily amount of municipal waste delivered for disposal in tonnes per day.
j Data refer to household waste only.
k Compositional figures relate to domestic waste.
l Compositional data refer to Tokyo only.
m Glass and ceramics.
n The category "Organic" refers to combustible components and water, and is derived from an average for six cities.
o Data include rubber.
p Data include metals.
q Compositional data refer to household waste in Vienna only.
r 1983.
s OECD estimates.
t Compositional data do not include Estonia, Kyrgyzstan, Latvia, Lithuania and Turkmenistan.

The composition of municipal wastes is described here in terms of the percentage contribution of paper and cardboard, plastics, glass, metals, and other materials to the total mass. These five categories total 100 per cent (although in some cases totals may not tally to precisely 100 per cent due to rounding). The category "Other" may include organic materials, kitchen waste, ash and grit, textiles, sweepings and garden refuse. Estimates of the percentage contribution of organic materials to the total mass of municipal wastes are also given, but as a separate category.

Data on municipal household solid wastes presented here represent official country statistics and have been extracted from State of the Environment-type reports. Definitions of the terms municipal solid waste and categories of municipal solid waste may vary between countries; inter-country comparisons should therefore be made with caution.

Sources:

1. Government of the Arab Republic of Egypt 1992 *Environmental Action Plan*, Egyptian Environmental Affairs Agency, Cairo.
2. Cook, D. and Kalbermatten, J. 1982 Prospects for resource recovery from urban solid wastes in developing countries. In: *Reuse of Solid Wastes, Proceedings of a Conference on the Practical Implications of the Reuse of Solid Waste*, London, 11-12 November 1981, Thomas Telford, London.
3. Ministry of Wildlife, Conservation and Tourism 1992 *Namibia's Green Plan* (Draft), Ministry of Wildlife, Conservation and Tourism, Windhoek.
4. Federal Environmental Protection Agency 1991 *Achieving Sustainable Development in Nigeria: National Report for the United Nations Conference on Environment and Development 1992*, Federal Environmental Protection Agency, Lagos.
5. National Environmental Protection Agency 1992 *Tunisia National Report to the United Nations Conference on Environment and Development*, National Environmental Protection Agency, Tunis.
6. OECD 1991 *OECD Environmental Data Compendium 1991*, Organisation for Economic Co-operation and Development, Paris.
7. World Resources Institute 1992 *World Resources 1992–93*, Oxford University Press, New York.
8. Cirillo, R.R., Jones, P.H., Faulstich, M.S., Loong, H., Barnes, D., Chiu, H.H. and Miller, W. 1988 *Hazardous Waste in the Pacific Basin*, Pacific Basin Consortium for Hazardous Waste Research, Honolulu.
9. Kaldjian, P. 1990 *Characterisation of Municipal Solid Waste in the United States: 1990 Update*. Environmental Protection Agency, Washington DC.
10. Environmental Protection Department 1991 *Environment Hong Kong 1991: A Review of 1990*, Environmental Protection Department, Hong Kong.
11. Ministry of the Environment 1992 *The Environment in Israel: National Report to the United Nations Conference on Environment and Development*, Ministry of the Environment, Jerusalem.
12. Jong-In, D. 1990 Hazardous waste management in Korea. Paper presented at the Hazardous Waste Workshop, Kyoto, Japan, 3–7 December 1990. Unpublished paper.
13. Federal Bureau of Statistics 1987 *Environment Statistics of Pakistan 1986*, Manager of Publications, Karachi.
14. Ministry of the Environment 1992 *Annual Report 1991*, Public Affairs Department, Ministry of the Environment, Singapore.
15. Anon 1990 Projected hazardous waste quantities by waste type. Paper presented at the Hazardous Waste Workshop, Kyoto, Japan, 3–7 December 1990. Unpublished paper.
16. Department of the Environment 1992 *Digest of Environmental Protection and Water Statistics No.14 1991*, HMSO, London.
17. Data supplied by the Institute of Global Climate and Ecology, Russian Academy of Sciences, Moscow.

✳ HMIP 1992 Seventeenth report, Incineration of Waste. Abercromby Liby 628.54 R

TABLE 8.3 Generation of hazardous wastes in selected countries by type of waste, most recent year available ($t\ a^{-1}$)

| Region/country | Year of data | Wastes consisting of, or contaminated with | | | | | | | | | | | Source(s) |
		Metals and metal compounds	Strong acids/alkalis	PCBs and halogenated hydrocarbons	Biocides and fine chemicals	Toxic inorganics	Solvents/oils	Toxic organics/intermediates	Resins, paints, sludges	Clinical	Other	Total	
AFRICA													
Côte d'Ivoire[a]	1991	100			200		20,000	3,000	200		2,900		1
Egypt	1990									13,000		63,000	2
Zambia[b]	1989	>50,150		>3	>10		>11,000[c]		>100				3
NORTH AMERICA													
Canada	1985	449,200[d]	19,100	120,000[e]	4,500		629,000		72,700[f]				4
USA	1986	8,762,561[g,h]	2,737,740[i]	5,015,060[h]	12,000		92,749,079[h]		630,000[f]				4
ASIA													
China	1987	114,000[j]			435,000[k]					56,814[m]		36,280,000[l]	5,6,7
China, Taiwan	1987											2,721,000	7
Hong Kong[n]	1988		55,000	1,300	1	806[o]	19,150	162	1,940	20	494		8,9
Israel[p]	1988											88,000	10
Korea[q]	1989		376,000				159,000[c]						11
Malaysia[r]	1987	58,185	29,905	676[s]	386,900[t]	109,268	26,131		36,885	3,476	34,348		7,12
Philippines		20–40[u]							80–150[v]				7,12
Singapore[r]	1989		1,415				18,280			12,780			7,12
Sri Lanka[r,w]		280		3		345	277	43			1,395		13
Thailand	1986	823,870	103,000	2,460		11,700	143,970	4,050		46,670	8,830		12,14
EUROPE													
Austria	1988	28,026	2,225	5,225	2,038.7		76,824		12,393	6,598	11,015		4,15
Czechoslovakia	1985											320,000	16
Finland	1987	2,070	185,400	109,293	388	766	110,663	950	16,389	2	496,203		4,15
Germany Fed. Rep.[x]	1987		1,207,040	147,740		241,230[y]	515,174		145,075	22,722	448,675		4,15,17
Italy	1991									140,800		3,246,000	4,18
Norway	1991	1,099	876	244	16	787	58,992	987	2,647	22	8	65,681	4,15,19
Portugal	1987		49	756			3,088		71.9				4,15
Switzerland	1989			1,190	500		190,000		11,000	1,520	6,600		15
USSR	1990	302,066,054[z]			566,696[aa]								16

Continued

TABLE 8.3 Continued

Region/country	Year of data	Metals and metal compounds	Strong acids/alkalis	PCBs and halogenated hydrocarbons	Biocides and fine chemicals	Toxic inorganics	Solvents/oils	Toxic organics/intermediates	Resins, paints, sludges	Clinical	Other	Total	Source(s)
Wastes consisting of, or contaminated with													
OCEANIA													
Australia	1984	1,700	33,990	1,717.5		30,000							7
Guam	1986									91		386[ab]	7

a Unquantified amounts of PCBs and hospital waste are also generated.
b Unquantified amounts of asbestos, solvents and heavy metal sludges are also generated.
c Data refer to oils only.
d Includes 263,000 t of mercury-contaminated wastes generated in 1989.
e Includes 6,500 t of high level PCB-contaminated wastes currently in storage and awaiting disposal.
f Data refer to waste paint only.
g Includes wastes containing silver and mercury only.
h Figure is based on a survey of hazardous waste generators and, therefore, is likely to be an underestimate of the total amount generated nationally.
i Data comprise concentrated acids only.
j Estimated arisings of chromium wastes at 18 plants in Beijing.
k Data refer to quantities of selected pesticides (including DDT) that have been placed in storage after restrictions were placed on their use.
l Estimate only.
m 1983.
n Excludes substances under MARPOL Annexes I and II.
o Includes 626 t of asbestos.
p Estimate based on the assumption that the total annual production of hazardous wastes is twice the amount received at the Ramat-Hovav site (i.e., 44,000 t a^{-1}).
q Data refer to all industrial waste arisings, not just hazardous wastes, within the categories given.
r Data are given in cubic metres per annum.
s Data are given in litres per annum.
t Data refer to pesticides only.
u Estimate refers to "toxic wastes" which are defined as acids, alkalis, heavy metals plus oxidizing and reducing agents. Data are given in millions of litres.
v Estimate refers to "hazardous waste" which are defined as chemical sludges, asbestos, oil, grease, paints, dyes, etc. Data are given in millions of litres.
w Data refer to wastes generated by industrial sources only.

x Provisional estimate only.
y Data refer to salt slags which contain aluminium.
z Figure is based on a survey of 10,322 enterprises and includes wastes containing cadmium, cobalt, arsenic, nickel, lead and mercury compounds.
aa Figure is based on a survey of 10,322 enterprises and refers to banned agrochemicals and pesticides, and benzopyrene.
ab Includes 91 t of clinical waste and a further 295 t of other hazardous wastes generated at military facilities on the island.

Sources:
1. Ministry of the Environment, Construction and Urbanisation 1992 *National Report on the State of the Environment in Côte d'Ivoire*, Ministry of the Environment, Construction and Urbanisation, Côte d'Ivoire (in French).
2. Government of the Arab Republic of Egypt 1992 *Environmental Action Plan*, Egyptian Environmental Affairs Agency, Cairo.
3. Uosukainen, J. 1990 Estimates of hazardous wastes in Zambia. Paper presented at the Hazardous Wastes Workshop, Kyoto, Japan, 3–7 December 1990. Unpublished paper.
4. OECD 1991 *Environmental Data Compendium 1991*, Organisation for Economic Co-operation and Development, Paris.
5. Shi Qing and Cheng Boxing 1987 Management of hazardous materials and wastes in China. Paper presented at the Workshop on Risk Assessment of Hazardous Chemicals in Developing Countries, East-West Center, Honolulu, Hawaii. Unpublished paper.
6. Cheng Boxing, Shi Qing and Wu Tianbac 1987 Research on management and strategies for chromium containing wastes in China. Paper presented at the Workshop on Risk Assessment of Hazardous Chemicals in Developing Countries, East-West Center, Honolulu, Hawaii. Unpublished paper.
7. Cirillo, R. R., Jones, P. H., Faulstich, M. S., Loong, H., Barnes, D., Chiu, H. H. and Miller, W. 1988 *Hazardous Waste in the Pacific Basin*, Pacific Basin Consortium for Hazardous Waste Research, Honolulu.
8. Environmental Protection Department 1989 ISWA Country Paper – Hong Kong. Paper presented at the ISWA Working Group Meeting on Hazardous Wastes, September 1989, Honolulu, Hawaii. Unpublished paper.
9. Environmental Protection Department 1991 *Environment Hong Kong 1991: A Review of 1990*, Environmental Protection Department, Hong Kong.
10. Ministry of the Environment 1992 *The Environment in Israel: National Report to the United Nations Conference on Environment and Development*, Ministry of the Environment, Jerusalem.
11. Jong-In, D. 1990 Hazardous waste management in Korea. Paper presented at the Hazardous Waste Workshop, Kyoto, Japan, 3–7 December 1990. Unpublished paper.
12. Anon 1989 Paper presented at the Workshop on Treatment and Storage of Hazardous Wastes, Bangkok, Thailand, 24–28 April 1989. Unpublished paper.
13. Andradi, A., Ellapola, R. and Mathes, H. 1990 Hazardous waste situation in Sri LAnka. Paper presented at the Hazardous Wastes Workshop, Kyoto, Japan, 3–7 December 1990. Unpublished paper.
14. Anon 1990 Projected hazardous waste quantities by waste type. Paper presented at the Hazardous Waste Workshop, Kyoto, Japan, 3–7 December, 1990. Unpublished paper.
15. UN ECE 1992 *The Environment in Europe and North America: Annotated statistics 1992*, United Nations, New York.
16. Ministry of Environment of the Czech Republic and Czechoslovak Academy of Sciences 1990 *Environment of the Czech Republic: Part II*, Czech Ministry of Environment, Prague.
17. Federal Environmental Agency 1991 *Facts and Figures on the Environment of Germany 1988/89*, Federal Environmental Agency, Berlin.
18. Ministry of the Environment 1992 *About the State of the Environment*, Ministry of the Environment, Italy (in Italian).
19. Central Bureau of Statistics 1992 *Natural Resources and the Environment 1991*, Central Bureau of Statistics, Oslo (in Norwegian with English titles).

TABLE 8.4 Known movements of hazardous wastes in selected countries (10^3 t a^{-1})

Region/country	Year	Imports	Exports	Source	Region/country	Year	Imports	Exports	Source
NORTH AMERICA					German Dem. Rep.	1988	1,150[d]		4
					Germany, Fed. Rep.	1988	20.2	805.4	1
Canada	1980	120	101	1	Ireland	1988		4,000	4
USA	1988	40	127	1	Italy			34.6[e]	5
					Luxembourg	1988		2.2	4
ASIA									
					Netherlands	1987		177	3
Hong Kong	1991		1,280[a]	2	Norway	1989		16.6	3
Japan	1985		0.04	1	Spain	1987		0.1	1
					Sweden	1988		30.2[f]	1
EUROPE					Switzerland	1989		112.3	3
					UK[g]	1990/91	44.1	0.525	6
Austria	1989	8.5	91.5	3					
Belgium	1989	54.7	15.1	4	**OCEANIA**				
Denmark[b]	1989	27	9	3					
Finland[c]	1990	16.3	19.2	3	Australia	1980		0.3	1
France	1988	250	45	1	New Zealand	1982		0.2	1

[a] Data refer to exports of non-hazardous waste for recycling.

[b] Import and export figures cover only hazardous waste covered by EEC Directive 84/631. Export figures do not include non-ferrous metals to other EEC countries (Article 17).

[c] Of the reported imports, 4,493 t, and of the exports, 19,133 t, are for reuse.

[d] Greenpeace reports that there is evidence of further substantial amounts of toxic waste imports.

[e] Quantity exported between 1 June 1989 and 31 December 1991.

[f] Data refer to selected hazardous wastes only.

[g] Data refer to the period 1 April 1990 to 31 March 1991 and are compiled under the 1988 Transfrontier Regulations. Imports do not include movements of hazardous non-ferrous metals wastes for recycling as the Regulations do not require these to be notified to central government.

Sources:

1. OECD 1991 *OECD Environmental Data Compendium 1991*, Organisation for Economic Co-operation and Development, Paris.
2. Environmental Protection Department 1991 *Environment Hong Kong 1991: A Review of 1990*, Environmental Protection Department, Hong Kong.
3. UN ECE 1992 *The Environment in Europe and North America: Annotated Statistics 1992*, United Nations, New York.
4. Vallette, J. and Spalding, H. (Eds) 1990 *The International Trade in Wastes: A Greenpeace Inventory*, (5th Edition), Washington DC.
5. Ministry of the Environment 1992 *About the State of the Environment*, Ministry of Environment, Italy (in Italian).
6. Department of the Environment 1992 *Digest of Environmental Protection and Water Statistics No. 14 1991*, HMSO, London.

TABLE 8.5 Treatment and disposal of municipal wastes in selected countries, most recent year available

Region/country	Annual arisings Year of data	Annual arisings (10^3 t)	Disposal Year of data	Landfill (10^3 t)	Incineration (10^3 t)	Incineration with energy recovery (%)	Composting/recovery (10^3 t)	Other (10^3 t)	Source(s)
AFRICA									
Egypt	1990	5,600		a	b		0.044[c]	1.7[d]	1
Tunisia	1991	1,200		a			e		2
NORTH AMERICA									
Canada	1989	16,000	1989	13,488	1,416[f]	7.1		242[f]	3
USA	1990	293,613	1988	130.6	25.5	96	23.5		4,5
ASIA									
Hong Kong	1991	7.4		5.2	2.2				6
Israel	1986	1.5[g]		98[h]					7
Japan	1988	48,283	1987	16,486	32,616	27.4	53	1,454	3
Korea	1989	78[i]		94[h]					8
Malaysia[j]			1984	19.4	4.2			80.5	9
EUROPE									
Austria	1988	2,700	1987	1,836	222	9	473	257	3
Belgium	1989	3,470[k]	1987	1,530	720	29.9		832	3
Denmark	1985	2,400	1985	1,260	540		100	500	3
Finland	1989	3,100	1987	2,000	50	100	l	450	3,10
France	1989	17,000	1989	7,684	6,970	67	1,207	1,139[m]	3
Germany, Fed Rep.	1987	19,483	1987	12,917	5,942		429	195	3
Greece	1989	3,147	1989	3,084	1[f]				3
Italy	1989	17,300	1989	5,286	2,794	21.2	834	5,052	3
Luxembourg	1990	170	1990	51	117	100	2		3
Netherlands	1988	6,900	1988	3,790	2,555	72	345	210	3
Norway	1989	2,000	1989	1,500	400	19	100		3
Portugal	1989	2,650	1989	742			280	1,656	3
Spain	1988	12,546	1988	9,713	604	60.8	2,229		3
Switzerland	1989	2,850	1989	460	2,270	80	90	30	3
OCEANIA									
Australia[n]	1989	12,796	1989	12,274	143	0	0		11

Continued

TABLE 8.5 Continued

a   Landfill practises are largely uncontrolled.
b   Although 80 incinerator plants with capacities ranging from 0.4–1.0 tons per hour exist, few are functioning.
c   Data refer to combined throughput in tons per hour for five composting plants.
d   Amount dumped or burnt on streets in Cairo each day, in tons.
e   About 200 families recover a small portion of the reclaimable material (glass, paper, metals, etc.) at one landfill site which also has a small composting plant.
f   1985.
g   Annual risings are in units of kilograms per person per day.
h   Data refer to the percentage of municipal wastes landfilled.
i   Data refer to the average daily arisings.
j   Disposal data represent the percentages of municipal wastes disposed of by "Landfill", "Incineration", and "Other" disposal options. The category "Other" refers to uncontrolled dumping, including open burning and dumping to rivers and coastal areas.
k   Data comprise 1982 figures for the Brussels region, 1987 figures for the Flanders region, and 1989 figures for the Walloon region.
l   In 1989, 18 per cent of municipal wastes were recovered.

m   "Other" disposal options include methanation and transfer to holding areas.
n   Data refer to the method of waste disposal by councils.

Sources:

1. Government of the Arab Republic of Egypt 1992 *Environmental Action Plan*, Egyptian Environmental Affairs Agency, Cairo.
2. National Environmental Protection Agency 1992 *Tunisia National Report to the United Nations Conference on Environment and Development*, National Environmental Protection Agency, Tunis.
3. OECD 1991 *OECD Environmental Data Compendium 1991*, Organisation for Economic Co-operation and Development, Paris.
4. Kaldjian, P. 1990 *Characterisation of Municipal Solid Waste in the United States: 1990 Update*, Environmental Protection Agency, Washington DC.
5. World Resources Institute 1992 *World Resources 1992–93*, Oxford University Press, New York.
6. Environmental Protection Department 1992 *Environment Hong Kong 1992: A Review of 1991*, Environmental Protection Department, Hong Kong.
7. Ministry of the Environment 1992 *The Environment in Israel - National Report to the United Nations Conference on Environment and Development*, Ministry of the Environment, Jerusalem.
8. Ministry of Environment 1991 *National Report of the Republic of Korea to UNCED 1992*, International Affairs Office, Ministry of Environment, Korea.
9. Cirillo, R. R., Jones, P. H., Faulstich, M. S., Loong, H., Barnes, D., Chiu, H. H. and Miller, W. 1988 *Hazardous Waste in the Pacific Basin*, Pacific Basin Consortium for Hazardous Waste Research, Honolulu.
10. Ministry for Foreign Affairs 1991 *Finland National Report to UNCED 1992*, Ministry of Environment, Helsinki.
11. Castles, I. 1992 *Australia's Environment: Issues and Facts*, Australian Bureau of Statistics, Belconnen.

TABLE 8.6 Treatment and disposal of hazardous wastes in selected countries, most recent year available (10^3 t a^{-1})

Region/country	Year of data	Physical/ chemical treatment	Incineration			Dumping at sea	Landfill	Storage/ containment	Other	Source(s)
			Land	At sea	Biological					
NORTH AMERICA										
USA	1985	254.9[a]	634[b]		68.9	37.1[c]	374	294[d]	84.7[e]	1
ASIA										
Hong Kong[f]	1991		12				8			2,3
Korea[g]	1989		1.9				17	4.5	34.3[h]	4
EUROPE										
Austria	1989	150	71.4		7[i]		2.5[j]			1
Denmark	1989	10	98.9							1
Finland	1987	205	31.4		0.2		44.9	1.3		1
Netherlands	1986	300	200			672[k]	275		35.3	1
Norway	1989	20.5	20.5							1
Portugal	1987		0.007				0.2	0.8		1
Switzerland	1989	122	177				67			1
UK[l]	1989	260	80				1,300			1
USSR	**1990**			5.5		160		564.1[m]	8.8[n]	5

a Includes 2,387 t sent for solidification.
b Includes 14,694 t sent for thermal treatment.
c Data refer to the neutralization of waste hydrochloric acid.
d Data refer to surface impoundment and underground injection.
e Includes recovered items (79,202 t).
f Data refer to clinical wastes only, and are in units of tonnes per day.
g Data are in units of tonnes per day.
h Includes recycled materials (31,064 t a^{-1}).
i 1988.
j Data refer to the landfilling of mercury and pesticide wastes only.
k Data refer to waste from foreign industries only, dumped at sea under Dutch jurisdiction in 1987.
l Data refer to 'Special' wastes which are defined as any substance which is dangerous to life by the Control of Pollution Act (Special Waste Regulations 1980). This includes drugs available only on prescription. Data on physical and chemical treatment include 17,000 t sent for solidification.
m Quantity includes 40 t sent to unregulated storage and 750 t stored in underground reservoirs.
n Comprises 6,792 t of reprocessed material and 2,049 t of material which has been "destroyed".

Sources:
1. UN ECE 1992 The Environment in Europe and North America: Annotated Statistics 1992, United Nations, New York.
2. Environmental Protection Department 1992 Environment Hong Kong 1992: A Review of 1991, Environmental Protection Department, Hong Kong.
3. Cirillo, R. R., Jones, P. H., Faulstich, M. S., Loong, H., Barnes, D., Chiu, H. H. and Miller, W. 1988 Hazardous Waste in the Pacific Basin, Pacific Basin Consortium for Hazardous Waste Research, Honolulu.
4. Jong-In, D. 1990 Hazardous waste management in Korea. Paper presented at the Hazardous Waste Workshop, Kyoto, Japan, 3–7 December 1990. Unpublished paper.
5. Data supplied by the Institute of Global Climate and Ecology, Russian Academy of Sciences, Moscow, Russia.

TABLE 8.7 Disposal of sewage sludge, 1980s (10^3 t a^{-1} dry matter[a])

| Region/country | Year of data | Total quantity generated | Disposal by | | | | | Source |
			Landfill	Incineration	Use in agriculture	Dumping at sea	Other	
NORTH AMERICA								
USA[b]	1988	5,570.7[c]	1,745.4[d]	759.8	2,336.7[e]	265.2	321.3[f]	1
EUROPE								
Austria	1989	200	67	74	57	0	0	2
Belgium	1984	29	15	6	8	0	0	2
Denmark	1987	131	39	35	57	0	0	2
Finland[g]	1988	1	0.25	0	0.5	0	0.25	1
France	1984	850	446	170	234	0	0	2
Germany, Fed. Rep.[g]	1987	51.7	3.4	1.9	6.4		5.9	1
Greece	1984	15	15	0	0	0	0	2
Ireland	1984	23	4	0	7	12	0	2
Italy	1984	800	440	90	270	0	0	2
Luxembourg	1984	15	3	0	12	0	0	2
Netherlands[g,h]	1987	3.9	0.36	0.06	1.5	1.47	0.51[i]	1
Norway[g]	1988	0.45	0.18		0.25		0.02	1
Poland[g]	1988	0.6	0.48		0.12	0	0	1
Spain	1984	280	28	0	173	79	0	2
Sweden[j]	1988	180	72	0	108	0	0	2
Switzerland	1987	250	80	57	113	0	0	2
UK[k]	1990	1,076	136	75	499	304	62	3

a Unless otherwise stated.
b Data refer to sewage sludges generated by publicly-owned treatment plants with secondary treatment or better.
c Data represent the lower end of the range.
d Comprises 1,124.3 t landfilled, 512.3 t sent for surface disposal, and 108.8 t monofilled.
e Includes some non-agricultural land applications.
f "Other" refers to distribution and marketing.
g Data are in units of million tonnes per year, wet matter.

h Data refer to outlet, not production, of sewage sludge.
i "Other" refers to use as compost and to use for sports grounds.
j Data are given in terms of wet weight. Dry matter content is about 20 per cent of the total amount generated.
k Data refer to amount of sludge disposed of, not raw sludge generated. Data on incineration refer to the amount of sludge fed to the incinerator, not the ash disposed of. Quantities of sewage sludge disposed of by the various disposal options may not necessarily sum to the total quantity generated. In such cases, the difference is unaccounted for.

Sources:
1. UN ECE 1992 *The Environment in Europe and North America: Annotated Statistics 1992*, United Nations, New York.
2. Matthews, P. J. 1992 Sewage sludge disposal in the UK: A new challenge for the next 20 years, *Journal of the Institution of Water and Environmental Management*, **6**(5), 551–559.
3. Department of the Environment 1992 *Digest of Environmental Protection and Water Statistics No. 14 1991*, HMSO, London.

TABLE 8.8 Litter in the marine environment (number of items, unless otherwise stated)

Region/country	Year	Glass	Metal	Wood	Textiles	Paper	Polystyrene	Rubber	Plastics	Miscellaneous	Fishing gear	Source
AFRICA												
Namibia[a]	1989	59.5	17			70.5[b]			52	57.0[c]	4.1[d]	1
	1990	333.9	44.5			63.6[b]			73.1	0.0[c]	3.2[d]	1
	1991	196.8	69.7			14.4[b]			49.2	32.8[c]	2.3[d]	1
NORTH AMERICA												
Ellesmere Island[e]	1990		136	3	10	6	2		6	35		2
ASIA												
Israel[f]		483	833	1,459	362	637[g]	603[h]	174[i]	12,317	487[j]	0	3
OCEANIA												
Australia[k]	1991	35,414	30,386	7,095	0	24,119[l]	23,592[m]	3,671	144,323	1,829	4,162	4
Australia[n]	1991	238	15		1		2	2	240.3[o]	p		5
Ducie Atoll[q]		201	109		27			1	392		223	6
Livingstone Island									50[r]			7

a   Data refer to kilograms of litter removed from the beach at Sandwich Harbour.
b   Paper and wood.
c   Litter from ships.
d   Lengths of fishing line found amounted to 16 km in 1989, 12.5 km in 1990 and 9 km in 1991.
e   Data refer to the number of items recorded on a 1,200 km ski traverse, conducted in the Spring of 1990.
f   Data are based on litter pieces counted in 472 samples taken from 5 m transects on six beaches between May 1988 and May 1989.
g   Cartons.
h   Styrofoam.
i   Foam rubber.
j   Ropes.
k   Data are based on the results of a one-day national survey conducted by volunteers as part of a Clean Up Australia Day (24 March, 1991). 4,651 completed forms were received from volunteers who collected a total of 20,000 t of litter on that day.
l   Paper and cardboard.
m   Polystyrene foam.
n   Data are based on the results of a survey conducted at Anxious Bay, Eyre Peninsula, South Australia, during 1–15 October, 1991.
o   Data are in kilograms and include various items of fishing gear.
p   Various items of 'land litter', including car wheels, and various cans and bottles were reported.
q   Data refer to the number of items noted in a 1.5 mile survey transect on the shore line. This survey was conducted during the Sir Peter Scott commemorative expedition to the Pitcairn Islands.
r   Data refer to the quantity of plastic (in pounds) found at Cape Shirreff Beach.

Sources:
1. Ministry of Wildlife, Conservation and Tourism 1992 *Namibia's Green Plan* (Draft), Ministry of Wildlife, Conservation and Tourism, Windhoek.
2. France, R. 1992 Garbage in paradise, *Nature* (London), **355**, 504.
3. Golik, A. and Gertner, Y. 1992 Litter on the Israeli coastline, *Marine Environmental Research*, **33**, 1–15.
4. Clean Up Australia Ltd 1991 *The Rubbish Report: Results of the First National Litter Survey Conducted on Clean-up Australia Day, 24 March 1991*, Clean-up Australia Ltd, Pyrmont.
5. Wace, N. 1991 Garbage in the Oceans, *Bogong*, **12**(1), 15–18.
6. Benton, T. 1991 Oceans of garbage, *Nature* (London), **352**, 113.
7. Anon. 1992 Plastic waste affects marine life, *The Environment Digest*, Number 59–60, 14.

TABLE 8.9   Recovery and recycling of selected materials, most recent available year

	Paper and cardboard			Glass			Coal ash		
Region/country	Year	Quantity recycled (10³ t)	Utilization rate[a] (%)	Year	Quantity recycled[b] (10³ t)	Recycling rate[c] (%)	Year	Quantity recycled (10³ t)	Utilization rate[d] (%)
AFRICA									
South Africa							1987	580	4
Tunisia	1990	30	38.5						
Zimbabwe	1990	40	44.4						
NORTH AMERICA									
Canada	1990	1,787	10.8	1985		12	1987		35
Costa Rica	1990	17	91.5						
Jamaica	1990	4							
Mexico	1990	2,168	74.6						
USA	1990	19,769	27.2				1989	15,894	24
SOUTH AMERICA									
Argentina	1990	332	37.3						
Brazil	1990	1,591	32.8						
Peru	1990	52	57.8						
ASIA									
China							1989	16,200	26
India							1991	6,750	17
Israel	1990	137	68.5						
Japan	1990	14,629	52.1				1989	1,918	49
Jordan	1990	13	86.3						
Korea	1990	3,342	73.9						
Myanmar	1990	2	13.6						
Philippines	1990	280	62.6						
Sri Lanka	1990	10	54.0						
Thailand	1990	730	77.1						
Turkey				1991	54	28			
EUROPE									
Austria	1990	1,143	39.0	1991	156	60			
Belgium				1991	223	55	1990	796	85
Czechoslovakia	1990	464	35.1	1987	302.4		1989	1,400	8
Denmark				1991	60	35	1990	880	90
Finland	1990	469	5.2	1991	15	31	1988	180	28
France	1990	3,295	46.7	1991	987	41	1987	130	59
Germany				1991	2,295	63			
German Dem. Rep.							1989	7,200	38
Germany, Fed. Rep.	1990	5,590	47.2				1989	6,464	56
Greece				1991	26	22			
Hungary	1990	288	64.9				1987	1,100	27
Ireland				1991	16	23			
Italy				1991	763	53	1988	900	69
Netherlands	1990	1,829	66.1	1991	360	70	1990	902	86

Continued

TABLE 8.9 Continued

Region/country	Paper and cardboard			Glass			Coal ash		
	Year	Quantity recycled (10^3 t)	Utilization rate[a] (%)	Year	Quantity recycled[b] (10^3 t)	Recycling rate[c] (%)	Year	Quantity recycled (10^3 t)	Utilization rate[d] (%)
Norway				1991	10	22			
Poland	1990	371		1991			1989	4,500	15
Portugal	1990	342	44.1	1991	50	30			
Spain	1990	2,209	64.1	1991	310	27	1987	1,217	14
Sweden	1990	1,034	12.3	1991	57	44	1988		50
Switzerland	1990	617	47.6	1991	199	71			
Yugoslavia				1987	308.3				
UK	1990	2,847	59.0	1991	385	21	1989	6,120	49
USSR	**1990**	**2,880[e]**		**1989**	**826**		**1989**	**11,500**	**9**
OCEANIA									
Australia				1983		57	1990	800	10
New Zealand	1990	53	6.9	1983		53			

[a] The "Utilization rate" represents the amount of waste paper reused for the production of paper and cardboard as a percentage of total paper and cardboard production.

[b] Quantities are exclusive of the manufacturer's cullet.

[c] The "Recycling rate" represents the amount of glass cullet used for glass production as a percentage of total glass consumption.

[d] The "Utilization rate" represents the amount of waste coal ash reused as a percentage of coal ash production.

[e] Data refer to recovery from municipal wastes only.

Sources:
Clarke, L. 1993 Utilization options for coal-residues: an international overview. Paper presented at the American Coal Ash Association's 10th International Coal Ash Utilization Symposium, Orlando, Florida, USA, January 1993. Unpublished paper.

FAO Advisory Committee on Pulp and Paper 1991 *Waste Paper Data 1988-90*, Food and Agricultural Organization of the United Nations, Rome.

FEVE 1992 *Glass Gazette*, **9**, 18.

OECD 1991 *OECD Environmental Data Compendium 1991*, Organisation for Economic Co-operation and Development, Paris.

Data for the USSR supplied by the Institute of Global Climate and Ecology, Russian Academy of Sciences, Moscow, Russia (glass and paper).

UN ECE 1992 *The Environment in Europe and North America: Annotated Statistics 1992*, United Nations, New York.

Part 9

Environmental Disasters

Significant advances in health, social and economic development throughout the world are being repeatedly interrupted by natural and man-made disasters. These events can have a devastating effect on communities, countries and even regions. The continuing deterioration of the environment and the increasing crowding of cities are increasing vulnerability to disasters. The scale of the setbacks caused by these events has highlighted that any long-term strategies towards sustainable development must be pursued with consideration of the global problem of disasters.

○ **On average, approximately 25,000 deaths are caused by natural disasters per year, and economic damage is estimated to be more than US$ 3,000 million a year.**

○ **The less developed countries suffered 97 per cent of the world's 825 major natural disasters between 1970 and 1985 and accounted for more than 99 per cent of all national disaster-related deaths.**

○ **Most energy-industry related accidents between 1969 and 1986 occurred as a result of coal mining, fires, oil and natural gas explosions, hydroelectric dam failure or core damage in nuclear power plants.**

○ **Between 1944 and 1987 there were 284 nuclear accidents world-wide, many of which were associated with mishandling of isotopes or inadvertent exposure to X-rays.**

It is important to recognize that an extreme natural event only becomes a natural disaster when it has a significant impact on human settlements and activities. Natural disasters may lead indirectly to environmental damage of a technological nature; for example, a flood damaging a dam containing toxic sludge or a chemical plant destroyed by an earthquake. While the events themselves cannot be prevented, their disastrous consequences can often be reduced by appropriate advance planning and the preparation of emergency measures on the part of the communities at risk.

During the last few years, the world has endured a wide diversity of disastrous events. Major earthquakes in Armenia, Iran, Egypt and India, severe flooding in Bangladesh, Pakistan and the Mississippi region of the USA, volcanic activity in the Pacific rim, particularly in the Philippines, and technological disasters such as the explosion in the city sewers

in Guadalajara, Mexico are just some of the events which have accounted for millions of deaths, countless homeless and unquantifiable socio-economic damage.

Such events also include those which have occurred in the wake of social and political changes and which have contributed to regional instability and humanitarian crises. The conflict over Kuwait is a notable example. In addition, widespread human suffering in places like the Kurdistan region and civil strife in countries such as Somalia, Mozambique and the former Yugoslavia have prompted major co-ordinated efforts working on disaster prevention, mitigation and preparedness, to take more account of disasters resulting from factors other than natural phenomena.

At the United Nations (UN), member states have created a new Department of Humanitarian Affairs (DHA) to co-ordinate, integrate and speed up the response of UN agencies to humanitarian crises (Tomblin, 1993). Absorbed into the DHA is the former Office of the United Nations Disaster Relief Co-ordinator (UNDRO) which, with its United Nations International Emergency Network (UNIENET) computer network, and considerable amount of experience and expertise in natural disasters, will be of significant value to the DHA.

In 1991, the United Nations Environment Programme (UNEP) Governing Council adopted a decision (16/9) that mandated the development, on an experimental basis, of a United Nations Centre for Urgent Environmental Assistance (UNCUEA). This centre would act in co-operation and in co-ordination with other UN agencies, focusing on assessment of, and response to, man-made environmental emergencies. In accordance with the Governing Council decision, the UNCUEA was established in Geneva for a period of 18 months. Since its establishment the UNCUEA has concentrated on the elaboration of proposals to improve the response of, and co-ordination within, the UN system to environmental aspects of different types of emergencies (Clerc, 1993). The establishment of the DHA and UNCUEA, along with the continuing efforts of the UNEP Industry and Environment Programme Activity Centre (IE/PAC) and particularly its Awareness and Preparedness for Emergencies at the Local Level (APELL) programme (see later), are indicative of the seriousness with which environmental disasters and emergencies are now considered by the UN system.

The concept of risk assessment and early warning of disasters is one which UNEP, in particular, has shown initiative and interest in developing. Warnings with long lead times, and increased confidence in such warnings, allow governments and organizations to prepare for hazardous events at many levels and allow populations to adjust their activities accordingly. In some cases natural disasters, such as tropical cyclones, are well served by co-ordinated early warning systems. Nevertheless, further development of early warning systems is necessary to cope with other potential disasters associated with weather, such as early warning systems for drought and famine, e.g., the Food and Agriculture Organization of the United Nations' (FAO) Famine Early Warning System (FEWS) (see Part 3: Natural Resources).

The greatest need for early warning systems is in developing countries. To become fully operational, such warning systems require the strengthening of the capacity within developing countries to assess their resources. This need was highlighted and requested by the United Nations Conference on Environment and Development (UNCED) held in Rio de Janeiro in June 1992 (see Part 10: International Co-operation).

As in previous editions of the 'UNEP Environmental Data Report', major categories of natural disasters are discussed. This edition, however, places greater emphasis on other environmental disasters, including industrial incidents and accidents, and the efforts of the international community to address disasters associated with hazardous substances.

References

Clerc, A. 1993 Personal communication, United Nations Centre for Urgent Environmental Assistance, Geneva.

Tomblin, J. 1993 Personal communication, United Nations Department of Humanitarian Affairs, Geneva.

Natural Disasters

It is now widely acknowledged that natural disasters are becoming increasingly significant in terms of numbers of events and magnitude of impact (UNEP, 1991). It has been estimated that over the last 20 years, more than three million people have been killed by natural disasters, while a further one billion had their lives adversely affected (WMO, 1990). On average, natural disasters are estimated to claim 25,000 lives and cause damage valued in excess of US$ 3,000 million per year (IE/PAC, 1992).

Natural disasters may occur very suddenly, as is the case with earthquakes, tsunamis, floods, volcanic eruptions, cyclones and landslides; or they may occur slowly like droughts and desertification. Few regions of the world have escaped the impact of one extreme natural phenomenon or another. However, by far the greatest burden caused by natural disasters falls on less developed countries and their most densely populated regions. Developing countries suffered 97 per cent of the world's 825 major natural disasters between 1970 and 1985, and accounted for more than 99 per cent of all natural disaster-related deaths (Abbott, 1991). This is because population growth, particularly in urban and coastal areas, and unsafe buildings are compounding existing risks from natural hazards in developing countries (UNEP, 1991). Although such natural hazards are integral features of the environment, the extent of their damage is largely a function of decisions made or postponed in the development process.

In recognition of the social and natural science component of natural disasters, the UN General Assembly proclaimed the International Decade for Natural Disaster Reduction (IDNDR) which began on 1 January, 1990. The main thrust of the IDNDR is an integrated approach to disaster reduction, with greater emphasis on pre-disaster planning, preparedness and prevention, while sustaining post-disaster relief capabilities. In short, the IDNDR aims to shift the focus from post-disaster reaction to pre-disaster action. It is also recognized that appropriate emergency planning for natural disasters, and its integration into national policies, could also be very helpful in preventing industrial or technological disasters. One of the first priorities towards achieving the aims of the IDNDR is the transfer of relevant knowledge to those countries recognizably most at risk from the effects of natural disasters and enhancing its application within them. The acquisition of relevant knowledge and assessments of risk depend heavily on existing data and information from previous natural disaster events.

As was reported in the previous edition of this report (UNEP, 1991), there are few global data bases for natural disasters. One of the most comprehensive global data bases is the EM-DAT data base on disasters which is managed by the Centre for Research on the Epidemiology of Disasters (CRED) at the Université Catholique in Brussels. Another global data base of major disasters since 1900 is held by the Office of United States Foreign Disaster Assistance (OFDA, 1989).

Within the UN system, DHA-Geneva administers UNIENET, which aims to place members of the international disaster management community in direct contact with each other and provide them instantaneously with background and operational disaster-related information. The data base of natural disasters within UNIENET was temporarily suspended in 1988, but is currently being re-established by DHA-Geneva.

Measurement of the impacts of natural disasters is a problem and subject to some debate. Almost all data bases concentrate on three main criteria: the number of people killed, the number of people affected and the damage caused, estimated in financial terms (usually in US$). The last of these three criteria has been subjected to much criticism because exchange and inflation rates can render the figures meaningless and irrelevant when listed against numbers of people killed, injured and homeless (Cruden, 1990). An attempt to quantify the risks to individual countries from natural disasters is described in Box 9.1.

As a result of the large numbers of disasters that occur, published listings are often limited to selected events based on, for example, numbers killed or financial damage reported. This approach often results in the omission of numerous events which occur in remote regions but can be recorded by global monitoring networks, such as those for seismic activity, and by satellite measurements. While these events may not significantly affect human settlements or livelihood, the study of them is equally important in terms of developing predictive capabilities for such events.

In previous editions of this report comprehensive data tables have been presented listing major events of earthquakes, tsunamis, volcanic eruptions, windstorms (i.e.,

BOX 9.1 Assessment of Disaster-Proneness

Perceptions of whether or not a given country is disaster-prone are usually influenced by knowledge of serious disasters in the recent past (often through media coverage). Such perceptions are therefore neither quantitative nor objective. Consequently, there is a need to establish a means of quantitatively determining national risk associated with the effects of disasters.

The overall effect of a disaster in a given country depends heavily on the economic health of that country. An event affecting a certain number of people and causing a certain amount of damage may have a disastrous effect on a country with a weak economy, whereas a disaster of similar magnitude in another country with a stronger economy may not be considered serious at all (Kerpelman, 1990). However, even in strong economies, the frequency and severity of disasters in terms of insured losses are rising, as shown in the table of property loss caused by disasters in the USA.

The most disaster-prone countries according to the UNDRO Disaster-Proneness Index

Total Index Rank[a]	Country	Average Index Rank[a]	Country
1	Montserrat	1	Montserrat
2	Vanuatu	2	Cook Is
3	Nicaragua	3	Yemen
4	Burkina	4	Vanuatu
5	Dominica	5	Tokelau
6	Cook Is	6	Burkina
7	Chad	7	Dominica
8	Bolivia	8	St Lucia
9	St Lucia	9	Antigua & Barbuda
10	Yemen	10	St Kitts & Nevis

[a] For explanation of Total and Average Index Rank see Table 9.2
Source: Summarized from Table 9.2

Insured property loss caused by disasters in the USA, 1986–92

Year	Insured loss (US$ 10^6)
1986	872
1987	945
1988	1,410
1989	7,640
1990	2,830
1991	14,220
1992 (estimated)	13,400

Source: McCarroll, 1992

In 1990, the United Nations Disaster Relief Organization (UNDRO) produced a 'Preliminary Study on the Identification of Disaster-prone Countries Based on Economic Impact' in which 195 countries were studied in order to assess the impact of disastrous events on their respective economies. Table 9.2 lists 73 of these countries and ranks them in order of the impact of disasters on their economies. The 10 most disaster-prone countries, derived from the study, are given in the summary table included here.

For several of the countries included in the study, the results (although very preliminary) have confirmed the assumption of disaster-proneness based on UNDRO's experience in those countries. Some results were, nevertheless, rather surprising. The degree to which disasters affected some countries was unexpected, as was the fact that other countries, subject to a long-continuing series of serious disasters of various types, were shown to have economies strong enough to support the effects of the disasters. For example, several small island countries (especially those island developing countries such as the Cook Islands, St Kitts & Nevis and Montserrat) had very high indices because they are particularly prone to tropical cyclones. A listing of significant natural disasters in Commonwealth island states was given in the third edition of this report (UNEP, 1991).

Due to the lack of a comprehensive, accurate and up-to-date global data base of historical information on disasters, the UNDRO Disaster-Proneness Index has produced some results which may be interpreted as distortions of the overall pattern of disasters around the world. For example, some countries had a relatively high index, based on a single disastrous event during the 20-year period which constituted the study (e.g., those island countries mentioned above, as well as Algeria and Liberia). India and the Philippines are examples of countries which continually suffer significant disasters, but whose indices suggest that they are not disaster-prone, at least with respect to the effect on their economies. Although the study and the index are preliminary they provide the basis for a more detailed continuation of the determination of a desired quantitative measure of disaster proneness.

References

Kerpelman, C. 1990 *Preliminary Study on the Identification of Disaster-prone Countries Based on Economic Impact*, Office of the United Nations Disaster Relief Coordinator, Geneva.

McCarroll, T. 1992 Through the roof, *Time Magazine*, **140**(15), 44–45.

UNEP 1991 *United Nations Environment Programme Environmental Data Report 1991/92*, Basil Blackwell, Oxford.

hurricanes, cyclones, typhoons, etc.), floods and landslides over recent decades, up to 1989 (UNEP 1987, 1989, 1991). In this edition the occurrences of all such major events from 1990–1992 have been listed together in Table 9.1. Such a listing is not comprehensive, due to inadequacies in the reporting of many events, and does not accurately identify so-called multiple hazards (i.e., the interaction of major hazards such as cyclones and floods, or floods and landslides) which cause some of the world's worst natural disasters. Some major natural hazards are discussed briefly below.

Earthquakes

Single earthquake events can cause devastation and loss of life on a scale far greater than any other individual natural hazard. However, because they are more infrequent than other natural disasters, long-term global comparisons indicate that the number of deaths and economic and environmental damage caused by earthquakes are far less than losses caused by events such as windstorms and floods.

Seismic activity is well monitored and documented throughout the world by a number of national and international recording networks. Comprehensive listings of seismic events are held by several agencies such as the International Seismological Centre (ISC) in Newbury, UK, and the National Earthquake Information Center at the World Data Center A in Boulder, Colorado, USA. Every year there are about 10^6 registerable seismic or microseismic tremors around the world. The vast majority of these are so small that they cannot be felt by people, and only 10–20 of them cause damage (IE/PAC, 1992). The ISC has approximately 36,000 episodes recorded in its data base for 1992 (Adams, 1993).

Detailed discussions of earthquakes and listings of major earthquakes from 1966 to 1990 have been included in previous editions of this report (UNEP 1987, 1989, 1991). Table 9.1 lists a further 16 earthquakes that have occurred since 1990.

Volcanic Eruptions

According to the Smithsonian Institute in Washington DC there are, at any time, about 20 to 30 volcanoes erupting in the world. Most go unreported because they are small eruptions or away from centres of population. However, the 1980s was the worst decade for volcanic disasters since the start of the century (Oppenheimer, 1990). Over 24,000 people were killed as a result of just two eruptions: Nevado del Ruiz in Colombia (1985) and El Chichon in Mexico (1982). A further 1,800 people died as a result of two lethal gas bursts from volcanic crater lakes in Cameroon in 1984 and 1986. Eruptions in the USA, Italy and Indonesia claimed more lives and caused extreme damage. Major volcanic eruptions have continued with similar frequency into the 1990s which has so far seen four major events including the worst volcanic disaster this century when Mount Pinatubo in the Philippines erupted on 2 April 1991, killing over 900 people.

The Mount Pinatubo eruption is thought by many experts to have significant and far-reaching effects on the global environment. These include global cooling during the next few years (see Part 2: Climate) and accelerated depletion of stratospheric ozone (see Part 1: Environmental Pollution, Atmosphere). About two to three times as much sulphur dioxide (SO_2) was released into the atmosphere compared with the El Chichon eruption in 1982. By mid-May, six weeks after the initial eruption, the volcano was emitting about 500 t of SO_2 per day; two weeks later this figure had increased to 5,000 t per day before falling to about 280 t per day in June 1991 (Bowler and Joyce, 1991). Satellites from the National Aeronautics and Space Administration (NASA) monitor volcanic material at altitudes between 17 and 40 km. The results received suggest the Mount Pinatubo eruption caused a 60–80 times increase in the amount of dust and aerosols in the stratosphere (Hecht, 1992). The effects of the Mount Pinatubo eruption on the global climate are discussed in more detail in Part 2: Climate.

Windstorms

Windstorms can influence precipitation and floods, and cause severe destruction to crops and property. Every country in the world is affected by either severe tropical cyclones (called hurricanes in the Atlantic, Caribbean and north-eastern Pacific; typhoons in the western Pacific; and cyclones in the Indian Ocean and in the sea around Australia), tornadoes, monsoons or thunderstorms.

Windstorms are the most frequently occurring natural hazards. Each year there are about 80 windstorms in which people are killed. Over the past 30 years or so the average annual death toll has been over 15,000 and the average annual damage has been estimated at about US$ 1,500 million (Smith, 1989). The total fatalities can be increased by drownings, especially through rare events such as the extreme tropical cyclones in Bangladesh in 1970 and 1991 which caused over 200,000 and 138,000 deaths respectively.

Bangladesh is highly prone to frequent destructive windstorms associated with tidal surges, particularly in the pre-monsoon months of April–May and the post-monsoon months of October–November. The magnitude of storm surges in Bangladesh is due to the shape of the coast, shallow waters and large time difference in the tides of the east and west coasts (BCAS, 1992). In the past, of the nine internationally recorded cases of extreme loss of life (40,000–200,000 deaths), seven cases were in the Bay of Bengal and the Arabian Sea.

It is now recognized that disaster mitigation with respect to windstorms should include an effective early warning system. Although cyclones do not respect political boundaries, the responsibility for provision of a warning system for each country rests with the national meteorological services. However, through national systems and internationally coordinated efforts such as the World Meteorological Organization's (WMO) World Weather Watch Programme (WWW) and its associated Tropical Cyclone Programme (TCP), there is now a warning system covering virtually every area directly affected by tropical cyclones (Vickers, 1991). The TCP was established in 1972 by WMO and the Economic and Social Commission for Asia and the Pacific (ESCAP). The initial members were Bangladesh, Myanmar (q.v. Burma), India, Pakistan, Sri Lanka and Thailand. Figure 9.1 shows the global occurrence (as percentages) of tropical cyclones and the areas covered by major international co-operative bodies concerned with provision of early warning systems.

The study of windstorms, particularly tropical cyclones, continues to attract the attention of scientists who regard their incidence as being a possible indicator of global climate change and who predict an increase in their frequency. However, there is little evidence at present to suggest that tropical cyclone frequency or intensity has increased with global changes throughout this century, perhaps because global warming has not yet made sufficient impact (WMO/UNEP, 1992).

Landslides

Landslides, which are defined as movements of large masses of rock, earth or debris down slopes (Cruden, 1990), are highly

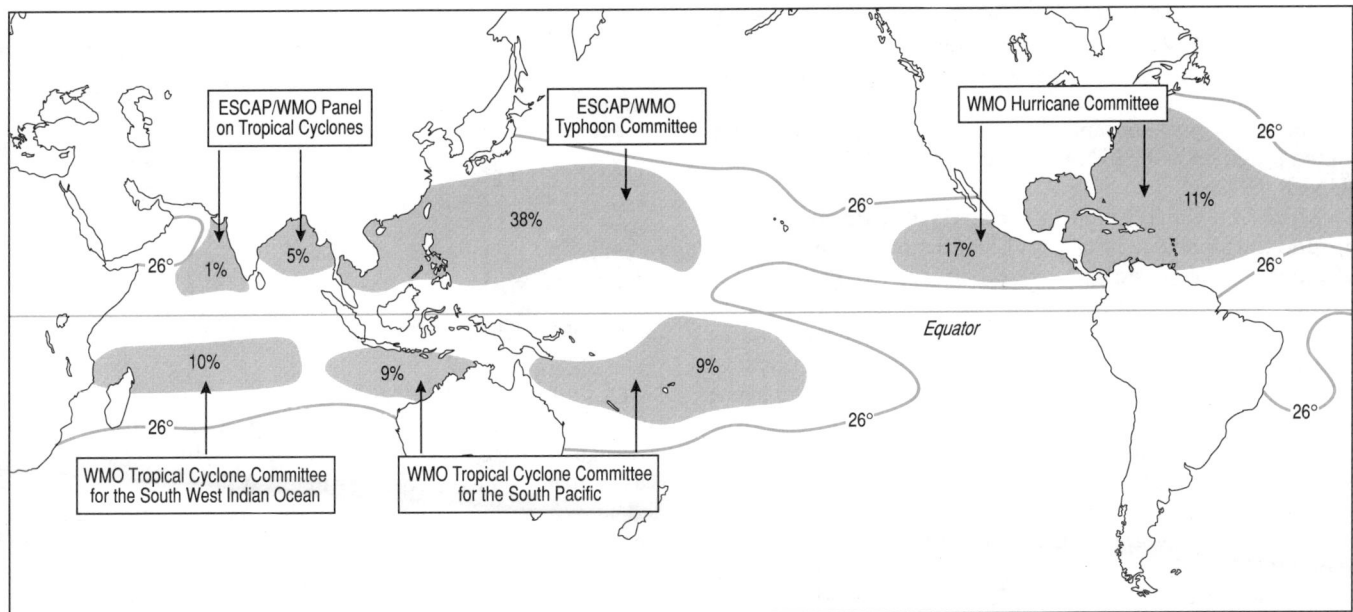

FIGURE 9.1 Global occurrence of tropical cyclones. Cyclones form only where sea temperatures exceed 26°C (dotted line). The names of tropical cyclone bodies and the basins covered by their programmes are also indicated, together with the percentage of cyclones occurring in that area
Source: After Smith, 1989

destructive and can affect every country that has topographic relief. At present, surprisingly little is known about the nature and environmental impacts of landslides or about the vulnerability of human communities to them. This is partly due to a lack of information from previous landslide events, the effects of which are often masked by association with more spectacular events such as earthquakes, volcanic eruptions, windstorms and floods from which they often result. Such an example occurred in Colombia in 1985 where, in the aftermath of the Nevado del Ruiz volcanic eruption, 22,000 people were killed by the landslide which ensued. There is also considerable inconsistency in the use of terminology and classifications in the reporting of events, which makes the compilation of data on the global occurrence of landslides more difficult than the documentation of other natural disasters.

Deficiencies in the information regarding landslides means that few people realize the extent of fatalities and economic damage which they may cause. Annual insured losses in the USA, Japan, Italy and India have been estimated at US$ 1 billion or more. In the USA alone, there are more than 25 landslide-related fatalities a year (Schuster and Fleming, 1986). Detailed national compilations of events suggest that in some countries landslides are the principal cause of death from natural disasters. Unfortunately, no systematic global comparisons of these data exist although some regional reports have been collected by Brabb and Harrod (1989).

It is now clear that an inventory of global landslide occurrence is needed to strengthen disaster management capabilities (Cruden and Brown, 1992). In view of this, and following successful landslide studies in a few countries, the International Geotechnical Societies/United Nations Educational, Scientific and Cultural Organization (UNESCO) Working Party on

World Landslide Inventories has undertaken to provide the means (using standardized terminology) to build a global data network based on local and national inventories.

Droughts

Drought may be defined as a period of two or more years with rainfall well below average (UNDP, 1992). Although the African continent has suffered the most dramatic impacts from drought during the past several decades, the vulnerability of many nations outside Africa to extended periods of water shortage has been increasing. In the past decade alone, droughts have occurred with considerable frequency and severity in most of the developed and developing world. Brazil, Argentina, Uruguay, Australia, the USA, Canada, India, China and most of the countries in south-east Asia are just a few examples of the countries seriously affected by drought.

Information on the impact of drought periods on a country basis is difficult to collate. International aid agencies, such as the International Committee of the Red Cross and the International Federation of Red Cross and Red Crescent Societies, make estimates of numbers of people who have died or have been displaced by droughts for some periods, but information comparable with that for other types of natural disasters is scarce (UNEP, 1989).

There is growing scientific speculation that occurrences of droughts in some regions of the world are significantly influenced by the "El Niño" phenomenon. Originally used to describe a regional oceanic current, and later the occasional warming of surface waters off the coasts of Ecuador, Peru and Chile, the term "El Niño" has recently become linked to the broader atmospheric process known as the Southern Oscillation. This is a fluctuation of periods of high and low atmospheric pressure between the south-eastern Pacific and Indian Ocean in the southern hemisphere. To describe more accurately the interaction of these two phenomena, their names have been joined together as the El Niño/Southern Oscillation or ENSO (UNEP, 1992a).

An ENSO event is usually associated with extreme weather around the globe and lasts from 12 to 18 months. Disturbance of global atmospheric circulation patterns by strong ENSO events frequently results in unusual weather in regions remote from the triggered phenomena and are described as "teleconnections" because of the absence of a direct physical relationship between the ENSO occurrence and the extreme weather events in other regions of the world. Many scientists are presently studying relationships between the ENSO phenomenon and possible changes in global climate (see Part 2: Climate).

The severe droughts in the Sahel and southern Africa are thought to have been influenced by ENSO events recorded in 1982–83 and 1986–87 (UNEP, 1992a). The social and economic consequences of severe droughts in the Sahel over the last 20 years have been devastating. The regular occurrences of droughts have contributed to repeated land degradation, crop failures and famine. It is now recognized that the ability to predict coming periods of drought in the Sahel region would be a fundamental component of an early warning system for environmental disasters. Climatological research which links the ENSO phenomenon with rainfall patterns in Africa is currently strengthening the possibility of the development of such a system. However, continued and improved programmes of monitoring of atmospheric and oceanic systems are necessary, along with regional studies in Africa that compare climatic data with changes in vegetation, soil and water.

Forest Fires

Data on forest fires in Europe and North America are collated by the Joint FAO/Economic Commission for Europe (ECE) Working Party on Forest Economics and Statistics, but there is a lack of information on wildfires in the tropics and sub-tropics. The largest recorded wildfire occurred in China in 1987, covering 10,000 km^2, killing nearly 200 people, destroying 12,000 homes and displacing 56,000 inhabitants. Australia is also frequently affected by forest fires. Table 9.3 presents time series data on number, area and average size of forest fires in Europe and North America. Data are presented separately for southern Europe because this region accounts for most of the fires and related damage within Europe.

Mediterranean countries have been the focus of continuous study as part of the UNEP Regional Seas' Mediterranean Action Plan (MAP). About 50,000 wildfires occur in the Mediterranean region each year, affecting areas of between 200,000 and 700,000 ha or approximately 0.6 per cent of the Mediterranean forest area (UNEP, 1992b).

The following table shows the number of forest fires, the percentage of the area burnt and the mean area burnt per fire in some Mediterranean countries:

	Number of fires (per 10,000 ha)	Area burnt (%)	Mean area burnt per fire (ha)
Cyprus	4	0.004	61.1
France	3	0.3	8.3
Greece	2	0.8	43.0
Israel	81	0.9	2.7
Italy	14	0.8	14.0
Spain	3	0.9	31.5
Turkey	1	0.1	8.8
Yugoslavia	1	0.9	22.2

Source: UNEP, 1992b

Droughts provide ideal conditions for forest fires to start and spread. In parts of southern Europe, especially Spain and Portugal, a recognized period of drought which began in 1989 has contributed to an increase in forest fires. This in turn has led to worsening economic conditions and environmental damage through loss of wood productivity, destruction of national parks and tourist sites, and increased desertification. In Portugal, the effects of successive dry and very dry years were felt by a series of major forest fires in 1990 and 1991. The combined effects considerably influenced defoliation of broadleaved and coniferous forests during that period (UN ECE/CEC, 1992) (for further discussion see Part 1: Environmental Pollution, Biological Monitoring).

References

Abbott, S. 1991 Courting Ruin, *Geographical Magazine*, **63**(8), 12–14.

Adams, J. 1993 Personal communication. International Siesmological Centre, Newbury, UK.

BCAS 1992 *Cyclone '91: An Environmental and Perceptional Study*, Bangladesh Centre for Advanced Studies, Dhaka.

Bowler, S. and Joyce, C. 1991 When sleeping giants wake, *New Scientist*, 6 July 1991.

Brabb, E. E. and Harrod, B. L. 1989 *Landslides: Extent and Economic Significance*, Balkema, Rotterdam.

Cruden, D. M. 1990 Personal communication, Department of Civil Engineering, University of Alberta, Edmonton, Canada.

Cruden, D. M. and Brown, W. M. 1992 *Progress Towards the World Landslide Inventory*, Proceedings of the sixth International Symposium on Landslides, Christchurch, New Zealand.

Hecht, J. 1992 Pinatubo cooling will test greenhouse models, *New Scientist*, 11 January 1992, 20.

IE/PAC 1992 *Hazard Identification and Evaluation in a Local Community*, Technical Report Series No. 12, United Nations Environment Programme, Industry and Environment Programme Activity Centre, Paris.

OFDA 1989 *Disaster History: Significant Data on Major Disasters Worldwide, 1900–Present*, Office of US Foreign Disaster Assistance, Washington DC.

Oppenheimer, C. 1990 Monitoring hot spots from space, *Geographical Magazine*, March 1990, 32–34.

Schuster, R. L. and Fleming, R. W. 1986 Economic Losses and Fatalities Due to Landslides, *Bulletin of the Association of Engineering Geologists*, **23**(1), 11–28.

Smith, D. K. 1989 *Natural Disaster Reduction: How Meteorological and Hydrological Services Can Help*, Report No. 722, World Meteorological Organization, Geneva.

UN ECE/CEC, 1992 *Forest Condition in Europe: 1992 Report*, United Nations Economic Commission for Europe, Geneva and Commission of the European Communities, Brussels.

UNDP 1992 *Assessment of Desertification and Drought in the Sudano-Sahelian Region, 1985–1991*, The United Nations Sudano-Sahelian Office, United Nations Development Programme, New York.

UNEP 1987 *United Nations Environment Programme Environmental Data Report 1987/88*, Basil Blackwell, Oxford.

UNEP 1989 *United Nations Environment Programme Environmental Data Report 1989/90*, Basil Blackwell, Oxford.

UNEP 1991 *United Nations Environment Programme Environmental Data Report 1991/92*, Basil Blackwell, Oxford.

UNEP 1992a *The El Niño Phenomenon*, UNEP/GEMS Library Series No. 8, United Nations Environment Programme, Nairobi.

UNEP 1992b Forest Fires in the Mediterranean, *Medwaves: News Bulletin of the Mediterranean Action Plan*, No. **25**, 1–5, United Nations Environment Programme, Athens.

Vickers, D. O. 1991 Tropical Cyclones, *Nature and Resources*, **27**(1), 31–37.

WMO 1990 *The Role of the World Meteorological Organization in the International Decade for Natural Disaster Reduction*, World Meteorological Organization, Geneva.

WMO/UNEP 1992 *Climate Change 1992: The Supplementary Report to the IPCC Scientific Assessment*, World Meteorological Organization, Geneva and United Nations Environment Programme, Nairobi.

Industrial Incidents and Accidents

The occurrence of accidents is a major contributor to the adverse effects on human health and the environment which are associated with industrialization and economic development. Such critical events most often occur at production facilities, or during the transport of hazardous materials (WHO, 1992a). Explosions, fires and collisions of moving vehicles can lead to the release of harmful chemical, physical or radioactive agents to the environment. New technological developments, or the improper management of existing technologies, can add to the growing complexity of risks which are unavoidably associated with industrial development.

Major and widely reported industrial accidents such as those in Seveso (Italy, 1976), Bhopal (India, 1985), Chernobyl (USSR, 1986), Schweitzerhalle (Switzerland, 1986) and Guadalajara (Mexico, 1991), have resulted in increased awareness of the potential hazards to the environment and human health. The result has been an increase in public interest in environmental issues and much public resistance to the location of industrial complexes in their neighbourhoods. The accident at Chernobyl is discussed in greater detail in Box 9.2.

Table 9.4 lists chronologically the major chemical or chemical-related accidents and incidents which occurred world-wide during the period 1986 to 1992. This data set is not comprehensive but presents a representative selection of major events and their effects. Incidents and accidents which occurred prior to 1986 are listed in previous editions of this report (UNEP 1987, 1989, 1991a).

More than 200 serious events occur in the Organisation for Economic Co-operation and Development (OECD) countries each year (UNEP, 1989). Available data suggest that most accidents are probably associated with the energy industries. A study of serious accidents in the energy sector between 1969 and 1986 by the Swiss Reinsurance Company found that the most severe accidents occurred as a result of coal-mining operations, fires, explosions of oil and natural gas, failure of hydroelectric dams and core damage with significant radionuclide release in nuclear power plants (WHO, 1992b). The Senior Expert Symposium on Electricity and the Environment also identified these as the energy-related industries from which the most potentially severe accidents may occur.

Oil Spills

In the aftermath of three of the world's largest oil spills from tankers in the last three years (Exxon Valdez, Aegean Sea and Braer), and the scale of the damage caused by releases of oil during and after the conflict over Kuwait in 1991, public awareness and concern for the ecological, economic and social consequences of such disasters has increased enormously. The impacts of such disasters are usually long-lived and the true extent of the damage often cannot be quantified for years following the event.

Oil spills from tankers and their resultant effects on the flora and fauna are generally well reported and documented. Although a major oil spill can seriously damage the environment, the effects are typically quite localized (UNEP, 1991a). Also, the extent of ecological damage does not necessarily relate to the quantity of oil spilt. In addition, oil spills account for only a small proportion of the total input of hydrocarbons to the marine environment; other sources include atmospheric deposition, river run-off and seepage of effluents.

The ecological effects on the marine environment from the oil spilt in the Prince William Sound, Alaska after the Exxon Valdez disaster in 1989 were extensive. Many of the ecological impacts have been indirect, such as increases in the abundance of some organisms (from intermediate trophic levels) because predators (such as birds and sea otters) were reduced in numbers. It has been estimated that the number of sea birds killed was between 100,000 and 300,000 and that only 800 were successfully cleaned and released to the wild. About US$ 18.3 million were spent on the capture and rehabilitation of sea otters; an estimated 4,000 died in the aftermath of the spill (Pain, 1993) and less than 200 were successfully cleaned and released (Shaw, 1992).

Estimates of the fate of the 38,000 t of crude oil lost by the Exxon Valdez are imprecise, but suggest that 30–40 per cent may have evaporated, 10–25 per cent was recovered, and the rest remains in the marine environment. Clean-up efforts are expected to aid natural recovery. The consequences of other oil spills affecting rocky shorelines, together with knowledge of previous natural and anthropogenic disturbances to Prince William Sound, suggest that most biotic communities and ecosystems will recover to approximately their pre-spill functional and structural characteristics within 5 to 25 years (Shaw, 1992).

Data on oil spills and tanker accidents are compiled by a number of organizations including the International Tanker Owners Pollution Federation Limited in London. These data suggest that the incidence of very large oil spills is relatively low. As detailed statistical analysis of the available data is rarely possible, the Federation places emphasis on identifying trends. Figure 9.2 shows that since 1970 the number of major spills each year (greater than 700 t) has

BOX 9.2 The Chernobyl Accident – An Update

In the early hours of Saturday, 26 April 1986, an accident occurred at Unit 4 of the Chernobyl nuclear power plant (RBMK-1000, a 3,200 MW light water-cooled graphite moderated power plant) in the Ukraine, which was to have global repercussions. Smoke, fumes and a large amount of radioactive material rose in a hot plume, almost 2 km high, and was carried throughout the western regions of the former USSR, including Belarus and the Russian Federation, to eastern and western Europe. It was also carried in much smaller amounts through the Northern Hemisphere (IAEA, 1991a). No radioactive materials from the accident appear to have reached the Southern Hemisphere because there is little transfer of air masses across the equator. The events of the accident and the fall-out of radioactive material have been detailed by the United Nations Scientific Committee on the Effects of Atomic Radiation (UNSCEAR) and the International Atomic Energy Agency (IAEA) (IAEA, 1991; UNEP, 1991).

It has been calculated by UNSCEAR that about 70 quadrillion Bq ^{137}Cs (about one quarter of the reactor core) was deposited over the Northern Hemisphere: 42 per cent was deposited in the former USSR, 37 per cent in Europe, 6 per cent in the oceans and the rest in various locations over the Northern Hemisphere. Similarly, 35 quadrillion Bq ^{134}Cs and 330 quadrillion Bq of ^{131}I may have been released. Whereas ^{131}I decays rapidly and is short-lived, ^{134}Cs will lead to elevated doses of radiation for years, while ^{137}Cs will lead to long-term doses of radiation for decades.

The Chernobyl accident resulted in severe radiation injuries and 31 immediate deaths. About 115,000 people were evacuated from a 30 km zone around the power plant where the worst contamination was measured.

Ten experts were sent to the area in June 1989 by WHO, and experts from the League of Red Cross and Red Crescent Societies went in early 1990. The aim of these missions was to assess the current health situation resulting from the event. In addition, in late 1989 the International Chernobyl Project was set up by the IAEA with the participation of the World Health Organization (WHO), WMO, FAO, International Labour Organization (ILO), the Commission of the European Communities (CEC) and UNSCEAR. The advisory committee comprised scientists from 10 countries, and altogether 200 experts from 25 countries took part in the project. Their task was not to undertake a totally new assessment, but to evaluate the quality and reliability of the existing assessment results. The radiological situation of approximately 825,000 people living in 2,225 settlements in the three affected republics was investigated. Key issues of concern were the true extent of contamination, radiation exposure of the population, health effects and protective measures.

The main concern of UNSCEAR has been to estimate the overall impact of the accident over the entire globe, leaving individual national assessments to the countries concerned. First year dose estimates show that the former USSR average effective dose (i.e., the equivalent dose weighted for the susceptibility to harm of different tissues in an individual) was 0.82 millisievert (mSv) per person compared with calculated values for south-eastern Europe (Bulgaria, Greece, Italy, former Yugoslavia) of 1.2 mSv, Scandinavia of 0.97 mSv and 0.94 mSv for central Europe. The low value obtained for the former USSR represents the dose averaged over the whole of the country, most of which was scarcely affected by the accident. Belarus received the highest average first year effective dose commitment of 2 mSv, about the same amount as received annually from natural radiation (i.e., the average annual effective dose of radiation from all sources is estimated at 2.4 mSv). This can be compared with medical test procedures of 0.4 to 1.0 mSv, weapons fallout of 0.01 mSv and nuclear power of 0.0002 mSv (UNEP, 1991).

Results of the dosimeter and health studies in the highly contaminated areas where local food restrictions were in force showed that the external radiation dose exceeded the background values in only 10 per cent of the inhabitants (IAEA, 1991b). A comparison of the health of people in seven "contaminated" and six "uncontaminated" control settlements with the same socio-economic structure showed that there were significant non-radiation related health disorders in the population of both the contaminated and uncontaminated settlements, but no health disorders that could be attributed directly to radiation (IAEA, 1991a,b). The former USSR data that were examined by the International Chernobyl Project Team did not indicate a substantial increase in incidence of leukaemia, cancer or hereditary effects. Nevertheless, future development of effects or increased numbers of cases cannot be excluded until longer time periods have elapsed. Based on estimates of absorbed thyroid dose in children, there is a possibility of a statistically-detectable increase in the incidence of thyroid tumours in the future. The International Chernobyl Project endorsed the proposal of the WHO Scientific Advisory Group on the Health Effects of Chernobyl that there should be long-term studies of selected high-risk populations. The accident had, and continues to have, considerable psychological consequences, such as anxiety and uncertainty, which have extended beyond the contaminated area. Other health indicators examined are documented in the appropriate reports (IAEA, 1991a,b).

References

IAEA 1991a *The International Chernobyl Project: An Overview, Assessment of Radiological Consequences and Evaluation of Protective Measures*, Report by International Advisory Committee, International Atomic Energy Agency, Geneva.

IAEA 1991b *The International Chernobyl Project: Summary Brochure, Assessment of Radiological Consequences and Evaluation of Protective Measures*, Report by International Advisory Committee, International Atomic Energy Agency, Vienna.

IAEA 1992 *The Chernobyl Accident: Updating of INSAG-1, INSAG-7*, A report by the International Nuclear Safety Advisory Group, Safety Series, Number 75-INSAG-7, International Atomic Energy Agency, Vienna.

UNEP 1991 *Radiation, Doses, Effects and Risks*, Second Edition, United Nations Environment Programme, Blackwell, Oxford.

decreased. By the end of the 1980s the average number of major oil spills each year had dropped to one-third of the average for the previous decade. The majority of oil spills are small, and for these the data on numbers and amounts are incomplete. However, reliable data on spills of over 7 t are collected, allowing an estimate of the total quantity spilt each year (Figure 9.2).

Major oil spills from tankers since 1979 are summarized in Figure 9.3 and the total incidence of spills by cause and magnitude, recorded during the period 1974–1992, are given below:

	Number of spills by magnitude of oil spilt			
	< 7 t	7–700 t	>700 t	Total
Operations				
Loading/discharging	2,708	254	13	2,975
Bunkering	540	23		563
Other operations	1,143	42		1,185
Accidents				
Collisions	134	177	71	382
Groundings	207	159	86	452
Hull failure	523	58	29	610
Fires and explosions	142	10	21	173
Other	2,181	157	43	2,381
TOTAL	7,578	880	263	8,721

Source: Data supplied by the International Tanker Owners Pollution Federation Limited, 1993

Nuclear Accidents

The large size of the nuclear industry in some countries, together with the number of radiation sources used for industrial and medical purposes, make accidents involving radiation inevitable (UNSCEAR, 1988). A number of the accidents involving radioactive materials that have already occurred have resulted in various degrees of human exposure and contamination of the environment. Between 1944 and 1987, a total of 284 nuclear accidents were recorded worldwide which, not including the Chernobyl disaster in the former USSR, affected 1,358 people of whom 33 died as a direct consequence. Most of these casualties were the result of mishandling individual radioisotope sources or inadvertent exposure of workers to X-rays (UNEP, 1991b). Data on occupational exposure to radiation were presented in the third edition of this report (UNEP, 1991a).

The selection of nuclear accidents given in Table 9.5 is not a comprehensive listing but illustrates the diversity of accidents. It includes the most significant and best documented events that have occurred since 1985. The table below shows the International Atomic Energy Agency (IAEA) levels of severity for nuclear events, together with the corresponding numbers of events of each level recorded in 1991:

Level	Severity class	Number of events
0	Below scale	22
1	Anomaly	25
2	Incident	25
3	Serious incident	6
4	Accident without offsite risks	0
5	Accident with offsite risks	0
6	Serious accident	0
7	Major accident	0

Source: IAEA, 1992

The Chernobyl disaster in 1986 was classed as a level seven event, and the Three Mile Island accident in the USA in 1979 was a level five event (IAEA, 1992). There were also four significant events presenting safety risks at nuclear power plants in 1991: one in Japan, where a steam generator tube was

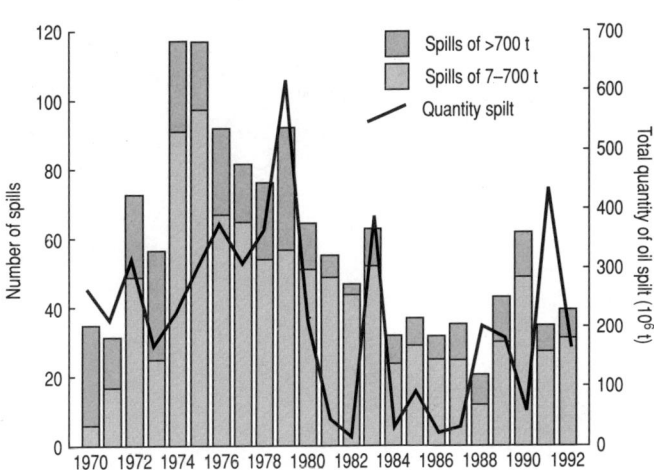

FIGURE 9.2 Number of oil spills and quantity spilt from accidents involving tankers, combination carriers and barges, 1970–1992
Source: Data supplied by the International Tanker Owners Pollution Federation Limited, 1993

ruptured; one in Ukraine, where there was an electrical fire at the Chernobyl-2 reactor; and two in the USA, a turbine failure at the Salem-2 reactor and a loss of coolant at the Oconee-3 reactor (IAEA, 1992). The impacts of the major Chernobyl disaster in 1986 are described in Box 9.2.

Hazardous Substances

Increasing concern over industrial accidents and the resultant severity of environmental deterioration has produced widespread consensus on the need to strengthen national and international capabilities for responding to emergencies involving hazardous substances. Recognition and identification of hazardous substances is a prerequisite for this.

There are a number of international inventories and guides for the identification of hazardous substances. The Environmental Chemical Data and Information Network (ECDIN) has a data base of over 60,000 chemicals. Within UNEP, the International Register of Potentially Toxic Chemicals (IRPTC) has data profiles containing physical and chemical properties, environmental and human toxicity, production and management data on about 800 chemicals, as well as international regulations and recommendations on over 8,000 chemicals.

One of the most established internationally co-ordinated efforts working towards improving prevention of chemical and other industrial accidents is the UNEP IE/PAC's APELL Programme, mentioned earlier. The main goal of APELL, which was launched in 1988, is to prevent technological accidents and their impacts by assisting decision-makers and technical personnel to improve community awareness of hazardous installations, and to prepare response plans for any unexpected events at these installations which may endanger life, property and the environment. Developed with the support of governments and the chemical industry, APELL includes training activities and the publication of various handbooks and documents; including the IE/PAC Technical Report No. 12 'Hazard Identification and Evaluation in a

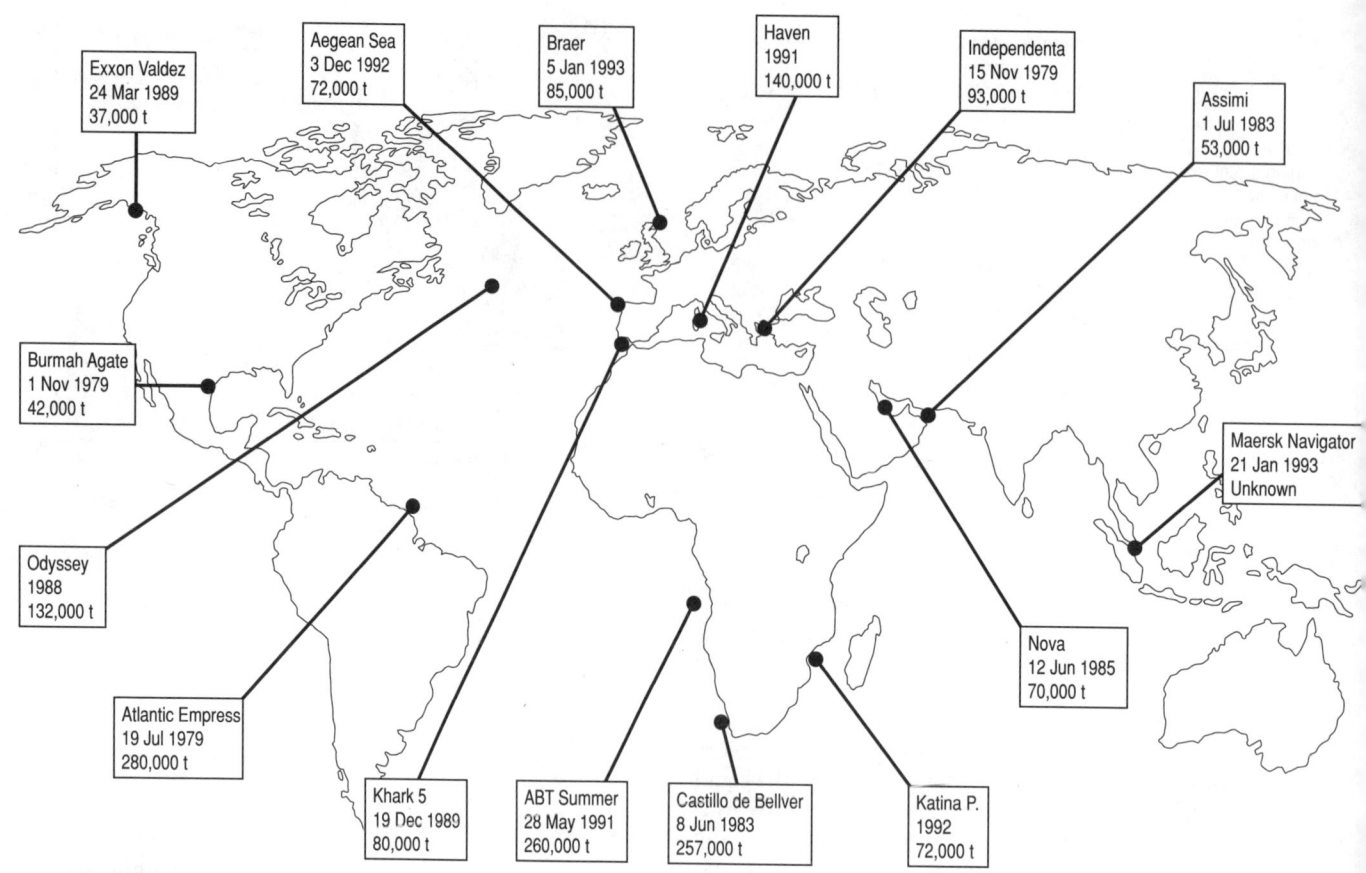

FIGURE 9.3 Location of major oil spills and quantities spilt since 1979
Source: Based on information supplied by the International Tanker Owners Pollution Federation Limited, 1993

Local Community' and the 'International Directory of Emergency Response Centres' compiled and published jointly with OECD (IE/PAC 1988, 1989–92, 1990, 1992).

Agenda 21, the action plan adopted at UNCED, supported APELL by specific reference to the need to strengthen the programme as a tool for the environmentally-sound management of toxic chemicals. The UN ECE Convention on the Transboundary Effects of Industrial Accidents mentions APELL as a tool for awareness and preparedness. In addition, APELL forms the awareness and preparedness component of the Interagency Risk Management Programme of the IAEA, UNEP, the United Nations Industrial Development Organization (UNIDO) and the World Health Organization (WHO) as well as undertaking work jointly with the International Labour Organization (ILO) and the International Programme on Chemical Safety (IPCS). The UNCUEA is building its capacity to respond to industrial accidents, taking into account relevant procedures and infrastructures created at the national level with the help of APELL.

References

IAEA 1992 *IAEA Yearbook*, International Atomic Energy Authority, Vienna.

IE/PAC 1988 *APELL Handbook*, United Nations Environment Programme, Industry and Environment Programme Activity Centre, Paris.

IE/PAC 1989–92 *APELL Newsletters*, **1–5**, United Nations Environment Programme, Industry and Environment Programme Activity Centre, Paris.

IE/PAC 1990 *Storage of Hazardous Material*, Technical Report No. 3, United Nations Environment Programme, Industry and Environment Programme Activity Centre, Paris.

IE/PAC 1992 *Hazard Identification and Evaluation in a Local Community*, Technical Report Series No. 12, United Nations Environment Programme, Industry and Environment Programme Activity Centre, Paris.

Pain, S. 1993 The two faces of the Exxon disaster, *New Scientist*, 22 May 1993, 11–13.

Shaw, D. G. 1992 The Exxon Valdez Oil-spill: Ecological and social consequences, *Environmental Conservation*, **19**(3), 253–258.

UNEP 1987 *United Nations Environment Programme Environmental Data Report 1987/88*, Basil Blackwell, Oxford.

UNEP 1989 *United Nations Environment Programme Environmental Data Report 1989/90*, Basil Blackwell, Oxford.

UNEP 1991a *United Nations Environment Programme Environmental Data Report 1991/92*, Basil Blackwell, Oxford.

UNEP 1991b *Radiation, Doses, Effects and Risks*, Second Edition, United Nations Environment Programme, Basil Blackwell, Oxford.

UNSCEAR 1988 *Sources, Effects and Risks of Ionizing Radiation*, United Nations Scientific Committee on the Effects of Atomic Radiation, Vienna.

WHO 1992a *WHO Commission on Health and Environment: Report of the Panel on Industry*, World Health Organization, Geneva.

WHO 1992b *Our Planet, Our Health: Report of the WHO Commission on Health and Environment*, World Health Organization, Geneva.

TABLE 9.1 Major natural disasters, 1990–1992

Region/country	Year	Month	Type of disaster	Number of persons Dead	Number of persons Injured	Number of persons Affected
AFRICA						
Benin	1991	July	Floods	30		
Chad	1991	August	Floods	39	14	
Egypt	1992	October	Earthquake	552	9,929	
Kenya	1990	April	Floods	44		
Madagascar	1991	February	Cyclone	36		125,000
Malawi	1991	March	Floods	1,172		218,000
	1991	March	Landslide	472		
Sudan	1990	May	Earthquake	31	0	10,000
	1991	December	Floods	2,000		
Tanzania	1990	April	Floods	283		142,000
NORTH AMERICA						
Canada	1991	January	Storm	33		
Costa Rica	1991	April	Earthquake	74	800	4,000
El Salvador	1991	June	Earthquake	1,000		
Mexico	1990	August	Hurricane	38		
	1990	September	Floods	85	100	17,000
Nicaragua	1992	September	Earthquake	105	489	13,587
USA	1990	December	Storm	79		
	1991	April	Floods	33		
SOUTH AMERICA						
Argentina	1992	January	Floods	104		
Bolivia	1992	December	Landslide	45		
Brazil	1991	January	Landslide	65		
	1992	February	Floods	41		
Chile	1991	June	Floods/storm	199	140	30,650
	1991	June	Landslide	61	750	48
Colombia	1991	July	Landslide	53		
Peru	1991	February	Floods	40		
	1991	February	Landslide	33		
	1991	April	Earthquake	38	750	
ASIA						
Afghanistan	1991	February	Earthquake	1,200	699	
	1991	February	Floods	567		16,000
	1991	May	Floods	700		
	1991	April	Floods	50		
	1992	September	Floods	450		500
Bangladesh	1990	March	Storms	242	200	
	1990	October	Tsunami	370		
	1990	December	Storm	250		
	1991	April	Floods	150,000		
	1991	April	Cyclone	138,866		15,000,000

Continued

TABLE 9.1 Continued

| Region/country | Year | Month | Type of disaster | Number of persons | | |
				Dead	Injured	Affected
Bangladesh	1991	May	Storm	33	300	
continued	1991	May	Floods	250		
	1991	June	Cyclone	125,720		
	1991	July	Floods	100		7,000,000
	1991	September	Floods	100		20,000,000
China	1990	April	Earthquake	126	2,049	24,574
	1990	June	Typhoon	56	148	
	1990	July	Typhoons	1,135	4,089	
	1990	July	Storm	33		
	1990	July	Typhoon	108		
	1990	August	Typhoons	559		20,000
	1991	January	Floods	3,000		
	1991	February	Earthquake	119	11	45,000
	1991	March	Storms	138	1,015	880,000
	1991	May	Floods	1,729	32,237	206,000,000
	1991	June	Floods	1,999	4,700	34,000,000
	1991	September	Landslide	200		
	1992	March	Floods	50		
	1992	March	Storm	35	362	3,300,000
	1992	April	Storm	94	3,796	
	1992	April	Floods	147	1,380	
India	1990	May	Typhoon	950		
	1990	May	Cyclone	514		6,500,000
	1990	August	Storms	620		
	1990	November	Cyclone	250		1,500,000
	1991	May	Floods	101		2,000,000
	1991	June	Floods	158		100,000
	1991	July	Floods	1,083		2,000,000
	1991	August	Floods	500		
	1991	August	Floods	57		
	1991	October	Earthquake	1,500	1,383	
	1991	October	Landslide	35		
	1991	November	Typhoon	125		
	1991	November	Floods	125		25,000
	1992	September	Floods	500		
	1992	October	Floods	51		100
Indonesia	1992	October	Landslide	75		
	1992	December	Earthquake/tsunami	2,500		
Iran	1990	June	Earthquake	40,000	105,000	
	1991	May	Earthquake	35,000		
Japan	1990	September	Typhoons	91		
	1991	May	Volcano	30		
	1991	May	Earthquake	37		
	1991	June	Volcano	43	20	10,000

Continued

TABLE 9.1 Continued

| Region/country | Year | Month | Type of disaster | Number of persons | | |
				Dead	Injured	Affected
Korea	1991	July	Floods	54	36	
	1991	August	Typhoon	72		
Pakistan	1991	February	Earthquake	504	574	247,000
	1992	August	Floods	112		
	1992	September	Floods	1,334		6,655,450
Philippines	1990	June	Typhoons	156	14	160,000
	1990	July	Earthquake	1,660	3,513	1,594,040
	1990	August	Typhoons	285		
	1990	November	Storm	100		
	1990	November	Typhoons	1,212	1,594	8,184,612
	1991	May	Storm	3,956	3,050	
	1991	June	Volcano	639		
	1991	August	Floods	50		
	1991	October	Typhoon	39	38	360,000
	1991	November	Typhoon/floods	8,840	6,149	996,908
	1992	September	Typhoon	50	13	942,764
Thailand	1990	October	Typhoon	36		
Turkey	1990	June	Storm	50		
	1991	May	Floods	30		16
	1992	February	Landslide	261	69	
	1992	March	Earthquake	547	2,000	
Viet Nam	1990	November	Cyclone	68		
	1991	December	Storm	100		
	1991	September	Floods	136		
	1991	December	Cyclone	150		
	1992	February	Cyclone	251		
	1992	October	Floods	55	10	44,109
EUROPE						
Romania	1991	July	Floods	73		
	1991	August	Floods	65		
USSR						
USSR	1990	July	Earthquake	43		
	1991	February	Earthquake	150		
	1991	May	Floods	1,700		
	1991	May	Landslide	64		
	1991	August	Storm	41		
Kyrgyzstan	1992	August	Earthquake	54	100	
OCEANIA						
Papua New Guinea	1991	February	Landslide	200		

Numbers of persons affected do not include numbers of homeless.

Only disasters for which 30 or more deaths were reported are included in the above table.

Sources:
Data extracted from the *EM-DAT Database on Disasters*, maintained by the Centre for Research on the Epidemiology of Disasters, Université Catholique de Louvain, Brussels, Belgium in February 1993.
Data supplied by the United Nations Department of Humanitarian Affairs, Geneva, Switzerland.

TABLE 9.2   UNDRO's Disaster-proneness index

Region/country	"Significant" disasters, 1970–89		GNP, 1980 values (10^6 US$)	Total Index		Average Index	
	Total damage, 1970–89 (10^3 US$)	No. of events 1970–89		Value (%)	Rank	Value (%)	Rank
AFRICA							
Algeria	5,200,000	1	36,650	14.19	37	14.19	17
Botswana	81,116	6	800	10.13	42	1.69	65
Burkina	2,390,396	4	1,250	191.23	4	47.81	6
Cape Verde	10,860	2	90	12.06	40	6.03	36
Chad	451,000	5	490	92.04	7	18.41	14
Comoros	61,171	3	100	61.18	12	20.39	13
Djibouti	15,708	5	160	9.82	43	1.96	60
Ethiopia	2,481,800	7	4,080	60.82	13	8.69	27
Gambia	31,000	2	210	14.79	35	7.37	32
Liberia	200,000	1	940	21.28	31	21.28	12
Madagascar	500,000	2	3,010	16.60	34	8.30	30
Malawi	27,400	1	1,160	2.36	66	2.36	55
Mali	277,000	3	1,230	22.52	27	7.51	31
Mauritania	247,340	4	600	41.15	18	10.28	23
Mauritius	436,059	4	1,070	40.68	19	10.17	24
Mozambique	86,500	2	3,262	2.65	64	1.32	70
Niger	359,500	3	1,670	21.53	30	7.17	35
Senegal	532,289	4	2,420	21.98	28	5.50	42
Sudan	375,000	1	6,740	5.56	53	5.56	40
Swaziland	54,152	1	430	12.60	39	12.60	21
Tunisia	90,000	1	8,020	1.37	70	1.37	69
NORTH AMERICA							
Antigua & Barbuda	38,000	1	100	38.00	20	38.00	9
Belize	10,000	2	140	7.15	48	3.58	48
Cuba	386,282	1	13,506	2.86	63	2.86	52
Dominica	70,650	3	50	141.30	5	47.10	7
Dominican Rep.	150,000	1	6,480	2.31	67	2.31	56
El Salvador	1,590,536	4	3,040	52.32	14	13.08	18
Guadeloupe	76,000	1	1,300	5.85	51	5.85	38
Guatemala	1,000,000	1	7,810	12.80	38	12.80	20
Haiti	131,286	2	1,420	9.21	44	4.61	43
Honduras	724,600	4	2,080	34.82	22	8.71	26
Jamaica	1,520,000	5	2,360	64.40	11	12.88	19
Martinique	155,000	2	1,410	11.00	41	5.50	41
Mexico	4,000,000	1	137,570	2.91	62	2.91	51
Montserrat	240,000	1	20	1,200.00	1	1,200.00	1
Nicaragua	4,200,988	8	2,030	206.95	3	25.87	11
Panama	135,000	2	3,180	4.25	57	2.12	58
St Kitts & Nevis	14,000	1	50	28.00	24	28.00	10
St Lucia	89,280	2	110	81.17	9	40.58	8
St Vincent & Grenadines	21,600	2	60	35.99	21	18.00	15

Continued

TABLE 9.2 Continued

Region/country	"Significant" disasters, 1970–89		GNP, 1980 values (10^6 US$)	Total Index		Average Index	
	Total damage, 1970–89 (10^3 US$)	No. of events 1970–89		Value (%)	Rank	Value (%)	Rank
SOUTH AMERICA							
Argentina	2,300,000	2	71,750	3.20	61	1.60	68
Bolivia	2,692,736	10	3,200	84.16	8	8.42	29
Brazil	8,200,000	2	255,070	3.21	60	1.61	67
Chile	1,500,000	1	25,410	5.90	50	5.90	37
Colombia	1,810,900	3	32,590	5.56	52	1.85	62
Ecuador	232,100	1	9,200	2.52	65	2.52	53
Paraguay	214,430	3	4,220	5.08	55	1.69	66
Peru	1,518,800	2	17,970	8.45	46	4.23	45
Uruguay	77,000	1	7,620	1.01	73	1.01	73
ASIA							
Afghanistan	734,220	6	2,710	27.10	25	4.52	44
Bangladesh	551,853	7	11,030	50.32	15	7.19	34
China	58,726,000	2	267,810	21.92	29	10.96	22
Japan	82,420,650	2	1,053,930	7.82	47	3.91	47
Jordan	42,000	1	3,470	1.21	72	1.21	72
Laos	43,146	1	240	17.98	32	17.98	16
Myanmar	96,000	1	5,630	1.71	69	1.71	64
Nepal	343,500	3	2,040	16.84	33	5.61	39
Pakistan	1,419,168	3	25,640	5.54	54	1.85	63
Sri Lanka	998,000	3	3,910	25.50	26	8.50	28
Thailand	400,000	1	31,550	1.27	71	1.27	71
Yemen	2,000,000	1	3,000	66.67	10	66.67	3
Yemen, Dem.	235,300	4	810	29.05	23	7.26	33
EUROPE							
Greece	1,812,000	2	39,910	4.54	56	2.27	57
Italy	25,470,153	2	359,210	7.08	49	3.54	49
Romania	4,594,000	3	50,870	9.03	45	3.01	50
Spain	3,900,000	1	195,670	1.99	68	1.99	59
OCEANIA							
Australia	6,000,000	1	147,140	4.08	58	4.08	46
Cook Is.	25,000	1	21	119.05	6	119.05	2
Fiji	163,000	6	1,110	14.68	36	2.45	54
French Polynesia	37,000	2	990	3.74	59	1.87	61
Tokelau	500	1	1	50.00	17	50.00	5
Tonga	25,100	5	50	50.20	16	10.04	25
Vanuatu	91,365	4	40	228.41	2	57.74	4

Continued

TABLE 9.2 Continued

The UNDRO Disaster-proneness Index for individual countries is calculated as follows (it is stressed that this work is preliminary and subject to further refinement):

Natural disasters occuring during the 20-year period, January 1970–July 1989 were assigned an individual index value corresponding to the damage caused by that disaster expressed as a percentage of national GNP:

$$\text{Index} = \frac{\text{Damage (in US\$)}}{\text{Annual GNP}} \times 100$$

Only "significant" disasters; i.e., disasters causing damage in financial terms amounting to 1 per cent or more of a country's total annual GNP were included in the calculation.

The Total Index for each country was then computed as the sum of the individual indices for each significant disaster. The Average Index is the arithmetic mean of the individual indices for each significant disaster during the period studied. The Total Index thus provides a measure of the total effect of disasters over 20 years, expressed as a percentage of total annual GNP. The Average Index measures the impact of the average significant disaster which struck the country during the 20-year period, as a percentage of its total annual GNP.

The selection of the period January 1970–July 1989 as the basis for the Disaster-proneness Index is driven by the availability of data on natural disasters; the data base of natural disaster information is considered to be less accurate prior to 1970. 1980 values for GNP were chosen on the assumption that 1980 data represent an average for the period studied.

Damage estimates in financial terms were unavailable for a number of disasters which occured during the study period. However, in some cases, estimates of the extent of damage in financial terms could be made from other information. Instances of this were disasters for which a figure was given for the number of persons affected by the disaster, but no corresponding figure on the damage caused. In such cases, where possible, existing data on similar types of disasters in the same country, and for which both figures were reported, was used to extrapolate the missing information as follows:

A unit damage figure per affected person was computed for that type of disaster in that country, by dividing total damage by total affected population for all such disasters for which both figures were reported. For each same type of disaster in that country for which no damage was reported, the affected population was multiplied by the unit damage to arrive at an estimated damage figure. This was then used to compute the index for that disaster.

Source:
Kerpelman, C. 1990 *Preliminary Study on the Identification of Disaster-prone Countries Based on Economic Impact*, United Nations Office of the Disaster Relief Co-ordinator, Geneva.

TABLE 9.3 Number, area and average size per year of fires on forest and other land in the UN ECE region, 1980–1988

Year	Europe			Southern Europe[a]			North America		
	Total no. of fires (10^3)	Total area burnt (10^3 ha)	Average area burnt per fire (ha)	Total no. of fires (10^3)	Total area burnt (10^3 ha)	Average area burnt per fire (ha)	Total no. of fires (10^3)	Total area burnt (10^3 ha)	Average area burnt per fire (ha)
1980	38.2	541.2	14.2	30.2	530.0	17.5	164.8	6,076.2	36.9
1981	46.3	768.5	16.6	41.0	760.0	18.6	199.3	7,105.1	35.6
1982	38.8	451.1	11.6	29.1	438.3	15.0	110.6	2,277.1	20.6
1983	34.6	504.7	14.6	26.3	495.1	18.8	104.1	1,987.7	19.1
1984	42.3	397.2	9.4	32.2	379.6	11.8	191.0	1,995.3	10.4
1985	54.3	1,058.3	19.5	47.6	1,049.8	22.1	147.9	2,882.8	19.5
1986	36.4	581.5	16.0	28.3	570.9	20.2	145.8	2,241.6	15.4
1987	39.8	447.6	11.2	34.5	440.1	12.7	158.4	3,102.7	19.6
1988	48.2	515.1	10.7	40.2	505.7	12.6	164.0	3,635.5	22.2
Average, 1980–88	42.1	585.0	13.8	34.4	574.4	16.6	154.0	3,478.2	22.1

[a] Includes Cyprus, France, Greece, Israel, Italy, Portugal, Spain, Turkey and Yugoslavia.

Source:
UN ECE 1992 *The Environment in Europe and North America: Annotated Statistics 1992*, United Nations Economic Commission for Europe, New York.

TABLE 9.4 Major chemical accidents/incidents, 1986–1992

Year	Country	Location	Type of accident/incident	Chemical(s) involved	Killed	Injured[a]	Evacuated[b]
1986	Canada	Hinton	Rail accident		40	90	
		Ontario	Transport accident	Petrol	0	0	5,000
	Italy	Naples		Petrol	5	150	2,000
	Mexico	Cardenas	Pipeline leakage	Gas	0	2	>20,000
	Switzerland	Basel	Warehouse fire	Various chemicals			
	UK	Hemel Hempstead	Road accident	Lead oxide	0	150	0
	USA	Lynchburg	Transport accident	Phosphorous oxychloride		125	
		Miamisburg	Rail accident (fire)	Phosphoric acid	0	140	40,000
		St Petersburg	Fire	Chloride	0	90	7,000
		St Petersburg	Leakage	Ammonia	0	52	8,000
	USSR	Chernobyl	Reactor explosion	Radionuclides	29	300	135,000
1987	Brazil	Goiana	Abandoned cancer treatment device	Radionuclides	2	241	0
	China	Guangxi Province		Methyl alcohol	55	3,600	
		Shanzi	Drinking water contamination	Ammonium bicarbonate	0	15,400	
	Egypt	Alexandria	Explosion	Smoke bombs	6	460	1,000
	France	Nantes	Fire	Fertilizers	0	24	25,000
	Guatemala	Puerto Camperiico	Clam contamination	Toxic waste	22		
	India	Bhopal	Industrial	Ammonia	0	0	200,000
	Mexico	Minatitlan	Process failure	Acrylonitrile		>200	10,000
	USA	Miamisburg	Rail accident	Red phosphorus		569	30,000
		Minot	Fire	Parathion	0	20	10,000
		Nantichoke	Fire	Sulphuric acid	0	0	18,000
		Salt Lake City	Industry	Ammonia	1	0	30,000
		Texas City	Process failure	Hydrofluoric acid	0	540	3,000
		Wilcon country	Rail accident	Sulphuric acid	0	0	3,000
	USSR	Annau	Rail accident	Chlorine	0	200	
1988	Canada	Quebec	Plant fire	Toxic smoke (PCBs)			3,500
	China	Liu Pan Shui	Explosion	Gas	45	5	
		Shanghai	Explosion at refinery	Petrochemicals	25		
		Shanghai	Explosion	Gas	45	23	
	France	St Avold	Explosion	Chemicals			
		Tours	Leakage	Chemicals	0	3	200,000
	India	Bombay	Fire at refinery	Oil	32		
	Mexico	Chihuahua	Explosion at storage site	Petrol	0	7	150,000
		Guadalupe	Explosion at storage site	Crude oil	20		200,000
		Mexico City	Explosion	Fireworks	62	87	
	Pakistan	Islamabad	Explosion at storage site	Explosives	>100	3,000	
	UK	North Sea	Explosion on platform	Oil, gas	167		
	USA	California	Chemical plant (leakage)	Chlorine gas			20,000
		Henderson	Explosion and fire	Perchlorinated ammonia	2	250	640
		Los Angeles	Process failure	Chlorine		37	20,000
		Springfield	Fire (leakage)	Chlorine	0	275	20,000
	USSR	Arzamas	Rail accident (explosion)	Explosives	73	720	90,000
		Shakhunya	Rail car fire	Herbicides			20,000
		Sverdlovsk	Rail accident (explosion)	Explosives	4	500	
	Yugoslavia	Sibanik	Process failure and fire	Fertilizers	0	0	60,000
1989	India	Bhatinda	Leakage	Ammonia		500	
		Brittania Chowk	Leakage	Chlorine		200	
	Pakistan	Garan Chashma	Explosion	Ammunition	40	>20	
	USA	Los Angeles	Release	Chlorine			11,000
		Pasadena	Explosion	Ethylene	23	125	1,300

Continued

TABLE 9.4 Continued

Year	Country	Location	Type of accident/incident	Chemical(s) involved	Killed	Injured[a]	Evacuated[b]
					\multicolumn{3}{	}{Number of people}	
	USSR	Acha Ufa	Explosion in pipeline	Natural gas	575	623	
		Ionava	Explosion and fire	Ammonia, fertilizers	6	53	30,000
		Yurga	Explosion	Ammunition	1	3	20,000
1990	Australia	Sydney	Fire and explosion	Gas			10,000
	Germany	Alfeld	Release from truck	Chlorine		200	
	India	Near Patna	Explosion on train	Gas	95	100	
		Uttar Pradesh	Poisoning in food factory		150		
	Nepal	Kathmandu	Polluted drinking water		100		
	Nigeria			Poisoned cough medicine	109		
	Thailand	Bangkok	Explosion	LPG	54	97	
	Turkey	Near Merzifon	Explosion in mine	Methane	68		
	USSR	Ufa	Leak from pesticide factory	Phenol			600,000
		Voronej-Rostov	Fire after multiple collision	Petroleum	55	14	
	Yugoslavia	Tozla	Explosion in mine	Gas	150		
1991	China	Shanxi Province	Explosion in mine	Gas	147		
		Maharashtra	Explosion		70		
	Ethiopia	Addis Ababa	Explosion	Ammunition	100	200	
	Italy	Livorno	Explosion (tanker in harbour)	Oil	140		
	Korea	Kumi	Release from electronics plant	Phenol		>100	
	Malaysia	Near Kuala Lumpur	Explosion	Fireworks	41	61	
	Mexico	Central Mexico	Road accident	Volatile industrial chemical	18	50	
	Thailand		Road accident (explosion)	Dynamite	171		
	Pakistan	Peshwar	Explosion	Ammunition	18	70	
	USA	Henderson	Leak in factory	Chlorine		55	15,000
		Sterlington	Explosion	Nitromethane	8	128	
1992	Mexico	Guadalajara	Explosion in city sewers	Gas	210	1,500	
	Turkey	Near Zonguldak	Explosion in mine	Gas	388	250	
	Sengal	Dakar	Peanut factory	Ammonia	40	300	

[a] In the case of poisonings, the data in the 'Injuries' column refer to the number of persons affected.

[b] Numbers of persons evacuated are generally approximate figures only.

LPG = Liquid pressurized gas.

Data included in the above table correspond to the period 1986 to the end of April, 1992.

Accidents/incidents listed are those which have caused at least 50 deaths or injuries and/or 1,000 people to be evacuated.

Accidents/incidents associated with the use of pesticides or drugs are not included. Oil spills are also excluded.

Figures for the numbers of injuries do not include numbers of deaths.

Incidents need not be of accidental origin and commonly result from ignorance or malpractice.

Sources:

OECD 1991 *OECD Environmental Data Compendium 1991*, Organisation for Economic Co-operation and Development, Paris.

OECD 1991 *The State of the Environment*, Organisation for Economic Co-operation and Development, Paris.

GEMS MARC 1993 *GEMS MARC Global Environmental Data Base*, Global Environment Monitoring System, Monitoring and Assessment Research Centre, London.

Information in the above table was supplemented by data extracted from the *EM-DAT Database on Disasters*, maintained by the Centre for Research on the Epidemiology of Disasters, Université Catholique de Louvain, Brussels, Belgium, in February 1993.

TABLE 9.5 Accidents involving nuclear installations and facilities since 1957

Region/country	Location	Year	Date	Comment
				Nuclear power plants
NORTH AMERICA				
USA	Michigan	1966	5 October	A sodium-cooling system failed at the Enrico Fermi fast breeder reactor near Monroe. About 20,000 Ci of fission products were released from the core but were almost completely contained. No exposure of the employees or the public occurred as a result of the incident[a].
	Three Mile Island	1979	28 March	The worst nuclear accident in the USA occurred at the Three Mile Island plant near Harrisburg, Pennsylvania. A partial meltdown of one of the reactors caused radioactive gas to be leaked into the atmosphere. However, there was no significant exposure of the public as the containment structures functioned as designed[a,b].
	Oklahoma	1986	6 January	One worker died and 100 were injured at a nuclear plant when a cylinder of nuclear material burst after being improperly treated[c].
ASIA				
Japan	Tsuruga	1981	25 April	Approximately 45 workers were exposed to radioactivity during repairs to a nuclear plant[c].
EUROPE				
France	Saint-Laurent	1969	17 October	A fuel-loading error sparked a partial meltdown at a gas-cooled power reactor[c].
USSR				
USSR	Shevchenko	1974		A reported explosion in a Soviet nuclear breeder plant[c].
	Chernobyl	1986	26 April	The worst nuclear accident on record. A fire at the Chernobyl nuclear power plant caused a huge release of radiation which spread over much of Europe. 35 quadrillion Bq ^{134}Cs and 330 quadrillion Bq ^{131}I may have been released. 31 people died in the immediate aftermath of the explosion and some 115,000 people were evacuated from a 30 km zone around the power plant where the highest levels of contamination were measured[a,b].
				Accidents at other nuclear installations
NORTH AMERICA				
USA	Tennessee	1959	November	A chemical explosion occurred at a radiochemical reprocessing plant at Oak Ridge National Laboratory. A small quantity of plutonium was released into the environment. Within the immediate vicinity of the plant, levels of radioactivity in excess of 50×10^{-9} Ci per 100 cm^2 were recorded in a number of places[a].
	Idaho	1961	3 January	An explosion of the Army Low-Power Reactor (SL-1) resulted in the death of three military personnel at the National Reactor Testing Station in Idaho Falls. Despite the fact that the explosion was a violent one, and that the reactor core contained about 10^6 Ci of medium-to-long-lived isotopes, it is thought that less than 10 Ci of ^{131}I were released to the environment[a].
EUROPE				
UK	Cumbria	1957	October	Fire destroyed the core of one of two plutonium-producing reactors at the Windscale (now Sellafield) nuclear complex. It has been estimated that between 17,000 and 22,000 Ci of radioactive material were released into the environment[a].
	Cumbria	1983	November	Radioactive waste was accidentally discharged into the Irish Sea from the Sellafield nuclear plant[b].

Continued

TABLE 9.5 Continued

Region/country	Location	Year	Date	Comment
USSR				
USSR	Kyshtym	1957	29 September	A serious accident occurred when a nuclear waste tank exploded releasing 100 quadrillion Bq of radiation over an area of 300 km^2. 1,150 people were evacuated from an area where the radiation levels exceeded 20,000 x 10^3 Bq km^{-2}; 280 from areas exceeding 2,400 x 10^3 Bq km^{-2}; 2,000 from areas exceeding 670 x 10^3 Bq km^{-2}; 4,200 from areas exceeding 330 x 10^3 Bq km^{-2}; and 3,100 from areas exceeding 120 x 10^3 Bq km^{-2} [c]. Another source reported that 74 quadrillion Bq were scattered over the provinces of Chelyabinsk, Sverdlovsk and Tyumensk[a,d].

Mishandling of industrial sources

Region/country	Location	Year	Date	Comment
AFRICA				
Morocco	Mohammedia	1984		A family of eight died after a passer-by picked up a source of ^{192}Ir used for construction and took it home[b].
NORTH AMERICA				
Mexico	Ciudad Juarez	1983		Between 300 and 500 people were exposed to radiation by an abandoned ^{60}Co source which was discovered in a shipment of scrap metal[b].
USA	Houston	1957	March	A minor accident occurred at a company licenced to encapsulate sources for gamma cameras. Two workers were opening a sealed can containing ^{192}Ir of about 135 Ci. Some of the ^{192}Ir escaped in dust form. The workers did not report the incident and it was not discovered until one month later that the laboratory had been contaminated. By this time, radiation had been spread to the outside community via workers' clothes[a].
SOUTH AMERICA				
Brazil	Goiania	1987		Four people died, 54 were hospitalized and 240 were contaminated when a source of ^{137}Cs was taken home and dismantled[b].
ASIA				
China	Shanxi Province	1992	September	Three people died and a further five people received excessive doses of radiation when a ^{60}Co source was taken home.

Satellites

Region/country	Location	Year	Date	Comment
NORTH AMERICA				
Canada	Northwest Terr.	1978	24 January	A Soviet satellite, Cosmos 954, powered by a nuclear reactor re-entered the atmosphere and spread radioactive debris over a 1,000 km path. The core of the reactor is estimated to have contained about 5,000 Ci of radioactive isotopes, of which 75 per cent is estimated to have remained in the upper atmosphere[a].
ASIA				
	Indian Ocean	1964	21 April	A newly-launched navigational satellite carrying a radioisotope power generator (SNAP 9-A) failed to reach orbital velocity and re-entered the atmosphere at about 50,000 m over the Indian Ocean. Although ^{238}Pu was known to be present in the upper atmosphere at the time (as a residue from earlier weapons testing), a new source of ^{238}Pu was detected four months after the abort. The isotopic power unit was known to contain about 17 x 10^3 Ci ^{238}Pu; by 1970 about 16 x 10^3 Ci ^{238}Pu was estimated to have been deposited at ground level[a].

Sources:
[a] Eisenbud, M. 1987 *Environmental Radioactivity: From Natural, Industrial and Military Sources* (third edition), Academic Press Inc., (London) Ltd, London.
[b] UNEP 1991 *Radiation, Doses, Effects and Risks*, (second edition), United

Nations Environment Programme, Nairobi.
[c] Data supplied by the Reuters Press Agency in 1992.
[d] Data supplied by B. G. Bennett, United Nations Scientific Committee on the Effects of Atomic Radiation, Vienna, in 1993.

Part 10

International Co-operation

Considerable progress towards international co-operation on environmental issues was achieved in June 1992 when the United Nations Conference on Environment and Development (UNCED) was held in Rio de Janeiro. This conference placed great emphasis on highlighting priority areas for action and co-operation, not only between governments, but also between governments and society as a whole. This chapter concentrates on the major themes associated with UNCED and progress towards international co-operation.

○ **Agenda 21, was the most significant product of UNCED, providing the framework for action and co-operation on environment and development leading into the next century.**

○ **The Global Environment Facility has nearly completed its pilot phase and has provided significant investment and technical assistance on global warming, biological diversity, international waters and the ozone layer.**

○ **New international agreements on climate change and biological diversity were opened for signature at UNCED.**

As a result of the important role of UNCED in raising governmental and public awareness of environmental issues, and its achievements towards sustainable development, most of this chapter is devoted to describing the processes and stages involved in UNCED and its major themes and results. Other aspects of international co-operation, such as information sources, and environmental awareness and public opinion have been described in earlier editions of the 'UNEP Environmental Data Report' (UNEP 1987, 1989, 1991).

A new register of environmental activities of the United Nations (UN) System has just been produced (UN, 1993a). This document draws on the data base compiled each year by the UN Advisory Committee on Co-ordination of Information Systems (ACCIS) for their Register of Development Activities in the UN System. The listings of activities are grouped according to UN agency, country and subject. It is estimated that, at present, the register probably covers only 50 per cent of actual environmental activities, although ways of improving this are being sought. It is anticipated that the register will be useful for reviewing the scope of UN activities in the environmental field.

The United Nations Conference on Environment and Development

The United Nations Conference on Environment and Development was held in Rio de Janeiro in Brazil on 3–14 June 1992. It was the largest conference ever organized by the UN and was the result of negotiation and co-operation of considerable complexity. It was attended by 178 governments and there were 120 heads of state present at the summit which concluded UNCED. The process of development and negotiation and the implications made at the conference are complex. The process of analysis and initial implementation of commitments of the conference are still being incorporated into national and international policies. The conference generated five formal documents:

○ The Rio Declaration on Environment and Development,
○ Agenda 21 – a programme outline for future action on environment and development,
○ The United Nations Framework Convention on Climate Change,
○ The Convention on Biological Diversity, and
○ The statement on principles for management, conservation and sustainable development of forests.

In addition to these documents, UNCED was very successful at raising awareness of the links between environment and development at the highest political level, bringing it to the forefront of international politics in the future.

The Conference dealt with a whole range of issues which contribute to the complex interactions of environment and development, in essence: the way in which human societies develop and change; their dependence on the environment; and their influence on the environment which, when subject to change, will have an influence in turn on development (see also Part 4: Population and Development). These issues include production and consumption of resources, population, economic growth, ecological systems, inequalities of resources, and sustainable development which takes account of the needs of succeeding generations.

The Preparatory Process and National Reporting

The main outputs and structure of UNCED were negotiated over the two years before the Conference. The primary instrument for the development of UNCED was the Preparatory Committee which received contributions from national governments and reports, international organizations, non-governmental organizations, sectoral conferences, and numerous other sources. The structure and detail of Agenda 21, the Rio Declaration and the logistics of implementation were all developed by the Preparatory Committee. Expert papers and sectoral reports were used to reach consensus on the issues in the 40 chapters of Agenda 21. For example, an International Conference on Water and the Environment, held in January 1992 in Dublin, brought together government-designated experts and representatives of international, intergovernmental and non-governmental organizations (NGOs). This Conference produced The Dublin Statement on Water and Sustainable Development (ICWE, 1992) which gave principles, an agenda and guidelines for implementation and follow-up for use in the preparatory process for UNCED.

There were also various regional preparatory conferences such as those for the African region held in Cairo in July 1991 and in Abidjan in November 1991. These conferences adopted a common position on the agenda for environment and development. As a result, Africa was able to present a draft text to the fourth Preparatory Committee in New York which became the basis for the chapter on combating desertification and drought in Agenda 21 (i.e., Chapter 12). The preparatory work in Africa also provided the basis for establishing the Intergovernmental Negotiating Committee for the Elaboration of an International Convention to Combat Desertification (INCD). Desertification is discussed in more detail in Part 3: Natural Resources.

National reports were requested by the Preparatory Committee for UNCED in order to evaluate national perspectives, experiences, policies, activities and issues. Guidelines were produced in an effort to make the reports more comparable (UN, 1990) and the United Nations Development Programme (UNDP) and other donors were requested to assist where necessary in the preparation of the reports. The guidelines suggested that the national reports should provide basic information on the interaction between environment and development and focus on strategic actions and policy implications: development plans, environment and natural resources problems and actions for their solution. It was suggested that a wide range of representatives from government agencies and NGOs (such as industry, trade unions, women and indigenous peoples) could be consulted in the preparation of the reports.

In the event, the national reports varied in scope and approach but retained structures to the common terms of reference. Some reports referred mainly to policy priorities while others almost took the form of 'State of the Environment' reports, with more or less detailed presentations of information and data on national environment and development. The Secretariat of UNCED took the task of summarizing the national reports in a consistent format, describing: the drafting process, problem areas, past and present capacity-building initiatives, recommendations and priorities on environment and development, financial arrangements and funding requirements, environmentally sound technologies, international co-operation, expectations

from UNCED, and a table of contents for the report. Despite the completion of this immense task, the summaries do not allow effective comparison between countries although they do give a good overview of national perceptions, problems and strategies. Table 10.1 gives a list of the reports submitted to the Secretariat of UNCED and indicates problem areas and capacity-building initiatives which were highlighted in the report summaries (UN 1992d, e, 1993b). A preliminary overview of the contents of the reports was prepared by the secretariat for the fourth session of the Preparatory Committee to highlight issues for the then forthcoming conference (UN, 1992f).

The Rio Declaration on Environment and Development

The Rio Declaration on Environment and Development had originally been proposed as an "Earth Charter". This Charter was to have contained the basic principles for the conduct of nations and peoples with respect to environment and development in order to ensure the future viability and integrity of the earth for human and other forms of life. The final form of the Earth Charter was originally agreed at the third Preparatory Committee session in Geneva, but at the fourth Preparatory Committee in New York the negotiations on the draft became difficult, with disagreements on the Charter becoming a distillation of the dichotomy of interests of the industrialized and industrializing countries. A compromise text was developed which became the Rio Declaration on Environment and Development (UN, 1992a). Although the text of the Declaration was a consensus, certain countries independently emphasized particular interpretations of issues such as the right to development, the responsibilities of developed countries and trade policies. The Rio Declaration has 27 principles which cover:

❍ The role of humans at the centre of concerns for sustainable development and their entitlement to a healthy and productive life in harmony with nature,
❍ National sovereignty on environment and development,
❍ The fulfilment of the right to development while ensuring sustainable development for future generations,
❍ The essential links between environment and sustainable development,
❍ The eradication of poverty,
❍ The need for co-operation of developed and developing countries on environment, development, finance and technology,
❍ Appropriate production and demographic policies,
❍ Public awareness and participation, and
❍ The role of the state in legislation, economy, liability and regulation.

Agenda 21

Agenda 21 is the most substantial product of UNCED and its preparatory process. It stands as the most comprehensive, far-reaching and, if implemented, the most effective programme of international action ever sanctioned by the international community. Agenda 21 is not a final and complete action programme, but is an evolving programme for co-operative action which integrates issues of environment and sustainable development. The intention is that national governments, international organizations and others review Agenda 21 in light of their own priorities and particular circumstances, and produce their own strategies for implementation of the global agenda both co-operatively and within their respective spheres of responsibility.

Agenda 21 itself is composed of 115 programme areas in 40 chapters, covering all aspects of environment and development (UN, 1992a,b,c). The chapters can be grouped into four themes; social and economic dimensions, conservation and management of resources for development, strengthening the role of major groups, and means of implementation. For each programme area, the basis for action is outlined (i.e., the basic premises for the need for action), then objectives to be achieved and activities to be undertaken and finally the means of implementation (particularly finance).

The underlying principle of Agenda 21 is the need for co-operation on environmental and developmental issues, based on a recognition of interdependence of different countries and peoples, the reliance of economic development on the environment, the impact of unsustainable development on the environment and the need for environmental protection as a part of the development process. These elements are part of a wider theme of the global nature of environmental problems and its links with the international nature of mutual economic dependence and development. The UNCED secretariat summarized Agenda 21 in a useful guide (UNCED, 1992) which identifies the main themes of:

○ Revitalization of growth with sustainability (economics, trade and debt),
○ Sustainable living (poverty, health and equity, consumption of resources),
○ Human settlements (urban issues, shelter, water supply and sanitation, wastes, transport and energy),
○ Efficient resource use (sustainable use of renewable resources, water, energy, biological diversity, minerals, forests, agriculture),
○ Global and regional resources (climate and weather, hydrological and carbon cycles, oceans and atmosphere),
○ Managing chemicals and wastes (reducing wastes, enhancing chemical safety, disposal and regulation), and
○ People's participation and responsibility (education, public awareness and training, accountability).

The implementation of Agenda 21 requires advances in many fields: data and information, capacity-building, technology transfer, legal and institutional strengthening and financial resources. The issue of data and information is one of particular relevance to the 'UNEP Environmental Data Report' and is mentioned with respect to many of the programme areas in Agenda 21. Chapter 40 of Agenda 21 deals, in detail, with "Information for Decision-Making" (UN, 1992c) and highlights a number of important issues under the programme areas of "bridging the data gap" and "improving information availability". In bridging the data gap, Agenda 21 aims to improve collection and use of data with greater relevance to users through: the use of indicators of sustainable development; improvements in monitoring and inventories of existing data; improvements in assessment and analysis; establishment of information frameworks to integrate environmental and development information; and strengthening the capacity to use traditional types of information. Improving the availability of information would require production of information usable for decision-making, establishing standards for handling information, and strengthening documentation on information and communication abilities.

Agenda 21 gave particular attention to the roles of the United Nations Environment Programme (UNEP) and UNDP. It gave a number of priority areas for UNEP: promoting international and UN agency co-operation on the environment; developing techniques such as natural resource accounting and environmental economics; environmental monitoring and assessment; scientific research for decision-making; dissemination of environmental information and data to governments and UN bodies; raising awareness; developing and implementing environmental law including co-ordination of convention secretariats; environmental impact assessment; information on environmentally-sound technologies; regional co-operation; technical, institutional and legal advice to governments; integration of environment and development in decision-making; and environmental emergencies. Since UNCED, these priority areas have been the focus of considerable discussion and action within UNEP and in consultation with co-operating governments and agencies.

The dissemination of information on the environment has been a priority of UNEP since it was established. This responsibility is fulfilled in part by the International Environmental Information System (INFOTERRA) which is an international network which puts enquirers directly in touch with experts and provides specific responses to particular queries by drawing together information from special sectoral sources, publications and on-line searches (UNEP, 1992). The INFOTERRA network links 155 countries through national focal points. In 1992 it processed 29,097 queries (INFOTERRA, 1993), meeting enquirers' needs by making information on environmental problems, policies and management available from governments, NGOs, industry, UN and intergovernmental organizations. Chapter 40 of Agenda 21 mentions the need for review and strengthening of INFOTERRA to ensure the effective dissemination of information (UN, 1992c para. 40.24).

The Commission on Sustainable Development

The establishment of a Commission on Sustainable Development (CSD) which would report to the UN Economic and Social Commission was specified in Chapter 38 of Agenda 21 (UN, 1992c). The role of the Commission was agreed as:
○ To monitor progress in implementation of Agenda 21,
○ To consider information and reports provided by governments,

Box 10.1 The UN Framework Convention on Climate Change

The aim of the UN Framework on Climate Change is to stabilize atmospheric concentrations of greenhouse gases at levels that will prevent human activities from interfering dangerously with the global climate system (UN, 1992). In effect, states agree to reduce emissions of greenhouse gases to earlier levels. Many countries and the European Community (but not the USA) have advocated that a voluntary goal is the reduction of carbon dioxide emissions to 1990 levels.

The Framework Convention has developed from a recognition of the need for co-operation on climate change, which arises from "common but differentiated responsibilities and respective capabilities" (UN, 1992). The increases measured in greenhouse gases have led to concern over climate change which could have consequences in the forms of impacts on ecosystems, agricultural production and sea level. Control of emissions, however, has serious implications for energy, economic and development policies.

The commitments made by Parties to the Framework Convention include:
○ Preparation of emissions and sinks inventories for greenhouse gases,
○ National programmes with measures to address climate change,
○ Development, use and transfer technologies to reduce emissions,
○ Sustainable management of sinks,
○ Preparation for impacts of climate change,
○ Technical and scientific co-operation,
○ Awareness, training and education,
○ Limiting greenhouse gas emissions and enhancing sinks, and
○ Reporting on activities and reviews of progress.

The convention places particular emphasis on the needs of developing countries and of countries where impacts of climate change are most likely to occur.

The Framework Convention provides for the establishment of a conference, secretariat and subsidiary body. When the Convention comes into force, financial co-operation, particularly for financial assistance to developing countries, will operate under the guidance of the conference. Until that time, financial assistance will be administered through the Global Environment Facility (UN, 1992). The Framework Convention will come into force when it has been ratified by 50 countries. By 30 April 1993 it had been ratified by 18 countries (Table 10.3).

Reference

UN 1992 *Convention on Climate Change* (non-official record). United Nations Department of Public Information, New York.

○ To review progress in the implementation of the commitments on the provision of financial resources and technology transfer,
○ To review progress toward the UN target of overseas development aid at 0.7 per cent of GNP of developed countries,
○ To review the adequacy of funding and mechanisms to reach the objectives stated in Chapter 33 of Agenda 21,
○ To receive and analyse information from NGOs,
○ To enhance dialogue with NGOs,
○ To consider progress in implementation of environmental agreements, and
○ To give recommendations to the UN General Assembly related to Agenda 21.

The Commission met for the first time in June 1993 and is made up of representatives of 53 states, observers from other nations, NGOs, UN agencies and other international organizations. The session reached agreement on a thematic programme of work for 1993–1997. In each year from 1994–1996, the following five cross-sectoral clusters of Agenda 21 will be reviewed:

○ Critical elements of sustainability, i.e., international economic policies, combating poverty, changing consumption patterns and demographic dynamics and sustainability,
○ Financial resources and mechanisms,
○ Education, science, transfer of environmentally-sound technologies, co-operation and capacity-building,
○ Decision-making structures, i.e., integrating environment and development in decision-making, international institutional arrangements, international legal instruments and mechanisms and information for decision-making, and
○ Roles of major groups such as women, youth, indigenous peoples, NGOs, local authorities, trade unions, business, scientists and farmers.

In each year the Commission will place particular emphasis on different chapters of Agenda 21 within each of the cross-sectoral clusters. The sectoral clusters will be reviewed in three phases, as follows:
1994 – Health, human settlements and fresh water; and toxic chemicals and hazardous wastes.
1995 – Land, desertification, forests and biodiversity.
1996 – Atmosphere, oceans and all kinds of seas.
The 1997 session will have an overall review and appraisal of Agenda 21 in preparation for the special session of the UN General Assembly to be held in the same year.

International Agreements at UNCED

In addition to the Conventions on Climate Change (Box 10.1) and Biological Diversity (Box 10.2) which were opened for signature at UNCED, a published document on forests was produced, the 'Non-Legally Binding Authoritative Statement of Principles for a Global Consensus on the Management, Conservation and Sustainable Development of All Types of Forests', otherwise known as the Forest Statement (UN, 1992c). This statement is a compromise which arose from efforts by some developed countries to produce a binding forest convention. This was opposed by a number of developing countries on grounds of national sovereignty and that they should be compensated in aid for any commitments made to forego the use of natural resources (IIED, 1992). The Forest Statement covers all aspects of use, information, management, policies, ecology, conservation, energy production, agro-forestry,

Box 10.2 The Convention on Biological Diversity

The UN Convention on Biological Diversity is based on the recognition of the value of biological diversity (both intrinsic value and value in terms of ecology) genetic diversity, social, economic and cultural systems, education, science and aesthetics. The preamble to the Convention raises concern over reductions in biological diversity caused by human activity, and mentions the lack of information and of relevant capacities to provide understanding of the issue. The Convention registers the need to prevent reductions in biological diversity through precautionary action, e.g., by conserving natural ecosystems and habitats. Partnerships, co-operation and communication are recognized as important in allowing groups such as indigenous peoples and women, together with the rest of society, to contribute fully to the conservation and sustainable use of biological diversity. Biological diversity in the Convention is taken to mean "the variability among living organisms from all sources including, *inter alia*, terrestrial, marine and other aquatic ecosystems and the ecological complexes of which they are part; this includes diversity within species of the ecosystems".

The Convention recognizes the right of states to exploit their own resources in following their environmental policies. It also asserts the responsibility of states to ensure that activities within their control do not cause damage to the environment of other states and highlights the need for states to co-operate with one another. The objectives of the Convention are:
○ The conservation of biological diversity,
○ The sustainable use of the elements which make up biological diversity, and
○ The just distribution of the benefits arising from the use of genetic resources.
The distribution of the benefits is intended to include funding, access to genetic resources and transfer of technologies.

There are a number of themes which are highlighted in the Convention:
○ Policy development: the development of national strategies for biological diversity and its integration into sectoral and cross-sectoral policies.
○ Monitoring and data: identification and monitoring of a) biological diversity for conservation and sustainable use and b) of human activities having an impact on biological diversity. Development and organization of the data from monitoring.
○ *In-situ* conservation: protected areas and their management for conservation of biological diversity, environmentally-sound and sustainable development near protected areas, restoration of degraded ecosystems, regulation of biotechnology, control of alien species, involvement of indigenous communities, legislation and regulation for protected species.
○ *Ex-situ* conservation: international co-operation on the conservation of organisms in their countries of origin, rehabilitation and reintroduction of threatened species, regulation of collection of biological resources in natural habitats.
○ Sustainable use and biological diversity: integration of biological diversity into decision-making, measures to avoid adverse impacts as a result of use, encouragement of traditional practices compatible with conservation and sustainable use, local participation in remedial action, co-operation between government and the private sector.
○ Genetic resources: the sovereign rights of states over natural resources mean that they determine access to genetic resources but should facilitate access by other states within the terms of the Convention. The Convention provides for close co-operation between those states providing and those acquiring genetic resources on access, research and sharing the results and benefits of research, development and use of genetic resources.
○ Impact assessments: introduction of environmental impact assessment with reference to biological diversity for projects and policies. Information exchange and consultation between countries on the likely adverse effects, emergency responses and liability.
○ Research and training: scientific and technical education to support monitoring and conservation. Research on biological diversity. Exchange of information and results of research.
○ Public education and awareness: promoting understanding of biological diversity and its importance.
○ Technology: access to, and transfer of, technology (including biotechnology) on fair terms to developing countries while respecting intellectual property rights. Human resources development and institution-building in science and technology.
○ Financial resources and mechanisms: financial support for national activities under the Convention. Provision of new financial resources by developed country parties to enable developing countries to implement the Convention. A mechanism for distribution of financial resources will be set up under the Conference of the Convention Parties.
○ Conference, Secretariat and subsidiary bodies: the Conference is to meet after the Convention enters into force at regular intervals and will establish the guidelines for detailed operation of the Convention and review implementation, amending the operation of the Convention through protocols and annexes as necessary. The Conference will designate a Secretariat from existing organizations and will establish a subsidiary body to provide scientific and technical assessments of biological diversity and information on technologies with actual or potential use in implementing the Convention.

At present there are two Annexes to the Convention. Annex I will contain information on ecosystems, habitats, species and genomes gathered under the identification and monitoring provisions of the Convention and Annex II provides a structure for arbitration.

The Convention was opened for signature at UNCED in July 1992 and will enter into force after 30 ratifications. By 30 April 1993, 15 countries had ratified the Convention (Table 10.3).

Reference

UN 1992 Convention on Biological Diversity (non-official record). UN Department of Public Information, New York.

economics, traditional uses, research, trade and co-operation.

Agenda 21 mentions the need for the establishment of an international convention to combat desertification (UN, 1992b para. 12.40). Following discussion at the UN General Assembly, the first session of the Intergovernmental Negotiating Committee for the elaboration of an international convention to combat desertification was held at UNEP in May–June 1993. Issues of links with trade, debt and poverty, regional focuses for action and financial support seem likely to continue to be topics for discussion. Agenda 21 suggested June 1994 as the date for finalizing the convention.

References

ICWE 1992 *The Dublin Statement on Water and Sustainable Development*, International Conference on Water and the Environment: Development Issues for the 21st Century, Dublin.

IIED 1992 Conventions on the environment: more to come, *Perspectives*, **9**, 9, International Institute for Environment and Development, London.

INFOTERRA 1993 Statistics for 1992 demonstrate increased growth, *Infoterra Bulletin*, **15**(2), 1.

UN 1990 *Guidelines for National Reports*, Preparatory Committee for the United Nations Conference on Environment and Development, First Session, 6–31 August 1990, A/CONF.151/PC/8 and A/CONF.151/PC/8/Add.1, United Nations General Assembly, New York.

UN 1992a *Report of the United Nations Conference on Environment and Development* (Rio de Janeiro, 3–14 June 1992) A/CONF.151/26 (Vol. I), United Nations, New York.

UN 1992b *Report of the United Nations Conference on Environment and Development* (Rio de Janeiro, 3–14 June 1992) A/CONF.151/26 (Vol. II), United Nations, New York.

UN 1992c *Report of the United Nations Conference on Environment and Development* (Rio de Janeiro, 3–14 June 1992) A/CONF.151/26 (Vol. III), United Nations, New York.

UN 1992d *Nations of the Earth Report – United Nations Conference on Environment and Development: National Reports Summaries*, Volume 1, United Nations Conference on Environment and Development, Geneva.

UN 1992e *Nations of the Earth Report – United Nations Conference on Environment and Development: National Reports Summaries*, Volume 2, United Nations Conference on Environment and Development, Geneva.

UN 1992f *Overview of National Reports*, Preparatory Committee for the United Nations Conference on Environment and Development, Fourth Session, 2 March–3 April 1992, A/CONF.151/PC/98, United Nations General Assembly, New York.

UN 1993a *Register of Environment Activities of the United Nations System, 1991*, United Nations, New York.

UN 1993b *Nations of the Earth Report – United Nations Conference on Environment and Development: National Reports Summaries*, Volume 3, United Nations Conference on Environment and Development, Geneva.

UNCED 1992 *The Global Partnership for Environment and Development: a Guide to Agenda 21* (draft September 1992) United Nations Conference on Environment and Development, Geneva.

UNEP 1992 *INFOTERRA Network*, United Nations Environment Programme, Nairobi.

The Global Environment Facility

The Global Environment Facility (GEF) started as a three-year (1991–1993) pilot programme jointly managed by UNDP, UNEP and the World Bank. The Facility was set up to assist developing countries deal with four main global environmental problems:

○ Reducing the atmospheric load of greenhouse gases which cause global warming,
○ Preserving the earth's biological diversity and its sustainable use and maintaining natural habitats,
○ Arresting the degradation of international waters, and
○ Protecting the stratospheric ozone layer from further depletion.

Global Environment Facility resources are used to explore ways of assisting developing countries to protect the global environment and to transfer environmentally benign technologies. This is done through the provision of grants and concessional funding for investment projects, technical assistance and, to a lesser extent, monitoring and research. All countries with a per capita income of less than US$ 4,000 a year and a UNDP programme in place are eligible for GEF funds (GEF, 1993a).

Twenty-four countries were present at the first meeting in Paris when the GEF was established in November 1990. From this initial group, composed equally of developed and developing nations, the number of participating countries rose to 46 by March 1993. A total of US$ 1.3 billion has been pledged for the pilot phase, most of which will have been committed by the end of 1993. The projects in the pilot phase have been submitted and approved as five portions (tranches) throughout the three-year period, representing a total of 113 projects worth about US$ 727 million (GEF, 1993b). Details of the GEF-financed projects on biological diversity are shown in Table 10.2. By May 1993, 32 projects worth US$ 250 million had received final approval (GEF, 1993c). The distribution of resources by May 1993 was as follows: 42 per cent for conserving biodiversity, 40 per cent for combating global warming, 17 per cent for protecting international waters, and 1 per cent for the reduction of ozone layer depletion (GEF, 1993b).

Global Environment Facility projects are identified through co-operation between UNEP, UNDP, the World Bank and the governments of the countries involved. Proposals for projects may also be made by NGOs, UN agencies or other organizations. Projects are screened by the Scientific and Technical Advisory Panel (STAP) before the Implementation Committee (made up of representatives of UNEP, UNDP and the World Bank) endorses submission to the GEF participants' meetings (GEF, 1993b). The Scientific and Technical Advisory Panel consists of 15 eminent scientists from industrial and developing countries. This independent group has formulated criteria and priorities for project selection, analytical frameworks for international waters, global warming and biodiversity conservation, and priorities for research which are currently under preparation. The members also review project proposals and advise on whether projects meet the established criteria and whether the priorities of the projects justify support (GEF, 1993a).

Responsibility for technical assistance activities is held by UNDP, which also helps to identify projects through pre-investment studies and runs the small grants programme (a fund that supports community-based activities by grassroots organizations and NGOs in developing countries). Ensuring that the GEF operates in line with international environmental conventions and agreements is the responsibility of UNEP, which also provides the secretariat, helps with environmental expertise for the GEF process, and is responsible for a number of convention support activities. The World Bank administers the Facility, acts as the repository of the Trust Fund, and is responsible for investment projects. The World Bank provides the GEF Chairman and the GEF Administrator's Office (GEF, 1993a).

As a result of UNCED, the GEF is being restructured. Chapter 33 of Agenda 21 (UN, 1992) states that it should be restructured to help to implement the global elements of Agenda 21, aiming to achieve transparent and democratic governance and universal participation, while at the same time ensuring predictability of and continued access to funding, paying particular attention to the needs of developing countries. The UNEP Governing Council adopted a decision in May 1993 to endorse further these principles of restructuring of the GEF. Replenishment of the GEF, expected to be decided in December 1993, will accompany the restructuring which will become operative in early 1994 (GEF, 1993a). A review of the Pilot Phase is currently in progress.

The GEF is the entity entrusted with the operation of the funding mechanism on an interim basis for the Conventions on Climate Change (Box 10.1) and Biological Diversity (Box 10.2), which are expected to enter into force in 1994–1995. The GEF also provides support funding to complement that from the Montreal Protocol to certain countries, particularly in Eastern Europe (GEF, 1993b).

References

GEF 1993a Information provided by Mikko Pyhala, Global Environment Facility Unit, UNEP, Nairobi.

GEF 1993b *Restructuring, Replenishment and the Record of the Pilot Phase.* Unpublished Information Sheet, Global Environment Facility, World Bank, Washington DC.

GEF 1993c *GEF Bulletin and Quarterly Operational Summary,* **8,** May 1993, Global Environment Facility, World Bank, Washington DC.

UN 1992 *Report of the United Nations Conference on Environment and Development* (Rio de Janeiro, 3–14 June 1992) A/CONF.151/26 (Vol. I) United Nations, New York.

International Agreements

A Register of International Treaties and Other Agreements in the Field of the Environment is maintained by UNEP. About 40 per cent of the agreements deal with marine systems and more than 25 per cent deal with nature conservation and terrestrial living resources (UNEP, 1991a). This register does not take account of bilateral agreements on such shared resources as fresh waters, but the proportions given do give an idea of the relative numbers of international agreements on different environmental themes which have been entered into by groups of states.

A complete list of treaties and agreements was given in the second edition of this report (UNEP, 1989) and the third edition (UNEP, 1991b) focused on the Montreal, Basel and UN

ECE Protocols (see below). The parties to a number of selected international agreements in natural resources, atmosphere, marine resources and hazardous substances are given in Table 10.3. The treaties and agreements are too numerous to discuss each, in detail, here. This edition, therefore, concentrates on selected agreements on conservation and natural resources, atmosphere and wastes and hazardous substances.

Conservation and Natural Resources

The Convention on Nature Protection and Wildlife Preservation in the Western Hemisphere (Washington, 1940) is an early agreement on species' protection and natural areas, with a primary focus on national parks, which is linked in implementation to the Organization of American States (OAS). The Convention is meant to aid uniform national park criteria in the region but its extension into national law has been uneven, a factor which has been linked to a lack of co-ordination by a specific secretariat and supporting administration (Forster and Osterwoldt, 1992). Although not specifically provided for, some assistance activities in training, research, technology transfer and management do take place. It seems possible that the role of the Convention may be extended; an OAS working group has recommended a comprehensive environmental programme for the region and one possibility would be to provide the Convention with greater capabilities for administration, co-ordination and the raising of finance (Forster and Osterwoldt, 1992).

The Antarctic Treaty is a focus for great interest because of its role in regulating activities in an almost untouched environment. Membership of the Treaty is open to any member state of the UN, and to others invited by the unanimous agreement of the Treaty Parties. Antarctica has been the subject of debate at every UN General Assembly since 1982, where some states which are not party to the Treaty have proposed that Antarctica should be regarded as a "Global Common" and governed by the UN. Treaty Parties have adopted an agreed approach of non-participation in such UN debates, communicating a consensus view through a co-ordinator. The Treaty is described in detail in Box 10.3.

The African Convention on the Conservation of Nature and Natural Resources (Algiers, 1968) aims to ensure effective conservation, use and development of natural resources. Development elements within the Convention acknowledge the importance of environmental management and provide an established structure for integration of environment and development, taking account of biological diversity and natural ecosystems (Forster and Osterwoldt, 1992). The Convention makes particular provisions for the establishment of protected areas for representative ecosystems (see also Part 3: Natural Resources for further discussion of designated areas). However, the level of activity associated with the Convention is low, because there is no provision for compliance monitoring or reporting. A draft amendment has been developed but is still being considered (Forster and Osterwoldt, 1992). The Association of South East Asian Nations' (ASEAN) Agreement on the Conservation of Nature and Natural Resources (Kuala Lumpur, 1985) is similar in outlook, being a regional agreement which aims to integrate environmental concerns into

Box 10.3 The Antarctic Treaty

The Antarctic Treaty is in fact an umbrella for a system of agreements. The Treaty itself regulates activities on the continent by banning military activity, nuclear explosions and disposal of radio-active wastes, guaranteeing freedom for scientific research and setting aside territorial disputes between certain of the Treaty Parties. Over 200 measures have been adopted on environmental protection and conservation, as well as others such as meteorology and telecommunications. There are five separate international agreements which, together with the main treaty, regulate activities in Antarctica. These are:

○ Agreed Measures for the Conservation of Antarctic Fauna and Flora (1964)
○ Convention for the Conservation of Antarctic Seals (1972)
○ Convention on the Conservation of Antarctic Marine Living Resources (1982)
○ Convention on the Regulation of Antarctic Mineral Resource Activities (1988)
○ Protocol on Environmental Protection to the Antarctic Treaty (1991)

Consultative (voting) status is open to all countries which have conducted significant scientific research in Antarctica (presently 26 countries). Other Acceding Parties (14 countries in all) are invited to attend and participate in the annual Antarctic Treaty Consultative Meetings.

The system of treaties is essentially regional, but research conducted in the region is of global importance; an important example is the discovery of holes in the earth's ozone layer. The agreements within the framework are based, to a greater or lesser extent, upon the anticipation of environmental problems. The Convention on the Conservation of Antarctic Marine Living Resources, which is based on an ecosystem approach, is a good example. This involves conservation and management of commercial stocks of fish, squid and krill to protect the stocks themselves and the predatory seabirds and seals which depend upon them.

Then Convention on Mineral Resource Activities has been superseded by the Environmental Protocol and is unlikely to come into force. It was negotiated with the aim of regulating any future mineral exploitation in Antarctica. The negotiations were complex and lasted for six years from 1982. However, pressures to prevent mineral exploitation ever taking place in Antarctica meant that certain countries felt unable to accede. The difficulties with the Convention stimulated initiatives for comprehensive environmental protection and the Environmental Protocol to the Antarctic Treaty was negotiated in less than one year. This Protocol was signed by 23 countries in 1991 and contains:

○ Designation of Antarctica as a natural reserve, devoted to science,
○ Principles for environmental protection, and
○ Mandatory rules in four Annexes on Environmental Impact Assessment, conservation of wildlife, waste disposal and management and prevention of marine pollution. A later fifth annex gives details for a protected areas system for Antarctica.

The Environmental Protocol makes an indefinite ban on mineral resource activities, with certain mechanisms for review. The Environmental Protocol enables mutual inspection and reporting to evaluate compliance and calls for annual reporting by each party on steps taken to implement the Protocol. It also allows for considerable flexibility in response to new environmental problems through the addition of new annexes. Debate continues among Parties on the possibility of a central secretariat to make communication and response more effective.

---

development planning. However, it has increased emphasis on species protection, genetic diversity and habitat preservation and addresses cross-sectoral environmental issues arising from development such as agricultural chemicals, land-use changes, and pollution control (Forster and Osterwoldt, 1992). The agreement has not yet entered into force.

The third regional agreement described in Table 10.3 is the Convention on the Conservation of European Wildlife and Their Natural Habitats (Berne, 1979) which also aims at species and habitat preservation, integrating conservation issues with a recognition of development effects. An interesting element of this Convention is its specific recognition of migratory species and attempts to involve developing countries within migratory ranges in North Africa in protection efforts under the Convention. This has partially succeeded, but little funding is available to make co-operation more effective (Forster and Osterwoldt, 1992). The Bonn Convention on the Conservation of Migratory Species of Wild Animals (Bonn, 1979) has global coverage, protecting specified migratory animals and conserving their habitat. Specific species or groups of species may be protected by the means of agreements between the states covering migratory ranges. So far a number of agreements are in place and others are being drafted, covering birds, marine mammals and bats.

The Convention on Biological Diversity (see Box 10.2) arose in 1987 when the Governing Council of UNEP requested the Executive Director to consult with governments and to consider the desirability and possible form of an international legal instrument to rationalize activities in the field of biological diversity and to address gaps in coverage (see UNEP Governing Council decision 14/26 17 June 1987). In 1989, the Governing Council requested the Executive Director to convene an Ad Hoc Working Group of Experts on Biological Diversity, the report of which formed the basis for a Working Group of Legal and Technical Experts (later renamed the Intergovernmental Negotiating Committee for a Convention on Biological Diversity) to negotiate a convention. The negotiations were successfully concluded by May 1992 and the Convention was open for signature at UNCED in June 1992. By the beginning of August 1993, 164 states and the European Community had signed the Convention and 25 States had ratified, accepted, approved or acceded to it. Certain issues arose during negotiations: financial support to developing countries to assist implementation; terms by which Contracting Parties (of particular relevance to developed countries) have access to genetic materials of other Contracting Parties (the majority of the genetic materials of interest are located in the tropical forests in developing countries); terms under which developing countries would have access to environmentally-sound technology and new biotechnologies developed from materials found in their tropical forests; the role of intellectual property rights in the transfer of technology; and the sharing of benefits from the exploitation of genetic resources (UN, 1992).

Atmosphere

International agreements on climate and atmosphere make up only 6 per cent of the total number listed in the Register of International Treaties (UNEP, 1991a). However, this field is presently receiving more attention because indications of atmospheric and climate change highlight the need for international agreements.

Europe and North America are the only areas covered by a regional agreement on air pollution but together they account for a large proportion of global air pollution (see Part 1: Environmental Pollution, Atmosphere). The Convention on Long-Range Transboundary Air Pollution (Geneva, 1979) aims to limit, reduce and prevent air pollution as far as possible, including transboundary air pollution (Benedick and Pronove, 1992). The Convention and protocols (on monitoring, sulphur emissions, and nitrogen oxides) were described in the previous 'UNEP Environmental Data Report' (UNEP, 1991b). An additional protocol was adopted in 1991, known as the Protocol concerning the Control of Emissions of Volatile Organic Compounds or their Transboundary Fluxes (Geneva, 1991). This latter Protocol gives a range of options for emission abatement but is not yet in force. Future priorities for control are heavy metals and persistent organic compounds (Benedick and Pronove, 1992).

The status of ratification for agreements on the protection of the ozone layer (the Vienna Convention and the Montreal Protocol) is shown in Table 10.3. The Montreal Protocol was discussed in the previous 'UNEP Environmental Data Report' (UNEP, 1991b). The Vienna Convention and the Montreal Protocol have the aims of promoting co-operation and research to protect the ozone layer and to avoid adverse effects of ozone depletion. The Montreal Protocol puts emphasis on the need to take account of the technical, economic and development needs of developing countries. Many developing countries were involved in the negotiations for the Montreal Protocol and the original 1987 Protocol allowed a 10-year period before developing countries were obliged to follow the reduction schedule for controlled substances (Benedick and Pronove, 1992) (see Part 1: Environmental Pollution, Atmosphere). An Amendment to the Protocol was negotiated in 1990 and came into force in 1992 and this makes a commitment to the transfer of technology to developing countries and financial co-operation. The Multilateral Ozone Fund was set up by the Parties to the Protocol in 1990 on an interim basis to meet agreed incremental costs to developing countries of implementing the control measures. In November 1992, the Parties to the protocol established the Fund on a permanent basis and, although the procedures of the agreements favour fair participation by developing countries, there is a continuing necessity for funds to be made available to finance participation by developing countries (Benedick and Pronove, 1992).

The UN Framework Convention on Climate Change was also opened for signature at UNCED in June 1992. The development process for the Framework Convention began in 1988 as a result of the UN General Assembly recognizing climate change as a common concern of humanity, in resolution 43/53. This resolution was followed by UNEP and the World Meteorological Organization (WMO) setting up the International Panel on Climate Change (IPCC). This Panel has produced detailed assessment reports on the problem of climate change (see Part 1: Environmental Pollution, Atmosphere and Part 2: Climate). The Framework Convention was negotiated by the Intergovernmental Negotiating Committee for a Framework Convention on Climate Change (INC) in the period leading up to UNCED. Details of the Convention are given in Box 10.1.

Wastes and Hazardous Substances

Since the publication of the third 'UNEP Environmental Data Report' (UNEP, 1991b), the Basel Convention on the Control of Transboundary Movements of Hazardous Wastes and Their Disposal (Basel, 1989) has come into force (on 5 May 1992) (UNEP, 1993). This Convention aims to reduce transboundary movements of specified hazardous and other wastes through environmentally sound and efficient management, to minimize the quantity of wastes generated and to support developing countries in waste management (Jones, 1992) (see also Part 8: Wastes). The Basel Convention arose from UNEP Governing Council recognizing the need to transfer environmental protection technology to developing countries (Jones, 1992). Parties to the Convention make particular commitments to promote environmentally-sound waste management, to communicate with one another on the transboundary movements of hazardous and other wastes subject to the Convention, and to co-operate including on the re-importation of wastes. The Convention requires co-operation between a number of international organizations which are involved in wastes and dumping regulation, principally the London Dumping Convention, the International Atomic Energy Agency (IAEA), the International Maritime Organization (IMO) and UNEP.

References

Benedick, R. E. and Pronove, R. 1992 Atmosphere and Outer Space. *Survey of Existing International Agreements and Instruments*, Research Paper No. 26, United Nations Conference on Environment and Development, Geneva.

Forster, M. J. and Osterwoldt, R. U. 1992 Nature Conservation and Terrestrial Living Resources. *Survey of Existing International Agreements and Instruments*, Research Paper No. 25, United Nations Conference on Environment and Development, Geneva.

Jones, W. F. 1992 Hazardous substances. *Survey of Existing International Agreements and Instruments*, Research Paper No. 29, United Nations Conference on Environment and Development, Geneva.

UN 1992 *Convention on Biological Diversity* (non-official record). United Nations Department of Public Information, New York.

UNEP 1989 *United Nations Environment Programme Environmental Data Report 1989/90*, Basil Blackwell, Oxford.

UNEP 1991a *Register of International Treaties and Other Agreements in the Field of the Environment*, UNEP/GC.16/Inf.4, United Nations Environment Programme, Nairobi.

UNEP 1991b *United Nations Environment Programme Environmental Data Report 1991/92*, Basil Blackwell, Oxford.

UNEP 1993 *Status of Global Environmental Conventions as at 30 April 1993*, Environmental Law and Institutions Programme Activity Centre, United Nations Environment Programme, Nairobi.

TABLE 10.1 Issues for concern and capacity-building initiatives identified by individual nations in their reports to UNCED, 1992

Region/country	Water resources	Water quality/supply	Marine environment	Air pollution	Climate change	Soils/desertification	Forests	Biodiversity	Agriculture	Mining	Wastes	Energy	Industry/agrochem.	Natural disasters	Population	Food security/quality	Health	Urban/housing	Economy/poverty	Legislation/institutions	Monitoring/research	Awareness/education	Infrastructure	Women
AFRICA																								
SADCC[a]	●○	●○			●	●○	●○	●○	●○					●	●				●○	●○	●	●○		
Algeria	●○	●○	●○		●○	○	○	●		●○		○		●	●		●			○				
Angola	●	●○	●○				●○	●	●○	●	○			●	●	●○	●○	●○	●	○		●○	●	
Benin	●	●○	●○			●○	●○	○	●					●		●○	●○	●○	●	○		○		○
Botswana	●	●○		●		●○	●	○	●○	●	●○	●		●○				●	●○	●○	●	●		
Burkina	●○	●○				●○	●	○	○	●	●○	○			●○	●	●		●	○		○	●○	○
Burundi	●○	●○			●	●	●○	●○	●○			●○						●○	○	●				
Cameroon	●	●○	●○	●		●○	●○		●○	○	●○		●		●○	●	●○	●○	●	●○	●			
Cape Verde	●○	●	●○			●○	●○	●○	●		●	●○		●		●	●	●○	●	●○	●			
Cent. African Rep.	○	●○		●		●○	●○	●	●○			●○	●		●	●	●	●○	●	●○	●	●○	●○	
Chad	●					●○	●○	○	●○			○	●			●		●	●○	●○	○	●○	●○	○
Congo		●	●		●	●○	●○		●○	●	○	○			●	●○	●○		●○			●○		●
Côte d'Ivoire	●	●○	●		●	●○	●○	●○		●○	○		●		●	●	●		●○	○	●○			○
Djibouti		●	●○			●	●○	●○				●			●		●		○			○		
Egypt	●○	●○	●○	●		●	○	○	●			●	●	●○		●	●○	●	●	○	○			
Equatorial Guinea		●○	●		●		●○	●○	●			●					○	●	○		●○	●○		
Ethiopia	●○	●○			○	●○	●○	○	●○					●○	●○	●○		○	●	●		●○	●○	○
Gabon		●	●○			●	●○	●○	○			●		●		●○	●○	○	●	○		
Gambia	●○	●○	○		●	●○	●○	●				●			●	●○	●		○				
Ghana	●○	●	●○	●○		●○	●○	●○	●○	●○		●○	●		●○	●	●○	●○	●○	●○	●	●○		
Guinea	●○	●○	●			●○	●○	●○	●○	●	●○	○		●○	●○	●	●○	●○	●○	○	●	●○		
Guinea-Bissau	●○	●	●○			●○	●○		●	○						●○	○		●			●		
Kenya	●○	●○	○	○		●○	●○	●○	●○		●	●		●○			●	●○	●○	●		●○		
Lesotho		●				●○	●	●○	●	●		○			●○		●○	●○	○	○	●		
Malawi	●○	●○				●○	●○	●○				○			●		●	○	●○		○		○	○
Mali	●○	●○		●○	●	●○	●	●○	●○		●○		○		●		○		●○	●○	○			
Mauritania	●○	●	●			●○	●○	●○	●○					●	●		●	●	○	●		○		
Mauritius	●	●○	●○	●		●	●○	●○	●	●○		●○		●○		●○		●○		○	●○	○		○
Morocco	●○	●○	○	●○		●○	●○	●○	●	●		●	●		●○	●○	●○	○	○	○	○	●		
Mozambique	●○	●○	●○			●○	●○	●○	●○			○		●		●○		●	●○	●		○		
Namibia	●					●	●○	●○	●○		●○		●○		●		●		●○	○			
Niger	●	●○			●	●○	●○	○	●○		●○	○			●○	●○	●○	●	●	○	○	○		
Nigeria	●	●○	●○			●	●○	●○			●		●		●○		●○			○		○		○
Rwanda		●				●○	●○	●	●○			●			●○	●	●○	●○	●○	○		●○		
São Tomé & Príncipe		●○				●	●○		●			●					●○	●	●○	●	●○			
Senegal	●○	○			○	●	●○	●○	●		●		●○			●○	●	○	●	●	●			
Seychelles	○	○	●○			●○	○	●		●○	○				●	●		●○		○		●		●
Sierra Leone		●	●○			●	●○		●○	●	●				●○		●○	●	●○	●○		●○		●○

Continued

TABLE 10.1 Continued

Region/country	Water resources	Water quality/supply	Marine environment	Air pollution	Climate change	Soils/desertification	Forests	Biodiversity	Agriculture	Mining	Wastes	Energy	Industry/agrochem.	Natural disasters	Population	Food security/quality	Health	Urban/housing	Economy/poverty	Legislation/institutions	Monitoring/research	Awareness/education	Infrastructure	Women
South Africa	●○	●○	●○	●○		●○	●	○	●	●	●○		●				●○			○	○	○		
Sudan	●	●	●			●○	●○	●○	●○		●○	●	●	●	●		●○	●	●○	●○	○	○		●○
Swaziland	○	●○		●○		●○	●○	●○	●○	○	●○		●	●○	●○	●	●○		●○	○		○		○
Tanzania	○	●○	●○		●	●○	●○	●○	●○	●	●		●			●	●○		●○	○	●○	○	●	
Togo		●○	●○	●		●○	●○	●○	●○		●○		●○			○	●○	●	●○	●○		○		
Tunisia	●○	●	●○			●○	○	●○			●○		●	●○			●○	●○	○	●			●○	○
Uganda	○	●			○	●	●○	●	●		●○	●○	●			●	●○	●	●○	●○	●	●	●○	
Zaire		●				●○	●○	●○	●			○	●○			●	●○	●	●		●	●	●○	
Zambia	●	●		●	●	●	●○	●○	●	●	●	○	●○			●	●	●	●	●○	○	○		
Zimbabwe	●○	●○		●		●○	●○	○	●○	○	●	○	●○	●	○		○		○	○	○	●○	○	
NORTH AMERICA																								
OECS[b]	●	●	●			●	●	●	●			●				●				●	●	●	●	
Antigua & Barbuda	●	●	●	●		●	●	●	●		●○		●○		●	●			●	○	○	●○	●○	
Bahamas	●○	●○	●○	○		●		●○	●○		●○			●○	●		●○	●○	●○	●○	●		●○	
Barbados	●	○	●○			●		○	●		●○	○	●		○	○	●○	●○	●○	●				
Belize	●	●○	●			●○	●○	●		●○	○		●○	●		●	●○	●	●○	●	●	●	●○	●○
Br. Virgin Is	●		●○		●		●○	●○	●○		●○			●○				●○		●○	●○	○	●	
Canada		●○	●○	○	●		●○	●○	●		○		●○				●		○	○	○	○		
Costa Rica	○	●○	●		●○	○	○	●	○		●	○	●			●	●		○	○	○			
Cuba[c]			○				○	○	○		○	○	○			○	○		○	○	○	○		
Dominican Rep.	●○	●○	●○			●	●○	●○	●○	●		●○	●○				○	●○	○	○	●			
El Salvador	○	●○	●○	●		●○	●○	●○			●		●		●	●	●	●	●○	●○	●○			
Guatemala		●○	●○	●		●	●○	●	●○			●	○		●	●○	●	●○	○	○	●○	●○		
Haiti	●	●	●○			●	●○	●	●		●		●○		●	●	●○	●	●○	●○	●○	●○		
Honduras	●○	●	●○			●○	●○	●	●○	●○	●		●				●○	●	●○	●○	●○	●○		●
Jamaica	●○	●	●○			●○	●○	●	●○		●○						●○	●	●○	○	○	○		
Mexico		●○	●○	●○			●○	●○	●○		●○									○	○			
Neth. Antilles	●	●	●	●		●○		●○			●		●			●		●	●○	○	●○	●○		
Nicaragua	●	●	●○			●	●○	●○	●	●		●○	●		●○	●	●○	●	●○	●○	●	○	○	
Panama	○	●	●			●○	○			●○	○		●			●	●		●○	○	○	●		
St Kitts & Nevis		○	●○			●	●	○	○		●○			○				○		○	○	●		
Trinidad & Tobago	●	●○	●○	●○	●○		○	○	●○		○	●○	●		○				●○	○	○	○		
Turks & Caicos Is							○	●												○		●		
USA	●○	●○	●○	●○	●○	○		●○			●○	○	●○			●	●○			●○	○	●○		
SOUTH AMERICA																								
Argentina	○	●	●	●○	●	●○	●	●○	○		●	●	●			●		●	●	○	●			
Bolivia	●○	●○		●		●	●○	●○	●	●	●	●	●			●	●	●	●	●○	●○	●		
Brazil	●	●○	●○	●○		●	●○	●○	●	●	●○	○	●	●		●	●	●	●○	○	●○	○		

Continued

TABLE 10.1 Continued

Region/country	Water resources	Water quality/supply	Marine environment	Air pollution	Climate change	Soils/desertification	Forests	Biodiversity	Agriculture	Mining	Wastes	Energy	Industry/agrochem.	Natural disasters	Population	Food security/quality	Health	Urban/housing	Economy/poverty	Legislation/institutions	Monitoring/research	Awareness/education	Infrastructure	Women
Chile	●○	●○	●○	●○		●	○	●○	●○	●○	●○		●○	●		●	●○	●	○	○	○	○	●	
Colombia		●		●		●	●○	●○		●○										○		○		
Ecuador	○	●○	●	●○		●○	●○	●○	●		●○				●		●	●	●	●○				
Falkland Is			●○					○	●○											○		○		●
Guyana		●○	●		●		●○	○	●○	○	●	○		●○		●	●○	●○	●○	○	●○	●○	●○	○
Paraguay	●	●○				●	●	●○	●○					●○			●○	●		○	○		○	
Peru	●	●				●	●	●○	●			●	●	●			●	●○	●	○		○		
Suriname	●	●	●				●○	●	●○	●○	●			●		●	●○	●○	●○	●		●○	●	
Uruguay	●○	●○	○		●	●	●○	●○	●○		●		●	●		●	●○		○	○		○		
Venezuela	●	●○	●		●○	●○	●○		●		●○		●○			○	●	●	●○	●○	○			
ASIA																								
Afghanistan		●					●	●○	●			●	●○	●	●	●	●	○	○	●○	●	●○	●○	●
Bahrain	●○	●○	●	●	●			●○	●○	●○		●			●		○		●○	●○	●○			
Bangladesh		●○		○		●○	●○	○	●○		●○	●		●○		●	●	●	○			●		
Bhutan	●○			●		●○	●○	●	●○		○	○	○	●○	●○				○	●○	●○	●		
China	●○					●○	●○	○	●○		○	○	○	●○	●	●	●○		●	●○	●○	●		
Cyprus	●	●	●○	●		●		○	●		○		●○						○	○	○			
Hong Kong		●○			●					●○		●							○	○	○			
India	●	●○	●○	●○		●	●○	●	●○		●○	●○		●○	●	●○	●○	●○		○	○	○		
Indonesia	●○	●○				●○	●○	●	●○		●○	●		●○		●○	●○		○	○				
Iran	○	●○	●	●○		●○	●○	●○	●○		●		●		●		●○		○	○	○	○		
Iraq	●○	●○				●○		●	●○		●○				●○	○		○	○		●	○		
Israel	●○	●○	●○	●○			●○	○	●		●○	○		●		●		○	○		○			
Japan	●○	●○	●○	●	●○	●○	●○	●○	●○		●○	●	●	●○		●	●○	○	○	●	○			
Jordan	●○	●○				●		●○		●○		●○			○		●○				●○			
Korea	○	●○	●		●○		●○	●	●○	●	●○	●○	●				●	●○	●○	○		○		
Korea, Dem.	●○	●○	○	○		○	●○	●○	○	○	●		●				○	○	●○					
Laos	●○	●○		●		●	●○	●○	●○	●			●○			○		●○	○	○				
Lebanon	●○	●○	●○	●○			○	○	●○		●		●		●		●○		●○	●	●○	●		
Malaysia	●○	●○	●○	●○	●○	●○	○	○	○		●		●						○	○	●○			
Maldives	●○	●				●			●○		●		●				●		○	●○	●○	●		
Mongolia	●○	●○		●○	●	●○	●○	●○	●		●		○		●			○	○	●○		●	○	
Myanmar			●○			○	●○	●○	○			○				●		○	○	●○	●○	○		
Nepal		●		●		●○	●		●○		●○	○	●	●		●	●○		○	○				
Oman	●○	●○	●	○			●○	●○	●○		●○	○			●		○	○					○
Pakistan	●○	●○	●	●○		●	●○	●○	●○		●○		●		●		●○		●○	○	○			
Palestine	●	●	●			●		●		●		●		●○	●		●				●			
Philippines	●○	●○	●○		●		●○	●○	●○	●○	●○	●	●○	●	●○	●	●○	●	●○	●○	●○	●○	●○	
Qatar	●○	●○	●		●		●	●○		●○			●○			●○			●○	●○	●			
Saudi Arabia	●○	●○	●○	●	●○	●○	○	●○	●○		●○		●○	●		●	●		○	○	○	●		

Continued

TABLE 10.1 Continued

Region/country	Water resources	Water quality/supply	Marine environment	Air pollution	Climate change	Soils/desertification	Forests	Biodiversity	Agriculture	Mining	Wastes	Energy	Industry/agrochem.	Natural disasters	Population	Food security/quality	Health	Urban/housing	Economy/poverty	Legislation/institutions	Monitoring/research	Awareness/education	Infrastructure	Women
Singapore	●○	●○	●○	○			○	●			○		○					●○		○		○		
Sri Lanka	●○	●○	●○	●		●			●○		●			●	●			●○	●	●○			●○	
Syria		●○		●○		●○	●○		●		○		●○	●○	●○	○	●○		○	○		○		
Thailand	●	●	●	●○	○	●	●○	●○	●		●○	○	●					●		○	○			
Turkey		●○	●○	●○		●○	●○	●	●○		●○	●○	●○			○		●○		○	○			
Viet Nam		●○	●	●		●○	●○	●○			●	●	●	●				○	●○	●○	●○	○		
Yemen	●○	●○	●○			●○		●						○			●		○	●○	●○	●		○

EUROPE

Region/country	Water resources	Water quality/supply	Marine environment	Air pollution	Climate change	Soils/desertification	Forests	Biodiversity	Agriculture	Mining	Wastes	Energy	Industry/agrochem.	Natural disasters	Population	Food security/quality	Health	Urban/housing	Economy/poverty	Legislation/institutions	Monitoring/research	Awareness/education	Infrastructure	Women
Arctic[d]		●	●										●						○	●○				
EEC[e]	●	●○	●○	●○	●○	●○	○	●○	●○		●○	●○	●○						○	○	○			
Austria		●○		●○		●○	●○	●○	●○		●○	○	●○					●		○	○	○		
Belgium		●○	●○	●○	●○	●	○	○	○		●○	○							○		○			
Bulgaria	●○	●○		●○		●○	○	○	●○			●○			●		●		○	●○	○	●○		
Croatia		●○	●○	●○		●	●○	●○	●○	●○	●○	●○	●○		●	○	●○		●○	●○	●○	○	●○	
Czech & Slovak Fed. Rep.	●○	●		●		●	●○	●○	●○	●○	●	●○	●				●		●○	●○	●○	●	●○	
Denmark	●	●○	○	●○		●○	●	●○	●○		●○	●	●○					○	○	○	○	○		
Finland		●○	●	●○	●○	●	●○	●○	●○		●	●	●○					○	○	○	○			
France	●	●○	●○	●○		●○	●○	○	●		○	●	●					○	○	○	○			
Germany	○	●○	●○	●○	●○	●○	●○	●	●○		●○		●○					●○		○		●○	●○	○
Greece	●○	●○	●○	●○		●	●	●○	●		●○		●○					●	●○	●○		●○	●○	○
Hungary	●	●		●		●		●○	●		●	●	●			●		●	●○	●○	●○	●		
Iceland		●	●○	●○		●○	●○	●○	●		○		●						○	○	○	○		
Ireland		●○	●	●○		●			●		●○		●○						○	●○		●○	●	
Italy	●○	●○	●			●○	●○	●	●○		●○	●○	●○	●○		○			●○	○	●○	○		
Malta	●○		○			●○	●	●	●○									●	○	○		○		
Monaco			●○	●○			●○	○			○			●					○	○	○			
Netherlands	●○	●○	●○	●○	●○	●	●	●	●○		●○	●○	●○						●○	○	○			
Norway		●○	●○	●○		○	●○	●○	○			○	●○		○				○	○				○
Poland	●○	●○	●○	●		●	●	●○	●		●		●		●	●			○	○	○			
Portugal		●○	●○	●○	○	●○	●○	●○			●○			●○		●○	●			○	○	○		
Romania	○	●○	●○	●○		●○	●○	○	●		●	○	●○						○	○				
Spain	●○	●○	●○	●○		●○	●○	●○	●○	●	●○	○	●						●○	○	○		○
Sweden		●○	●○	●○	○	●	●○	●○	●○		●○		●○			●			○	○				
Switzerland		●○	●○	●○		○	●○	●○	○		○		●○		○			○	○					
UK	●○	●○	○	●○	●○		●○	●○	○		●○	○	●○		●			○	○	○	○	○		
Yugoslavia		●○	●	●	●	●	●○	●			●		●		○	●○	●		○	○	●			

FORMER USSR[f]

Region/country	Water resources	Water quality/supply	Marine environment	Air pollution	Climate change	Soils/desertification	Forests	Biodiversity	Agriculture	Mining	Wastes	Energy	Industry/agrochem.	Natural disasters	Population	Food security/quality	Health	Urban/housing	Economy/poverty	Legislation/institutions	Monitoring/research	Awareness/education	Infrastructure	Women
		●○	●	●○	●	●	●○		●○				●					●	●	●○	●○	○		
Azerbaijan	●	●○	●	●		●○	●○	●○	●○	●	●	●	●○	●		○		●	○	○	○	●		

Continued

TABLE 10.1 Continued

Region/country	Water resources	Water quality/supply	Marine environment	Air pollution	Climate change	Soils/desertification	Forests	Biodiversity	Agriculture	Mining	Wastes	Energy	Industry/agrochem.	Natural disasters	Population	Food security/quality	Health	Urban/housing	Economy/poverty	Legislation/institutions	Monitoring/research	Awareness/education	Infrastructure	Women
Belarus	●	●○		●○		●○	●○	●○			●○		●				●		●	○	○	○		
Estonia	●	●		●		●		●○	●		●		●				●		●○	●○			●○	
Kazakhstan	●	●○	●	●○	●		●	●○	○		●		●	●○			●		●○	●○		○	○	
Latvia	○	●○		●			●	●○	●		●○		●○			●	●	●	●○	●○	○	●○		
Lithuania	●	●	●	●○			●	●○	○		●	●○	●		●	●	●	●	●○	○		●○		●
Russia	●	●○	●○	●○		●	●	●○	●○		●	●	●○	●		●	●○	●○		○		●○	●	
Ukraine	●○	●○	●	●○		●○	●○	●○	●		○		●○				●			○	○	○	○	

OCEANIA

Region/country	Water resources	Water quality/supply	Marine environment	Air pollution	Climate change	Soils/desertification	Forests	Biodiversity	Agriculture	Mining	Wastes	Energy	Industry/agrochem.	Natural disasters	Population	Food security/quality	Health	Urban/housing	Economy/poverty	Legislation/institutions	Monitoring/research	Awareness/education	Infrastructure	Women
PIDC[g]	●	●	●		●		●	●○	●		●	●	●			●	●	●		●○	●		●	●
Australia	●○	●○	●○	●○	●○	●○	○	●○	●○	●○	●○	●○	●○	●			●○	●○		○	●○			○
Fiji	●○	●○	●○		●		●○	●	●		○	●○		●	●		●	●		●○		●		
Marshall Is	●	●	●○		●			●○	●	○	●	●		●			●			○	○	●		
New Zealand	●○	●○	●○	●○	●○	●○	○	●○	●○			●○	●○				●○			●○	○	○		○
Niue	●	●○	●○			●○	●○	●○	●○		●			●○	●○	●○	○	○		●○	○	○		●○
Papua New Guinea		●	●○		●	●	●○	●○	●○	●○			○	●○		●○	●○	●○	●○	●	●○	●		●○
Solomon Is		●○	●○		●	●	●○	●○	●○	●○		●	●○				●		●	●○	●○	●○		
Tokelau		●	●○		●			●○				○	●○	●		●○	○	○	●○	●○	●○	●○	●○	
Tonga	●○	●○	●○			●	●○	●○			●		●○	●		●	●	●	●○	●○	●○	●○	●○	
Vanuatu	●○	●○	●○		●		●○	●○	●		●			●		●	●		●○	●○	●○	○		
Western Samoa	●○	●○	●○		●	●	●○	●○	○		●		●○				●		●○	●○	○	○		

a The Southern African Development Coordination Conference (SADCC) covers Angola, Botswana, Lesotho, Malawi, Mozambique, Namibia, Swaziland, Tanzania, Zambia and Zimbabwe.

b The Organization of Eastern Caribbean States (OECS) covers Antigua and Barbuda, The Commonwealth of Dominica, Grenada, Montserrat, St Kitts and Nevis, St Lucia, St Vincent and the Grenadines and The British Virgin Islands.

c No specific problem areas mentioned.

d Presented formally by the Government of Finland. Countries participating in the Arctic Environmental Protection Strategy in 1991 comprise Canada, Denmark, Finland, Iceland, Norway, Sweden, USA and the former USSR.

e The European Economic Community (EEC) comprises Belgium, Denmark, France, Germany, Greece, Ireland, Italy, Luxembourg, Netherlands, Portugal, Spain and UK.

f Report for the USSR was presented in 1991, before it ceased to exist as a political entity.

g The Pacific Islands Developing Countries (PIDC) are Cook Is, Fiji, Kiribati, Marshall Is, Micronesia, Niue, Palau, Papua New Guinea, Solomon Is, Tokelau, Tonga, Tuvalu, Vanuatu and Western Samoa.

The table gives a summary of the problem areas and the capacity building initiatives which were highlighted in the summaries of national reports submitted to the UN Conference on Environment and Development. There was no common set of criteria for description of problem areas and capacity building so the above table uses some discretion in allocating problems to particular categories.

Sources:

UN 1992 *Nations of the Earth Report – United Nations Conference on Environment and Development: National Reports Summaries*, Volume 1, United Nations Conference on Environment and Development, Geneva.

UN 1992 *Nations of the Earth Report – United Nations Conference on Environment and Development: National Reports Summaries*, Volume 2, United Nations Conference on Environment and Development, Geneva.

UN 1993 *Nations of the Earth Report – United Nations Conference on Environment and Development: National Reports Summaries*, Volume 3, United Nations Conference on Environment and Development, Geneva.

● – Problem area.
○ – Past and present capacity-building initiative.

TABLE 10.2 Global Environment Facility biodiversity projects

Project	GEF funding (10⁶ US$)	Lead agency[a]	Other funding (10⁶ US$)	NGO involvement	STAP areas[b]	Species rich area	Ecologically diverse area	High level of endemism	Important gene pools/relatives	Sustainable management	Threatened area	Impact on other ecosystem	Migratory habitat	International laws/agreements	Ecosystems approach	Protected areas	Sustainable use of biota	Education/training	Inventories/research	Institutional strengthening	Public awareness
FIRST TRANCHE																					
Congo	10	WB		●	1	●	●				●				●	●					
Kenya	6.2	WB	50	●	1,2						●				●	●		●			
Uganda	4	WB	30	●	1	●	●	●			●					●				●	
Bhutan	10	WB	10	●	5	●	●	●			●	●				●				●	
Laos	5	WB	78	●	1						●					●					
Philippines	20	WB	224	●	1						●		●		●	●				●	
Algeria	10	WB		●	2		●				●		●		●	●			●		
Poland	4.5	WB	3	●	1						●					●			●	●	
Brazil – biodiversity	30	WB	117	●	1						●					●		●		●	
Mexico	25	WB	50	●	1,4	●	●	●		●	●					●				●	
East Africa	10	UNDP		●	3	●	●	●	●		●					●				●	
West/Central Africa	1	UNDP		●	3					●	●		●			●			●		
South Pacific – biodiversity	8.2	UNDP	2.5	●	3	●		●			●					●	●			●	
Viet Nam	3	UNDP		●	1,4	●		●								●			●		
Amazon – regional	4.5	UNDP	1.8	●	1,3	●	●	●						●		●			●	●	
Colombia	9	UNDP	1	●	1,4	●		●		●	●				●	●			●	●	
Guyana	3	UNDP	0.5		1	●				●						●			●	●	
SECOND TRANCHE																					
Ghana	7.2	WB	20		2	●					●	●	●	●		●	●			●	●
Malawi	4	WB	8.8	●	5	●		●							●	●	●	●			
Seychelles	1.8	WB	10	●	4			●						●		●				●	
Czech Rep.	2.3	WB	170		5	●		●	●				●	●		●				●	
Slovakia	2	WB			5	●		●					●			●				●	
Peru	4	WB			1,5	●	●	●			●	●				●					●
Nepal	3.8	UNDP		●	5		●	●			●				●	●	●				●
Papua New Guinea	5	UNDP		●	1	●	●	●			●				●	●	●				
Sri Lanka	4.1	UNDP		●	1,4	●		●		●		●			●	●				●	
Argentina	2.8	UNDP		●	4	●			●		●	●			●	●				●	
Belize	3	UNDP	1.2	●	4	●	●	●							●	●	●			●	
Costa Rica	8	UNDP	6.8	●	1	●	●	●		●					●	●	●			●	
Cuba	2	UNDP			4	●		●						●	●	●				●	
Biodiv. country studies (1)	7	UNEP	2.5		3	●								●		●			●	●	
Africa – Lake Tanganyika	10			●	5,4	●		●			●	●				●			●	●	●
THIRD TRANCHE																					
Central Africa – LANDSAT	1.8	WB		●	3														●		
Zimbabwe	5	WB	13	●	6			●	●			●			●	●	●				●
Indonesia	12	WB	22	●	4,2	●	●			●	●				●	●					

Continued

TABLE 10.2 Continued

Project	GEF funding (10^6 US$)	Lead agency[a]	Other funding (10^6 US$)	NGO involvement	STAP areas[b]	Species rich area	Ecologically diverse area	High level of endemism	Important gene pools/relatives	Sustainable management	Threatened area	Impact on other ecosystem	Migratory habitat	International laws/agreements	Ecosystems approach	Protected areas	Sustainable use of biota	Education/training	Inventories/research	Institutional strengthening	Public awareness
Turkey	5	WB	100		4	●	●	●	●							●	●				
Egypt – Red Sea	4.8	WB			6	●				●	●	●	●			●	●				
Danube Delta	6	WB	242	●	6,4					●	●		●		●		●		●	●	
Bolivia	4.5	WB	4.5	●	4	●				●	●					●	●		●	●	●
Ecuador	6	WB	30	●	1	●	●	●		●	●				●	●	●		●	●	
Yemen – Red Sea	2.8	UNDP			1	●				●	●	●	●			●	●				
Jordan	6.3	UNDP		●	1	●	●				●	●	●			●	●				
Dominican Rep.	3	UNDP		●	1,4	●				●	●					●	●				
Uruguay	3	UNDP		●	4	●	●				●					●			●		
E. Caribbean – ENCORE	7	USAID	4		1,4	●		●			●		●	●		●		●		●	
El Salvador	9.5	USAID	20		4						●	●				●			●	●	
Jamaica	6.6	USAID	6.2		1,4	●	●	●			●					●	●		●		●
Nicaragua	5.4	USAID	2.6		1	●	●			●		●				●				●	●
S. Pacific – Env Protection	1.7	USAID			3	●	●	●		●	●			●		●	●				

FOURTH TRANCHE

Project	GEF funding (10^6 US$)	Lead agency[a]	Other funding (10^6 US$)	NGO involvement	STAP areas[b]	Species rich area	Ecologically diverse area	High level of endemism	Important gene pools/relatives	Sustainable management	Threatened area	Impact on other ecosystem	Migratory habitat	International laws/agreements	Ecosystems approach	Protected areas	Sustainable use of biota	Education/training	Inventories/research	Institutional strengthening	Public awareness
Mozambique	5	WB			1,3	●	●			●	●		●			●	●			●	
W. Africa – game ranching	7	WB			3					●	●						●				
Ethiopia	2.5	UNDP			3	●	●	●	●								●		●	●	●
Côte d'Ivoire	3	UNDP			1.4	●				●	●										
Burkina	2.5	UNDP	2.5		6					●	●					●		●	●	●	
Alternatives to slash and burn	3	UNDP			1	●	●	●	●										●		
Biodiv. country studies (2)	2	UNEP			5									●					●	●	
Biodiv. data management	4	UNEP		●	5									●			●		●		
Global Biodiv. Assessment	2	UNEP		●	5									●					●		●
Philippines	25	USAID		●	1	●		●		●	●				●	●				●	
Uganda – action programme	20	USAID		●	1	●					●						●				

FIFTH TRANCHE

Project	GEF funding (10^6 US$)	Lead agency[a]	Other funding (10^6 US$)	NGO involvement	STAP areas[b]	Species rich area	Ecologically diverse area	High level of endemism	Important gene pools/relatives	Sustainable management	Threatened area	Impact on other ecosystem	Migratory habitat	International laws/agreements	Ecosystems approach	Protected areas	Sustainable use of biota	Education/training	Inventories/research	Institutional strengthening	Public awareness
Cameroon	5	WB	2		1	●	●	●		●	●					●	●		●	●	
Mauritius	0.2	UNDP	0		1	●		●			●					●	●		●		●
Mongolia	1.5	UNDP	1.5		5	●	●	●		●	●					●	●		●	●	
Indonesia/Malaysia – rhino	2	UNDP	6		1						●			●		●	●		●	●	

[a] The main lead agencies are the United Nations Development Programme (UNDP), the United Nations Environment Programme (UNEP) and the World Bank (WB). USAID is a Parallel Co-financing agency.

[b] The Scientific and Technical Panel (STAP) has defined the following classifications for biodiversity projects supported by GEF:
1 Tropical forest and Mediterranean.
2 Wetlands.
3 Regional.
4 Marine and coastal.
5 Other.
6 Arid and semi-arid.

Sources:
Data supplied by GEF Unit, Nairobi.
GEF 1993 Bulletin and Quarterly Operational Summary, No. 8, May 1993, Global Environment Facility, World Bank, Washington DC.

TABLE 10.3 Selected international agreements relating to the environment

Region/country	Conservation and natural resources												Atmosphere					Marine						Waste	
	1. Washington Nature Protection	2. Antarctic Treaty	3. African Nature and Natural Res.	4. Ramsar Convention - Wetlands	5. World Cultural & Natural Heritage	6. CITES – Endangered Species	7. Bonn – Migratory Species	8. Berne – European Wildlife	9. International Tropical Timber	10. Plant Genetic Resources FAO	11. ASEAN Nature Conservation	12. Biological Diversity	13. Transboundary Air Pollution ECE	14. Vienna Convention – Ozone Layer	15. Montreal Protocol – Ozone	16. Amendment Montreal Protocol	17. UN Climate Change	18. London Dumping Convention	19. MARPOL 1973/78 – Ship Pollution	20. Protection of Mediterranean – MAP	21. Wider Caribbean – Cartagena	22. UN Law of the Sea	23. Regulation of Whaling – IWC	24. Basel Convention	25. IAEA Notification Convention

AFRICA

Region/country	1	2	3	4	5	6	7	8	9	10	11	12	13	14	15	16	17	18	19	20	21	22	23	24	25
Algeria	r	r	r	r								s		r	r	r	s		r	r					
Angola			r									s					s						r		
Benin			r	r		r				r		s					s								
Botswana			r									s		r	r		s						r		
Burkina	r	r	r	r		r	r			r		s		sr	sr		s								
Burundi			r	r								s					s								
Cameroon	r		r	r		r			r	r		s		r	r	r	s						r		
Cape Verde			r							r		s					s				r		r		
Cent. African Rep.	r		r	r						r		s		r	r		s								
Chad		r		r						r		s		r			s								
Congo	r		r	r			r			r		s			s		s	r	r			r			
Côte d'Ivoire	r		r				r			r		s		r	r		s	r	r				r		
Djibouti	r		r							r		s					s				r		r		
Egypt	r	r	r	r	sr	r	r			r		s		sr	sr	r	s		r	r		r	r	r	r
Equatorial Guinea			r							r			r												
Ethiopia			r	r						r		s					s								
Gabon		r	r	r			r			r		s					s	r	r						
Gambia			r	r						r		s		sr	r		s				r		r		
Ghana	r	r	r	r		r	r			r		s		r	sr	r	s				r		r		
Guinea			r	r						r		s		r	r	r	s						r		
Guinea-Bissau		r	r							r		s					s						r		
Kenya	r	r	r	r						r		s		r	sr		s	r						r	r
Lesotho												s					s								
Liberia	r		r				r	r				s					s				r				
Libya			r							r		s		r	r		s		r	r					
Madagascar	r		r	r	s					r		s					s								
Malawi	r		r	r						r		s		r	r		s								
Mali	r	r	r		r					r		s					s						r		
Mauritania		r	r							r		s					s								
Mauritius			r							r		sr		r	r	r	sr						r		r
Morocco	r	r	r	r	s					r		s		s	s		s	r	r						
Mozambique	r		r	r						r		s					s								

Continued

TABLE 10.3 Continued

Region/country	Conservation and natural resources												Atmosphere					Marine						Waste	
	1.	2.	3.	4.	5.	6.	7.	8.	9.	10.	11.	12.	13.	14.	15.	16.	17.	18.	19.	20.	21.	22.	23.	24.	25.
Namibia						r						s					s					r			
Niger			r	r	r	r	sr			r		s	r	r			s								
Nigeria			r		r	r	r					s	r	r			s	r				r		sr	r
Rwanda			r			r				r		s					s								
São Tomé & Príncipe												s					s					r			
Senegal			r	r	r	r	r	r		r		s	r	s			s					r	r	r	
Seychelles			r		r	r						sr	r	r		r	sr	r	r			r	r		
Sierra Leone									r								s								
Somalia						r	sr															r			
South Africa		c		r		r	r			r			r	r	r			r	r			r			r
Sudan			r		r	r				r		s	r	r			s					r			
Swaziland			r									s	r	r			s								
Tanzania			r		r	r				r		s	r	r		r	s					r		r	
Togo			r			r	s		r	r		s	r	sr			s			r		r			
Tunisia			r	r	r	r	r			r		s	r	r			s	r	r	r		r			r
Uganda			r	r	r	r	s					s	r	sr			s					r			
Zaire			r		r	r	r					s					s	r				r			
Zambia			r	r	r	r				r		s	r	r			s					r			
Zimbabwe					r	r				r		s	r	r			sr								
NORTH AMERICA																									
Antigua & Barbuda						r				r		sr	r	r		r	sr	r	r		r	r	r	r	
Bahamas						r						s	r				s		r			r		r	
Barbados						r				r		s	r	r			s				r				
Belize					r	r				r		s					s					r	r		
Canada				r	r	r		r				sr	r	sr	sr	r	sr	r						r	r
Costa Rica	r				r	r				r		s	r	r			s	r			r		r		r
Cuba		a			r	r				r		s	r	r			s	r			r	r			r
Dominica										r			r	r		r					r	r			
Dominican Rep.	r				r	r				r		s					s	r							
El Salvador	r				r	r				r		s	r	r			s							sr	
Grenada										r		s	r	r			s				r	r			
Guatemala	r	a		r	r	r						s	r	r			s	r				r		s	r
Haiti	r					r				r		s					s	r						s	
Honduras						r	r		r	r		s					s	r							
Jamaica						r		s		r		s	r	r		r	s	r	r			r	r		
Mexico	r			r	r	r				r		sr		sr	sr	r	sr	r			r	r	r	sr	r
Nicaragua	r				r	r				r		s	r	r			s								
Panama	r			r	r	r	r		r	r		s	r	sr			s	r	r			r		sr	
St Kitts & Nevis						r						sr	r	r			sr								
St Lucia						r	r											r			r	r	r		
St Vincent & Grenadines						r													r		r		r		
Trinidad & Tobago	r					r			r	r		s	r	r			s					r	r		
USA	r	c		r	r	r			r				r	sr	sr	r	sr	r	r			r	r	s	r

Continued

TABLE 10.3 Continued

Region/country	Conservation and natural resources												Atmosphere					Marine						Waste	
	1.	2.	3.	4.	5.	6.	7.	8.	9.	10.	11.	12.	13.	14.	15.	16.	17.	18.	19.	20.	21.	22.	23.	24.	25.
SOUTH AMERICA																									
Argentina	r	c			r	r	r			r		s	sr	sr	r	s		r				r		sr	r
Bolivia			r	r	r				r	r		s				s								s	
Brazil	r	c			r	r			r			s	r	r		s		r	r			r	r	r	r
Chile	r	c	r		r	r	r			r		s	sr	sr	r	s		r					r	sr	
Colombia		a			r	r			r	r		s	r			s				r		r		s	
Ecuador	r			r	r	r			r	r		sr	r	r	r	sr			r				r	sr	
Guyana					r	r						s				s									
Paraguay	r				r	r		s		r		s	r	r	r	s					r				
Peru	r	c			r	r			r	r		s	sr	r	r	s				r		r			
Suriname	r			r		r						s				s		r	r						
Uruguay	r	c		r	r	r	r					s	r	r		s			r				r	sr	r
ASIA																									
Afghanistan					r	r						s				s		r						s	
Bahrain					r					r		s	r	r	r	s						r		sr	
Bangladesh					r	r				r		s	r	r		s								r	r
Bhutan												s				s									
Brunei						r							r						r						
Cambodia					r																				
China		c			r	r			r			sr	r	r	r	sr		r	r				r	sr	r
Cyprus					r	r		r		r		s	r	r	r	s		r	r	r		r		sr	r
India		c		r	r	r	sr		r	r		s	r	r	r	s			r				r	sr	r
Indonesia					r	r			r			s	r	sr	r	s			r			r			
Iran				r	r	r				r		s	r	r		s								r	
Iraq					r					r												r			r
Israel						r	r			r		s	r	sr	r	s			r	r				s	r
Japan		c		r		r			r			s	r	sr	r	s		r	r			r			r
Jordan				r	r	r						s	r	r		s		r						sr	r
Korea		c			r				r	r		s	r	r	r	s			r				r		r
Korea, Dem.		a								r		s				s			r						
Kuwait										r		s	r	r								r		s	
Laos					r																				
Lebanon					r					r		s	r	r	r	s		r	r					s	
Malaysia					r	r			r			s	r	r											r
Maldives					r							sr	r	sr	r	sr								r	
Mongolia					r							s				s									r
Myanmar												s				s			r						
Nepal				r	r	r			r	r		s				s									
Oman					r							s				s		r	r			r	r		
Pakistan				r	r	r	r					s	r	r	r	s									r
Philippines					r	r	s		r	r		s	r	sr		s		r				r	r	s	
Qatar					r							s													
Saudi Arabia					r		r						r	r										sr	r

Continued

TABLE 10.3 Continued

Region/country	Conservation and natural resources												Atmosphere					Marine						Waste	
	1.	2.	3.	4.	5.	6.	7.	8.	9.	10.	11.	12.	13.	14.	15.	16.	17.	18.	19.	20.	21.	22.	23.	24.	25.
Singapore						r						s		r	r	r	s	r							
Sri Lanka			r	r	r	sr				r		s		r	r		s							r	r
Syria					r					r		s		r	r					r	r			sr	
Thailand					r	r			r			s		r	sr	r	s							s	r
Turkey					r			r		r		s	r	r	r					r	r			s	r
United Arab Em.						r						s		r	r			r						sr	r
Viet Nam				r	r												s		r						r
Yemen					r							s					s					r			
EUROPE																									
EEC								r					r							r					
Albania					r															r					
Austria		a		r		r		r	r	r		s	r	sr	sr	r	s		r					sr	r
Belgium		c		r		r	r	r	r	r		s	r	sr	sr		s	r	r					s	
Bulgaria		a	r	r	r		r			r		s	r	r	r		s		r						
Croatia[a]												s		r	r		s								
Czech and Slovak Fed. Rep.[a]		a		r	r								r						r						r
Czech Republic[a]						r																			
Denmark		a		r	r	r	sr	r	r	r		s	r	sr	sr	r	s	r	r			r		s	r
Finland		c		r	r	r	r	r	r	r		s	r	sr	sr	r	s	r	r			r		sr	r
France		c		r	r	r	sr	r	r	r		s	r	sr	sr	r	s	r	r	r	r	r		sr	r
Germany		c		r	r	r	sr		r	r		s	r	sr	sr	r	s	r	r			r		s	r
Greece		a		r	r	r	s	r	r	r		s	r	sr	sr		s	r	r	r				s	r
Holy See				r																					
Hungary		a		r	r	r	r	r		r		s	r	r	r		s	r	r					sr	r
Iceland				r						r		s	r	r	r		s	r	r		r	r			
Ireland				r	r		r	r	r	r		s	r	r	sr	r	s	r				r		s	r
Italy		c		r	r	r	sr	r	r	r		s	r	sr	sr	r	s	r	r	r				s	r
Liechtenstein						r		r				s	r	r	r		s							sr	
Luxembourg						r	r	r				s	r	r	sr	r	s	r	r					s	
Malta				r	r	r						s		s	sr		s	r	r	r					
Monaco				r	r							sr		r	r	r	sr	r		r			r	r	r
Netherlands		c		r		r	r	r	r	r		s	r	sr	sr	r	s	r	r		r		r	sr	r
Norway		c		r	r	r	sr	r	r	r		s	r	sr	sr	r	s	r	r				r	sr	r
Poland		c		r	r	r				r		s	r	r	r		s	r	r					sr	r
Portugal				r	r	r	sr	r	r	r		s	r	r	sr	r	s	r	r					s	
Romania		a		r	r							s	r	r	r	r	s							r	r
San Marino												s					s								
Slovakia[a]						r																			
Slovenia[a]												s		r	r	r	s								
Spain		c		r	r	r	sr	r	r	r		s	r	r	sr	r	s	r	r	r			r	s	r
Sweden		c		r	r	r	sr	r	r	r		s	r	sr	sr	r	s	r	r				r	sr	r
Switzerland		a		r	r	r		r	r	r		s	r	sr	sr	r	s	r	r				r	sr	r

Continued

TABLE 10.3 Continued

Region/country	Conservation and natural resources 1.	2.	3.	4.	5.	6.	7.	8.	9.	10.	11.	12.	Atmosphere 13.	14.	15.	16.	17.	Marine 18.	19.	20.	21.	22.	23.	Waste 24.	25.
UK		c		r	r	r	sr	r	r	r		s	r	sr	sr	r	s	r	r		r		r	s	r
Yugoslavia				r	r					r		s	r	r	r		s	r	r	r		r			r
FORMER USSR[a]				r	r				r	r			r					r	r			r			r
Armenia													s				s								
Azerbaijan													s				s								
Belarus				r									s	r	sr	sr	s	r							r
Estonia					r								s				s		r					r	
Kazakhstan													s				s								
Latvia													s				s							r	
Lithuania													s				s		r						
Moldova													s				s								
Russia		c			r								s	sr	sr	r	s							s	
Ukraine													s	r	sr	sr	s	r							r
OCEANIA																									
Australia		c		r	r	r	r		r				s	r	sr	r	sr	r	r			r		r	r
Comoros													s				s								
Cook Is													sr				sr								
Fiji				r						r			sr	r	r		sr				r				
Kiribati														r	r		s	r							
Marshall Is													sr	r	r	r	sr		r		r				
Micronesia													s				s				r				
Nauru										r			s				s	r							
New Zealand		c		r	r	r				r			s	sr	sr	r	s	r				r		s	r
Papua New Guinea		a			r				r				sr	r	r		sr	r							
Samoa										r			s	r	r		s								
Solomon Is										r			s				s	r				r			
Tonga										r															
Tuvalu													s				s			r					
Vanuatu					r								sr				sr		r						

[a] The new countries which formerly made up the Czech and Slovak Federal Republic, Yugoslavia and the USSR have ratified a number of agreements on their own behalf. The status of these new countries with respect to agreements ratified by the Czech and Slovak Federal Republic, Yugoslavia and the USSR has not been confirmed in the sources available at the time of writing.

r – Ratified.
s – Signatory.
c – Consultative party.
a – Acceding party.

For most of the agreements shown, only ratifications are indicated. Where agreements have only recently been adopted - particularly for those opened for signature at the UN Conference on Environment and Development in June 1992 – signatories are shown. Agreements are only integrated into national law when ratified.

Key to Agreements
Place and date in brackets are those of adoption of the agreement.
1. Convention on Nature Protection and Wildlife Preservation in the Western Hemisphere (Washington, 12 October 1940). Entry into force 1 May 1942. Status as at May 1991.
2. The Antarctic Treaty (Washington, 1 December 1959), and related agreements and protocols. Entry into force 23 June 1961. Status as at 30 June 1992.
3. African Convention on the Conservation of Nature and Natural Resources (Algiers, 15 September 1968). Entry into force 16 June 1969. Status as at 30 June 1992.
4. Convention on Wetlands of International Importance Especially as Waterfowl Habitat (Ramsar, 2 February 1971). Entry into force 21 December 1975. Status as at 6 March 1993.
5. Convention concerning the Protection of the World Cultural and Natural Heritage (Paris, 12 November 1972). Entry into force 17 December 1975. Status as at 30 June 1992.

TABLE 10.3   Continued

6. Convention on International Trade in Endangered species of Wild Fauna and Flora (Washington, 3 March 1973). Entry into force 1 July 1975. Status as at 30 April 1993.
7. Convention on the Conservation of Migratory Species of Wild Animals (Bonn, 23 June 1979). Entry into force 1 November 1983. Status as at 30 April 1993.
8. Convention on the Conservation of European Wildlife and Their Natural Habitats (Berne, 19 September 1979). Entry into force 1June 1982. Status as at 30 June 1992.
9. The International Tropical Timber Agreement (Geneva, 18 November 1983). Entry into force 1 April 1985. Status as at 30 June 1992.
10. International Undertaking on Plant Genetic Resources (Rome, 23 November 1983) as supplemented. Status as at 1 January 1992.
11. ASEAN Agreement on the Conservation of Nature and Natural Resources (Kuala Lumpur, 9 July 1985). Not yet in force. Status as at 30 June 1992.
12. Convention on Biological Diversity (Rio de Janeiro, 5 June 1992). Not yet in force. Status as at 30 April 1993.
13. Convention on Long-Range Transboundary Air Pollution (Geneva, 13 November 1979), and related protocols. Entry into force 16 March 1983. Status as at 1 January 1992.
14. Vienna Convention for the Protection of the Ozone Layer (Vienna, 22 March 1985). Entry into force 22 September 1988. Status as at 30 April 1993.
15. Montreal Protocol on Substances that Deplete the Ozone Layer (Montreal, 16 September 1987). Entry into force 1 January 1989. Status as at 30 April 1993.
16. Amendment to the Montreal Protocol on Substances that Deplete the Ozone Layer, (London, 29 June 1990). Entry into force 10 August 1992. Status as at 30 April 1993.
17. United Nations Framework Convention on Climate Change (New York, 1992). Not yet in force. Status as at 30 April 1993.
18. Convention on the Prevention of Marine Pollution by Dumping of Wastes and Other Matter (London, 29 December 1972), as amended. Original entry into force 30 August 1975. Amendments not yet in force. Status as at May 1991.
19. International Convention for the Prevention of Pollution from Ships (London, 2 November 1973), as amended. Entry into force (1978 protocol) 2 October 1983. Status as at 30 June 1992.
20. Convention for the Protection of the Mediterranean Sea Against Pollution (Barcelona, 16 February 1976) and related protocols. Entry into force 12 February 1978. Status as at 1 January 1992.
21. Convention for the Protection and Development of the Marine Environment of the Wider Caribbean Region (Cartagena de Indias, 24 March 1983) and related protocols. Entry into force 11 October 1986. Status as at 1 January 1992.
22. United Nations convention on the Law of the Sea (Montego Bay, 10 December 1982). Not yet in force. Status as at 30 June 1992.
23. International Convention for the Regulation of Whaling (Washington, 2 December 1946), as amended. Entry into force 10 November 1948, amendment 4 April 1959. Status as at 30 June 1992.
24. Basel Convention on the Control of Transboundary Movements of Hazardous Wastes and their Disposal (Basel, 22 March 1989). Entry into force 5 May 1992. Status as at 30 April 1993.
25. Convention on Early Notification of a Nuclear Accident (Vienna, 26 September 1986). Entry into force 27 October 1986. Status as at 30 June 1992.

Sources:

Benedick, R. E. and Pronove, R. 1992 *Atmosphere and Outer Space*. Survey of Existing International Agreements and Instruments, Research Paper No. 26, United Nations Conference on Environment and Development, Geneva.

Boyle, A. E., Freestone, D. A. C., Kummer, K. and Ong, D. M. 1992 *Marine Environment and Marine Pollution*. Survey of Existing International Agreements and Instruments, Research Paper No. 27, United Nations Conference on Environment and Development, Geneva.

Forster, M. J. and Osterwoldt, R. U. 1992 *Nature Conservation and Terrestrial Living Resources*. Survey of Existing International Agreements and Instruments, Research Paper No. 25, United Nations Conference on Environment and Development, Geneva.

Mazzanti, M. R. 1992 *Nuclear Safety*. Survey of Existing International Agreements and Instruments, Research Paper No. 30, United Nations Conference on Environment and Development, Geneva.

Szekely, A. and Kwiatkowska, B. 1992 *Marine Living Resources*. Survey of Existing International Agreements and Instruments, Research Paper No. 28, United Nations Conference on Environment and Development, Geneva.

UNEP 1991 *Register of International Treaties and other Agreements in the Field of the Environment*. UNEP/GC.16/Inf.4, United Nations Environment Programme, Nairobi.

UNEP 1993 *Status of Global Environmental Conventions as at 30 April 1993*. Environmental Law and Institutions Programme Activity Centre, United Nations Environment Programme, Nairobi.

Data supplied by the Environmental Law and Institutions Programme Activity Centre, United Nations Environment Programme, Nairobi in 1993.

The 'UNEP Environmental Data Report' is produced on behalf of UNEP by the staff of the GEMS Monitoring and Assessment Research Centre in London with the co-operation of the World Resources Institute (Washington DC) and the UK Department of the Environment (London). This is the fourth 'UNEP Environmental Data Report'. Previous issues were published in 1987, 1989, 1991. Each edition draws upon data gathered by GEMS and other major UNEP programmes, including those concerned with health, atmosphere, climate, water, soils, oceans, deforestation, desertification, biological diversity, ecosystems and hazardous chemicals. The report includes material from many other sources, in particular from UNEP's UN partners such as the WHO, WMO, FAO, UNDP, UNSTAT, UNSCEAR, UNDP, UNFPA, DESIPA, UNICEF, UNESCO, UNCUEA, UNIDO, DHA, UN ECE, and IAEA. Material from international sources, for example, the OECD, the World Bank, IMO, CEC, IRF and WEC has also substantiated sections of the report.

Compilation of this report has been further assisted by the invaluable comments, suggestions and critical information which have been willingly provided by numerous national agencies and federal/state/provincial organizations. National "State of the Environment"-type reports, as well as reports submitted to the World Commission on Environment and Development, have provided previously unpublished information. Staff of NGOs around the world and colleagues within King's College London, University of London have also provided comments on draft versions of the report.

Finally we would wish to thank UNEP's own staff, particularly within the Programme Activity Centres of GEMS, IE, IRPTC, INFOTERRA, ELI, OCA and GRID and within UNEP associated centres such as HEM, NWRI and RISØ for their objective reviews and contributions to the identification of the pressing environmental issues of today.

Associate Editors
A. D. Willcocks, GEMS MARC, London
P. J. Peterson, UNEP and GEMS MARC, London

Steering Committee
M. D. Gwynne, UNEP GEMS/PAC, Nairobi
A. L. Hammond, WRI, Washington DC
P. S. MacCormack, UK DoE, London
C. Ogden, UK DoE, London
P. J. Peterson, UNEP/GEMS MARC, London
A. D. Willcocks, GEMS MARC, London

Compilers
P. S. Burgess, GEMS MARC
D. V. Chapman, Consultant, GEMS MARC
N. Henninger, WRI
J. A. Jackson, GEMS MARC
J. D. D. Lee, GEMS MARC
C. Ogden, UK DoE
P. J. Peterson, UNEP/GEMS MARC
E. Rodenberg, WRI
A. D. Webster, GEMS MARC
E. Wil de Gose, GEMS MARC
A. D. Willcocks, GEMS MARC

Editorial and Production Assistance
D. V. Chapman, Consultant, GEMS MARC
J. A. Jackson, GEMS MARC
C. J. Meads, GEMS MARC
F. Preston, GEMS MARC

Contributors/Reviewers
H. Abaza, UNEP/EEU, Nairobi, Kenya
R. D. Adams, International Siesmological Centre, Newbury, UK
Y. Adebayo, UNEP/TEB, Nairobi, Kenya
A. L. Aggarwal, National Environmental Engineering Research Institute, Nagpur, India
J. Allen, EUROSTAT, Luxembourg
J. Aloisi de Larderel, UNEP IE/PAC, Paris, France
A. L. Alusa, UNEP/Climate Unit, Nairobi, Kenya
J. K. Angell, Air Resources Laboratory, NOAA, Silver Spring, Maryland, USA
A. Ayoub, UNEP/TEB, Nairobi, Kenya
F. Balkau, UNEP IE/PAC, Paris, France
G. M. Bankobeza, UNEP/Ozone Secretariat, Nairobi, Kenya
A. Bárcena, Earth Council, San José, Costa Rica
P. Bartelmus, UNSTAT, New York, USA
A. Beg, Pakistan Agro Chemicals, Karachi, Pakistan
B. Bender, UNEP IRPTC/PAC, Geneva, Switzerland
B. Bennett, UNSCEAR, Vienna, Austria
J. Berney, CITES Secretariat, Lausanne, Switzerland
R. Bisson, NWRI/CCIW, Burlington, Ontario, Canada
M. Bjorklund, UNEP/EM (Wildlife and Protected Areas), Nairobi, Kenya
C. Boelcke, UNEP GEMS/PAC, Nairobi, Kenya
V. G. Boldirev, WMO, Geneva, Switzerland
J. Bower, Warren Spring Laboratory, Stevenage, UK
D. A. Bryant, WRI, Washington DC, USA
D. Calamari, Institute of Agricultural Entomology, University of Milan, Milan, Italy
M. E. Cheatle, UNEP GEMS/PAC, Nairobi, Kenya
Chen Changjie, Institute of Environmental Health Monitoring, Chinese Academy of Sciences, Beijing, People's Republic of China
D. Chibanda, Environmental Health Services, Harare, Zimbabwe
R. Christ, UNEP/Climate Unit, Nairobi, Kenya
J. Christensen, UNEP Collaborating Centre on Energy and Environment, Risø National Laboratory, Roskilde, Denmark
L. Clarke, IEA Coal Research Ltd, London, UK
A. Clerc, UNCUEA, Geneva, Switzerland
D. Cocks, Division of Wildlife and Ecology, CSIRO, Lyneham, Australia
T. Conway, NOAA/CMDL, Boulder, Colorado, USA
M. Collins, WCMC, Cambridge, UK
D. M. Cruden, University of Alberta, Edmonton, Canada
A. Dahl, UNEP/Earthwatch, Geneva, Switzerland
B. E. Davies, University of Bradford, Bradford, UK
J. Delcambre, Consultant to IE/PAC, Paris, France
B. Dimitriades, US EPA, Research Triangle Park, North Carolina, USA
F. Dixon, World Energy Council, London, UK
E. J. Dlugokencky, NOAA/CMDL, Boulder, Colorado, USA
J. Drako, Hydrometeorological Institute, Bratislava, Slovakia

R. Duffield, NWRI/CCIW, Burlington, Ontario, Canada

T. Dutkiewitz, WHO Collaborating Centre for Occupational Health, Institute of Occupational Medicine, Lodz, Poland

E. G. Dutton, NOAA/CMDL, Boulder, Colorado, USA

H. El-Habr, UNEP/TEB, Nairobi, Kenya

J. W. Elkins, NOAA/CMDL, Boulder, Colorado, USA

G. Evans, Environmental Monitoring Systems Laboratory, US EPA, Research Triangle Park, North Carolina, USA

R. Gambell, IWC, Cambridge, UK

N. Gebremedhin, UNEP/EM, Nairobi, Kenya

G. Goldstein, WHO/EHE, Geneva, Switzerland

R. Gonzalez-Garcia, Urban Development and Ecology, Mexico DF, Mexico

H. Gopalan, UNEP/TEB, Nairobi, Kenya

C. Grey, International Tanker Owners Pollution Federation Ltd, London

D. Guha-Sapir, Centre for Research on the Epidemiology of Disease, Brussels, Belgium

M. D. Gwynne, UNEP GEMS/PAC, Nairobi, Kenya

W. Haeberli, WGMS, Zurich, Switzerland

J. Harrison, WCMC, Cambridge, UK

D. L. Hawksworth, International Mycological Institute, Egham, UK

L. Heligman, Population Division, UNDESIPA, New York, USA

R. Helmer, WHO/PEP, Geneva, Switzerland

J. R. Hickman, Health and Welfare Canada, Ottawa, Canada

K. Hill, United Nations Conference on Environment and Development Secretariat, Geneva, Switzerland

J. Huismans, UNEP/Earthwatch Co-ordination and Environmental Assessment, Nairobi, Kenya

M. Johannessen, NIVA, Oslo, Norway

K. C. Jones, University Of Lancaster, Lancaster, UK

P. D. Jones, University of East Anglia, Norwich, UK

A. Kahnert, Statistical Division, UN ECE, Geneva, Switzerland

C. Kerpelman, UNDRO, Geneva, Switzerland

I. Keshavjee, UNEP/TEB, Nairobi, Kenya

T. Kjellström, WHO/PEP, Geneva, Switzerland

S. Kleemola, Data Processing Unit, Environmental Data Centre, Helsinki,

W. D. Komhyr, NOAA/CMDL, Boulder, Colorado, USA

J. G. Kretzschmar, VITO, Mol, Belgium

L. Kromkatchev, UNEP/DC, Nairobi, Kenya

P. M. Lang, NOAA/CMDL, Boulder, Colorado, USA

L. Laugeri, WHO/EHE, Geneva, Switzerland

Liang Xiyan, Beijing Municipal Monitoring Centre, Beijing, People's Republic of China

M. Laurijssen, UNEP IRPTC/PAC, Geneva, Switzerland

M. Lorenz, Federal Research Centre for Forestry and Forest Products, Hamburg

D. Mage, WHO/PEP, US EPA, Research Triangle Park, N. Carolina, USA

M. D. Mazzola, Direccion de Relaciones Sanitarias Internacionales, Buenos Aires, Argentina

B. Mendonca, NOAA/CMDL, Boulder, Colorado, USA

J. Miller, Environment Division, WMO, Geneva, Switzerland

D. Mitchell, UNEP GEMS/PAC, Nairobi, Kenya

B. M. Moore, King's College London, London UK

A. B. Murray, UNEP/HEM Office, Munich, Germany

F. Murray, Murdoch University, Murdoch, Australia

W. Musani, UNEP/ELI, Nairobi, Kenya

M. Nagai, UNEP/ELI, Nairobi, Kenya

M. M. Nasralla, Cairo National Research Centre, Cairo, Egypt

P. Novelli, NOAA/CMDL, Boulder, Colorado, USA

T. O'Connor, Coastal Monitoring Branch, NOAA National Ocean Service, Rockville, Maryland, USA

E. D. Ongley, NWRI/CCIW, Burlington, Ontario, Canada

M. Opelz, IAEA, Geneva, Switzerland

G. Ozolins, WHO/PEP, Geneva, Switzerland

K. F. Panzer, Bundesforschungsanstalt für Forst-und Holzwirtschaft, Hamburg, Germany

F. G. Pariboni, FAO, Rome, Italy

N. Parkinson, King's College London, London, UK

D. B. Peakall, Canadian Wildlife Service, Environment Canada, Ottawa, Ontario, Canada

J. K. Piotrowski, Medical Academy, Lodz, Poland

K. Pond, Farnborough College of Technology, Farnborough, UK

N. Previsich, Health and Welfare Canada, Ottawa, Canada

R. Price, King's College London, London, UK

M. Pyhala, UNEP/GEF Unit, Nairobi, Kenya

W. Rast, UNEP, Water and Lithosphere Unit, Nairobi, Kenya

M. Raizenne, Health and Welfare Canada, Ottawa, Canada

F. Ramade, University of Paris-South, Orsay, France

A. A. Razak, Al-Azhar University, Cairo, Egypt

D-G. Rhee, Ministry of the Environment, Seoul, Korea

D. Robinson, Rutgers University, Newark New Jersey, USA

A. Ross, IMO, London, UK

Å. Rühling, Swedish Environmental Research Institute and Department of Plant Ecology, University of Lund, Sweden

I. Rummel-Bulska, SBC, Geneva, Switzerland

F. Ya. Rovinsky, Institute of Global Climate and Ecology, Academy of Sciences, Moscow, Russia

N. Sabogal, UNEP/Ozone Secretariat, Nairobi, Kenya

K. M. Sarma, UNEP/Ozone Secretariat, Nairobi, Kenya

W. Scherer, IOC, Paris, France

M. Schomaker, UNEP GEMS/PAC, Nairobi, Kenya

P. Schroder, UNEP OCA/PAC, Nairobi, Kenya

W. Seltzer, UNSTAT, New York, USA

R. M. Shende, UNEP IE/PAC, Paris, France

J. S. Singh, UNFPA, New York, USA

J. Siriswasdi, Office of the National Environment Board, Bangkok, Thailand

T. Spence, Joint Planning Office for the GCOS, WMO, Geneva, Switzerland

L. Spencer, UNEP/INFOTERRA, Nairobi, Kenya

A. van Strein, Netherlands Central Bureau of Statistics, Voorburg, Netherlands

J. Stronkherst, Ministry of Transport, Tidal Waters Division, Middelberg, Netherlands

N. Sundararaman, IPCC, c/o WMO, Geneva, Switzerland

B. M. Taal, UNEP/TEB(Forests), Nairobi, Kenya

S. Tamplin, WHO Regional Office for Western Pacific, Manila, Philippines

C. Tavera, UNEP/GEF Unit, Nairobi, Kenya

R. Thakre, National Environmental Engineering Research Institute, Nagpur, India

A. Tolkachev, IOC, Paris, France

J. Tomblin, UN DHA-Geneva, Switzerland

L. Truppi, US EPA, Research Triangle Park, North Carolina, USA

E. B. Tutuwan, Ministry of Planning and Regional Development, Yaounde, Cameroon

V. Vandeweerd, UNEP GEMS/PAC, Nairobi, Kenya

G. Varallyay, Research Institute of Soil Science and Agricultural Chemistry, Hungarian Academy of Sciences, Budapest, Hungary

N. Wace, The Australian National University, Canberra ACT, Australia

C. C. Wallén, Consultant to UNEP, Nairobi, Kenya

P. Wardle, FAO, Rome, Italy

P.J. Whitfield, King's College London, London, UK

G. B. Wiersma, College of Forest Resources, University of Maine, USA

W. Wooster, Department of Marine Sciences, University of Washington, Seattle, USA

H. Yakowitz, OECD, Paris, France

A. S. Yarnatovsky, Institute of Global Climate and Ecology, Academy of Sciences, Moscow, Russia

Zhang Jinhua, UNEP/TEB, Nairobi, Kenya

Zhao Dianwu, Research Centre for Environmental Sciences, Chinese Academy of Sciences, Beijing, People's Republic of China

Zhao Ji, Department of Geography, Beijing Normal University, Beijing, People's Republic of China

Typesetting and Layout

I. Bertin and C. Ketch, Cork

Chapman Bounford Associates, London (figures)

Upper Case Ltd, Cork

List of Country Names

Countries for which data are provided in this report are listed below. Both the short name (i.e., the abbreviated form used within the body of this report) and the full official state title are given. This listing includes independent sovereign nations and selected dependent territories.

Countries are listed in alphabetical order according to their current status, i.e., status as of June 1993. Recent changes (within the last five years) in country nomenclature and designation are, however, noted. Since the publication of the previous edition of this report in May 1991, the following major changes have occurred: By the end of 1991 the Union of Soviet Socialist Republics (USSR) finally dissolved; the 15 member republics have since become separate independent nations. On 1 January 1993 the federation of Czechoslovakia formally split into two states, the Czech Republic and the Slovak Republic (Slovakia). Following a period of unrest commencing in 1991 the federation of Yugoslavia has split into several new entities; these comprise the Republic of Croatia, the Republic of Bosnia and Herzegovina, the Republic of Slovenia, the Former Yugoslavian Republic of Macedonia and the Federal Republic of Yugoslavia (Serbia and Montenegro). In May 1993 Eritrea split from Ethiopia and became an independent state.

Within the data tables included in this report, countries have been listed in alphabetical order under the following continental regional groupings; Africa, North America, South America, Asia, Europe, (Former) USSR and Oceania. Although there is no internationally agreed regional grouping scheme, the above classification is widely used in international statistical compilations and thus for ease of reference has been adopted here. In some cases, however, alternative regional groupings have been retained from the original source data bases or documents.

In the absence of international standardization on the terms 'developed' and 'developing' countries, designation of countries as developed or developing in the present publication is as assigned in the source data bases or documents. For the most part 'developed' countries include all the European nations (including those of Eastern Europe) the former USSR, USA, Canada, Japan, Australia and New Zealand. Deviations from this usage are indicated in the footnotes to individual tables. Please note that designations such as 'developed' and 'developing' are intended for statistical convenience only and do not necessarily express a judgement about a stage reached by a particular country in the development process.

In cases where geographical changes have taken place, countries are listed in data tables and figures according to their status at the time of data collection. For example, Germany is listed separately as the Federal Republic of Germany (Germany, Fed. Rep.) and the German Democratic Republic (German Dem. Rep.) when data relate to the time before their unification on 3 October 1990. Similarly the former USSR is listed as 'USSR' in data tables and figures reporting pre-1992 data. Elsewhere in the report countries are referred to by their current name. In the case of Yemen, where the short form of the unified state (Yemen) is the same as that of the former Yemen Arab Republic, a footnote is used in individual tables to identify the unified state. Deviations from this convention are detailed in the notes to individual tables.

The description and classification of countries and territories in this study and the arrangement of the material do not imply the expression of any opinion whatsoever of the Secretariat of the United Nations concerning the legal status of any country, territory, city or area, or of its authorities, or concerning the delimitation of its frontiers or boundaries, or regarding its economic system or degree of development.

Short Name	Full State Title/Notes
Afghanistan	Islamic State of Afghanistan
Albania	Republic of Albania
Algeria	Democratic and Popular Republic of Algeria
American Samoa	American Samoa (US Dependency)
Andorra	Principality of Andorra
Angola	People's Republic of Angola
Anguilla	Anguilla (UK Dependency)
Antigua & Barbuda	Antigua and Barbuda
Argentina	Argentine Republic
Armenia	Republic of Armenia (formerly part of the USSR)
Aruba	Aruba (Netherlands Dependency, formerly part of the Netherlands Antilles)
Australia	Commonwealth of Australia
Austria	Republic of Austria
Azerbaijan	Azerbaijani Republic (formerly part of the USSR)
Bahamas	Commonwealth of the Bahamas
Bahrain	State of Bahrain
Bangladesh	People's Republic of Bangladesh
Barbados	Barbados
Belarus	Republic of Belarus (formerly part of the USSR)
Belgium	Kingdom of Belgium
Belize	Belize

Benin	Republic of Benin	French Polynesia	French Polynesia (French Dependency)
Bermuda	Bermuda (UK Dependency)	French Southern Tr.	French Southern Territory (French Dependency)
Bhutan	Kingdom of Bhutan		
Bolivia	Republic of Bolivia	Gabon	Gabonese Republic
Bosnia and Herzegovina	Republic of Bosnia and Herzegovina (formerly part of Yugoslavia)	Gambia	Republic of The Gambia
		Gaza Strip	Gaza Strip
Botswana	Republic of Botswana	Georgia	Republic of Georgia (formerly part of the USSR)
Brazil	Federative Republic of Brazil		
Br. Ind. Oc. Tr.	British Indian Ocean Territory (UK Dependency)	Germany	Federal Republic of Germany (formerly German Dem. Rep. and Germany, Fed. Rep.)
Br. Virgin Is	British Virgin Islands (UK Dependency)		
Brunei	Brunei Darussalam	Ghana	Republic of Ghana
Bulgaria	Republic of Bulgaria	Gibraltar	Gibraltar (UK Dependency)
Burkina	People's Democratic Republic of Burkina	Greece	Hellenic Republic
		Greenland	Greenland (Danish Dependency)
Burundi	Republic of Burundi	Grenada	Grenada
Cameroon	United Republic of Cameroon	Guadeloupe	Guadeloupe (French Dependency)
Cambodia	Cambodia (formerly Kampuchea)	Guam	Guam (US Dependency)
Canada	Canada	Guatemala	Republic of Guatemala
Cape Verde	Republic of Cape Verde	Guinea	Republic of Guinea
Cayman Is	Cayman Islands (UK Dependency)	Guinea-Bissau	Republic of Guinea-Bissau
Cent. African Rep.	Central African Republic	Guyana	Co-operative Republic of Guyana
Chad	Republic of Chad	Haiti	Republic of Haiti
Chile	Republic of Chile	Honduras	Republic of Honduras
China	People's Republic of China	Hong Kong	Hong Kong (UK Dependency)
China, Taiwan	China (Taiwan)	Holy See	Holy See
Christmas Is.	Christmas Island (Kiribati Dependency)	Hungary	Republic of Hungary
		Iceland	Republic of Iceland
Cocos Is	Cocos (Keeling) Islands (Mauritius Dependency)	India	Republic of India
		Indonesia	Republic of Indonesia
Colombia	Republic of Colombia	Iran	Islamic Republic of Iran
Comoros	Federal and Islamic Republic of Comoros	Iraq	Republic of Iraq
		Ireland	Republic of Ireland
Congo	Republic of the Congo	Israel	State of Israel
Cook Is.	Cook Island (New Zealand Dependency)	Italy	Italian Republic
		Jamaica	Jamaica
Costa Rica	Republic of Costa Rica	Japan	Japan
Côte d'Ivoire	Republic of Côte d'Ivoire	Jordan	Hashemite Kingdom of Jordan
Croatia	Republic of Croatia (formerly part of Yugoslavia)	Kazakhstan	Republic of Kazakhstan (formerly part of the USSR)
Cuba	Republic of Cuba	Kenya	Republic of Kenya
Cyprus	Republic of Cyprus	Kiribati	Republic of Kiribati
Czech Republic	The Czech Republic (formerly part of Czechoslovakia)	Korea	Republic of Korea
		Korea, Dem.	Democratic People's Republic of Korea
Denmark	Kingdom of Denmark		
Djibouti	Republic of Djibouti	Kuwait	State of Kuwait
Dominica	Commonwealth of Dominica	Kyrgyzstan	Republic of Kyrgyzstan (formerly part of the USSR)
Dominican Rep.	Dominican Republic		
East Timor	East Timor	Laos	Lao People's Democratic Republic
Ecuador	Republic of Ecuador	Latvia	Republic of Latvia (formerly part of the USSR)
Egypt	Arab Republic of Egypt		
El Salvador	Republic of El Salvador	Lebanon	Lebanese Republic
Equatorial Guinea	Republic of Equatorial Guinea	Lesotho	Kingdom of Lesotho
Eritrea	Eritrea (formerly part of Ethiopia)	Liberia	Republic of Liberia
Estonia	Republic of Estonia (formerly part of the USSR)	Libya	Great Socialist People's Libyan Arab Jamahiriya
Ethiopia	People's Democratic Republic of Ethiopia	Liechtenstein	Principality of Liechtenstein
		Lithuania	Republic of Lithuania (formerly part of the USSR)
Faeroe Is	Faeroe Islands (Danish Dependency)		
Falkland Is	Falkland Islands (UK Dependency)	Luxembourg	Grand Duchy of Luxembourg
Fiji	The Republic of Fiji	Macau	Macau (Portuguese Dependency)
Finland	Republic of Finland	Macedonia	The former Yugoslavian Republic of Macedonia (formerly part of Yugoslavia)
France	French Republic		
French Guiana	French Guiana (French Dependency)		

Madagascar	Republic of Madagascar	Singapore	Republic of Singapore
Malawi	Malawi	Slovakia	Slovak Republic (formerly part of Czechoslovakia)
Malaysia	Malaysia		
Maldives	Republic of Maldives	Slovenia	Republic of Slovenia (formerly part of Yugoslavia)
Mali	Republic of Mali		
Malta	Republic of Malta	Solomon Is	Solomon Islands
Marshall Is	Republic of the Marshall Islands	Somalia	Somali Democratic Republic
Martinique	Martinique (French Dependency)	South Africa	Republic of South Africa
Mauritania	Islamic Republic of Mauritania	Spain	Kingdom of Spain
Mauritius	Republic of Mauritius	Sri Lanka	Democratic Socialist Republic of Sri Lanka
Mexico	United Mexican States		
Micronesia	Federated States of Micronesia	St Kitts & Nevis	Federation of Saint Christopher and Nevis
Moldova	Republic of Moldova (formerly part of the USSR)		
		St Helena	Saint Helena (UK Dependency)
Monaco	Principality of Monaco	St Lucia	Saint Lucia
Mongolia	Mongolian People's Republic	St Pierre & Miquelon	Saint Pierre and Miquelon (French Dependency)
Montserrat	Montserrat (UK Dependency)		
Morocco	Kingdom of Morocco	St Vincent & Grenadines	Saint Vincent and the Grenadines
Mozambique	Republic of Mozambique		
Myanmar	Union of Myanmar (formerly Burma)	Sudan	Republic of the Sudan
		Suriname	Republic of Suriname
Namibia	Republic of Namibia	Swaziland	Kingdom of Swaziland
Nauru	Republic of Nauru	Sweden	Kingdom of Sweden
Nepal	Kingdom of Nepal	Switzerland	Swiss Confederation
Neth. Antilles	Netherlands Antilles (Netherlands Dependency)	Syria	Syrian Arab Republic
		Tajikistan	Republic of Tajikistan (formerly part of the USSR)
Netherlands	Kingdom of the Netherlands		
New Caledonia	New Caledonia (French Dependency)	Tanzania	United Republic of Tanzania
		Thailand	Kingdom of Thailand
New Zealand	New Zealand	Togo	Republic of Togo
Nicaragua	Republic of Nicaragua	Tokelau	Tokelau (New Zealand Dependency)
Niger	The Republic of Niger	Tonga	Kingdom of Tonga
Nigeria	Federal Republic of Nigeria	Trinidad & Tobago	Republic of Trinidad and Tobago
Niue	Niue (New Zealand Dependency)	Tunisia	Tunisian Republic
Norfolk Is.	Norfolk Island (Australian Dependency)	Turkey	Republic of Turkey
		Turkmenistan	Turkmenistan (formerly part of the USSR)
Northern Marianas Is	Northern Marianas Islands, Commonwealth of Islands		
		Turks & Caicos Is	Turks and Caicos Islands (UK Dependency)
Norway	Kingdom of Norway		
Oman	Sultanate of Oman	Tuvalu	Tuvalu
Pacific Is. Tr. Tr.	Trust Territory of the Pacific Islands	Uganda	Republic of Uganda
Pakistan	Islamic Republic of Pakistan	Ukraine	Ukraine (formerly part of the USSR)
Panama	Republic of Panama	UK	United Kingdom of Great Britain and Northern Ireland
Papua New Guinea	Independent State of Papua New Guinea		
		United Arab Em.	United Arab Emirates
Paraguay	Republic of Paraguay	Uruguay	Oriental Republic of Uruguay
Peru	Republic of Peru	USA	United States of America
Philippines	Republic of the Philippines	US Virgin Is	United States Virgin Islands
Poland	Republic of Poland	Uzbekistan	Republic of Uzbekistan (formerly part of the USSR)
Portugal	Portuguese Republic		
Puerto Rico	Puerto Rico (US Dependency)	Vanuatu	Republic of Vanuatu
Qatar	State of Qatar	Venezuela	Republic of Venezuela
Réunion	Réunion (French Dependency)	Viet Nam	Socialist Republic of Viet Nam
Romania	Romania	Wallis Is	Wallis and Futuna Islands (French Dependency)
Russia	Russian Federation (formerly part of the USSR)		
		Western Sahara	Western Sahara
Rwanda	Republic of Rwandese	Yemen	Republic of Yemen (formerly Yemen, Dem. and Yemen)
Samoa	Independent State of Western Samoa		
San Marino	Republic of San Marino	Yugoslavia (Serbia & Montenegro)	Federal Republic of Yugoslavia (formerly Yugoslavia)
São Tomé & Príncipe	Democratic Republic of São Tomé and Príncipe		
		Zaire	Republic of Zaire
Saudi Arabia	Kingdom of Saudi Arabia	Zambia	Republic of Zambia
Senegal	Republic of Senegal	Zimbabwe	Republic of Zimbabwe
Seychelles	Republic of Seychelles		
Sierra Leone	Republic of Sierra Leone		

List of Abbreviations

APELL — Awareness and Preparedness for Emergencies at Local Level (IE/PAC)

ASEAN — Association of South East Asian Nations

BAPMoN — Background Air Pollution Monitoring Network (WMO/US EPA/UNEP(GEMS))

CDIAC — Carbon Dioxide Information Analysis Center

CEC — Commission of the European Communities

CIPEL — International Surveillance Commission of Lake Léman

CITES — Convention on International Trade in Endangered Species of Wild Fauna and Flora

CMA — Chemical Manufacturers Association

CMDL — Climate Monitoring and Diagnostics Laboratory (formerly Climate Monitoring for Climate Change)

CORINE — European Community Co-ordinated Information System on the State of the Environment and Natural Resources

CRED — Centre for Research on the Epidemiology of Disasters

CSD — Commission for Sustainable Development

DHA — Department of Humanitarian Affairs

EEC — European Economic Community

ELI/PAC — Environmental Law and Institutions Programme Activity Centre (UNEP)

EMEP — European Monitoring and Evaluation Programme (UN ECE/UNEP/WMO)

EPI — Expanded Programme on Immunization (WHO)

ESCAP — Economic and Social Commission for Asia and the Pacific

EUROSTAT — Statistical Office of the European Communities

FAO — Food and Agriculture Organization of the United Nations

FCCC — Framework Convention on Climate Change

GAW — Global Atmosphere Watch (WMO)

GCOS — Global Climate Observing System (WMO/IOC/UNEP/ICSU)

GEF — Global Environment Facility (UNDP/UNEP/World Bank)

GEMS/PAC — Global Environment Monitoring System (UNEP)

GESAMP — IMO/FAO/UNESCO/WMO/WHO/IAEA/UN/UNEP Group of Experts on Scientific Aspects of Marine Pollution

GLASOD — Global Assessment of Human-induced Soil Degradation (ISRIC/UNEP)

GOOS — Global Ocean Observing System (IOC/ICSU/WMO/UNEP)

GO$_3$OS — Global Ozone Observing System (WMO)

GRID/PAC — Global Resource Information Database (UNEP)

GTOS — Global Terrestrial Observing System (UNEP/UNESCO/FAO/WMO/ICSU)

GTS — Global Telecommunications System (WWW(WMO))

HEALs — Human Exposure Assessment Locations (WHO/UNEP)

HEM — Harmonization of Environmental Measurement (UNEP)

IAEA — International Atomic Energy Agency

IARC — International Agency for Research on Cancer

IATA — International Air Transport Association

ICAO — International Civil Aviation Organization

ICES — International Council for the Exploration of the Sea

ICP — International Co-operative Programme

ICPIC — International Cleaner Production Information Computer System (IE/PAC)

ICSI — International Commission on Snow and Ice

ICSU — International Council for Scientific Unions

IDNDR — International Decade for Natural Disaster Reduction

IEA — International Energy Agency

IE/PAC — Industry and Environment Programme Activity Centre (UNEP)

IGBP — International Geosphere-Biosphere Programme (ICSU)

ILO — International Labour Organization

IMO — International Maritime Organization

INC — International Negotiating Committee

INCD — International Convention to Combat Desertification

INFOTERRA/PAC — International Environmental Information System (UNEP)

IOC — Intergovernmental Oceanographic Commission (UNESCO)

IPCC — Intergovernmental Panel on Climate Change (WMO/UNEP)

IPCS — International Programme on Chemical Safety (UNEP/WHO/ILO)

IRF — International Road Federation

IRPTC/PAC — International Register of Potentially Toxic Chemicals (UNEP)

ISC — International Seismological Centre

ISO — International Standards Organization

ISRIC — International Soil Reference and Information Centre

ITTO — International Tropical Timber Organization

IUCN	The World Conservation Union (formerly the International Union for Conservation of Nature and Natural Resources)
IWC	International Whaling Commission
JECFA	Joint FAO/WHO Expert Committee on Food Additives
JMPR	Joint FAO/WHO Meeting on Pesticide Residues
LDC	London Dumping Convention
MAP	Mediterranean Action Plan (UNEP OCA/PAC)
MED POL	Monitoring and Research Programme of the Mediterranean Action Plan (UNEP OCA/PAC)
MPAP	Mar del Plata Action Plan
MSC	Meteorological Synthesizing Centre
NAS	United States National Academy of Sciences
NASA	National Aeronautics and Space Administration
NGO	Non-governmental Organization
NILU	Norwegian Institute for Air Research
NIVA	Norwegian Institute for Water Research
NOAA	National Oceanic and Atmospheric Administration
NSWS	National Surface Water Survey
NWRI	National Water Research Institute of Canada
OCA/PAC	Oceans and Coastal Areas Programme Activity Centre (UNEP)
OECD	Organisation for Economic Co-operation and Development
OFDA	Office of the US Foreign Disaster Assistance
PACD	Plan of Action to Combat Desertification
PSMSL	Permanent Service for Mean Sea Level
SBC	Secretariat of the Basel Convention
SEI	Stockholm Environment Institute
SOTOR	World Soils and Terrain Digital Database (ISRIC/UNEP/International Society of Soil Science)
UIC	Union Internationale des Chemins de Fer
UN	United Nations
UNCED	United Nations Conference on Environment and Development
UNCHS	United Nations Centre for Human Settlements (Habitat)
UNCTAD	United Nations Conference on Trade and Development
UNCUEA	United Nations Centre for Urgent Environmental Assistance
UNDESIPA	United Nations Department for Economic and Social Information and Policy Analysis (formerly the United Nations Department of Economic and Social Development)
UNDP	United Nations Development Programme
UNDRO	United Nations Disaster Relief Organization
UN ECE	United Nations Economic Commission for Europe
UNEP	United Nations Environment Programme
UNESCO	United Nations Educational, Scientific and Cultural Organization
UNFPA	United Nations Population Fund (formerly United Nations Fund for Population Activities)
UNICEF	United Nations Children's Fund
UNIDO	United Nations Industrial Development Organization
UNIENET	The United Nations International Emergency Network
UNSCEAR	United Nations Scientific Committee on the Effects of Atomic Radiation
UNSTAT	United Nations Statistical Office
US AID	United States Agency for International Development
US EPA	United States Environmental Protection Agency
WCDMP	World Climate Data and Monitoring Programme (WMO)
WCMC	World Conservation Monitoring Centre
WCP	World Climate Programme (WMO)
WCRP	World Climate Research Programme (WMO/ICSU/UNESCO)
WEC	World Energy Council (formerly World Energy Conference)
WGMS	World Glacier Monitoring Service
WHO	World Health Organization
WMO	World Meteorological Organization
WWF	World Wide Fund for Nature (formerly World Wildlife Fund)
WWW	World Weather Watch (WMO)